Pseudo-Differential Operators
Theory and Applications
Vol. 3

Pseudo-Differential Operators: Theory and Applications is a series of moderately priced graduate-level textbooks and monographs appealing to students and experts alike. Pseudo-differential operators are understood in a very broad sense and include such topics as harmonic analysis, PDE, geometry, mathematical physics, microlocal analysis, time-frequency analysis, imaging and computations. Modern trends and novel applications in mathematics, natural sciences, medicine, scientific computing, and engineering are highlighted.

Nicolas Lerner

Metrics on the Phase Space and Non-Selfadjoint Pseudo-Differential Operators

Birkhäuser
Basel · Boston · Berlin

Author:

Nicolas Lerner
Projet Analyse fonctionnelle
Institut de Mathématique de Jussieu
Université Pierre et Marie Curie (Paris VI)
4, Place Jussieu
75252 Paris cedex 05
France
e-mail: lerner@math.jussieu.fr

2010 Mathematics Subject Classification: 35S05, 35A05, 47G30

Library of Congress Control Number: 2009940363

Bibliographic information published by Die Deutsche Bibliothek
Die Deutsche Bibliothek lists this publication in the Deutsche Nationalbibliografie;
detailed bibliographic data is available in the Internet at <http://dnb.ddb.de>.

ISBN 978-3-7643-8509-5 Birkhäuser Verlag AG, Basel · Boston · Berlin

© 2010 Birkhäuser Verlag AG
Basel · Boston · Berlin
P.O. Box 133, CH-4010 Basel, Switzerland
Part of Springer Science+Business Media
Printed on acid-free paper produced of chlorine-free pulp. TCF∞
Printed in Germany

ISBN 978-3-7643-8509-5   e-ISBN 978-3-7643-8510-1

9 8 7 6 5 4 3 2 1   www.birkhauser.ch

Contents

Preface   ix

1 Basic Notions of Phase Space Analysis   1
   1.1   Introduction to pseudo-differential operators   1
      1.1.1   Prolegomena   1
      1.1.2   Quantization formulas   9
      1.1.3   The $S_{1,0}^m$ class of symbols   11
      1.1.4   The semi-classical calculus   22
      1.1.5   Other classes of symbols   27
   1.2   Pseudo-differential operators on an open subset of $\mathbb{R}^n$ 28
      1.2.1   Introduction   28
      1.2.2   Inversion of (micro)elliptic operators   32
      1.2.3   Propagation of singularities   37
      1.2.4   Local solvability   42
   1.3   Pseudo-differential operators in harmonic analysis   51
      1.3.1   Singular integrals, examples   51
      1.3.2   Remarks on the Calderón-Zygmund theory and classical
          pseudo-differential operators   54

2 Metrics on the Phase Space   57
   2.1   The structure of the phase space   57
      2.1.1   Symplectic algebra   57
      2.1.2   The Wigner function   58
      2.1.3   Quantization formulas   58
      2.1.4   The metaplectic group   60
      2.1.5   Composition formula   62
   2.2   Admissible metrics   67
      2.2.1   A short review of examples of pseudo-differential calculi . .   67
      2.2.2   Slowly varying metrics on $\mathbb{R}^{2n}$ 68
      2.2.3   The uncertainty principle for metrics   72
      2.2.4   Temperate metrics   74
      2.2.5   Admissible metric and weights   76

	2.2.6	The main distance function	80
2.3		General principles of pseudo-differential calculus	83
	2.3.1	Confinement estimates	84
	2.3.2	Biconfinement estimates	84
	2.3.3	Symbolic calculus	91
	2.3.4	Additional remarks	94
	2.3.5	Changing the quantization	100
2.4		The Wick calculus of pseudo-differential operators	100
	2.4.1	Wick quantization	100
	2.4.2	Fock-Bargmann spaces	104
	2.4.3	On the composition formula for the Wick quantization	106
2.5		Basic estimates for pseudo-differential operators	110
	2.5.1	L^2 estimates	110
	2.5.2	The Gårding inequality with gain of one derivative	113
	2.5.3	The Fefferman-Phong inequality	115
	2.5.4	Analytic functional calculus	134
2.6		Sobolev spaces attached to a pseudo-differential calculus	137
	2.6.1	Introduction	137
	2.6.2	Definition of the Sobolev spaces	138
	2.6.3	Characterization of pseudo-differential operators	140
	2.6.4	One-parameter group of elliptic operators	146
	2.6.5	An additional hypothesis for the Wiener lemma: the geodesic temperance	152

3 Estimates for Non-Selfadjoint Operators   **161**
3.1		Introduction	161
	3.1.1	Examples	161
	3.1.2	First-bracket analysis	171
	3.1.3	Heuristics on condition (Ψ)	174
3.2		The geometry of condition (Ψ)	177
	3.2.1	Definitions and examples	177
	3.2.2	Condition (P)	179
	3.2.3	Condition (Ψ) for semi-classical families of functions	181
	3.2.4	Some lemmas on C^3 functions	190
	3.2.5	Inequalities for symbols	194
	3.2.6	Quasi-convexity	200
3.3		The necessity of condition (Ψ)	203
3.4		Estimates with loss of $k/k+1$ derivative	205
	3.4.1	Introduction	205
	3.4.2	The main result on subellipticity	207
	3.4.3	Simplifications under a more stringent condition on the symbol	207
3.5		Estimates with loss of one derivative	209
	3.5.1	Local solvability under condition (P)	209

		3.5.2	The two-dimensional case, the oblique derivative problem	.	216
		3.5.3	Transversal sign changes		220
		3.5.4	Semi-global solvability under condition (P)		225
	3.6	(Ψ) does not imply solvability with loss of one derivative			226
		3.6.1	Introduction		226
		3.6.2	Construction of a counterexample		232
		3.6.3	More on the structure of the counterexample		246
	3.7	Condition (Ψ) does imply solvability with loss of 3/2 derivatives	.	250
		3.7.1	Introduction		250
		3.7.2	Energy estimates		251
		3.7.3	From semi-classical to local estimates		263
	3.8	Concluding remarks			283
		3.8.1	A short historical account of solvability questions		283
		3.8.2	Open problems		284
		3.8.3	Pseudo-spectrum and solvability		285
4	**Appendix**				**287**
	4.1	Some elements of Fourier analysis			287
		4.1.1	Basics		287
		4.1.2	The logarithm of a non-singular symmetric matrix		289
		4.1.3	Fourier transform of Gaussian functions		291
		4.1.4	Some standard examples of Fourier transform		295
		4.1.5	The Hardy Operator		299
	4.2	Some remarks on algebra			300
		4.2.1	On simultaneous diagonalization of quadratic forms		300
		4.2.2	Some remarks on commutative algebra		301
	4.3	Lemmas of classical analysis			303
		4.3.1	On the Faà di Bruno formula		303
		4.3.2	On Leibniz formulas		305
		4.3.3	On Sobolev norms		306
		4.3.4	On partitions of unity		308
		4.3.5	On non-negative functions		310
		4.3.6	From discrete sums to finite sums		317
		4.3.7	On families of rapidly decreasing functions		319
		4.3.8	Abstract lemma for the propagation of singularities		322
	4.4	On the symplectic and metaplectic groups			324
		4.4.1	The symplectic structure of the phase space		324
		4.4.2	The metaplectic group		334
		4.4.3	A remark on the Feynman quantization		337
		4.4.4	Positive quadratic forms in a symplectic vector space		338
	4.5	Symplectic geometry			344
		4.5.1	Symplectic manifolds		344
		4.5.2	Normal forms of functions		345
	4.6	Composing a large number of symbols			346

4.7   A few elements of operator theory 356
    4.7.1   A selfadjoint operator 356
    4.7.2   Cotlar's lemma 357
    4.7.3   Semi-classical Fourier integral operators 361
4.8   On the Sjöstrand algebra 366
4.9   More on symbolic calculus 367
    4.9.1   Properties of some metrics 367
    4.9.2   Proof of Lemma 3.2.12 on the proper class 368
    4.9.3   More elements of Wick calculus 370
    4.9.4   Some lemmas on symbolic calculus 374
    4.9.5   The Beals-Fefferman reduction 376
    4.9.6   On tensor products of homogeneous functions 378
    4.9.7   On the composition of some symbols 379

Bibliography   **383**

Index   **395**

Preface

This book is devoted to the study of pseudo-differential operators, with special emphasis on non-selfadjoint operators, a priori estimates and localization in the phase space. We have tried here to expose the most recent developments of the theory with its applications to local solvability and semi-classical estimates for non-selfadjoint operators.

The first chapter, *Basic Notions of Phase Space Analysis*, is introductory and gives a presentation of very classical classes of pseudo-differential operators, along with some basic properties. As an illustration of the power of these methods, we give a proof of propagation of singularities for real-principal type operators (using a priori estimates, and not Fourier integral operators), and we introduce the reader to local solvability problems. That chapter should be useful for a reader, say at the graduate level in analysis, eager to learn some basics on pseudo-differential operators.

The second chapter, *Metrics on the Phase Space* begins with a review of symplectic algebra, Wigner functions, quantization formulas, metaplectic group and is intended to set the basic study of the phase space. We move forward to the more general setting of metrics on the phase space, following essentially the basic assumptions of L. Hörmander (Chapter 18 in the book [73]) on this topic. We use the notion of confinement, introduced by J.-M. Bony and the author and we follow the initial part of the paper [20] on these topics. We expose as well some elements of the so-called Wick calculus. We present some key examples related to the Calderón-Zygmund decompositions such that the Fefferman-Phong inequality and we prove that the analytic functional calculus works for admissible metrics. We give a description of the construction of Sobolev spaces attached to a pseudo-differential calculus, following the paper by J.-M. Bony and J.-Y. Chemin [19]; this construction of Sobolev spaces has been discussed in the aforementioned paper and also in several articles of R. Beals such as [6] (see also the paper [7] for a key lemma of characterization of pseudo-differential operators).

The third and last chapter, *Estimates for Non-Selfadjoint Operators*, is devoted to the more difficult and less classical topic of non-selfadjoint pseudo-differential operators. We discuss the details of the various types of estimates that can be proved or disproved, depending on the geometry of the symbols. We start with a rather elementary section containing examples and various classical models such as the Hans Lewy example. Next, we move forward with a quite easy discussion on the analysis of the first Poisson bracket of the imaginary and real part. The following sections are more involved; in particular we start a discussion on the geometry of condition (Ψ), with some known facts on flow-invariant sets, but we expose also the contribution of N. Dencker in the understanding of that geometric condition, with various inequalities satisfied by symbols. The next two sections are concerned respectively with the proof of the necessity of condition (Ψ) for local solvability and also with subelliptic estimates: on these two topics, we refer essentially to the existing literature, but we mention the results to hopefully provide the reader with some continuous overview of the subject. Then we enter into the discussion of estimates with loss of one derivative; we start with a detailed proof of the Beals-Fefferman result on local solvability with loss of one derivative under condition (P). Although this proof is classical, it seems useful to review its main arguments based on Calderón-Zygmund decompositions to understand how this type of cutting and stopping procedure works in a rather simple setting (at any rate simpler than in the section devoted to condition (Ψ)). We show, following the author's counterexample, that an estimate with loss of one derivative is not a consequence of condition (Ψ). Finally, we give a proof of an estimate with loss of $3/2$ derivatives under condition (Ψ), following the articles of N. Dencker [35] and the author's [98]. We end that chapter with a short historical account of solvability questions and also with a list of open questions.

There is also a lengthy appendix to this book. Some topics of this appendix are simply very classical material whose re-exposition might benefit to the reader by providing an immediate access to a reference for some calculations or formulas: it is the case of the first two sections of that appendix *Some elements of Fourier analysis, Some remarks of algebra* and also of the fourth one *On the symplectic and metaplectic groups*. Other parts of the appendix are devoted to technical questions, which would have impeded the reader in his progression: this is the case in particular of the very last section *More on symbolic calculus*.

It is our hope that the first two parts of the book are accessible to graduate students with a decent background in Analysis. The third chapter is directed more to researchers but should also be accessible to the readers able to get some good familiarity with the first two chapters, in which the main tools for the proofs of Chapter 3 are provided.

Acknowledgements. I wish to express my thanks to Jean-Michel Bony for numerous discussions on Sobolev spaces attached to a pseudo-differential calculus given by a metric on the phase space and also for many helpful indications on related matters; the entry [18] refers to these private discussions and files. For several months, the author had the privilege of exchanging several letters and files with Lars Hörmander on the topic of solvability. The author is most grateful for the help generously provided. These personal communications are referred to in the text as [78] and are important in all subsections of Section 3.7. It is my pleasure to thank the editors of this Birkhäuser series, Luigi Rodino and Man Wah Wong, for inviting me to write this book. Finally I wish to acknowledge gratefully the authorization of the *Annals of Mathematics* and of the *Publications of RIMS* to reproduce some parts of the articles [92] and [100].

Chapter 1

Basic Notions of Phase Space Analysis

1.1 Introduction to pseudo-differential operators

1.1.1 Prolegomena

A differential operator of order m on $\mathbb{R}^n$ can be written as

$$a(x, D) = \sum_{|\alpha| \le m} a_\alpha(x) D_x^\alpha,$$

where we have used the notation (4.1.4) for the multi-indices. Its *symbol* is a polynomial in the variable ξ and is defined as

$$a(x, \xi) = \sum_{|\alpha| \le m} a_\alpha(x) \xi^\alpha, \qquad \xi^\alpha = \xi_1^{\alpha_1} \dots \xi_n^{\alpha_n}.$$

We have the formula

$$(a(x, D)u)(x) = \int_{\mathbb{R}^n} e^{2i\pi x \cdot \xi} a(x, \xi) \hat{u}(\xi) d\xi, \qquad (1.1.1)$$

where $\hat{u}$ is the Fourier transform as defined in (4.1.1). It is possible to generalize the previous formula to the case where a is a tempered distribution on $\mathbb{R}^{2n}$.

Let u, v be in the Schwartz class $\mathscr{S}(\mathbb{R}^n)$. Then the function

$$\mathbb{R}^n \times \mathbb{R}^n \ni (x, \xi) \mapsto \hat{u}(\xi) \bar{v}(x) e^{2i\pi x \cdot \xi} = \Omega_{u,v}(x, \xi) \qquad (1.1.2)$$

belongs to $\mathscr{S}(\mathbb{R}^{2n})$ and the mapping $(u, v) \mapsto \Omega_{u,v}$ is sesquilinear continuous. Using this notation, we can provide the following definition.

Definition 1.1.1. Let $a \in \mathscr{S}'(\mathbb{R}^{2n})$ be a tempered distribution. We define the operator $a(x, D) : \mathscr{S}(\mathbb{R}^n) \longrightarrow \mathscr{S}^*(\mathbb{R}^n)$ by the formula

$$\langle a(x,D)u, v\rangle_{\mathscr{S}^*(\mathbb{R}^n), \mathscr{S}(\mathbb{R}^n)} = \prec a, \Omega_{u,v} \succ_{\mathscr{S}'(\mathbb{R}^{2n}), \mathscr{S}(\mathbb{R}^{2n})},$$

where $\mathscr{S}^*(\mathbb{R}^n)$ is the antidual of $\mathscr{S}(\mathbb{R}^n)$ (continuous antilinear forms). The distribution a is called the symbol of the operator $a(x, D)$.

N.B. The duality product $\langle u, v\rangle_{\mathscr{S}^*(\mathbb{R}^{2n}), \mathscr{S}(\mathbb{R}^{2n})}$, is linear in the variable u and antilinear in the variable v. We shall use the same notation for the dot product in the *complex* Hilbert space L^2 with the notation

$$\langle u, v\rangle_{L^2} = \int u(x)\overline{v(x)}dx.$$

The general rule that we shall follow is to always use the sesquilinear duality as above, except if specified otherwise. For the real duality, as in the left-hand side of the formula in Definition 1.1.1, we shall use the notation $\prec u, v \succ = \int u(x)v(x)dx$, e.g., for $u, v \in \mathscr{S}(\mathbb{R}^n)$.

Although the previous formula is quite general, since it allows us to *quantize*[1] any tempered distribution on $\mathbb{R}^{2n}$, it is not very useful, since we cannot compose this type of operators. We are in fact looking for an algebra of operators and the following theorem provides a simple example.

In the sequel we shall denote by $C_b^\infty(\mathbb{R}^{2n})$ the (Fréchet) space of C^∞ functions on $\mathbb{R}^{2n}$ which are bounded as well as all their derivatives.

Theorem 1.1.2. *Let $a \in C_b^\infty(\mathbb{R}^{2n})$. Then the operator $a(x, D)$ is continuous from $\mathscr{S}(\mathbb{R}^n)$ into itself.*

Proof. Using Definition 1.1.1, we have for $u, v \in \mathscr{S}(\mathbb{R}^n), a \in C_b^\infty(\mathbb{R}^{2n})$,

$$\langle a(x,D)u, v\rangle_{\mathscr{S}^*(\mathbb{R}^n), \mathscr{S}(\mathbb{R}^n)} = \iint e^{2i\pi x \cdot \xi} a(x, \xi)\hat{u}(\xi)\bar{v}(x)dx d\xi.$$

On the other hand the function $U(x) = \int e^{2i\pi x \cdot \xi} a(x, \xi)\hat{u}(\xi)d\xi$ is smooth and such that, for any multi-indices α, β,

$$x^\beta D_x^\alpha U(x) = (-1)^{|\beta|} \sum_{\alpha' + \alpha'' = \alpha} \frac{\alpha!}{\alpha'!\alpha''!} \int e^{2i\pi x \cdot \xi} D_\xi^\beta \big(\xi^{\alpha'} (D_x^{\alpha''} a)(x, \xi)\hat{u}(\xi)\big)d\xi$$

$$= (-1)^{|\beta|} \sum_{\alpha' + \alpha'' = \alpha} \frac{\alpha!}{\alpha'!\alpha''!} \int e^{2i\pi x \cdot \xi} D_\xi^\beta \big((D_x^{\alpha''} a)(x, \xi)\widehat{D^{\alpha'} u}(\xi)\big)d\xi$$

and thus

$$\sup_{x \in \mathbb{R}^n} |x^\beta D_x^\alpha U(x)| \leq \sum_{\substack{\alpha' + \alpha'' = \alpha \\ \beta' + \beta'' = \beta}} \frac{\alpha!}{\alpha'!\alpha''!} \frac{\beta!}{\beta'!\beta''!} \|D_\xi^{\beta'} D_x^{\alpha''} a\|_{L^\infty(\mathbb{R}^{2n})} \|D^{\beta''} \widehat{D^\alpha u}\|_{L^1(\mathbb{R}^n)}.$$

---

[1] We mean simply here that we are able to define a linear mapping from $\mathscr{S}'(\mathbb{R}^{2n})$ to the set of continuous operators from $\mathscr{S}(\mathbb{R}^n)$ to $\mathscr{S}'(\mathbb{R}^n)$.

Since the Fourier transform and ∂_{x_j} are continuous on $\mathscr{S}(\mathbb{R}^n)$, we get that the mapping $u \mapsto U$ is continuous from $\mathscr{S}(\mathbb{R}^n)$ into itself. The above defining formula for $a(x, D)$ ensures that $a(x, D)u = U$. $\qquad\square$

The Schwartz space $\mathscr{S}(\mathbb{R}^{2n})$ is not dense in the Fréchet space $C_b^\infty(\mathbb{R}^{2n})$ (e.g., $\forall \varphi \in \mathscr{S}(\mathbb{R}^{2n}), \sup_{x \in \mathbb{R}^{2n}} |1 - \varphi(x)| \geq 1$) but, in somewhat pedantic terms, one may say that this density is true for the bornology on $C_b^\infty(\mathbb{R}^{2n})$; in simpler terms, let a be a function in $C_b^\infty(\mathbb{R}^{2n})$ and take for instance

$$a_k(x, \xi) = a(x, \xi)e^{-(|x|^2 + |\xi|^2)k^{-2}}.$$

It is easy to see that each a_k belongs to $\mathscr{S}(\mathbb{R}^{2n})$, that the sequence (a_k) is bounded in $C_b^\infty(\mathbb{R}^{2n})$ and converges in $C^\infty(\mathbb{R}^{2n})$ to a. This type of density will be enough for the next lemma.

Lemma 1.1.3. *Let (a_k) be a sequence in $\mathscr{S}(\mathbb{R}^{2n})$ such that (a_k) is bounded in the Fréchet space $C_b^\infty(\mathbb{R}^{2n})$ and (a_k) converges in $C^\infty(\mathbb{R}^{2n})$ to a function a. Then a belongs to $C_b^\infty(\mathbb{R}^{2n})$ and for any $u \in \mathscr{S}(\mathbb{R}^n)$, the sequence $(a_k(x, D)u)$ converges to $a(x, D)u$ in $\mathscr{S}(\mathbb{R}^n)$.*

Proof. The fact that a belongs to $C_b^\infty(\mathbb{R}^{2n})$ is obvious. Using the identities in the proof of Theorem 1.1.2 we see that

$$x^\beta D_x^\alpha \big(a_k(x, D)u - a(x, D)u\big) = x^\beta D_x^\alpha \big((a_k - a)(x, D)u\big)$$

$$= (-1)^{|\beta|} \sum_{\substack{\alpha' + \alpha'' = \alpha \\ \beta' + \beta'' = \beta}} \frac{\alpha!}{\alpha'!\alpha''!} \frac{\beta!}{\beta'!\beta''!} \int e^{2i\pi x \cdot \xi} \big(D_\xi^{\beta'} D_x^{\alpha''}(a_k - a)\big)(x, \xi) D_\xi^{\beta''} \widehat{D^{\alpha'}u}(\xi) d\xi$$

$$= \sum_{\substack{\alpha' + \alpha'' = \alpha \\ \beta' + \beta'' = \beta}} \frac{\alpha!}{\alpha'!\alpha''!} \frac{\beta!}{\beta'!\beta''!} (1 + |x|^2)^{-1}$$

$$\times \int (1 + |D_\xi|^2) \big(e^{2i\pi x \cdot \xi}\big) \big(D_\xi^{\beta'} D_x^{\alpha''}(a_k - a)\big)(x, \xi) D_\xi^{\beta''} \widehat{D^{\alpha'}u}(\xi) d\xi$$

which is a (finite) sum of terms of type $V_k(x) = (1 + |x|^2)^{-1} \int e^{2i\pi x \cdot \xi} b_k(x, \xi) w_u(\xi) d\xi$ with the sequence (b_k) bounded in $C_b^\infty(\mathbb{R}^{2n})$ and converging to 0 in $C^\infty(\mathbb{R}^{2n})$, $u \mapsto w_u$ linear continuous from $\mathscr{S}(\mathbb{R}^n)$ into itself. As a consequence we get that, with R_1, R_2 positive parameters,

$$|V_k(x)| \leq \sup_{\substack{|x| \leq R_1 \\ |\xi| \leq R_2}} |b_k(x, \xi)| \int_{|\xi| \leq R_2} |w_u(\xi)| d\xi \mathbf{1}\{|x| \leq R_1\}$$

$$+ \int_{|\xi| \geq R_2} |w_u(\xi)| d\xi \sup_{k \in \mathbb{N}} \|b_k\|_{L^\infty(\mathbb{R}^{2n})} \mathbf{1}\{|x| \leq R_1\}$$

$$+ R_1^{-2} \mathbf{1}\{|x| \geq R_1\} \sup_{k \in \mathbb{N}} \|b_k\|_{L^\infty(\mathbb{R}^{2n})} \int |w_u(\xi)| d\xi,$$

implying

$$|V_k(x)| \leq \varepsilon_k(R_1, R_2) \int |w_u(\xi)| d\xi + \eta(R_2) \sup_{k \in \mathbb{N}} \|b_k\|_{L^\infty(\mathbb{R}^{2n})}$$

$$+ \theta(R_1) \sup_{k \in \mathbb{N}} \|b_k\|_{L^\infty(\mathbb{R}^{2n})} \int |w_u(\xi)| d\xi,$$

with $\lim_{k \to +\infty} \varepsilon_k(R_1, R_2) = 0, \lim_{R \to +\infty} \eta(R) = \lim_{R \to +\infty} \theta(R) = 0$. Thus we have for all positive R_1, R_2,

$$\limsup_{k \to +\infty} \|V_k\|_{L^\infty} \leq \eta(R_2) \sup_{k \in \mathbb{N}} \|b_k\|_{L^\infty(\mathbb{R}^{2n})} + \theta(R_1) \sup_{k \in \mathbb{N}} \|b_k\|_{L^\infty(\mathbb{R}^{2n})} \int |w_u(\xi)| d\xi,$$

entailing (by taking the limit when R_1, R_2 go to infinity) that $\lim_{k \to +\infty} \|V_k\|_{L^\infty} = 0$ which gives the result of the lemma. $\qquad\square$

Theorem 1.1.4. *Let $a \in C_b^\infty(\mathbb{R}^{2n})$: the operator $a(x, D)$ is bounded on $L^2(\mathbb{R}^n)$.*

Proof. Since $\mathscr{S}(\mathbb{R}^n)$ is dense in $L^2(\mathbb{R}^n)$, it is enough to prove that there exists a constant C such that for all $u, v \in \mathscr{S}(\mathbb{R}^n)$,

$$|\langle a(x, D)u, v \rangle_{\mathscr{S}^*(\mathbb{R}^n), \mathscr{S}(\mathbb{R}^n)}| \leq C \|u\|_{L^2(\mathbb{R}^n)} \|v\|_{L^2(\mathbb{R}^n)}.$$

We introduce the polynomial on $\mathbb{R}^n$ defined by $P_k(t) = (1 + |t|^2)^{k/2}$, where $k \in 2\mathbb{N}$, and the function

$$W_u(x, \xi) = \int u(y) P_k(x - y)^{-1} e^{-2i\pi y \cdot \xi} dy.$$

The function W_u is the partial Fourier transform of the function $\mathbb{R}^n \times \mathbb{R}^n \ni (x, y) \mapsto u(y) P_k(x - y)^{-1}$ and if $k > n/2$ (we assume this in the sequel), we obtain that $\|W_u\|_{L^2(\mathbb{R}^{2n})} = c_k \|u\|_{L^2(\mathbb{R}^n)}$. Moreover, since $u \in \mathscr{S}(\mathbb{R}^n)$, the function W_u belongs to $C^\infty(\mathbb{R}^{2n})$ and satisfies, for all multi-indices α, β, γ,

$$\sup_{(x, \xi) \in \mathbb{R}^{2n}} P_k(x) \xi^\gamma |(\partial_x^\alpha \partial_\xi^\beta W_u)(x, \xi)| < \infty.$$

In fact we have

$$\xi^\gamma(\partial_x^\alpha \partial_\xi^\beta W_u)(x, \xi) = \int \overbrace{u(y)(-2i\pi y)^\beta}^{\in \mathscr{S}(\mathbb{R}^n)} \partial^\alpha(1/P_k)(x - y)(-1)^{|\gamma|} D_y^\gamma(e^{-2i\pi y \cdot \xi}) dy$$

$$= \sum_{\gamma' + \gamma'' = \gamma} \frac{\gamma!}{\gamma'! \gamma''!} \int D_y^{\gamma'}\left(u(y)(-2i\pi y)^\beta\right)(-2i\pi)^{-|\gamma''|}$$

$$\times \partial^{\gamma'' + \alpha}(1/P_k)(x - y)(e^{-2i\pi y \cdot \xi}) dy$$

and

$$|\partial^\alpha(1/P_k)(x - y)| \leq C_{\alpha,k}(1 + |x - y|)^{-k} \leq C_{\alpha,k}(1 + |x|)^{-k}(1 + |y|)^k.$$

From Definition 1.1.1, we have

$$\langle a(x,D)u,v\rangle_{\mathscr{S}^*(\mathbb{R}^n),\mathscr{S}(\mathbb{R}^n)} = \iint_{\mathbb{R}^n\times\mathbb{R}^n} e^{2i\pi x\cdot\xi}a(x,\xi)\hat{u}(\xi)\bar{v}(x)dxd\xi,$$

and we obtain, using an integration by parts justified by the regularity and decay of the functions W above,

$$\langle a(x,D)u,v\rangle$$
$$= \iint a(x,\xi)P_k(D_\xi)\left(\int u(y)P_k(x-y)^{-1}e^{2i\pi(x-y)\cdot\xi}dy\right)\bar{v}(x)dxd\xi$$
$$= \iint a(x,\xi)P_k(D_\xi)\Big(\underbrace{e^{2i\pi x\cdot\xi}W_u(x,\xi)\bar{v}(x)}_{\in\mathscr{S}(\mathbb{R}^{2n})}\Big)dxd\xi$$
$$= \iint (P_k(D_\xi)a)(x,\xi)W_u(x,\xi)P_k(D_x)\left(\int e^{2i\pi x\cdot(\xi-\eta)}P_k(\xi-\eta)^{-1}\overline{\hat{v}(\eta)}d\eta\right)dxd\xi$$
$$= \iint (P_k(D_\xi)a)(x,\xi)W_u(x,\xi)P_k(D_x)\left(W_{\bar{\hat{v}}}(\xi,x)e^{2i\pi x\cdot\xi}\right)dxd\xi$$
$$= \sum_{0\le l\le k/2} C^l_{k/2}\iint |D_x|^{2l}\Big((P_k(D_\xi)a)(x,\xi)W_u(x,\xi)\Big)W_{\bar{\hat{v}}}(\xi,x)e^{2i\pi x\cdot\xi}dxd\xi$$
$$= \sum_{\substack{|\alpha|\le k\\ |\beta|+|\gamma|\le k}} c_{\alpha\beta\gamma}\iint \underbrace{(D_\xi^\alpha D_x^\beta a)(x,\xi)}_{\text{bounded}} D_x^\gamma(W_u)(x,\xi)\ \underbrace{W_{\bar{\hat{v}}}(\xi,x)}_{\substack{\in L^2(\mathbb{R}^{2n})\text{ with norm}\\ c_k\|v\|_{L^2(\mathbb{R}^n)}}}\ e^{2i\pi x\cdot\xi}dxd\xi.$$

Checking now the x-derivatives of W_u, we see that

$$D_x^\gamma(W_u)(x,\xi) = \int u(y)D^\gamma(1/P_k)(x-y)e^{-2i\pi y\cdot\xi}dy,$$

and since $D^\gamma(1/P_k)$ belongs to $L^2(\mathbb{R}^n)$ (since $k>n/2$), we get that the $L^2(\mathbb{R}^{2n})$ norm of $D_x^\gamma(W_u)$ is bounded above by $c_\gamma\|u\|_{L^2(\mathbb{R}^n)}$. Using the Cauchy-Schwarz inequality, we obtain that

$$|\langle a(x,D)u,v\rangle| \le \sum_{\substack{|\alpha|\le k\\ |\beta|+|\gamma|\le k}} c_{\alpha\beta\gamma}\|\partial_\xi^\alpha\partial_x^\beta a\|_{L^\infty(\mathbb{R}^{2n})}\|D_x^\gamma W_u\|_{L^2(\mathbb{R}^{2n})}\|W_{\bar{\hat{v}}}\|_{L^2(\mathbb{R}^{2n})}$$
$$\le C_n\|u\|_{L^2(\mathbb{R}^n)}\|v\|_{L^2(\mathbb{R}^n)}\sup_{\substack{|\alpha|\le k\\ |\beta|\le k}}\|\partial_\xi^\alpha\partial_x^\beta a\|_{L^\infty(\mathbb{R}^{2n})},$$

where C_n depends only on n and $2\mathbb{N}\ni k>n/2$, which is the sought result. $\qquad\square$

N.B. The number of derivatives needed to control the $\mathscr{L}(L^2(\mathbb{R}^n))$ (the linear continuous operators on $L^2(\mathbb{R}^n)$) norm of $a(x,D)$ can be slightly improved (cf. [22], [28]). The elementary proof above is due to I.L. Hwang ([79]).

The next theorem gives us our first algebra of pseudo-differential operators.

Theorem 1.1.5. *Let a, b be in $C_b^\infty(\mathbb{R}^{2n})$. Then the composition $a(x, D)b(x, D)$ makes sense as a bounded operator on $L^2(\mathbb{R}^n)$ (also as a continuous operator from $\mathscr{S}(\mathbb{R}^n)$ into itself), and $a(x, D)b(x, D) = (a \diamond b)(x, D)$ where $a \diamond b$ belongs to $C_b^\infty(\mathbb{R}^{2n})$ and is given by the formula*

$$(a \diamond b)(x, \xi) = (\exp 2i\pi D_y \cdot D_\eta)(a(x, \xi + \eta)b(y + x, \xi))_{|y=0, \eta=0}, \qquad (1.1.3)$$

$$(a \diamond b)(x, \xi) = \iint e^{-2i\pi y \cdot \eta} a(x, \xi + \eta) b(y + x, \xi) dy d\eta, \qquad (1.1.4)$$

when a and b belong to $\mathscr{S}(\mathbb{R}^{2n})$. The mapping $a, b \mapsto a \diamond b$ is continuous for the topology of Fréchet space of $C_b^\infty(\mathbb{R}^{2n})$. Also if $(a_k), (b_k)$ are sequences of functions in $\mathscr{S}(\mathbb{R}^{2n})$, bounded in $C_b^\infty(\mathbb{R}^{2n})$, converging in $C^\infty(\mathbb{R}^{2n})$ respectively to a, b, then a and b belong to $C_b^\infty(\mathbb{R}^{2n})$, the sequence $(a_k \diamond b_k)$ is bounded in $C_b^\infty(\mathbb{R}^{2n})$ and converges in $C^\infty(\mathbb{R}^{2n})$ to $a \diamond b$.

Remark 1.1.6. From Lemma 4.1.2, we know that the operator $e^{2i\pi D_y \cdot D_\eta}$ is an isomorphism of $C_b^\infty(\mathbb{R}^{2n})$, which gives a meaning to the formula (1.1.3), since for $a, b \in C_b^\infty(\mathbb{R}^{2n})$, (x, ξ) given in $\mathbb{R}^{2n}$, the function $(y, \eta) \mapsto a(x, \xi + \eta)b(y + x, \xi) = C_{x, \xi}(y, \eta)$ belongs to $C_b^\infty(\mathbb{R}^{2n})$ as well as $JC_{x, \xi}$ and we can take the value of the latter at $(y, \eta) = (0, 0)$.

Proof. Let us first assume that $a, b \in \mathscr{S}(\mathbb{R}^{2n})$. The kernels k_a, k_b of the operators $a(x, D), b(x, D)$ belong also to $\mathscr{S}(\mathbb{R}^{2n})$ and the kernel k_c of $a(x, D)b(x, D)$ is given by (we use Fubini's theorem)

$$k(x, y) = \int k_a(x, z) k_b(z, y) dz = \iiint a(x, \xi) e^{2i\pi(x-z) \cdot \xi} b(z, \zeta) e^{2i\pi(z-y) \cdot \zeta} d\zeta d\xi dz.$$

The function k belongs also to $\mathscr{S}(\mathbb{R}^{2n})$ and we get, for $u, v \in \mathscr{S}(\mathbb{R}^n)$,

$$\langle a(x, D)b(x, D)u, v \rangle_{L^2(\mathbb{R}^n)}$$

$$= \iiiint\!\!\int a(x, \xi) e^{2i\pi(x-z) \cdot \xi} b(z, \zeta) e^{2i\pi(z-y) \cdot \zeta} u(y) \bar{v}(x) d\zeta d\xi dz dy dx.$$

$$= \iiint\!\!\int a(x, \xi) e^{2i\pi(x-z) \cdot \xi} b(z, \zeta) e^{2i\pi z \cdot \zeta} \hat{u}(\zeta) d\zeta d\xi dz \bar{v}(x) dx.$$

$$= \iiint\!\!\int a(x, \xi) e^{2i\pi(x-z) \cdot \xi} b(z, \zeta) e^{2i\pi(z-x) \cdot \zeta} d\xi dz e^{2i\pi x \cdot \zeta} \hat{u}(\zeta) d\zeta \bar{v}(x) dx.$$

$$= \iint c(x, \zeta) e^{2i\pi x \cdot \zeta} \hat{u}(\zeta) d\zeta \bar{v}(x) dx,$$

with

$$c(x, \zeta) = \iint a(x, \xi) e^{2i\pi(x-z) \cdot (\xi-\zeta)} b(z, \zeta) d\xi dz$$

$$= \iint a(x, \xi + \zeta) e^{-2i\pi z \cdot \xi} b(z + x, \zeta) d\xi dz, \qquad (1.1.5)$$

which is indeed (1.1.4). With $c = a \diamond b$ given by (1.1.4), using that $a, b \in \mathscr{S}(\mathbb{R}^{2n})$ we get, using the notation (4.1.4) and $P_k(t) = (1 + |t|^2)^{1/2}, k \in 2\mathbb{N}$,

$$c(x, \xi) = \iint P_k(D_\eta)\left(e^{-2i\pi y \cdot \eta}\right) P_k(y)^{-1} a(x, \xi + \eta) b(y + x, \xi) dy d\eta$$

$$= \iint e^{-2i\pi y \cdot \eta} P_k(y)^{-1} (P_k(D_2)a)(x, \xi + \eta) b(y + x, \xi) dy d\eta$$

$$= \iint P_k(D_y)\left(e^{-2i\pi y \cdot \eta}\right) P_k(\eta)^{-1} P_k(y)^{-1} (P_k(D_2)a)(x, \xi + \eta) b(y + x, \xi) dy d\eta$$

$$= \sum_{0 \le l \le k/2} C_{k/2}^l \iint e^{-2i\pi y \cdot \eta} |D_y|^{2l} \left(P_k(y)^{-1} b(y + x, \xi)\right)$$

$$\times P_k(\eta)^{-1} (P_k(D_2)a)(x, \xi + \eta) dy d\eta. \qquad (1.1.6)$$

We denote by $a \tilde{\diamond} b$ the right-hand side of the previous formula and we note that, when $k > n$, it makes sense as well for $a, b \in C_b^\infty(\mathbb{R}^{2n})$, since $|\partial_t^\alpha (1/P_k)(t)| \le C_{\alpha,k}(1 + |t|)^{-k}$. We already know that $a \diamond b = a \tilde{\diamond} b$ for a, b in the Schwartz class and we want to prove that it is also true for $a, b \in C_b^\infty(\mathbb{R}^{2n})$. Choosing an even $k > n$ (take $k = n + 1$ or $n + 2$), we also get

$$\|a \tilde{\diamond} b\|_{L^\infty(\mathbb{R}^{2n})} \le C_n \sup_{|\alpha| \le n+2} \|\partial_\xi^\alpha a\|_{L^\infty(\mathbb{R}^{2n})} \sup_{|\beta| \le n+2} \|\partial_x^\beta b\|_{L^\infty(\mathbb{R}^{2n})}.$$

Moreover, we note from (1.1.6) that

$$\partial_{\xi_j}(a \tilde{\diamond} b) = (\partial_{\xi_j} a) \tilde{\diamond} b + a \tilde{\diamond} (\partial_{\xi_j} b), \partial_{x_j}(a \tilde{\diamond} b) = (\partial_{x_j} a) \tilde{\diamond} b + a \tilde{\diamond} (\partial_{x_j} b)$$

and as a result

$$\|\partial_\xi^\alpha \partial_x^\beta (a \tilde{\diamond} b)\|_{L^\infty(\mathbb{R}^{2n})}$$

$$\le C_{n,\alpha,\beta} \sup_{\substack{|\alpha'| \le n+2, |\beta'| \le n+2 \\ \alpha'' + \alpha''' = \alpha, \ \beta'' + \beta''' = \beta}} \|\partial_\xi^{\alpha' + \alpha''} \partial_x^{\beta''} a\|_{L^\infty(\mathbb{R}^{2n})} \|\partial_x^{\beta' + \beta'''} \partial_\xi^{\alpha'''} b\|_{L^\infty(\mathbb{R}^{2n})},$$

$$(1.1.7)$$

which gives also the continuity of the bilinear mapping $C_b^\infty(\mathbb{R}^{2n}) \times C_b^\infty(\mathbb{R}^{2n}) \ni (a, b) \mapsto a \tilde{\diamond} b \in C_b^\infty(\mathbb{R}^{2n})$. We have for $u, v \in \mathscr{S}(\mathbb{R}^n), a, b \in C_b^\infty(\mathbb{R}^{2n})$,

$$a_k(x, \xi) = e^{-(|x|^2 + |\xi|^2)/k^2} a(x, \xi), \quad b_k(x, \xi) = e^{-(|x|^2 + |\xi|^2)/k^2} b(x, \xi),$$

from Lemma 1.1.3 and Theorem 1.1.2, with limits in $\mathscr{S}(\mathbb{R}^n)$,

$$a(x, D)b(x, D)u = \lim_k a_k(x, D)b(x, D)u = \lim_k \left(\lim_l a_k(x, D)b_l(x, D)u\right),$$

and thus, with $\Omega_{u,v}(x, \xi) = e^{2i\pi x \cdot \xi} \hat{u}(\xi) \bar{v}(x)$ (which belongs to $\mathscr{S}(\mathbb{R}^{2n})$),

$$\langle a(x, D)b(x, D)u, v \rangle_{L^2} = \lim_k \left(\lim_l \langle (a_k \diamond b_l)(x, D)u, v \rangle\right)$$

$$= \lim_k \left(\lim_l \iint (a_k \diamond b_l)(x, \xi) \Omega_{u,v}(x, \xi) dx d\xi\right) = \iint (a \tilde{\diamond} b)(x, \xi) \Omega_{u,v}(x, \xi) dx d\xi,$$

which gives indeed $a(x, D)b(x, D) = (a\tilde{\diamond}b)(x, D)$. This property gives at once the continuity properties stated at the end of the theorem, since the weak continuity property follows immediately from (1.1.6) and the Lebesgue dominated convergence theorem, whereas the Fréchet continuity follows from (1.1.7). Moreover, with the same notation as above, we have with

$$C_{x,\xi}^{(a,b)}(y, \eta) = a(x, \xi + \eta)b(y + x, \xi)$$

(see Remark 1.1.6) for each $(x, \xi) \in \mathbb{R}^{2n}$,

$$(JC_{x,\xi}^{(a,b)})(0,0) = \lim_k(JC_{x,\xi}^{(a_k,b_k)})(0,0) = \lim_k\big((a_k \diamond b_k)(x,\xi)\big) = (a\tilde{\diamond}b)(x,\xi)$$

which proves (1.1.3). The proof of the theorem is complete.   □

Definition 1.1.7. Let $A : \mathscr{S}(\mathbb{R}^n) \longrightarrow \mathscr{S}'(\mathbb{R}^n)$ be a linear operator. The adjoint operator $A^* : \mathscr{S}(\mathbb{R}^n) \longrightarrow \mathscr{S}'(\mathbb{R}^n)$ is defined by

$$\langle A^*u, v\rangle_{\mathscr{S}^*(\mathbb{R}^n),\mathscr{S}(\mathbb{R}^n)} = \overline{\langle Av, u\rangle}_{\mathscr{S}^*(\mathbb{R}^n),\mathscr{S}(\mathbb{R}^n)},$$

where $\mathscr{S}^*(\mathbb{R}^n)$ is the antidual of $\mathscr{S}(\mathbb{R}^n)$ (continuous antilinear forms).

Theorem 1.1.8. *Let $a \in \mathscr{S}'(\mathbb{R}^{2n})$ and $A = a(x, D)$ be given by Definition 1.1.1. Then the operator A^* is equal to $a^*(x, D)$, where $a^* = J\bar{a}$ (J is given in Lemma 4.1.2). If a belongs to $C_b^\infty(\mathbb{R}^{2n})$, $a^* = J\bar{a} \in C_b^\infty(\mathbb{R}^{2n})$ and the mapping $a \mapsto a^*$ is continuous from $C_b^\infty(\mathbb{R}^{2n})$ into itself.*

Proof. According to Definitions 1.1.7 and 1.1.1, we have for $u, v \in \mathscr{S}(\mathbb{R}^n)$, with $\Omega_{v,u}(x,\xi) = e^{2i\pi x\cdot\xi}\hat{v}(\xi)\bar{u}(x)$,

$$\langle A^*u, v\rangle_{\mathscr{S}^*(\mathbb{R}^n),\mathscr{S}(\mathbb{R}^n)} = \overline{\langle Av, u\rangle}_{\mathscr{S}^*(\mathbb{R}^n),\mathscr{S}(\mathbb{R}^n)} = \overline{\prec a, \Omega_{v,u} \succ}_{\mathscr{S}'(\mathbb{R}^{2n}),\mathscr{S}(\mathbb{R}^{2n})}$$

$$= \prec \bar{a}, \overline{\Omega}_{v,u} \succ_{\mathscr{S}'(\mathbb{R}^{2n}),\mathscr{S}(\mathbb{R}^{2n})}.$$

On the other hand, we have

$$\big(J^{-1}(\overline{\Omega}_{v,u})\big)(x,\xi) = \iint e^{2i\pi(x-y)\cdot(\xi-\eta)}e^{-2i\pi y\cdot\eta}\overline{\hat{v}}(\eta)u(y)dyd\eta$$

$$= \bar{v}(x)e^{2i\pi x\cdot\xi}\hat{u}(\xi) = \Omega_{u,v}(x,\xi),$$

so that, using (4.1.16), we get

$$\langle A^*u, v\rangle_{\mathscr{S}^*(\mathbb{R}^n),\mathscr{S}(\mathbb{R}^n)} = \prec \bar{a}, J\Omega_{u,v} \succ_{\mathscr{S}'(\mathbb{R}^{2n}),\mathscr{S}(\mathbb{R}^{2n})} = \prec J\bar{a}, \Omega_{u,v} \succ_{\mathscr{S}'(\mathbb{R}^{2n}),\mathscr{S}(\mathbb{R}^{2n})}$$

and finally $A^* = (J\bar{a})(x, D)$. The last statement in the theorem follows from Lemma 4.1.2.   □

Comment. In this introductory section, we have seen a very general definition of quantization (Definition 1.1.1), an easy $\mathscr{S}$ continuity theorem (Theorem 1.1.2), a trickier L^2-boundedness result (Theorem 1.1.4), a composition formula (Theorem 1.1.5) and an expression for the adjoint (Definition 1.1.7). These five steps are somewhat typical of the construction of a pseudo-differential calculus and we shall see many different examples of this situation. The above prolegomena provide a quite explicit and elementary approach to the construction of an algebra of pseudo-differential operators in a rather difficult framework, since we did not use any asymptotic calculus and did not have at our disposal a "small parameter". The proofs and simple methods that we used here will be useful later as well as many of the results.

1.1.2 Quantization formulas

We have already seen in Definition 1.1.1 and in the formula (1.1.1) a way to associate to a tempered distribution $a \in \mathscr{S}'(\mathbb{R}^{2n})$ an operator from $\mathscr{S}(\mathbb{R}^n)$ to $\mathscr{S}'(\mathbb{R}^n)$. This question of quantization has of course many links with quantum mechanics and we want here to study some properties of various quantization formulas, such as the Weyl quantization and the Feynman formula along with several variations around these examples. We are given a function a defined on the phase space $\mathbb{R}^n \times \mathbb{R}^n$ (a is a "Hamiltonian") and we wish to associate to this function an operator. For instance, we may introduce the one-parameter formulas, for $t \in \mathbb{R}$,

$$(\mathrm{op}_t \, a)u(x) = \iint e^{2i\pi(x-y)\cdot\xi} a\big((1-t)x + ty, \xi\big)u(y)dyd\xi. \tag{1.1.8}$$

When $t = 0$, we recognize the standard quantization introduced in Definition 1.1.1, quantizing $a(x)\xi_j$ in $a(x)D_{x_j}$ (see (4.1.4)). However, one may wish to multiply first and take the derivatives afterwards: this is what the choice $t = 1$ does, quantizing $a(x)\xi_j$ in $D_{x_j}a(x)$. The more symmetrical choice $t = 1/2$ was made by Hermann Weyl [150]: we have

$$(\mathrm{op}_{\frac{1}{2}} \, a)u(x) = \iint e^{2i\pi(x-y)\cdot\xi} a\big(\frac{x+y}{2}, \xi\big)u(y)dyd\xi, \tag{1.1.9}$$

and thus $\mathrm{op}_{\frac{1}{2}}(a(x)\xi_j) = \frac{1}{2}\big(a(x)D_{x_j} + D_{x_j}a(x)\big)$. This quantization is widely used in quantum mechanics, because a real-valued Hamiltonian gets quantized by a (formally) selfadjoint operator. We shall see that the most important property of that quantization remains its symplectic invariance, which will be studied in detail in Chapter 2; a different symmetrical choice was made by Richard Feynman who used the formula

$$\iint e^{2i\pi(x-y)\cdot\xi} \, \big(a(x,\xi) + a(y,\xi)\big)\frac{1}{2}u(y)dyd\xi, \tag{1.1.10}$$

keeping the selfadjointness of real Hamiltonians, but losing the symplectic invariance (see Section 4.4.3). The reader may be confused by the fact that we did not bother about the convergence of the integrals above. Before providing a definition, we may assume that $a \in \mathscr{S}(\mathbb{R}^{2n}), u, v \in \mathscr{S}(\mathbb{R}^n), t \in \mathbb{R}$ and compute

$$\langle (\mathrm{op}_t a)u, v \rangle = \iiint a\big((1-t)x + ty, \xi\big)e^{2i\pi(x-y)\cdot\xi}u(y)\bar{v}(x)dyd\xi dx$$

$$= \iiint a(z,\xi)e^{-2i\pi s\cdot\xi}u(z + (1-t)s)\bar{v}(z - ts)dzd\xi ds$$

$$= \iiint a(x,\xi)e^{-2i\pi z\cdot\xi}u(x + (1-t)z)\bar{v}(x - tz)dxd\xi dz,$$

so that with

$$\Omega_{u,v}(t)(x,\xi) = \int e^{-2i\pi z\cdot\xi}u(x + (1-t)z)\bar{v}(x - tz)dz, \qquad (1.1.11)$$

which is easily seen[2] to be in $\mathscr{S}(\mathbb{R}^{2n})$ when $u, v \in \mathscr{S}(\mathbb{R}^n)$, we can give the following definition.[3]

Definition 1.1.9. Let $a \in \mathscr{S}'(\mathbb{R}^{2n})$ be a tempered distribution and $t \in \mathbb{R}$. We define the operator $\mathrm{op}_t a : \mathscr{S}(\mathbb{R}^n) \longrightarrow \mathscr{S}^*(\mathbb{R}^n)$ by the formula

$$\langle (\mathrm{op}_t a)u, v \rangle_{\mathscr{S}^*(\mathbb{R}^n),\mathscr{S}(\mathbb{R}^n)} = \prec a, \Omega_{u,v}(t) \succ_{\mathscr{S}'(\mathbb{R}^{2n}),\mathscr{S}(\mathbb{R}^{2n})},$$

where $\mathscr{S}^*(\mathbb{R}^n)$ is the antidual of $\mathscr{S}(\mathbb{R}^n)$ (continuous antilinear forms).

Proposition 1.1.10. *Let $a \in \mathscr{S}'(\mathbb{R}^{2n})$ be a tempered distribution and $t \in \mathbb{R}$. We have*

$$\mathrm{op}_t a = \mathrm{op}_0(J^t a) = (J^t a)(x, D),$$

with J^t defined in Lemma 4.1.2.

Proof. Let $u, v \in \mathscr{S}(\mathbb{R}^n)$. With the $\mathscr{S}(\mathbb{R}^{2n})$ function $\Omega_{u,v}(t)$ given above, we have for $t \neq 0$,

$$\big(J^t\Omega_{u,v}(0)\big)(x,\xi) = |t|^{-n}\iint e^{-2i\pi t^{-1}(x-y)\cdot(\xi-\eta)}\Omega_{u,v}(0)(y,\eta)dyd\eta$$

$$= |t|^{-n}\iint e^{-2i\pi t^{-1}(x-y)\cdot(\xi-\eta)}\hat{u}(\eta)\bar{v}(y)e^{2i\pi y\cdot\eta}dyd\eta$$

$$= \iint e^{-2i\pi z\cdot(\xi-\eta)}\hat{u}(\eta)\bar{v}(x - tz)e^{2i\pi(x-tz)\cdot\eta}dzd\eta$$

---

[2]In fact the linear mapping $\mathbb{R}^n \times \mathbb{R}^n \ni (x, z) \mapsto (x - tz, x + (1-t)z)$ has determinant 1 and $\Omega_{u,v}(t)$ appears as the partial Fourier transform of the function $\mathbb{R}^n \times \mathbb{R}^n \ni (x, z) \mapsto \bar{v}(x - tz)u(x + (1-t)z)$, which is in the Schwartz class.

[3]The reader can check that this is consistent with Definition 1.1.1.

$$= \int e^{-2i\pi z \cdot \xi} u(x + (1-t)z) \bar{v}(x - tz) dz = \Omega_{u,v}(t)(x, \xi), \qquad (1.1.12)$$

so that

$$\begin{aligned}
\langle (\mathrm{op}_t a) u, v \rangle_{\mathscr{S}^*(\mathbb{R}^n), \mathscr{S}(\mathbb{R}^n)} &= \prec a, \Omega_{u,v}(t) \succ_{\mathscr{S}'(\mathbb{R}^{2n}), \mathscr{S}(\mathbb{R}^{2n})} && \text{(Definition 1.1.9)} \\
&= \prec a, J^t \Omega_{u,v}(0) \succ_{\mathscr{S}'(\mathbb{R}^{2n}), \mathscr{S}(\mathbb{R}^{2n})} && \text{(property (1.1.12))} \\
&= \prec J^t a, \Omega_{u,v}(0) \succ_{\mathscr{S}'(\mathbb{R}^{2n}), \mathscr{S}(\mathbb{R}^{2n})} && \text{(easy identity for } J^t) \\
&= \langle (J^t a)(x, D) u, v \rangle_{\mathscr{S}^*(\mathbb{R}^n), \mathscr{S}(\mathbb{R}^n)} && \text{(Definition 1.1.1),}
\end{aligned}$$

completing the proof. $\qquad \square$

Remark 1.1.11. Theorem 1.1.8 and the previous proposition give in particular that $a(x, D)^* = \mathrm{op}_1(\bar{a}) = (J\bar{a})(x, D)$, a formula which in fact motivates the study of the group J^t. On the other hand, using the Weyl quantization simplifies somewhat the matter of taking adjoints since we have, using Remark 4.1.3,

$$\left(\mathrm{op}_{1/2}(a) \right)^* = \left(\mathrm{op}_0(J^{1/2} a) \right)^* = \mathrm{op}_0(J(\overline{J^{1/2} a})) = \mathrm{op}_0(J^{1/2} \bar{a}) = \mathrm{op}_{1/2}(\bar{a})$$

and in particular if a is real-valued, $\mathrm{op}_{1/2}(a)$ is formally selfadjoint. The Feynman formula as displayed in (1.1.10) amounts to quantizing the Hamiltonian a by

$$\frac{1}{2} \mathrm{op}_0(a + Ja)$$

and we see that $\left(\mathrm{op}_0(a + Ja) \right)^* = \mathrm{op}_0(J\bar{a} + J(\overline{Ja})) = \mathrm{op}_0(J\bar{a} + \bar{a})$, which also provides selfadjointness for real-valued Hamiltonians.

Lemma 1.1.12. *Let* $a \in \mathscr{S}(\mathbb{R}^{2n})$. *Then for all* $t \in \mathbb{R}$, $\mathrm{op}_t(a)$ *is a continuous mapping from* $\mathscr{S}'(\mathbb{R}^n)$ *into* $\mathscr{S}(\mathbb{R}^n)$.

Proof. Let $a \in \mathscr{S}(\mathbb{R}^{2n})$: we have for $u \in \mathscr{S}'(\mathbb{R}^n)$, $A = a(x, D)$,

$$x^\beta (D_x^\alpha A u)(x) = \sum_{\alpha' + \alpha'' = \alpha} \frac{1}{\alpha'! \alpha''!} \langle \hat{u}(\xi), e^{2i\pi x \cdot \xi} \xi^{\alpha'} x^\beta (D_x^{\alpha''} a)(x, \xi) \rangle_{\mathscr{S}'(\mathbb{R}^n_\xi), \mathscr{S}(\mathbb{R}^n_\xi)},$$

so that $A u \in \mathscr{S}(\mathbb{R}^n)$ and the same property holds for $\mathrm{op}_t(a)$ since J^t is an isomorphism of $\mathscr{S}'(\mathbb{R}^{2n})$. $\qquad \square$

1.1.3 The $S_{1,0}^m$ class of symbols

Differential operators on $\mathbb{R}^n$ with smooth coefficients are given by a formula (see (1.1.1))

$$a(x, D) u = \sum_{|\alpha| \le m} a_\alpha(x) D_x^\alpha$$

where the a_α are smooth functions. Assuming some behaviour at infinity for the a_α, we may require that they are $C_b^\infty(\mathbb{R}^n)$ (see page 2) and a natural generalization is to consider operators $a(x, D)$ with a symbol a of type $S_{1,0}^m$, i.e., smooth functions on $\mathbb{R}^{2n}$ satisfying

$$|(\partial_\xi^\alpha \partial_x^\beta a)(x, \xi)| \leq C_{\alpha\beta} \langle\xi\rangle^{m-|\alpha|}, \quad \langle\xi\rangle = (1 + |\xi|^2)^{1/2}. \tag{1.1.13}$$

The best constants $C_{\alpha\beta}$ in (1.1.13) are the semi-norms of a in the Fréchet space $S_{1,0}^m$. We can define, for $a \in S_{1,0}^m, k \in \mathbb{N}$,

$$\gamma_{k,m}(a) = \sup_{(x,\xi)\in\mathbb{R}^{2n}, |\alpha|+|\beta|\leq k} |(\partial_\xi^\alpha \partial_x^\beta a)(x, \xi)| \langle\xi\rangle^{-m+|\alpha|}. \tag{1.1.14}$$

Example. The function $\langle\xi\rangle^m$ belongs to $S_{1,0}^m$: the function $\mathbb{R} \times \mathbb{R}^n \ni (\tau, \xi) \mapsto (\tau^2 + |\xi|^2)^{m/2}$ is (positively) homogeneous of degree m on $\mathbb{R}^{n+1}\backslash\{0\}$, and thus $\partial_\xi^\alpha((\tau^2 + |\xi|^2)^{m/2})$ is homogeneous of degree $m - |\alpha|$ and bounded above by

$$C_\alpha(\tau^2 + |\xi|^2)^{\frac{m-|\alpha|}{2}}.$$

Since the restriction to $\tau = 1$ and the derivation with respect to ξ commute, it gives the answer.

We shall see that the class of operators $\mathrm{op}(S_{1,0}^m)$ is suitable ($\mathrm{op}(b)$ is $\mathrm{op}_0 b$, see Proposition 1.1.10) to invert elliptic operators, and useful for the study of singularities of solutions of PDE. We see that the elements of $S_{1,0}^m$ are temperate distributions, so that the operator $a(x, D)$ makes sense, according to Definition 1.1.1. We have also the following result.

Theorem 1.1.13. *Let $m \in \mathbb{R}$ and $a \in S_{1,0}^m$. Then the operator $a(x, D)$ is continuous from $\mathscr{S}(\mathbb{R}^n)$ into itself.*

Proof. With $\langle D\rangle = \mathrm{op}(\langle\xi\rangle)$, we have $a(x, D) = \mathrm{op}(a(x, \xi)\langle\xi\rangle^{-m})\langle D\rangle^m$. The function $a(x, \xi)\langle\xi\rangle^{-m}$ belongs to $C_b^\infty(\mathbb{R}^{2n})$ so that we can use Theorem 1.1.2 and the fact that $\langle D\rangle^m$ is continuous on $\mathscr{S}(\mathbb{R}^n)$ to get the result. □

Theorem 1.1.14. *Let $a \in S_{1,0}^0$. Then the operator $a(x, D)$ is bounded on $L^2(\mathbb{R}^n)$.*

Proof. Since $S_{1,0}^0 \subset C_b^\infty(\mathbb{R}^{2n})$, it follows from Theorem 1.1.4. □

Theorem 1.1.15. *Let m_1, m_2 be real numbers and $a_1 \in S_{1,0}^{m_1}, a_2 \in S_{1,0}^{m_2}$. Then the composition $a_1(x, D)a_2(x, D)$ makes sense as a continuous operator from $\mathscr{S}(\mathbb{R}^n)$ into itself and $a_1(x, D)a_2(x, D) = (a_1 \diamond a_2)(x, D)$ where $a_1 \diamond a_2$ belongs to $S_{1,0}^{m_1+m_2}$ and is given by the formula*

$$(a_1 \diamond a_2)(x, \xi) = (\exp 2i\pi D_y \cdot D_\eta)\Big(a_1(x, \xi + \eta)a_2(y + x, \xi)\Big)_{|y=0,\eta=0}. \tag{1.1.15}$$

N.B. From Lemma 4.1.5, we know that the operator $e^{2i\pi D_y \cdot D_\eta}$ is an isomorphism of $S_{1,0}^m(\mathbb{R}^{2n})$, which gives a meaning to the formula (1.1.15), since for $a_j \in S_{1,0}^{m_j}(\mathbb{R}^{2n})$, (x, ξ) given in $\mathbb{R}^{2n}$, the function $(y, \eta) \mapsto a_1(x, \xi + \eta)a_2(y + x, \xi) = C_{x,\xi}(y, \eta)$ belongs to $S_{1,0}^{m_1}(\mathbb{R}^{2n})$ as well as $JC_{x,\xi}$ and we can take the value of the latter at $(y, \eta) = (0, 0)$.

Proof. We assume first that both a_j belong to $\mathscr{S}(\mathbb{R}^{2n})$. The formula (1.1.4) provides the answer. Now, rewriting the formula (1.1.6) for an even integer k, we get

$$(a_1 \diamond a_2)(x,\xi) = \sum_{0 \leq l \leq k/2} C_{k/2}^l \iint e^{-2i\pi y \cdot \eta} |D_y|^{2l} \left(\langle y \rangle^{-k} a_2(y+x,\xi) \right)$$

$$\times \langle \eta \rangle^{-k} ((\langle D_\eta \rangle^k a_1)(x,\xi+\eta) dy d\eta. \quad (1.1.16)$$

We denote by $a_1 \tilde\diamond a_2$ the right-hand side of (1.1.16) and we note that, when $k > n + |m_1|$, it makes sense (and it does not depend on k) as well for $a_j \in S_{1,0}^{m_j}$, since

$$|\partial_y^\alpha \langle y \rangle^{-k}| \leq C_{\alpha,k} \langle y \rangle^{-k}, \quad |\partial_y^\beta a_2(y+x,\xi)| \leq C_\beta \langle \xi \rangle^{m_2}, \quad |\partial_\eta^\gamma a_1(x,\xi+\eta)| \leq C_\gamma \langle \xi+\eta \rangle^{m_1}$$

so that the absolute value of the integrand above is[4][5]

$$\lesssim \langle y \rangle^{-k} \langle \eta \rangle^{-k} \langle \xi \rangle^{m_2} \langle \xi+\eta \rangle^{m_1} \lesssim \langle y \rangle^{-k} \langle \eta \rangle^{-k+|m_1|} \langle \xi \rangle^{m_1+m_2}.$$

Remark 1.1.16. Note that this proves that the mapping

$$S_{1,0}^{m_1} \times S_{1,0}^{m_2} \ni (a_1,a_2) \mapsto a_1 \tilde\diamond a_2 \in S_{1,0}^{m_1+m_2}$$

is bilinear continuous. In fact, we have already proven that

$$|(a_1 \tilde\diamond a_2)(x,\xi)| \leq C \langle \xi \rangle^{m_1+m_2},$$

and we can check directly that $a_1 \tilde\diamond a_2$ is smooth and satisfies

$$\partial_{\xi_j} (a_1 \tilde\diamond a_2) = (\partial_{\xi_j} a_1) \tilde\diamond a_2 + a_1 \tilde\diamond (\partial_{\xi_j} a_2)$$

so that $|\partial_{\xi_j}(a_1 \tilde\diamond a_2)(x,\xi)| \leq C \langle \xi \rangle^{m_1+m_2-1}$, and similar formulas for higher order derivatives.

Remark 1.1.17. Let (c_k) be a bounded sequence in the Fréchet space $S_{1,0}^m$ converging in $C^\infty(\mathbb{R}^{2n})$ to c. Then c belongs to $S_{1,0}^m$ and for all $u \in \mathscr{S}(\mathbb{R}^n)$, the sequence $(c_k(x,D)u)$ converges to $c(x,D)u$ in $\mathscr{S}(\mathbb{R}^n)$. In fact, the sequence of functions $(c_k(x,\xi)\langle\xi\rangle^{-m})$ is bounded in $C_b^\infty(\mathbb{R}^{2n})$ and we can apply Lemma 1.1.3 to get that $\lim_k \mathrm{op}(c_k(x,\xi)\langle\xi\rangle^{-m})\langle D\rangle^m u = \mathrm{op}(c(x,\xi)\langle\xi\rangle^{-m})\langle D\rangle^m u = \mathrm{op}(c)u$ in $\mathscr{S}(\mathbb{R}^n)$.

The remaining part of the argument is the same as in the proof of Theorem 1.1.5, after (1.1.7). □

---

[4]We use $\langle \xi+\eta \rangle \leq 2^{1/2}\langle\xi\rangle\langle\eta\rangle$ so that,

$$\forall s \in \mathbb{R}, \forall \xi, \eta \in \mathbb{R}^n, \quad \langle\xi+\eta\rangle^s \leq 2^{|s|/2}\langle\xi\rangle^s\langle\eta\rangle^{|s|}, \quad (1.1.17)$$

a convenient inequality (to get it for $s \geq 0$, raise the first inequality to the power s, and for $s < 0$, replace ξ by $-\xi - \eta$) a.k.a. Peetre's inequality.

[5]We use here the notation $a \lesssim b$ for the inequality $a \leq Cb$, where C is a "controlled" constant (here C depends only on k, m_1, m_2).

Theorem 1.1.18. *Let s, m be real numbers and $a \in S^m_{1,0}$. Then the operator $a(x, D)$ is bounded from $H^{s+m}(\mathbb{R}^n)$ to $H^s(\mathbb{R}^n)$.*

Proof. Let us recall that $H^s(\mathbb{R}^n) = \{u \in \mathscr{S}'(\mathbb{R}^n), \langle\xi\rangle^s \hat{u}(\xi) \in L^2(\mathbb{R}^n)\}$. From Theorem 1.1.15, the operator $\langle D\rangle^s a(x, D)\langle D\rangle^{-m-s}$ can be written as $b(x, D)$ with $b \in S^0_{1,0}$ and so from Theorem 1.1.14, it is a bounded operator on $L^2(\mathbb{R}^n)$. Since $\langle D\rangle^\sigma$ is an isomorphism of $H^\sigma(\mathbb{R}^n)$ onto $L^2(\mathbb{R}^n)$ with inverse $\langle D\rangle^{-\sigma}$, it gives the result. $\square$

Corollary 1.1.19. *Let r be a symbol in $S^{-\infty}_{1,0} = \cap_m S^m_{1,0}$. Then $r(x, D)$ sends $\mathscr{E}'(\mathbb{R}^n)$ into $\mathscr{S}(\mathbb{R}^n)$.*

Proof. We have for $v \in \mathscr{E}'$ and $\psi \in C^\infty_c(\mathbb{R}^n)$ equal to 1 on a neighborhood of the support of v, iterating

$$x_j D^\beta r(x, D)v = [x_j, D^\beta r(x, D)]\psi v + D^\beta r(x, D)\psi x_j v = r_j(x, D)v, \quad r_j \in S^{-\infty}_{1,0},$$

that $x^\alpha D^\beta r(x, D)v = r_{\alpha\beta}(x, D)v, r_{\alpha\beta} \in S^{-\infty}_{1,0}$, and thus

$$x^\alpha D^\beta r(x, D)v \in \cap_s H^s(\mathbb{R}^n) \subset C^\infty_b(\mathbb{R}^n),$$

completing the proof. $\square$

Theorem 1.1.20. *Let m_1, m_2 be real numbers and $a_1 \in S^{m_1}_{1,0}, a_2 \in S^{m_2}_{1,0}$. Then $a_1(x, D)a_2(x, D) = (a_1 \diamond a_2)(x, D)$, the symbol $a_1 \diamond a_2$ belongs to $S^{m_1+m_2}_{1,0}$ and we have the asymptotic expansion, for all $N \in \mathbb{N}$,*

$$a_1 \diamond a_2 = \sum_{|\alpha| < N} \frac{1}{\alpha!} D^\alpha_\xi a_1 \partial^\alpha_x a_2 + r_N(a_1, a_2), \tag{1.1.18}$$

with $r_N(a_1, a_2) \in S^{m_1+m_2-N}_{1,0}$. Note that $D^\alpha_\xi a_1 \partial^\alpha_x a_2$ belong to $S^{m_1+m_2-|\alpha|}_{1,0}$.

Proof. The proof is very similar to the proof of Lemma 4.1.5. We can use the formula (1.1.15) and apply that lemma to get the desired formula with

$$r_N(a_1, a_2)(x, \xi)$$
$$= \int_0^1 \frac{(1-\theta)^{N-1}}{(N-1)!} e^{2i\pi\theta D_z \cdot D_\zeta}(2i\pi D_z \cdot D_\zeta)^N (a_1(x, \zeta)a_2(z, \xi))d\theta_{|z=x, \zeta=\xi}. \tag{1.1.19}$$

The function $(z, \zeta) \mapsto b_{x,\xi}(z, \zeta) = \langle\xi\rangle^{-m_2}(2i\pi D_z \cdot D_\zeta)^N a_1(x, \zeta)a_2(z, \xi)$ belongs to $S^{m_1-N}_{1,0}(\mathbb{R}^{2n}_{z,\zeta})$ uniformly with respect to the parameters $(x, \xi) \in \mathbb{R}^{2n}$: it satisfies, using the notation (1.1.14), for $\max(|\alpha|, |\beta|) \leq k$,

$$|\partial^\alpha_\zeta \partial^\beta_z b_{x,\xi}(z, \zeta)| \leq \gamma_{k,m_1}(a_1)\gamma_{k,m_2}(a_2)\langle\zeta\rangle^{m_1-N-|\alpha|}.$$

Applying Lemma 4.1.5, we obtain that the function

$$\rho_{x,\xi}(z, \zeta) = \int_0^1 \frac{(1-\theta)^{N-1}}{(N-1)!}(J^\theta b_{x,\xi})(z, \zeta)d\theta$$

belongs to $S_{1,0}^{m_1-N}(\mathbb{R}_{z,\zeta}^{2n})$ uniformly with respect to x,ξ, so that in particular

$$\sup_{(x,\xi,z,\zeta)\in\mathbb{R}^{4n}} |\rho_{x,\xi}(z,\zeta)\langle\zeta\rangle^{-m_1+N}| = C_0 < +\infty.$$

Since $r_N(a_1,a_2)(x,\xi)\langle\xi\rangle^{-m_2} = \rho_{x,\xi}(x,\xi)$, we obtain

$$|r_N(a_1,a_2)(x,\xi)| \leq C_0\langle\xi\rangle^{m_1+m_2-N}. \tag{1.1.20}$$

Using the formula (1.1.19) above gives as well the smoothness of $r_N(a_1,a_2)$ and with the identities (consequences of $\partial_{x_j}(a_1 \diamond a_2) = (\partial_{x_j}a_1) \diamond a_2 + a_1 \diamond (\partial_{x_j}a_2)$)

$$\partial_{x_j}\big(r_N(a_1,a_2)\big) = r_N(\partial_{x_j}a_1,a_2) + r_N(a_1,\partial_{x_j}a_2)$$
$$\partial_{\xi_j}\big(r_N(a_1,a_2)\big) = r_N(\partial_{\xi_j}a_1,a_2) + r_N(a_1,\partial_{\xi_j}a_2),$$

it is enough to reapply (1.1.20) to get the result $r_N \in S_{1,0}^{m_1+m_2-N}$. $\qquad\square$

We have already seen in Theorem 1.1.8 that the adjoint (in the sense of Definition 1.1.7) of the operator $a(x,D)$ is equal to $a^*(x,D)$, where $a^* = J\bar{a}$ (J is given in Lemma 4.1.2). Lemma 4.1.5 gives the following result.

Theorem 1.1.21. *Let $a \in S_{1,0}^m$. Then $a^* = J\bar{a}$ and the mapping $a \mapsto a^*$ is continuous from $S_{1,0}^m$ into itself. Moreover, for all integers N, we have*

$$a^* = \sum_{|\alpha|<N}\frac{1}{\alpha!}D_\xi^\alpha\partial_x^\alpha\bar{a} + r_N(a), \quad r_N(a) \in S_{1,0}^{m-N}.$$

A consequence of the above results is the following.

Corollary 1.1.22. *Let $a_j \in S_{1,0}^{m_j}, j = 1,2$. Then we have*

$$a_1 \diamond a_2 \equiv a_1 a_2 \mod S_{1,0}^{m_1+m_2-1}, \tag{1.1.21}$$

$$a_1 \diamond a_2 - a_2 \diamond a_1 \equiv \frac{1}{2i\pi}\{a_1,a_2\} \mod S_{1,0}^{m_1+m_2-2}, \tag{1.1.22}$$

*where the **Poisson bracket*** $\{a_1,a_2\} = \displaystyle\sum_{1\leq j\leq n}\frac{\partial a_1}{\partial\xi_j}\frac{\partial a_2}{\partial x_j} - \frac{\partial a_1}{\partial x_j}\frac{\partial a_2}{\partial\xi_j}. \tag{1.1.23}$

For $a \in S_{1,0}^m$, $a^* \equiv \bar{a} \mod S_{1,0}^{m-1}$. \tag{1.1.24}

Theorem 1.1.23. *Let a be a symbol in $S_{1,0}^m$ such that $\inf_{(x,\xi)\in\mathbb{R}^{2n}} |a(x,\xi)|\langle\xi\rangle^{-m} > 0$. Then there exists $b \in S_{1,0}^{-m}$ such that*

$$b(x,D)a(x,D) = \mathrm{Id} + l(x,D),$$
$$a(x,D)b(x,D) = \mathrm{Id} + r(x,D), \qquad r,l \in S_{1,0}^{-\infty} = \cap_\nu S_{1,0}^\nu.$$

Proof. We remark first that the smooth function $1/a$ belongs to $S_{1,0}^{-m}$: it follows from the Faà de Bruno formula (see, e.g., our appendix, Section 4.3.1), or more elementarily, from the fact that, for $|\alpha| + |\beta| \geq 1$, $\partial_\xi^\alpha \partial_x^\beta(\frac{1}{a}a) = 0$, entailing the Leibniz formula

$$a\partial_\xi^\alpha \partial_x^\beta(1/a) = \sum_{\substack{\alpha'+\alpha''=\alpha, \beta'+\beta''=\beta \\ |\alpha'|+|\beta'|<|\alpha|+|\beta|}} \partial_\xi^{\alpha'}\partial_x^{\beta'}(1/a)\partial_\xi^{\alpha''}\partial_x^{\beta''}(a)c(\alpha',\beta'),$$

with constants $c(\alpha', \beta')$. Arguing by induction on $|\alpha| + |\beta|$, we get

$$|a\partial_\xi^\alpha \partial_x^\beta(1/a)| \lesssim \sum_{\alpha'+\alpha''=\alpha} \langle\xi\rangle^{-m-|\alpha'|}\langle\xi\rangle^{m-|\alpha''|} \lesssim \langle\xi\rangle^{-|\alpha|}$$

and from $|a| \gtrsim \langle\xi\rangle^m$, we get $1/a \in S_{1,0}^{-m}$. Now, we can compute, using Theorem 1.1.20,

$$\frac{1}{a} \diamond a = 1 + l_1, \quad l_1 \in S_{1,0}^{-1}.$$

Inductively, we can assume that there exist $(b_0, \ldots, b_N)$ with $b_j \in S^{-m-j}$ such that

$$(b_0 + \cdots + b_N) \diamond a = 1 + l_{N+1}, \quad l_{N+1} \in S_{1,0}^{-N-1}. \tag{1.1.25}$$

We can now take $b_{N+1} = -l_{N+1}/a$ which belongs to S^{-m-N-1} and this gives

$$(b_0 + \cdots + b_N + b_{N+1}) \diamond a = 1 + l_{N+1} - l_{N+1} + l_{N+2}, \quad l_{N+2} \in S_{1,0}^{-N-2}.$$

Lemma 1.1.24. *Let $\mu \in \mathbb{R}$ and $(c_j)_{j\in\mathbb{N}}$ be a sequence of symbols such that $c_j \in S_{1,0}^{\mu-j}$. Then there exists $c \in S_{1,0}^\mu$ such that*

$$c \sim \sum_j c_j, \quad i.e., \quad \forall N \in \mathbb{N}, \quad c - \sum_{0\leq j<N} c_j \in S_{1,0}^{\mu-N}.$$

Proof. The proof is based on a Borel-type argument similar to the one used to construct a C^∞ function with an arbitrary Taylor expansion. Let $\omega \in C_b^\infty(\mathbb{R}^n)$ such that $\omega(\xi) = 0$ for $|\xi| \leq 1$ and $\omega(\xi) = 1$ for $|\xi| \geq 2$. Let $(\lambda_j)_{j\in\mathbb{N}}$ be a sequence of numbers ≥ 1. We want to define

$$c(x,\xi) = \sum_{j\geq 0} c_j(x,\xi)\omega(\xi\lambda_j^{-1}), \tag{1.1.26}$$

and we shall show that a suitable choice of λ_j will provide the answer. We note that, since $\lambda_j \geq 1$, the functions $\xi \mapsto \omega(\xi\lambda_j^{-1})$ make a bounded set in the Fréchet space $S_{1,0}^0$. Multiplying the c_j by $\langle\xi\rangle^{-\mu}$, we may assume that $\mu = 0$. We have then, using the notation (1.1.14) (in which we drop the second index),

$$|c_j(x,\xi)|\omega(\xi\lambda_j^{-1}) \leq \gamma_0(c_j)\langle\xi\rangle^{-j}\mathbf{1}\{|\xi| \geq \lambda_j\} \leq \gamma_0(c_j)\lambda_j^{-j/2}\langle\xi\rangle^{-j/2},$$

so that,

$$\forall j \geq 1, \ \lambda_j \geq 2^2 \gamma_0(c_j)^{\frac{2}{j}} = \mu_j^{(0)} \implies \forall j \geq 1, \ |c_j(x,\xi)| \omega(\xi \lambda_j^{-1}) \leq 2^{-j} \langle \xi \rangle^{-j/2},$$

showing that the function c can be defined as above in (1.1.26) and is a continuous bounded function. Let $1 \leq k \in \mathbb{N}$ be given. Calculating (with $\omega_j(\xi) = \omega(\xi \lambda_j^{-1})$) the derivatives $\partial_\xi^\alpha \partial_x^\beta (c_j \omega_j)$ for $|\alpha| + |\beta| = k$, we get

$$|\partial_\xi^\alpha \partial_x^\beta (c_j \omega_j)| \leq \gamma_k(c_j \omega_j) \langle \xi \rangle^{-j-|\alpha|} \mathbf{1}\{|\xi| \geq \lambda_j\} \leq \tilde{\gamma}_k(c_j) \lambda_j^{-j/2} \langle \xi \rangle^{-|\alpha|-\frac{j}{2}},$$

so that

$$\forall j \geq k, \ \lambda_j \geq 2^2 \big(\tilde{\gamma}_k(c_j)\big)^{\frac{2}{j}} = \mu_j^{(k)} \implies \forall j \geq k, \ |\partial_\xi^\alpha \partial_x^\beta (c_j \omega_j)| \leq 2^{-j} \langle \xi \rangle^{-|\alpha|-\frac{j}{2}},$$
$$\tag{1.1.27}$$

showing that the function c can be defined as above in (1.1.26) and is a C^k function such that

$$|(\partial_\xi^\alpha \partial_x^\beta c)(x,\xi)| \leq \sum_{0 \leq j < k} \tilde{\gamma}_k(c_j) \langle \xi \rangle^{-j-|\alpha|} + \sum_{j \geq k} 2^{-j} \langle \xi \rangle^{-|\alpha|} \leq C_k \langle \xi \rangle^{-|\alpha|}.$$

It is possible to fulfill the conditions on the λ_j above for all $k \in \mathbb{N}$: just take

$$\lambda_j \geq \sup_{0 \leq k \leq j} \mu_j^{(k)}.$$

The function c belongs to $S_{1,0}^0$ and

$$r_N = c - \sum_{0 \leq j < N} c_j = \sum_{0 \leq j < N} \underbrace{(\omega_j - 1) c_j}_{\in S_{1,0}^{-\infty}} + \sum_{j \geq N} c_j \omega_j,$$

and for $|\alpha| + |\beta| = k$, using the estimates (1.1.27), we obtain

$$\sum_{j \geq N} |\partial_\xi^\alpha \partial_x^\beta (c_j \omega_j)(x,\xi)| \leq \sum_{N \leq j < \max(2N,k)} \overbrace{|\partial_\xi^\alpha \partial_x^\beta (c_j \omega_j)(x,\xi)|}^{\lesssim \langle \xi \rangle^{-|\alpha|-j} \lesssim \langle \xi \rangle^{-|\alpha|-N}}$$
$$+ \sum_{j \geq \max(2N,k)} \underbrace{|\partial_\xi^\alpha \partial_x^\beta (c_j \omega_j)(x,\xi)|}_{\lesssim 2^{-j} \langle \xi \rangle^{-|\alpha|-\frac{j}{2}} \lesssim 2^{-j} \langle \xi \rangle^{-|\alpha|-N}},$$

proving that $r_N \in S_{1,0}^{-N}$. The proof of the lemma is complete. $\qquad \square$

Going back to the proof of the theorem, we can take, using Lemma 1.1.24, $S_{1,0}^{-m} \ni b \sim \sum_{j \geq 0} b_j$, and for all $N \in \mathbb{N}$,

$$b \diamond a \in \sum_{0 \leq j < N} b_j \diamond a + S_{1,0}^{-N-m} \diamond a = 1 + S_{1,0}^{-N},$$

providing the first equality in Theorem 1.1.23. To construct a right approximate inverse, i.e., to obtain the second equality in this theorem with an a priori different b, follows the same lines (or can be seen as a direct consequence of the previous identity by applying it to the adjoint a^*); however we are left with the proof that the right and the left approximate inverse could be taken as the same. We have proven that there exists $b^{(1)}, b^{(2)} \in S_{1,0}^{-m}$ such that

$$b^{(1)} \diamond a \in 1 + S_{1,0}^{-\infty}, \quad a \diamond b^{(2)} \in 1 + S_{1,0}^{-\infty}.$$

Now we calculate, using[6] Theorem 1.1.5, $(b^{(1)} \diamond a) \diamond b^{(2)} = b^{(2)} \mod S_{1,0}^{-\infty}$ which is also $b^{(1)} \diamond (a \diamond b^{(2)}) = b^{(1)} \mod S_{1,0}^{-\infty}$ so that $b^{(1)} - b^{(2)} \in S_{1,0}^{-\infty}$, providing the result and completing the proof of the theorem. $\qquad\square$

An important consequence of the proof of the previous theorem is the possible microlocalization of this result.

Theorem 1.1.25. *Let χ be a symbol in $S_{1,0}^0$ and let a be a symbol in $S_{1,0}^m$ such that $\inf_{(x,\xi) \in \text{supp}\,\chi} |a(x,\xi)| \langle \xi \rangle^{-m} > 0$. Let ψ be a symbol in $S_{1,0}^0$ such that $\text{supp}\,\psi \subset \text{int}\,\{\chi = 1\}$. Then there exists $b \in S_{1,0}^{-m}$ such that*

$$b(x,D)a(x,D) = \psi(x,D) + l(x,D), \quad l \in S_{1,0}^{-\infty}.$$

Proof. We consider the symbol $b_0 = \chi/a$, which belongs obviously to $S_{1,0}^{-m}$. We have

$$b_0 \diamond a = \chi + l_1, \quad l_1 \in S_{1,0}^{-1}, \quad \left(-\frac{\chi l_1}{a} + \frac{\chi}{a}\right) \diamond a = \chi + l_1(1 - \chi) + l_2, \ l_2 \in S_{1,0}^{-2}.$$

Inductively, we may assume that there exists $(b_0, \ldots, b_N)$ with $b_j \in S^{-m-j}$ such that

$$(b_0 + b_1 + \cdots + b_N) \diamond a = \chi + \sum_{1 \le j \le N} l_j(1 - \chi) + l_{N+1}, \quad l_{N+1} \in S_{1,0}^{-1-N}.$$

Choosing $b_{N+1} = -\chi l_{N+1}/a$, we get

$$(b_0 + b_1 + \cdots + b_N + b_{N+1}) \diamond a = \chi + \sum_{1 \le j \le N+1} l_j(1 - \chi) + l_{N+2}, \quad l_{N+2} \in S_{1,0}^{-2-N}.$$

Taking now a symbol $\psi \in S_{1,0}^0$ such that $\text{supp}\,\psi \subset \chi^{-1}(\{1\})$, we obtain for all $N \in \mathbb{N}$, the existence of symbols $b_0, \ldots, b_N$ with $b_j \in S^{-m-j}$ such that

$$\psi \diamond (b_0 + b_1 + \ldots + b_N) \diamond a = \psi \diamond \chi + \psi \diamond \sum_{1 \le j \le N} l_j(1 - \chi) + \psi \diamond l_{N+1} \quad (l_{N+1} \in S_{1,0}^{-1-N})$$

$$= \psi + r_{N+1}, \quad r_{N+1} \in S_{1,0}^{-1-N}.$$

---

[6]A consequence of Theorem 1.1.5 is the associativity of the "law" $\diamond$ since

$$\text{op}(a \diamond (b \diamond c)) = \text{op}(a)\big(\text{op}(b)\text{op}(c)\big) = \big(\text{op}(a)\text{op}(b)\big)\text{op}(c) = \text{op}((a \diamond b) \diamond c)$$

so that the injectivity property of Remark 4.1.6 gives the answer.

Using now Lemma 1.1.24, we find a symbol $b \in S_{1,0}^{-m}$ such that, for all $N \in \mathbb{N}$, $\psi \diamond b \diamond a \in \psi + S_{1,0}^{-1-N}$, i.e., we find $\tilde{b} \in S_{1,0}^{-m}$ such that $\tilde{b} \diamond a \equiv \psi$ (mod $S_{1,0}^{-\infty}$). $\square$

We end this introduction with the so-called *sharp Gårding inequality*, a result proven in 1966 by L. Hörmander [64] and extended to systems the same year by P. Lax and L. Nirenberg [86].

Theorem 1.1.26. *Let a be a non-negative symbol in $S_{1,0}^m$. Then there exists a constant C such that, for all $u \in \mathscr{S}(\mathbb{R}^n)$,*

$$\mathrm{Re}\langle a(x, D)u, u \rangle + C\|u\|^2_{H^{\frac{m-1}{2}}(\mathbb{R}^n)} \geq 0. \tag{1.1.28}$$

Proof. First reductions. We may assume that $m = 1$: in fact, the statement for $m = 1$ implies the result by considering, for a non-negative $a \in S_{1,0}^m$, the operator $\langle D \rangle^{\frac{1-m}{2}} a(x, D) \langle D \rangle^{\frac{1-m}{2}}$ which, according to Theorem 1.1.20 has a symbol in $S_{1,0}^1$, which belongs to $\langle \xi \rangle^{1-m} a(x, \xi) + S_{1,0}^0$. Applying the result for $m = 1$, and the L^2-boundedness of operators with symbols in $S_{1,0}^0$, we get for all $u \in \mathscr{S}(\mathbb{R}^n)$,

$$\mathrm{Re}\langle \langle D \rangle^{\frac{1-m}{2}} a(x, D) \langle D \rangle^{\frac{1-m}{2}} u, u \rangle + C\|u\|^2_{L^2(\mathbb{R}^n)} \geq 0,$$

which gives the sought result when applied to $u = \langle D \rangle^{\frac{m-1}{2}} v$. We may also replace $a(x, D)$ by a^w, where a^w is the operator with Weyl symbol a . In fact, according to Lemma 4.1.2, $J^{1/2}a - a \in S_{1,0}^0$ and $\mathrm{op}(S_{1,0}^0)$ is L^2-bounded.
Main step: a result with a small parameter. We consider a non-negative $a \in S_{1,0}^1$ and

$$\varphi \in C_c^\infty((0, +\infty); \mathbb{R}_+) \text{ such that } \int_0^{+\infty} \varphi(h) \frac{dh}{h} = 1. \tag{1.1.29}$$

This implies

$$a(x, \xi) = \int_0^{+\infty} \underbrace{\varphi(\langle \xi \rangle h) a(x, \xi)}_{=a_h(x,\xi)} \frac{dh}{h}. \tag{1.1.30}$$

We have, with $\Gamma_h(x, \xi) = 2^n \exp -2\pi(h^{-1}|x|^2 + h|\xi|^2)$ and $X = (x, \xi)$,

$$(a_h * \Gamma_h)(X) = a_h(X) + \int_0^1 (1 - \theta) a_h''(X + \theta Y) Y^2 \Gamma_h(Y) dY \, d\theta$$

$$= a_h(X) + r_h(X). \tag{1.1.31}$$

The main step of the proof is that $(a_h * \Gamma_h)^w \geq 0$, a result following from the next calculation (for $u \in \mathscr{S}(\mathbb{R}^n)$), due to Definition 1.1.9. We have, with $\Omega_{u,u}$ defined in (1.1.11),

$$\langle (a_h * \Gamma_h)^w u, u \rangle = \iint (a_h * \Gamma_h)(x, \xi) \left(\int e^{-2i\pi z \cdot \xi} u(x + \frac{z}{2}) \bar{u}(x - \frac{z}{2}) dz \right) dx d\xi$$

$$= \iint a(y, \eta) (\Omega_{u,u}(1/2) * \Gamma_h)(y, \eta) dy d\eta,$$

and since

$$(\Omega_{u,u}(1/2) * \Gamma_h)(x,\xi)$$

$$= \iiint e^{-2i\pi z \cdot (\xi-\eta)} u(x-y+\frac{z}{2})\bar{u}(x-y-\frac{z}{2})2^n \exp{-2\pi(h^{-1}|y|^2 + h|\eta|^2)}dz\,dy\,d\eta$$

$$= \iint e^{-2i\pi z \cdot \xi} u(x-y+\frac{z}{2})\bar{u}(x-y-\frac{z}{2})2^{n/2} e^{-2\pi h^{-1}|y|^2} h^{-n/2} e^{-\frac{\pi}{2h}|z|^2} dz\,dy$$

$$= \iint u(x-y_1)\bar{u}(x-y_2) e^{-2i\pi(y_2-y_1)\cdot\xi} 2^{n/2} h^{-n/2} e^{-\frac{\pi}{2h}|y_1+y_2|^2} e^{-\frac{\pi}{2h}|y_1-y_2|^2} dy_1\,dy_2$$

$$= 2^{n/2} h^{-n/2} \left| \int u(x-y_1) e^{2i\pi y_1 \cdot \xi} e^{-\pi h^{-1}|y_1|^2} dy_1 \right|^2 \geq 0,$$

we get indeed $(a_h * \Gamma_h)^w \geq 0$. From (1.1.30) and (1.1.31), we get

$$a^w = \int_0^{+\infty} a_h^w h^{-1} dh = \int_0^{+\infty} (a_h * \Gamma_h)^w h^{-1} dh - \int_0^{+\infty} r_h^w h^{-1} dh$$

$$\geq -\int_0^{+\infty} r_h^w h^{-1} dh.$$

Last step: $\int_0^{+\infty} r_h^w h^{-1} dh$ is L^2-bounded. This is a technical point, where the main difficulty is coming from the integration in h. We have from (1.1.31) and the fact that Γ_h is an even function,

$$r_h(X) = \frac{1}{8\pi} \operatorname{trace}_h a_h''(X) + \frac{1}{3!} \iint_0^1 (1-\theta)^3 a_h^{(4)}(X+\theta Y)Y^4 \Gamma_h(Y)dY\,d\theta,$$

with $\operatorname{trace}_h a_h''(X) = h \operatorname{trace} \partial_x^2 a_h + h^{-1} \operatorname{trace} \partial_\xi^2 a_h$. Since $\varphi \in C_c^\infty((0,+\infty))$, we have

$$\int_0^{+\infty} h \operatorname{trace} \partial_x^2 a_h h^{-1} dh = \operatorname{trace} \partial_x^2 a(x,\xi) \int_0^{+\infty} \varphi(\langle\xi\rangle h)dh = c \operatorname{trace} \partial_x^2 a \langle\xi\rangle^{-1},$$

with $c = \int_0^{+\infty} \varphi(t)dt$. The symbol $c \operatorname{trace} \partial_x^2 a \langle\xi\rangle^{-1}$ belongs to $S_{1,0}^0$ as well as the other term $\int_0^{+\infty} h^{-1} \operatorname{trace} \partial_\xi^2 a_h(x,\xi) h^{-1} dh$: we have

$$(\partial_\xi a_h)(x,\xi) = (\partial_\xi a)(x,\xi)\varphi(h\langle\xi\rangle) + a(x,\xi)\varphi'(h\langle\xi\rangle)h\langle\xi\rangle^{-1}\xi,$$

$$(\partial_\xi^2 a_h)(x,\xi) = (\partial_\xi^2 a)(x,\xi)\varphi(h\langle\xi\rangle) + 2\partial_\xi a(x,\xi)\varphi'(h\langle\xi\rangle)h$$

$$+ a(x,\xi)\varphi''(h\langle\xi\rangle)h^2\langle\xi\rangle^{-2}\xi^2 + a(x,\xi)\varphi'(h\langle\xi\rangle)h\partial_\xi(\xi\langle\xi\rangle^{-1}),$$

and checking for instance the term $\int_0^{+\infty} h^{-1}(\partial_\xi^2 a)(x,\xi)\varphi(h\langle\xi\rangle)\frac{dh}{h}$, we see that it is equal to

$$(\partial_\xi^2 a)(x,\xi) \int_0^{+\infty} h^{-1}\varphi(h\langle\xi\rangle)\frac{dh}{h} = (\partial_\xi^2 a)(x,\xi)\langle\xi\rangle \int_0^{+\infty} h^{-1}\varphi(h)\frac{dh}{h}$$

$$= c_1(\partial_\xi^2 a)(x,\xi)\langle\xi\rangle \in S_{1,0}^0,$$

whereas the other terms are analogous. We are finally left with the term

$$\rho(X) = \frac{1}{3!} \iiint_0^1 (1-\theta)^3 a_h^{(4)}(X+\theta Y)Y^4\Gamma_h(Y)dYh^{-1}dhd\theta,$$

and we note that on the integrand of (1.1.30), the product $h\langle\xi\rangle$ is bounded above and below by fixed constants and that the integral can in fact be written as

$$a(x,\xi) = \int_{\kappa_0\langle\xi\rangle^{-1}}^{\kappa_1\langle\xi\rangle^{-1}} \varphi(\langle\xi\rangle h)a(x,\xi)dh/h$$

with $0 < \kappa_0 = \min \operatorname{supp} \varphi < \kappa_1 = \max \operatorname{supp} \varphi$. Consequently the symbol a_h satisfies the following estimates:

$$|\partial_\xi^\alpha \partial_x^\beta a_h| \leq C_{\alpha\beta} h^{-1+|\alpha|}$$

where the $C_{\alpha\beta}$ are some semi-norms of a (and thus independent of h). As a result, the above estimates can be written in a more concise and convenient way, using the multilinear forms defined by the derivatives. We have, with $T = (t,\tau) \in \mathbb{R}^n \times \mathbb{R}^n$,

$$|a_h^{(l)}(X)T^l| \leq C_l h^{-1} g_h(T)^{l/2}, \quad \text{with } g_h(t,\tau) = |t|^2 + h^2|\tau|^2.$$

We calculate

$$\rho^{(k)}(X)T^k = \frac{1}{3!} \iiint_0^1 (1-\theta)^3 a_h^{(4+k)}(X+\theta Y)Y^4 T^k \Gamma_h(Y)dYh^{-1}dhd\theta,$$

which satisfies with $\omega_h(t,\tau) = h^{-1}g_h(t,\tau)$,

$$|\rho^{(k)}(X)T^k|$$
$$\leq \frac{C_{4+k}}{4!} \iint 1\{h \leq \kappa_1\}h^{-1}g_h(T)^{k/2} \underbrace{g_h(Y)^2}_{=h^2\omega_h(Y)^2} 2^n e^{-2\pi\omega_h(Y)}dYh^{-1}dh$$
$$\leq \frac{C_{4+k}}{4!}g_h(T)^{k/2} \iint \omega_h(Y)^2 1\{h \leq \kappa_1\}2^n e^{-2\pi\omega_h(Y)}dYdh \leq \tilde{C}_k(|t|+|\tau|)^k$$

and this proves that the function ρ belongs to $C_b^\infty(\mathbb{R}^{2n})$, as well as $J^{1/2}\rho$ (Lemma 4.1.2) and thus $\rho^w = (J^{1/2}\rho)(x,D)$ is bounded on L^2 (Theorem 1.1.4). The proof is complete. $\qquad\square$

Remark 1.1.27. Theorem 1.1.26 remains valid for systems, even in infinite dimension. For definiteness, let us assume simply that $a(x,\xi)$ is an $N \times N$ Hermitian non-negative matrix of symbols in $S_{1,0}^1$. Then for all $u \in \mathscr{S}(\mathbb{R}^n; \mathbb{C}^N)$, the inequality (1.1.28) holds. The vector space $\mathbb{C}^N$ can be replaced in the above statement by an infinite-dimensional complex Hilbert space H with a valued in $\mathscr{L}(H)$ and the proof above requires essentially no change.

Remark 1.1.28 (**The Hörmander classes** $S^m_{\rho,\delta}$). There are many other examples of classes of symbols, such as the $S^m_{\rho,\delta}$ class: smooth functions a on $\mathbb{R}^{2n}$ such that

$$|(\partial^\alpha_\xi \partial^\beta_x a)(x,\xi)| \leq C_{\alpha\beta}\langle\xi\rangle^{m-\rho|\alpha|+\delta|\beta|}. \qquad (1.1.32)$$

We see that it corresponds to the notation $S^m_{1,0}$ (see (1.1.13)) and also $C^\infty_b(\mathbb{R}^{2n}) = S^0_{0,0}$. An important motivation for studying this class of operators comes from Hörmander's characterization of hypoelliptic operators with constant coefficients (see Theorem 11.1.3 in [72]) in which it appeared that the parametrices of these operators were in this type of class.

It is not difficult to prove that op($S^m_{\rho,\delta}$) is continuous from $\mathscr{S}(\mathbb{R}^n)$ into itself. Assuming $1 \geq \rho \geq \delta \geq 0$, $\delta < 1$, one can prove that op($S^0_{\rho,\delta}$) $\subset \mathscr{L}(L^2(\mathbb{R}^n))$, but it is not so easy; we shall see in the second chapter a very general approach to this type of L^2-boundedness questions, where that class will be a particular case (see Theorem 2.5.1). Theorem 1.1.18 of Sobolev continuity still holds true. Composition formulas given by Theorem 1.1.20 should be modified so that $r_N(a_1,a_2)$ belongs to $S^{m_1+m_2-(\rho-\delta)N}_{\rho,\delta}$, as well as in Theorem 1.1.21 where $r_N(a)$ belongs to $S^{m-(\rho-\delta)N}_{\rho,\delta}$. In particular, we note that when $\rho = \delta$, no asymptotic expansions exist and for instance the Poisson bracket of symbols in $S^0_{\delta,\delta}$ is still in $S^0_{\delta,\delta}$. Theorems 1.1.23, 1.1.25 of inversion of elliptic operators do not require any change, provided $\rho > \delta$ (note that $S^{-\infty}_{\rho,\delta} = S^{-\infty}_{1,0}$), whereas in Theorem 1.1.26, the $\frac{m-1}{2}$ should be replaced by $\frac{m-(\rho-\delta)}{2}$. Except for the L^2-boundedness, all these results can be obtained by simple inspection of the previous proofs; the reader eager for a complete and more general description will find in the second chapter a thorough justification.

1.1.4   The semi-classical calculus

The books by M. Dimassi and J. Sjöstrand [37], by A. Grigis and J. Sjöstrand [49], by B. Helffer [55, 56, 57, 58], by A. Martinez [103] and by D. Robert [125], were based on a long tradition in quantum mechanics (see, e.g., the book by V. Maslov and M. Fedoriuk [104]) of semi-classical approximation, in which one tries to look at the behaviour of various physical quantities when the Planck constant goes to zero.

A semi-classical symbol of order m is defined as a family of smooth functions $a(\cdot,\cdot,h)$ defined on the phase space $\mathbb{R}^{2n}$, depending on a parameter $h \in (0,1]$, such that, for all multi-indices α, β,

$$\sup_{(x,\xi,h)\in\mathbb{R}^n\times\mathbb{R}^n\times(0,1]} |(\partial^\alpha_\xi \partial^\beta_x a)(x,\xi,h)|h^{m-|\alpha|} < +\infty. \qquad (1.1.33)$$

The set of semi-classical symbols of order m will be denoted by S^m_{scl}. A typical example of such a symbol of order 0 is a function $a_1(x,h\xi)$ where a_1 belongs to $C^\infty_b(\mathbb{R}^{2n})$: we have indeed $\partial^\alpha_\xi \partial^\beta_x (a_1(x,h\xi)) = (\partial^\alpha_\xi \partial^\beta_x a_1)(x,h\xi)h^{|\alpha|}$. It turns out that this version of the semi-classical calculus is certainly the easiest to understand

and that Theorem 1.1.4 implies the main continuity result for these symbols. The reader has also to keep in mind that we are not dealing here with a single function defined on the phase space, but with a family of symbols depending on a (small) parameter h, a way to express that the constants occurring in (1.1.33) are "independent of h". We shall review the results of the section on the $S_{1,0}^m$ class of symbols and show how they can be transferred to the semi-classical framework, *mutatis mutandis* and almost without any new argument. To understand the correspondence between symbols in $S_{1,0}^m$ and semi-classical symbols, it is essentially enough to think of the $S_{1,0}$ calculus as a semi-classical calculus with small parameter $\langle \xi \rangle^{-1}$.

We can define, for $a \in S_{scl}^m, k \in \mathbb{N}$,

$$\gamma_{k,m}(a) = \sup_{(x,\xi,h) \in \mathbb{R}^{2n} \times (0,1], |\alpha|+|\beta| \leq k} |(\partial_\xi^\alpha \partial_x^\beta a)(x,\xi,h)| h^{m-|\alpha|}. \tag{1.1.34}$$

Theorem 1.1.29. *Let $a \in S_{scl}^m$. Then the operator $a(x,D,h)h^m$ is continuous from $\mathscr{S}(\mathbb{R}^n)$ into itself with constants independent of $h \in (0,1]$.*

Proof. We have $a(x,D,h) = \mathrm{op}(a(x,\xi,h))$. The set $\{a(x,\xi,h)h^m\}_{h \in (0,1]}$ is bounded in $C_b^\infty(\mathbb{R}^{2n})$, so that we can use Theorem 1.1.2 to get the result. □

Theorem 1.1.30. *Let $a \in S_{scl}^m$. Then the operator $a(x,D,h)h^m$ is bounded on $L^2(\mathbb{R}^n)$ with a norm bounded above independently of $h \in (0,1]$.*

Proof. The set $\{a(x,\xi,h)h^m\}_{h \in (0,1]}$ being bounded in $C_b^\infty(\mathbb{R}^{2n})$, it follows from Theorem 1.1.4. □

Theorem 1.1.31. *Let m_1, m_2 be real numbers and $a_1 \in S_{scl}^{m_1}, a_2 \in S_{scl}^{m_2}$. Then the composition $a_1(x,D,h)a_2(x,D,h)$ makes sense as a continuous operator from $\mathscr{S}(\mathbb{R}^n)$ into itself, as well as a bounded operator on $L^2(\mathbb{R}^n)$ and*

$$a_1(x,D,h)a_2(x,D,h) = (a_1 \diamond a_2)(x,D,h)$$

where $a_1 \diamond a_2$ belongs to $S_{scl}^{m_1+m_2}$ and is given by the formula

$$(a_1 \diamond a_2)(x,\xi,h) = (\exp 2i\pi D_y \cdot D_\eta)\Big(a_1(x,\xi+\eta,h)a_2(y+x,\xi,h)\Big)\Big|_{y=0,\eta=0}. \tag{1.1.35}$$

Proof. This is a direct consequence of Theorem 1.1.5 since

$$\cup_{j=1,2}\{h^{m_j}a_j(x,\xi,h)\}_{h \in (0,1]} \quad \text{is bounded in } C_b^\infty(\mathbb{R}^{2n}).$$
□

Theorem 1.1.32. *Let m_1, m_2 be real numbers and $a_1 \in S_{scl}^{m_1}, a_2 \in S_{scl}^{m_2}$. Then $a_1(x,D,h)a_2(x,D,h) = (a_1 \diamond a_2)(x,D,h)$, the symbol $a_1 \diamond a_2$ belongs to $S_{scl}^{m_1+m_2}$ and we have the asymptotic expansion, for all $N \in \mathbb{N}$,*

$$a_1 \diamond a_2 = \sum_{|\alpha|<N} \frac{1}{\alpha!} D_\xi^\alpha a_1 \partial_x^\alpha a_2 + r_N(a_1,a_2), \tag{1.1.36}$$

with $r_N(a_1,a_2) \in S_{scl}^{m_1+m_2-N}$. Note that $D_\xi^\alpha a_1 \partial_x^\alpha a_2$ belongs to $S_{scl}^{m_1+m_2-|\alpha|}$.

Proof. Since $h^{m_j} a_j(x, \xi, h), j = 1, 2$, belongs to S^0_{scl}, we may assume that $m_1 = m_2 = 0$. We can use the formula (1.1.15) and apply the formula (4.1.20) to get the desired formula with

$$r_N(a_1, a_2)(x, \xi, h) = \int_0^1 \frac{(1-\theta)^{N-1}}{(N-1)!} e^{2i\pi\theta D_z \cdot D_\zeta}$$

$$\times (2i\pi D_z \cdot D_\zeta)^N \big(a_1(x, \zeta, h) a_2(z, \xi, h)\big) d\theta_{|_{z=x, \zeta=\xi}}. \quad (1.1.37)$$

The function $(z, \zeta) \mapsto b_{x,\xi,h}(z, \zeta) = (2i\pi D_z \cdot D_\zeta)^N a_1(x, \zeta, h) a_2(z, \xi, h)$ belongs to $S^{-N}_{scl}(\mathbb{R}^{2n}_{z,\zeta})$ uniformly with respect to the parameters $(x, \xi) \in \mathbb{R}^{2n}$: it satisfies, using the notation (1.1.34), for $\max(|\alpha|, |\beta|) \leq k$,

$$|\partial_\zeta^\alpha \partial_z^\beta b_{x,\xi,h}(z, \zeta)| \leq \gamma_{k,m_1}(a_1) \gamma_{k,m_2}(a_2) h^{N+|\alpha|}.$$

Applying Lemma 4.1.2, we obtain that the function

$$\rho_{x,\xi,h}(z, \zeta) = \int_0^1 \frac{(1-\theta)^{N-1}}{(N-1)!} (J^\theta b_{x,\xi,h})(z, \zeta) d\theta$$

belongs to $S^{-N}_{scl}(\mathbb{R}^{2n}_{z,\zeta})$ uniformly with respect to x, ξ, h, so that in particular

$$\sup_{(x,\xi,z,\zeta) \in \mathbb{R}^{4n}, h \in (0,1]} |\rho_{x,\xi,h}(z, \zeta) h^{-N}| = C_0 < +\infty.$$

Since $r_N(a_1, a_2)(x, \xi) = \rho_{x,\xi,h}(x, \xi)$, we obtain

$$|r_N(a_1, a_2)(x, \xi)| \leq C_0 h^N. \quad (1.1.38)$$

Using the formula (1.1.37) above gives as well the smoothness of $r_N(a_1, a_2)$ and with the identities (consequences of $\partial_{x_j}(a_1 \diamond a_2) = (\partial_{x_j} a_1) \diamond a_2 + a_1 \diamond (\partial_{x_j} a_2)$)

$$\partial_{x_j}\big(r_N(a_1, a_2)\big) = r_N(\partial_{x_j} a_1, a_2) + r_N(a_1, \partial_{x_j} a_2)$$

$$\partial_{\xi_j}\big(r_N(a_1, a_2)\big) = r_N(\partial_{\xi_j} a_1, a_2) + r_N(a_1, \partial_{\xi_j} a_2),$$

it is enough to reapply (1.1.38) to get the result $r_N \in S^{-N}_{scl}$. $\square$

Lemma 4.1.2 and Taylor's expansion (1.1.37) give the following result.

Theorem 1.1.33. *Let $a \in S^m_{scl}$. Then $a^* = J\bar{a}$ and the mapping $a \mapsto a^*$ is continuous from S^m_{scl} into itself. Moreover, for all integers N, we have*

$$a^* = \sum_{|\alpha| < N} \frac{1}{\alpha!} D_\xi^\alpha \partial_x^\alpha \bar{a} + r_N(a), \quad r_N(a) \in S^{m-N}_{scl}.$$

Corollary 1.1.34. *Let $a_j \in S^{m_j}_{scl}, j = 1, 2$. Then we have*

$$a_1 \diamond a_2 \equiv a_1 a_2 \mod S^{m_1+m_2-1}_{scl}, \quad (1.1.39)$$

$$a_1 \diamond a_2 - a_2 \diamond a_1 \equiv \frac{1}{2i\pi} \{a_1, a_2\} \mod S^{m_1+m_2-2}_{scl}, \quad (1.1.40)$$

$$for \; a \in S^m_{scl}, \quad a^* \equiv \bar{a} \mod S^{m-1}_{scl}. \quad (1.1.41)$$

Lemma 1.1.35. *Let $\mu \in \mathbb{R}$ and $(c_j)_{j\in\mathbb{N}}$ be a sequence of symbols such that $c_j \in S_{scl}^{\mu-j}$. Then there exists $c \in S_{scl}^{\mu}$ such that*

$$c \sim \sum_j c_j, \qquad i.e., \quad \forall N \in \mathbb{N}, \quad c - \sum_{0\le j<N} c_j \in S_{scl}^{\mu-N}.$$

Proof. The proof is almost identical to the proof of Lemma 1.1.24. Let $\omega \in C_b^\infty(\mathbb{R};\mathbb{R}_+)$ such that $\omega(t) = 0$ for $t \le 1$ and $\omega(t) = 1$ for $t \ge 2$. Let $(\lambda_j)_{j\in\mathbb{N}}$ be a sequence of numbers ≥ 1. We want to define

$$c(x,\xi,h) = \sum_{j\ge 0} c_j(x,\xi,h)\omega(h^{-1}\lambda_j^{-1}), \qquad (1.1.42)$$

and we shall show that a suitable choice of λ_j will provide the answer. Multiplying the c_j by h^μ, we may assume that $\mu = 0$. We have then

$$|c_j(x,\xi,h)|\omega(h^{-1}\lambda_j^{-1}) \le \gamma_0(c_j)h^j \mathbf{1}\{1 \ge h\lambda_j\} \le \gamma_0(c_j)\lambda_j^{-j},$$

so that,

$$\forall j \ge 1, \ \lambda_j \ge 2\gamma_0(c_j)^{\frac{1}{j}} = \mu_j^{(0)} \implies \forall j \ge 1, \ |c_j(x,\xi,h)|\omega(h^{-1}\lambda_j^{-1}) \le 2^{-j},$$

showing that the function c can be defined as above in (1.1.42) and is a continuous bounded function. Let $1 \le k \in \mathbb{N}$ be given. Calculating (with $\omega_j = \omega(h^{-1}\lambda_j^{-1})$) the derivatives $\omega_j\partial_\xi^\alpha\partial_x^\beta(c_j)$ for $|\alpha| + |\beta| = k$, we get

$$\omega_j|\partial_\xi^\alpha\partial_x^\beta(c_j)| \le \gamma_k(c_j)h^{j+|\alpha|}\mathbf{1}\{1 \ge h\lambda_j\} \le \gamma_k(c_j)\lambda_j^{-j/2}h^{|\alpha|+\frac{j}{2}},$$

so that

$$\forall j \ge k, \ \lambda_j \ge 2^2\big(\gamma_k(c_j)\big)^{\frac{2}{j}} = \mu_j^{(k)} \implies \forall j \ge k, \ |\partial_\xi^\alpha\partial_x^\beta(c_j\omega_j)| \le 2^{-j}h^{|\alpha|+\frac{j}{2}},$$
$$(1.1.43)$$

showing that the function c can be defined as above in (1.1.42) and is a C^k function such that

$$|(\partial_\xi^\alpha\partial_x^\beta c)(x,\xi,h)| \le \sum_{0\le j<k} \gamma_k(c_j)h^{j+|\alpha|} + \sum_{j\ge k} 2^{-j}h^{|\alpha|} \le C_k h^{|\alpha|}.$$

It is possible to fulfill the conditions on the λ_j above for all $k \in \mathbb{N}$: just take $\lambda_j \ge \sup_{0\le k\le j} \mu_j^{(k)}$. The function c belongs to S_{scl}^0 and, with $S_{scl}^{-\infty} = \cap_{m\in\mathbb{R}}S_{scl}^m$,

$$r_N = c - \sum_{0\le j<N} c_j = \sum_{0\le j<N} \underbrace{(\omega_j - 1)c_j}_{\in S_{scl}^{-\infty}} + \sum_{j\ge N} c_j\omega_j,$$

and for $|\alpha| + |\beta| = k$, using the estimates (1.1.43), we obtain

$$\sum_{j\geq N} |\partial_\xi^\alpha \partial_x^\beta (c_j\omega_j)(x,\xi,h)| \leq \underbrace{\sum_{N\leq j<\max(2N,k)} |\partial_\xi^\alpha \partial_x^\beta(c_j\omega_j)(x,\xi,h)|}_{\lesssim h^{|\alpha|+j}\lesssim h^{|\alpha|+N}}$$

$$+ \sum_{j\geq\max(2N,k)} \underbrace{|\partial_\xi^\alpha \partial_x^\beta (c_j\omega_j)(x,\xi,h)|}_{\lesssim 2^{-j}h^{|\alpha|+\frac{j}{2}}\lesssim 2^{-j}h^{|\alpha|+N}},$$

proving that $r_N \in S_{scl}^{-N}$. The proof of the lemma is complete. $\qquad\square$

Remark 1.1.36. These asymptotic results (as well as the example $a_1(x,h\xi)$ with $a_1 \in C_b^\infty(\mathbb{R}^{2n})$ see page 22) led many authors to set a slightly different framework for the semi-classical calculus; instead of dealing with a family of symbols $a(x,\xi,h)$ satisfying the estimates (1.1.33), one deals with a function $a \in C_b^\infty(\mathbb{R}^{2n})$ and considers the operator $a(x,hD_x)$ or the operator $a(x,h\xi)^w$; another way to express this is to modify the quantization formula and to define for instance

$$(a^{w_h}u)(x) = \iint e^{\frac{2i\pi}{h}\langle x-y,\xi\rangle} a(\frac{x+y}{2},\xi)u(y)dyd\xi h^{-n}, \text{ i.e., } a^{w_h} = a(x,h\xi)^w. \tag{1.1.44}$$

Then, using Lemma 1.1.35, given a sequence $(a_j)_{j\geq 0}$ in $C_b^\infty(\mathbb{R}^{2n})$, it is possible to consider $a(x,\xi,h) \in S_{scl}^0$ with

$$a(x,\xi,h) \sim \sum_{j\geq 0} h^j a_j(x,h\xi), \quad \text{i.e., } \forall N, \ a(x,\xi,h) - \sum_{0\leq j<N} h^j a_j(x,h\xi) \in S_{scl}^{-N}.$$

The symbol a_0 is the *principal symbol* and

$$a(x,\xi,h)^w \sim \sum_{j\geq 0} h^j a_j^{w_h}, \quad \text{i.e., } \forall N, \ a(x,\xi,h)^w - \sum_{0\leq j<N} h^j a_j^{w_h} = h^N r_{N,h}^{w_h},$$

where $\{r_{N,h}\}_{0<h\leq 1}$ is bounded in $C_b^\infty(\mathbb{R}^{2n})$: in fact we have from Theorem 1.1.32,

$$h^N r_{N,h}(x,h\xi) = s_N(x,\xi,h), \quad s_N \in S_{scl}^{-N}, \quad \text{i.e., } r_{N,h}(x,\xi) = h^{-N}s_N(x,h^{-1}\xi,h),$$

and thus

$$|(\partial_\xi^\alpha \partial_x^\beta r_{N,h})(x,\xi) = h^{-N-|\alpha|}(\partial_\xi^\alpha \partial_x^\beta s_N)(x,h^{-1}\xi,h)| \leq h^{-N-|\alpha|}\gamma_{\alpha,\beta,N}h^{N+|\alpha|}.$$

If $a,b \in S_{scl}^0$ and $a \sim \sum_{j\geq 0} h^j a_j(x,h\xi), b \sim \sum_{j\geq 0} h^j b_j(x,h\xi)$ as above, then one can prove, using Corollary 1.1.34 and Lemma 4.1.2

$$a^w b^w \equiv (a_0 b_0)^{w_h} \mod h(S_{scl}^0)^w, \tag{1.1.45}$$

$$[a^w, b^w] \equiv \frac{h}{2i\pi}\{a_0,b_0\}^{w_h} \mod h^2(S_{scl}^0)^w. \tag{1.1.46}$$

There are many variations on this theme, and in particular, one can replace the space $C_b^\infty(\mathbb{R}^{2n})$ by a more general one, involving some weight functions, for instance with polynomial growth at infinity. At this point, we are leaving an introduction to the pseudo-differential calculus and will use our more general approach of Chapter 2, involving metrics on the phase space, which incorporate all these variations. Expecting these generalizations, we shall not use the w_h quantization in this book, except for the present remark.

Theorem 1.1.37. *Let a be a symbol in S_{scl}^0 such that*

$$\inf_{(x,\xi)\in\mathbb{R}^{2n},h\in(0,1]} |a(x,\xi,h)| > 0.$$

Then there exists $b \in S_{scl}^0$ such that

$$\begin{aligned} b(x,D,h)a(x,D,h) &= \mathrm{Id} + l(x,D,h), \\ a(x,D,h)b(x,D,h) &= \mathrm{Id} + r(x,D,h), \end{aligned} \quad r,l \in S_{scl}^{-\infty} = \cap_\nu S_{scl}^\nu.$$

Proof. The only change to perform in the proof of Theorem 1.1.23 to get this result is to replace everywhere $S_{1,0}$ by S_{scl}. □

Theorem 1.1.38. *Let χ be a symbol in S_{scl}^0 and let a be a symbol in S_{scl}^0 such that $\inf_{h\in(0,1],(x,\xi)\in\text{supp}\,\chi(\cdot,\cdot,h)} |a(x,\xi,h)| > 0$. Let ψ be a symbol in S_{scl}^0 such that $\text{supp}\,\psi(\cdot,\cdot,h) \subset \{(x,\xi), \chi(x,\xi,h) = 1\}$. Then there exists $b \in S_{scl}^0$ such that*

$$b(x,D,h)a(x,D,h) = \psi(x,D,h) + l(x,D,h), \quad l \in S_{scl}^{-\infty}.$$

Proof. Here also we have only to follow the proof of Theorem 1.1.25 and use Lemma 1.1.35 instead of Lemma 1.1.24 in the course of the proof. □

Theorem 1.1.39. *Let a be a non-negative symbol in S_{scl}^0. Then there exists a constant C such that, for all $u \in \mathscr{S}(\mathbb{R}^n)$,*

$$\mathrm{Re}\langle a(x,D,h)u, u\rangle + hC\|u\|_{L^2(\mathbb{R}^n)}^2 \geq 0. \tag{1.1.47}$$

Equivalently, there exists $C \geq 0$ such that $a^w + Ch \geq 0$.

Proof. The proof of Theorem 1.1.26 contains a proof of this result: noticing that it is harmless to replace the standard quantization by the Weyl quantization for this result, since $J^{1/2}a - a$ belongs to S_{scl}^{-1} (see the formula (4.1.20) and Lemma 4.1.5), we use the formula (1.1.31) to obtain $(a * \Gamma_h)^w \geq 0$. The difference $a * \Gamma_h - a$ is $\int_0^1 (1-\theta) \int_{\mathbb{R}^{2n}} a''(X + \theta Y, h)Y^2 \Gamma_h(Y) dY d\theta$, which belongs to S_{scl}^0. □

1.1.5 Other classes of symbols

Shubin's classes of pseudo-differential operators

If one considers a polynomial $p \in \mathbb{C}[\mathbb{R}^{2n}]$ with degree $2m$, it satisfies the following estimates, using the notation $\langle X \rangle = (1 + |x|^2 + |\xi|^2)^{1/2}$ for $X = (x,\xi) \in \mathbb{R}^n \times \mathbb{R}^n$,

$$|(\partial_\xi^\alpha \partial_x^\beta p)(x,\xi)| \leq C_{\alpha\beta} \langle X \rangle^{2m-|\alpha|-|\beta|}. \tag{1.1.48}$$

It is then natural to consider the class Σ^m of smooth symbols defined by (1.1.48). Since $\Sigma^0 \subset C_b^\infty(\mathbb{R}^{2n})$, Theorem 1.1.4 gives the $L^2(\mathbb{R}^n)$-boundedness of $\operatorname{op}(\Sigma^0)$. It would be tedious to repeat the arguments of the two previous sections and we refer the reader to Chapter 2 and in particular to the introductory Section 2.2.1 to see how that class fits in the general framework of classes defined by a metric on the phase space.

Let us simply justify our notation Σ^m: if $a_j \in \Sigma^{m_j}, j = 1, 2$, the Poisson bracket $\{a_1, a_2\} = \partial_\xi a_1 \cdot \partial_x a_2 - \partial_\xi a_2 \cdot \partial_x a_1 \in$ belongs to $\Sigma^{m_1+m_2-1}$ and in particular satisfies the estimate $|\{a_1, a_2\}| \le C\langle X\rangle^{2m_1-1+2m_2-1}$. To keep the statements of Corollaries 1.1.22, 1.1.34, we need to consider the symbols satisfying (1.1.48) as symbols of "order m".

1.2   Pseudo-differential operators on an open subset of $\mathbb{R}^n$

1.2.1   Introduction

The main reason for studying the class $S_{1,0}^m$ of pseudo-differential operators as introduced in the second subsection of Section 1.1.3 is that the parametrix of an elliptic differential operator of order m has a symbol in the class $S_{1,0}^{-m}$. More specifically, we have the following result.

Proposition 1.2.1. *Let m be a non-negative integer, Ω an open set of $\mathbb{R}^n$ and let $A = \sum_{|\alpha|\le m} a_\alpha(x)D_x^\alpha$ be a differential [7] operator with C^∞ coefficients on Ω (i.e., $a_\alpha \in C^\infty(\Omega)$). We assume that A is elliptic, i.e.,*

$$\forall (x, \xi) \in \Omega \times (\mathbb{R}^n\backslash\{0\}), \quad \sum_{|\alpha|=m} a_\alpha(x)\xi^\alpha \ne 0.$$

Then, if u is a distribution on Ω such that Au belongs to $H_{loc}^s(\Omega)$, we obtain that u belongs to $H_{loc}^{s+m}(\Omega)$, implying that singsupp $u =$ singsupp Au *(for the C^∞ singular supports[8]).*

This result will be proven in the next subsection in a far greater generality; first we shall use the notion of wave-front-set which microlocalizes the notion of singular support and also we shall prove this result for a microelliptic pseudo-differential operator. In fact, the proof relies essentially on Theorem 1.1.23 which allows the invertibility of an operator of the same type as A above. Nevertheless, one should note that the function $(x, \xi) \mapsto \sum_{|\alpha|\le m} a_\alpha(x)\xi^\alpha$ does not belong to $S_{1,0}^m$ since in the first place it is not defined on $\mathbb{R}^{2n}$ when $\Omega \ne \mathbb{R}^n$, and even if Ω were equal to $\mathbb{R}^n$, we do not have any control on the growth of the a_α at infinity. Also we see that the ellipticity condition concerns only the *principal symbol*, i.e.,

---

[7]We use the notation (4.1.4) for the D_x^α.

[8]For $v \in \mathscr{D}'(\Omega)$, $(\text{singsupp } v)^c$ is the union of the open subsets ω of Ω such that $v_{|\omega} \in C^\infty(\omega)$.

the function $\sum_{|\alpha|=m} a_\alpha(x)\xi^\alpha$. To get a good understanding (and a simple proof) of the previous result, we have to introduce the notion of pseudo-differential operator on an open set of $\mathbb{R}^n$, as well as the proper notion of ellipticity. The elliptic regularity theorem will be a simple consequence of the calculus of pseudo-differential operators on an open set of $\mathbb{R}^n$. One of the most important results of this theory is that pseudo-differential operators are geometrical objects that can be defined on a smooth manifold without reference to a coordinate chart; this invariance by change of coordinates has had a tremendous influence on the success of microlocal methods in geometrical problems such as the index theorem.

Definition 1.2.2. Let Ω be an open subset of $\mathbb{R}^n$ and $m \in \mathbb{R}$. $S_{loc}^m(\Omega \times \mathbb{R}^n)$ is defined as the set of $a \in C^\infty(\Omega \times \mathbb{R}^{2n})$ such that for any compact subset K of Ω, for all multi-indices $\alpha, \beta \in \mathbb{N}^n$, there exists $C_{K\alpha\beta}$ such that, for $(x, \xi) \in K \times \mathbb{R}^n$,

$$|(\partial_\xi^\alpha \partial_x^\beta a)(x, \xi)| \leq C_{K\alpha\beta}\langle\xi\rangle^{m-|\alpha|}. \tag{1.2.1}$$

We note in particular that the differential operators of order m with C^∞ coefficients in Ω have a symbol in $S_{loc}^m(\Omega \times \mathbb{R}^n)$, i.e., can be written as

$$(Au)(x) = \int e^{2i\pi x \cdot \xi} a(x, \xi)\hat{u}(\xi)d\xi, \quad \text{for } u \in C_c^\infty(\Omega), \tag{1.2.2}$$

with $a(x, \xi) = \sum_{|\alpha|\leq m} a_\alpha(x)\xi^\alpha, a_\alpha \in C^\infty(\Omega)$.

Theorem 1.2.3. *Let Ω be an open set of $\mathbb{R}^n$ and let a be a symbol in $S_{loc}^m(\Omega \times \mathbb{R}^n)$. Then the formula (1.2.2) defines a continuous linear operator (denoted also by $a(x, D)$) from $C_c^\infty(\Omega)$ into $C^\infty(\Omega)$, from $\mathscr{E}'(\Omega)$ into $\mathscr{D}'(\Omega)$, and from $H_{comp}^{s+m}(\Omega)$ to $H_{loc}^s(\Omega)$ for all $s \in \mathbb{R}$.*

Proof. To obtain the last result, we note that for $\chi \in C_c^\infty(\Omega)$, the operator $\chi(x)a(x, D)$ has the symbol $\chi(x)a(x, \xi)$ which belongs to $S_{1,0}^m$ and thus, from Theorem 1.1.18, $\chi(x)a(x, D)$ sends continuously $H^{s+m}(\mathbb{R}^n)$ into $H^s(\mathbb{R}^n)$, which gives that $a(x, D)$ sends continuously $H_{comp}^{s+m}(\Omega)$ into $H_{loc}^s(\Omega)$. This implies also that the formula (1.2.2) defines an operator from $\mathscr{S}(\mathbb{R}^n)$ into $C^\infty(\Omega)$. Moreover the formula (1.2.2) defines a mapping from $\mathscr{E}'(\Omega)$ into $\mathscr{D}'(\Omega)$, via the identity[9]

$$\prec a(x, D)u, \varphi \succ_{\mathscr{D}'(\Omega), \mathscr{D}(\Omega)} = \prec \hat{u}(\xi), \int \varphi(x)a(x, \xi)e^{2i\pi x \cdot \xi}dx \succ_{\mathscr{S}'(\mathbb{R}^n), \mathscr{S}(\mathbb{R}^n)}. \quad \square$$

---

[9] Using Theorem 1.1.2, we see that for $\varphi \in C_c^\infty(\Omega)$, the function

$$\xi \mapsto V_\varphi(\xi) = \int_{\mathbb{R}^n} a(x, \xi)\varphi(x)e^{2i\pi x \cdot \xi}dx$$

belongs to $\mathscr{S}(\mathbb{R}^n)$: for $\chi \in C_c^\infty(\Omega)$ equal to 1 on the support of φ, we consider the symbol $b(x, \xi) = \chi(\xi)a(\xi, x)\langle x\rangle^{-m}$ which belongs to $C_b^\infty(\mathbb{R}^{2n})$ and we have

$$V_\varphi(\xi) = \langle\xi\rangle^m \int_{\mathbb{R}^n} \chi(x)a(x, \xi)\langle\xi\rangle^{-m}\varphi(x)e^{2i\pi x \cdot \xi}dx = \langle\xi\rangle^m (\text{op}(b)\hat{\varphi})(\xi).$$

Definition 1.2.4. Let Ω be an open set of $\mathbb{R}^n$ and $m \in \mathbb{R}$. The set of operators $\{a(x,D)\}_{a \in S^m_{loc}(\Omega \times \mathbb{R}^n)}$ as given by the formula (1.2.2) is defined as $\Psi^m(\Omega)$, the set of pseudo-differential operators of order m on Ω.

We have to modify slightly the quantization of our symbols to get an algebra of operators, sending for instance $C^\infty_c(\Omega)$ into itself. Let us consider a locally finite partition of unity (see Section 4.3.4 in the appendix) in Ω, $\mathbf{1}_\Omega(x) = \sum_{j \in \mathbb{N}} \varphi_j(x)$ where each φ_j belongs to $C^\infty_c(\Omega)$. Let a be a symbol in $S^m_{loc}(\Omega \times \mathbb{R}^n)$ and A be the operator defined by the formula (1.2.2). We consider the operator

$$\widetilde{A} = \sum_{\substack{j,k \\ \text{supp}\,\varphi_j \cap \text{supp}\,\varphi_k \neq \emptyset}} \varphi_j A \varphi_k. \tag{1.2.3}$$

The operator $\varphi_j A \varphi_k$ has a symbol in $S^m_{1,0}$ which is given by $\varphi_j a \diamond \varphi_k$. We consider the finite sets

$$J_j = \{k \in \mathbb{N}, \text{supp}\,\varphi_j \cap \text{supp}\,\varphi_k \neq \emptyset\} \tag{1.2.4}$$

and the function

$$\Phi_j = \sum_{k \in J_j} \varphi_k \in C^\infty_c(\Omega), \quad \Phi_j = 1 \text{ on a neighborhood of supp}\,\varphi_j, \tag{1.2.5}$$

(see Lemma 4.3.6) so that for all multi-indices α

$$\varphi_j(x) \partial^\alpha_x (1 - \Phi_j)(x) = 0. \tag{1.2.6}$$

We check now the symbol $\tilde{a} = \sum_j \varphi_j a \diamond \Phi_j$ of $\widetilde{A}$. Given a compact subset of Ω it meets only finitely many supp φ_j and thus $\tilde{a}$ belongs to $S^m_{loc}(\Omega \times \mathbb{R}^n)$. On the other hand, we have on $\Omega \times \mathbb{R}^n$,

$$a - \tilde{a} = \sum_j \varphi_j a - \varphi_j a \diamond \Phi_j = \sum_j \varphi_j a \diamond (1 - \Phi_j)$$

and we get from Theorem 1.1.20 and (1.2.6) that that each $\varphi_j a \diamond (1 - \Phi_j)$ belongs to $S^{-\infty}_{1,0}$; moreover the sum is locally finite, so that $a - \tilde{a} \in S^{-\infty}_{loc}(\Omega \times \mathbb{R}^n) = \cap_{m \in \mathbb{R}} S^m_{loc}(\Omega \times \mathbb{R}^n)$.

Proposition 1.2.5. *Let Ω be an open set of $\mathbb{R}^n$, let a be a symbol in $S^m_{loc}(\mathbb{R}^n)$. There exists a symbol $\tilde{a} \in S^m_{loc}(\Omega \times \mathbb{R}^n)$ such that*

(i) *$a - \tilde{a} \in S^{-\infty}_{loc}(\Omega \times \mathbb{R}^n)$, $a(x,D) - \tilde{a}(x,D)$ sends $\mathscr{E}'(\Omega)$ into $C^\infty(\Omega)$,*

(ii) *the operator $\tilde{a}(x,D)$ is properly supported[10], and sends continuously $C^\infty_c(\Omega)$ into itself, $C^\infty(\Omega)$ into itself, $\mathscr{E}'(\Omega)$ into itself, $\mathscr{D}'(\Omega)$ into itself,*

---

[10]A continuous linear operator $A : \mathscr{D}(V) \longrightarrow \mathscr{D}'(U)$ is said to be properly supported when both projections of the support of the kernel k from supp k in U, V are proper, i.e., for every compact $L \subset V$, there exists a compact $K \subset U$ such that supp $v \subset L \Longrightarrow$ supp $Av \subset K$ and for every compact $K \subset U$, there exists a compact $L \subset V$ such that supp $v \subset L^c \Longrightarrow$ supp $Av \subset K^c$.

(iii) $\tilde{a}(x,D)$ *defines a continuous linear operator from* $H^{s+m}_{comp}(\Omega)$ *to* $H^s_{comp}(\Omega)$, *from* $H^{s+m}_{loc}(\Omega)$ *to* $H^s_{loc}(\Omega)$.

Proof. We have already proven (i), using $\mathscr{E}'(\Omega) = \cup_s H^s_{comp}(\Omega)$ and Theorem 1.2.3. Using the above notation, we get for $u \in C^\infty_c(\Omega)$,

$$\tilde{a}(x,D)u = \sum_j \varphi_j a(x,D) \Phi_j u \qquad (1.2.7)$$

with a finite sum of $C^\infty_c(\Omega)$ functions since $\operatorname{supp} u$ meets only finitely many $\operatorname{supp}\Phi_j$. If $u \in C^\infty(\Omega)$, we have $\Phi_j u \in C^\infty_c(\Omega) \subset C^\infty_c(\mathbb{R}^n)$ and $\sum_j \varphi_j a(x,D)\Phi_j u$ is a locally finite sum of $C^\infty_c(\Omega)$ functions, thus a $C^\infty(\Omega)$ function. For $u \in \mathscr{E}'(\Omega)$ with a (compact) support $K \subset \Omega$, the $\Phi_j u$ are all zero, except for a finite set of indices J_K and then $\sum_{j \in J_K} \varphi_j a(x,D)\Phi_j u$ belongs to $\mathscr{E}'(\Omega)$. If $u \in \mathscr{D}'(\Omega)$, we have $\Phi_j u \in \mathscr{E}'(\Omega)$ and $\sum_j \varphi_j a(x,D)\Phi_j u$ is a locally finite sum of distributions in Ω and thus a distribution on Ω, proving (ii). The assertion (iii) and the continuity properties are direct consequences of (ii) and of Theorem 1.2.3. $\qquad\square$

Remark 1.2.6. Let us now consider a symbol a belonging to $S^m_{loc}(\Omega \times \mathbb{R}^n)$. We can quantify this symbol into a properly supported operator, say $\operatorname{Op}_\Omega(a)$, given by the formula (1.2.7), which has the properties of the operator $\tilde{a}(x,D)$ in Proposition 1.2.5. This quantization defines a linear mapping from $S^m_{loc}(\Omega \times \mathbb{R}^n)$ to the quotient $\Psi^m_{ps}(\Omega)/\Psi^{-\infty}_{ps}(\Omega)$, where $\Psi^m_{ps}(\Omega)$ stands for the properly supported pseudo-differential operators of order m on Ω, and $\Psi^{-\infty}_{ps}(\Omega) = \cap_{m \in \mathbb{R}} \Psi^m_{ps}(\Omega)$. A change in the choice of the partition of unity (φ_j) will not change this mapping. From Proposition 1.2.5, we see that the operators of $\Psi^m_{ps}(\Omega)$ are continuous from $C^\infty_c(\Omega)$ into itself, from $C^\infty(\Omega)$ into itself, from $\mathscr{E}'(\Omega)$ into itself, from $\mathscr{D}'(\Omega)$ into itself, from $H^{s+m}_{comp}(\Omega)$ into $H^s_{comp}(\Omega)$ and from $H^{s+m}_{loc}(\Omega)$ into $H^s_{loc}(\Omega)$. Note also that if $A \in \Psi^{-\infty}_{ps}(\Omega)$ and $u \in \mathscr{D}'(\Omega)$, if ω is a relatively compact open subset of Ω, u belongs to $H^s_{loc}(\omega)$ for some s and thus $Au \in H^{+\infty}_{loc}(\omega)$ so that $Au \in C^\infty(\omega)$, proving that $\Psi^{-\infty}_{ps}(\Omega)$ sends $\mathscr{D}'(\Omega)$ into $C^\infty(\Omega)$.

Theorem 1.2.7. *Let* Ω *be an open set of* $\mathbb{R}^n$, $m_1, m_2 \in \mathbb{R}$. *Let* $a_j \in S^{m_j}_{loc}(\Omega \times \mathbb{R}^n)$. *Then the operator* $\operatorname{Op}_\Omega(a_1)\operatorname{Op}_\Omega(a_2)$ *belongs to* $\Psi^{m_1+m_2}_{ps}(\Omega)$ *and is such that,*

$$\operatorname{Op}_\Omega(a_1)\operatorname{Op}_\Omega(a_2) = \operatorname{Op}_\Omega(a_1 a_2) \quad \mathrm{mod}\ \Psi^{m_1+m_2-1}_{ps}(\Omega),$$

$$\operatorname{Op}_\Omega(a_1)\operatorname{Op}_\Omega(a_2) = \operatorname{Op}_\Omega(a_1 a_2 + D_\xi a_1 \cdot \partial_x a_2) \quad \mathrm{mod}\ \Psi^{m_1+m_2-2}_{ps}(\Omega),$$

and more generally, for all $N \in \mathbb{N}$,

$$\operatorname{Op}_\Omega(a_1)\operatorname{Op}_\Omega(a_2) = \operatorname{Op}_\Omega\Big(\sum_{|\alpha| < N} \frac{1}{\alpha!} D^\alpha_\xi a_1 \partial^\alpha_x a_2 \Big) \quad \mathrm{mod}\ \Psi^{m_1+m_2-N}_{ps}(\Omega).$$

Proof. Let $\psi_1, \psi_2 \in C^\infty_c(\Omega)$ with $\psi_2 = 1$ on a neighborhood of the support of ψ_1.

From Theorem 1.1.20, Proposition 1.2.5 and Remark 1.2.6, we have

$$\psi_1 \text{Op}_\Omega(a_1)\psi_2 \text{Op}_\Omega(a_2) = (\psi_1 a_1)(x,D)(\psi_2 a_2)(x,D) \mod \Psi^{-\infty}(\Omega)$$

$$= \psi_1 \big(\sum_{|\alpha|<N} \frac{1}{\alpha!} D_\xi^\alpha a_1 \partial_x^\alpha a_2 \big)(x,D) \mod \Psi^{m_1+m_2-N}(\Omega)$$

$$= \psi_1 \text{Op}_\Omega \big(\sum_{|\alpha|<N} \frac{1}{\alpha!} D_\xi^\alpha a_1 \partial_x^\alpha a_2 \big) \mod \Psi^{m_1+m_2-N}(\Omega),$$

and since the lhs and the first term in the rhs are both properly supported, the equality holds $\mod \Psi_{ps}^{m_1+m_2-N}(\Omega)$. It means that for all $\psi_1 \in C_c^\infty(\Omega)$, we have

$$\psi_1 \text{Op}_\Omega(a_1)\text{Op}_\Omega(a_2) = \psi_1 \text{Op}_\Omega \big(\sum_{|\alpha|<N} \frac{1}{\alpha!} D_\xi^\alpha a_1 \partial_x^\alpha a_2 \big) \mod \Psi_{ps}^{m_1+m_2-N}(\Omega).$$

Since the operator $\text{Op}_\Omega(a)$ is properly supported and completely determined mod $\Psi_{ps}^{-\infty}(\Omega)$ by its definition on $C_c^\infty(\Omega)$, it concludes the proof. $\square$

Let $a \in S_{loc}^m(\Omega \times \mathbb{R}^n)$ and let us consider as above the operator $\text{Op}_\Omega(a) = \sum_{j \sim k} \varphi_j a(x,D)\varphi_k$, $j \sim k$ meaning $\text{supp}\,\varphi_j \cap \text{supp}\,\varphi_k \neq \emptyset$. With Φ_j given by (1.2.5), we have $\text{Op}_\Omega(a) = \sum_{j \sim k} \varphi_j \Phi_j a(x,D)\varphi_k$ and thus the adjoint operator is

$$\sum_{j \sim k} \varphi_k J(\overline{\Phi_j a})(x,D)\varphi_j.$$

Since $J(\overline{\Phi_j a}) = \sum_{|\alpha|<N} \frac{1}{\alpha!} D_\xi^\alpha \partial_x^\alpha (\Phi_j \bar{a}) + r_{N,j}$ with $r_{N,j} \in S_{1,0}^{m-N}$, we get

$$(\text{Op}_\Omega(a))^* = \sum_{j \sim k} \varphi_k \sum_{|\alpha|<N} \frac{1}{\alpha!} D_\xi^\alpha \partial_x^\alpha (\Phi_j \bar{a})(x,D)\varphi_j + \sum_{j \sim k} \varphi_k r_{N,j}(x,D)\varphi_j.$$

Let $a^* \in S_{loc}^m(\Omega \times \mathbb{R}^n)$ such that for all N,

$$a^* - \sum_{|\alpha|<N} \frac{1}{\alpha!} D_\xi^\alpha \partial_x^\alpha \bar{a} \in S_{loc}^{m-N}(\Omega \times \mathbb{R}^n).$$

Since Φ_j is 1 near the support of φ_j, we obtain $(\text{Op}_\Omega(a))^* = (\text{Op}_\Omega(a^*))$ modulo $\Psi_{ps}^{-\infty}(\Omega)$.

1.2.2   Inversion of (micro)elliptic operators

Definitions

Let Ω be an open subset of $\mathbb{R}^n$ and $(x_0, \xi_0) \in \Omega \times (\mathbb{R}^n\backslash\{0\}) = \dot{T}^*(\Omega)$; a *conic-neighborhood* of (x_0, ξ_0) is defined as a subset of $\Omega \times \mathbb{R}^n\backslash\{0\}$ containing for some positive r the set

$$W_{x_0,\xi_0}(r) = \{(x,\xi) \in \mathbb{R}^n \times \mathbb{R}^n\backslash\{0\}, |x-x_0| < r, \left| \frac{\xi}{|\xi|} - \frac{\xi_0}{|\xi_0|} \right| < r, |\xi| > \frac{1}{r} \}. \quad (1.2.8)$$

Definition 1.2.8. Let $a \in S_{loc}^m(\Omega \times \mathbb{R}^n)$ and $(x_0, \xi_0) \in \Omega \times \mathbb{R}^n \backslash \{0\}$. The symbol a is said to be *elliptic at* (x_0, ξ_0), when there exists a conic-neighborhood W of (x_0, ξ_0) such that

$$\inf_{(x,\xi) \in W} |a(x,\xi)||\xi|^{-m} > 0. \tag{1.2.9}$$

The points of $\dot{T}^*(\Omega)$ where a is not elliptic are called characteristic points.

Let us give an example of an elliptic symbol of order 0 at (x_0, ξ_0). Considering a function $\chi_0 \in C_c^\infty(\mathbb{R}), \chi_0(t) = 1$ for $t \leq 1$, $\chi_0(t) = 0$ for $t \geq 2$, we define on $\mathbb{R}^{2n}$ for $r > 0$,

$$\theta_{r,x_0,\xi_0}(x,\xi) = \chi_0(r^{-2}|x - x_0|^2)\chi_0\left(r^{-2}\left|\frac{\xi}{|\xi|} - \frac{\xi_0}{|\xi_0|}\right|^2\right)(1 - \chi_0(2r^2|\xi|^2)). \tag{1.2.10}$$

It is easy to check that θ_{r,x_0,ξ_0} belongs to $S_{1,0}^0$, is elliptic at (x_0, ξ_0) (note that $\theta_{r,x_0,\xi_0} \equiv 1$ on $W_{x_0,\xi_0}(r)$ and $\mathrm{supp}\, \theta_{r,x_0,\xi_0} \subset W_{x_0,\xi_0}(2r)$).

Definition 1.2.9. A function a defined on $\Omega \times \mathbb{R}^n$ will be called *positively-homogeneous of degree* m when for all $\xi \in \mathbb{R}^n$ with $|\xi| \geq 1$ and all $t \geq 1$, $a_m(x, t\xi) = t^m a_m(x, \xi)$. A function a defined on $\Omega \times \mathbb{R}^n \backslash \{0\}$ will be called *positively-homogeneous of degree* m when for all $\xi \in \mathbb{R}^n \backslash \{0\}$ and all $t > 0$, $a_m(x, t\xi) = t^m a_m(x, \xi)$.

Lemma 1.2.10. *Let $a \in S_{loc}^m(\Omega \times \mathbb{R}^n)$ and $(x_0, \xi_0) \in \Omega \times \mathbb{R}^n \backslash \{0\}$ such that the symbol a is elliptic at (x_0, ξ_0). Then for $b \in S_{loc}^{m'}(\Omega \times \mathbb{R}^n)$ with $m' < m$, the symbol $a + b$ is elliptic at (x_0, ξ_0). In particular, if there exists $a_m \in C^\infty(\Omega \times \mathbb{R}^n)$ positively-homogeneous of degree m such that*

$$a_m(x_0, \xi_0/|\xi_0|) \neq 0, \quad a - a_m \in S_{loc}^{m-1}(\Omega \times \mathbb{R}^n),$$

then the symbol a is elliptic at (x_0, ξ_0). This is the case in particular of a differential operator with $C^\infty(\Omega)$ coefficients $\sum_{|\alpha| \leq m} a_\alpha(x) D_x^\alpha$ such that

$$0 \neq a_m(x_0, \xi_0)(= \sum_{|\alpha|=m} a_\alpha(x_0)\xi_0^\alpha).$$

Proof. The first part of the lemma is obvious since, for K a compact subset of Ω, $\lim_{|\xi| \to +\infty}(\sup_{x \in K} |b(x,\xi)|)|\xi|^{-m} = 0$. The second part is due to the fact that the property of homogeneity and the smoothness of a imply[11] that $a_m \in S_{loc}^m(\Omega \times \mathbb{R}^n)$. $\square$

Remark 1.2.11. Note that if $\mathrm{Op}_\Omega(a_1) = \mathrm{Op}_\Omega(a_2)$ with $a_j \in S_{loc}^m(\Omega \times \mathbb{R}^n)$, then $\mathrm{Op}_\Omega(a_1 - a_2) \in \Psi_{ps}^{-\infty}(\Omega)$ and thus

$$(a_1 - a_2)(x, D) = r(x, D) \text{ with } r \in S_{loc}^{-\infty}(\Omega \times \mathbb{R}^n).$$

---

[11] For $\omega \in C_b^\infty(\mathbb{R}^n)$ vanishing for $|\xi| \leq 1/2$ and equal to 1 for $|\xi| \geq 1$, we have in fact

$$a_m(x,\xi) = \omega(\xi)a_m(x,\xi/|\xi|)|\xi|^m + (1 - \omega(\xi))a_m(x,\xi) \in S_{loc}^m(\Omega \times \mathbb{R}^n) + S_{loc}^{-\infty}(\Omega \times \mathbb{R}^n).$$

A consequence of Remark 4.1.6 is that, for all $\chi \in C_c^\infty(\Omega)$, $\chi(x)(a_1 - a_2)(x, \xi) = \chi(x)r(x, \xi)$ which gives $a_1 - a_2 = r$ (as functions of $S_{loc}^{-\infty}(\Omega \times \mathbb{R}^n)$). As a result, the characteristic points of a_1 and a_2 are the same, and one may define char $\mathrm{Op}_\Omega(a)$ as the characteristic points of a.

Lemma 1.2.12. *Let $a \in S_{loc}^m(\Omega \times \mathbb{R}^n)$ and $(x_0, \xi_0) \in \Omega \times \mathbb{R}^n \backslash \{0\}$ such that the symbol a is elliptic at (x_0, ξ_0). Then there exists $r > 0$ and $b \in S_{loc}^{-m}(\Omega \times \mathbb{R}^n)$, elliptic at (x_0, ξ_0) such that*

$$\mathrm{Op}_\Omega(b)\mathrm{Op}_\Omega(a) = \mathrm{Op}_\Omega(\theta_{r,x_0,\xi_0}) + \mathrm{Op}_\Omega(\rho),$$

with $\rho \in S_{loc}^{-\infty}(\Omega \times \mathbb{R}^n)$ and θ_{r,x_0,ξ_0} is given by (1.2.10).

Proof. Since a is elliptic at (x_0, ξ_0), we may assume that (1.2.9) holds for some conic-neighborhood $W_{x_0,\xi_0}(r_0)$. Let us consider the symbol $\theta_{r_1,x_0,\xi_0} \in S_{1,0}^0$ with $r_1 = r_0/2$ so that $\mathrm{supp}\,\theta_{r_1,x_0,\xi_0} \subset W_{x_0,\xi_0}(r_0)$. The assumption of Theorem 1.1.25 is verified with $\chi = \theta_{r_1,x_0,\xi_0}$. Considering $r_2 = r_0/4$ so that $\mathrm{supp}\,\theta_{r_2,x_0,\xi_0} \subset W_{x_0,\xi_0}(r_1) \subset \{\theta_{r_1,x_0,\xi_0} = 1\}$ we can find $b_1 \in S_{1,0}^{-m}$ such that, omitting the subscripts x_0, ξ_0,

$$b_1(x, D)\big(\theta_{r_1}a\big)(x, D) = \theta_{r_2}(x, D) + \rho(x, D), \quad \rho \in S_{1,0}^{-\infty},$$

implying with $r = r_0/8$,

$$\theta_r(x, D)b_1(x, D)\big(\theta_{r_1}a\big)(x, D) = \theta_r(x, D) + \tilde{\rho}(x, D), \quad \tilde{\rho} \in S_{1,0}^{-\infty},$$

and thus, with $b = \theta_r \diamond b_1$, which belongs to $S_{1,0}^m$, we have modulo $\Psi_{ps}^{-\infty}(\Omega)$,

$$\mathrm{Op}_\Omega(b)\mathrm{Op}_\Omega(a) \equiv \mathrm{Op}_\Omega(b)\mathrm{Op}_\Omega(\theta_{r_1}a) + \mathrm{Op}_\Omega(\theta_r \diamond b_1)\mathrm{Op}_\Omega((1 - \theta_{r_1})a)$$
$$\equiv \mathrm{Op}_\Omega(b)\mathrm{Op}_\Omega(\theta_{r_1}a)$$
$$\equiv \mathrm{Op}_\Omega(\theta_r). \qquad \square$$

Definition 1.2.13. Let Ω be an open set of $\mathbb{R}^n$, $a \in S_{loc}^m(\Omega \times \mathbb{R}^n)$, $A = \mathrm{Op}_\Omega(a)$. We define the essential support of A, denoted by $\mathrm{essupp}\,A$, as the complement in $\Omega \times \mathbb{R}^n \backslash \{0\}$ of the points (x_0, ξ_0) for which there exists a conic-neighborhood W so that a is of order $-\infty$ in W, i.e.,

$$\forall (N, \alpha, \beta) \in \mathbb{N} \times \mathbb{N}^n \times \mathbb{N}^n, \quad \sup_{(x,\xi) \in W} |(\partial_\xi^\alpha \partial_x^\beta a)(x, \xi)||\xi|^N < \infty.$$

Note that from Remark 1.2.11, this definition depends only on A and the essential support is a closed conic subset of $\dot{T}^*(\Omega)$. Thanks to Lemma 1.2.10, if $A = \mathrm{Op}_\Omega(a_m + b)$ with $a_m \in C^\infty(\Omega \times \mathbb{R}^n)$ positively-homogeneous of degree m and $b \in S_{loc}^{m-1}(\Omega \times \mathbb{R}^n)$, then char $A = \{(x, \xi) \in \dot{T}^*(\Omega), a_m(x, \xi) = 0\}$.

Remark 1.2.14. Let Ω be an open set of $\mathbb{R}^n$ and $(A, B) \in \Psi_{ps}^{m_1}(\Omega) \times \Psi_{ps}^{m_2}(\Omega)$. Then we have

$$\mathrm{essupp}(AB) \subset \mathrm{essupp}\,A \cap \mathrm{essupp}\,B. \qquad (1.2.11)$$

In fact if $\dot{T}^*(\Omega) \ni (x_0, \xi_0)$ belongs to $(\mathrm{essupp}\,A)^c \cup (\mathrm{essupp}\,B)^c$, the composition formula of Theorem 1.2.7 shows that (x_0, ξ_0) is in $(\mathrm{essupp}\,AB)^c$.

Theorem 1.2.15. *Let Ω be an open set of $\mathbb{R}^n$, $A \in \Psi_{ps}^m(\Omega)$. Let (x_0, ξ_0) be an elliptic point for A, i.e., $(x_0, \xi_0) \notin \operatorname{char} A$. Then there exist $B \in \Psi_{ps}^{-m}(\Omega), R, S \in \Psi_{ps}^0(\Omega)$, such that*

$$BA = \operatorname{Id} + R, \quad AB = \operatorname{Id} + S, \quad (x_0, \xi_0) \notin \operatorname{essupp} R, \quad (x_0, \xi_0) \notin \operatorname{essupp} S. \tag{1.2.12}$$

Proof. Lemma 1.2.12 implies the first result. On the other hand we can prove similarly that there exists $B_1 \in \Psi_{ps}^{-m}(\Omega)$ such that $AB_1 = \operatorname{Id} + S_1, (x_0, \xi_0) \notin \operatorname{essupp} S_1$. Now we see that

$$B = B(AB_1 - S_1) = (\operatorname{Id} + R)B_1 - BS_1 = B_1 + RB_1 - BS_1,$$

so that $AB = \operatorname{Id} + S, (x_0, \xi_0) \notin \operatorname{essupp} S$, (using (1.2.11)). The proof is complete. $\square$

The wave-front-set of a distribution

Definition 1.2.16. *Let Ω be an open set of $\mathbb{R}^n$ and $u \in \mathscr{D}'(\Omega)$. The wave-front-set of u, denoted by WFu, is the subset of $\dot{T}^*(\Omega)$ whose complement is given by*

$$(WFu)^c = \{(x, \xi) \in \dot{T}^*(\Omega), \exists W \text{conic-neighborhood of } (x, \xi) \text{ such that}$$
$$\forall a \in S_{loc}^m(\Omega \times \mathbb{R}^n) \text{ with } \operatorname{supp} a \subset W, \text{ we have } \operatorname{Op}_\Omega(a)u \in C^\infty(\Omega)\}. \tag{1.2.13}$$

Proposition 1.2.17. *Let Ω be an open set of $\mathbb{R}^n$ and $u \in \mathscr{D}'(\Omega)$. The wave-front-set of u is a closed conic subset of $\dot{T}^*(\Omega)$ whose canonical projection[12] on Ω is singsupp u. Moreover, we have*

$$(WFu)^c = \{(x, \xi) \in \dot{T}^*(\Omega), \exists a \in S_{loc}^0(\Omega \times \mathbb{R}^n) \text{ elliptic at } (x, \xi)$$
$$\text{with } Op_\Omega(a)u \in C^\infty(\Omega)\}. \tag{1.2.14}$$

Proof. The first assertion (closed conic) follows immediately from the definition. Now if $x_0 \notin \operatorname{singsupp} u$, there exists $r_0 > 0$ such that $u_{|B(x_0,r_0)}$ is C^∞; here, $B(x, r)$ stands for the open Euclidean ball of $\mathbb{R}^n$ with center x and radius r. As a result if $\xi_0 \in \mathbb{S}^{n-1}$, $a \in S_{loc}^m(\Omega \times \mathbb{R}^n)$ with $\operatorname{supp} a \subset W_{x_0,\xi_0}(r_1), r_1 = r_0/2$, $\chi_0 \in C_c^\infty(B(x_0, r_0)), \chi_0 = 1$ on $B(x_0, r_1)$,

$$\operatorname{Op}_\Omega(a)u = \overbrace{\operatorname{Op}_\Omega(a)}^{\in C_c^\infty(\Omega)} \underbrace{\chi_0 u}_{\in C_c^\infty(\Omega)} + \overbrace{\operatorname{Op}_\Omega(a)(1 - \chi_0)}^{\in C^\infty(\Omega)} u}_{\in \Psi_{ps}^{-\infty}(\Omega)}$$

since $A \in \Psi_{ps}^{-\infty}(\Omega)$ sends $\mathscr{D}'(\Omega)$ into $C^\infty(\Omega)$, proving that $\{x_0\} \times \mathbb{S}^{n-1} \subset (WFu)^c$. Conversely, if $x_0 \in \operatorname{singsupp} u$, there must exist some $\xi_0 \in \mathbb{S}^{n-1}$ such that

---

[12]This is the mapping $\dot{T}^*(\Omega) \ni (x, \xi) \mapsto x \in \Omega$.

$(x_0, \xi_0) \in WFu$, otherwise $\{x_0\} \times \mathbb{S}^{n-1} \subset (WFu)^c$ and using the compactness of $\mathbb{S}^{n-1}$, we could find an open neighborhood ω of x_0 in Ω, such that for all $a \in S^0_{loc}(\Omega \times \mathbb{R}^n), \operatorname{supp} a \subset \omega \times \mathbb{R}^n, \operatorname{Op}_\Omega(a)u \in C^\infty(\Omega)$; taking $a(x, \xi) = \chi(x)$ where $\chi \in C^\infty_c(\omega)$ would give $u \in C^\infty(\omega)$, contradicting $x_0 \in \operatorname{singsupp} u$. Calling N_u the complement of the set defined by (1.2.14), we see immediately that $(WFu)^c \subset N^c_u$; conversely, if $(x_0, \xi_0) \in N^c_u$, we can find A such that

$$A \in \Psi^0_{ps}(\Omega), \quad Au \in C^\infty(\Omega), \quad (x_0, \xi_0) \notin \operatorname{char} A.$$

Applying Theorem 1.2.15, we find $B \in \Psi^0_{ps}(\Omega)$ so that (1.2.12) holds and this implies, for $c \in S^m_{loc}(\Omega \times \mathbb{R}^n)$,

$$BAu = u + Ru \Longrightarrow \operatorname{Op}_\Omega(c)u = \underbrace{\operatorname{Op}_\Omega(c)BAu}_{\in C^\infty(\Omega)} - \operatorname{Op}_\Omega(c)Ru$$

and since $(x_0, \xi_0) \notin \operatorname{essupp} R$, there exists a conic-neighborhood W of (x_0, ξ_0) such that R is of order $-\infty$ in W so that, taking c supported in W will imply $\operatorname{Op}_\Omega(c)R \in \Psi^{-\infty}_{ps}(\Omega)$ and $\operatorname{Op}_\Omega(c)Ru \in C^\infty(\Omega)$, proving that $(x_0, \xi_0) \notin WFu$. The proof of the proposition is complete. □

Lemma 1.2.18. *Let Ω be an open set of $\mathbb{R}^n$ and $u \in \mathscr{D}'(\Omega)$. Then*

$$(WFu)^c = \{(x, \xi) \in \dot{T}^*(\Omega), \exists W \text{ conic-neighborhood of } (x, \xi) \text{ such that}$$
$$\forall A \in \Psi^m_{ps}(\Omega), \text{with } \operatorname{essupp} A \subset W, \text{we have } Au \in C^\infty(\Omega)\}. \quad (1.2.15)$$

Proof. Calling M_u the complement of the set defined by (1.2.15), we have obviously $M^c_u \subset (WFu)^c$ and conversely if $(x_0, \xi_0) \notin WFu$, there exists $r_0 > 0$ such that for all $a \in S^m_{loc}(\Omega \times \mathbb{R}^n)$ supported in $W_{x_0, \xi_0}(r_0)$, $\operatorname{Op}_\Omega(a)u \in C^\infty(\Omega)$. Now if $B \in \Psi^m_{ps}(\Omega)$, with $\operatorname{essupp} B \subset W_{x_0, \xi_0}(r_0/2)$, we have

$$B = \operatorname{Op}_\Omega(b) = \operatorname{Op}_\Omega(\underbrace{b\theta_{r_0/2}}_{\substack{\text{supported} \\ \text{in } W(r_0)}}) \mod \psi^{-\infty}_{ps}(\Omega)$$

and thus $Bu \in C^\infty(\Omega)$, completing the proof of the lemma. □

The elliptic regularity theorem

Theorem 1.2.19. *Let Ω be an open set of $\mathbb{R}^n$ and $A \in \Psi^m_{ps}(\Omega)$. Then for $u \in \mathscr{D}'(\Omega)$,*

$$WF(Au) \subset WFu \subset \operatorname{char} A \cup WF(Au).$$

Proof. If $(x_0, \xi_0) \notin WFu$, there exists a conic-neighborhood W of (x_0, ξ_0) such that (1.2.15) holds and taking $C \in \Psi^m_{ps}(\Omega)$ with $\operatorname{essupp} C \subset W$, we get from (1.2.11) that $\operatorname{essupp} CA \subset W$, and Lemma 1.2.18 implies that $(x_0, \xi_0) \notin WF(Au)$.

On the other hand, if $(x_0, \xi_0) \notin \operatorname{char} A$ and $(x_0, \xi_0) \notin WF(Au)$, Theorem 1.2.15 provides $B \in \Psi_{ps}^{-m}(\Omega)$ satisfying (1.2.12): we get

$$u = BAu - Ru, \quad (x_0, \xi_0) \notin \operatorname{essupp} R \quad \text{i.e., } R \text{ of order } -\infty \text{ on } W,$$

where W is a conic-neighborhood of (x_0, ξ_0). Taking $C \in \Psi_{ps}^m(\Omega)$ with $\operatorname{essupp} C \subset W$ we have

$$Cu = CBAu - \underbrace{CR}_{\in \Psi_{ps}^{-\infty}(\Omega)} u,$$

so that $CRu \in C^\infty(\Omega)$. On the other hand, since $(x_0, \xi_0) \notin WF(Au)$, thanks to Lemma 1.2.18, there exists a conic-neighborhood W_1 of (x_0, ξ_0) such that, for all $P \in \Psi_{ps}^m(\Omega)$ with $\operatorname{essupp} P \subset W_1$, we have $PAu \in C^\infty(\Omega)$. This proves that $CBAu \in C^\infty(\Omega)$, provided $\operatorname{essupp} C \subset W_1$ and with $\operatorname{essupp} C \subset W_1 \cap W$ we get $Cu \in C^\infty(\Omega)$, which implies $(x_0, \xi_0) \notin WFu$, using Lemma 1.2.18. $\qquad \square$

Corollary 1.2.20. *Let Ω be an open set of $\mathbb{R}^n$, $A \in \Psi_{ps}^m(\Omega)$. Then for $u \in \mathscr{D}'(\Omega)$, $\operatorname{singsupp}(Au) \subset \operatorname{singsupp} u \subset \operatorname{singsupp}(Au) \cup \operatorname{pr}(\operatorname{char} A)$ and in particular, if A is elliptic on Ω, i.e., $\operatorname{char} A = \emptyset$, we obtain that $\operatorname{singsupp} u = \operatorname{singsupp}(Au)$.*

Definition 1.2.21 (H^s **wave-front-set**). Let Ω be an open set of $\mathbb{R}^n$, $s \in \mathbb{R}$ and $u \in \mathscr{D}'(\Omega)$. The H^s wave-front-set of u, denoted by $WF_s u$, is the subset of $\dot{T}^*(\Omega)$ whose complement is given by

$$(WF_s u)^c = \{(x, \xi) \in \dot{T}^*(\Omega), \exists W \text{ conic-neighborhood of } (x, \xi) \text{ such that}$$
$$\forall A \in \psi^0(\Omega) \text{ with } \operatorname{essupp} A \subset W, \text{we have } Au \in H_{loc}^s(\Omega)\}. \quad (1.2.16)$$

When $(x, \xi) \notin WF_s u$, we shall say that u is H^s at (x, ξ) and write $u \in H_{(x,\xi)}^s$.

    The proof of the following theorem is a simple adaptation of the proof of Theorem 1.2.19.

Theorem 1.2.22. *Let Ω be an open set of $\mathbb{R}^n$, $s, m \in \mathbb{R}$ and $A \in \Psi_{ps}^m(\Omega)$. Then for $u \in \mathscr{D}'(\Omega)$,*

$$WF_s(Au) \subset WF_{s+m} u \subset \operatorname{char} A \cup WF_s(Au). \quad (1.2.17)$$

1.2.3 Propagation of singularities

Let Ω be an open subset of $\mathbb{R}^n$, $m \in \mathbb{R}$ and $P \in \Psi_{ps}^m(\Omega)$ a pseudo-differential operator with symbol p such that

$$p(x, \xi) = p_m(x, \xi)\omega(\xi) + p_{m-1}(x, \xi), \quad (1.2.18)$$

with p_m positively-homogeneous of degree m and $p_{m-1} \in S_{loc}^{m-1}(\Omega \times \mathbb{R}^n)$ and

$$\omega \in C^\infty(\mathbb{R}^n), \begin{cases} \omega(\xi) = 0 & \text{for } |\xi| \leq 1/2, \\ \omega(\xi) = 1 & \text{for } |\xi| \geq 1. \end{cases} \quad (1.2.19)$$

We shall say that p_m is the[13] principal symbol of P. Note also that the function $p_m(x,\xi)\omega(\xi)$ is positively-homogeneous with degree m, according to the terminology of Definition 1.2.9. In the sequel, we shall mainly consider operators of that type.

Theorem 1.2.23. *Let P be as above, $t_0 < t_1 \in \mathbb{R}$ and $I = [t_0, t_1] \ni t \mapsto \gamma(t) \in \dot{T}^*(\Omega)$ be a null bicharacteristic[14] curve of $\operatorname{Re} p_m$. Let us assume that $\operatorname{Im} p_m \geq 0$ on a conic-neighborhood of $\gamma(I)$. Let $s \in \mathbb{R}$ and $u \in \mathscr{D}'(\Omega)$ such that $Pu \in H^s$ at $\gamma(I)$ (i.e., $WF_s Pu \cap \gamma(I) = \emptyset$). Then*

$$\gamma(t_0) \in WF_{s+m-1}u \Longrightarrow \gamma(t_1) \in WF_{s+m-1}u. \tag{1.2.20}$$

Remark 1.2.24. The property (1.2.20) means that the singularities are propagating *forward* when the imaginary part of p_m is non-negative (see the discussion in Section 4.3.8). A statement equivalent to (1.2.20) is

$$\gamma(t_1) \notin WF_{s+m-1}u \Longrightarrow \gamma(t_0) \notin WF_{s+m-1}u, \tag{1.2.21}$$

meaning that the regularity is propagating backward in that case. If we change the sign condition on $\operatorname{Im} p_m$, we have to reverse the direction of propagation of singularities and we have, under $\operatorname{Im} p_m \leq 0$ near $\gamma(I)$,

$$\gamma(t_1) \in WF_{s+m-1}u \Longrightarrow \gamma(t_0) \in WF_{s+m-1}u. \tag{1.2.22}$$

When the imaginary part of p_m is identically 0, the propagation takes place in both directions and $WF_{s+m-1}u \backslash WF_s(Pu)$ is invariant by the Hamiltonian flow of p_m; this implies in particular for a real-valued p_m that $WFu \backslash WF(Pu)$ is invariant by the Hamiltonian flow of p_m.

Proof of Theorem 1.2.23. Multiplying by an elliptic operator of order $1 - m$, we are reduced to the case $m = 1$. We have to prove that

$$u \in H^s_{\gamma(t_1)}, Pu \in H^s_{\gamma(I)} \Longrightarrow u \in H^s_{\gamma(t_0)}. \tag{1.2.23}$$

It is enough to prove that

$$u \in H^s_{\gamma(t_1)}, Pu \in H^s_{\gamma(I)}, u \in H^{s-\frac{1}{2}}_{\gamma(I)} \Longrightarrow u \in H^s_{\gamma(t_0)}. \tag{1.2.24}$$

---

[13]If p_m, q_m are positively-homogeneous of degree m on $\Omega \times \mathbb{R}^n \backslash \{0\}$ such that for $|\xi| \geq R > 0$ $|p_m(x,\xi) - q_m(x,\xi)| \leq C|\xi|^{m-1}$, this implies $|p_m(x,\xi/|\xi|) - q_m(x,\xi/|\xi|)| \leq C|\xi|^{-1}$ and thus $p_m = q_m$ on $\Omega \times \mathbb{R}^n \backslash \{0\}$.

[14]For a C^1 on $\Omega \times \mathbb{R}^n$, the Hamiltonian vector field of a is

$$H_a = \sum_{1 \leq j \leq n} \left(\frac{\partial a}{\partial \xi_j} \frac{\partial}{\partial x_j} - \frac{\partial a}{\partial x_j} \frac{\partial}{\partial \xi_j} \right).$$

The integral curves of H_a are called the bicharacteristic curves of a. Since $H_a(a) = 0$, a is constant along its bicharacteristic curves; the *null* bicharacteristic curves are those on which a vanishes.

In fact, since $\gamma(I)$ is compact, we may assume that $u \in H_{\gamma(I)}^{s_0-\frac{1}{2}}$ for some s_0. The property $(1.2.24)_{s_0}$ is identical to $(1.2.23)_{s_0}$. Assume now that $u \in H_{\gamma(t_1)}^{s_0+\frac{1}{2}}, Pu \in H_{\gamma(I)}^{s_0+\frac{1}{2}}$: this implies that $u \in H_{\gamma(t_1)}^{s_0}, Pu \in H_{\gamma(I)}^{s_0}$ and since $u \in H_{\gamma(I)}^{s_0-\frac{1}{2}}$, the property $(1.2.24)_{s_0}$ gives eventually $u \in H_{\gamma(I)}^{s_0}$ so that the property $(1.2.24)_{s_0+\frac{1}{2}}$ gives $u \in H_{\gamma(I)}^{s_0+\frac{1}{2}}$. Inductively, we assume that for some $k \in \mathbb{N}^*$,

$$u \in H_{\gamma(t_1)}^{s_0+\frac{k}{2}}, Pu \in H_{\gamma(I)}^{s_0+\frac{k}{2}} \implies u \in H_{\gamma(I)}^{s_0+\frac{k}{2}}. \tag{1.2.25}$$

Then if $u \in H_{\gamma(t_1)}^{s_0+\frac{k+1}{2}}, Pu \in H_{\gamma(I)}^{s_0+\frac{k+1}{2}}$, (1.2.25) gives $u \in H_{\gamma(I)}^{s_0+\frac{k}{2}}$ and the property $(1.2.24)_{s_0+\frac{k+1}{2}}$ gives $u \in H_{\gamma(I)}^{s_0+\frac{k+1}{2}}$, so that (1.2.24) implies (1.2.25) for all $k \in \mathbb{N}$ and all s_0 such that $u \in H_{\gamma(I)}^{s_0-\frac{1}{2}}$, meaning that(1.2.24) implies (1.2.23). Now to prove (1.2.24), it is enough to get it for $s = 0$: assuming (1.2.24) for $s = 0$ and considering a properly supported pseudo-differential operator E_s, E_{-s} of order $s, -s$, elliptic on a neighborhood of $\gamma(I)$, whose symbols have an asymptotic expansion $\sum_{j\in\mathbb{N}} c_{\pm s-j}, c_{\pm s-j}$ positively-homogeneous of degree $\pm s - j$ and such that

$$E_{-s}E_s = \mathrm{Id} + R, \quad \gamma(I) \subset (\mathrm{essupp}\, R)^c,$$

we get under the hypothesis of (1.2.24) that

$$E_s u \in H_{\gamma(t_1)}^0, E_s P E_{-s} E_s u \in H_{\gamma(I)}^0, E_s u \in H_{\gamma(I)}^{-1/2},$$

and since the operator $E_s P E_{-s}$ is of order 1 with the same principal symbol as P, we can then apply (1.2.24) for $s = 0$, entailing $E_s u \in H_{\gamma(t_0)}^0$ which gives $u \in H_{\gamma(t_0)}^s$ using the ellipticity of E_s. The remaining part of the proof is devoted to establishing (1.2.24) for $s = 0$. As a last preliminary remark we note that

$$J = \{t \in [t_0, t_1], u \in H_{\gamma(s)}^0 \text{ for } s \in [t, t_1]\}$$

is a non-empty open interval of $[t_0, t_1]$; if $\inf J$ belongs to J it is also closed and thus equal to $[t_0, t_1]$; as a result, we may assume that $J =]t_0, t_1]$. Of course there is no loss of generality setting $t_0 = 0, t_1 = 1$. Summing-up, we have to prove

$$u \in H_{\gamma(t)}^0 \text{ for } t \in]0, 1], Pu \in H_{\gamma([0,1])}^0, u \in H_{\gamma([0,1])}^{-1/2} \implies u \in H_{\gamma(0)}^0. \tag{1.2.26}$$

We may also assume that u is compactly supported: if $\varphi \in C_c^\infty(\Omega)$ is 1 near the first projection of $\gamma([0, 1])$, we have $P\varphi u = [P, \varphi]u + \varphi Pu$ and since

$$(\mathrm{essupp}[P, \varphi])^c \supset \gamma([0, 1])$$

we get that φu satisfies as well the assumptions of (1.2.26). On the other hand, if p_1 is the real part of the principal symbol of P, we may assume that at $\gamma(0)$,

$dp_1 \wedge \xi \cdot dx \neq 0$, otherwise $\partial_\xi p_1(\gamma(0)) = 0, \partial_x p_1(\gamma(0)) = \lambda \xi(0)$ and the solution $\gamma(t) = (x(t), \xi(t))$ of

$$\dot{x}(t) = \partial_\xi p_1(x(t), \xi(t)), \quad \dot{\xi}(t) = -\partial_x p_1(x(t), \xi(t)), \quad (x(0), \xi(0)) = \gamma(0),$$

is $x(t) = x(0)$, $\xi(t) = e^{-\lambda t}\xi(0)$; since the wave-front-set is conic, (1.2.26) is obvious. Let us consider W_0 a conic-neighborhood of $\gamma([0,1])$ such that in W_0 $Pu \in H^0, u \in H^{-1/2}$. Let $m_0 \in S^0$ real-valued. The symbol of P is $p_1 + iq_1 + p_0 + iq_0$ with $p_j, q_j \in S^j$ real-valued, p_1, q_1 positively-homogeneous of degree 1 and $q_1 \geq 0$. We calculate with $M = m_0(x, D), A = \frac{1}{2}(P + P^*), B = \frac{1}{2i}(P - P^*)$, for $v \in C_c^\infty(\Omega)$,

$$2\operatorname{Re}\langle Pv, iM^*Mv \rangle$$
$$= \langle [A, iM^*M]v, v \rangle + 2\operatorname{Re}\langle M^*[M, B]v, v \rangle + 2\operatorname{Re}\langle BMv, Mv \rangle. \quad (1.2.27)$$

From the Gårding inequality (Theorem 1.1.26) we have

$$2\operatorname{Re}\langle BMv, Mv \rangle + \alpha_0 \|Mv\|_0^2 \geq 0, \quad \alpha_0 \text{ a semi-norm of } q_1. \quad (1.2.28)$$

On the other hand, the principal symbol of $M^*[M, B]$ is purely imaginary, belongs to S^0, and so that

$$2\operatorname{Re}\langle M^*[M, B]v, v \rangle + C_1 \|v\|_{-1/2}^2 \geq 0. \quad (1.2.29)$$

We have also

$$\langle [A, iM^*M]v, v \rangle + C_2 \|v\|_{-1/2}^2 \geq \langle \{p_1, m^2\}(x, D)v, v \rangle \frac{1}{2\pi}. \quad (1.2.30)$$

As a result, we have

$$\|MPv\|_0^2 + (1 + \alpha_0)\|Mv\|_0^2 + (C_1 + C_2)\|v\|_{-1/2}^2 \geq \langle \{p_1, m^2\}(x, D)v, v \rangle \frac{1}{2\pi}. \quad (1.2.31)$$

We can find $(t, y, \tau, \eta) \in \mathbb{R} \times \mathbb{R}^{n-1} \times \mathbb{R} \times \mathbb{R}^{n-1}$ as C^∞ local symplectic coordinates near $\gamma(0)$, $(x, \xi) \mapsto (t, y)$ homogeneous functions of degree 0 with respect to ξ, $(x, \xi) \mapsto (\tau, \eta)$ homogeneous functions of degree 1 with respect to ξ, so that $\partial_t = H_{p_1}$. We choose $\theta_0 \in C_c^\infty(\mathbb{R})$ supported on $[-\varepsilon_0, \varepsilon_0]$ with $\varepsilon_0 > 0$, positive on $(-\varepsilon_0, \varepsilon_0)$, with L^2-norm 1 and consider $\theta_1(t) = \theta_0(t - 3\varepsilon_0)$: with

$$\kappa(t) = \int_{-\varepsilon_0}^t (\theta_0(s)^2 - \theta_1(s)^2)ds = \int_{t-3\varepsilon_0}^t \theta_0(s)^2 ds,$$

imitating the multiplier method of Section 4.3.8, we have for the C^∞ function κ supported in $[-\varepsilon_0, 4\varepsilon_0]$,

$$0 \leq \kappa \leq 1, \quad \dot{\kappa} = \theta_0^2 - \theta_1^2, \quad [-\varepsilon_0, 2\varepsilon_0] \subset \{\dot{\kappa} \geq 0\}, \quad \operatorname{supp}\kappa \subset [-\varepsilon_0, 4\varepsilon_0],$$

and the following variation table.

t	$-\varepsilon_0$		0		ε_0		$2\varepsilon_0$		$3\varepsilon_0$		$4\varepsilon_0$
$\dot{\kappa}(t)$	0	$+$	$\dot{\kappa}(0)>0$	$+$	0	0	0	$-$	$\dot{\kappa}(3\varepsilon_0)<0$	$-$	0
$\kappa(t)$	0	$\nearrow$	$\kappa(0)>0$	$\nearrow$	1	1	1	$\searrow$	$\kappa(3\varepsilon_0)>0$	$\searrow$	0

We multiply now the function κ by ν^2 with $\nu \in C^\infty(\mathbb{R}^{2n-1}_{y,\tau,\eta};[0,1]), \nu(0)=1, \nu$ homogeneous with degree 0 with respect to τ, η, and we get that

$$0 \leq \kappa(t)\nu^2(Y) \leq 1, \quad \partial_t\big(\kappa(t)\nu^2(Y)\big) = \big(\theta_0^2(t) - \theta_1^2(t)\big)\nu^2(Y),$$

$$[-\varepsilon_0, 2\varepsilon_0] \times \operatorname{supp}\nu \subset \{\dot{\kappa} \otimes \nu^2 \geq 0\}, \quad \operatorname{supp}\kappa \otimes \nu^2 \subset [-\varepsilon_0, 4\varepsilon_0] \times \operatorname{supp}\nu,$$

$$\kappa(0)\nu^2(0) > 0, \quad \dot{\kappa}(0)\nu^2(0) > 0, \quad \{\dot{\kappa} \otimes \nu^2 < 0\} \subset [2\varepsilon_0, 4\varepsilon_0] \times \operatorname{supp}\nu.$$

The mapping

$$\mathbb{R} \times \mathbb{R}^{n-1} \times \mathbb{R} \times \mathbb{R}^{n-1} \ni (t,y,\tau,\eta) \mapsto x(t,y,\tau,\eta), \xi(t,y,\tau,\eta) \in \mathbb{R}^n \times \mathbb{R}^n$$

is a local symplectomorphism Θ, with x homogeneous of degree 0, ξ homogeneous of degree 1, and the push-forward μ of $\kappa \otimes \nu^2$ by Θ given by $\mu \circ \Theta = \kappa \otimes \nu^2$ is homogeneous of degree 0 with respect to τ, η and satisfies

$$\mu \in C_c^\infty(\dot{T}^*(\Omega);[0,1]), \quad \operatorname{supp}\mu \subset \Theta([-\varepsilon_0, 4\varepsilon_0] \times \operatorname{supp}\nu), \quad \mu(\gamma(0)) > 0,$$

$$H_{p_1}(\mu) = \chi_0^2 - \chi_1^2, \quad \chi_0, \chi_1 \in C_c^\infty(\dot{T}^*(\Omega)), \quad \chi_0(\gamma(0)) > 0,$$

$$\operatorname{supp}\chi_0 = \Theta([-\varepsilon_0, \varepsilon_0] \times \operatorname{supp}\nu), \quad \operatorname{supp}\chi_1 = \Theta([2\varepsilon_0, 4\varepsilon_0] \times \operatorname{supp}\nu),$$

$$\text{and thus } \operatorname{supp}\chi_0 \cap \operatorname{supp}\chi_1 = \emptyset.$$

The function μ is homogeneous of degree 0 such that

$$\operatorname{supp}\mu \subset W, \quad \text{a conic-neighborhood of } \gamma([-\varepsilon_0, 4\varepsilon_0]) \subset W_0,$$

$$H_{p_1}(\mu) = \chi_0^2 - \chi_1^2, \quad \text{and} \quad \operatorname{supp}\chi_1 \subset W_1,$$

where $W_1 \subset W_0$ is a conic-neighborhood of $\gamma(3\varepsilon_0)$ with $u \in H^0$ on W_1.

We consider now, with $T = t \circ \Theta^{-1}$ so that $H_{p_1}(T) = 1$, and T is homogeneous with degree 0, the symbol

$$m = \mu e^{\lambda T}$$

where λ is such that $\lambda \geq (1 + \alpha_0)2\pi$ and α_0 is given in (1.2.28). Checking:

$$H_{p_1}(m^2) - 2\pi(1 + \alpha_0)m^2 = 2\mu e^{\lambda T}\big(H_{p_1}(\mu)e^{\lambda T} + \lambda e^{\lambda T}\mu\big) - 2\pi(1 + \alpha_0)\mu^2 e^{2\lambda T}$$

$$= 2\mu e^{2\lambda T}H_{p_1}(\mu) + \mu^2 e^{2\lambda T}\big(2\lambda - 2\pi(1 + \alpha_0)\big) \geq 2\mu e^{2\lambda T}H_{p_1}(\mu).$$

Since $\mu e^{2\lambda T}H_{p_1}(\mu)$ is supported in W, positive at $\gamma(0)$ and non-negative except on a neighborhood of $\gamma(3\varepsilon_0)$, in fact such that

$$\mu e^{2\lambda T}H_{p_1}(\mu) = \mu(e^{\lambda T}\chi_0)^2 - \mu(e^{\lambda T}\chi_1)^2,$$

the inequality (1.2.31) gives with $A_0, A_1 \in \Psi_{ps}^0(\Omega)$, A_0 elliptic at $\gamma(0)$, essupp $A_1 \subset W_1$,

$$\|A_0 v\|_0^2 \leq \|A_1 v\|_0^2 + \|MPv\|_0^2 + (C_1 + C_2)\|v\|_{-1/2}^2. \tag{1.2.32}$$

Replacing in that inequality v by Nv where $N \in \psi_{ps}^0(\Omega)$, essupp $N \subset W_0$, with a symbol equal to 1 on W gives, with $R \in \psi_{ps}^{-\infty}(\Omega)$,

$$\|A_0 v\|_0^2 \leq \|A_1 v\|_0^2 + \|MPv\|_0^2 + (C_1 + C_2)\|Nv\|_{-1/2}^2 + \|Rv\|_0^2. \tag{1.2.33}$$

Since $u \in \mathscr{E}'(\Omega)$, we can apply (1.2.33) to $v_\varepsilon = u * \rho_\varepsilon$ where $\rho_\varepsilon(x) = \rho(x/\varepsilon)\varepsilon^{-n}$, with $\rho \in C_c^\infty(\mathbb{R}^n)$ of integral 1 and ε small enough so that $v_\varepsilon \in C_c^\infty(\Omega)$. Since u is H^0 on W_1 and also $H^{-1/2}$ on W_0, Pu is H^0 on W_0, we get that the $\|A_0 v_\varepsilon\|_0^2$ is bounded for $\varepsilon \to 0_+$, implying that the weak limit $A_0 u$ in $\mathscr{E}'(\Omega)$ belongs to H^0, proving that u is H^0 at $\gamma(0)$. The proof of Theorem 1.2.23 is complete. $\square$

1.2.4   Local solvability

Functional analysis arguments

Definition 1.2.25. Let Ω be an open subset of $\mathbb{R}^n$, $x_0 \in \Omega$, $m \in \mathbb{R}$ and $P \in \Psi_{ps}^m(\Omega)$ a properly supported pseudo-differential operator. We shall say that P is locally solvable at x_0 if there exists an open neighborhood $V \subset \Omega$ of x_0 such that

$$\forall f \in C^\infty(\Omega), \quad \exists u \in \mathscr{D}'(\Omega) \quad \text{with } Pu = f \text{ in } V. \tag{1.2.34}$$

Note that this definition makes sense since P is properly supported (in particular P is an endomorphism of $\mathscr{D}'(\Omega)$) and we can actually restrict a distribution to an open set. Moreover the set of points $x \in \Omega$ such that P is locally solvable at x is open.

Definition 1.2.26. Let Ω, x_0, P be as above and let $\mu \geq 0$. We shall say that P is locally solvable at x_0 with loss of μ derivatives if, for every $s \in \mathbb{R}$, there exists an open neighborhood $V \subset \Omega$ of x_0 such that

$$\forall f \in H_{loc}^s(\Omega), \quad \exists u \in H_{loc}^{s+m-\mu}(\Omega) \quad \text{with } Pu = f \text{ in } V. \tag{1.2.35}$$

Remark 1.2.27. Note that the neighborhood V above may depend on s.

Lemma 1.2.28. *Let Ω be an open subset of $\mathbb{R}^n$, $x_0 \in \Omega$, $m \in \mathbb{R}$ and let $P \in \Psi_{ps}^m(\Omega)$ be a pseudo-differential operator solvable at x_0. Then there exists a neighborhood $V \subset \Omega$ of x_0, $N \in \mathbb{N}$, $C > 0$ such that*

$$\forall v \in C_c^\infty(V), \quad C\|P^* v\|_N \geq \|v\|_{-N}. \tag{1.2.36}$$

Proof. The solvability of P at x_0 gives the existence of a neighborhood V of x_0 such that (1.2.34) holds. We consider now $v_0 \in C_c^\infty(V)$ such that $P^* v_0 = 0$: Then, for all $\varphi \in C_c^\infty(V)$ the solvability property implies the existence of $u \in \mathscr{D}'(\Omega)$ with $Pu = \varphi$ in V. As a result, we have, for all $\varphi \in C_c^\infty(V)$,

$$\int \varphi(x)\overline{v_0(x)}dx = \langle (Pu)_{|V}, v_0\rangle_{\mathscr{D}_*(V),\mathscr{D}(V)} = \langle Pu, v_0\rangle_{\mathscr{D}_*(\Omega),\mathscr{D}(\Omega)}$$

$$= \langle u, P^*v_0\rangle_{\mathscr{D}_*(\Omega),\mathscr{D}(\Omega)} = 0, \quad (1.2.37)$$

which gives $v_0 = 0$. Then for any compact subset K of V, the space $C_K^\infty(V)$ is a metrizable topological space for the topology given by the countable family of norms $\{\|P^*v\|_{H^r}\}_{r\in\mathbb{N}}$. Let $\tilde{K}$ be a compact subset of V, a neighborhood of a given compact subset K of V and consider the space $C_{\tilde{K}}^\infty(V)$, equipped with its standard Fréchet topology, where the semi-norms may be given by the family $\{\|\varphi\|_{H^s}\}_{s\in\mathbb{N}}$. For a fixed $v \in C_K^\infty(V)$, the mapping

$$C_{\tilde{K}}^\infty(V) \ni \varphi \mapsto \int \varphi(x)\overline{v(x)}dx$$

is obviously continuous since $C_{\tilde{K}}^\infty(V)$ is equipped with its standard Fréchet topology. For a fixed $\varphi \in C_{\tilde{K}}^\infty(V)$, the mapping

$$C_K^\infty(V) \ni v \mapsto \int \varphi(x)\overline{v(x)}dx$$

is continuous for the topology on $C_K^\infty(V)$ given by the family $\{\|P^*v\|_{H^r}\}_{r\in\mathbb{N}}$ since $\varphi = Pu$ on V with $u \in \mathscr{D}'(\Omega)$ and thus, as in (1.2.37),

$$\left| \int \varphi(x)\overline{v(x)}dx = \langle u, P^*v\rangle_{\mathscr{D}_*(\Omega),\mathscr{D}(\Omega)} \right| \le C_u\|P^*v\|_r.$$

A separately continuous bilinear form on the product of a Fréchet space with a metrizable space is in fact continous so that

$$\exists C > 0, \exists N \in \mathbb{N}, \forall v \in C_K^\infty(V), \forall \varphi \in C_{\tilde{K}}^\infty(V),$$

$$\left| \int \varphi(x)\overline{v(x)}dx \right| \le C\|P^*v\|_N\|\varphi\|_N,$$

and since $\tilde{K}$ is a neighborhood of K, it gives the lemma. $\qquad\qquad\square$

Lemma 1.2.29. *Let Ω be an open subset of $\mathbb{R}^n$, $x_0 \in \Omega$, $m, s, \mu \in \mathbb{R}$ and let $P \in \Psi_{ps}^m(\Omega)$ be a pseudo-differential operator such that there exists an open neighborhood V of x_0 such that*

$$\forall f \in H_{loc}^s(\Omega), \quad \exists u \in H_{loc}^{s+m-\mu}(\Omega) \quad with\ Pu = f\ in\ V. \quad (1.2.38)$$

Then there exists a neighborhood $W \subset \Omega$ of x_0, $C > 0$ such that

$$\forall v \in C_c^\infty(W), \quad C\|P^*v\|_{-s-m+\mu} \ge \|v\|_{-s}. \quad (1.2.39)$$

Proof. We consider $v_0 \in C_c^\infty(V)$ such that $P^* v_0 = 0$: Then, for all $\varphi \in C_c^\infty(V)$ the solvability property (1.2.38) implies (1.2.34) and the proof of Lemma 1.2.28 gives $v_0 = 0$. Then for any compact subset K of V, the space $C_K^\infty(V)$ is a normed space with the norm $\|P^* v\|_{-s-m+\mu}$. Let $\tilde{K}$ be a compact subset of V, a neighborhood of a given compact subset K of V and consider the Hilbert space $H_{\tilde{K}}^s(V)$. For a fixed $v \in C_K^\infty(V)$, the mapping $H_{\tilde{K}}^s(V) \ni \varphi \mapsto \langle \varphi, v \rangle$ is obviously continuous since $|\langle \varphi, v \rangle| \leq \|\varphi\|_s \|v\|_{-s}$. For a fixed $\varphi \in H_{\tilde{K}}^s(V)$, the mapping

$$C_K^\infty(V) \ni v \mapsto \langle \varphi, v \rangle$$

is continuous for the topology on $C_K^\infty(V)$ given by the norm $\|P^* v\|_{-s-m+\mu}$ since $\varphi = Pu$ on V with $u \in H_{loc}^{s+m-\mu}(\Omega)$ and thus, as in (1.2.37), with $\chi \in C_c^\infty(\Omega)$, $\chi = 1$ near the support of $P^*(C_K^\infty(V))$,

$$\left| \langle \varphi, v \rangle = \langle u, P^* v \rangle_{\mathscr{D}^*(\Omega), \mathscr{D}(\Omega)} = \langle \chi u, P^* v \rangle_{H^{s+m-\mu}, \mathscr{D}(\Omega)} \right| \leq \|\chi u\|_{s+m-\mu} \|P^* v\|_{-s-m+\mu}.$$

As before, this bilinear form is continuous,

$$\exists C > 0, \forall v \in C_K^\infty(V), \forall \varphi \in H_{\tilde{K}}^s(V), |\langle \varphi, v \rangle| \leq C \|P^* v\|_{-s-m+\mu} \|\varphi\|_s,$$

and since $\tilde{K}$ is a neighborhood of K, it gives $\|v\|_{-s} \leq C \|P^* v\|_{-s-m+\mu}$, for all $v \in C_K^\infty(V)$. Choosing K as a compact neighborhood of x_0 included in V, we can take $W = \operatorname{int} K$ to obtain the lemma. $\qquad\square$

Lemma 1.2.30. *Let Ω be an open subset of $\mathbb{R}^n$, $x_0 \in \Omega$, $m \in \mathbb{R}$ and $P \in \Psi_{ps}^m(\Omega)$ a properly supported pseudo-differential operator. Let $s \in \mathbb{R}$. Assume that there exists $\mu \geq 0$ and an open neighborhood $V \subset \Omega$ of x_0 such that,*

$$\exists C > 0, \forall v \in C_c^\infty(V), \quad \|v\|_{-s} \leq C \|P^* v\|_{-s-m+\mu}. \qquad (1.2.40)$$

Then, for all $f \in H_{loc}^s(\Omega)$, there exists $u \in H^{s+m-\mu}(\Omega)$ such that $Pu = f$ in V.

Proof. Let $f_0 \in H_{loc}^s(\Omega)$. The inequality (1.2.40) implies the injectivity of P^* on $C_c^\infty(V)$. Assuming as we may that $V \Subset \Omega$, we get that the space $P^*(C_c^\infty(V))$ is a subspace of $C_K^\infty(\Omega)$, where K is a compact subset of Ω. We consider $P^*(C_c^\infty(V))$ as a subspace of $H_0^{-s-m+\mu}(\Omega)$ and we can define the linear form

$$P^*(C_c^\infty(V)) \ni P^* v \mapsto \langle v, f_0 \rangle$$

which satisfies the following estimate: with $\chi \in C_c^\infty(\Omega)$ equal to 1 on $\overline{V}$, we have

$$|\langle v, f_0 \rangle| \leq \|v\|_{-s} \|\chi f_0\|_s \leq C \|P^* v\|_{-s-m+\mu} \|\chi f_0\|_s.$$

We can extend this linear form to the whole $H_0^{-s-m+\mu}(\Omega)$ to a linear form ξ with norm $\leq C \|\chi f_0\|_s$ by the Hahn-Banach theorem. This means that there exists $u_0 \in H^{s+m-\mu}(\Omega)$ such that

$$\forall g \in H_0^{-s-m+\mu}(\Omega), \quad \langle g, u_0 \rangle = \xi(g), \quad \|u_0\|_{H^{s+m-\mu}(\Omega)} \leq C \|\chi f_0\|_s,$$

and in particular for all $v \in C_c^\infty(V)$,

$$\langle v, f_0 \rangle = \xi(P^*v) = \langle P^*v, u_0 \rangle = \langle v, Pu_0 \rangle$$

and thus $Pu_0 = f_0$ on V. $\qquad\square$

Remark 1.2.31. Note in particular that if the estimate

$$\|v\|_{\sigma+m-\mu} \le C\|P^*v\|_\sigma$$

is proven true for any $\sigma \in \mathbb{R}$, for $v \in C_c^\infty(V_\sigma)$, where V_σ is a neighborhood of x_0 (which may depend on σ), then the result of the lemma holds and P is locally solvable at x_0 with loss of μ derivatives (see Remark 1.2.27). The estimate above can be true for $\mu = 0$ if and only if P is elliptic at x_0. Moreover the two previous lemmas show that the local solvability questions for a properly supported operator are equivalent to proving an a priori estimate of the type (1.2.36), (1.2.40).

Lemma 1.2.32. *Let Ω be an open subset of $\mathbb{R}^n$, $x_0 \in \Omega$, $m \in \mathbb{R}$ and $P \in \Psi_{ps}^m(\Omega)$ a properly supported pseudo-differential operator. If P is solvable at x_0, then there exists a neighborhood V of x_0 and an integer N such that*

$$\forall f \in C^N(\Omega), \quad \exists u \in \mathscr{E}'^N(\Omega), \quad \text{with } Pu = f \text{ in } V. \qquad (1.2.41)$$

Proof. In fact, from (1.2.36), the estimate

$$C\|P^*v\|_N \ge \|v\|_{-N} \qquad (1.2.42)$$

holds for all $v \in C_c^\infty(V)$ for some neighborhood V of x_0 and some $N \in \mathbb{N}$. We may assume $V \Subset \Omega$. Let $f_0 \in C^N(\Omega)$. The inequality (1.2.42) implies the injectivity of P^* on $C_c^\infty(V)$. We consider the space $P^*(C_c^\infty(V))$ as a subspace of $C_K^N(\Omega)$, where K is a compact subset of Ω and we can define the linear form

$$P^*(C_c^\infty(V)) \ni P^*v \mapsto \langle v, f_0 \rangle$$

which satisfies the following estimate: with $\chi \in C_c^\infty(\Omega)$, $\chi = 1$ near $\overline{V}$, we have

$$|\langle v, f_0 \rangle| \le \|v\|_{-N} \|\chi f_0\|_N \le C\|P^*v\|_N \|\chi f_0\|_N \le C_1 \|\chi f_0\|_N \sup_{\substack{|\alpha| \le N \\ x \in K}} |\partial^\alpha(P^*v)(x)|.$$

By the Hahn-Banach theorem, we can extend this linear form to a linear form ξ defined on $C^N(\Omega)$ such that

$$\forall g \in C^N(\Omega), \quad |\xi(g)| \le C_1 \|\chi f_0\|_N \sup_{\substack{|\alpha| \le N \\ x \in K}} |\partial^\alpha g(x)|.$$

This means that there exists $u_0 \in \mathscr{E}'^N(\Omega)$ such that $\forall g \in C^N(\Omega)$, $\langle g, u_0 \rangle = \xi(g)$ and in particular for all $v \in C_c^\infty(V)$,

$$\langle v, f_0 \rangle = \xi(P^*v) = \langle P^*v, u_0 \rangle = \langle v, Pu_0 \rangle$$

and thus $Pu_0 = f_0$ on V. $\qquad\square$

Remarks on solvability with loss of μ derivative(s)

To establish local solvability at x_0 with loss of μ derivatives, it is enough to prove (see Lemma 1.2.30 and Remark 1.2.31) that for every $s \in \mathbb{R}$, there exists $r > 0$ and $C > 0$ such that

$$\forall v \in C_c^\infty(B(x_0, r)), \quad C\|P^*v\|_{-s-m+\mu} \geq \|v\|_{-s}. \tag{1.2.43}$$

However, we may be able to prove a weaker estimate only for some s. The next lemma establishes local solvability as a consequence of a weak estimate.

Lemma 1.2.33. *Let Ω be an open subset of $\mathbb{R}^n$, $x_0 \in \Omega, m \in \mathbb{R}, \mu \geq 0, s < n/2$ and $P \in \Psi_{ps}^m(\Omega)$ a properly supported pseudo-differential operator. Let us assume that there exists $r > 0, C > 0$ such that*

$$\forall v \in C_c^\infty(B(x_0, r)), \quad C\|v\|_{-s-1} + C\|P^*v\|_{-s-m+\mu} \geq \|v\|_{-s}. \tag{1.2.44}$$

Then, there exists $r > 0, C > 0$ such that (1.2.43) holds and for all $f \in H_{loc}^s(\Omega)$, there exists $u \in H^{s+m-\mu}(\Omega)$ such that $Pu = f$ in V, where V is some neighborhood of x_0. In particular P is locally solvable at x_0.

Proof. Lemma 4.3.5 implies that $\|u\|_{-s-1} \leq \phi(r)\|u\|_{-s}$ with $\lim_{r\to 0} \phi(r) = 0$, provided $-s > -n/2$ and this proves that shrinking r leads to (1.2.43) which implies the lemma by applying Lemma 1.2.30. $\qquad\square$

On the other hand, we may also want to stick on our Definition 1.2.26 of solvability with loss of μ derivatives for which we need to prove an estimate for every $s \in \mathbb{R}$.

Lemma 1.2.34. *Let Ω be an open subset of $\mathbb{R}^n$, $x_0 \in \Omega$, $m \in \mathbb{R}, \mu \geq 0$ and $P \in \Psi_{ps}^m(\Omega)$ a properly supported pseudo-differential operator, with homogeneous principal symbol p_m such that*

$$\mathbb{S}^{n-1} \ni \xi \mapsto p_m(x_0, \xi) \text{ is not identically } 0. \tag{1.2.45}$$

Let us assume that for every $s \in \mathbb{R}$, there exists $r > 0, C_1, C_2$ such that

$$\forall v \in C_c^\infty(B(x_0, r)), \quad C_2\|v\|_{-s-1} + C_1\|P^*v\|_{-s-m+\mu} \geq \|v\|_{-s}. \tag{1.2.46}$$

Then, for every $s \in \mathbb{R}$, there exists $r > 0, C > 0$ such that (1.2.43) holds.

Proof. If (1.2.43) were not true, we could find a sequence $(v_k)_{k\geq 1}$ such that $v_k \in C_c^\infty(B(x_0, k^{-1}))$ with $\|v_k\|_{-s} = 1$ and $\lim_k \|P^*v_k\|_{-s-m+\mu} = 0$. The estimate (1.2.46) implies that $C_2\|v_k\|_{-s-1} \geq 1/2$ for k large enough and since the sequence (v_k) is compact in H^{-s-1}, it has a subsequence strongly convergent in H^{-s-1} towards $v_0 \neq 0$, which is a weak limit in H^{-s}. We have

$$P^*v_0 = 0, \quad \mathrm{supp}\, v_0 = \{x_0\}, \quad \text{so that } v_0 = Q(D)\delta_{x_0}$$

where Q is a non-zero polynomial. As a consequence, we have for $u \in C_c^\infty(\Omega)$,

$$\langle \bar{Q}(D)Pu, \delta_{x_0} \rangle = \langle u, P^* v_0 \rangle = 0 \Longrightarrow (\bar{Q}(D)Pu)(x_0) = 0, \text{ for all } u \in C_c^\infty(\Omega),$$

and that, for all $\xi \in \mathbb{R}^n \backslash \{0\}$, $\bar{q}(\xi)p(x_0, \xi) = 0$, where q is the principal part of Q (homogeneous with degree ν) and p the principal symbol of P, since

$$\bar{q}(\xi)p(x_0, \xi)u(x) = \lim_{t \to +\infty} \left((\bar{Q}(D)P)(e^{2i\pi t \langle \cdot, \xi \rangle}u(\cdot)) \right)(x_0) t^{-m-\nu} e^{-2i\pi t \langle x_0, \xi \rangle}.$$

Since q is a non-zero polynomial[15], this implies that $\xi \mapsto p(x_0, \xi)$ is identically zero, contradicting the assumption. The proof of the lemma is complete. $\qquad \square$

Lemma 1.2.35. *Let Ω be an open subset of $\mathbb{R}^n$, $x_0 \in \Omega$, $m \in \mathbb{R}$ and $P \in \Psi_{ps}^m(\Omega)$ a properly supported pseudo-differential operator. To get $(1.2.44)_s$ it suffices to prove that for every $\xi_0 \in \mathbb{S}^{n-1}$, there exists some $\varphi \in S_{1,0}^0$ non-characteristic at (x_0, ξ_0), such that*

$$\exists C, \exists r_0 > 0, \forall r \in]0, r_0], \exists A(r), \exists \varepsilon(r) \text{ with } \lim_{r \to 0} \varepsilon(r) = 0, \forall v \in C_c^\infty(B(x_0, r)),$$

$$\varepsilon(r)\|v\|_{\sigma+m-\mu} + A(r)\|v\|_{\sigma+m-\mu-1} + C\|P^* v\|_\sigma \geq \|\varphi^w v\|_{\sigma+m-\mu}, \qquad (1.2.47)$$

with $\sigma = -s - m + \mu$.

Proof. To get $(1.2.44)_s$ with $s = -\sigma - m + \mu$, it is enough to prove that, for every $\xi_0 \in \mathbb{S}^{n-1}$, there exists some $\varphi \in S_{1,0}^0$ non-characteristic at (x_0, ξ_0), $r > 0$, $C > 0$, such that

$$\forall v \in C_c^\infty(B(x_0, r)), \quad C\|v\|_{\sigma+m-\mu-1} + C\|P^* v\|_\sigma \geq \|\varphi^w v\|_{\sigma+m-\mu}. \qquad (1.2.48)$$

In fact, if $(1.2.48)$ holds, one can find finitely many $\varphi_1, \ldots, \varphi_\nu$ such that

$$\sum_{1 \leq j \leq \nu} |\varphi_j|^2 \quad \text{is elliptic at } x_0$$

and for all $v \in C_c^\infty(B(x_0, r_0))$ (r_0 is the minimum of the $r_j > 0$ corresponding to each φ_j),

$$\|v\|_{\sigma+m-\mu}^2 \leq C_1 \sum_{1 \leq j \leq \nu} \|\varphi_j^w v\|_{\sigma+m-\mu}^2 + C_2 \|v\|_{\sigma+m-\mu-1}^2$$

$$\leq C_1 \nu 2C^2 \|P^* v\|_\sigma^2 + (C_2 + C_1 \nu 2C^2)\|v\|_{\sigma+m-\mu-1}^2,$$

which gives $C\|v\|_{\sigma+m-\mu-1} + C\|P^* v\|_\sigma \geq \|v\|_{\sigma+m-\mu}$. The same argument as above gives the implication $(1.2.47)_\sigma \Longrightarrow (1.2.44)_s$. $\qquad \square$

---

[15]The open set $\{\xi \in \mathbb{R}^n, q(\xi) \neq 0\}$ is dense since the closed set $\{\xi \in \mathbb{R}^n, q(\xi) = 0\}$ cannot have interior points because q is a non-zero polynomial.

Remark 1.2.36. Assume in particular that $m = 1$ and that, on a conic-neighborhood of $(0; e_n)$ the principal symbol of p is

$$\xi_1 + q(x_1, x', \xi'), \quad q \text{ complex-valued.} \tag{1.2.49}$$

Considering a positively-homogeneous function of n variables χ_0 supported in

$$\{\xi \in \mathbb{R}^n, |\xi_1| \leq C_1 |\xi'|\},$$

equal to 1 on $\{\xi \in \mathbb{R}^n, |\xi_1| \leq C_0 |\xi'|, |\xi| \geq 1\}$, and ψ_0 a positively-homogeneous function of $n-1$ variables supported in a conic-neigborhood of $\xi_0' = (0, \ldots, 0, 1) \in \mathbb{R}^{n-1}$ and equal to one on a conic-neighborhood of ξ_0', the symbol

$$l_1(x, \xi) = \xi_1 + q(x_1, x', \xi') \psi_0(\xi') \chi_0(\xi) \in S^1_{1,0}$$

coincides with p on some conic-neighborhood of $(0, e_n)$ and we have

$$l_1^w = D_1 + \underbrace{\big(q(x_1, x', \xi') \psi_0(\xi') \big)^w}_{q_1} - \big(q(x_1, x', \xi') \psi_0(\xi')(1 - \chi_0(\xi)) \big)^w.$$

The symbol q_1 does not belong to $S^1_{1,0}(\mathbb{R}^{2n})$, but only to $S^1_{1,0}(\mathbb{R}^{2n-2}_{x',\xi'})$, uniformly with respect to x_1. Let us assume that for $v \in \mathscr{S}(\mathbb{R}^n)$, $\operatorname{supp} v \subset \{|x_1| \leq T\}$,

$$C \|D_1 v + q_1^w v\|_0 \geq T^{-1} \|v\|_0. \tag{1.2.50}$$

We consider χ_1 a positively-homogeneous function of n variables supported in $\{\xi \in \mathbb{R}^n, |\xi_1| \leq C_0 |\xi'|\}$, and we apply (1.2.50) to $\rho(x_1 T^{-1}) \chi_1^w u$ where $u \in \mathscr{S}(\mathbb{R}^n)$, supported in $|x_1| \leq T/2$ and $\rho_1 \in C^\infty_{[-2,2]}(\mathbb{R})$, equal to 1 on $[-1, 1]$: we get

$$2C \|l_1^w \rho(x_1 T^{-1}) \chi_1^w u + \big((1 - \chi_0) q_1\big)^w \rho(x_1 T^{-1}) \chi_1^w u\| \geq T^{-1} \|\rho(x_1 T^{-1}) \chi_1^w u\|_0.$$

The term $\big((1 - \chi_0) q_1\big)^w \rho(x_1 T^{-1}) \chi_1^w$ has a symbol in $S^{-\infty}_{1,0}(\mathbb{R}^{2n})$ (to be checked directly by the composition formula whose expansion is 0, or see in the appendix Lemma 4.9.17). The term

$$[l_1^w, \rho(x_1 T^{-1})] \chi_1^w u = (2i\pi T)^{-1} \underbrace{\rho'(x_1 T^{-1})}_{\substack{= 0 \text{ on } [-T, T]}} \chi_1^w \underbrace{\rho(2x_1 T^{-1})}_{\substack{\text{supported in} \\ [-T, T]}} u = r_T^w u,$$

$r_T \in S^{-\infty}_{1,0}(\mathbb{R}^{2n})$. The term $[l_1^w, \chi_1^w]$ is L^2-bounded and we have thus

$$\|\chi_1^w l_1^w u\|_0 + \|u\|_0 + \alpha(T) \|u\|_{-1} \geq \frac{c_0}{T} \|\chi_1^w u\|$$

which gives (1.2.47) for $m = 1, \sigma = 0, \mu = 1$. Staying with the case $\mu = 1$ (loss 1), the argument is not different for other values of m, σ. We can of course replace the assumption (1.2.49) by $e(x, \xi)(\xi_1 + q(x_1, x', \xi'))$ where e is elliptic on a conic-neighborhood of $(0; e_n)$.

Remark 1.2.37. On the other hand, when $\mu > 1$, the rhs in the estimate (1.2.50) has to be replaced by $\|v\|_{1-\mu}$ and the fact that this suffices to prove local solvability requires some particular care. This point is discussed in details in Subsection 3.7.3 of Section 3.7 where an estimate with loss of $3/2$ derivatives is established.

Operators of real principal type

Theorem 1.2.38. *Let Ω be an open subset of $\mathbb{R}^n$, $m \in \mathbb{R}$ and $P \in \Psi_{ps}^m(\Omega)$ a pseudo-differential operator with symbol p such that (1.2.18) holds with p_m real-valued and positively-homogeneous of degree m. We assume that P is of principal type, i.e.,*

$$\forall (x, \xi) \in \dot{T}^*(\Omega), \quad p_m(x, \xi) = 0 \Longrightarrow \partial_\xi p_m(x, \xi) \neq 0. \tag{1.2.51}$$

Then the operator P is locally solvable at every point of Ω with loss of one derivative.

Proof. One could of course use the estimate (1.2.33), but the argument for solvability alone is so simple in that case that it may be worthy to look at it. We consider a point $x_0 \in \Omega$. If p_m is elliptic at x_0, the estimate

$$C\|v\|_{\sigma+m-1} + C\|P^*v\|_\sigma \geq \|v\|_{\sigma+m}$$

follows from Theorem 1.2.15 for v supported in $B(x_0, r_0)$ with $r_0 > 0$ small enough. If there exists $\xi_0 \neq 0$ such that $p_m(x_0, \xi_0) = 0$, we may choose the coordinates so that $x_0 = 0, \xi_0 = e_n$ and

$$p_m(x, \xi) = (\xi_1 + a(x_1, x', \xi'))e(x, \xi)$$

with a homogeneous of degree 1 with respect to ξ', e elliptic with degree $m - 1$ on a conic-neighborhood of $(0, e_n)$. According to Remark 1.2.36 the question reduces to proving an estimate for the operator $L = D_{x_1} + a(x_1, x', \xi')^w$ where $a \in C^\infty(\mathbb{R}^{2n-1}; \mathbb{R})$ such that, for all α, β,

$$\sup_{(x, \xi') \in \mathbb{R}^n \times \mathbb{R}^{n-1}} |(\partial_{\xi'}^\alpha \partial_x^\beta a)(x, \xi')| \langle \xi' \rangle^{-1+|\alpha|} < \infty.$$

We find

$$2 \operatorname{Re} \langle D_{x_1} u + a(x_1, x', \xi')^w u, i x_1 u \rangle_{L^2} = \frac{1}{2\pi} \|u\|_0^2, \tag{1.2.52}$$

so that for $u \in C_c^\infty(\mathbb{R}^n)$, $u = 0$ on $|x_1| \geq T$, we have $2\|Lu\|_0 T\|u\|_0 \geq \frac{1}{2\pi}\|u\|_0^2$ and thus $\|Lu\|_0 \geq \frac{1}{4\pi T}\|u\|_0$. $\quad\square$

Operators of principal type, complex symbols with a non-negative imaginary part

Theorem 1.2.39. *Let Ω be an open subset of $\mathbb{R}^n$, $m \in \mathbb{R}$ and $P \in \Psi_{ps}^m(\Omega)$ a pseudo-differential operator with symbol p such that (1.2.18) holds with p_m complex-valued and positively-homogeneous of degree m. We assume that P is of principal type (see (1.2.51)) such that the function $\operatorname{Im} p_m$ is non-negative (resp. non-positive). Then the operator P is locally solvable at every point of Ω with loss of one derivative.*

Proof. To handle the complex-valued case, we see that the principal type condition (1.2.51) implies that at a non-elliptic point,

$$p_m(x_0, \xi_0) = 0, \quad d(\operatorname{Im} p_m)(x_0, \xi_0) = 0, \quad \partial_\xi(\operatorname{Re} p_m)(x_0, \xi_0) \neq 0.$$

The first thing that we can do is to use the estimate (1.2.32): assuming $m = 1$, the bicharacteristic curve $\dot{\gamma} = H_{\operatorname{Re} p_m}(\gamma)$ starting at $\gamma(0) = (x_0, \xi_0)$ we find $A_0, A_1 \in \Psi_{ps}^0(\Omega)$, A_0 elliptic at $\gamma(0)$, $\operatorname{essupp} A_1 \subset W_1$, W_1 conic-neighborhood of $\gamma(3\varepsilon_0)$ where $\varepsilon_0 > 0$, $\gamma(0) \notin W_1$, $M \in \Psi_{ps}^0(\Omega)$,

$$\|A_0 v\|_0^2 \le \|A_1 v\|_0^2 + \|M P v\|_0^2 + (C_1 + C_2)\|v\|_{-1/2}^2. \qquad (1.2.53)$$

Applying this to $v \in C_c^\infty(B(x_0, r))$ with r small enough, the principal-type assumption $\partial_\xi \operatorname{Re} p_m(x_0, \xi_0) \ne 0$ implies, with χ_0 supported in the unit ball of $\mathbb{R}^n$, that $A_1 v = A_1 \chi_0((\cdot - x_0)/r) v$ with $\operatorname{essupp} A_1 \chi_0((\cdot - x_0)/r) = \emptyset$, so that (1.2.47) holds with $\mu = 1, m = 1, \sigma = 0$. The other cases are analogous.

On the other hand, it is also interesting to find directly a multiplier method, as in the real-principal type case. We need only to handle

$$L_\pm = D_{x_1} u + a(x_1, x', \xi')^w u \pm i b(x, \xi)^w + r_0(x, \xi)^w$$

with $b \in S_{1,0}^1, b \ge 0$, $a(x_1, \cdot, \cdot)$ real-valued in $S_{1,0}^1(\mathbb{R}_{x',\xi'}^{2n-2})$ uniformly in x_1 and $r_0 \in S_{1,0}^0$. With $\theta \in C^\infty(\mathbb{R}; \mathbb{R})$, we calculate

$$2\operatorname{Re}\langle D_{x_1} u + a(x_1, x', \xi')^w u + i b(x, \xi)^w u, i\theta(x_1)^2 u\rangle_{L^2} = \frac{1}{\pi}\langle \theta\theta' u, u\rangle + 2\operatorname{Re}\langle b^w u, \theta^2 u\rangle.$$

We have

$$\theta^2 b^w + b^w \theta^2 = \theta\theta b^w + b^w \theta\theta = \theta[\theta, b^w] + 2\theta b^w \theta + [b^w, \theta]\theta = 2\theta b^w \theta + [[b^w, \theta], \theta]$$

and thus, from Gårding inequality (Theorem 1.1.26), we find

$$2\operatorname{Re}\langle L_+ u, i\theta^2 u\rangle \ge \frac{1}{\pi}\langle \theta\theta' u, u\rangle + \langle [[b^w, \theta], \theta]u, u\rangle - C_0\|\theta u\|^2 - \|[r_0, \theta]u\|^2$$

where C_0 depends on semi-norms of b, r_0; to handle the term r_0, we have used

$$\langle r_0^w u, \theta^2 u\rangle = \langle [\theta, r_0^w]u, \theta u\rangle + \langle r_0^w \theta u, \theta u\rangle.$$

We have with $\lambda > 0$, $\theta(x_1) = e^{\lambda x_1}$, for $u \in C_c^\infty(\mathbb{R}^n)$ vanishing at $|x_1| \ge 1/\lambda$ (so that with $\chi_0 \in C_c^\infty(\mathbb{R})$ equal to 1 on $[-1, 1]$, $u = \chi_0(\lambda x_1)u$),

$$2\operatorname{Re}\langle L_+ u, ie^{2\lambda x_1} u\rangle \ge (\pi^{-1}\lambda - C_0)\|e^{\lambda x_1} u\|^2 - C(\lambda)\|u\|_{-1/2}^2$$

implying $2\pi\lambda^{-1}\|\chi_0(\lambda x_1)e^{\lambda x_1} L_+ u\|_0^2 + C(\lambda)\|u\|_{-1/2}^2 \ge (\frac{\lambda}{2\pi} - C_0)\|e^{\lambda x_1} u\|^2$, and assuming χ_0 valued in $[0, 1]$, vanishing on $(-2, 2)^c$, and $\lambda \ge 4\pi C_0$,

$$2\pi\lambda^{-1}e^4\|L_+ u\|_0^2 + C(\lambda)\|u\|_{-1/2}^2 \ge \frac{\lambda}{4\pi e^2}\|u\|^2.$$

Choosing $\lambda = 1 + 4\pi C_0$, we find that there exists $r_0 > 0$ such that, for $u \in C_c^\infty(\mathbb{R}^n)$, with $\operatorname{diameter}(\operatorname{supp} u) \le r_0$,

$$C_1\|L_+ u\|_0 + C_1\|u\|_{-1/2} \ge \|u\|_0,$$

proving the local solvability of L_+^*; the case of L_-^* is analogous. $\qquad\square$

1.3 Pseudo-differential operators in harmonic analysis

1.3.1 Singular integrals, examples

The Hilbert transform

A basic object in the classical theory of harmonic analysis is the Hilbert transform, given by the one-dimensional convolution with $pv(1/\pi x) = \frac{d}{\pi dx}(\ln|x|)$, where we consider here the distribution derivative of the $L^1_{\text{loc}}(\mathbb{R})$ function $\ln|x|$. We can also compute the Fourier transform of $pv(1/\pi x)$, which is given by $-i\operatorname{sign}\xi$ (see, e.g., (4.1.21)). As a result the Hilbert transform $\mathscr{H}$ is a unitary operator on $L^2(\mathbb{R})$ defined by

$$\widehat{\mathscr{H}u}(\xi) = -i\operatorname{sign}\xi\,\hat{u}(\xi). \tag{1.3.1}$$

It is also given by the formula

$$(\mathscr{H}u)(x) = \lim_{\epsilon\to 0_+}\frac{1}{\pi}\int_{|x-y|\geq\epsilon}\frac{u(y)}{x-y}dy.$$

The Hilbert transform is certainly the first known example of a *Fourier multiplier* ($\mathscr{H}u = F^{-1}(a\hat{u})$ with a bounded a). Since the sign function is bounded, it is obviously bounded on $L^2(\mathbb{R})$, but is is tempting to relate that result to Theorem 1.1.14 of L^2-boundedness of the $S^0_{1,0}$ class; naturally the singularity at 0 of the sign function prevents it to be a symbol in that class.

The Riesz operators, the Leray-Hopf projection

The Riesz operators are the natural multi-dimensional generalization of the Hilbert transform. We define for $u \in L^2(\mathbb{R}^n)$,

$$\widehat{R_j u}(\xi) = \frac{\xi_j}{|\xi|}\hat{u}(\xi), \quad \text{so that } R_j = D_j/|D| = (-\Delta)^{-1/2}\frac{\partial}{i\partial x_j}. \tag{1.3.2}$$

The R_j are selfadjoint bounded operators on $L^2(\mathbb{R}^n)$ with norm 1.

We can also consider the $n \times n$ matrix of operators given by $Q = R \otimes R = (R_j R_k)_{1\leq j,k\leq n}$ sending the vector space of $L^2(\mathbb{R}^n)$ vector fields into itself. The operator Q is selfadjoint and is a projection since $\sum_l R_l^2 = \operatorname{Id}$ so that $Q^2 = (\sum_l R_j R_l R_l R_k)_{j,k} = Q$. As a result the operator

$$\mathbb{P} = \operatorname{Id} - R \otimes R = \operatorname{Id} - |D|^{-2}(D \otimes D) = \operatorname{Id} - \Delta^{-1}(\nabla \otimes \nabla) \tag{1.3.3}$$

is also an orthogonal projection, the Leray-Hopf projector (a.k.a. the Helmholtz-Weyl projector); the operator $\mathbb{P}$ is in fact the orthogonal projection onto the closed subspace of L^2 vector fields with null divergence. We have for a vector field $u = \sum_j u_j\partial_j$, the identities $\operatorname{grad}\operatorname{div} u = \nabla(\nabla\cdot u)$, $\operatorname{grad}\operatorname{div} = \nabla \otimes \nabla = (-\Delta)(iR\otimes iR)$, so that

$$Q = R \otimes R = \Delta^{-1}\operatorname{grad}\operatorname{div}, \quad \operatorname{div} R \otimes R = \operatorname{div},$$

which implies $\operatorname{div}\mathbb{P}u = \operatorname{div}u - \operatorname{div}(R\otimes R)u = 0$, and if $\operatorname{div}u = 0$, $\mathbb{P}u = u$. The Leray-Hopf projector is in fact the $(n\times n)$-matrix-valued Fourier multiplier given by $\operatorname{Id}-|\xi|^{-2}(\xi\otimes\xi)$. This operator plays an important role in fluid mechanics since the Navier-Stokes system for incompressible fluids can be written for a given divergence-free v_0,

$$\begin{cases} \partial_t v + \mathbb{P}\big((v\cdot\nabla)v\big) - \nu\triangle v = 0, \\ \mathbb{P}v = v, \\ v_{|t=0} = v_0. \end{cases}$$

As already said for the Riesz operators, $\mathbb{P}$ is *not* a classical pseudo-differential operator, because of the singularity at the origin: however it is indeed a Fourier multiplier with the same functional properties as those of R.

In three dimensions the curl operator is given by the matrix

$$\operatorname{curl} = \begin{pmatrix} 0 & -\partial_3 & \partial_2 \\ \partial_3 & 0 & -\partial_1 \\ -\partial_2 & \partial_1 & 0 \end{pmatrix} = \operatorname{curl}^*, \tag{1.3.4}$$

since we can note that the matrix

$$C(\xi) = 2\pi\begin{pmatrix} 0 & -i\xi_3 & i\xi_2 \\ i\xi_3 & 0 & -i\xi_1 \\ -i\xi_2 & i\xi_1 & 0 \end{pmatrix}$$

is purely imaginary and anti-symmetric, a feature that could not happen for scalar Fourier multiplier. We get also $\operatorname{curl}^2 = -\triangle\operatorname{Id} + \operatorname{grad}\operatorname{div}$ and (the Biot-Savard law)

$$\operatorname{Id} = (-\triangle)^{-1}\operatorname{curl}^2 + \triangle^{-1}\operatorname{grad}\operatorname{div}, \quad \text{also equal to} \quad (-\triangle)^{-1}\operatorname{curl}^2 + \operatorname{Id} - \mathbb{P},$$

which gives $\operatorname{curl}^2 = -\triangle\mathbb{P}$, so that

$$[\mathbb{P}, \operatorname{curl}] = \triangle^{-1}\big(\triangle\mathbb{P}\operatorname{curl} - \triangle\operatorname{curl}\mathbb{P}\big) = \triangle^{-1}\big(-\operatorname{curl}^3 + \operatorname{curl}(-\triangle\mathbb{P})\big) = 0,$$
$$\mathbb{P}\operatorname{curl} = \operatorname{curl}\mathbb{P} = \operatorname{curl}(-\triangle)^{-1}\operatorname{curl}^2 = \operatorname{curl}\big(\operatorname{Id} - \triangle^{-1}\operatorname{grad}\operatorname{div}\big) = \operatorname{curl}$$

since $\operatorname{curl}\operatorname{grad} = 0$ (note also the adjoint equality $\operatorname{div}\operatorname{curl} = 0$).

These examples show that some interesting cases of Fourier multipliers are quite close to pseudo-differential operators, with respect to the homogeneity and behaviour for large frequencies, although the singularities at the origin in the momentum space make them slightly different. They belong to the family of singular integrals that we shall review briefly.

Theorem 1.3.1. *Let Ω be a function in $L^1(\mathbb{S}^{n-1})$ such that $\int_{\mathbb{S}^{n-1}}\Omega(\omega)d\sigma(\omega) = 0$. Then the following formula defines a tempered distribution T:*

$$\langle T, \varphi\rangle = \lim_{\epsilon\to 0+}\int_{|x|\geq\epsilon}\Omega\Big(\frac{x}{|x|}\Big)|x|^{-n}\varphi(x)dx = -\int(x\cdot\partial_x\varphi(x))\Omega\Big(\frac{x}{|x|}\Big)|x|^{-n}\ln|x|dx.$$

The distribution T is homogeneous of degree $-n$ on $\mathbb{R}^n$ and, if Ω is odd, the Fourier transform of T is a bounded function.

N.B. We shall use the principal-value notation $T = pv\big(|x|^{-n}\Omega(\frac{x}{|x|})\big)$; when $n = 1$ and $\Omega = \text{sign}$, we recover the principal value $pv(1/x) = \frac{d}{dx}(\ln|x|)$ which is odd, homogeneous of degree -1, and whose Fourier transform is $-i\pi\,\text{sign}\,\xi$ (see, e.g., (4.1.21)).

Proof. Let φ be in $\mathscr{S}(\mathbb{R}^n)$ and $\epsilon > 0$. Using polar coordinates, we check

$$\int_{\mathbb{S}^{n-1}} \Omega(\omega) \int_\epsilon^{+\infty} \varphi(r\omega) \frac{dr}{r} d\sigma(\omega)$$

$$= \int_{\mathbb{S}^{n-1}} \Omega(\omega)\Big[\varphi(\epsilon\omega)\ln(\epsilon^{-1}) - \int_\epsilon^{+\infty} \omega \cdot d\varphi(r\omega)\ln r\,dr\Big]d\sigma(\omega).$$

Since the mean value of Ω is 0, we get the first statement of the theorem, noticing that the function $x \mapsto \Omega(x/|x|)|x|^{-n+1}\ln(|x|)(1+|x|)^{-2}$ is in $L^1(\mathbb{R}^n)$. We have

$$\langle x \cdot \partial_x T, \varphi \rangle = -\langle T, x \cdot \partial_x \varphi \rangle - n\langle T, \varphi \rangle \tag{1.3.5}$$

and we see that

$$\langle T, x \cdot \partial_x \varphi \rangle = \lim_{\epsilon \to 0_+} \int_{\mathbb{S}^{n-1}} \Omega(\omega) \int_\epsilon^{+\infty} r\omega \cdot (d\varphi)(r\omega) \frac{dr}{r} d\sigma(\omega)$$

$$= \int_{\mathbb{S}^{n-1}} \Omega(\omega) \int_0^{+\infty} \omega \cdot (d\varphi)(r\omega)dr d\sigma(\omega)$$

$$= \int_{\mathbb{S}^{n-1}} \Omega(\omega) \int_0^{+\infty} \frac{d}{dr}(\varphi(r\omega))dr d\sigma(\omega) = -\varphi(0)\int_{\mathbb{S}^{n-1}} \Omega(\omega)d\sigma(\omega) = 0$$

so that (1.3.5) implies that $x \cdot \partial_x T = -nT$ which is the homogeneity of degree $-n$ of T. As a result the Fourier transform of T is a homogeneous distribution with degree 0.

N.B. Note that the formula $-\int(x \cdot \partial_x\varphi(x))\Omega(\frac{x}{|x|})|x|^{-n}\ln|x|dx$ makes sense for $\Omega \in L^1(\mathbb{S}^{n-1}), \varphi \in \mathscr{S}(\mathbb{R}^n)$ and defines a tempered distribution. For instance, if $n = 1$ and $\Omega = 1$, we get the distribution derivative $\frac{d}{dx}(\text{sign}\,x\ln|x|)$. However, the condition of mean value 0 for Ω on the sphere is necessary to obtain T as a principal value, since in the discussion above, the term factored out by $\ln(1/\epsilon)$ is $\int_{\mathbb{S}^{n-1}} \Omega(\omega)\varphi(\epsilon\omega)d\sigma(\omega)$ which has the limit $\varphi(0)\int_{\mathbb{S}^{n-1}} \Omega(\omega)d\sigma(\omega)$. On the other hand, from the defining formula of T, we get with $\Omega_j(\omega) = \frac{1}{2}(\Omega(\omega)+(-1)^j\Omega(-\omega))$ (Ω_1(resp.Ω_2) is the odd (resp. even) part of Ω),

$$\langle T, \varphi \rangle = \int_{\mathbb{S}^{n-1}} \Omega(\omega)\langle\frac{d}{dt}(H(t)\ln t), \varphi(t\omega)\rangle_{\mathscr{S}'(\mathbb{R}_t),\mathscr{S}(\mathbb{R}_t)}d\sigma(\omega)$$

$$= \int_{\mathbb{S}^{n-1}} \Omega_1(\omega)\langle pv(\frac{1}{2t}), \varphi(t\omega)\rangle_{\mathscr{S}'(\mathbb{R}_t),\mathscr{S}(\mathbb{R}_t)}d\sigma(\omega)$$

$$+ \int_{\mathbb{S}^{n-1}} \Omega_2(\omega)\langle\frac{d}{dt}(H(t)\ln t), \varphi(t\omega)\rangle_{\mathscr{S}'(\mathbb{R}_t),\mathscr{S}(\mathbb{R}_t)}d\sigma(\omega), \tag{1.3.6}$$

since

$$A_1 = \int_{\mathbb{S}^{n-1}} \Omega_1(\omega)\langle pv(\frac{1}{2t}), \varphi(t\omega)\rangle d\sigma(\omega) = -\frac{1}{2}\int_{\mathbb{S}^{n-1}} \Omega_1(\omega)\langle \ln|t|, \omega \cdot d\varphi(t\omega)\rangle d\sigma(\omega)$$

and $\langle \ln|t|, \omega \cdot d\varphi(t\omega)\rangle = \int_0^{+\infty} \omega \cdot d\varphi(t\omega) \ln t\, dt + \int_0^{+\infty} \omega \cdot d\varphi(-s\omega)(\ln s) ds$ so that

$$A_1 = \int_{\mathbb{S}^{n-1}} \Omega_1(\omega)\langle H(t)\ln t, \underbrace{-\frac{1}{2}\omega \cdot (d\varphi(t\omega) + d\varphi(-t\omega))}_{-\frac{1}{2}\frac{d}{dt}(\varphi(t\omega) - \varphi(-t\omega))}\rangle d\sigma(\omega)$$

and thus since Ω_1 is odd,

$$A_1 = \int_{\mathbb{S}^{n-1}} \Omega_1(\omega)\langle \frac{d}{dt}(H(t)\ln t), \frac{1}{2}(\varphi(t\omega) - \varphi(-t\omega))\rangle d\sigma(\omega)$$

$$= \int_{\mathbb{S}^{n-1}} \Omega_1(\omega)\langle \frac{d}{dt}(H(t)\ln t), \varphi(t\omega)\rangle d\sigma(\omega).$$

Let us show that, when Ω is odd, the Fourier transform of T is bounded. Using (1.3.6) and (4.1.21) we get

$$\langle \hat{T}, \psi \rangle = \int_{\mathbb{S}^{n-1}} \Omega(\omega)\langle pv(\frac{1}{2t}), \widehat{\psi}(t\omega)\rangle d\sigma(\omega)$$

$$= -\frac{i\pi}{2}\int_{\mathbb{R}^n} \int_{\mathbb{S}^{n-1}} \Omega(\omega)\,\mathrm{sign}(\omega \cdot \xi)\varphi(\xi)d\xi d\sigma(\omega)$$

proving that

$$\hat{T}(\xi) = -\frac{i\pi}{2}\int_{\mathbb{S}^{n-1}} \Omega(\omega)\,\mathrm{sign}(\omega \cdot \xi)d\sigma(\omega) \tag{1.3.7}$$

which is indeed a bounded function since $\Omega \in L^1(\mathbb{S}^{n-1})$.   □

1.3.2 Remarks on the Calderón-Zygmund theory and classical pseudo-differential operators

It is possible to generalize Theorem 1.3.1 in several directions. In particular the L^p-boundedness ($1 < p < \infty$) of these homogeneous singular integrals can be established, provided some regularity assumptions are made on $\hat{T}$ (see, e.g., Theorem 7.9.5 in [71], the reference books on harmonic analysis by E.M. Stein [132] and J. Duoandikoetxea [40]).

Also a Calderón-Zygmund theory of singular integrals with "variable coefficients", given by some kernel $k(x,y)$ satisfying some conditions analogous to homogeneous functions of degree $-n$ of $x - y$, has reached a high level of refinement (see, e.g., the book by R. Coifman & Y. Meyer [28] and the developments in [40]). Although that theory is not independent of the theory of classical pseudo-differential operators, the fact that the symbols do have a singularity at $\xi = 0$

make them quite different ; the constrast is even more conspicuous for the L^p theory of Calderón-Zygmund operators, which is very well understood although its analogue for general pseudo-differential operators (see, e.g., [111]) has not reached the same level of understanding. We have seen above that the classes $S^0_{\rho,\delta}$ (see Remark 1.1.28) give rise to L^2-bounded operators provided $0 \leq \delta \leq \rho \leq 1, \delta < 1$, and it is possible to prove that the operators with symbol in the class $S^0_{1,0}$ are L^p-bounded, $1 < p < \infty$. The method of proof of that result is not significantly different from the proof of the Calderón-Zygmund theorem of L^p-boundedness for standard homogeneous singular integrals and is based on the weak $(1,1)$ regularity and the Marcinkiewicz interpolation theorem. However, some operators with symbol in the class $S^0_{1,\delta}$ with $0 < \delta < 1$ are not L^p-bounded for $p \neq 2$.

The present book is almost entirely devoted to the developments of the L^2 theory of pseudo-differential operators, but it is certainly useful to keep in mind that some very natural and useful examples of singular integrals are not pseudo-differential operators. For the very important topic of L^p-theory of pseudo-differential operators, we refer the reader to [110].

Chapter 2

Metrics on the Phase Space

In this chapter, we describe a general version of the pseudo-differential calculus, due to L. Hörmander ([69], Chapter 18 in [73]). That version followed some earlier generalizations due to R. Beals and C. Fefferman ([8]) and to R. Beals ([6]). It was followed by some other generalizations due to A. Unterberger [145] and to a joint work of J.-M. Bony and N. Lerner ([20]), whose presentation we follow. We also give a precised version of the Fefferman-Phong inequality, following [100] where we provide an upper bound for the number of derivatives sufficient to obtain that inequality. Finally, we study the Sobolev spaces naturally attached to an admissible metric on the phase space, essentially following the paper by J.-M. Bony and J.-Y. Chemin [19].

2.1 The structure of the phase space

2.1.1 Symplectic algebra

Let us consider a finite-dimensional real vector space E (the configuration[1] space $\mathbb{R}^n_x$) and its dual space E^* (the momentum space $\mathbb{R}^n_\xi$). The phase space is defined as $\mathcal{P} = E \oplus E^*$; its running point will be denoted in general by an upper case letter $(X = (x, \xi), Y = (y, \eta))$. The symplectic form on $\mathcal{P}$ is given by

$$[(x, \xi), (y, \eta)] = \langle \xi, y \rangle_{E^*, E} - \langle \eta, x \rangle_{E^*, E} \qquad (2.1.1)$$

where $\langle \cdot, \cdot \rangle_{E^*, E}$ stands for the bracket of duality. The symplectic group $Sp(2n)$ is the subgroup of the linear group of $\mathcal{P}$ preserving (2.1.1). With

$$\sigma = \begin{pmatrix} 0 & \mathrm{Id}(E^*) \\ -\mathrm{Id}(E) & 0 \end{pmatrix}, \qquad (2.1.2)$$

---

[1]We shall identify the configuration space E with $\mathbb{R}^n$, but the reader may keep in mind that all the theory can be developed without this identification.

we have for $X, Y \in \mathcal{P}$, $[X, Y] = \langle \sigma X, Y \rangle_{\mathcal{P}^*, \mathcal{P}}$, so that the equation of the symplectic group is

$$A^* \sigma A = \sigma.$$

One can describe (see Section 4.4.1 in the appendix) a set of generators for the symplectic group $Sp(2n)$, identifying $\mathcal{P}$ with $\mathbb{R}^n_x \times \mathbb{R}^n_\xi$: the following mappings generate the symplectic group,

$$(x, \xi) \mapsto (Tx, {}^t T^{-1}\xi), \text{ where } T \text{ is an automorphism of } E, \qquad \text{(i)}$$

$$(x_k, \xi_k) \mapsto (\xi_k , -x_k), \text{ and the other coordinates fixed,} \qquad \text{(ii)}$$

$$(x, \xi) \mapsto (x, \xi + Sx), \text{ where } S \text{ is symmetric from } E \text{ to } E^*. \qquad \text{(iii)}$$

2.1.2   The Wigner function

For $u, v \in \mathscr{S}(\mathbb{R}^n)$, we define the Wigner function $\mathcal{H}(u, v)$ on the phase space by the formula

$$\mathcal{H}(u, v)(x, \xi) = \int u(x + \frac{y}{2})\bar{v}(x - \frac{y}{2})e^{-2i\pi y \cdot \xi} dy. \qquad (2.1.3)$$

The mapping $(u, v) \mapsto \mathcal{H}(u, v)$ is sesquilinear continuous from $\mathscr{S}(\mathbb{R}^n) \times \mathscr{S}(\mathbb{R}^n)$ to $\mathscr{S}(\mathbb{R}^{2n})$. Since $\mathcal{H}(u, v)$ is the partial Fourier transform of the function $(x, y) \mapsto u(x + y/2)\bar{v}(x - y/2)$, the Wigner function also satisfies,

$$\|\mathcal{H}(u, v)\|_{L^2(\mathbb{R}^{2n})} = \|u\|_{L^2(\mathbb{R}^n)}\|v\|_{L^2(\mathbb{R}^n)}, \qquad (2.1.4)$$

$$\mathcal{H}(u, v)(x, \xi) = 2^n \langle \sigma_{x, \xi} u, v \rangle_{L^2(\mathbb{R}^n)}, \qquad (2.1.5)$$

$$\text{with} \quad (\sigma_{x, \xi} u)(y) = u(2x - y) \exp -4i\pi(x - y) \cdot \xi \qquad (2.1.6)$$

and the so-called "phase symmetries" σ_X are unitary and selfadjoint on $L^2(\mathbb{R}^n)$.

2.1.3   Quantization formulas

Let a be temperate distribution on $\mathbb{R}^{2n}$. We have seen in Section 1.1.2 and in Definition 1.1.9 that the Weyl quantization rule associates to this distribution the operator a^w defined formally on functions $u \in \mathscr{S}(\mathbb{R}^n)$ by

$$(a^w u)(x) = \iint e^{2i\pi(x-y) \cdot \xi} \, a(\frac{x + y}{2}, \xi) \, u(y) dy d\xi. \qquad (2.1.7)$$

Since the previous formula does not obviously make sense, we use the Wigner function (2.1.3) to provide a weak definition,

$$\langle a^w u, v \rangle_{\mathscr{S}^*(\mathbb{R}^n), \mathscr{S}(\mathbb{R}^n)} = \prec a, \mathcal{H}(u, v) \succ_{\mathscr{S}'(\mathbb{R}^{2n}), \mathscr{S}(\mathbb{R}^{2n})}, \qquad (2.1.8)$$

where $\mathscr{S}^*(\mathbb{R}^n)$ is the antidual of $\mathscr{S}(\mathbb{R}^n)$ and the second bracket is for real duality. For instance we have

$$(x \cdot \xi)^w = (x \cdot D_x + D_x \cdot x)/2,$$

with $D_x = \frac{1}{2i\pi}\frac{\partial}{\partial x}$ whereas the *classical* quantization rule would map the Hamiltonian $x \cdot \xi$ to the operator $x \cdot D_x$. The formula (2.1.8) can be written as

$$(a^w u, v) = \iint a(x, \xi)\mathcal{H}(u, v)(x, \xi)dxd\xi,$$

where the Wigner function $\mathcal{H}$ is defined by (2.1.3). We have indeed

$$\langle a^w u, v\rangle = \iiint a\left(\frac{x+y}{2}, \xi\right)e^{2i\pi(x-y)\xi}u(y)\bar{v}(x)dydxd\xi$$

$$= \iiint a(x', \xi)e^{2i\pi x''\xi}u\left(x' - \frac{x''}{2}\right)\bar{v}\left(x' + \frac{x''}{2}\right)dx'dx''d\xi$$

$$= \iint a(x, \xi)\mathcal{H}(u, v)(x, \xi)dxd\xi,$$

as claimed in (2.1.8). Now, we have also from that formula

$$\mathcal{H}(u, v)(x, \xi) = 2^n \int u(2x - z)\bar{v}(z)e^{-4i\pi(x-z)\xi}dz = 2^n \langle \sigma_{x,\xi}u, v\rangle_{L^2(\mathbb{R}^n)}$$

which can be written ([145], [150]), as

$$a^w = \int_{\mathbb{R}^{2n}} a(X)2^n \sigma_X dX = \int_{\mathbb{R}^{2n}} \hat{a}(\Xi)\exp(2i\pi\Xi \cdot M)d\Xi, \qquad (2.1.9)$$

where $\Xi \cdot M = \hat{x} \cdot x + \hat{\xi} \cdot D_x$ (here $\Xi = (\hat{x}, \hat{\xi})$). The last formula can be proven by applying Plancherel's formula to (2.1.8): for $u, v \in \mathcal{S}(\mathbb{R}^n)$, we have

$$\prec a, \mathcal{H}(u, v) \succ_{\mathscr{S}'(\mathbb{R}^{2n}),\mathscr{S}(\mathbb{R}^{2n})} = \prec \hat{a}, \check{\widehat{\mathcal{H}(u, v)}} \succ_{\mathscr{S}'(\mathbb{R}^{2n}),\mathscr{S}(\mathbb{R}^{2n})}$$

$$\in\mathscr{S}(\mathbb{R}^{2n})$$

and $\check{\widehat{\mathcal{H}(u, v)}}(\hat{x}, \hat{\xi}) = \iint e^{2i\pi(x\cdot\hat{x}+\xi\cdot\hat{\xi})}\mathcal{H}(u, v)(x, \xi)\, dxd\xi$ is also, according to (2.1.8), equal to $\langle b^w u, v\rangle$ with $b(x, \xi) = e^{2i\pi(x\cdot\hat{x}+\xi\cdot\hat{\xi})}$, that is the exponential of a purely imaginary linear form.

Lemma 2.1.1. *If L is a linear form on the phase space $\mathbb{R}^{2n}$, $a \in \mathscr{S}'(\mathbb{R}^{2n})$, we have*

$$L^w L^w = (L^2)^w, \quad \forall k \in \mathbb{N}, \ (L^w)^k = (L^k)^w, \qquad (2.1.10)$$

$$L^w a^w = \left(La + \frac{1}{4i\pi}\{L, a\}\right)^w, \quad a^w L^w = \left(La - \frac{1}{4i\pi}\{L, a\}\right)^w \qquad (2.1.11)$$

where the Poisson bracket $\{L, a\} = H_L(a)$ is given by (1.1.23). If L is real-valued, L^w is an (unbounded) selfadjoint operator on $L^2(\mathbb{R}^n)$ with domain $\{u \in L^2(\mathbb{R}^n), L^w u \in L^2(\mathbb{R}^n)\}$ and we have $\exp(iL^w) = (e^{iL})^w$.

Proof. Using (2.1.7) for the linear forms ξ_j and x_j, one gets at once (2.1.11) by linearity and this implies readily (2.1.10). When the linear form L is real-valued, the function e^{iL} belongs to $\mathscr{S}'(\mathbb{R}^{2n})$ and for $t \in \mathbb{R}$ (using (2.1.11) for the second equation),

$$\frac{d}{dt}(e^{itL})^w = (iLe^{itL})^w = iL^w(e^{itL})^w = (e^{itL})^w iL^w$$

which gives $(e^{itL})^w = \exp itL^w$ and the last assertion of the lemma. $\square$

Note also that for a linear form L given by $L(X) = [X, T]$, (2.1.1) and (2.1.11) give

$$[L^w, a^w] = \frac{1}{2i\pi}(a' \cdot T)^w. \tag{2.1.12}$$

The operator with Weyl symbol e^{iL} is an isomorphism of $\mathscr{S}(\mathbb{R}^n)$ whose inverse has Weyl symbol e^{-iL} since

$$\frac{d}{dt}\left((e^{itL})^w(e^{-itL})^w\right) = \left((e^{itL})^w iL^w(e^{-itL})^w - (e^{itL})^w iL^w(e^{-itL})^w\right) = 0$$

so that $(e^{itL})^w(e^{-itL})^w = \mathrm{Id}$. Moreover $(e^{iL})^w$ can be extended to a unitary operator of $L^2(\mathbb{R}^n)$ since, according to Remark 1.1.11, for $u, v \in \mathcal{S}(\mathbb{R}^n)$,

$$\langle (e^{iL})^w u, (e^{iL})^w v\rangle_{L^2(\mathbb{R}^n)} = \langle (e^{-iL})^w(e^{iL})^w u, v\rangle_{L^2(\mathbb{R}^n)} = \langle u, v\rangle_{L^2(\mathbb{R}^n)},$$

and the rôle of $L, -L$ can be exchanged so that $(e^{-iL})^w(e^{iL})^w = (e^{iL})^w(e^{-iL})^w = \mathrm{Id}_{L^2(\mathbb{R}^n)}$. On the other hand, one may also consider the unbounded operator L^w on $L^2(\mathbb{R}^n)$ with domain $D = \{u \in L^2(\mathbb{R}^n), L^w u \in L^2(\mathbb{R}^n)\}$ and show directly that, for L a real-valued linear form, L^w is a selfadjoint operator. (cf., e.g., Section 4.7 in the appendix). The formulas (2.1.9) give in particular

$$\|a^w\|_{\mathcal{L}(L^2)} \le \min(2^n\|a\|_{L^1(\mathbb{R}^{2n})}, \|\widehat{a}\|_{L^1(\mathbb{R}^{2n})}), \tag{2.1.13}$$

where $\mathcal{L}(L^2)$ stands for the space of bounded linear maps from $L^2(\mathbb{R}^n)$ into itself.

2.1.4   The metaplectic group

The metaplectic group $Mp(n)$ is the subgroup of the group of unitary transformations of $L^2(\mathbb{R}^n)$ generated by the multiplication by complex numbers with modulus 1 and

$(M_T u)(x) = |\det T|^{-1/2}u(T^{-1}x)$, where T is an automorphism of E,   (j)

partial Fourier transformation, with respect to x_k for $k = 1, \dots, n$,   (jj)

multiplication by $\exp(i\pi\langle Sx, x\rangle)$, where S is symmetric from E to E^*.   (jjj)

A detailed study of the symplectic and metaplectic groups is given in the appendix, Section 4.4.1. In particular, one can define a surjective homomorphism of groups

$$\Pi : Mp(n) \to Sp(2n),$$

with kernel $\mathbb{S}^1 \mathrm{Id}_{L^2(\mathbb{R}^n)}$ and such that, if $\chi = \Pi(M)$ and u, v are in $L^2(\mathbb{R}^n)$, $\mathcal{H}(u, v)$ is their Wigner function,

$$\mathcal{H}(Mu, Mv) = \mathcal{H}(u, v) \circ \chi^{-1}.$$

This is Segal's formula [128] which could be rephrased as follows.

Theorem 2.1.2. *Let $a \in \mathscr{S}'(\mathbb{R}^{2n})$ and $\chi \in Sp(2n)$. There exists M in the fiber of χ (M is a unitary transformation which is also an automorphism of $\mathscr{S}(\mathbb{R}^n)$ and $\mathscr{S}'(\mathbb{R}^n)$) such that*

$$(a \circ \chi)^w = M^* a^w M. \tag{2.1.14}$$

In particular, the images by Π of the transformations (j), (jj), (jjj) *are respectively* (i), (ii), (iii) *in page 58 of Section 2.1. Moreover, if χ is the phase translation, $\chi(x, \xi) = (x + x_0, \xi + \xi_0)$, (2.1.14) is fulfilled with $M = \tau_{x_0, \xi_0}$, the phase translation given by*

$$(\tau_{x_0, \xi_0} u)(y) = u(y - x_0) \, e^{2i\pi \langle y - \frac{x_0}{2}, \xi_0 \rangle}. \tag{2.1.15}$$

If $\chi(x, \xi) = (2x - x_0, 2\xi - \xi_0)$ is the symmetry with respect to (x_0, ξ_0), M in (2.1.14) is, up to a unit factor, the phase symmetry σ_{x_0, ξ_0} defined in (2.1.6).

Proof. The proof is a consequence of Remark 4.4.19 of the appendix. $\square$

The following lemma describes the natural link between the phase translations and the phase symmetries.

Lemma 2.1.3. *The phase symmetry σ_X given in (2.1.6) is unitary and selfadjoint on $L^2(\mathbb{R}^n)$, the phase translation τ_X given in (2.1.15) is unitary and $\tau_X^* = \tau_{-X}$. Moreover, we have for $Y, Z \in \mathbb{R}^{2n}$,*

$$\sigma_Y \sigma_Z = \tau_{2Y - 2Z} e^{4i\pi[Z, Y]} = \int_{\mathbb{R}^{2n}} e^{-4i\pi[X - Y, X - Z]} 2^n \sigma_X \, dX, \tag{2.1.16}$$

$$\tau_{2Z} \sigma_Y = \sigma_{Y + Z} e^{4i\pi[Z, Y]}, \quad \sigma_Y \tau_{2Z} = \sigma_{Y - Z} e^{4i\pi[Z, Y]}, \quad \tau_Y \tau_Z = \tau_{Y + Z} e^{i\pi[Y, Z]}. \tag{2.1.17}$$

Proof. The first statements are already proven. Let us note that

$$T \mapsto 2Z - T \mapsto 2Y - (2Z - T)$$

is the composition of the symmetry w.r.t. Z followed by the symmetry w.r.t. Y, which gives the translation by $2Y - 2Z$, so that the first formula in (2.1.16) is natural; as well as (2.1.17), it can be checked directly by an elementary computation left to the reader. Let us verify the second formula in (2.1.16). For $u \in \mathscr{S}(\mathbb{R}^n)$, we have, with $Y = (y, \eta)$, $Z = (z, \zeta)$, both in $\mathbb{R}^n \times \mathbb{R}^n$,

$$(\sigma_{y,\eta} \sigma_{z,\zeta} u)(x) = e^{4i\pi(x-y)\cdot\eta}(\sigma_{z,\zeta} u)(2y - x) = e^{4i\pi(x-y)\cdot\eta} e^{4i\pi(2y-x-z)\cdot\zeta} u(2z - 2y + x).$$

Since we have

$$[X - Y, X - Z] = [X, Y - Z] + [Y, Z],$$

the following integral can be considered as a Fourier transform of the tempered distribution (depending on the parameter x) $X' \mapsto (\sigma_{X'}u)(x)$. We calculate

$$\int_{\mathbb{R}^{2n}} e^{-4i\pi[X'-Y,X'-Z]} 2^n (\sigma_{X'}u)(x) dX'$$

$$= \iint e^{-4i\pi(\xi'-\eta)\cdot(x'-z)+4i\pi(x'-y)\cdot(\xi'-\zeta)} 2^n u(2x'-x) e^{4i\pi(x-x')\cdot\xi'} dx' d\xi'$$

$$= \iint e^{-4i\pi(\xi'-\eta)\cdot(\frac{x}{2}+\frac{x''}{2}-z)+4i\pi(\frac{x}{2}+\frac{x''}{2}-y)\cdot(\xi'-\zeta)} u(x'') e^{4i\pi(\frac{x}{2}-\frac{x''}{2})\cdot\xi'} dx'' d\xi'$$

$$= \iint e^{4i\pi\xi'\cdot(z-y+\frac{x-x''}{2})} e^{4i\pi\eta\cdot(\frac{x}{2}+\frac{x''}{2}-z)+4i\pi(\frac{x}{2}+\frac{x''}{2}-y)\cdot(-\zeta)} u(x'') dx'' d\xi'$$

$$= \int \delta(2z-2y+x-x'') e^{4i\pi\eta\cdot(\frac{x}{2}+\frac{x''}{2}-z)+4i\pi(\frac{x}{2}+\frac{x''}{2}-y)\cdot(-\zeta)} u(x'') dx''$$

$$= u(2z-2y+x) e^{4i\pi\eta\cdot(x-y)+4i\pi(x+z-2y)\cdot(-\zeta)}. \qquad \square$$

2.1.5 Composition formula

According to Theorem 1.1.2 and to Lemma 4.1.2, the operator a^w is continuous from $\mathscr{S}(\mathbb{R}^n)$ into itself whenever $a \in C_b^\infty(\mathbb{R}^{2n})$ and thus in particular when $a \in \mathscr{S}(\mathbb{R}^{2n})$. Using for $a, b \in \mathscr{S}(\mathbb{R}^{2n})$ the first formula in (2.1.9), we obtain

$$a^w b^w = \iint_{\mathbb{R}^{2n} \times \mathbb{R}^{2n}} a(Y) b(Z) 2^{2n} \sigma_Y \sigma_Z dY dZ.$$

Using the last formula of (2.1.16) we get $a^w b^w = (a \sharp b)^w$ with

$$(a \sharp b)(X) = 2^{2n} \iint_{\mathbb{R}^{2n} \times \mathbb{R}^{2n}} e^{-4i\pi[X-Y,X-Z]} a(Y) b(Z) dY dZ. \qquad (2.1.18)$$

We can compare this with the classical composition formula, as studied in Theorem 1.1.5, namely $\mathrm{op}(a)\mathrm{op}(b) = \mathrm{op}(a \diamond b)$ with

$$(a \diamond b)(x, \xi) = \iint_{\mathbb{R}^n \times \mathbb{R}^n} e^{-2i\pi y \cdot \eta} a(x, \xi+\eta) b(y+x, \xi) dy d\eta.$$

Another method to perform that calculation would be to use the kernels of the operators a^w, b^w, in the spirit of the proof of Theorem 1.1.5. For future reference, we note that the distribution kernel k_a of the operator $a(x, D)$ (for $a \in \mathscr{S}'(\mathbb{R}^{2n})$) as defined by (1.1.1) is

$$k_a(x, y) = \int e^{2i\pi(x-y)\cdot\xi} a(x, \xi) d\xi = \hat{a}^2(x, y-x) \qquad (2.1.19)$$

so that $\hat{a}^2(x, y) = k_a(x, y+x)$ and in the distribution sense

$$a(x, \xi) = \int e^{2i\pi y \cdot \xi} k_a(x, y+x) dy. \qquad (2.1.20)$$

The distribution kernel κ_a of the operator a^w (for $a \in \mathscr{S}'(\mathbb{R}^{2n})$) as defined by (2.1.8) is (in the distribution sense)

$$\kappa_a(x, y) = \int e^{2i\pi(x-y)\cdot\xi} a\left(\frac{x+y}{2}, \xi\right) d\xi \tag{2.1.21}$$

so that $\kappa_a(x - \frac{t}{2}, x + \frac{t}{2}) = \int e^{-2i\pi t\cdot\xi} a(x, \xi) d\xi = \hat{a}^2(x, t)$ and thus

$$a(x, \xi) = \int \kappa_a\left(x - \frac{t}{2}, x + \frac{t}{2}\right) e^{2i\pi t\xi} dt. \tag{2.1.22}$$

The formulas (2.1.19), (2.1.20), (2.1.21) and (2.1.22) are proved above for a general distribution symbol $a \in \mathscr{S}'(\mathbb{R}^{2n})$ whereas the composition formula (2.1.18) (as well as (1.1.4)) require some more hypotheses to make sense (even as "oscillatory" integrals as in Section 7.8 of [71] or in a weak distribution sense), which is quite natural since we need to compose the operators a^w, b^w; at this point, we have proven the composition formula (2.1.18) for a, b both in $\mathscr{S}(\mathbb{R}^{2n})$, a case in which a^w and b^w send (continuously) $\mathscr{S}(\mathbb{R}^n)$ into itself; note also that Theorem 1.1.20 has already provided a composition formula along with an asymptotic expansion in the classical quantization framework for the class $S_{1,0}^m$ of symbols studied in Section 1.1.3. One of the goals of this chapter is to show that it is possible to construct various algebras of pseudo-differential operators while considerably relaxing the assumptions on the symbol classes.

Remark 2.1.4. The symplectic invariance of the Weyl quantization given by the formula (2.1.14) is its most important property. The fact that the Weyl formula quantizes real Hamiltonians into formally selfadjoint operators is shared with other quantizations such as the Feynman quantization, given by $a^F = \frac{1}{2}(\mathrm{op}(a) + \mathrm{op}(Ja))$ (see Remark 1.1.11). The reader can check Section 4.4.3 of our appendix to see that the Feynman quantization does not satisfy the symplectic invariance property.

For $a_1, a_2 \in \mathscr{S}(\mathbb{R}^{2n})$, the formula (2.1.18) can be written as well as (we use the symplectic form σ given in (2.1.2))

$$(a_1 \natural a_2)(X) = \iint e^{-2i\pi\langle\sigma Y_1, Y_2\rangle} a_1(2^{-1/2}Y_1 + X) a_2(2^{-1/2}Y_2 + X) dY_1 dY_2$$

so that using (see Proposition 4.1.1) that the Fourier transform of

$$\mathbb{R}^{2n} \times \mathbb{R}^{2n} \ni (X_1, X_2) \mapsto \exp -2i\pi\langle\sigma X_1, X_2\rangle = \exp -i\pi \begin{pmatrix} X_1 & X_2 \end{pmatrix} \begin{pmatrix} 0 & -\sigma \\ \sigma & 0 \end{pmatrix} \begin{pmatrix} X_1 \\ X_2 \end{pmatrix}$$

is $\exp i\pi \begin{pmatrix} \Xi_1 & \Xi_2 \end{pmatrix} \begin{pmatrix} 0 & \sigma^{-1} \\ -\sigma^{-1} & 0 \end{pmatrix} \begin{pmatrix} \Xi_1 \\ \Xi_2 \end{pmatrix} = \exp 2i\pi\langle\Xi_1, \sigma^{-1}\Xi_2\rangle$: in fact the $4n \times 4n$ matrix

$$\begin{pmatrix} 0 & -\sigma \\ \sigma & 0 \end{pmatrix} = \begin{pmatrix} 0 & 0 & 0 & -I_E \\ 0 & 0 & I_{E^*} & 0 \\ 0 & I_E & 0 & 0 \\ -I_{E^*} & 0 & 0 & 0 \end{pmatrix}$$

is symmetric with determinant 1 and signature 0 and its inverse is

$$
\begin{pmatrix}
0 & 0 & 0 & -I_{E^*} \\
0 & 0 & I_E & 0 \\
0 & I_{E^*} & 0 & 0 \\
-I_E & 0 & 0 & 0
\end{pmatrix}.
$$

We obtain, after applying Plancherel's formula and performing a simple affine change of variables,

$$
(a_1 \natural a_2)(X) = \iint \exp i\pi \langle \Xi_1, \sigma^{-1}\Xi_2 \rangle
$$

$$
\times \left(\iint \exp 2i\pi \big((X-Y_1)\cdot \Xi_1 + (X-Y_2)\cdot \Xi_2\big) a_1(Y_1)a_2(Y_2)dY_1 dY_2 \right) d\Xi_1 d\Xi_2
$$

which is

$$
(a_1 \natural a_2)(X) = \exp(i\pi[D_{X_1}, D_{X_2}])\big(a_1(X_1)a_2(X_2)\big)_{|X_1=X_2=X}, \tag{2.1.23}
$$

where the symplectic form on the dual of the phase space is naturally given by $[\Xi_1, \Xi_2] = \langle -\sigma^{-1}\Xi_1, \Xi_2 \rangle$ so that

$$
[D_{X_1}, D_{X_2}] = [(D_{x_1}, D_{\xi_1}), (D_{x_2}, D_{\xi_2})] = D_{\xi_1}\cdot D_{x_2} - D_{x_1}\cdot D_{\xi_2}. \tag{2.1.24}
$$

Remark 2.1.5. For future reference, it will be useful to notice that

$$
\big(i\pi[D_{X_1}, D_{X_2}]\big)^j (a_1 \otimes a_2)_{|X_1=X_2=X} = 2^{-j}a_1^{(j)}(X)\big((\sigma^{-1}D_X)^j a_2\big)(X).
$$

In fact, the lhs is equal to

$$
\iint \big(i\pi\langle -\sigma^{-1}\Xi_1, \Xi_2 \rangle\big)^j \widehat{a_1}(\Xi_1)\widehat{a_2}(\Xi_2)e^{2i\pi(X\cdot(\Xi_1+\Xi_2))}d\Xi_1 d\Xi_2
$$

$$
= \iint \big(i\pi\langle \Xi_1, \sigma^{-1}\Xi_2 \rangle\big)^j \widehat{a_1}(\Xi_1)\widehat{a_2}(\Xi_2)e^{2i\pi(X\cdot(\Xi_1+\Xi_2))}d\Xi_1 d\Xi_2
$$

$$
= \iint \big(i\pi\mathtt{Fourier}(\langle D, \sigma^{-1}\Xi_2 \rangle)^j a_1\big)(\Xi_1)\widehat{a_2}(\Xi_2)e^{2i\pi(X\cdot(\Xi_1+\Xi_2))}d\Xi_1 d\Xi_2
$$

$$
= \int \big(\tfrac{1}{2}(\langle \partial, \sigma^{-1}\Xi_2 \rangle)^j a_1\big)(X)\widehat{a_2}(\Xi_2)e^{2i\pi X\cdot\Xi_2}d\Xi_2
$$

$$
= 2^{-j}\int a_1^{(j)}(X)(\sigma^{-1}\Xi)^j \widehat{a_2}(\Xi)e^{2i\pi X\cdot\Xi}d\Xi \quad \text{which is the rhs.}
$$

Remark 2.1.6. We can observe that the expression

$$
a_1 \boxtimes a_2 = \exp(i\pi[D_{X_1}, D_{X_2}])(a_1 \otimes a_2) = \exp \frac{1}{4i\pi}[\partial_{X_1}, \partial_{X_2}](a_1 \otimes a_2) \tag{2.1.25}
$$

makes sense for all tempered distributions $a_j \in \mathscr{S}'(\mathbb{R}^{2n})$ and provides a distribution in $\mathscr{S}'(\mathbb{R}^{4n})$ although the formula (2.1.23) requires a restriction to the diagonal of $\mathbb{R}^{2n} \times \mathbb{R}^{2n}$ of the distribution $a_1 \boxtimes a_2$, restriction which is naturally subjected to some conditions on the symbols a_1, a_2.

It is convenient to give right now an "asymptotic" version of the composition formulas (2.1.18) and (1.1.4), as it is already done with (1.1.18). We use

$$
\begin{aligned}
\frac{1}{k!}\left(\frac{1}{4i\pi}[\partial_X,\partial_Y]\right)^k &= \frac{1}{k!}\left(\frac{1}{4i\pi}\sum_{1\leq j\leq n}(\partial_{\xi_j}\partial_{y_j} - \partial_{x_j}\partial_{\eta_j})\right)^k \\
&= \frac{1}{2^k k!}\left(\sum_{1\leq j\leq n}(D_{\xi_j}\partial_{y_j} - \partial_{x_j}D_{\eta_j})\right)^k \\
&= \frac{1}{2^k}\sum_{\gamma_1+\cdots+\gamma_n=k}\prod_{1\leq j\leq n}\frac{1}{\gamma_j!}(D_{\xi_j}\partial_{y_j} - \partial_{x_j}D_{\eta_j})^{\gamma_j} \\
&= \frac{1}{2^k}\sum_{\substack{\gamma_1+\cdots+\gamma_n=k \\ \gamma_j=\alpha_j+\beta_j}}\prod_{1\leq j\leq n}\frac{1}{\alpha_j!\beta_j!}(D_{\xi_j}\partial_{y_j})^{\alpha_j}(-1)^{|\beta_j|}(\partial_{x_j}D_{\eta_j})^{\beta_j} \\
&= \frac{1}{2^k}\sum_{|\alpha|+|\beta|=k}\frac{1}{\alpha!\beta!}D_\xi^\alpha\partial_y^\alpha(-1)^{|\beta|}\partial_x^\beta D_\eta^\beta,
\end{aligned}
$$

and we get from (2.1.25)

$$
(a\sharp b)(x,\xi) = \sum_{0\leq k<\nu} 2^{-k}\sum_{|\alpha|+|\beta|=k}\frac{(-1)^{|\beta|}}{\alpha!\beta!}D_\xi^\alpha\partial_x^\beta a\, D_\xi^\beta\partial_x^\alpha b + r_\nu(a,b), \tag{2.1.26}
$$

$$
\text{with}\quad r_\nu(a,b)(X) = R_\nu\big(a(X)\otimes b(Y)\big)_{|X=Y}, \tag{2.1.27}
$$

$$
R_\nu = \int_0^1 \frac{(1-\theta)^{\nu-1}}{(\nu-1)!}\exp\frac{\theta}{4i\pi}[\partial_X,\partial_Y]d\theta\left(\frac{1}{4i\pi}[\partial_X,\partial_Y]\right)^\nu. \tag{2.1.28}
$$

Defining

$$
\omega_k(a,b) = 2^{-k}\sum_{|\alpha|+|\beta|=k}\frac{(-1)^{|\beta|}}{\alpha!\beta!}D_\xi^\alpha\partial_x^\beta a\, D_\xi^\beta\partial_x^\alpha b, \tag{2.1.29}
$$

we get that

$$
\omega_0 = ab, \quad \omega_1 = \frac{1}{4i\pi}\{a,b\},
$$

so that the beginning of this expansion is thus

$$
ab + \frac{1}{4i\pi}\{a,b\},
$$

where the Poisson bracket $\{a,b\}$ is given by (1.1.23). The $\omega_k(a,b)$ with k even are symmetric in a,b and skew-symmetric for k odd: this is obvious from the above expression coming from $[\partial_X,\partial_Y]^k$. Also, when a,b are real-valued the $\omega_k(a,b)$ with k even are real and purely imaginary for k odd. This can be compared to the classical expansion formula (1.1.18), which is less symmetrical. We can prove as well, using the Taylor expansion of the exponential function, another expression of R_ν in (2.1.33),

$$\overbrace{}^{\omega_j(a_1,a_2)}$$

$$(a_1 \sharp a_2)(X) = \sum_{0 \le j < \nu} \frac{(i\pi)^j}{j!} [D_{X_1}, D_{X_2}]^j (a_1 \otimes a_2)_{|X_1=X_2=X}$$

$$+ \int_0^1 \frac{(1-\theta)^{\nu-1}}{(\nu-1)!} 2^{2n} \theta^{-2n} \iint e^{-\frac{4i\pi}{\theta}[X-Y_1, X-Y_2]} (i\pi)^\nu [D_{Y_1}, D_{Y_2}]^\nu (a_1 \otimes a_2)(Y_1, Y_2)$$

$$dY_1 dY_2 d\theta. \quad (2.1.30)$$

Remark 2.1.7. The expansion formula above are proven true at this point only for $a, b \in \mathscr{S}(\mathbb{R}^{2n})$, but we shall see later on that for rather general symbols, these formulas can actually be used as true asymptotic expansions. To build-up this from the previous sections, we give the result for the class of symbols (1.1.13) introduced in the first section (a proof for very general classes of symbols will be given later on, so we omit it here). For $a_j \in S_{1,0}^{m_j}, j = 1, 2$ we have

$$a_1 \sharp a_2 = a_1 a_2 + \frac{1}{4i\pi} \{a_1, a_2\} \quad \text{mod } S_{1,0}^{m_1+m_2-2}, \quad (2.1.31)$$

$$a_1 \sharp a_2 + a_2 \sharp a_1 = 2a_1 a_2 \quad \text{mod } S_{1,0}^{m_1+m_2-2}, \quad (2.1.32)$$

$$a_1 \sharp a_2 - a_2 \sharp a_1 = \frac{1}{2i\pi} \{a_1, a_2\} \quad \text{mod } S_{1,0}^{m_1+m_2-3}. \quad (2.1.33)$$

Let us give a more explicit expression for the third term ω_2 in the expansion (2.1.26). We have

$$(4i\pi)^2 \omega_2(a, b) = \sum_{|\alpha|=2} \frac{1}{\alpha!} \left(\partial_\xi^\alpha a \partial_x^\alpha b + \partial_x^\alpha a \partial_\xi^\alpha b \right) - \sum_{|\alpha|=|\beta|=1} \partial_\xi^\alpha \partial_x^\beta a \, \partial_x^\alpha \partial_\xi^\beta b.$$

Remark 2.1.8. We have implicitly considered that our symbols were scalar-valued, but it is of course interesting, e.g., for the study of systems of PDE, to consider the case where a is a tempered distribution on $\mathbb{R}^{2n}$ valued in some Banach algebra of operators. For instance a may be a smooth function, bounded as well as all its derivatives, from $\mathbb{R}^{2n}$ into $\mathscr{B}(\mathbb{H})$ where $\mathbb{H}$ is a Hilbert space and $\mathscr{B}(\mathbb{H})$ stands for the bounded operators on $\mathbb{H}$. Even the finite-dimensional case is of interest, but one has to pay careful attention to the composition formula, because $\mathscr{B}(\mathbb{H})$ is not commutative. We see that part of the discussion above remains valid but that the product $a(Y)b(Z)$ in (2.1.18) cannot be replaced by $b(Z)a(Y)$. In particular, the formula (2.1.31) remains true in the non-commutative case, provided the Poisson bracket is defined as in (1.1.23), respecting the order of the products; however, the formulas following (2.1.31) must be modified in the non-commutative case: in particular we have, say for finite-dimensional square systems a_1, a_2 with entries in $S_{1,0}^{m_j}$,

$$a_1 \sharp a_2 - a_2 \sharp a_1 = a_1 a_2 - a_2 a_1 + \frac{1}{4i\pi} \{a_1, a_2\} - \frac{1}{4i\pi} \{a_2, a_1\} \quad \text{mod } S_{1,0}^{m_1+m_2-2}$$

so that the symbol of the commutator $[a_1^w, a_2^w]$ is, modulo $S_{1,0}^{m_1+m_2-2}$,

$$\underbrace{\llcorner a_1(x,\xi), a_2(x,\xi) \lrcorner}_{\substack{\text{commutator of the matrices} \\ a_1(x,\xi), a_2(x,\xi)}} + \frac{1}{4i\pi} \sum_{1 \le j \le n} \frac{\partial a_1}{\partial \xi_j} \frac{\partial a_2}{\partial x_j} + \frac{\partial a_2}{\partial x_j} \frac{\partial a_1}{\partial \xi_j} - \frac{\partial a_1}{\partial x_j} \frac{\partial a_2}{\partial \xi_j} - \frac{\partial a_2}{\partial \xi_j} \frac{\partial a_1}{\partial x_j}.$$

2.2 Admissible metrics

2.2.1 A short review of examples of pseudo-differential calculi

We have already seen some examples of symbol classes, such as

- $S_{\rho,\delta}^m$: C^∞ functions a defined on $\mathbb{R}^{2n}$ satisfying (1.1.32). The conditions

$$\delta < 1, 0 \le \delta \le \rho \le 1$$

  are required. The classes $S_{1,0}^m$ provide a good framework for the analysis of PDE in an open set and in particular are useful to construct parametrices of (micro)elliptic operators and to prove some results of propagation of singularities for real-principal type operators (see, e.g., our Section 1.2.3). The classes $S_{\rho,\delta}^m$ were introduced in [65] as the natural space containing the parametrices of hypoelliptic operators with constant coefficients.

- S_{scl}^m: C^∞ functions a defined on $\mathbb{R}^{2n} \times (0,1]$ satisfying (1.1.33). These classes constitute the natural framework of semi-classical analysis, where the operators depend on a small parameter h, likened to the Planck constant.

- The classes Σ^m: C^∞ functions a defined on $\mathbb{R}^{2n}$ satisfying (1.1.48). These classes are extensively used in the book [129] to study the spectral properties of globally defined pseudo-differential operators on $\mathbb{R}^n$.

Further steps in the generalization, essentially motivated by the proof of solvability of principal-type differential operators satisfying Nirenberg-Treves' condition (P) in [8], were taken by R. Beals and C. Fefferman in their 1973 paper [9], in which the authors introduced symbols a which are C^∞ functions on $\mathbb{R}^{2n}$ such that

$$|(\partial_\xi^\alpha \partial_x^\beta a)(x,\xi)| \le C_{\alpha\beta} \Phi(x,\xi)^{M-|\alpha|} \varphi(x,\xi)^{m-|\beta|} \tag{2.2.1}$$

where the positive functions Φ, φ should satisfy some conditions. The article [69] by L. Hörmander in 1979 (see also Chapter 18 in [73]), introduced the more flexible notion of *metric* on the phase space, a notion also used in the paper [145] of A. Unterberger and in a more general context by J.-M. Bony and N. Lerner in [20]. It is easy to see that the conditions (2.2.1) provide a more general framework than the three examples at the beginning of this section and one can consider the following Riemannian structure on the phase space $\mathbb{R}^{2n}$:

$$g_{x,\xi} = \frac{|dx|^2}{\varphi(x,\xi)^2} + \frac{|d\xi|^2}{\Phi(x,\xi)^2} : \text{ for } (t,\tau) \in \mathbb{R}^n \times \mathbb{R}^n, \quad g_{x,\xi}(t,\tau) = \frac{|t|^2}{\varphi(x,\xi)^2} + \frac{|\tau|^2}{\Phi(x,\xi)^2}.$$

The estimates (2.2.1) for $|\alpha| + |\beta| = 1$ (when $M = m = 0$), can be written as well as $|(\nabla a)(X) \cdot T| \leq C g_X(T)^{1/2}$ ($X = (x,\xi), T = (t,\tau)$) and for higher-order derivatives as

$$|(\nabla^k a)(X)(T_1, \ldots, T_k)| \leq C g_X(T_1)^{1/2} \cdots g_X(T_k)^{1/2}.$$

As a particular case, the class $S^m_{\rho,\delta}$ defined above is related to the following metric on the phase space,

$$g_{x,\xi} = \frac{|dx|^2}{\langle \xi \rangle^{-2\delta}} + \frac{|d\xi|^2}{\langle \xi \rangle^{2\rho}} : \quad \text{for } (t,\tau) \in \mathbb{R}^n \times \mathbb{R}^n, \ g_{x,\xi}(t,\tau) = \frac{|t|^2}{\langle \xi \rangle^{-2\delta}} + \frac{|\tau|^2}{\langle \xi \rangle^{2\rho}}, \quad (2.2.2)$$

with $\langle \xi \rangle = (1 + |\xi|^2)^{1/2}$. These examples will provide a guideline for the definitions in the sequel. A pseudo-differential calculus will be essentially the datum of a metric g that should satisfy some local and some global conditions. The symplectic structure and the uncertainty principle will also play a natural rôle in the constraints imposed on the metric g.

2.2.2   Slowly varying metrics on $\mathbb{R}^{2n}$

Definition 2.2.1. Let g be a (measurable) mapping from $\mathbb{R}^{2n}$ into $\mathcal{C}$, the cone of positive-definite quadratic forms on $\mathbb{R}^{2n}$: for each $X \in \mathbb{R}^{2n}$, g_X is a positive-definite quadratic form on $\mathbb{R}^{2n}$. We shall say that g is a slowly varying metric on $\mathbb{R}^{2n}$ if

$$\exists C > 0, \exists r > 0, \forall X, Y, T \in \mathbb{R}^{2n},$$

$$g_X(Y - X) \leq r^2 \Longrightarrow C^{-1} g_Y(T) \leq g_X(T) \leq C g_Y(T). \quad (2.2.3)$$

Remark 2.2.2. The previous property will be satisfied if we ask only that

$$\exists C > 0, \exists r > 0, \forall X, Y, T \in \mathbb{R}^{2n}, \quad g_X(Y-X) \leq r^2 \Longrightarrow g_Y(T) \leq C g_X(T). \quad (2.2.4)$$

In fact, assuming (2.2.4) gives that whenever $g_X(Y - X) \leq C^{-1} r^2$(which is $\leq r^2$ since $C \geq 1$ from (2.2.4) with $X = Y$) this implies $g_Y(X - Y) \leq r^2$, and thus $g_X(T) \leq C g_Y(T)$, so that (2.2.3) is satisfied with C_0, r_0 such that $r_0 = C^{-1/2} r, C_0 = C$.

Examples 2.2.3. If we consider for a positive Lipschitz-continuous function ρ defined on $\mathbb{R}^{2n}$, and a Euclidean norm $\|\|$ on $\mathbb{R}^{2n}$, the metric defined by

$$g_X(T) = \frac{\|T\|^2}{\rho(X)^2},$$

one can check easily that g is slowly varying: assuming $g_X(Y - X) \leq r^2$ means that $\|Y - X\| \leq r\rho(X)$ so that, if L is the Lipschitz constant of ρ, we have

$$\rho(X) \leq \rho(Y) + L\|X - Y\| \leq \rho(Y) + Lr\rho(X),$$

so that choosing $0 < r < \frac{1}{L}$ gives the desired property (2.2.4) with $C = (1 - Lr)^{-2}$.
Let us verify that the metric (2.2.2) is slowly varying when $\rho \leq 1$. We assume that
$g_{x,\xi}((y, \eta) - (x, \xi)) \leq r^2$ so that $|\eta - \xi|^2 \leq r^2(1 + |\xi|^2)^{\rho}$ and we obtain

$$|\xi|^2 \leq 2r^2(1 + |\xi|^2) + 2|\eta|^2$$

and thus $(1 + |\xi|^2)(1 - 2r^2) \leq 2(1 + |\eta|^2)$ which proves (2.2.4) if $r < 2^{-1/2}$. Note
that the metric (2.2.2) with $\rho > 1$ is not slowly varying since

$$\frac{\langle \eta \rangle^{\rho}}{\langle \xi \rangle^{\rho}} \leq \sup_{(t,\tau)} \frac{g_{x,\xi}(t, \tau)^{1/2}}{g_{y,\eta}(t, \tau)^{1/2}}$$

and for $k > 1$ with $|\xi| = k, |\eta| = k^{\rho}/\ln k$, we have for $k \to +\infty$, $|\xi - \eta| \sim k^{\rho}/\ln k \ll k^{\rho} \sim \langle \xi \rangle^{\rho}$, so that

$$|\xi - \eta| \langle \xi \rangle^{-\rho} = g_{x,\xi}(x - x, \xi - \eta)^{1/2} \to 0, \text{ but } \langle \eta \rangle \langle \xi \rangle^{-1} \sim k^{\rho-1}/\ln k \to +\infty.$$

Definition 2.2.4. Let g be a slowly varying metric on $\mathbb{R}^{2n}$. A function $M : \mathbb{R}^{2n} \to$
$]0, +\infty)$ is a g-slowly varying weight if

$$\exists C > 0, \exists r > 0, \forall X, Y \in \mathbb{R}^{2n}, g_X(Y - X) \leq r^2 \implies C^{-1} \leq \frac{M(X)}{M(Y)} \leq C. \quad (2.2.5)$$

Remark 2.2.5. It is easy to verify that $\langle \xi \rangle$ is a slowly varying weight for the metrics
(2.2.2) when $\rho \leq 1$: we have

$$\frac{\langle \xi \rangle^2}{\langle \eta \rangle^2} \leq \frac{1 + 2|\xi - \eta|^2}{1 + |\eta|^2} + 2 \leq 3 + 2\frac{|\xi - \eta|^2}{\langle \xi \rangle^{2\rho}} \frac{\langle \xi \rangle^{2\rho}}{\langle \eta \rangle^2} \leq 3 + 2g_{x,\xi}((y, \eta) - (x, \xi)) \frac{\langle \xi \rangle^2}{\langle \eta \rangle^2}$$

which gives the answer provided $g_X(Y - X) < 1/2$. Note also that products
of slowly varying weights for g are still slowly varying weights for g, and more
generally that M^{θ} is a slowly varying weight if M is a slowly varying weight and
$\theta \in \mathbb{R}$.

Definition 2.2.6. Let g be a slowly varying metric[2] on $\mathbb{R}^{2n}$ and let M be a g-slowly
varying weight. The space of symbols $S(M, g)$ is defined as the set of C^{∞} functions
on $\mathbb{R}^{2n}$ such that $\forall k \in \mathbb{N}, \exists C_k, \forall X, T_1, \ldots, T_k \in \mathbb{R}^{2n}$,

$$|a^{(k)}(X)(T_1, \ldots, T_k)| \leq C_k M(X) \prod_{1 \leq j \leq k} g_X(T_j)^{1/2}. \quad (2.2.6)$$

For $a \in S(M, g), l \in \mathbb{N}$, we shall note that

$$\|a\|_{S(M,g)}^{(l)} = \max_{0 \leq k \leq l} \sup_{\substack{X, T_j \\ g_X(T_j)=1}} |a^{(k)}(X)(T_1, \ldots, T_k)| M(X)^{-1}. \quad (2.2.7)$$

The space $S(M, g)$, with the countable family of semi-norms $(\| \cdot \|_{S(M,g)}^{(l)})_{l \in \mathbb{N}}$ is a
Fréchet space.

---

[2]The reader may notice that these slow variations properties are not needed for setting the
definition which makes sense for a metric g and a positive function M.

Looking back at the couple of examples above, we can check readily that

$$S_{\rho,\delta}^m = S\big(\langle\xi\rangle^m, \frac{|dx|^2}{\langle\xi\rangle^{-2\delta}} + \frac{|d\xi|^2}{\langle\xi\rangle^{2\rho}}\big) \quad (\text{see } (1.1.32)),$$

$$S_{scl}^m = S\big(h^{-m}, |dx|^2 + h^2|d\xi|^2\big) \quad (\text{see } (1.1.33)),$$

$$\Sigma^m = S\big((1 + |x|^2 + |\xi|^2)^m, \frac{|dx|^2 + |d\xi|^2}{1 + |x|^2 + |\xi|^2}\big) \quad (\text{see } (1.1.48)),$$

and (2.2.1) defines $S(\Phi^M\varphi^m, \frac{|dx|^2}{\varphi^2} + \frac{|d\xi|^2}{\Phi^2})$. It is not difficult to illustrate the meaning of the localization via a slowly varying metric: around each point X, one may draw the ball

$$U_{X,r} = \{Y, g_X(Y - X) \le r^2\} \tag{2.2.8}$$

and if r is chosen small enough, *uniformly with respect to* X, the shape and size of $U_{Y,r}$ will be equivalent to the shape and size of $U_{X,r}$ for $Y \in U_{X,r}$. R. Beals and C. Fefferman in [9] preferred to speak about "boxes", since their metrics were defined by (2.2.1) and they would consider instead of the Euclidean ball U the product of (Euclidean) balls,

$$B_{x,\xi,r} = \Big\{(y,\eta) \in \mathbb{R}^n \times \mathbb{R}^n, \max\Big(\frac{|y - x|}{\varphi(x,\xi)}, \frac{|\eta - \xi|}{\Phi(x,\xi)}\Big) \le r\Big\}.$$

The main feature of a slowly varying metric is that it is possible to introduce some partitions of unity related to the metric in a way made precise in the following theorem. The reader will find in the appendix, Section 4.3.4 a construction of a partition of unity for a metric defined only on some open subset of $\mathbb{R}^N$.

Theorem 2.2.7. *Let g be a slowly varying metric on $\mathbb{R}^{2n}$. Then there exists $r_0 > 0$ such that, for all $r \in (0, r_0]$, there exists a family $(\varphi_Y)_{Y \in \mathbb{R}^{2n}}$ of functions uniformly in $S(1, g)$ (i.e., $\forall k, \sup_Y \|\varphi_Y\|_{S(1,g)}^{(k)} < \infty$) supported in $U_{Y,r}$ such that, for all $X \in \mathbb{R}^{2n}$,*

$$\int_{\mathbb{R}^{2n}} \varphi_Y(X)|g_Y|^{1/2}dY = 1,$$

where $|g_Y|$ is the determinant of g_Y (with respect to the standard Euclidean norm).

Proof. Let $\chi_0 \in C^\infty(\mathbb{R}; [0, 1])$ non-increasing such that $\chi_0(t) = 1$ on $t \le 1/2$, $\chi_0(t) = 0$ on $t \ge 1$. Let us assume that the property (2.2.3) is satisfied with some positive constants $C_0, r_0 > 0$. We define, for $r \in (0, r_0]$, the function

$$\omega(X, r) = \int_{\mathbb{R}^{2n}} \chi_0\big(r^{-2}g_Y(X - Y)\big)|g_Y|^{1/2}dY.$$

Since χ_0 is non-increasing and $X \mapsto \chi_0\big(r^{-2}g_Y(X - Y)\big)$ is supported in $U_{Y,r}$, we have

$$\omega(X,r) \geq \int_{\mathbb{R}^{2n}} \chi_0\big(r^{-2}C_0 g_X(X-Y)\big)|g_X|^{1/2}C_0^{-n}dY = \underbrace{\int_{\mathbb{R}^{2n}} \chi_0(|Z|^2)dZ C_0^{-2n} r^{2n}}_{\text{a positive constant}},$$

(2.2.9)

and an estimate from above of the same type. Also we see that, using the notation $\|T\|_Y = g_Y(T)^{1/2}$ and $\langle,\rangle_Y$ for the associated dot product, we have

$$\omega'(X,r)T = \int_{\mathbb{R}^{2n}} \chi_0'\big(r^{-2}\|X-Y\|_Y^2\big)r^{-2}2\langle(X-Y),T\rangle_Y|g_Y|^{1/2}dY,$$

and by induction, for $k \in \mathbb{N}^*, T \in \mathbb{R}^{2n}$, $\omega^{(k)}(X,r)T^k$ is a finite sum of terms of type

$$c_{p,k} \int_{\mathbb{R}^{2n}} \chi_0^{(p)}\big(r^{-2}\|X-Y\|_Y^2\big)r^{-2p}\langle X-Y,T\rangle_Y^{2p-k}g_Y(T)^{k-p}|g_Y|^{1/2}dY, \quad (2.2.10)$$

where $c_{p,k}$ is a constant depending only on p,k and $p \in [k/2,k] \cap \mathbb{N}$. The support of $\chi_0^{(p)}$, the condition $0 < r \leq r_0$, shows as above that (2.2.10) is bounded above by

$$c(p,k,r,C_0)\|\chi_0^{(p)}\|_{L^\infty} \int_{|Z|\leq 1} |Z|^{2p-k}dZ g_X(T)^{k/2},$$

so that for all k, $|\omega^{(k)}(X,r)T^k| \leq C_k g_X(T)^{k/2}$. According to Section 4.2.2, this implies that the multilinear forms $\omega^{(k)}(X,r)$ have a bounded norm with respect to g_X. Thus we obtain that $\omega(\cdot,r)$ belongs to $S(1,g)$ and the estimate from below (2.2.9) shows that $1/\omega(\cdot,r)$ belongs as well to $S(1,g)$. As a consequence, we get that

$$1 = \int_{\mathbb{R}^{2n}} \underbrace{\chi_0\big(r^{-2}g_Y(X-Y)\big)\omega(X,r)^{-1}}_{=\varphi_Y(X)}|g_Y|^{1/2}dY,$$

where the family $(\varphi_Y)_{Y\in\mathbb{R}^{2n}}$ satisfies the requirements of the theorem. $\qquad\square$

N.B. A discrete version of Theorem 2.2.7 is given in the appendix with Theorem 4.3.7.

Remark 2.2.8. The slow variation property does not require any local regularity, not even continuity[3]. Once a slowly varying metric g is given, the partition of unity allows us to find a smooth metric $\widetilde{g}$ somewhat equivalent to g: we define

$$\widetilde{g}_X(T) = \int_{\mathbb{R}^{2n}} \varphi_Y(X)g_Y(T)|g_Y|^{1/2}dY$$

which is a positive-definite quadratic form such that for $T \neq 0$,

$$\frac{\widetilde{g}_X(T)}{g_X(T)} = \int_{\mathbb{R}^{2n}} \varphi_Y(X)\frac{g_Y(T)}{g_X(T)}|g_Y|^{1/2}dY \in [C^{-1},C]$$

---

[3]Note that the integral version of the partition requires some measurability assumption.

since for all $X \in \operatorname{supp} \varphi_Y, \frac{g_Y(T)}{g_X(T)} \in [C^{-1}, C]$. The same can be done with a g-slowly varying weight M (Definition 2.2.4): we define

$$\widetilde{M}(X) = \int_{\mathbb{R}^{2n}} \varphi_Y(X) M(Y) |g_Y|^{1/2} dY$$

and we see that the ratio $\widetilde{M}(X)/M(X)$ belongs to $[C^{-1}, C]$ where C is the fixed constant given in (2.2.5). Moreover, using the notation of the proof of Theorem 2.2.7 and $\chi_1 = 1$ on the support of χ_0, χ_1 supported in $(-\infty, 2]$, we get that $\widetilde{M}$ is smooth with $|\widetilde{M}^{(k)}(X) T^k| \leq M(X) g_X(T)^{k/2} C_k \int_{\mathbb{R}^{2n}} \chi_1(r^{-2} g_Y(X - Y)) |g_Y|^{1/2} dY$, so that

$$\widetilde{M} \in S(M, g), \quad \forall X \in \mathbb{R}^{2n}, \quad \frac{\widetilde{M}(X)}{M(X)} \in [C^{-1}, C]. \tag{2.2.11}$$

An immediate consequence of Theorem 2.2.7 is the following

Proposition 2.2.9. *Let g be a slowly varying metric on $\mathbb{R}^{2n}$, M be a g-slowly varying weight. and a be a symbol in $S(M, g)$. Then there exists $r_0 > 0$ such that, for all $0 < r \leq r_0$,*

$$a(X) = \int_{\mathbb{R}^{2n}} a_Y(X) |g_Y|^{1/2} dY$$

where $a_Y \in S(M(Y), g_Y)$ uniformly and $\operatorname{supp} a_Y \subset U_{Y,r}$.

2.2.3   The uncertainty principle for metrics

The slow variation properties exposed in the previous section do not use the symplectic structure of the phase space $\mathbb{R}^{2n}$. It is quite natural to expect to have some contraints linking the metric g with the symplectic form (2.1.1). Let us first consider a slightly more abstract framework with the configuration space E given as an n-dimensional real vector space and the phase space $\mathcal{P} = E \oplus E^*$ equipped with the symplectic form σ given by (4.4.1). We refer the reader to the appendix, Section 4.4.1 for some basic facts on symplectic algebra.

Let us consider a metric on $\mathcal{P}$, i.e., a mapping $\mathcal{P} \ni X \mapsto g_X \in \mathcal{C}$, where $\mathcal{C}$ is the cone of positive-definite quadratic forms on $\mathcal{P}$. We shall always assume that

$$\forall X \in \mathcal{P}, \quad g_X \leq g_X^\sigma, \tag{2.2.12}$$

where g_X^σ is given by (4.4.18). Moreover we define, for $X \in \mathcal{P}$,

$$\lambda_g(X) = \inf_{T \neq 0} \left(\frac{g_X^\sigma(T)}{g_X(T)} \right)^{1/2} \tag{2.2.13}$$

and we note from the previous hypotheses that

$$g_X \leq \lambda_g(X)^{-2} g_X^\sigma, \quad \lambda_g(X) \geq 1. \tag{2.2.14}$$

Examples 2.2.10. Let us review the couple of examples above; for the example (2.2.2), which we write

$$g_{x,\xi} = \langle\xi\rangle^{2\delta}|dx|^2 + \langle\xi\rangle^{-2\rho}|d\xi|^2,$$

Lemma 4.4.25 shows that $g_{x,\xi}^\sigma = \langle\xi\rangle^{2\rho}|dx|^2 + \langle\xi\rangle^{-2\delta}|d\xi|^2$, $\lambda_g(x,\xi) = \langle\xi\rangle^{\rho-\delta}$, so that the condition (2.2.12) is equivalent to $\rho \geq \delta$. The semi-classical metric satisfies

$$g_{x,\xi} = |dx|^2 + h^2|d\xi|^2, \quad g_{x,\xi}^\sigma = h^{-2}|dx|^2 + |d\xi|^2, \quad \lambda = h^{-1},$$

and the condition $h \in (0,1]$ appears to be the expression of (2.2.12). For the global metric defined for $X \in \mathbb{R}^{2n}$ by

$$g_X = \langle X\rangle^{-2}g_0, \quad g_0 = g_0^\sigma, \quad g_X^\sigma = \langle X\rangle^2 g_0, \quad \lambda_g(X) = \langle X\rangle^2.$$

For the metric related to the estimates (2.2.1) we have

$$g_{x,\xi} = \varphi(x,\xi)^{-2}|dx|^2 + \Phi(x,\xi)^{-2}|d\xi|^2, \quad g_{x,\xi}^\sigma = \Phi(x,\xi)^2|dx|^2 + \varphi(x,\xi)^2|d\xi|^2, \quad \lambda = \Phi\varphi.$$

Looking now at Lemma 4.4.25, we see that for each $(x,\xi) \in \mathcal{P}$, there exists some symplectic coordinates (y,η) such that

$$g_{x,\xi}(y,\eta) = \sum_{1\leq j\leq n} \lambda_j^{-1}(y_j^2 + \eta_j^2), \quad g_{x,\xi}^\sigma(y,\eta) = \sum_{1\leq j\leq n} \lambda_j(y_j^2 + \eta_j^2), \qquad (2.2.15)$$

with $\lambda(x,\xi) = \min_{1\leq j\leq n}\lambda_j$, so that (2.2.12) can be expressed as

$$\min_{1\leq j\leq n} \lambda_j \geq 1.$$

Comment. Checking the ellipsoid $\{(y,\eta), g_{x,\xi}(y,\eta) \leq 1\}$, we see that the conjugate axes have size $\lambda_j^{1/2}$, and we can write symbolically that $\Delta y_j \Delta \eta_j \sim \lambda_j$. Now the Heisenberg uncertainty principle requires a lower bound on these products, so that it is natural to require as well a lower bound for the λ_j, that is to set a rule which prevents the localization from being too sharp on conjugate axes. Although the analogy with this physical principle is only heuristical at this point, the next lemma provides a mathematical reason to require (2.2.12).

Lemma 2.2.11. *Let g be a metric on $\mathbb{R}^{2n}$, m_1, m_2 be two positive functions on $\mathbb{R}^{2n}$ and $a_j \in S(m_j, g)$, $j = 1,2$ (cf. Definition 2.2.6). We have*

$$\omega_k(a_1, a_2) \in S(m_1 m_2 \lambda_g^{-k}, g)$$

where $\omega_k(a_1, a_2)$ is defined in (2.1.29).

Proof. Using Lemma 4.4.25, we may assume that for a given $X \in \mathbb{R}^{2n}$, the metric g_X has the form (2.2.15) and since the linear change of variables used to achieve that form is symplectic, it leaves unchanged the brackets $[\partial_{X_1}, \partial_{X_2}]$ involved in the formula defining ω_k. We check for $X, T \in \mathbb{R}^{2n}$,

$$\frac{1}{l!}\omega_k(a_1, a_2)^{(l)}(X)T^l$$

$$= \frac{1}{k!}(\frac{1}{4i\pi})^k \sum_{l_1+l_2=l} \frac{1}{l_1!l_2!}[\partial_{X_1}, \partial_{X_2}]^k \big(a_1^{(l_1)}(X_1)T^{l_1} \otimes a_2^{(l_2)}(X_2)T^{l_2}\big)_{X_1=X_2=X},$$

and we see that we have a linear combination of terms

$$a_1^{(l_1+k)}(X)T^{l_1}S_1' \ldots S_k' a_2^{(l_2+k)}(X)T^{l_2}S_1'' \ldots S_k''$$

with $\Gamma(S_{j'}') = \Gamma(S_{j''}'') = 1, \Gamma = \sum_{1 \leq j \leq n} |dy_j|^2 + |d\eta_j|^2$. As a result, we get

$$|\omega_k(a_1, a_2)^{(l)}(X)T^l| \leq C g_X(T)^{l/2} \left(\sup_S \frac{g_X(S)}{\Gamma(S)}\right)^k$$

and since

$$\sup_S \frac{g_X(S)}{\Gamma(S)} = \max(\lambda_j^{-1}) = (\min \lambda_j)^{-1} = \lambda_g(X)^{-1},$$

this gives the result. $\square$

Comment. Although we have not proven yet a composition formula, it is not difficult to guess that we expect that (2.1.26) provides some asymptotic expansion and for instance that the weight of the product $a_1 a_2$ (say $m_1 m_2$) should be larger than the weight of the Poisson bracket $\{a_1, a_2\}$ which is $m_1 m_2 \lambda_g^{-1}$, which requires that $\lambda_g \geq 1$. It will be important for our calculi to deal with large regions of the phase space where $\lambda_g = 1$: at these places no asymptotic expansion is valid (e.g., for the example (2.2.2) when $\rho = \delta$), since all terms in (2.1.26) are potentially of the same size but we should certainly avoid having regions where λ_g is not bounded from below since the divergence of (2.1.26) will get out of control.

2.2.4   Temperate metrics

We begin with a remark, stressing the non-locality of the composition formulas of pseudo-differential operators.

Remark 2.2.12. Looking at the composition formula (2.1.18), we see that, whenever a, b are compactly supported (and say in $L^1(\mathbb{R}^{2n})$), $a \sharp b$ is an entire function; if in addition, the supports of a and b are disjoint with a, b smooth, all the terms $\omega_k(a, b)$ (given in (2.1.29)) are identically 0. Nevertheless, taking $a \in C_c^\infty(\mathbb{R}^{2n})$(not identically 0), and b defined by

$$b(X) = \overline{\hat{a}(-2\sigma X)}\chi(X),$$

where χ is a non-negative C_c^∞ function with $\mathrm{supp}\,\chi \cap \mathrm{supp}\,a = \emptyset, \mathrm{supp}\,\chi \neq \emptyset$ (so that $a, b \in C_c^\infty(\mathbb{R}^{2n})$ with disjoint supports), we get from (2.1.18)

$$(a \sharp b)(0) = 2^{2n} \int \hat{a}(-2\sigma Z) b(Z) dZ = 2^{2n} \int |\hat{a}(-2\sigma Z)|^2 \chi(Z) dZ > 0$$

since the function $\hat{a}(-2\sigma \cdot)$ is analytic and cannot vanish on the non-empty open set $\mathrm{int}(\mathrm{supp}\,\chi)$. As a consequence, the composition formula (2.1.18) is not local, and the values of $a \sharp b$ at a given point may depend on the values of a, b far away from that point. This indicates that, for a given metric g defined on $\mathbb{R}^{2n}$, some assumption should be made to relate g_X to g_Y for any $X, Y \in \mathbb{R}^{2n}$.

Definition 2.2.13. Let g be a metric on $\mathbb{R}^{2n}$. We shall say that g is temperate if

$$\exists C > 0, \exists N \geq 0, \forall X, Y, T \in \mathbb{R}^{2n}, \quad \frac{g_X(T)}{g_Y(T)} \leq C\big(1 + (g_X^\sigma \wedge g_Y^\sigma)(X - Y)\big)^N, \quad (2.2.16)$$

where g^σ is given by (4.4.18) and $\wedge$ in (4.4.23).

The definition in [73] looked less symmetrical and weaker, since the rhs of the inequality in (2.2.16) was replaced by $(1 + g_X^\sigma(X - Y))^N$. Definition 2.2.13 was introduced in [20] to handle pseudo-differential calculus on an open set of $\mathbb{R}^{2n}$. Let us show now that these conditions are equivalent when considering metrics defined on the whole[4] $\mathbb{R}^{2n}$.

Lemma 2.2.14. *Let g be a metric on $\mathbb{R}^{2n}$ such that*

$$\exists C > 0, \exists N \geq 0, \forall X, Y, T \in \mathbb{R}^{2n}, \quad \frac{g_X(T)}{g_Y(T)} \leq C\big(1 + g_X^\sigma(X - Y)\big)^N. \quad (2.2.17)$$

Then g is temperate, i.e., satisfies (2.2.16).

Proof. For $X, Y \in \mathbb{R}^{2n}$, according to (4.4.27) there exists $Z \in \mathbb{R}^{2n}$ such that

$$(g_X^\sigma \wedge g_Y^\sigma)(X - Y) = 2g_X^\sigma(X - Z) + 2g_Y^\sigma(Z - Y).$$

Now from (2.2.17), we know that

$$g_X \leq C_1 g_Z \big(1 + g_X^\sigma(X - Z)\big)^{N_1}, \quad g_Z^\sigma \leq C_1 g_Y^\sigma \big(1 + g_Y^\sigma(Y - Z)\big)^{N_1}.$$

Then we get $g_Z^\sigma(Z - Y) \leq C_1 \big(1 + g_Y^\sigma(Y - Z)\big)^{N_1 + 1}$ and thus

$$g_Z \leq C_2 g_Y \big(1 + g_Y^\sigma(Z - Y)\big)^{N_1(N_1 + 1)}.$$

---

[4]This equivalence does not hold in general when the metric is defined only on a proper open subset of $\mathbb{R}^{2n}$ and that was the reason for introducing the formulation above.

We obtain

$$
\begin{aligned}
\frac{g_X(T)}{g_Y(T)} &\leq C_1 \frac{g_Z(T)}{g_Y(T)} \big(1 + g_X^\sigma(X - Z)\big)^{N_1} \\
&\leq C_1 C_2 \big(1 + g_Y^\sigma(Z - Y)\big)^{N_1(N_1+1)} \big(1 + g_X^\sigma(X - Z)\big)^{N_1} \\
&\leq C_3 \big(1 + 2g_X^\sigma(X - Z) + 2g_Y^\sigma(Z - Y)\big)^{N_1(N_1+1)} \\
&= C_3 \big(1 + (g_X^\sigma \wedge g_Y^\sigma)(X - Y)\big)^{N_1(N_1+1)}. \qquad\qquad \square
\end{aligned}
$$

2.2.5 Admissible metric and weights

Definition 2.2.15. Let g be a metric on $\mathbb{R}^{2n}$, which is slowly varying (see (2.2.3)), satisfies the uncertainty principle (see (2.2.12)) and is temperate (see (2.2.16)). We shall say then that g is an admissible metric. Let M be a slowly varying weight for g (see (2.2.5)) such that

$$
\exists C, N, \forall X, Y \in \mathbb{R}^{2n}, \quad \frac{M(X)}{M(Y)} \leq C\big(1 + (g_X^\sigma \wedge g_Y^\sigma)(X - Y)\big)^N, \qquad (2.2.18)
$$

then we shall say that M is an admissible weight for g.

Remark 2.2.16. The previous property is equivalent to the weaker

$$
\exists C, N, \forall X, Y \in \mathbb{R}^{2n}, \quad \frac{M(X)}{M(Y)} \leq C\big(1 + g_X^\sigma(X - Y)\big)^N. \qquad (2.2.19)
$$

Let X, Y be given in $\mathbb{R}^{2n}$ and Z such that $(g_X^\sigma \wedge g_Y^\sigma)(X - Y) = 2g_X^\sigma(X - Z) + 2g_Y^\sigma(Z - Y)$. We have

$$
\frac{M(X)}{M(Y)} = \frac{M(X)}{M(Z)} \frac{M(Z)}{M(Y)} \leq C^2 \big(1 + g_X^\sigma(X - Z)\big)^N \big(1 + g_Z^\sigma(Y - Z)\big)^N,
$$

and since g satisfies (2.2.16), we can estimate from above the $\big(1 + g_Z^\sigma(Y - Z)\big)^N$ by $\big(1 + g_Y^\sigma(Y - Z)\big)^{N_1}$ and this gives the property of Definition 2.2.15.

Remark 2.2.17. Note that, for an admissible metric g, the function λ_g defined in (2.2.13) is an admissible weight. First λ_g is a slowly varying weight for g, since for $r > 0$ small enough, $g_X(Y - X) \leq r^2$ implies[5] $g_X \sim g_Y$ and thus $g_X^\sigma \sim g_Y^\sigma$, so that

$$
\frac{g_X^\sigma(T)}{g_X(T)} \sim \frac{g_Y^\sigma(T)}{g_Y(T)} \implies \lambda_g(X) \sim \lambda_g(Y).
$$

---

[5]Here and thereafter, for a, b positive quadratic forms, we shall use the abuse of language $a \lesssim b$ (resp. $a \gtrsim b$, $a \sim b$) to mean that the ratio $a(T)/b(T)$ is bounded above (resp. below, resp. above and below) by "fixed" positive constants which depend only on the "structure" constants occurring in (2.2.3), (2.2.16), (2.2.5).

Moreover we have

$$\frac{g_X^\sigma(T)}{g_X(T)} \le C_1 \big(1 + (g_X^\sigma \wedge g_Y^\sigma)(X - Y)\big)^{N_1} \frac{g_Y^\sigma(T)}{g_Y(T)}$$

$$\Longrightarrow \lambda_g(X)^2 \le \lambda_g(Y)^2 C_1 \big(1 + (g_X^\sigma \wedge g_Y^\sigma)(X - Y)\big)^{N_1}.$$

Naturally, the powers of λ_g are also admissible weights. The weights are multiplicative whereas the "orders" of standard classes are additive ; although we shall not use this terminology, it would be natural to say that the symbols of order μ related to the metric g are those in $S(\lambda_g^\mu, g)$.

Lemma 2.2.18. *The metric (2.2.2) is admissible when $0 \le \delta \le \rho \le 1, \delta < 1$. The metric (2.2.2) is not admissible when $\rho = \delta = 1$. The metrics $|dx|^2 + h^2|d\xi|^2$, (see (1.1.33), $h \in (0, 1]$), $\frac{|dx|^2 + |d\xi|^2}{1 + |x|^2 + |\xi|^2}$ (see (1.1.48)) are admissible.*

Proof. We have already seen with Examples 2.2.3 that the slow variation of (2.2.2) is equivalent to $\rho \le 1$. Also the study of Examples 2.2.10 shows that the uncertainty principle for these metrics is equivalent to $\delta \le \rho$. The metric of the class $S_{1,1}$ is not temperate since that would imply

$$\exists C, \ \exists N, \ \forall \xi, \forall \eta, \quad \langle \xi \rangle \le C \langle \eta \rangle \left(1 + |\xi - \eta|/\langle \xi \rangle \right)^N,$$

which cannot hold for $\eta = 0$ and unbounded ξ. Let us show that the metric of the class $S_{\rho,\delta}$ is temperate whenever $\delta < 1, \delta \le \rho \le 1$: we check

$$\frac{g_{x,\xi}(t, \tau)}{g_{y,\eta}(t, \tau)} = \frac{\langle \xi \rangle^{2\delta}|t|^2 + \langle \xi \rangle^{-2\rho}|\tau|^2}{\langle \eta \rangle^{2\delta}|t|^2 + \langle \eta \rangle^{-2\rho}|\tau|^2} \le \max\left(\langle \xi \rangle^{2\delta} \langle \eta \rangle^{-2\delta}, \langle \xi \rangle^{-2\rho} \langle \eta \rangle^{2\rho}\right)$$

and we also have $g_{x,\xi}^\sigma(x - y, \xi - \eta) = \langle \xi \rangle^{2\rho}|x - y|^2 + \langle \xi \rangle^{-2\delta}|\xi - \eta|^2 \ge \langle \xi \rangle^{-2\delta}|\xi - \eta|^2$. According to Lemma 2.2.14, it is enough to check

$$\frac{1 + |\xi|}{1 + |\eta|} + \frac{1 + |\eta|}{1 + |\xi|} \le 2C \big(1 + (1 + |\xi|)^{-\delta}|\xi - \eta|\big)^{\frac{1}{1-\delta}},$$

which is a consequence of the symmetrical

$$\frac{1 + |\eta|}{1 + |\xi|} \le C \big(1 + (1 + |\xi| + |\eta|)^{-\delta}|\xi - \eta|\big)^{\frac{1}{1-\delta}}, \tag{2.2.20}$$

which we prove now. The estimate (2.2.20) is obvious for $|\eta| \le 2|\xi|$ and for $|\eta| \le 1$. If $|\eta| \ge 2|\xi|, |\eta| \ge 1$, the rhs of (2.2.20) is bounded below by $C'|\eta|$, implying (2.2.20). The metric $|dx|^2 + h^2|d\xi|^2$ is constant and the condition (2.2.12) means $h \le 1$. The admissibility of the last metric of the lemma is an easy exercise left to the reader. $\qquad \square$

Definition 2.2.19. Let g be a metric on $\mathbb{R}^{2n}$. The metric g is said to be *symplectic* whenever for all $X \in \mathbb{R}^{2n}$, $g_X = g_X^\sigma$, where g_X^σ is defined in (4.4.18). We define the *symplectic intermediate metric* $g^\natural$ by

$$g_X^\natural = \sqrt{g_X \cdot g_X^\sigma}. \qquad (2.2.21)$$

The metric $g^\natural$ is symplectic, according to (4.4.28) and, if (2.2.12) holds, we have $g_X \le g_X^\natural = (g_X^\natural)^\sigma \le g_X^\sigma$.

Assuming (2.2.12) for g we get from Lemma 4.4.25 that

$$g_X \le \lambda_g(X)^{-1} g_X^\natural \le g_X^\natural = \left(g_X^\natural\right)^\sigma \le \lambda_g(X) g_X^\natural \le g_X^\sigma. \qquad (2.2.22)$$

The following very useful observation is due to J. Toft [137].

Proposition 2.2.20. *Let g be an admissible metric on $\mathbb{R}^{2n}$. Then the metric $g^\natural$ is admissible and there exists C, N such that, for all $X, Y \in \mathbb{R}^{2n}$,*

$$g_X \le C g_Y \left(1 + (g_X^\natural \wedge g_Y^\natural)(Y - X)\right)^N. \qquad (2.2.23)$$

Moreover if M is an admissible weight for g, it is also an admissible weight for $g^\natural$.

Proof. The properties (2.2.3) and (2.2.17) are satisfied for g: the constants occurring there will be denoted by r_0, C_0, N_0. Let X be given in $\mathbb{R}^{2n}$. According to Lemma 4.4.25, there exists a basis $(e_1, \ldots, e_n, \epsilon_1, \ldots, \epsilon_n)$ of $\mathbb{R}^{2n}$ such that

the matrix of the symplectic form σ is $\begin{pmatrix} 0 & I_n \\ -I_n & 0 \end{pmatrix}$,

the matrix of g_X is $\begin{pmatrix} h & 0 \\ 0 & h \end{pmatrix}$ with $h = \operatorname{diag}(h_1, \ldots, h_n)$, $(h_j \in (0, 1])$,

so that the matrix of g_X^σ is $\begin{pmatrix} h^{-1} & 0 \\ 0 & h^{-1} \end{pmatrix}$ and the matrix of $g_X^\natural$ is I_{2n}.

For $T \in \mathbb{R}^{2n}$, we shall write $T = \sum_{1 \le j \le n}(t_j e_j + \tau_j \epsilon_j)$, $T_j = t_j$ for $1 \le j \le n$ and $T_j = \tau_{j-n}$ for $n + 1 \le j \le 2n$. For $Y \in \mathbb{R}^{2n}$, $r > 0$ we define

$$J = \{j \in \{1, \ldots, 2n\}, h_j |X_j - Y_j|^2 \le r^2\}, \quad Z_j = \begin{cases} Y_j, & \text{if } j \in J, \\ X_j, & \text{if } j \notin J. \end{cases}$$

We have $g_X(X - Z) = \sum_{j \in J} h_j |X_j - Y_j|^2 \le 2nr^2$ and if $2nr^2 \le r_0^2$, we obtain that the ratios $g_X(S)/g_Z(S)$ (and thus from Remark 4.4.24, $g_X^\sigma(S)/g_Z^\sigma(S)$) belong to $[C_0^{-1}, C_0]$. Using the temperance of g, we get

$$g_X \le C_0 g_Z \le C_0^2 g_Y \left(1 + g_Z^\sigma(Y - Z)\right)^{N_0}$$

$$\le C_0^2 g_Y \left(1 + C_0 g_X^\sigma(Y - Z)\right)^{N_0} = C_0^2 g_Y \left(1 + C_0 \sum_{j \notin J} h_j^{-1} |Y_j - X_j|^2\right)^{N_0}$$

$$\le C_0^2 g_Y \left(1 + C_0 \sum_{j \notin J} r^{-2} |Y_j - X_j|^4\right)^{N_0}$$

$$\le C_0^2 g_Y \left(1 + C_0 r^{-2} g_X^\natural (Y - X)^2\right)^{N_0},$$

that is

$$g_X \le C_1 g_Y \left(1 + g_X^\natural (Y - X)\right)^{2N_0}. \tag{2.2.24}$$

We can repeat the previous argument, replacing g by g^σ (and also replacing g by a weight M, admissible for g):

$$g_X^\sigma \le C_0 g_Z^\sigma \le C_0^2 g_Y^\sigma \left(1 + g_Z^\sigma (Y - Z)\right)^{N_0}$$
$$\le C_0^2 g_Y^\sigma \left(1 + C_0 g_X^\sigma (Y - Z)\right)^{N_0} = C_0^2 g_Y^\sigma \left(1 + C_0 \sum_{j \notin J} h_j^{-1} |Y_j - X_j|^2\right)^{N_0}$$
$$\le C_0^2 g_Y^\sigma \left(1 + C_0 \sum_{j \notin J} r^{-2} |Y_j - X_j|^4\right)^{N_0}$$
$$\le C_0^2 g_Y^\sigma \left(1 + C_0 r^{-2} g_X^\natural (Y - X)^2\right)^{N_0},$$

which implies

$$g_Y \le C_1 g_X \left(1 + g_X^\natural (Y - X)\right)^{2N_0}. \tag{2.2.25}$$

Let X, Y be given in $\mathbb{R}^{2n}$. There exists $T \in \mathbb{R}^{2n}$ such that

$$(g_X^\natural \wedge g_Y^\natural)(X - Y) = 2g_X^\natural (X - T) + 2g_Y^\natural (T - Y)$$

and we have thus from (2.2.24), (2.2.25),

$$g_X \le C_1 g_T \left(1 + g_X^\natural (X - T)\right)^{2N_0}, \quad g_T \le C_1 g_Y \left(1 + g_Y^\natural (T - Y)\right)^{2N_0},$$

providing $g_X \le C_2 g_Y \left(1 + g_X^\natural (X - T) + g_Y^\natural (T - Y)\right)^{4N_0}$, which is (2.2.23). On the other hand, the metric $g^\natural$ is slowly varying since it is larger than the slowly varying g and $g^\natural = \sqrt{g \cdot g^\sigma}$; the latter formula and the now proven (2.2.23) give the temperance of $g^\natural$ which also satisfies trivially (2.2.12). The proof of the proposition is complete. □

The following lemma will be useful in the sequel.

Lemma 2.2.21. *Let g be an admissible metric on $\mathbb{R}^{2n}$ and let G be a slowly varying metric on $\mathbb{R}^{2n}$. We assume that $G = \mu g$ is conformal to g, with $\mu \le \lambda_g$. Then the metric G is admissible.*

Proof. We have $G^\sigma = \mu^{-1} g^\sigma \ge \mu^{-1} \lambda_g^2 g \ge \mu g = G$, so that G satisfies (2.2.12) and we get also $G^\natural = g^\natural$. Since we have assumed that G is slowly varying, we may assume, to prove (2.2.17), that for X, Y given in $\mathbb{R}^{2n}$, $G_Y(Y - X) \ge r_0^2$ with some positive r_0. Using now (2.2.22) for G (for which (2.2.12) holds), we get, since $G \le \lambda_G^{-1} G^\natural$,

$$g_Y^\natural (Y - X) = G_Y^\natural (Y - X) \ge \lambda_G(Y) r_0^2 = r_0^2 \lambda_g(Y) \mu(Y)^{-1}.$$

Then for $T \in \mathbb{R}^{2n}$, using the temperance of g (which implies (2.2.23)), we get

$$\frac{G_X(T)}{G_Y(T)} = \frac{\mu(X)}{\mu(Y)} \frac{g_X(T)}{g_Y(T)} \le \frac{\lambda_g(X)}{r_0^2 \lambda_g(Y)} g_Y^\natural (Y - X) C \left(1 + (g_X^\natural \wedge g_Y^\natural)(Y - X)\right)^N,$$

and the rhs of this inequality is bounded above by a power of $1 + g_X^\flat(X - Y)$, yielding (2.2.17) and the lemma. $\square$

Lemma 2.2.22. *Let g be an admissible metric on $\mathbb{R}^{2n}$ and m be an admissible weight for g. Let $f \in C^\infty(\mathbb{R}_+^*; \mathbb{R}_+^*)$ such that*

$$\forall r \in \mathbb{N}, \quad \sup_{t>0} |f^{(r)}(t)| f(t)^{-1} t^r < \infty. \tag{2.2.26}$$

Then there exists $C > 0$ and $\tilde{m} \in S(m, g)$ such that $\forall X$, $C^{-1}m(X) \leq \tilde{m}(X) \leq Cm(X)$, $f(\tilde{m})$ is an admissible weight for g and $f(\tilde{m}) \in S(f(\tilde{m}), g)$.

N.B. The assumption on f is satisfied in particular for the function $t \mapsto t^\theta$ where $\theta \in \mathbb{R}$.

Proof. Remark 2.2.8 gives the existence of $\tilde{m}$ and $C > 0$ satisfying $\tilde{m} \in S(m, g)$ and (2.2.11), which gives also that $\tilde{m}$ is admissible for g. For $T \in \mathbb{R}^{2n}$, we check with $M = f(\tilde{m})$, using the Faà di Bruno formula (see, e.g., Section 4.3.1),

$$M^{(k)}(X)T^k = k! \sum_{\substack{k_1+\cdots+k_r=k \\ 1\leq r\leq k,\ k_j\geq 1}} \frac{1}{r!} f^{(r)}(\tilde{m}) \frac{1}{k_1!\ldots k_r!} \tilde{m}^{(k_1)}T^{k_1}\ldots m^{(k_r)}T^{k_r},$$

which entails, thanks to (2.2.26),

$$|M^{(k)}(X)T^k| \leq C \sum_{1\leq r\leq k} f(\tilde{m})\tilde{m}(X)^{-r}m(X)^r g_X(T)^{k/2} \leq kCg_X(T)^{k/2}M(X),$$

which is the sought result. $\square$

2.2.6   The main distance function

Definition 2.2.23. Let g be an admissible metric on $\mathbb{R}^{2n}$ (see Definition 2.2.15). We define, for $X, Y \in \mathbb{R}^{2n}$ and $r > 0$, using the notation in (4.4.18), (4.4.23), (2.2.8),

$$\delta_r(X, Y) = 1 + (g_X^\sigma \wedge g_Y^\sigma)(U_{X,r} - U_{Y,r}). \tag{2.2.27}$$

Lemma 2.2.24. *Let g and δ_r be as above, with (2.2.16) satisfied with N_0. Then, there exist some positive constants $(C_j)_{1\leq j\leq 5}, (N_j)_{1\leq j\leq 2}$ and $r_0 > 0$ such that, for $0 < r \leq r_0$, $\forall X, Y, T \in \mathbb{R}^{2n}$,*

$$\delta_r(X, Y) \leq C_1\Big(1 + \min\big(g_X^\sigma(U_{X,r} - U_{Y,r}), g_Y^\sigma(U_{X,r} - U_{Y,r})\big)\Big)$$
$$\leq C_1\Big(1 + \max\big(g_X^\sigma(U_{X,r} - U_{Y,r}), g_Y^\sigma(U_{X,r} - U_{Y,r})\big)\Big)$$
$$\leq C_2\delta_r(X, Y)^{N_0+1}, \tag{2.2.28}$$

$$\frac{g_X(T)}{g_Y(T)} \leq C_3\delta_r(X, Y)^{N_0}, \quad \sup_{X\in\mathbb{R}^{2n}} \int_{\mathbb{R}^{2n}} \delta_r(X, Y)^{-N_1}|g_Y|^{1/2}dY < \infty, \tag{2.2.29}$$

and for $g_X(Y - X) \geq C_4r^2$, $\lambda_g(X) + \lambda_g(Y) \leq C_5\delta_r(X, Y)^{N_2}$. $(2.2.30)$

Proof. We note that from the second inequality in (4.4.26),

$$\delta_r(X,Y) \le 1 + 2\min\big(g_X^\sigma(U_{X,r} - U_{Y,r}), g_Y^\sigma(U_{X,r} - U_{Y,r})\big),$$

and since g is temperate, for $X' \in U_{X,r}, Y' \in U_{Y,r}$, we have, for r small enough,

$$g_X^\sigma \wedge g_Y^\sigma \sim g_{X'}^\sigma \wedge g_{Y'}^\sigma \gtrsim g_{X'}^\sigma \big(1 + (g_{X'}^\sigma \wedge g_{Y'}^\sigma)(X' - Y')\big)^{-N_0}$$
$$\sim g_X^\sigma \big(1 + (g_X^\sigma \wedge g_Y^\sigma)(X' - Y')\big)^{-N_0}$$

which implies $g_X^\sigma \wedge g_Y^\sigma \gtrsim g_X^\sigma \delta_r(X,Y)^{-N_0}$ and thus $g_X^\sigma(U_{X,r} - U_{Y,r}) \lesssim \delta_r(X,Y)^{N_0+1}$, yielding (2.2.28); moreover, with the same notation, we have

$$\frac{g_X}{g_Y} \sim \frac{g_{X'}}{g_{Y'}} \lesssim \big(1 + (g_{X'}^\sigma \wedge g_{Y'}^\sigma)(X' - Y')\big)^{N_0} \sim \big(1 + (g_X^\sigma \wedge g_Y^\sigma)(X' - Y')\big)^{N_0},$$

which gives the first inequality in (2.2.29). Still with the same notation, we have

$$\begin{aligned}
1 + g_X(X - Y) &\le 1 + 3g_X(X - X') + 3g_X(X' - Y') + 3g_X(Y' - Y)\\
&\lesssim \delta_r(X,Y)^{N_0}\big(1 + g_X(X - X') + g_X(X' - Y') + g_Y(Y' - Y)\big)\\
&\lesssim \delta_r(X,Y)^{N_0}\big(1 + g_X^\sigma(X' - Y')\big)\\
&\lesssim \delta_r(X,Y)^{2N_0}\big(1 + (g_X^\sigma \wedge g_Y^\sigma)(X' - Y')\big)\\
&\lesssim \delta_r(X,Y)^{2N_0+1},
\end{aligned}$$

and since $\frac{|g_Y|^{1/2}}{|g_X|^{1/2}} \lesssim \delta_r(X,Y)^{nN_0}$ we obtain the integrability property in (2.2.29). We have also from (2.2.14)

$$\lambda_g(X)^2 g_X(X' - Y') \le g_X^\sigma(X' - Y') \lesssim \delta_r(X,Y)^{N_0}(g_X^\sigma \wedge g_Y^\sigma)(X' - Y') \le \delta_r(X,Y)^{N_0+1}.$$

As a consequence, for C large enough, if $g_X(X - Y) \ge Cr^2$ we get $g_X(X' - Y') \ge r^2$ (otherwise $g_X(X' - Y') < r^2$ and $g_X \sim g_Y$ with $g_X(X - Y) \le C_0 r^2$). The previous inequality gives $\lambda_g(X) \lesssim \delta_r(X,Y)^{N_1}$ and (2.2.30). $\qquad\square$

Lemma 2.2.25. *Let g be an admissible metric on $\mathbb{R}^{2n}$ and M be an admissible weight for g (cf. Definition 2.2.15). There exist $r_0 > 0, C > 0, N_0 \ge 0$, such that for $0 < r \le r_0$, $X, Y \in \mathbb{R}^{2n}$,*

$$\frac{M(X)}{M(Y)} \le C\delta_r(X,Y)^{N_0}. \tag{2.2.31}$$

Proof. Since M is admissible for g, it satisfies (2.2.18) and using that M is slowly varying for g and g is slowly varying, for r small enough, we have for $X' \in U_{X,r}, Y' \in U_{Y,r}$,

$$\frac{M(X)}{M(Y)} \sim \frac{M(X')}{M(Y')} \le C\big(1 + (g_{X'}^\sigma \wedge g_{Y'}^\sigma)(X' - Y')\big)^N \sim \delta_r(X,Y)^N,$$

which completes the proof. $\qquad\square$

The remaining part of that section is a small variation on Definition 2.2.23, and Lemmas 2.2.24, 2.2.25, based upon the property given by Proposition 2.2.20: that result allows us to replace $\delta_r(X, Y)$ by a smaller quantity $\Delta_r(X, Y)$ defined below while keeping the same properties (2.2.29).

Definition 2.2.26. Let g be an admissible metric on $\mathbb{R}^{2n}$ (see Definition 2.2.15). We define, for $X, Y \in \mathbb{R}^{2n}$ and $r > 0$, using the notation in (2.2.21), (4.4.18), (4.4.23), (2.2.8),

$$\Delta_r(X, Y) = 1 + (g_X^\flat \wedge g_Y^\flat)(U_{X,r} - U_{Y,r}). \tag{2.2.32}$$

Lemma 2.2.27. *Let g and Δ_r be as above, with (2.2.23) satisfied with N_0. Then, there exist some positive constants $(C_j)_{1 \leq j \leq 5}, (N_j)_{1 \leq j \leq 2}$ and $r_0 > 0$ such that, for $0 < r \leq r_0, \forall X, Y, T \in \mathbb{R}^{2n}$,*

$$\Delta_r(X, Y) \leq C_1\Big(1 + \min\big(g_X^\flat(U_{X,r} - U_{Y,r}), g_Y^\flat(U_{X,r} - U_{Y,r})\big)\Big)$$

$$\leq C_1\Big(1 + \max\big(g_X^\flat(U_{X,r} - U_{Y,r}), g_Y^\flat(U_{X,r} - U_{Y,r})\big)\Big)$$

$$\leq C_2\Delta_r(X, Y)^{N_0+1}, \tag{2.2.33}$$

$$\frac{g_X(T)}{g_Y(T)} \leq C_3\Delta_r(X, Y)^{N_0}, \quad \sup_{X \in \mathbb{R}^{2n}} \int_{\mathbb{R}^{2n}} \Delta_r(X, Y)^{-N_1} |g_Y|^{1/2} dY < \infty, \tag{2.2.34}$$

and for $g_X(Y - X) \geq C_4 r^2, \quad \lambda_g(X) + \lambda_g(Y) \leq C_5\Delta_r(X, Y)^{N_2}. \tag{2.2.35}$

Proof. We note that from the second inequality in (4.4.26),

$$\Delta_r(X, Y) \leq 1 + 2\min\big(g_X^\flat(U_{X,r} - U_{Y,r}), g_Y^\flat(U_{X,r} - U_{Y,r})\big),$$

and since g is slowly varying and satisfies (2.2.23), for $X' \in U_{X,r}, Y' \in U_{Y,r}$, we have, for r small enough,

$$g_X^\flat \wedge g_Y^\flat \sim g_{X'}^\flat \wedge g_{Y'}^\flat \gtrsim g_{X'}^\flat\big(1 + (g_{X'}^\flat \wedge g_{Y'}^\flat)(X' - Y')\big)^{-N_0}$$

$$\sim g_X^\flat\big(1 + (g_X^\flat \wedge g_Y^\flat)(X' - Y')\big)^{-N_0}$$

which implies $g_X^\flat \wedge g_Y^\flat \gtrsim g_X^\flat \Delta_r(X, Y)^{-N_0}$ and thus

$$g_X^\flat(U_{X,r} - U_{Y,r}) \lesssim \Delta_r(X, Y)^{N_0+1},$$

yielding (2.2.33); moreover, with the same notation, we have

$$\frac{g_X}{g_Y} \sim \frac{g_{X'}}{g_{Y'}} \lesssim \big(1 + (g_{X'}^\flat \wedge g_{Y'}^\flat)(X' - Y')\big)^{N_0} \sim \big(1 + (g_X^\flat \wedge g_Y^\flat)(X' - Y')\big)^{N_0},$$

which gives the first inequality in (2.2.34). Still with the same notation, we have

$$1 + g_X(X - Y) \leq 1 + 3g_X(X - X') + 3g_X(X' - Y') + 3g_X(Y' - Y)$$
$$\lesssim \Delta_r(X,Y)^{N_0}\left(1 + g_X(X - X') + g_X(X' - Y') + g_Y(Y' - Y)\right)$$
$$\lesssim \Delta_r(X,Y)^{N_0}\left(1 + g_X^\natural(X' - Y')\right)$$
$$\lesssim \Delta_r(X,Y)^{2N_0}\left(1 + (g_X^\natural \wedge g_Y^\natural)(X' - Y')\right)$$
$$\lesssim \Delta_r(X,Y)^{2N_0+1},$$

and since $\frac{|g_Y|^{1/2}}{|g_X|^{1/2}} \lesssim \Delta_r(X,Y)^{nN_0}$ we obtain the integrability property in (2.2.34). We have also from (2.2.22)

$$\lambda_g(X)g_X(X'-Y') \leq g_X^\natural(X'-Y') \lesssim \Delta_r(X,Y)^{N_0}(g_X^\natural \wedge g_Y^\natural)(X'-Y') \leq \Delta_r(X,Y)^{N_0+1}.$$

As a consequence, for C large enough, if $g_X(X-Y) \geq Cr^2$ we get $g_X(X'-Y') \geq r^2$ (otherwise $g_X(X'-Y') < r^2$ and $g_X \sim g_Y$ with $g_X(X-Y) \leq C_0 r^2$). The previous inequality gives $\lambda_g(X) \lesssim \Delta_r(X,Y)^{N_1}$ and (2.2.35). □

Lemma 2.2.28. *Let g be an admissible metric on $\mathbb{R}^{2n}$ and M be an admissible weight for g (cf. Definition 2.2.15). There exist $r_0 > 0, C > 0, N_0 \geq 0$, such that for $0 < r \leq r_0, X, Y \in \mathbb{R}^{2n}$,*

$$\frac{M(X)}{M(Y)} \leq C\Delta_r(X,Y)^{N_0}. \tag{2.2.36}$$

Proof. Since M is admissible for g, it is an admissible weight for $g^\natural$, thanks to Proposition 2.2.20, and using that M is slowly varying for g and g is slowly varying, for r small enough, we have for $X' \in U_{X,r}, Y' \in U_{Y,r}$,

$$\frac{M(X)}{M(Y)} \sim \frac{M(X')}{M(Y')} \leq C\left(1 + (g_{X'}^\natural \wedge g_{Y'}^\natural)(X' - Y')\right)^N \sim \Delta_r(X,Y)^N,$$

which completes the proof. □

2.3 General principles of pseudo-differential calculus

We need now to attack the hardest part of our task, namely, we should provide an argument to prove that the composition formula (2.1.23) holds for $a \in S(m_1, g), b \in S(m_2, g)$ and that $a \sharp b \in S(m_1 m_2, g)$. Our idea is to use the partition of unity of Proposition 2.2.9 so that the problem is reduced to handling

$$\iint_{\mathbb{R}^{2n} \times \mathbb{R}^{2n}} (a_Y \sharp b_Z) |g_Y|^{1/2} |g_Z|^{1/2} dY \, dZ.$$

In fact, to show a convergence result for that integral, it will be enough to study carefully the compositions $a_Y \sharp b_Z$. The first thing that we realize is that $a_Y \sharp b_Z$ is not compactly supported, since it is an entire function on $\mathbb{C}^{2n}$ (see Remark 2.2.12). Nevertheless, we should expect some localization properties of some sort for that symbol. In particular, if $Y = Z$, that symbol should be "small" far away from Y. In the next section, we give a precise quantitative version of these assertions.

2.3.1   Confinement estimates

Definition 2.3.1 (Confined symbols). Let g be a positive-definite quadratic form on $\mathbb{R}^{2n}$ such that $g \leq g^\sigma$. Let a be a smooth function in $\mathbb{R}^{2n}$ and $U \subset \mathbb{R}^{2n}$. We shall say that a is g-confined in U if $\forall k \in \mathbb{N}, \forall N \geq 0, \exists C_{k,N}$ such that, for all $X, T \in \mathbb{R}^{2n}$,

$$|a^{(k)}(X)T^k| \leq C_{k,N} g(T)^{k/2}\big(1 + g^\sigma(X-U)\big)^{-N/2}. \qquad (2.3.1)$$

We shall set

$$\|a\|_{g,U}^{(k,N)} = \sup_{\substack{X,T\in\mathbb{R}^{2n} \\ g(T)=1}} |a^{(k)}(X)T^k|\big(1 + g^\sigma(X-U)\big)^{N/2} \qquad (2.3.2)$$

$$\text{and} \quad \|a\|_{g,U}^{(l)} = \max_{k\leq l} \|a\|_{g,U}^{(k,l)}. \qquad (2.3.3)$$

When U is compact, any smooth function supported in U is confined in U. The set of the g-confined symbols in U will be denoted by $\mathrm{Conf}(g, U)$.

N.B. Let us note that it is an easy point of commutative algebra to check that $T_1 \otimes \cdots \otimes T_k$ is a linear combination of powers $S \otimes \cdots \otimes S$ with k factors (see 4.2.4 in the appendix); this means in particular that the estimates above are equivalent to estimates controlling $a^{(k)}(X)T_1 \otimes \cdots \otimes T_k$. We see also that, as a set, $\mathrm{Conf}(g, U)$ coincides with the Schwartz class on $\mathbb{R}^{2n}$. Naturally, the set of semi-norms (2.3.3) will be most important for our purpose as well as uniform estimates with respect to g.

2.3.2   Biconfinement estimates

Our most important technical result is the following.

Theorem 2.3.2. *Let g_1, g_2 be two positive-definite quadratic forms on $\mathbb{R}^{2n}$ such that $g_j \leq g_j^\sigma$ and let $a_j, j = 1, 2$ be g_j-confined in U_j, a g_j-ball of radius ≤ 1. Then, for all k, l, for all $X, T \in \mathbb{R}^{2n}$,*

$$|(a_1 \sharp a_2)^{(l)}T^l| \leq A_{k,l}(g_1+g_2)(T)^{l/2}\Big(1+(g_1^\sigma \wedge g_2^\sigma)(X-U_1)+(g_1^\sigma \wedge g_2^\sigma)(X-U_2)\Big)^{-k/2}, \qquad (2.3.4)$$

with $A_{k,l} = \gamma(n, k, l)\|a_1\|_{g_1,U_1}^{(m)}\|a_2\|_{g_2,U_2}^{(m)}, \quad m = 2n + 1 + k + l.$

Corollary 2.3.3. *Under the assumptions of the previous theorem, for all non-negative integers (p, N), there exists $q = q(p, N)$ and $C = C(p, N)$, such that*

$$\|a_1 \sharp a_2\|_{g_1+g_2,U_1}^{(p)} + \|a_1 \sharp a_2\|_{g_1+g_2,U_2}^{(p)}$$

$$\leq C \|a_1\|_{g_1,U_1}^{(q)} \|a_2\|_{g_2,U_2}^{(q)} \left(1 + (g_1^\sigma \wedge g_2^\sigma)(U_1 - U_2)\right)^{-N}, \quad (2.3.5)$$

$$\|a_1 \sharp a_2\|_{g_1,U_1}^{(p)} + \|a_1 \sharp a_2\|_{g_2,U_2}^{(p)} \leq C \|a_1\|_{g_1,U_1}^{(q)} \|a_2\|_{g_2,U_2}^{(q)}$$

$$\times \left(1 + \sup_{g_1(T)=1} g_2(T) + \sup_{g_2(T)=1} g_1(T)\right)^p \left(1 + (g_1^\sigma \wedge g_2^\sigma)(U_1 - U_2)\right)^{-N}. \quad (2.3.6)$$

Proof that the theorem implies the corollary. The first inequality follows from the previous theorem by applying the triangle inequality to the quadratic form $g_1^\sigma \wedge g_2^\sigma$. We note that (2.3.4) implies

$$|(a_1 \sharp a_2)^{(l)} T^l| \leq g_1(T)^{l/2} A_{k,l} \left(1 + \sup_{g_1(T)=1} g_2(T)\right)^{l/2}$$

$$\times \left(1 + (g_1^\sigma \wedge g_2^\sigma)(X - U_1) + (g_1^\sigma \wedge g_2^\sigma)(X - U_2)\right)^{-k/2},$$

and thus, using the inequality (4.4.37), we get

$$|(a_1 \sharp a_2)^{(l)} T^l| \leq g_1(T)^{l/2} A_{k_1+k_2,l} \left(1 + \sup_{g_1(T)=1} g_2(T)\right)^{l/2}$$

$$\times \left(1 + (g_1^\sigma \wedge g_2^\sigma)(U_1 - U_2)\right)^{-k_1/2} \left(1 + g_1^\sigma(X - U_1)\right)^{-k_2/2} 2^{k_2/2} \left(1 + \sup_{g_1^\sigma(T)=1} g_2^\sigma(T)\right)^{k_2/2},$$

and[6]

$$|(a_1 \sharp a_2)^{(l)} T^l| g_1(T)^{-l/2} \left(1 + g_1^\sigma(X - U_1)\right)^{k_2/2} \leq A_{k_1+k_2,l} 2^{k_2/2}$$

$$\times \left(1 + \sup_{g_1(T)=1} g_2(T)\right)^{l/2} \left(1 + (g_1^\sigma \wedge g_2^\sigma)(U_1 - U_2)\right)^{-k_1/2} \left(1 + \sup_{g_2(T)=1} g_1(T)\right)^{k_2/2},$$

yielding

$$\|a_1 \sharp a_2\|_{g_1,U_1}^{(p)} \leq \max_{l \leq p} A_{k_1+p,l} 2^{p/2}$$

$$\times \left(1 + (g_1^\sigma \wedge g_2^\sigma)(U_1 - U_2)\right)^{-k_1/2} \left(1 + \sup_{g_1(T)=1} g_2(T) + \sup_{g_2(T)=1} g_1(T)\right)^p$$

which gives the second inequality of the corollary, completing its proof. $\qquad \square$

---

[6] We use here the implication (with $\alpha_{12} > 0$) $g_1 \leq \alpha_{12} g_2 \implies g_2^\sigma \leq \alpha_{12} g_1^\sigma$, which follows from Remark 4.4.24.

Comment. If for instance each a_j is supported in the ball U_j, the composition $a_1 \natural a_2$ is "confined" in both balls U_1, U_2 in the precise quantitative sense given by the decay estimates (2.3.4). In particular, when $a_1 = a_2 = a$ is supported in the g-ball U, the composition $a \natural a$ is no longer supported in U but is "confined" in U. Obviously the class of functions supported in U is not stable by composition (the operation $\natural$), but the class of confined functions (in fact the Schwartz class) is indeed stable by composition.

Proof of Theorem 2.3.2. We start from the formula (2.1.18)

$$(a_1 \natural a_2)(X) = \iint a_1(Y_1) a_2(Y_2) e^{-4i\pi[Y_1 - X, Y_2 - X]} dY_1 dY_2 2^{2n}$$

and for Y_1, X given in $\mathbb{R}^{2n}$ we define $\theta = \theta_{Y_1, X} \in \mathbb{R}^{2n}$ such that
$$g_2(\theta) = 1, \quad [\theta, Y_1 - X] = g_2^\sigma(Y_1 - X)^{1/2} \qquad \text{(see (4.4.21))}.$$
Then for $k \in \mathbb{N}$, since θ does not depend on Y_2,

$$\left(\frac{1}{2} \langle \theta, D_{Y_2} \rangle + 1 \right)^k (\exp -4i\pi[Y_1 - X, Y_2 - X])$$
$$= \left(1 + g_2^\sigma(Y_1 - X)^{1/2} \right)^k \exp -4i\pi[Y_1 - X, Y_2 - X].$$

Consequently, we have

$$(a_1 \natural a_2)(X)$$
$$= \iint a_1(Y_1) \sum_{0 \le l \le k} a_2^{[l]}(Y_2) e^{-4i\pi[Y_1 - X, Y_2 - X]} \left(1 + g_2^\sigma(Y_1 - X)^{1/2} \right)^{-k} dY_1 dY_2 2^{2n}$$

with $a_2^{[l]} = (-4i\pi)^{-l} \binom{k}{l} a_2^{(l)} \theta^l$. As a consequence, for all k, k_1, N, we get

$$|(a_1 \natural a_2)(X)| \le \iint \|a_1\|_{g_1, U_1}^{(0, k_1)} \left(1 + g_1^\sigma(Y_1 - U_1) \right)^{-k_1/2} \left(1 + g_2^\sigma(Y_1 - X) \right)^{-k/2}$$
$$\times \|a_2\|_{g_2, U_2}^{(k, N)} \left(1 + g_2^\sigma(Y_2 - U_2) \right)^{-N/2} dY_1 dY_2 2^{2n+k}. \quad (2.3.7)$$

We notice then that
$$\left(1 + g_1^\sigma(Y_1 - U_1) \right)^{k_1/2} \left(1 + g_2^\sigma(Y_1 - X) \right)^{k_1/2} \ge \left(1 + g_1^\sigma(Y_1 - U_1) + g_2^\sigma(Y_1 - X) \right)^{k_1/2}$$
$$\ge 2^{-k_1/2} \left(1 + (g_1^\sigma \wedge g_2^\sigma)(X - U_1) \right)^{k_1/2},$$

so that for $k = k_1 + k_2$, the estimate (2.3.7) gives

$$|(a_1 \natural a_2)(X)| \le \|a_1\|_{g_1, U_1}^{(0, k_1)} \|a_2\|_{g_2, U_2}^{(k, N)} \left(1 + (g_1^\sigma \wedge g_2^\sigma)(X - U_1) \right)^{-k_1/2}$$
$$\times \iint \left(1 + g_2^\sigma(Y_1 - X) \right)^{-k_2/2} \left(1 + g_2^\sigma(Y_2 - U_2) \right)^{-N/2} dY_1 dY_2 2^{2n+k+\frac{k_1}{2}}.$$

Since U_2 is a g_2-ball of radius ≤ 1 with center X_2, we have for $Z \in U_2$,

$$1 + g_2(Y_2 - X_2) \leq 1 + 2g_2(Y_2 - Z) + 2g_2(Z - X_2) \leq 3\big(1 + g_2(Y_2 - Z)\big)$$
$$\leq 3\big(1 + g_2^\sigma(Y_2 - Z)\big)$$

which implies $1 + g_2(Y_2 - X_2) \leq 3\big(1 + g_2^\sigma(Y_2 - U_2)\big)$. As a consequence, we get

$$|(a_1 \sharp a_2)(X)| \leq \|a_1\|_{g_1,U_1}^{(0,k_1)} \|a_2\|_{g_2,U_2}^{(k,N)} \big(1 + (g_1^\sigma \wedge g_2^\sigma)(X - U_1)\big)^{-k_1/2}$$
$$\times \iint \big(1 + g_2^\sigma(Y_1 - X)\big)^{-k_2/2} \big(1 + g_2(Y_2 - X_2)\big)^{-N/2} dY_1 dY_2 2^{2n+k+\frac{k_1}{2}} 3^{N/2}.$$

Since the product of the determinants of g_2 and g_2^σ is 1, choosing $k_2 = N = 2n+1$, we obtain

$$|(a_1 \sharp a_2)(X)| \leq \|a_1\|_{g_1,U_1}^{(0,k_1)} \|a_2\|_{g_2,U_2}^{(2n+1+k_1,2n+1)} \big(1 + (g_1^\sigma \wedge g_2^\sigma)(X - U_1)\big)^{-k_1/2} C(k_1,n).$$

Analogously, we get

$$|(a_1 \sharp a_2)(X)| \leq \|a_2\|_{g_1,U_1}^{(0,k_1)} \|a_1\|_{g_2,U_2}^{(2n+1+k_1,2n+1)} \big(1 + (g_1^\sigma \wedge g_2^\sigma)(X - U_2)\big)^{-k_1/2} C(k_1,n).$$

Taking the minimum of the rhs in the two previous inequalities, we obtain with $k_1 = k$,

$$|(a_1 \sharp a_2)(X)| \leq C(k,n) 2^{k/2} \|a_1\|_{g_1,U_1}^{(2n+1+k)} \|a_2\|_{g_2,U_2}^{(2n+1+k)}$$
$$\times \big(1 + (g_1^\sigma \wedge g_2^\sigma)(X - U_1) + (g_1^\sigma \wedge g_2^\sigma)(X - U_2)\big)^{-k/2}, \quad (2.3.8)$$

which is indeed (2.3.4) for $l = 0$. Writing (2.1.18) as

$$(a_1 \sharp a_2)(X) = 2^{2n} \iint e^{-4i\pi[Y_1,Y_2]} a_1(Y_1 + X) a_2(Y_2 + X) dY_1 dY_2,$$

we realize that if $D = \partial_{X_j}$, we have

$$D(a_1 \sharp a_2) = Da_1 \sharp a_2 + a_1 \sharp Da_2, \quad (2.3.9)$$

so that applying (2.3.8) to the previous formula gives (2.3.4) for $l \geq 1$. The proof of Theorem 2.3.2 is complete. $\qquad\square$

We need now to get the analogous result for the expansion of $a_1 \sharp a_2$.

Theorem 2.3.4. *Let g_1, g_2, a_1, a_2 as in Theorem 2.3.2. For $\nu \in \mathbb{N}$, we recall from Definitions (2.1.26), (2.1.29),*

$$r_\nu(a_1, a_2)(X) = a_1 \sharp a_2 - \sum_{0 \leq j < \nu} \frac{1}{j!} \big(i\pi[D_{X_1}, D_{X_2}]\big)^j (a_1 \otimes a_2)|_{diagonal}.$$

Then, for all k, l, ν, for all $X, T \in \mathbb{R}^{2n}$, we have

$$\left|\left(r_\nu(a_1, a_2)\right)^{(l)} T^l\right| \le A_{k,l,\nu}(g_1 + g_2)(T)^{l/2} \Lambda_{1,2}^{-\nu}$$
$$\times \left(1 + (g_1^\sigma \wedge g_2^\sigma)(X - U_1) + (g_1^\sigma \wedge g_2^\sigma)(X - U_2)\right)^{-k/2}, \quad (2.3.10)$$

with $A_{k,l,\nu} = C(k, n, l, \nu) \|a_1\|_{g_1, U_1}^{(2n+1+k+l+\nu)} \|a_2\|_{g_2, U_2}^{(2n+1+k+l+\nu)}$ and

$$\Lambda_{1,2} = \inf_T \left(\frac{g_1^\sigma(T)}{g_2(T)}\right)^{1/2} = \inf_T \left(\frac{g_2^\sigma(T)}{g_1(T)}\right)^{1/2}. \quad (2.3.11)$$

Remark 2.3.5. First of all, note that we can get a result similar to Corollary 2.3.3: for all (p, N, ν), there exists $q = q(p, N, \nu)$ and $C = C(p, N, \nu)$, such that

$$\|R_\nu(a_1, a_2)\|_{g_1+g_2, U_1}^{(p)} + \|R_\nu(a_1, a_2)\|_{g_1+g_2, U_2}^{(p)}$$
$$\le C\|a_1\|_{g_1, U_1}^{(q)} \|a_2\|_{g_2, U_2}^{(q)} \Lambda_{1,2}^{-\nu} \left(1 + (g_1^\sigma \wedge g_2^\sigma)(U_1 - U_2)\right)^{-N}. \quad (2.3.12)$$

Next, let us give a proof of the identity (2.3.11). We have, using (4.4.22),

$$\frac{g_1^\sigma(T)}{g_2(T)} = \frac{[T, U_T]^2}{g_1(U_T)g_2(T)} \le \frac{g_2(T)g_2^\sigma(U_T)}{g_1(U_T)g_2(T)} = \frac{g_2^\sigma(U_T)}{g_1(U_T)}, \quad \text{with } U_T = g_1^{-1}\sigma T,$$

so that $\inf_T \frac{g_1^\sigma(T)}{g_2(T)} \le \inf_T \frac{g_2^\sigma(U_T)}{g_1(U_T)}$ and since the mapping $T \mapsto U_T$ is an isomorphism, we get (2.3.11).

Proof of the theorem. We have from (2.1.18),

$$(a_1 \sharp a_2)(X) = \iint a_1(X + Y_1)a_2(Y_2)e^{-4i\pi\langle\sigma Y_1, Y_2 - X\rangle} dY_1 dY_2 2^{2n}$$
$$= \iint a_1(X + \frac{1}{2}\sigma^{-1}\Xi)a_2(Y_2)e^{-2i\pi\langle\Xi, Y_2 - X\rangle} d\Xi dY_2$$
$$= \int a_1(X + \frac{1}{2}\sigma^{-1}\Xi)\widehat{a_2}(\Xi)e^{2i\pi\langle\Xi, X\rangle} d\Xi, \quad (2.3.13)$$

and[7] consequently,

---

[7]The expression (2.3.13) is interesting on its own and can be written as

$$(a_1 \sharp a_2)(X) = (a_1(X + \frac{1}{2}\sigma^{-1}D_X)a_2)(X). \quad (2.3.14)$$

$$(a_1 \sharp a_2)(X) = \sum_{0 \le j < \nu} a_1^{(j)}(X) \frac{1}{j! 2^j} \int (\sigma^{-1}\Xi)^j \widehat{a}_2(\Xi) e^{2i\pi\langle\Xi,X\rangle} d\Xi$$

$$+ \int_0^1 \frac{(1-\theta)^{\nu-1}}{(\nu-1)! 2^\nu} \int a_1^{(\nu)}(X + \frac{\theta}{2}\sigma^{-1}\Xi)(\sigma^{-1}\Xi)^\nu \widehat{a}_2(\Xi) e^{2i\pi\langle\Xi,X\rangle} d\Xi d\theta$$

$$= \sum_{0 \le j < \nu} a_1^{(j)}(X) \frac{1}{j! 2^j} (\sigma^{-1}D_X)^j a_2(X) + r_\nu(a_1, a_2)(X),$$

which corresponds to the expansion of the theorem according to Remark 2.1.5 with

$$r_\nu(a_1, a_2)(X) = \int_0^1 \frac{(1-\theta)^{\nu-1}}{(\nu-1)! 2^\nu} \int a_1^{(\nu)}(X + \frac{\theta}{2}\sigma^{-1}\Xi)(\sigma^{-1}\Xi)^\nu \widehat{a}_2(\Xi) e^{2i\pi\langle\Xi,X\rangle} d\Xi d\theta,$$

so that

$$r_\nu(a_1, a_2)(X)$$
$$= \int_0^1 \frac{(1-\theta)^{\nu-1}}{(\nu-1)!} \int a_1^{(\nu)}(X + \theta Z) Z^\nu \widehat{a}_2(2\sigma Z) e^{2i\pi\langle 2\sigma Z, X\rangle} dZ d\theta 2^{2n}$$
$$= \int_0^1 \frac{(1-\theta)^{\nu-1}}{(\nu-1)!} \iint a_1^{(\nu)}(X + \theta Z) Z^\nu a_2(Y_2) e^{2i\pi\langle 2\sigma Z, X\rangle - 2i\pi\langle Y_2, 2\sigma Z\rangle} dZ dY_2 d\theta 2^{2n}$$
$$= \int_0^1 \frac{(1-\theta)^{\nu-1}}{(\nu-1)!} \iint a_1^{(\nu)}(X + \theta(Y_1 - X))(Y_1 - X)^\nu a_2(Y_2)$$
$$\times e^{-4i\pi[Y_1 - X, Y_2 - X]} dY_1 dY_2 2^{2n} d\theta.$$

We consider now (using 4.4.22) $T = T_{Y_1, X}$ such that $g_2(T) = 1$ and $[T, Y_1 - X] = g_2^\sigma(Y_1 - X)^{1/2}$. Then for $k \in \mathbb{N}$,

$$\left(\frac{1}{2}\langle T, D_{Y_2}\rangle + 1\right)^k e^{-4i\pi[Y_1 - X, Y_2 - X]} = e^{-4i\pi[Y_1 - X, Y_2 - X]} (1 + g_2^\sigma(Y_1 - X)^{1/2})^k.$$

As a consequence we get

$$|r_\nu(a_1, a_2)(X)|$$
$$\le \iiint_0^1 \frac{(1-\theta)^{\nu-1}}{(\nu-1)!} 2^{2n+k} \|a_1^{(\nu)}\|_{g_1, U_1}^{(0, k_1)} \left(1 + g_1^\sigma(X + \theta(Y_1 - X) - U_1)\right)^{-k_1/2}$$
$$\times g_1(Y_1 - X)^{\nu/2} (1 + g_2^\sigma(Y_1 - X))^{-k/2}$$
$$\times \|a_2\|_{g_2, U_2}^{(k, N)} \left(1 + g_2^\sigma(Y_2 - U_2)\right)^{-N/2} dY_1 dY_2 d\theta.$$

Since $\theta \in [0, 1]$, we have $g_2^\sigma(Y_1 - X) \ge g_2^\sigma(\theta(Y_1 - X))$ and for $k = k_1 + k_2 + \nu$ and $N = 2n + 1 = k_2$,

$$|r_\nu(a_1, a_2)(X)|$$

$$\leq \iiint_0^1 \frac{(1-\theta)^{\nu-1}}{(\nu-1)!} 2^{2n+k} \|a_1\|_{g_1, U_1}^{(\nu, k_1)} \left(1 + g_1^\sigma\big(X + \theta(Y_1 - X) - U_1\big)\right)^{-k_1/2}$$

$$\times \Lambda_{12}^{-\nu} g_2^\sigma (Y_1 - X)^{\nu/2} \big(1 + g_2^\sigma\big(\theta(Y_1 - X)\big)\big)^{-\frac{k_1}{2}} (1 + g_2^\sigma(Y_1 - X))^{-\frac{k_2+\nu}{2}}$$

$$\times \|a_2\|_{g_2, U_2}^{(k, N)} \big(1 + g_2^\sigma(Y_2 - U_2)\big)^{-(2n+1)/2} dY_1 dY_2 d\theta.$$

We use now that

$$2g_1^\sigma(X + \theta(Y_1 - X) - U_1) + 2g_2^\sigma(\theta(X - Y_1)) \geq (g_1^\sigma \wedge g_2^\sigma)(X - U_1)$$

and we obtain

$$|r_\nu(a_1, a_2)(X)|$$

$$\leq C(k, n, \nu) \|a_1\|_{g_1, U_1}^{(\nu, k_1)} \|a_2\|_{g_2, U_2}^{(k, 2n+1)} \Lambda_{12}^{-\nu} \big(1 + (g_1^\sigma \wedge g_2^\sigma)(X - U_1)\big)^{-k_1/2}$$

$$\times \iint (1 + g_2^\sigma(Y_1 - X))^{-\frac{k_2}{2}} \big(1 + g_2^\sigma(Y_2 - U_2)\big)^{-(2n+1)/2} dY_1 dY_2. \qquad (2.3.15)$$

Reasoning as in the proof of Theorem 2.3.2, we get (2.3.10) for $l = 0$, which is easy to generalize to $l \geq 1$. The proof of Theorem 2.3.4 is complete. $\square$

The next lemma is an important remark which will be useful in the sequel.

Lemma 2.3.6. *Let g be a positive-definite quadratic form on $\mathbb{R}^{2n}$ such that $g \leq g^\sigma$, let $c \in S(1, g)$ and $\psi \in \mathrm{Conf}(g, U_r)$ where U_r is a g-ball with radius $r \leq 1$. Then $\psi \sharp c$ belongs to $\mathrm{Conf}(g, U_r)$, and the mapping $\mathrm{Conf}(g, U_r) \times S(1, g) \ni (\psi, c) \mapsto \psi \sharp c \in \mathrm{Conf}(g, U_r)$ is continuous.*

Proof. Using a partition of unity $1 = \int \phi_T |g|^{1/2} dT$ for the constant metric g, with $\phi_T \in S(1, g)$ supported in a g-ball with center T and radius r, we get, assuming that U_r has center 0, $\psi \sharp c = \int_{\mathbb{R}^{2n}} \psi \sharp (c \phi_T) |g|^{1/2} dT$, and the biconfinement estimates of Theorem 2.3.2 give

$$|(\psi \sharp c)^{(l)}(X) S^l| \leq C_{k, l} g(S)^{l/2} \int_{\mathbb{R}^{2n}} \big(1 + g^\sigma(X - U_r) + g^\sigma(X - U_r - T)\big)^{-k/2} |g|^{1/2} dT,$$

with $C_{k, l}$ semi-norms of ψ, c, the inequalities

$$\sup_{g(S) \leq 1} |(\psi \sharp c)^{(l)}(X) S^l| \big(1 + g^\sigma(X - U_r)\big)^{N/2}$$

$$\leq C_{k+N, l} \int_{\mathbb{R}^{2n}} \big(1 + g^\sigma(X - U_r - T)\big)^{-k/2} |g|^{1/2} dT.$$

For $g(T') \leq r^2 \leq 1$, we have

$$g(X - T) \leq 2g(X - T - T') + 2g(T') \leq 2g^\sigma(X - T - T') + 2r^2$$

and thus $1 + g(X - T) \leq 3 + 2g^\sigma(X - T - U_r)$ so that for $k = 2n + 1$,

$$\sup_{g(S) \leq 1} |(\psi \sharp c)^{(l)}(X)S^l|(1 + g^\sigma(X - U_r))^{N/2}$$

$$\leq C_{2n+1+N,l}3^{n+1/2} \int_{\mathbb{R}^{2n}} (1 + g(X - T))^{-n-\frac{1}{2}}|g|^{1/2}dT = C_{2n+1+N,l}3^{n+1/2}\alpha(n),$$

completing the proof of the lemma. $\qquad\square$

2.3.3 Symbolic calculus

The previous section dealt with the study of the composition formula (2.1.18) for two symbols in the Schwartz class attached to two constant metrics. That will be essentially enough to tackle the more interesting case where each symbol belongs to a symbol class attached to a general admissible metric.

Theorem 2.3.7. *Let g be an admissible metric on $\mathbb{R}^{2n}$ (see Definition 2.2.15), m_1, m_2 two admissible weights for g, $a_j \in S(m_j, g), j = 1, 2$. Then, for all $\nu \in \mathbb{N}$,*

$$a_1 \sharp a_2 - \sum_{0 \leq j < \nu} \frac{1}{j!} (i\pi[D_{X_1}, D_{X_2}])^j (a_1 \otimes a_2)_{|\text{diagonal}} \in S(m_1 m_2 \lambda_g^{-\nu}, g). \qquad (2.3.16)$$

In particular, we have $a_1 \sharp a_2 = a_1 a_2 + \frac{1}{4i\pi}\{a_1, a_2\} + r_2$, $r_2 \in S(m_1 m_2 \lambda_g^{-2}, g)$. The mappings $S(m_1, g) \times S(m_2, g) \ni (a_1, a_2) \mapsto a_1 \sharp a_2 \in S(m_1 m_2, g)$ as well as (2.3.16) are continuous for the Fréchet space structure given by the semi-norms (2.2.7).

Proof. We shall first use the partition of unity of Proposition 2.2.9 and consider

$$\iint (a_{1,Y_1} \sharp a_{2,Y_2})(X)|g_{Y_1}|^{1/2}|g_{Y_2}|^{1/2}dY_1 dY_2$$

and notice that, from Theorem 2.3.2, we get

$$|(a_{1,Y_1} \sharp a_{2,Y_2})(X)|$$
$$\leq C_{N,M}\delta_r(Y_1, Y_2)^{-N} (1 + (g_{Y_1}^\sigma \wedge g_{Y_2}^\sigma)(X - U_{Y_1,r}))^{-M} m_1(Y_1)m_2(Y_2).$$

On the other hand, Lemmas 2.2.25–2.2.24 give

$$\frac{m_2(Y_2)}{m_2(Y_1)} \leq C_0 \delta_r(Y_1, Y_2)^{N_0}, \quad \frac{g_{Y_1}^\sigma}{g_{Y_2}^\sigma} \leq C_0 \delta_r(Y_1, Y_2)^{N_0},$$

so that

$$|(a_{1,Y_1} \sharp a_{2,Y_2})(X)| \lesssim \delta_r(Y_1, Y_2)^{-N+N_0} (1 + (g_{Y_1}^\sigma \wedge g_{Y_2}^\sigma)(X - U_{Y_1,r}))^{-M} (m_1 m_2)(Y_1)$$
$$\lesssim (m_1 m_2)(Y_1)(1 + g_{Y_1}^\sigma(X - U_{Y_1,r}))^{-M} \delta_r(Y_1, Y_2)^{-N+N_0+N_0 M}$$
$$\lesssim (m_1 m_2)(Y_1)\delta_r(Y_1, X)^{-M} \delta_r(Y_1, Y_2)^{-N+N_0+N_0 M}$$
$$\lesssim (m_1 m_2)(X)\delta_r(Y_1, X)^{-M+N_0} \delta_r(Y_1, Y_2)^{-N+N_0+N_0 M}.$$

Choosing now M so that $M - N_0$ is large enough, then choosing N so that $N - N_0 - N_0 M$ is large enough, we can use the integrability condition in (2.2.29) to secure $|a_1 \sharp a_2| \leq m_1 m_2$. To obtain the estimate for the derivatives is analogous (using (2.3.9)) as well as the estimate for the expansion, using Theorem 2.3.4. □

In fact the previous theorem can be extended to the following statement, quite useful in the third chapter. The reader may wait to hit that chapter to pay more attention to these results, which look more technical than the previous one.

Theorem 2.3.8. *Let g_1, g_2 be two admissible metrics on $\mathbb{R}^{2n}$, m_j an admissible weight for g_j, $a_j \in S(m_j, g_j), j = 1, 2$. According to the identity (2.3.11), we define*

$$\Lambda_{12}(X) = \inf_T \left(\frac{g^\sigma_{1,X}(T)}{g_{2,X}(T)} \right)^{1/2} = \inf_T \left(\frac{g^\sigma_{2,X}(T)}{g_{1,X}(T)} \right)^{1/2},$$

and we assume that $\inf_{X \in \mathbb{R}^{2n}} \Lambda_{1,2}(X) \geq 1$. We also assume that there exist $C > 0, N \geq 0$ such that for all $X, Y, T \in \mathbb{R}^{2n}$,

$$\frac{g_{1,X}(T) + g_{2,X}(T)}{g_{1,Y}(T) + g_{2,Y}(T)} \leq C\big(1 + (g^\sigma_{1,X} \wedge g^\sigma_{2,X})(X - Y)\big)^N, \qquad (2.3.17)$$

$$\frac{m_j(X)}{m_j(Y)} \leq C\big(1 + (g^\sigma_{1,X} \wedge g^\sigma_{2,X})(X - Y)\big)^N, \quad j = 1, 2. \qquad (2.3.18)$$

Then the metric $g_1 \vee g_2 = \frac{1}{2}(g_1 + g_2)$ is admissible, the product $m_1 m_2$ is an admissible weight for $g_1 \vee g_2$ and, for all $\nu \in \mathbb{N}$,

$$a_1 \sharp a_2 - \sum_{0 \leq j < \nu} \frac{1}{j!} \big(i\pi[D_{X_1}, D_{X_2}]\big)^j (a_1 \otimes a_2)_{|\text{diagonal}} \in S(m_1 m_2 \Lambda_{12}^{-\nu}, g_1 \vee g_2).$$

Moreover, the term $\omega_j = \frac{1}{j!}\big(i\pi[D_{X_1}, D_{X_2}]\big)^j (a_1 \otimes a_2)_{|\text{diagonal}} \in S(m_1 m_2 \Lambda_{12}^{-j}, g_1 \vee g_2)$.

Proof. The metric $g_1 \vee g_2$ is obviously slowly varying since the condition

$$(g_{1,X} \vee g_{2,X})(Y - X) \leq r^2$$

implies that for both $j = 1, 2$ we have $g_{j,X}(Y - X) \leq 2r^2$, so that the slow variation of the g_j implies $g_{j,Y} \leq C g_{j,X}$ and $g_{1,Y} \vee g_{2,Y} \leq C(g_{1,X} \vee g_{2,X})$ which gives the slow variation of $g_1 \vee g_2$, according to Remark 2.2.2. We have also with $\lambda_j = \lambda_{g_j}$, the inequalities

$$g^\sigma_j \geq \lambda_j^2 g_j, \quad g^\sigma_1 \geq \Lambda_{12}^2 g_2, \quad g^\sigma_2 \geq \Lambda_{12}^2 g_1,$$

so that

$$g^\sigma_1 \wedge g^\sigma_2 \geq \frac{1}{2} \min(\Lambda_{12}^2, \lambda_1^2) g_1 + \frac{1}{2} \min(\Lambda_{12}^2, \lambda_2^2) g_2 \geq \min(\Lambda_{12}^2, \lambda_1^2, \lambda_2^2)(g_1 \vee g_2),$$

and the metric $g_1 \vee g_2$ satisfies the uncertainty principle (2.2.12) since $(g_1 \vee g_2)^\sigma = g^\sigma_1 \wedge g^\sigma_2$ and $\min(\Lambda_{12}^2, \lambda_1^2, \lambda_2^2) \geq 1$. The temperance of $g_1 \vee g_2$ follows from Lemma

2.2.14 since our hypothesis (2.3.17) is precisely (2.2.17) for $g_1 \vee g_2$, which is now proven admissible. The weight m_j is slowly varying for g_j and a fortiori for $g_1 \vee g_2 \geq g_j/2$; it is also admissible for $g_1 \vee g_2$ since our hypothesis (2.3.18) is (2.2.19). Since $a_j \in S(m_j, g_j) \subset S(m_j, g_1 \vee g_2)$, $g_1 \vee g_2$ is admissible, m_j is an admissible weight for $g_1 \vee g_2$, we can apply the Theorem 2.3.7 to obtain $a_1 \sharp a_2 \in S(m_1 m_2, g_1 \vee g_2)$. To obtain the last statements, we have to consider the integral formula provided by Proposition 2.2.9,

$$(a_1 \sharp a_2)(X) = \iint (a_{1,Y_1} \sharp a_{2,Y_2})(X) |g_{Y_1}|^{1/2} |g_{Y_2}|^{1/2} dY_1 dY_2$$

and to apply Theorem 2.3.4 to $a_{1,Y_1} \sharp a_{2,Y_2}$ with the same proof as Theorem 2.3.7. The very last assertion of the theorem follows from the previous one since we have

$$\underbrace{\omega_\nu + a_1 \sharp a_2 - \sum_{0 \leq j \leq \nu} \omega_\nu}_{S(m_1 m_2 \Lambda_{12}^{-\nu-1}, g_1 \vee g_2)} = a_1 \sharp a_2 - \sum_{0 \leq j < \nu} \omega_j \in S(m_1 m_2 \Lambda_{12}^{-\nu}, g_1 \vee g_2),$$

which completes the proof. $\qquad\square$

Remark 2.3.9. One important point of this theorem is that the quantity Λ_{12} can be much larger than $\min(\lambda_1, \lambda_2)$, as it is the case when $g_1 = g_1^\sigma$ so that $\lambda_1 = 1$ and $g_2 = \lambda_2^{-1} g_1$. In this situation, $\min(\lambda_1, \lambda_2) = 1$ whereas $\Lambda_{12} = \lambda_2^{1/2}$. The next corollary says a little bit more.

Corollary 2.3.10. *Let* g_1, g_2 *be two admissible metrics on* $\mathbb{R}^{2n}$ *such that* $g_1^\flat = g_2^\flat$ *(note that this is the case when* g_2 *and* g_1 *are conformal, cf. Remark 4.4.30). Let* m_j *be an admissible weight for* g_j, $a_j \in S(m_j, g_j), j = 1, 2$. *With the notation of the previous theorem, we have*

$$\Lambda_{12} \geq \sqrt{\lambda_1 \lambda_2}, \quad \text{(with equality when } g_2 \text{ and } g_1 \text{ are conformal)}.$$

The metric $g_1 \vee g_2 = \frac{1}{2}(g_1 + g_2)$ *is admissible, the product* $m_1 m_2$ *is an admissible weight for* $g_1 \vee g_2$ *and, for all* $\nu \in \mathbb{N}$, *the last statement of the previous theorem holds.*

Proof. We have, according to Lemma 4.4.25,

$$\lambda_j(X) g_{j,X} \leq g_{j,X}^\flat = \Gamma_X = \Gamma_X^\sigma \leq \lambda_j(X)^{-1} g_{j,X}^\sigma$$

so[8] that $\Lambda_{12} \geq \sqrt{\lambda_1 \lambda_2}$. Using now Proposition 2.2.20, we check that

$$\frac{g_{1,X}(T) + g_{2,X}(T)}{g_{1,Y}(T) + g_{2,Y}(T)} \leq C \big(1 + (\Gamma_X \wedge \Gamma_Y)(X - Y)\big)^N$$

and we have a similar estimate for the ratios $m_j(X)/m_j(Y)$. All the assumptions of Theorem 2.3.8 are satisfied and we obtain the corollary. $\qquad\square$

---

[8]When the g_j are conformal, $g_2 = \mu g_1$, we have $\Lambda_{12}^2 = \mu^{-1} \lambda_1^2 = \mu \lambda_2^2 \implies \Lambda_{12} = \sqrt{\lambda_1 \lambda_2}$.

2.3.4   Additional remarks

In this section, we prove some technical results on confined symbols which shall be useful in the study of Sobolev spaces.

Lemma 2.3.11. *Let g be a positive-definite quadratic form on $\mathbb{R}^{2n}$ such that $g \leq g^{\sigma}$, U be a closed g-ball of radius r, and χ be a linear symplectic mapping of $\mathbb{R}^{2n}$. Then if $a \in \mathrm{Conf}(g, U)$ (cf. Definition 2.3.1), the function $a \circ \chi$ belongs to $\mathrm{Conf}(\tilde{g}, \tilde{U})$ with*

$$\tilde{g} = \chi^{*} g \chi, \quad \tilde{U} = \chi^{-1} U, \quad \|a\|_{g,U}^{(k,N)} = \|a \circ \chi\|_{\tilde{g},\tilde{U}}^{(k,N)}.$$

Note that $\tilde{g}$ is the positive-definite quadratic form given by the pull-back by χ of g and that $\tilde{U}$ is the $\tilde{g}$-ball of radius r.

Proof. We define $b(Y) = a(\chi Y)$ and we get

$$|b^{(k)}(Y) S^{k}| = |a^{(k)}(\chi Y)(\chi S)^{k}| \leq \|a\|_{g,U}^{k,N} g(\chi S)^{k/2} \Big(1 + g^{\sigma}\big(\chi(Y - \tilde{U})\big)\Big)^{-N/2},$$

so that using (4.4.18) and $\chi^{*} \sigma \chi = \sigma$ (cf. (4.4.2)), we infer

$$(\tilde{g})^{\sigma} = \sigma^{*}(\tilde{g})^{-1}\sigma = \sigma^{*}\chi^{-1}g^{-1}(\chi^{*})^{-1}\sigma = \chi^{*}\sigma^{*}g^{-1}\sigma\chi = \chi^{*}g^{\sigma}\chi,$$

so that $g^{\sigma}\big(\chi(Y - \tilde{U})\big) = (\tilde{g})^{\sigma}(Y - \tilde{U})$, implying that $b \in \mathrm{Conf}(\tilde{g}, \tilde{U})$ and the inequality $\|a \circ \chi\|_{\tilde{g},\tilde{U}}^{k,N} \leq \|a\|_{g,U}^{k,N}$; the reverse inequality follows from the same proof with a replaced by b and χ by χ^{-1}. $\quad\square$

Lemma 2.3.12. *Let g be a positive-definite quadratic form on $\mathbb{R}_{x}^{n} \times \mathbb{R}_{\xi}^{n}$ such that $g \leq g^{\sigma}$ and $g(x, \xi) = g(x, -\xi)$. Let U_{r} be a g-ball with radius $r \leq 1$. Then for all $t \in \mathbb{R}$, the operator $J^{t} = \exp 2i\pi t D_{x} \cdot D_{\xi}$ (see Lemma 4.1.2) is continuous from $\mathrm{Conf}(g, U_{r})$ into itself and more precisely, for all $l, \nu \in \mathbb{N}$, there exists $l'(l, \nu) \in \mathbb{N}, C(l, \nu) > 0, \mu(l, \nu) \geq 0$, such that for all $t \in \mathbb{R}$,*

$$\Big\| J^{t}a - \sum_{0 \leq j < \nu} (2i\pi t D_{x} \cdot D_{\xi})^{j} a \Big\|_{g,U_{r}}^{(l)} \leq C(l, \nu) \|a\|_{g,U_{r}}^{(l')} \lambda^{-\nu}(1 + |t|)^{\mu(l,\nu)}. \quad (2.3.19)$$

Proof. Since $g(x, \xi) = g(x, -\xi)$, we have $g(x, \xi) = \gamma_{1}(x) + \gamma_{2}(\xi)$ where the γ_{j} are positive-definite quadratic forms on $\mathbb{R}^{n}$, identified with $n \times n$ symmetric matrices with positive eigenvalues. Then there exists $V \in O(n)$, i.e., such that $V^{*}V = \mathrm{Id}$, with

$$g(Vy, V^{*-1}\eta) = g(Vy, V\eta) = \langle \gamma_{1}Vy, Vy \rangle + \langle \gamma_{2}V\eta, V\eta \rangle = \sum_{1 \leq j \leq n} \mu_{j} y_{j}^{2} + \langle \gamma_{2}V\eta, V\eta \rangle.$$

With $M = \mathrm{diag}(\mu_{j}^{-1/2})$, we get $g(VMy, VM^{-1}\eta) = \|y\|^{2} + \langle \gamma\eta, \eta \rangle$ where γ is a positive symmetric matrix and $\|\cdot\|$ is the standard Euclidean norm on $\mathbb{R}^{n}$. Let $U \in O(n)$ such that $U^{*}\gamma U = \mathrm{diag}(\nu_{j})$: we obtain

$$g(VMUy, VM^{-1}U\eta) = \|Uy\|^{2} + \sum_{1 \leq j \leq n} \nu_{j}\eta_{j}^{2} = \sum_{1 \leq j \leq n} (y_{j}^{2} + \nu_{j}\eta_{j}^{2}).$$

We note also that $\left((VMU)^*\right)^{-1} = \left(U^*M\,V^*\right)^{-1} = VM^{-1}U$ so that the mapping $(y,\eta) \mapsto (VMUy, VM^{-1}U\eta)$ is symplectic (cf. Theorem 4.4.3). Furthermore, we can use the symplectic mapping $(y_j,\eta_j) \mapsto (\nu_j^{1/4}y_j, \nu_j^{-1/4}\eta_j)$ to reach the identity $\sum_{1\le j\le n} h_j(y_j^2 + \eta_j^2) = g(B^{-1}y, B^*\eta)$ with some $B \in Gl(n,\mathbb{R})$ (and positive h_j); the condition $g \le g^\sigma$ is expressed by $\max_{1\le j\le n} h_j \le 1$. Moreover that type of symplectic mapping leaves invariant[9] the quadratic form $x \cdot \xi$, so that, with $b(x,\xi) = a(B^{-1}x, B^*\xi)$, we have

$$(J^t b)(x,\xi) = (J^t a)(B^{-1}x, B^*\xi).$$

Using Lemma 2.3.11, we may indeed assume that

$$g = \sum_{1\le j\le n} h_j(|dx_j|^2 + |d\xi_j|^2), \quad \lambda^{-1} = \max_{1\le j\le n} h_j \le 1.$$

It is also enough to check the result only for $J_1^t = \exp itD_{x_1}D_{\xi_1}$ since $h_1 \le \lambda^{-1}$ and J^t is the commutative product of $\exp itD_{x_j}D_{\xi_j}$. We have then with $s \in \mathbb{R}$, $h_1 = \lambda_1^{-1}$, using the formula of Lemma 4.1.2,

$$(J_1^{s^2} a)(x,\xi) = \int_{\mathbb{R}^2} e^{-2i\pi y_1 \eta_1} a(sy_1 + x_1, x', s\eta_1 + \xi_1, \xi')\,dy_1 d\eta_1.$$

We use the identity $(i + \lambda_1^{1/2}y_1)^{-1}(i - \lambda_1^{1/2}D_{\eta_1})e^{-2i\pi y_1\eta_1} = e^{-2i\pi y_1\eta_1}$, which gives for $k \in \mathbb{N}$,

$$(J_1^{s^2} a)(x,\xi)$$
$$= \int_{\mathbb{R}^2} e^{-2i\pi y_1\eta_1}(i+\lambda_1^{1/2}y_1)^{-k}\left((i+s\lambda_1^{1/2}D_{\xi_1})^k a\right)(sy_1+x_1, x', s\eta_1+\xi_1, \xi')\,dy_1 d\eta_1.$$

Similarly, the identity $(i + \lambda_1^{1/2}\eta_1)^{-1}(i - \lambda_1^{1/2}D_{y_1})e^{-2i\pi y_1\eta_1} = e^{-2i\pi y_1\eta_1}$, gives for $l \in \mathbb{N}$,

$$(J_1^{s^2} a)(x,\xi) = \int_{\mathbb{R}^2} e^{-2i\pi y_1\eta_1}(i+\lambda_1^{1/2}\eta_1)^{-l}$$
$$\times (i+\lambda_1^{1/2}D_{y_1})^l\left[(i+\lambda_1^{1/2}y_1)^{-k}\left((i+s\lambda_1^{1/2}D_{\xi_1})^k a\right)(sy_1+x_1, x', s\eta_1+\xi_1, \xi')\right]dy_1 d\eta_1.$$

As a consequence, $(J_1^{s^2} a)(x,\xi)$ is a finite sum[10] of terms of type

$$\alpha_{l',l''}(x,\xi) = \int_{\mathbb{R}^2} e^{-2i\pi y_1\eta_1}(i+\lambda_1^{1/2}\eta_1)^{-l}(i+\lambda_1^{1/2}y_1)^{-k-l''}\lambda_1^{l''}$$
$$\times \left((i+s\lambda_1^{1/2}D_{x_1})^{l'}(i+s\lambda_1^{1/2}D_{\xi_1})^k a\right)(sy_1 + x_1, x', s\eta_1 + \xi_1, \xi')\,dy_1 d\eta_1.$$

---

[9]This is not the case for all symplectic mappings: the mappings (iii), (jjj) in Proposition 4.4.4 do *not* leave invariant $x \cdot \xi$.

[10]The number of terms depends only on l and the coefficient of each term is bounded above by a constant depending only on k, l.

with $l', l'' \in \mathbb{N}$ with $l' + l'' \leq l$. This implies that, assuming that 0 is the center of the ball U_r, for $q \in \mathbb{N}$,

$$|(J_1^{s^2} a)(x, \xi)| \leq C_{k,l} \lambda_1^{N(k,l)} \|a\|_{g,U_r}^{(k+l,q)} \langle s \rangle^{k+l}$$
$$\times \int_{\mathbb{R}^2} \Big[1 + \inf_{g(z,\zeta) \leq r^2} \big(\lambda_1^{1/2} |s y_1 + x_1 - z_1| + \lambda_1^{1/2} |s \eta_1 + \xi_1 - \zeta_1|$$
$$+ \sum_{2 \leq j \leq n} \lambda_j^{1/2} (|x_j - z_j| + |\xi_j - \zeta_j|) \big) \Big]^{-q}$$
$$\times (1 + \lambda_1^{1/2} |\eta_1|)^{-l} (1 + \lambda_1^{1/2} |y_1|)^{-k} dy_1 d\eta_1.$$

As a result, for $k, l \geq \max(2, q)$, if $|s| \leq 1$, we get with $X = (x, \xi)$,

$$|(J_1^{s^2} a)(x, \xi)| \leq C'_{k,l} \lambda_1^{N(k,l)} \|a\|_{g,U_r}^{(k+l,q)} \big(1 + g^\sigma (X - U_r) \big)^{-q/2}. \tag{2.3.20}$$

If $|s| > 1$ (and $k, l \geq \max(2, q)$), we have

$$(1 + \lambda_1^{1/2} |\eta_1|)^{-l} (1 + \lambda_1^{1/2} |y_1|)^{-k}$$
$$\leq (1 + \lambda_1^{1/2} |\eta_1|)^{-2} (1 + \lambda_1^{1/2} |y_1|)^{-2} (1 + \lambda_1^{1/2} |\eta_1|)^{-l+2} (1 + \lambda_1^{1/2} |y_1|)^{-k+2}$$
$$\leq (1 + \lambda_1^{1/2} |\eta_1|)^{-2} (1 + \lambda_1^{1/2} |y_1|)^{-2} (1 + \lambda_1^{1/2} |\eta_1| + \lambda_1^{1/2} |y_1|)^{-\min(k-2,l-2)}$$
$$\leq (1 + \lambda_1^{1/2} |\eta_1|)^{-2} (1 + \lambda_1^{1/2} |y_1|)^{-2} (1 + |s| \lambda_1^{1/2} |\eta_1| + |s| \lambda_1^{1/2} |y_1|)^{-\min(k-2,l-2)}$$
$$\times |s|^{\min(k-2,l-2)},$$

entailing for $k, l \geq \max(2, q) + 2, |s| > 1$,

$$|(J_1^{s^2} a)(x, \xi)| \leq C''_{k,l} \lambda_1^{N(k,l)} \|a\|_{g,U_r}^{(k+l,q)} \big(1 + g^\sigma (X - U_r) \big)^{-q/2} \langle s \rangle^{2k+2l}. \tag{2.3.21}$$

Moreover we have

$$J_1^t a = \sum_{0 \leq j < \nu} (2 i \pi t D_{x_1} \cdot D_{\xi_1})^j a + \int_0^1 \frac{(1 - \theta)^{\nu-1}}{(\nu - 1)!} J_1^{\theta t} d\theta (2 i \pi t D_{x_1} \cdot D_{\xi_1})^\nu a, \tag{2.3.22}$$

and applying (2.3.20) or (2.3.21), we get for $k = l = \max(2, q) + 2$,

$$|(J_1^t a)(X)| \leq \big(1 + g^\sigma (X - U_r) \big)^{-q/2}$$
$$\times \Big(\sum_{0 \leq j < \nu} \|a\|_{g,U_r}^{(2j,q)} (2\pi |t|)^j + \frac{(2\pi |t|)^\nu}{\nu!} \|(\partial_{x_1} \cdot \partial_{\xi_1})^\nu a\|_{g,U_r}^{(k+l,q)} \langle t \rangle^{k+l} C'''_{k,l} \lambda_1^{N(k,l)} \Big).$$

Since $\|(\partial_{x_1} \cdot \partial_{\xi_1})^\nu a\|_{g,U_r}^{(p,q)} \leq \lambda_1^{-\nu} \|a\|_{g,U_r}^{(2\nu+p,q)}$, for a given q, with $k = l = \max(2, q) + 2$, we may choose $\nu \geq N(k, l)$ and obtain

$$|(J_1^t a)(X)| \leq \big(1 + g^\sigma (X - U_r) \big)^{-q/2} C_q \langle t \rangle^{\mu_q} \|a\|_{g,U_r}^{(m_q,q)}. \tag{2.3.23}$$

Since J_1^t commutes with the derivations, we get from (2.3.23), for $k \in \mathbb{N}, T \in \mathbb{R}^{2n}$,

$$|(J_1^t a)^{(k)}(X)T^k| = |(J_1^t a^{(k)}T^k)(X)|$$

$$\leq \left(1 + g^\sigma(X - U_r)\right)^{-q/2} C_q \langle t \rangle^{\mu_q} \|a\|_{g,U_r}^{(m_q+k,q)} g(T)^{k/2}, \quad (2.3.24)$$

and using $J^t = J_1^t \ldots J_n^t$ we get (2.3.19) for $\nu = 0$. To obtain the result for $\nu > 0$, we use the formula (2.3.22) for J^t instead of J_1^t and the estimate

$$\|(\partial_x \cdot \partial_\xi)^\nu a\|_{g,U_r}^{(p,q)} \leq \lambda^{-\nu} C_\nu \|a\|_{g,U_r}^{(2\nu+p,q)}.$$

The proof of the lemma is complete.   □

Lemma 2.3.13. *Let g be a positive quadratic form on $\mathbb{R}^{2n}$ such that $g \leq g^\sigma$, U_r be a closed g-ball with radius $r \leq 1$ and $a \in \mathrm{Conf}(g, U_r)$. Then for all $R > r$, there exist $b, c \in \mathrm{Conf}(g, U_R)$ such that $a = b \sharp c$, where U_R is the g-ball with radius R and the same center as U_r. The semi-norms of b, c depend only on R, r and the semi-norms of a.*

Proof. Using Lemma 4.4.25 and Lemma 2.3.11, we see that it is enough to prove the lemma assuming

$$g = \sum_{1 \leq j \leq n} h_j(|dx_j|^2 + |d\xi_j|^2), \quad \max_{1 \leq j \leq n} h_j \leq 1,$$

where $\{(x_j, \xi_j)\}$ is a set of linear symplectic coordinates. From Lemma 2.3.12, we see that is is enough to prove the result for the ordinary quantization $\mathrm{op}_0(a) = a(x, D)$ as given by Proposition 1.1.10. Also we may assume that the center of U_r is 0. We can now apply Lemma 4.3.20 to get that

$$a(x, \xi) = a_1(x)a_2(x, \xi), \quad \text{with } a_2 \in \mathrm{Conf}(g, U_r)$$

and with $\gamma = \sum_{1 \leq j \leq n} h_j|dx_j|^2$, $\mathbf{b}_R$ the γ-ball with radius R and center 0,

$$\forall k, N \in \mathbb{N}, \quad \sup_{t, \gamma(t) \leq 1} |a_1^{(k)}(x)t^k|\left(1 + \gamma^{-1}(x - \mathbf{b}_R)\right)^N < +\infty. \quad (2.3.25)$$

Defining $a_3 = J\overline{a_2}$, we get that $a_3 \in \mathrm{Conf}(g, U_r)$ and applying again Lemma 4.3.20, we obtain $a_3(x, \xi) = a_4(x, \xi)a_5(\xi)$ with a_5 satisfying (2.3.25). As a result we have

$$a(x, D) = a_1(x)a_2(x, D), \quad a_2(x, D)^* = a_3(x, D) = a_4(x, D)a_5(D),$$
$$\text{and thus } a(x, D) = a_1(x)\overline{a_5}(D)(J\overline{a_4})(x, D).$$

The symbol $J\overline{a_4}$ belongs to $\mathrm{Conf}(g, U_r)$ whereas the classical symbol of the operator $a_1(x)\overline{a_5}(D)$ is $a_1(x)\overline{a_5}(\xi)$, which is g-confined in the "box" $\mathbf{b}_R \times \mathbf{b}_R$, which gives only the g-confinement in $U_{R\sqrt{2}}$. However, we can replace a by $J^{-1/2}a$ and get

$$a^w = (J^{-1/2}(b_1(x)b_2(\xi)))^w c^w, \quad b_1, b_2 \text{ satisfying } (2.3.25), \ c \in \mathrm{Conf}(g, U_r).$$

We may consider the symplectic rotation χ_θ, for a real parameter θ, $(x_j, \xi_j) \mapsto (x_j \cos\theta - \xi_j \sin\theta, x_j \sin\theta + \xi_j \cos\theta)$, which leaves invariant g, and we may prove a factorization

$$(c \circ \chi_\theta)^w = b_3(x)b_4(D)d^w, \quad b_3, b_4 \text{ satisfying } (2.3.25), \ d \in \mathrm{Conf}(g, U_r),$$

so that $a^w = \left(J^{-1/2}(b_1 \otimes b_2)\right)^w \left(J^{-1/2}(b_3 \otimes b_4)\right)^w d^w$. Now the symbol

$$J^{-1/2}(b_1 \otimes b_2))\sharp J^{-1/2}(b_3 \otimes b_4)$$

is g-confined in $\mathbf{b}_R \times \mathbf{b}_R \cap \chi_\theta(\mathbf{b}_R \times \mathbf{b}_R)$. Factorizing again d^w, and varying θ, we can get the result of the lemma with R replaced by any $R' > R$, which gives the same result. $\square$

Definition 2.3.14 (Uniformly confined family of symbols). Let g be an admissible metric on $\mathbb{R}^{2n}$ (see Definition 2.2.15), r_0 be the positive constant given by Theorem 2.2.7, and let $(\psi_Y)_{Y \in \mathbb{R}^{2n}}$ be a family of functions on $\mathbb{R}^{2n}$. We shall say that this family is a *uniformly confined family of symbols* if there exists $r \in]0, r_0[$ such that $\forall k, N \in \mathbb{N}$,

$$\sup_{Y \in \mathbb{R}^{2n}, T \neq 0} |\psi_Y^{(k)}(X)T^k| g_Y(T)^{-k/2}\left(1 + g_Y^\sigma(X - U_{Y,r})\right)^{N/2} < +\infty, \qquad (2.3.26)$$

where $U_{Y,r}$ is the closed g_Y-ball with radius r. In other words, each ψ_Y is g_Y-confined in $U_{Y,r}$ (according to Definition 2.3.1) and

$$\forall l \in \mathbb{N}, \quad \omega_l = \sup_{Y \in \mathbb{R}^{2n}} \|\psi_Y\|_{g_Y, U_{Y,r}}^{(l)} < +\infty. \qquad (2.3.27)$$

The constants $(\omega_l)_{l \in \mathbb{N}}$ are the semi-norms of the family $(\psi_Y)_{Y \in \mathbb{R}^{2n}}$.

The next theorem is an immediate consequence of Lemma 2.3.13 and of Definition 2.3.14.

Theorem 2.3.15. *Let g be an admissible metric on $\mathbb{R}^{2n}$ (see Definition 2.2.15), and let $(\phi_Y)_{Y \in \mathbb{R}^{2n}}$ be a uniformly confined family of symbols. Then there exists $(\psi_Y)_{Y \in \mathbb{R}^{2n}}, (\theta_Y)_{Y \in \mathbb{R}^{2n}}$ two uniformly confined family of symbols such that*

$$\phi_Y = \psi_Y \sharp \theta_Y.$$

The semi-norms of the families $(\psi_Y)_{Y \in \mathbb{R}^{2n}}, (\theta_Y)_{Y \in \mathbb{R}^{2n}}$ are bounded above by functions of the semi-norms of $(\phi_Y)_{Y \in \mathbb{R}^{2n}}$.

Proposition 2.3.16. *Let g be an admissible metric on $\mathbb{R}^{2n}$, m be a g-admissible weight, and $(\phi_Y)_{Y \in \mathbb{R}^{2n}}$ be a uniformly confined family of symbols. Then the function a defined by*

$$a(X) = \int_{\mathbb{R}^{2n}} m(Y)\phi_Y(X)|g_Y|^{1/2}dY$$

belongs to $S(m,g)$; if $b \in S(m,g)$, the family $(m(Y)^{-1}\phi_Y \sharp b)_{Y \in \mathbb{R}^{2n}}$ is a uniformly confined family of symbols. Moreover, using the notation (2.1.29), defining

$$\rho_Y = m(Y)^{-1}\lambda_g(Y)^{\nu}\Big(\phi_Y \sharp b - \sum_{0 \le j < \nu} \omega_j(\phi_Y, b)\Big),$$

the family $(\rho_Y)_{Y \in \mathbb{R}^{2n}}$ is a uniformly confined family of symbols.

Remark 2.3.17. *For b as above, the family $\big(m(Y)^{-1}\lambda_g(Y)^j\omega_j(\varphi_Y, b)\big)_{Y \in \mathbb{R}^{2n}}$ is trivially a uniformly confined family of symbols.*

Proof. We have for all k, N,

$$|a^{(k)}(X)T^k|$$

$$\left| \int_{\mathbb{R}^{2n}} m(Y)\phi_Y^{(k)}(X)T^k\big(1 + g_Y^\sigma(X - U_{Y,r})\big)^{N/2}\big(1 + g_Y^\sigma(X - U_{Y,r})\big)^{-N/2}|g_Y|^{1/2}dY \right|$$

$$\le g_X(T)^{k/2}m(X) \int_{\mathbb{R}^{2n}} \frac{m(Y)}{m(X)}\|\phi_Y\|_{g_Y,U_{Y,r}}^{(\max(k,N))} \frac{g_Y(T)^{k/2}}{g_X(T)^{k/2}}\delta_r(X,Y)^{-N/2}|g_Y|^{1/2}dY,$$

so that using (2.2.29) and (2.2.31), choosing N large enough, we get $a \in S(m,g)$. On the other hand, if $b \in S(m,g)$, the family $(m(Y)^{-1}\phi_Y \sharp b)_{Y \in \mathbb{R}^{2n}}$ is a uniformly confined family of symbols since

$$m(Y)^{-1}(\phi_Y \sharp b)^{(k)}(X)S^k = \int m(Y)^{-1}\big(\phi_Y \sharp(b\varphi_T)\big)^{(k)}(X)S^k|g_T|^{1/2}dT$$

which is bounded above by (see Theorem 2.3.2, Lemmas 2.2.24, 2.2.25)

$$(1 + g_Y^\sigma(X - U_{Y,r}))^{-N}g_Y(S)^{k/2} \int m(Y)^{-1}m(T)\frac{(g_Y + g_T)(S)^{k/2}}{g_Y(S)^{k/2}}\delta_r(Y,T)^{NN_0}$$

$$\times C_M\big(1 + (g_Y^\sigma \wedge g_T^\sigma)(X - U_{Y,r}) + (g_Y^\sigma \wedge g_T^\sigma)(X - U_{T,r})\big)^{-M+N}|g_T|^{1/2}dT$$

$$\le C' \int \delta_r(Y,T)^{-M/2}|g_T|^{1/2}dT < \infty$$

for M large enough (N, k are given, N_0 is a constant given by (2.2.29), M can be chosen as large as we wish). To get the very last statement, we repeat the previous proof, using Theorem 2.3.4 instead of Theorem 2.3.2. $\qquad\square$

2.3.5   Changing the quantization

Theorem 2.3.18. *Let g be an admissible metric on $\mathbb{R}^{2n}$ (see Definition 2.2.15) such that*

$$\forall (x, \xi) \in \mathbb{R}^{2n}, \forall (y, \eta) \in \mathbb{R}^{2n}, \quad g_{x,\xi}(y, \eta) = g_{x,\xi}(y, -\eta). \tag{2.3.28}$$

Let m be an admissible weight for g and $a \in S(m, g)$. Then for all $t \in \mathbb{R}$, the symbol $J^t a = \exp 2i\pi D_x \cdot D_\xi a$ belongs to $S(m, g)$ and the mapping $a \mapsto J^t a$ is an isomorphism of the Fréchet space $S(m, g)$, with polynomial bounds in the real variable t. Moreover, with the weight λ_g defined in (2.2.13), we have

$$J^t a - \sum_{0 \le j < \nu} \frac{1}{j!} (2i\pi t D_x \cdot D_\xi)^j a \in S(m\lambda_g^{-\nu}, g). \tag{2.3.29}$$

Proof. Using a partition of unity, as given by Proposition 2.2.9, Lemma 2.3.12 gives the result. □

Theorem 2.3.19. *Let g be an admissible metric on $\mathbb{R}^{2n}$ (see Definition 2.2.15) such that (2.3.28) is satisfied, let m_1, m_2 be two admissible weights for g, $a_j \in S(m_j, g), j = 1, 2$. With $a(x, D) = (J^{-1/2} a)^w$ (see also (1.1.1)), we have*

$$a_1(x, D) a_2(x, D) = (a_1 \diamond a_2)(x, D), \quad a_1 \diamond a_2 \in S(m_1 m_2, g), \tag{2.3.30}$$

$$\forall \nu \in \mathbb{N}, \quad a_1 \diamond a_2 = \sum_{|\alpha| < \nu} \frac{1}{\alpha!} D_\xi^\alpha a_1 \partial_x^\alpha a_2 + r_\nu(a_1, a_2), \tag{2.3.31}$$

with $r_\nu(a_1, a_2) \in S(m_1 m_2 \lambda_g^{-\nu}, g)$. In particular, we have

$$a_1 \diamond a_2 = a_1 a_2 + \frac{1}{2i\pi} \partial_\xi a_1 \cdot \partial_x a_2 + r_2, \quad r_2 \in S(m_1 m_2 \lambda_g^{-2}, g).$$

Proof. We have $a_1 \diamond a_2 = (J^{-1/2} a_1) \sharp (J^{-1/2} a_2)$ so that the result follows from Theorem 2.3.7 and Theorem 2.3.18. □

2.4   The Wick calculus of pseudo-differential operators

2.4.1   Wick quantization

We recall here some facts on the so-called Wick quantization, as used in [93], [94], [95].

Definition 2.4.1. Let $Y = (y, \eta)$ be a point in $\mathbb{R}^n \times \mathbb{R}^n$. The operator Σ_Y is defined as $\left[2^n e^{-2\pi |\cdot - Y|^2} \right]^w$. Let a be in $L^\infty(\mathbb{R}^{2n})$. The Wick quantization of a is defined as

$$a^{\text{Wick}} = \int_{\mathbb{R}^{2n}} a(Y) \Sigma_Y dY. \tag{2.4.1}$$

Remark 2.4.2. The operator Σ_Y is a rank-one orthogonal projection: we have

$$\Sigma_Y u = (Wu)(Y)\tau_Y\varphi_0 \quad \text{with } (Wu)(Y) = \langle u, \tau_Y\varphi_0\rangle_{L^2(\mathbb{R}^n)}, \tag{2.4.2}$$

where $\varphi_0(x) = 2^{n/4}e^{-\pi|x|^2}$ and $(\tau_{y,\eta}\varphi_0)(x) = \varphi_0(x-y)e^{2i\pi\langle x-\frac{y}{2},\eta\rangle}$. $\quad$ (2.4.3)

In fact we get from the definition of Σ_Y that, for $u \in \mathscr{S}(\mathbb{R}^n)$,

$$\begin{aligned}
(\Sigma_{y,\eta}u)(x) &= \iint u(z)e^{2i\pi(x-z)\cdot\xi}2^n e^{-2\pi|\frac{x+z}{2}-y|^2}e^{-2\pi|\xi-\eta|^2}\,dz\,d\xi \\
&= \int u(z)e^{2i\pi(x-z)\cdot\eta}2^{n/2}e^{-2\pi|\frac{x+z}{2}-y|^2}e^{-\frac{\pi}{2}|x-z|^2}\,dz \\
&= \int u(z)e^{-2i\pi(z-\frac{y}{2})\cdot\eta}2^{n/4}e^{-\pi|z-y|^2}\,dz\, 2^{n/4}e^{-\pi|x-y|^2}e^{2i\pi(x-\frac{y}{2})\cdot\eta} \\
&= \langle u, \tau_{y,\eta}\varphi_0\rangle\tau_{y,\eta}\varphi_0.
\end{aligned}$$

Proposition 2.4.3. (1) *Let a be in $L^\infty(\mathbb{R}^{2n})$. Then $a^{Wick} = W^*a^\mu W$ and $1^{Wick} = Id_{L^2(\mathbb{R}^n)}$ where W is the isometric mapping from $L^2(\mathbb{R}^n)$ to $L^2(\mathbb{R}^{2n})$ given above, and a^μ the operator of multiplication by a in $L^2(\mathbb{R}^{2n})$. The operator $\pi_{\mathcal{H}} = WW^*$ is the orthogonal projection on a closed proper subspace $\mathcal{H}$ of $L^2(\mathbb{R}^{2n})$ and has the kernel*

$$\Pi(X,Y) = e^{-\frac{\pi}{2}|X-Y|^2}e^{-i\pi[X,Y]}, \tag{2.4.4}$$

where $[,]$ is the symplectic form (2.1.1). Moreover, we have

$$\|a^{Wick}\|_{\mathcal{L}(L^2(\mathbb{R}^n))} \le \|a\|_{L^\infty(\mathbb{R}^{2n})}, \tag{2.4.5}$$

$$a(X) \ge 0 \text{ for all } X \text{ implies } a^{Wick} \ge 0. \tag{2.4.6}$$

(2) *Let m be a real number, and $p \in S(\Lambda^m, \Lambda^{-1}\Gamma)$, where Γ is the Euclidean norm on $\mathbb{R}^{2n}$. Then $p^{Wick} = p^w + r(p)^w$, with $r(p) \in S(\Lambda^{m-1}, \Lambda^{-1}\Gamma)$ so that the mapping $p \mapsto r(p)$ is continuous. More precisely, one has*

$$r(p)(X) = \int_0^1 \int_{\mathbb{R}^{2n}} (1-\theta)p''(X+\theta Y)Y^2 e^{-2\pi\Gamma(Y)}2^n\,dY\,d\theta.$$

Note that $r(p) = 0$ if p is affine and $r(p) = \frac{1}{8\pi}\operatorname{trace} p''$ if p is a polynomial with degree ≤ 2.

(3) *For $a \in L^\infty(\mathbb{R}^{2n})$, the Weyl symbol of a^{Wick} is*

$$a * 2^n \exp -2\pi\Gamma, \text{ which belongs to } S(1,\Gamma) \text{ with } k^{th}\text{-semi-norm } c(k)\|a\|_{L^\infty}. \tag{2.4.7}$$

(4) *Let $\mathbb{R} \ni t \mapsto a(t,X) \in \mathbb{R}$ such that, for $t \le s$, $a(t,X) \le a(s,X)$. Then, for $u \in C_c^1(\mathbb{R}_t, L^2(\mathbb{R}^n))$, assuming $a(t,\cdot) \in L^\infty(\mathbb{R}^{2n})$,*

$$\int_{\mathbb{R}} \operatorname{Re}\langle D_t u(t), ia(t)^{Wick}u(t)\rangle_{L^2(\mathbb{R}^n)}\,dt \ge 0. \tag{2.4.8}$$

(5) *With the operator Σ_Y given in Definition 2.4.1, we have the estimate*

$$\|\Sigma_Y\Sigma_Z\|_{\mathcal{L}(L^2(\mathbb{R}^n))} \leq 2^n e^{-\frac{\pi}{2}\Gamma(Y-Z)}. \tag{2.4.9}$$

(6) *More precisely, the Weyl symbol of $\Sigma_Y\Sigma_Z$ is, as a function of the variable*
$X \in \mathbb{R}^{2n}$, *setting $\Gamma(T) = |T|^2$,*

$$e^{-\frac{\pi}{2}|Y-Z|^2} e^{-2i\pi[X-Y,X-Z]} 2^n e^{-2\pi|X-\frac{Y+Z}{2}|^2}. \tag{2.4.10}$$

Comment. Part of this proposition is well summarized by the following diagram:

$$
\begin{array}{ccc}
L^2(\mathbb{R}^{2n}) & \xrightarrow[\text{(multiplication by }a)]{a} & L^2(\mathbb{R}^{2n}) \\[2mm]
W \uparrow & & \downarrow W^* \\[2mm]
L^2(\mathbb{R}^n) & \xrightarrow[a^{\text{Wick}}]{} & L^2(\mathbb{R}^n)
\end{array}
$$

Proof. For $u, v \in \mathscr{S}(\mathbb{R}^n)$, we have

$$\langle a^{\text{Wick}}u, v \rangle = \int_{\mathbb{R}^{2n}} a(Y)\langle \Sigma_Y u, v \rangle_{L^2(\mathbb{R}^n)} dY = \int_{\mathbb{R}^{2n}} a(Y)(Wu)(Y)\overline{(Wv)(Y)} dY,$$

which gives

$$a^{\text{Wick}} = W^* a^\mu W. \tag{2.4.11}$$

Also we have from (2.4.1) that $1^{\text{Wick}} = \text{Id}$, since

$$1^{\text{Wick}} = \int_{\mathbb{R}^{2n}} \Sigma_Y dY \quad \text{has Weyl symbol} \int_{\mathbb{R}^{2n}} 2^n e^{-2\pi|X-Y|^2} dY = 1.$$

This implies that

$$W^* W = \text{Id},$$

i.e., W is isometric from $L^2(\mathbb{R}^n)$ into $L^2(\mathbb{R}^{2n})$. The operator WW^* is bounded selfadjoint and is a projection since $WW^*WW^* = WW^*$. Defining $\mathcal{H}$ as ran W, we get that WW^* is the orthogonal projection onto $\mathcal{H}$, since the range of WW^* is included in the range of W, and for $\Phi \in \mathcal{H}$, we have

$$\Phi = Wu = WW^*Wu \in \text{ran}(WW^*).$$

Moreover ran W is closed since W is isometric, that latter property implying also, using (2.4.11), the property (2.4.5), whereas (2.4.6) follows from (2.4.1) and $\Sigma_Y \geq 0$ as an orthogonal projection. The kernel of the operator WW^* is, from (2.4.2),

(2.4.3), with $X = (x, \xi), Y = (y, \eta)$,

$$\Pi(X, Y) = \langle \tau_Y \varphi_0, \tau_X \varphi_0 \rangle_{L^2(\mathbb{R}^n)}$$

$$= 2^{n/2} \int_{\mathbb{R}^n} e^{-\pi|t-x|^2} e^{-\pi|t-y|^2} e^{2i\pi(t-\frac{y}{2})\cdot\eta} e^{-2i\pi(t-\frac{x}{2})\cdot\xi} dt$$

$$= e^{-\frac{\pi}{2}|x-y|^2} 2^{n/2} \int_{\mathbb{R}^n} e^{-\frac{\pi}{2}|2t-x-y|^2} e^{2i\pi t \cdot (\eta-\xi)} dt e^{i\pi(x\cdot\xi-y\cdot\eta)}$$

$$= e^{-\frac{\pi}{2}|x-y|^2} 2^{n/2} \int_{\mathbb{R}^n} e^{-2\pi|t|^2} e^{2i\pi(t+\frac{x+y}{2})\cdot(\eta-\xi)} dt e^{i\pi(x\cdot\xi-y\cdot\eta)}$$

$$= e^{-\frac{\pi}{2}|x-y|^2} e^{-\frac{\pi}{2}|\xi-\eta|^2} e^{i\pi(x+y)\cdot(\eta-\xi)} e^{i\pi(x\cdot\xi-y\cdot\eta)}$$

$$= e^{-\frac{\pi}{2}|x-y|^2} e^{-\frac{\pi}{2}|\xi-\eta|^2} e^{i\pi(x\eta-y\xi)} = e^{-\frac{\pi}{2}|X-Y|^2} e^{-i\pi[X,Y]},$$

which is (2.4.4). Postponing the proof of $\mathcal{H} \neq L^2(\mathbb{R}^{2n})$ until after the proof of (2), we have proven (1). To obtain (2), we note that (2.4.1) gives directly that

$$a^{\text{Wick}} = (a * 2^n \exp -2\pi\Gamma)^w$$

and the second-order Taylor expansion gives (2) while (3) is obvious from the convolution formula. Note also that $u \in \mathscr{S}(\mathbb{R}^n)$ implies $Wu \in \mathscr{S}(\mathbb{R}^{2n})$ since $e^{-i\pi y \cdot \eta}(Wu)(y, \eta)$ is the partial Fourier transform with respect to x of $\mathbb{R}^n \times \mathbb{R}^n \ni (x, y) \mapsto u(x) 2^{n/4} e^{-\pi|x-y|^2}$: this gives also another proof that W is isometric since

$$\iint |u(x)|^2 2^{n/2} e^{-2\pi|x-y|^2} dx dy = \|u\|_{L^2(\mathbb{R}^n)}^2.$$

We calculate now, for $u \in \mathscr{S}(\mathbb{R}^n)$ with L^2 norm 1, using the already proven (2) on the Wick quantization of linear forms,

$$2\operatorname{Re}\langle \pi_{\mathcal{H}}\xi_1 Wu, ix_1 Wu \rangle_{L^2(\mathbb{R}^{2n})} = 2\operatorname{Re}\langle W^* \xi_1 Wu, iW^* x_1 Wu \rangle_{L^2(\mathbb{R}^n)}$$

$$= 2\operatorname{Re}\langle \xi_1^{\text{Wick}} u, ix_1^{\text{Wick}} u \rangle_{L^2(\mathbb{R}^n)} = 2\operatorname{Re}\langle D_1 u, ix_1 u \rangle_{L^2(\mathbb{R}^n)} = 1/2\pi.$$

If $\mathcal{H}$ were the whole $L^2(\mathbb{R}^{2n})$, the projection $\pi_{\mathcal{H}}$ would be the identity and we would have

$$0 = 2\operatorname{Re}\langle \xi_1 Wu, ix_1 Wu \rangle_{L^2(\mathbb{R}^{2n})} = 2\operatorname{Re}\langle \pi_{\mathcal{H}}\xi_1 Wu, ix_1 Wu \rangle_{L^2(\mathbb{R}^{2n})} = 1/2\pi.$$

Let us prove (4). We have from the Lebesgue dominated convergence theorem,

$$\alpha = \int_{\mathbb{R}} \operatorname{Re}\langle D_t u(t), ia(t)^{\text{Wick}} u(t) \rangle_{L^2(\mathbb{R}^n)} dt$$

$$= -\lim_{h \to 0_+} \int_{\mathbb{R}} \frac{1}{2\pi h} \operatorname{Re}\langle u(t+h) - u(t), a(t)^{\text{Wick}} u(t+h) \rangle_{L^2(\mathbb{R}^n)} dt$$

$$= \lim_{h \to 0_+} \frac{1}{2\pi h} \left(-\int_{\mathbb{R}} \operatorname{Re}\langle u(t), a(t-h)^{\text{Wick}} u(t) \rangle_{L^2(\mathbb{R}^n)} dt \right.$$

$$\left. + \int_{\mathbb{R}} \operatorname{Re}\langle u(t), a(t)^{\text{Wick}} u(t+h) \rangle_{L^2(\mathbb{R}^n)} dt \right)$$

$$= \lim_{h \to 0_+} \Big\{ \underbrace{\frac{1}{2\pi h} \int_{\mathbb{R}} \mathrm{Re}\langle (a(t) - a(t-h))^{\mathrm{Wick}} u(t), u(t)\rangle_{L^2(\mathbb{R}^n)} dt}_{=\beta(h)}$$

$$+ \underbrace{\int_{\mathbb{R}} \mathrm{Re}\langle \frac{-1}{2\pi hi}(u(t+h) - u(t)), ia(t)^{\mathrm{Wick}} u(t)\rangle_{L^2(\mathbb{R}^n)} dt}_{\text{with limit } -\alpha} \Big\}.$$

The previous calculation shows that $\beta(h)$ has a limit when $h \to 0_+$ and $2\alpha = \lim_{h \to 0_+} \beta(h)$. Since the function $a(t) - a(t-h)$ is non-negative, the already proven (2.4.6) implies that the operator $(a(t) - a(t-h))^{\mathrm{Wick}}$ is also non-negative, implying $\beta(h) \geq 0$ which gives $\alpha \geq 0$, i.e., (2.4.8)[11]. Since for the Weyl quantization, one has from (2.1.13), $\|a^w\|_{\mathcal{L}(L^2(\mathbb{R}^n))} \leq 2^n \|a\|_{L^1(\mathbb{R}^{2n})}$, we get the result (2.4.9) from (2.4.10). Let us finally prove the latter formula. From the composition formula (2.1.18), we obtain that the Weyl symbol ω of $\Sigma_Y \Sigma_Z$ is

$$\omega(X) = 2^{2n} \iint e^{-4i\pi[X-X_1, X-X_2]} 2^{2n} e^{-2\pi|X_1-Y|^2} e^{-2\pi|X_2-Z|^2} dX_1 dX_2$$

$$= 2^{4n} \iint e^{-4i\pi[X-Y, X-X_2]} e^{-2i\pi\langle X_1, 2\sigma(X-X_2)\rangle} e^{-2\pi|X_1|^2} e^{-2\pi|X_2-Z|^2} dX_1 dX_2$$

$$= 2^{3n} \int e^{-4i\pi[X-Y, X-X_2]} e^{-2\pi|X-X_2|^2} e^{-2\pi|X_2-Z|^2} dX_2$$

$$= 2^{3n} e^{-\pi|X-Z|^2} \int e^{-4i\pi[X-Y, X-X_2]} e^{-\pi|X+Z-2X_2|^2} dX_2$$

$$= 2^{3n} e^{-\pi|X-Z|^2} e^{-2i\pi[X-Y, X-Z]} \int e^{-4i\pi[X-Y, -X_2]} e^{-4\pi|X_2|^2} dX_2$$

$$= 2^n e^{-\pi|X-Z|^2} e^{-2i\pi[X-Y, X-Z]} e^{-\pi|X-Y|^2}$$

$$= 2^n e^{-2i\pi[X-Y, X-Z]} e^{-2\pi|X-\frac{Y+Z}{2}|^2} e^{-\frac{\pi}{2}|Y-Z|^2}. \qquad \square$$

2.4.2 Fock-Bargmann spaces

There are also several links with the so-called Fock-Bargmann spaces (the space $\mathcal{H}$ above), that we can summarize with the following definitions and properties.

Proposition 2.4.4. *With $\mathcal{H}$ defined in Proposition 2.4.3 we have*

$$\mathcal{H} = \{\Phi \in L^2(\mathbb{R}^{2n}_{y,\eta}), \quad \Phi = f(z) \exp -\frac{\pi}{2}|z|^2, \ z = \eta + iy \ , \ f \ \text{entire}\}, \qquad (2.4.12)$$

i.e., $\mathcal{H} = \mathrm{ran} W = L^2(\mathbb{R}^{2n}) \cap \ker(\bar{\partial} + \frac{\pi}{2}z).$

---

[11] Note that (2.4.8) is simply a way of writing that $\frac{d}{dt}(a(t)^{\mathrm{Wick}}) \geq 0$, which is a consequence of (2.4.6) and of the non-decreasing assumption made on $t \mapsto a(t, X)$.

Proof. For $v \in L^2(\mathbb{R}^n)$, we have, with the notation $z^2 = \sum_{1 \le j \le n} z_j^2$ for $z \in \mathbb{C}^n$,

$$(Wv)(y, \eta) = \int_{\mathbb{R}^n} v(x) 2^{n/4} e^{-\pi(x-y)^2} e^{-2i\pi(x-\frac{y}{2})\eta} dx$$

$$= \int_{\mathbb{R}^n} v(x) 2^{n/4} e^{-\pi(x-y+i\eta)^2} dx e^{-\frac{\pi}{2}(y^2+\eta^2)} e^{-\frac{\pi}{2}(\eta+iy)^2} \qquad (2.4.13)$$

and we see that $Wv \in L^2(\mathbb{R}^{2n}) \cap \ker(\bar\partial + \frac{\pi}{2}z)$. Conversely, if $\Phi \in L^2(\mathbb{R}^{2n}) \cap \ker(\bar\partial + \frac{\pi}{2}z)$, we have $\Phi(x, \xi) = e^{-\frac{\pi}{2}(x^2+\xi^2)} f(\xi + ix)$ with $\Phi \in L^2(\mathbb{R}^{2n})$ and f entire. This gives

$$(WW^*\Phi)(x, \xi) = \iint e^{-\frac{\pi}{2}\left((\xi-\eta)^2 + (x-y)^2 + 2i\xi y - 2i\eta x\right)} \Phi(y, \eta) dy d\eta$$

$$= e^{-\frac{\pi}{2}(\xi^2+x^2)} \iint e^{-\frac{\pi}{2}(\eta^2 - 2\xi\eta + y^2 - 2xy + 2i\xi y - 2i\eta x)} \Phi(y, \eta) dy d\eta$$

$$= e^{-\frac{\pi}{2}(\xi^2+x^2)} \iint e^{-\frac{\pi}{2}\left(\eta^2 + y^2 + 2iy(\xi+ix) - 2\eta(\xi+ix)\right)} \Phi(y, \eta) dy d\eta$$

$$= e^{-\frac{\pi}{2}(\xi^2+x^2)} \iint e^{-\pi(y^2+\eta^2)} e^{\pi(\eta-iy)(\xi+ix)} f(\eta + iy) dy d\eta$$

$$= e^{-\frac{\pi}{2}|z|^2} \iint e^{-\pi|\zeta|^2} e^{\pi\bar\zeta z} f(\zeta) dy d\eta \quad (\zeta = \eta + iy, \ z = \xi + ix)$$

$$= e^{-\frac{\pi}{2}|z|^2} \iint f(\zeta) \prod_{1 \le j \le n} \frac{1}{\pi(z_j - \zeta_j)} \frac{\partial}{\partial\bar\zeta_j} \left(e^{-\pi|\zeta|^2} e^{\pi\bar\zeta z}\right) dy d\eta$$

$$= e^{-\frac{\pi}{2}|z|^2} \langle f(\zeta) \prod_{1 \le j \le n} \frac{\partial}{\partial\bar\zeta_j} \left(\frac{1}{\pi(\zeta_j - z_j)}\right), e^{-\pi|\zeta|^2} e^{\pi\bar\zeta z} \rangle_{\mathscr{S}'(\mathbb{R}^{2n}), \mathscr{S}(\mathbb{R}^{2n})}$$

$$= e^{-\frac{\pi}{2}|z|^2} f(z),$$

since f is entire. This implies $WW^*\Phi = \Phi$ and $\Phi \in \operatorname{ran} W$, completing the proof of the proposition. $\qquad\square$

Proposition 2.4.5. *Defining*

$$\mathscr{H} = \ker(\bar\partial + \frac{\pi}{2}z) \cap \mathscr{S}'(\mathbb{R}^{2n}), \qquad (2.4.14)$$

the operator W given by (2.4.2) can be extended as a continuous mapping from $\mathscr{S}'(\mathbb{R}^n)$ onto $\mathscr{H}$ (the $L^2(\mathbb{R}^n)$ dot product is replaced by a bracket of (anti)duality). The operator $\widetilde\Pi$ with kernel Π given by (2.4.4) defines a continuous mapping from $\mathscr{S}(\mathbb{R}^{2n})$ into itself and can be extended as a continuous mapping from $\mathscr{S}'(\mathbb{R}^{2n})$ onto $\mathscr{H}$. It verifies

$$\widetilde\Pi^2 = \widetilde\Pi, \quad \widetilde\Pi_{|\mathscr{H}} = \operatorname{Id}_{\mathscr{H}}. \qquad (2.4.15)$$

Proof. As above we use that $e^{-i\pi y\eta}(Wv)(y, \eta)$ is the partial Fourier transform w.r.t. x of the tempered distribution on $\mathbb{R}^{2n}_{x,y}$,

$$v(x)2^{n/4}e^{-\pi(x-y)^2}.$$

Since $e^{\pm i\pi y\eta}$ are in the space $\mathscr{O}_M(\mathbb{R}^{2n})$ of multipliers of $\mathscr{S}(\mathbb{R}^{2n})$, that transformation is continuous and injective from $\mathscr{S}'(\mathbb{R}^n)$ into $\mathscr{S}'(\mathbb{R}^{2n})$. Replacing in (2.4.13) the integrals by brackets of duality, we see that $W(\mathscr{S}'(\mathbb{R}^n)) \subset \mathscr{H}$. Conversely, if $\Phi \in \mathscr{H}$, the same calculations as above give (2.4.15) and (2.4.14). $\square$

2.4.3   On the composition formula for the Wick quantization

In this section, we prove some formulas of composition for operators with very irregular Wick symbols.

Lemma 2.4.6. *For $p, q \in L^\infty(\mathbb{R}^{2n})$ real-valued with $p'' \in L^\infty(\mathbb{R}^{2n})$, we have*

$$\mathrm{Re}\big(p^{Wick}q^{Wick}\big) = \Big(pq - \frac{1}{4\pi}\nabla p \cdot \nabla q\Big)^{Wick} + R, \qquad (2.4.16)$$

$$\|R\|_{\mathcal{L}(L^2(\mathbb{R}^n))} \le C(n)\|p''\|_{L^\infty}\|q\|_{L^\infty}. \qquad (2.4.17)$$

The product $\nabla p \cdot \nabla q$ above makes sense (see our appendix Section 4.3.2) as a tempered distribution since ∇p is a Lipschitz-continuous function and ∇q is the derivative of an L^∞ function: in fact, we shall use as a definition (see Section 4.3.2) $\nabla p \cdot \nabla q = \nabla \cdot (\underbrace{q}_{L^\infty} \underbrace{\nabla p}_{Lip.}) - \underbrace{q}_{L^\infty} \underbrace{\Delta p}_{L^\infty} .$

Proof. Using Definition (2.4.1) , we see that

$$\begin{aligned}
p^{\mathrm{Wick}}q^{\mathrm{Wick}} &= \iint_{\mathbb{R}^{2n}\times\mathbb{R}^{2n}} p(Y)q(Z)\Sigma_Y\Sigma_Z dY dZ \\
&= \iint \Big(p(Z) + p'(Z)(Y - Z) + p_2(Z,Y)(Y - Z)^2\Big)q(Z)\Sigma_Y\Sigma_Z dY dZ \\
&= \int (pq)(Z)\Sigma_Z dZ + \iint p'(Z)(Y - Z)\Sigma_Y dY q(Z)\Sigma_Z dZ + R_0,
\end{aligned}$$

with $R_0 = \iiint_0^1 (1 - \theta)p''(Z + \theta(Y - Z))(Y - Z)^2 q(Z)\Sigma_Y\Sigma_Z dY dZ d\theta$.

Remark 2.4.7. *Let ω be a measurable function defined on $\mathbb{R}^{2n} \times \mathbb{R}^{2n}$ such that*

$$|\omega(Y, Z)| \le \gamma_0\big(1 + |Y - Z|\big)^{N_0}.$$

Then the operator $\iint \omega(Y, Z)\Sigma_Y\Sigma_Z dY dZ$ is bounded on $L^2(\mathbb{R}^n)$ with $\mathcal{L}(L^2(\mathbb{R}^n))$ norm bounded above by a constant depending on γ_0, N_0. This is an immediate consequence of Cotlar's Lemma 4.7.1 and of the estimate (2.4.9).

Using that remark, we obtain that

$$\|R_0\|_{\mathcal{L}(L^2(\mathbb{R}^n))} \le C_1(n)\|p''\|_{L^\infty(\mathbb{R}^{2n})}\|q\|_{L^\infty(\mathbb{R}^{2n})}. \qquad (2.4.18)$$

We check now $\int (Y - Z)\Sigma_Y dY$ whose Weyl symbol is, as a function of X,

$$\int (Y - Z)2^n e^{-2\pi|X-Y|^2} dY = \int (X - Z)2^n e^{-2\pi|X-Y|^2} dY = X - Z.$$

So with $L_Z(X) = X - Z$, we have $\int (Y - Z)\Sigma_Y dY \Sigma_Z = (X - Z)^w \Sigma_Z = L_Z^w \Sigma_Z$ and thus

$$\text{Re} \int (Y - Z)\Sigma_Y dY \Sigma_Z = \text{Re}(L_Z^w \Sigma_Z) = \left((X - Z)2^n e^{-2\pi|X-Z|^2}\right)^w$$

$$= \frac{1}{4\pi}\partial_Z (2^n e^{-2\pi|X-Z|^2})^w, \qquad (2.4.19)$$

so that

$$\text{Re} \int (Y - Z)\Sigma_Y dY \Sigma_Z = \frac{1}{4\pi}\partial_Z(\Sigma_Z). \qquad (2.4.20)$$

Using that p and q are real-valued, the formula for $\text{Re}(p^{\text{Wick}}q^{\text{Wick}})$ becomes

$$\text{Re}(p^{\text{Wick}}q^{\text{Wick}}) = \int (pq)(Z)\Sigma_Z dZ + \int p'(Z)q(Z)\frac{1}{4\pi}\partial_Z \Sigma_Z dZ + \text{Re} R_0$$

$$= \int \left((pq)(Z) - \frac{1}{4\pi}p'(Z) \cdot q'(Z)\right)\Sigma_Z dZ - \int \frac{1}{4\pi}\,\text{trace}\,p''(Z)q(Z)\Sigma_Z dZ + \text{Re} R_0,$$

that is the result of the lemma, using (2.4.18) and (2.4.5) for the penultimate term on the line above. $\qquad \square$

The next lemma is more involved.

Lemma 2.4.8. *For p a measurable real-valued function such that p'', $(p'p'')'$, $(pp'')'' \in L^\infty$, we have*

$$p^{\text{Wick}}p^{\text{Wick}} = \int \left[p(Z)^2 - \frac{1}{4\pi}|\nabla p(Z)|^2\right]\Sigma_Z dZ + S, \qquad (2.4.21)$$

$$\|S\|_{\mathcal{L}(L^2(\mathbb{R}^n))} \leq C(n) \left(\|p''\|_{L^\infty}^2 + \|(p''p')'\|_{L^\infty} + \|(pp'')''\|_{L^\infty}\right). \qquad (2.4.22)$$

Here p'' stands for the vector (tensor) with components $(\partial_X^\alpha p)_{|\alpha|=2}$, whereas the components of $(p''p')'$ are $\partial_X^\alpha(\partial_X^\beta \partial_X^\gamma p)_{|\alpha|=1,|\beta|=2}$, and those of $(pp'')''$ are
$$\quad\quad\quad\quad\quad\quad\quad\quad\quad\quad\quad {}_{|\gamma|=1}$$

$$\partial_X^\alpha(p\partial_X^\beta p)_{|\alpha|=|\beta|=2}.$$

Proof. We have

$$p^{\text{Wick}}p^{\text{Wick}} = \iint p(Y)p(Z)\Sigma_Y \Sigma_Z dY dZ$$

$$= \iint p(Y)\big(p(Y) + p'(Y)(Z - Y)\big)\Sigma_Y \Sigma_Z dY dZ$$

$$+ \iiint_0^1 p(Y)(1 - \theta)p''(Y + \theta(Z - Y))d\theta(Z - Y)^2 \Sigma_Y \Sigma_Z dY dZ d\theta$$

so that, using (2.4.20) for the terms pp' in the double integral above, we get, noting $\text{trace}(p'') = \Delta p$,

$$p^{\text{Wick}} p^{\text{Wick}} = \left(p^2 - \frac{1}{4\pi}|\nabla p|^2 - \frac{1}{4\pi}p\Delta p\right)^{\text{Wick}} + \text{Re}(\Omega_0 + \Omega_1 + \Omega_2), \qquad (2.4.23)$$

with

$$\Omega_0 = \iiint_0^1 p(Y + \theta(Z - Y))p''(Y + \theta(Z - Y))(Z - Y)^2 \Sigma_Y \Sigma_Z dY\, dZ(1 - \theta)d\theta, \qquad (2.4.24)$$

$$\Omega_1 = \iiint_0^1 p'(Y + \theta(Z - Y))\theta(Y - Z)$$
$$\times\, p''(Y + \theta(Z - Y))(Z - Y)^2 \Sigma_Y \Sigma_Z dY\, dZ(1 - \theta)d\theta \qquad (2.4.25)$$

and from Remark 2.4.7,

$$\|\Omega_2\|_{\mathcal{L}(L^2(\mathbb{R}^n))} \le C_1(n)\|p''\|_{L^\infty}^2. \qquad (2.4.26)$$

We write now $\Omega_0 = \Omega_{00} + \Omega_{01}$, $\Omega_1 = \Omega_{10} + \Omega_{11}$ with

$$\Omega_{00} = \frac{1}{2}\iint p(Y)p''(Y)(Z - Y)^2 \Sigma_Y \Sigma_Z dY\, dZ,$$

$$\Omega_{01} = \iiint_0^1 \Big((pp'')(Y + \theta(Z - Y)) - (pp'')(Y)\Big)(Z - Y)^2 \Sigma_Y \Sigma_Z dY\, dZ(1 - \theta)d\theta,$$

$$\Omega_{10} = -\frac{1}{6}\iiint_0^1 p'(Y)(Z - Y)p''(Y)(Z - Y)^2 \Sigma_Y \Sigma_Z dY\, dZ,$$

$$\|\Omega_{11}\|_{\mathcal{L}(L^2(\mathbb{R}^n))} \le C_2(n)\|(p'p'')'\|_{L^\infty}. \qquad (2.4.27)$$

We have also $\Omega_{01} = \Omega_{010} + \Omega_{011}$ with

$$\Omega_{010} = \frac{1}{6}\iint (pp'')'(Y)(Z - Y)(Z - Y)^2 \Sigma_Y \Sigma_Z dY\, dZ,$$

$$\|\Omega_{011}\|_{\mathcal{L}(L^2(\mathbb{R}^n))} \le C_3(n)\|(pp'')''\|_{L^\infty}. \qquad (2.4.28)$$

From (2.4.23)–(2.4.28), it suffices to check that the following term is a remainder satisfying the estimate (2.4.22) to get the result of Lemma 2.4.8:

$$\tilde{\Omega} = -\frac{1}{4\pi}\int p(Y)\,\text{trace}\,p''(Y)\Sigma_Y dY + \frac{1}{2}\,\text{Re}\iint (pp'')(Y)(Z - Y)^2 \Sigma_Y \Sigma_Z dY\, dZ$$
$$+ \frac{1}{6}\,\text{Re}\iint (pp'')'(Y)(Z - Y)(Z - Y)^2 \Sigma_Y \Sigma_Z dY\, dZ$$
$$- \frac{1}{6}\,\text{Re}\iiint_0^1 p'(Y)(Z - Y)p''(Y)(Z - Y)^2 \Sigma_Y \Sigma_Z dY\, dZ. \qquad (2.4.29)$$

The real part of the Weyl symbol of $\int (Z_j - Y_j)(Z_k - Y_k)(Z_l - Y_l)\Sigma_Y \Sigma_Z dZ$ is (see (2.4.10))

$$\int (Z_j - Y_j)(Z_k - Y_k)(Z_l - Y_l)e^{-\frac{\pi}{2}|Y-Z|^2}$$

$$\times \cos(2\pi[X-Y, X-Z])2^n e^{-2\pi|X-\frac{Y+Z}{2}|^2} dZ$$

$$= \int T_j T_k T_l e^{-2\pi|T/2|^2} \cos(2\pi[X-Y, T])2^n e^{-2\pi|X-Y-\frac{T}{2}|^2} dT$$

$$= \int T_j T_k T_l \cos(2\pi[X-Y, T])e^{-\pi|X-Y-T|^2} dT 2^n e^{-\pi|X-Y|^2} = \nu_{jkl}(X-Y)$$

with

$$\nu_{jkl}(S) = \int T_j T_k T_l \cos(2\pi[S, T])e^{-\pi|S-T|^2} dT 2^n e^{-\pi|S|^2} \tag{2.4.30}$$

$$= 2^n e^{-\pi|S|^2} \int (T_j + S_j)(T_k + S_k)(T_l + S_l) \cos(2\pi[S, T])e^{-\pi|T|^2} dT \tag{2.4.31}$$

$$= 2^n e^{-\pi|S|^2} \int (T_j T_k S_l + T_k T_l S_j + T_l T_j S_k + S_j S_k S_l) \cos(2\pi[S, T])e^{-\pi|T|^2} dT. \tag{2.4.32}$$

We notice that the function $S \mapsto \int_{\mathbb{R}^{2n}} T_j T_k \exp(2i\pi[S, T])e^{-\pi|T|^2} dT$ is a second-order derivative of $S \mapsto \int_{\mathbb{R}^{2n}} \exp(2i\pi[S, T])e^{-\pi|T|^2} dT = e^{-\pi|S|^2}$ so that

$$2^n e^{-\pi|S|^2} S_l \int_{\mathbb{R}^{2n}} T_j T_k \cos(2\pi[S, T])e^{-\pi|T|^2} dT = e^{-2\pi|S|^2} S_l P_{jk}(S),$$

with P_{jk} even, second-order and real polynomial. The function $S_{l_1} S_{l_2} S_{l_3} e^{-2\pi|S|^2}$ is always a linear combination of derivatives of Schwartz functions on $\mathbb{R}^{2n}$, since

- if $l_1 < l_2 \leq l_3$, it is the derivative with respect to S_{l_1} of $S_{l_2} S_{l_3} e^{-2\pi|S|^2}(-4\pi)^{-1}$,
- if $l_1 = l_2 < l_3$, it is the derivative with respect to S_{l_3} of $S_{l_1} S_{l_2} e^{-2\pi|S|^2}(-4\pi)^{-1}$,
- if $l_1 = l_2 = l_3 = l$, it is a linear combination of the third and first derivative with respect to S_l of $e^{-2\pi|S|^2}$, since

$$(e^{t^2})''' = (12t + 8t^3)e^{t^2}, \qquad t^3 e^{t^2} = \frac{1}{8}(e^{t^2})''' - \frac{3}{4}(e^{t^2})'.$$

As a result the function ν_{jkl} defined by (2.4.32) is a linear combination of derivatives with respect to S_j, S_k or S_l of Schwartz functions on $\mathbb{R}^{2n}$. Integrating by parts in the last two terms of (2.4.29), we see that their $\mathcal{L}(L^2)$ norm is bounded from above by $C_4(n)(\|(pp'')''\|_{L^\infty} + \|(p'p'')'\|_{L^\infty})$. Looking at (2.4.29), we see that we are left with

$$\tilde{\Omega}_0 = -\frac{1}{4\pi} \int p(Y) \operatorname{trace} p''(Y)\Sigma_Y dY + \frac{1}{2} \operatorname{Re} \iint (pp'')(Y)(Z-Y)^2 \Sigma_Y \Sigma_Z dY dZ. \tag{2.4.33}$$

The real part of the operator $\int (Z_j - Y_j)(Z_k - Y_k)\Sigma_Y \Sigma_Z dZ$ has the Weyl symbol (function of X)

$$\int T_j T_k e^{-\pi|X-Y-T|^2} \cos(2\pi[X-Y,T]) dT 2^n e^{-\pi|X-Y|^2} \tag{2.4.34}$$

$$= \int \left((X_j - Y_j)(X_k - Y_k) + T_j T_k \right) e^{-\pi|T|^2} \cos(2\pi[X-Y,T]) dT 2^n e^{-\pi|X-Y|^2} \tag{2.4.35}$$

$$= \int (S_j S_k + T_j T_k) e^{-\pi|T|^2} \cos(2\pi[S,T]) dT 2^n e^{-\pi|S|^2}, \quad S = X - Y. \tag{2.4.36}$$

• If $j \neq k$, both terms in (2.4.36) are second-order derivatives with respect to Y of a Schwartz function in $\mathbb{R}^{2n}$. In fact the first term is

$$S_j S_k 2^n e^{-2\pi|S|^2} = \partial_{S_j} \partial_{S_k} \left(2^n e^{-2\pi|S|^2} / 16\pi^2 \right) = \partial_{Y_j} \partial_{Y_k} \left(2^n e^{-2\pi|S|^2} / 16\pi^2 \right)$$

and the second term is equal to $-S_{j'} S_{k'} 2^n e^{-2\pi|S|^2}$, with $j' \neq k'$, also a second-order derivative. The contribution of these terms in (2.4.33) is then, after integration by parts, an L^2 bounded operator with norm $\leq C_5(n) \|(pp'')''\|_{L^\infty}$.

• If $j = k$, with $j' = j \pm n$ (in fact $j' = j + n$ if $1 \leq j \leq n$ and $j' = j - n$ if $1 + n \leq j \leq 2n$), we note that (2.4.36) is equal to

$$S_j^2 2^n e^{-2\pi|S|^2} - \frac{1}{4\pi^2} e^{-\pi|S|^2} \partial_{S_{j'}}^2 \left(2^n e^{-\pi|S|^2} \right) = 2^n e^{-2\pi|S|^2} \left(S_j^2 - \frac{1}{4\pi^2}(4\pi^2 S_{j'}^2 - 2\pi) \right).$$

Taking into account the contribution of all these terms in (2.4.33), we see that we are left with

$$-\frac{1}{4\pi} \int p(Y) \operatorname{trace} p''(Y) \Sigma_Y dY + \frac{1}{2} \iint \frac{1}{2\pi} \operatorname{trace}(pp'')(Y) \Sigma_Y dY = 0.$$

The proof of Lemma 2.4.8 is complete.   □

2.5   Basic estimates for pseudo-differential operators

2.5.1 L^2 estimates

Theorem 2.5.1. *Let g be an admissible metric on $\mathbb{R}^{2n}$ (see Definition 2.2.15), $a \in S(1,g)$ (cf. Definition 2.2.6). Then a^w is bounded on $L^2(\mathbb{R}^n)$ and $\|a^w\|_{\mathcal{L}(L^2)}$ can be estimated by a semi-norm of the symbol a.*

Before proving this theorem, we need a lemma.

Lemma 2.5.2. *Let g be a positive-definite quadratic form on $\mathbb{R}^{2n}$ such that $g \leq g^\sigma$. Let a be a smooth function in $\mathbb{R}^{2n}$ and U be a g-ball with radius ≤ 1 such that a is g-confined in U (see Definition 2.3.1). Then*

$$\|a^w\|_{\mathcal{L}(L^2(\mathbb{R}^n))} \leq C(n) \|a\|_{g,U}^{(2n+1)}. \tag{2.5.1}$$

Proof of the lemma. For $k \in \mathbb{N}$, we have

$$\hat{a}(-\Xi) = \int \left(\left(1 - \langle g^{-1}D_X, \Xi \rangle \right)^k a \right)(X) e^{2i\pi X \cdot \Xi} dX \left(1 + \langle g^{-1}\Xi, \Xi \rangle \right)^{-k},$$

so that, with $k = 2n + 1$,

$$|\hat{a}(-\Xi)| \leq \left(1 + \langle g^{-1}\Xi, \Xi \rangle \right)^{-2n-1} \int \sum_{0 \leq l \leq 2n+1} C_{2n+1}^l |a^{(l)}(X)(g^{-1}\Xi)^l| dX,$$

entailing

$$|\hat{a}(-\Xi)| \leq \left(1 + \langle g^{-1}\Xi, \Xi \rangle \right)^{-2n-1}$$
$$\times \int \sum_{0 \leq l \leq 2n+1} C_{2n+1}^l \|a\|_{g,U}^{(2n+1,2n+1)} \langle gg^{-1}\Xi, g^{-1}\Xi \rangle^{l/2} \left(1 + g^\sigma(X - U) \right)^{-n-\frac{1}{2}} dX.$$

This gives

$$|\hat{a}(-\Xi)| \leq \left(1 + \langle g^{-1}\Xi, \Xi \rangle \right)^{-2n-1} \left(1 + \langle g^{-1}\Xi, \Xi \rangle^{1/2} \right)^{2n+1} \|a\|_{g,U}^{(2n+1,2n+1)}$$
$$\times \int \left(1 + g^\sigma(X - U) \right)^{-n-\frac{1}{2}} dX.$$

With X_0 standing for the center of U, we have, with $X' \in U$,

$$1 + g(X - X_0) \leq 1 + 2g(X - X') + 2g(X' - X_0) \leq 3 + 2g^\sigma(X - X')$$

and consequently $1 + g(X - X_0) \leq 3\left(1 + g^\sigma(X - U) \right)$, so that

$$|\hat{a}(-\Xi)|$$
$$\leq \left(1 + \langle g^{-1}\Xi, \Xi \rangle \right)^{-n-\frac{1}{2}} 2^{n+\frac{1}{2}} \|a\|_{g,U}^{(2n+1,2n+1)} \int \left(1 + g(X - X_0) \right)^{-n-\frac{1}{2}} dX 3^{n+\frac{1}{2}}$$
$$\leq \left(1 + \langle g^{-1}\Xi, \Xi \rangle \right)^{-n-\frac{1}{2}} \|a\|_{g,U}^{(2n+1,2n+1)} |g|^{-1/2} c(n).$$

As a consequence, the L^1-norm of $\hat{a}$ is bounded above by the rhs of (2.5.1), and using the estimate (2.1.13) we get the result of the lemma. $\qquad\square$

Proof of Theorem 2.5.1. It is now a simple matter to obtain the L^2-boundedness of Theorem 2.5.1. We write, using the partition of unity given by Theorem 2.2.7,

$$a^w = \int a_Y^w |g_Y|^{1/2} dY$$

and we note that from the previous lemma, $\sup_Y \|a_Y^w\|_{\mathcal{L}(L^2)} < \infty$. Moreover, the estimates (2.3.4), (2.2.29) imply that, for all $l, k = k_1 + k_2$,

$$|(\bar{a}_Y \sharp a_Z)^{(l)} T^l|$$

$$\leq A_{k,l}(g_Y + g_Z)(T)^{l/2} \left(1 + (g_Y^\sigma \wedge g_Z^\sigma)(X - U_Y) + (g_Y^\sigma \wedge g_Z^\sigma)(X - U_Z)\right)^{-k/2}$$

$$\leq A_{k,l} g_Y(T)^{l/2} \delta_r(Y,Z)^{N_0 l} \delta_r(Y,Z)^{-k_1} \left(1 + g_Y^\sigma(X - U_Y)\right)^{-k_2/2} \delta_r(Y,Z)^{N_0 k_2}$$

$$\leq A_{k,l} g_Y(T)^{l/2} \left(1 + g_Y^\sigma(X - U_Y)\right)^{-k_2/2} \delta_r(Y,Z)^{N_0 k_2 + N_0 l - k_1},$$

so that, for l, N, M given, one can choose

$$k_2 = N, \quad k_1 - N_0 N - N_0 l \geq M$$

and get that $\bar{a}_Y \sharp a_Z$ is g_Y-confined in U_Y, so that with the previous lemma, we obtain

$$\max(\|\bar{a}_Y^w a_Z^w\|_{\mathcal{L}(L^2)}^{1/2}, \|a_Y^w \bar{a}_Z^w\|_{\mathcal{L}(L^2)}^{1/2}) \leq C_M \delta(Y,Z)^{-M} \qquad (2.5.2)$$

for all M. To conclude, we note that (2.5.2) implies that the assumptions of (Cotlar's) Lemma 4.7.1 are fulfilled. The proof of Theorem 2.5.1 is complete. $\qquad \square$

The following lemma will be useful later on.

Lemma 2.5.3. *Let g be an admissible metric on $\mathbb{R}^{2n}$ and $\int \varphi_Y |g_Y|^{1/2} dY = 1$ be a partition of unity related to g (cf. Theorem 2.2.7). There exists a positive constant C such that for all $u \in L^2(\mathbb{R}^n)$,*

$$C^{-1} \|u\|_{L^2(\mathbb{R}^n)}^2 \leq \int \|\varphi_Y^w u\|_{L^2(\mathbb{R}^n)}^2 |g_Y|^{1/2} dY \leq C\|u\|_{L^2(\mathbb{R}^n)}^2. \qquad (2.5.3)$$

Proof of the lemma. The right inequality is indeed a direct consequence of Cotlar's Lemma 4.7.1 so we leave it to the reader and provide a proof of the more involved left inequality. The confinement estimates of Corollary 2.3.3 imply that for all $N \geq 1$, there exists C_N such that

$$\|\varphi_Y^w \varphi_Z^w\|_{\mathcal{L}(L^2)} \leq C_N \delta_r(Y,Z)^{-N}.$$

Now since the φ_Y are real-valued and a partition of unity, we obtain with $L^2(\mathbb{R}^n)$ norms and dot products

$$\|u\|^2 = \iint \langle \varphi_Y^w u, \varphi_Z^w u \rangle |g_Y|^{1/2} |g_Z|^{1/2} dY \, dZ$$

$$\leq \iint_{\delta_r(Y,Z) \leq \alpha} \left(\frac{\alpha}{\delta_r(Y,Z)}\right)^{N_0} \|\varphi_Y^w u\| \|\varphi_Z^w u\| |g_Y|^{1/2} |g_Z|^{1/2} dY \, dZ$$

$$+ \underbrace{\iint_{\delta_r(Y,Z) > \alpha} \langle \varphi_Y^w \varphi_Z^w u, u \rangle |g_Y|^{1/2} |g_Z|^{1/2} dY \, dZ,}_{\langle Ru, u \rangle} \qquad (2.5.4)$$

where α is a positive parameter to be chosen later. We check now the selfadjoint operator R defined above: from Cotlar's lemma, we have

$$\|R\|_{\mathcal{L}(L^2)} \le \sup_{\substack{Y_0, Z_0 \\ \text{with } \delta_r(Y_0, Z_0) > \alpha}} \left[\iint_{\delta_r(Y,Z) > \alpha} \|\varphi_{Y_o}^w \varphi_{Z_o}^w \varphi_Y^w \varphi_Z^w\|_{\mathcal{L}(L^2)}^{1/2} |g_Y|^{1/2} |g_Z|^{1/2} dY dZ \right],$$

and since

$$\|\varphi_{Y_o}^w \varphi_{Z_o}^w \varphi_Y^w \varphi_Z^w\|_{\mathcal{L}(L^2)}^2 \le \|\varphi_{Y_o}^w \varphi_{Z_o}^w\| \; \|\varphi_Y^w \varphi_Z^w\| \; \|\varphi_{Y_o}^w\| \; \|\varphi_{Z_o}^w \varphi_Y^w\| \; \|\varphi_Z^w\|$$

$$\le C_N \delta_r(Z_o, Y_o)^{-N} C_N \delta_r(Z, Y)^{-N} C_N \delta_r(Z_o, Y)^{-N} \sup_Z \|\varphi_Z^w\|^2,$$

we get that for $N/4 = N_1$, with N_1 such that (2.2.29) holds,

$$\|R\|_{\mathcal{L}(L^2)} \le C_N^{3/4} \sup_Y \|\varphi_Y^w\|^{1/2}$$

$$\times \sup_{\substack{Y_o, Z_o \\ \text{with } \delta_r(Y_o, Z_o) > \alpha}} \delta_r(Y_o, Z_o)^{-N/4} \iint \delta_r(Y, Z)^{-N/4} \delta_r(Z_o, Y)^{-N/4} |g_Y|^{1/2} |g_Z|^{1/2} dY dZ$$

$$\le C_N^{3/4} \sup_Y \|\varphi_Y^w\|^{1/2} \alpha^{-N/4} \int \delta_r(Z_o, Y)^{-N/4} \int \delta_r(Z, Y)^{-N/4} |g_Z|^{1/2} dZ |g_Y|^{1/2} dY$$

$$\le C_{4N_1}^{3/4} \sup_Y \|\varphi_Y^w\|^{1/2} \; \alpha^{-N_1} C_0^2. \tag{2.5.5}$$

We choose now the parameter α so that $C_{4N_1}^{3/4} \sup_Y \|\varphi_Y^w\|^{1/2} \alpha^{-N_1} C_0^2 = 1/2$ and we get from (2.5.4) and (2.5.5) that

$$\|u\|^2 \le \iint_{\delta_r(Y,Z) \le \alpha} \left(\frac{\alpha}{\delta_r(Y, Z)} \right)^{N_1} \|\varphi_Y^w u\| \|\varphi_Z^w u\| |g_Y|^{1/2} |g_Z|^{1/2} dZ dY + \frac{1}{2} \|u\|^2.$$

The kernel $\delta_r(Y, Z)^{-N_1}$ is of Schur type (Remark 4.7.3) on $L^2(\mathbb{R}^{2n}, |g_Z|^{1/2} dZ)$ from (2.2.29) and thus we obtain

$$\|u\|^2 \le 2\alpha^{N_1} C_0 \int \|\varphi_Y^w u\|^2 |g_Y|^{1/2} dY,$$

which is the statement of the lemma whose proof is now complete. $\qquad\square$

2.5.2 The Gårding inequality with gain of one derivative

We want to prove in this section a generalization of Theorems 1.1.26, 1.1.39; it says that a non-negative symbol of order 1, related to an admissible metric, is quantized by an operator which is semi-bounded from below. According to Remark 2.2.17, to be of order 1 for a symbol a means that $a \in S(\lambda_g, g)$. The main point in this generalization is that the non-negativity for the operator as a consequence of the non-negativity of its symbol holds true as well for any admissible metric g.

However, we want also to deal with systems, including infinite-dimensional systems and prove our inequality in that framework. So far we have dealt only with scalar-valued (complex-valued) symbols ; let us first consider a symbol a defined on $\mathbb{R}^{2n}$ but valued in the algebra of $N \times N$ matrices. It means simply that $a = (a_{jk})_{1 \leq j,k \leq N}$ where each a_{jk} belongs to $S(m, g)$ for some g-admissible weight m. In Remark 2.1.8, we have already considered that case, including the infinite-dimensional case where the $N \times N$ matrices are replaced by the algebra $\mathscr{B}(\mathbb{H})$ of bounded operator on some Hilbert space. Although many results can be extended without much change to this "matrix-valued" case, it is very important to keep in mind that $\mathscr{B}(\mathbb{H})$ is not commutative when $\dim \mathbb{H} > 1$ and that the composition formula and the Poisson bracket should be given the proper definition, taking into account the position of the various terms. Anyhow, we shall skip checking all the details of that calculus for systems of pseudo-differential operators and take advantage of the very simple proof using the Wick calculus to extend the result to that case.

Theorem 2.5.4. *Let g be an admissible metric on $\mathbb{R}^{2n}$, $\mathbb{H}$ be a Hilbert space, a be a symbol in $S(\lambda_g, g)$ valued in the non-negative symmetric bounded operators on $\mathbb{H}$. Then the operator a^w is semi-bounded from below, and more precisely, there exists $l \in \mathbb{N}$ and C depending only on n such that*

$$\forall u \in \mathscr{S}(\mathbb{R}^n; \mathbb{H}), \quad \langle a^w u, u \rangle + C \|a\|_{S(\lambda_g, g)}^{(l)} \|u\|_{L^2(\mathbb{R}^n; \mathbb{H})}^2 \geq 0. \tag{2.5.6}$$

Under the same hypothesis, the same result is true with $\langle a^w u, u \rangle$ replaced by $\operatorname{Re}\langle a(x, D)u, u \rangle$.

Proof. Using Theorem 2.2.7, one can find a family $(\varphi_Y)_{Y \in \mathbb{R}^{2n}}$ of functions uniformly in $S(1, g)$ supported in $U_{Y,r}$, non-negative, such that $\int \varphi_Y |g_Y|^{1/2} dY = 1$. With $(\psi_Y)_{Y \in \mathbb{R}^{2n}}$ uniformly in $S(1, g)$ and real-valued, supported in $U_{Y,2r}$, equal to 1 on $U_{Y,r}$, we have

$$\psi_Y \sharp \varphi_Y a \sharp \psi_Y = \varphi_Y a + r_Y, \tag{2.5.7}$$

and Proposition 2.3.16 implies that $(r_Y)_{Y \in \mathbb{R}^{2n}}$ is a uniformly confined family of symbols, so that

$$a^w \equiv \int_{\mathbb{R}^{2n}} \psi_Y^w (\varphi_Y a)^w \psi_Y^w |g_Y|^{1/2} dY, \qquad \mod \mathscr{L}(L^2(\mathbb{R}^n)). \tag{2.5.8}$$

The symbol $\varphi_Y a$ belongs uniformly to $S(\lambda_g(Y), g_Y) \subset S(\lambda_g(Y), \lambda_g(Y)^{-1} g_Y^\sharp)$, thanks to (2.2.22) and $g_Y^\sharp = (g_Y^\sharp)^\sigma$. Using a linear symplectic mapping and Segal's formula of Theorem 2.1.2, we get that $(\varphi_Y a)^w$ is unitary equivalent to some α^w with $0 \leq \alpha \in S(\mu, \mu^{-1}|dX|^2)$ with semi-norms bounded above independently of Y and $\mu = \lambda_g(Y)$. Proposition 2.4.3(1)(2) imply that $\alpha^w + C \geq 0$, where C is a semi-norm of α and thus of a, so that $(a\varphi_Y)^w + C \geq 0$. Plugging this in (2.5.8), we get the result since

$$\int \psi_Y^w \psi_Y^w |g_Y|^{1/2} dY \in \mathscr{L}(L^2(\mathbb{R}^n)), \tag{2.5.9}$$

thanks to Cotlar's lemma. $\qquad\square$

Comment. The reader may think that we did not pay much attention to the fact that the symbol was valued in $\mathscr{B}(\mathbb{H})$; in fact, since the ψ_Y, χ_Y are scalar-valued, the formulas (2.5.7), (2.5.8) hold without change (except that $L^2(\mathbb{R}^n)$ becomes $L^2(\mathbb{R}^n; \mathbb{H})$) and it is a simple matter to check that the $\mathscr{B}(\mathbb{H})$-valued version of Proposition 2.4.3 holds true, with the non-negativity condition $a(X) \geq 0$ meaning $a(X)$ is a non-negative symmetric bounded operator in $\mathbb{H}$.

2.5.3 The Fefferman-Phong inequality

First versions

Let us consider a classical second-order symbol $a(x,\xi)$, i.e., a smooth function defined on $\mathbb{R}^n \times \mathbb{R}^n$ such that, for all multi-indices α, β,

$$|(\partial_\xi^\alpha \partial_x^\beta a)(x,\xi)| \leq C_{\alpha\beta}(1 + |\xi|)^{2-|\alpha|}. \tag{2.5.10}$$

The Fefferman-Phong inequality states that, if a satisfies (2.5.10) and is a non-negative function, there exists C such that, for all $u \in \mathcal{S}(\mathbb{R}^n)$,

$$\mathrm{Re}\langle a(x,D)u, u\rangle_{L^2(\mathbb{R}^n)} + C\|u\|_{L^2(\mathbb{R}^n)}^2 \geq 0, \tag{2.5.11}$$

or equivalently (with an a priori different constant C),

$$a^w + C \geq 0. \tag{2.5.12}$$

The constant C in (2.5.11), (2.5.12) depends only on a finite number of $C_{\alpha\beta}$ in (2.5.10). More generally, we shall prove the following extension of Theorem 2.5.4, an extension which works only in the scalar case (i.e., for scalar-valued symbols).

Theorem 2.5.5. *Let g be an admissible metric on $\mathbb{R}^{2n}$, a be a non-negative symbol in $S(\lambda_g^2, g)$. Then the operator a^w is semi-bounded from below, and more precisely, there exists $l \in \mathbb{N}$ and C depending only on n such that*

$$\forall u \in \mathscr{S}(\mathbb{R}^n), \quad \langle a^w u, u\rangle + C\|a\|_{S(\lambda_g^2,g)}^{(l)} \|u\|_{L^2(\mathbb{R}^n)}^2 \geq 0. \tag{2.5.13}$$

Under the same hypothesis, the same result is true with $\langle a^w u, u\rangle$ replaced by $\mathrm{Re}\langle a(x,D)u, u\rangle$.

Reduction to the constant-metric case. This theorem is a consequence of the same statement where g is replaced by a constant metric, i.e., a positive-definite quadratic form on $\mathbb{R}^{2n}$ such that $g \leq g^\sigma$. Using the partition of unity $(\varphi_Y)_{Y \in \mathbb{R}^{2n}}$ given by Theorem 2.2.7, of functions uniformly in $S(1, g)$ supported in $U_{Y,r}$, non-negative, such that $\int \varphi_Y |g_Y|^{1/2} dY = 1$. With $(\psi_Y)_{Y \in \mathbb{R}^{2n}}$ uniformly in $S(1, g)$, supported in $U_{Y,2r}$, equal to 1 on $U_{Y,r}$, we have

$$\psi_Y \sharp \varphi_Y a \sharp \psi_Y = \varphi_Y a + r_Y, \tag{2.5.14}$$

with $(r_Y)_{Y \in \mathbb{R}^{2n}}$ a uniformly confined family of symbols (see Proposition 2.3.16), implying that a^w is equal modulo $\mathscr{L}(L^2(\mathbb{R}^n))$ to $\int \psi_Y^w (\varphi_Y a)^w \psi_Y^w |g_Y|^{1/2} dY$. Now the symbol $\varphi_Y a$ is non-negative and uniformly in $S(\lambda_g(Y)^2, g_Y)$ and the result for a constant metric g_Y will give the result, applying (2.5.9). *We are left with the proof of Theorem 2.5.5 in the case g is a constant metric.* $\square$

The papers [25], [118] contain some counterexamples to the generalization of the above statement to systems, and we shall see that the proof relies heavily on a Calderón-Zygmund decomposition for a non-negative function. We are also concerned with the number of derivatives necessary to handle (2.5.13), i.e., to give an explicit bound from above for the number l occurring there. We have chosen to prove a more precise result involving an algebra of pseudo-differential operators introduced by J.Sjöstrand in [130] and [131]. To formulate our result, we need first to introduce that algebra.

Sjöstrand algebra of pseudo-differential operators

In [130] and [131], J. Sjöstrand introduced a Wiener-type algebra of pseudo-differential operators as follows. Let $\mathbb{Z}^{2n}$ be the standard lattice in $\mathbb{R}_X^{2n}$ and let $1 = \sum_{j \in \mathbb{Z}^{2n}} \chi_0(X - j), \chi_0 \in C_c^\infty(\mathbb{R}^{2n})$, be a partition of unity. We note $\chi_j(X) = \chi_0(X - j)$.

Proposition 2.5.6. *Let a be a tempered distribution on $\mathbb{R}^{2n}$. We shall say that a belongs to the class $\mathcal{A}$ if $\omega_a \in L^1(\mathbb{R}^{2n})$, with $\omega_a(\Xi) = \sup_{j \in \mathbb{Z}^{2n}} |\mathcal{F}(\chi_j a)(\Xi)|$, where $\mathcal{F}$ is the Fourier transform. Moreover, we have*

$$S_{0,0}^0 \subset S_{0,0;2n+1}^0 \subset \mathcal{A} \subset C^0(\mathbb{R}^{2n}) \cap L^\infty(\mathbb{R}^{2n}), \tag{2.5.15}$$

where $S_{0,0;2n+1}$ is the set of functions defined on $\mathbb{R}^{2n}$ such that $|(\partial_\xi^\alpha \partial_x^\beta a)(x, \xi)| \leq C_{\alpha\beta}$ for $|\alpha| + |\beta| \leq 2n+1$. $\mathcal{A}$ is a Banach algebra for multiplication with the norm $\|a\|_{\mathcal{A}} = \|\omega_a\|_{L^1(\mathbb{R}^{2n})}$.

Proof. In fact, we have the implications $a \in \mathcal{A} \implies \mathcal{F}(\chi_j a) \in L^1(\mathbb{R}^{2n}) \implies \chi_j a \in C^0 \cap L^\infty$, and, since the sum is locally finite with a fixed overlap[12], we get $a \in C^0 \cap L^\infty$. Moreover, if $a \in S_{0,0;2n+1}^0$, i.e., is bounded as well as all its derivatives

---

[12]If $\cap_{j \in J} \operatorname{supp} \chi_j \neq \emptyset$ then $\operatorname{card} J \leq N_0$, where N_0 depends only on the compact set $\operatorname{supp} \chi_0$.

of order $\leq 2n + 1$, we have, with $P(\Xi) = (1 + \|\Xi\|^2)^n$ the formula $\mathcal{F}(\chi_j a)(\Xi) = P(\Xi)^{-1}\mathcal{F}\big(P(D_X)(\chi_j a)\big)$. We get the identity

$$\mathcal{F}(\chi_j a)(\Xi) = P(\Xi)^{-1}(\Xi_1 + i)^{-1}\mathcal{F}\big((D_{X_1} + i)P(D_X)(\chi_j a)\big).$$

This entails, in the cone $\{\Xi \in \mathbb{R}^{2n}, 2n|\Xi_1| \geq \|\Xi\|\}$ and thus everywhere[13]

$$|\mathcal{F}(\chi_j a)(\Xi)| \leq \underbrace{P(\Xi)^{-1}(1 + \|\Xi\|)^{-1}}_{\in L^1(\mathbb{R}^{2n})} \mathrm{mes}(\mathrm{supp}\,\chi_0) \sup_{0 \leq k \leq 2n+1} \|a^{(l)}\|_{L^\infty} C_n,$$

yielding the result. □

Remark 2.5.7. Since $1 \in \mathcal{A}$, $\mathcal{A}$ is not included in $\mathcal{F}(L^1(\mathbb{R}^{2n}))$. Moreover $\mathcal{A}$ contains $\mathcal{F}(L^1)$: let a be a function in $\mathcal{F}(L^1)$. With the above notation, we have

$$|\mathcal{F}(\chi_j a)(\Xi)| = \left| \int \hat{\chi}_0(\Xi - N)\hat{a}(N)e^{2i\pi j(N-\Xi)}dN \right| \leq \int |\hat{\chi}_0(\Xi - N)||\hat{a}(N)|dN,$$

and thus $\int |\omega_a(\Xi)|d\Xi \leq \|\hat{a}\|_{L^1}\|\widehat{\chi_0}\|_{L^1}$, which gives the inclusion. Moreover, $\mathcal{A}$ is a Banach commutative algebra for multiplication.

Proposition 2.5.8. *The algebra $\mathcal{A}$ is stable by change of quantization, i.e., for all t real, $a \in \mathcal{A} \Longleftrightarrow J^t a = \exp(2i\pi t D_x \cdot D_\xi)a \in \mathcal{A}$. The bilinear map $a_1, a_2 \mapsto a_1 \natural a_2$ is defined on $\mathcal{A} \times \mathcal{A}$ and continuous-valued in $\mathcal{A}$, which is a (non-commutative) Banach algebra for $\natural$. The maps $a \mapsto a^w, a(x, D)$ are continuous from $\mathcal{A}$ to $\mathcal{L}(L^2(\mathbb{R}^n))$.*

Proof. The proof is given in [130]. A. Boulkhemair established a lot more results on this algebra in his paper [21]. □

Remark 2.5.9. The standard Wiener's Lemma states that if $a \in \ell^1(\mathbb{Z}^d)$ is such that $u \mapsto a * u = C_a u$ is invertible as an operator on $\ell^2(\mathbb{Z}^d)$, then the inverse operator is of the form C_b for some $b \in \ell^1(\mathbb{Z}^d)$. In [131] the author proves several types of Wiener lemma for $\mathcal{A}$. First a commutative version, saying that if $a \in \mathcal{A}$ and $1/a$ is a bounded function, then $1/a$ belongs to $\mathcal{A}$. Next, Theorem 4.1 of [131] provides a non-commutative version of the Wiener lemma for the algebra $\mathcal{A}$: if an operator a^w with $a \in \mathcal{A}$ is invertible as a continuous operator on L^2, then the inverse operator is b^w with $b \in \mathcal{A}$. In the paper [51], K. Gröchenig and M. Leinert prove several versions of the non-commutative Wiener lemma, and their definition of the twisted convolution ((1.1) in [51]) is indeed very close to (a discrete version of) the composition formula (1.2.2) above.

Also J. Sjöstrand proved in Proposition 5.1 of [131] the standard Gårding inequality with gain of one derivative for his class, in the semi-classical setting, where h is a small parameter in $(0, 1]$:

$$a \geq 0, a'' \in \mathcal{A} \Longrightarrow a(x, h\xi)^w + Ch \geq 0. \tag{2.5.16}$$

---

[13] $\mathbb{R}^{2n} = \cup_{1 \leq k \leq 2n}\{\Xi \in \mathbb{R}^{2n}, 2n|\Xi_k| \geq \|\Xi\|\}$ since the complement of that union is empty: it is not possible to find Ξ so that $\max_{1 \leq k \leq 2n} 2n|\Xi_k| < \|\Xi\| \leq 2n \max_{1 \leq k \leq 2n} |\Xi_k|$.

A consequence of the result of [21] is that[14]

$$a \geq 0, a^{(4)} \in S_{0,0}^0 \implies a(x, h\xi)^w + Ch^2 \geq 0. \tag{2.5.17}$$

We shall improve that statement, keeping the conclusion and weakening the hypothesis:

$$a \geq 0, a^{(4)} \in \mathcal{A} \implies a(x, h\xi)^w + Ch^2 \geq 0. \tag{2.5.18}$$

From the first two inclusions in (2.5.15), we see that (2.5.18) implies (2.5.17). Moreover the constant C in (2.5.18) will depend only on the dimension and on the norm of $a^{(4)}$ in $\mathcal{A}$, which is much more precise than the dependence of C in (2.5.17), which depends on a finite number of semi-norms of a in $S_{0,0}^0$. Although (2.5.18) looks stronger than (2.5.16) since $h^2 \ll h$, it is not obvious to actually *deduce* (2.5.16) from (2.5.18). Anyhow we shall see that they are both true and that the proof of (2.5.16) is an immediate consequence of the most elementary properties of the so-called Wick quantization exposed in our Section 2.4.1. Note also that a version of the Hörmander-Melin inequality with gain of 6/5 of derivatives was given, in the semi-classical setting, by F. Hérau ([61]): this author used the assumption (6.4) of Theorem 6.2 of [73], but with a limited regularity on the symbol a, which is only such that $a^{(3)} \in \mathcal{A}$.

Fefferman-Phong inequalities

Theorem 2.5.10. *Let n be a positive integer. There exists a constant C_n such that, for all non-negative functions a defined on $\mathbb{R}^{2n}$ satisfying $a^{(4)} \in \mathcal{A}$, the operator a^w is semi-bounded from below and, more precisely, satisfies*

$$a^w + C_n \|a^{(4)}\|_{\mathcal{A}} \geq 0. \tag{2.5.19}$$

The Banach algebra $\mathcal{A}$ is defined in Proposition 2.5.6. Note that the constant C_n depends only on the dimension n.

Proof. The proof is given in the subsection starting on page 122. □

Corollary 2.5.11. *Let n be a positive integer.*

(i) *Let $a(x, \xi)$ be a non-negative function defined on $\mathbb{R}^n \times \mathbb{R}^n$ such that (2.5.10) is satisfied for $|\alpha| + |\beta| \leq 2n + 5$. Then (2.5.11) and (2.5.12) hold with a constant C depending only on n and on $\max_{|\alpha|+|\beta| \leq 2n+5} C_{\alpha\beta}$.*

(ii) *Let $a(x, \xi, h)$ be a non-negative function defined on $\mathbb{R}^n \times \mathbb{R}^n \times (0, 1]$ such that*

$$|(\partial_\xi^\alpha \partial_x^\beta a)(x, \xi, h)| \leq h^{|\alpha|} C_{\alpha\beta}, \quad \text{for } 4 \leq |\alpha| + |\beta| \leq 2n + 5.$$

Then $a^w + Ch^2 \geq 0$ and $\operatorname{Re} a(x, D) + Ch^2 \geq 0$ hold with a constant C depending only on n and on $\max_{4 \leq |\alpha|+|\beta| \leq 2n+5} C_{\alpha\beta}$.

---

[14]In fact the operator $h^{-2} a(x, h\xi)^w$ is unitarily equivalent to $h^{-2} a(h^{1/2}x, h^{1/2}\xi)^w$ and the function $b(x, \xi) = h^{-2} a(h^{1/2}x, h^{1/2}\xi)$ is non-negative and satisfies $b^{(4)}(x, \xi) = a^{(4)}(h^{1/2}x, h^{1/2}\xi)$ which is uniformly in $S_{0,0}^0$ whenever h is bounded and $a^{(4)} \in S_{0,0}^0$.

(iii) Let $a(x, \xi)$ be a non-negative function defined on $\mathbb{R}^n \times \mathbb{R}^n$ such that $a^{(4)}$ belongs to $\mathcal{A}$. Then $a(x, h\xi)^w + C \|a^{(4)}\|_{\mathcal{A}} h^2 \geq 0$ and $\operatorname{Re} a(x, hD) + C \|a^{(4)}\|_{\mathcal{A}} h^2 \geq 0$ hold with a constant C depending only on n.

(iv) Let $a(x, \xi, h)$ be a non-negative function defined on $\mathbb{R}^n \times \mathbb{R}^n \times (0, 1]$ such that, for $|\alpha| + |\beta| = 4$, the functions $(x, \xi) \mapsto (\partial_1^\beta \partial_2^\alpha a)(xh^{1/2}, \xi h^{-1/2}, h) h^{-|\alpha|}$ belong to $\mathcal{A}$ with a norm bounded above by ν_0 for all $h \in (0, 1]$. Then $a^w + C\nu_0 h^2 \geq 0$ and $\operatorname{Re} a(x, D) + C\nu_0 h^2 \geq 0$ hold with a constant C depending only on n.

Proof. That corollary is proven in the subsection starting on page 129. $\qquad\square$

Remark 2.5.12. The recent paper [23] by A. Boulkhemair cuts significantly the requirement on the number of derivatives down to $n + 4 + \varepsilon$.

Remark 2.5.13. Theorem 2.5.10 implies Theorem 2.5.5 since it implies the constant metric particular case of the latter statement. In fact repeating the argument before (2.5.9), we see that for a non-negative in $S(\lambda_g^2, g)$ where g is a constant admissible metric, i.e., a positive-definite quadratic form such that $g \leq g^\sigma$, the operator a^w is unitarily equivalent to α^w where α is non-negative in $S(\mu^2, \mu^{-1}|dX|^2)$ with the same semi-norms and $\mu \geq 1$; in particular we have that $\alpha^{(4)}$ belongs to $S_{0,0}^0$ which is a subset of $\mathcal{A}$. Applying Theorem 2.5.10 gives readily the constant-metric case.

Remark 2.5.14. We have paid attention to counting precisely the number of derivatives only for the standard metric $|dx|^2 + \langle \xi \rangle^{-2} |d\xi|^2$ and not for any admissible metric. However, we do not have any necessary condition supporting the sharpness of the requirements of the best results of [23]. It is certainly possible to obtain a rather explicit estimate of l in Theorem 2.5.5, but the reader will see in the proof that to obtain that the operator $\int r_Y^w |g_Y|^{1/2} dY$ is a bounded operator in $L^2(\mathbb{R}^n)$ with r_Y defined by (2.5.14), requires some effort, even in the standard case. In the case of a general admissible metric, it is quite likely that Corollary 2.5.11(i) can be generalized, and maybe improved as in [23], but we have not pursued these improvements. The number of derivatives required to have the Fefferman-Phong inequality for a general admissible metric will be of course larger than for the constant metric (say $4+n+\epsilon$ as in [23]), but that number seems also to depend on the temperance properties of the metric, namely on the number N_1 occurring in (2.2.29). The notions introduced in our Definition 4.9.10 may play a rôle in formulating a statement on this topic.

The following improvement of the Gårding inequality with gain of one derivative is due to J. Sjöstrand in [131].

Theorem 2.5.15. *Let a be a non-negative function defined on $\mathbb{R}^{2n}$ such that the second derivatives a'' belong to $\mathcal{A}$. Then we have*

$$a^w + C_n \|a''\|_{\mathcal{A}} \geq 0. \tag{2.5.20}$$

Proof. Although a proof of this result is given in [131] (*Proposition* 5.1), it is a nice and simple introduction to our more complicated argument of the sequel. From

Proposition 2.4.3, we have

$$a^w = a^{\text{Wick}} - r(a)^w \geq -r(a)^w,$$

with $r(a)(X) = \int_0^1 \int_{\mathbb{R}^{2n}} (1-\theta) a''(X + \theta Y) Y^2 e^{-2\pi|Y|^2} 2^n dY d\theta$. Since $\mathcal{A}$ is stable by translation (see Lemma 4.8.1), we see that $r(a) \in \mathcal{A}$ and thus $r(a)^w$ is bounded on $L^2(\mathbb{R}^n)$ from Proposition 2.5.8.   □

Remark 2.5.16. This theorem implies as well the following semi-classical version; let a be function satisfying the assumption of Theorem 2.5.15. For $h \in (0,1]$, we define $A_h(x, \xi) = h^{-1} a(x h^{1/2}, \xi h^{1/2})$. The function A_h is non-negative with a second derivative bounded in $\mathcal{A}$ by cst $\times \|a''\|_{\mathcal{A}}$ (see Lemma 4.8.1), so that the previous theorem implies, with C depending only on the dimension, that $A_h^w + C\|a''\|_{\mathcal{A}} \geq 0$. Since A_h^w is unitarily equivalent to $h^{-1} a(x, h\xi)^w$, this gives

$$a(x, h\xi)^w + hC\|a''\|_{\mathcal{A}} \geq 0. \tag{2.5.21}$$

Proof of Theorem 2.5.10, Step 1: Sharp estimates for the remainders

Property (2.4.6) falls short of providing a proof for the Fefferman-Phong inequality, which gains two derivatives. However, we shall be able to use some of the properties of that quantization to handle an improved version of the Fefferman-Phong inequality.

Lemma 2.5.17. *Let a be a function defined on $\mathbb{R}^{2n}$ such that the fourth derivatives $a^{(4)}$ belong to $\mathcal{A}$. Then we have*

$$a^w = \left(a - \frac{1}{8\pi} \text{trace } a''\right)^{\text{Wick}} + \rho_0(a^{(4)})^w,$$

with $\rho_0(a^{(4)}) \in \mathcal{A}$ and more precisely $\|\rho_0(a^{(4)})\|_{\mathcal{A}} \leq C_n \|a^{(4)}\|_{\mathcal{A}}$.

Proof. The Weyl symbol σ_a of a^{Wick} is

$$\sigma_a(X) = \int a(X + Y) 2^n e^{-2\pi|Y|^2} dY$$

$$= a(X) + \int \frac{1}{2} a''(X) Y^2 2^n e^{-2\pi|Y|^2} dY$$

$$\quad + \frac{1}{3!} \iint_0^1 (1-\theta)^3 a^{(4)}(X + \theta Y) Y^4 2^n e^{-2\pi|Y|^2} dY d\theta$$

$$= a(X) + \frac{1}{8\pi} \text{trace } a''(X)$$

$$\quad + \frac{1}{3!} \iint_0^1 (1-\theta)^3 a^{(4)}(X + \theta Y) Y^4 2^n e^{-2\pi|Y|^2} dY d\theta.$$

Moreover the Weyl symbol θ_a of $(\text{trace } a'')^{\text{Wick}}$ is, from Proposition 2.4.3,

$$\theta_a(X) = \operatorname{trace} a''(X) + \int_0^1 \!\!\! \int_{\mathbb{R}^{2n}} (1-\theta)(\operatorname{trace} a'')''(X+\theta Y)Y^2 e^{-2\pi|Y|^2} 2^n dY d\theta.$$

As a result, the Weyl symbol of the operator $\left(a - \frac{1}{8\pi}\operatorname{trace} a''\right)^{\mathrm{Wick}}$ is

$$a + \frac{1}{3!}\int\!\!\!\int_0^1 (1-\theta)^3 a^{(4)}(X+\theta Y)Y^4 2^n e^{-2\pi|Y|^2} dY d\theta$$

$$- \frac{1}{8\pi}\int_0^1 \!\!\! \int_{\mathbb{R}^{2n}} (1-\theta)(\operatorname{trace} a'')''(X+\theta Y)Y^2 e^{-2\pi|Y|^2} 2^n dY d\theta.$$

We get the equality in the lemma with

$$\rho_0(a^{(4)})(X) = \frac{1}{8\pi}\int_0^1 \!\!\! \int_{\mathbb{R}^{2n}} (1-\theta)(\operatorname{trace} a'')''(X+\theta Y)Y^2 e^{-2\pi|Y|^2} 2^n dY d\theta$$

$$- \frac{1}{3!}\int\!\!\!\int_0^1 (1-\theta)^3 a^{(4)}(X+\theta Y)Y^4 2^n e^{-2\pi|Y|^2} dY d\theta. \quad (2.5.22)$$

We note now that ρ_0 depends linearly on $a^{(4)}$ and that

$$\rho_0(a^{(4)})(X) = \int\!\!\!\int_0^1 a^{(4)}(X+\theta Y)\underbrace{M(\theta,Y)}_{\substack{\text{polynomial} \\ \text{in } Y, \theta.}} e^{-2\pi|Y|^2} dY d\theta. \quad (2.5.23)$$

Looking now at the formula (2.5.23) and applying Lemma 4.8.1, we get

$$\|\rho_0(a^{(4)})\|_{\mathcal{A}} \le \int\!\!\!\int_0^1 M(\theta,Y)e^{-2\pi|Y|^2} dY d\theta C_0 \|a^{(4)}\|_{\mathcal{A}} = C_1 \|a^{(4)}\|_{\mathcal{A}}.$$

The proof of Lemma 2.5.17 is complete. $\qquad\qquad\qquad\qquad\qquad\square$

Remark 2.5.18. We note that, from Lemma 2.5.17 and the L^2-boundedness of operators with symbols in $\mathcal{A}$, Theorem 2.5.19 is reduced to proving

$$a \ge 0, a^{(4)} \in \mathcal{A} \Longrightarrow \left(a - \frac{1}{8\pi}\operatorname{trace} a''\right)^{\mathrm{Wick}} \text{is semi-bounded from below.} \quad (2.5.24)$$

Naturally, one should not expect the quantity $a - \frac{1}{8\pi}\operatorname{trace} a''$ to be non-negative: this quantity will take negative values even in the simplest case $a(x,\xi) = x^2 + \xi^2$, so that the positivity of the quantization expressed by (2.4.6) is far from enough to get our result. We shall prove below a stronger version of (2.5.24). In particular, one may certainly weaken the assumption to require only the implicit-looking $(a^{(4)})^w \in \mathscr{L}(L^2(\mathbb{R}^n))$.

The proof, Step 2: non-negative $C^{3,1}$ functions are sums of squares of $C^{1,1}$ functions

The key element in the proof is the following result on the decomposition of a $C^{3,1}$ non-negative function as a sum of squares of functions in $C^{1,1}$ with controlled bounds .

Theorem 2.5.19. *Let m be a non-negative integer. There exists an integer N and a positive constant C such that the following property holds. Let a be a non-negative $C^{3,1}$ function[15] defined on $\mathbb{R}^m$ such that $a^{(4)} \in L^\infty$; then we can write*

$$a = \sum_{1 \le j \le N} b_j^2, \tag{2.5.25}$$

where the b_j are $C^{1,1}$ functions such that b_j'', $(b_j' b_j'')'$, $(b_j b_j'')'' \in L^\infty$. More precisely, we have

$$\|b_j''\|_{L^\infty}^2 + \|(b_j' b_j'')'\|_{L^\infty} + \|(b_j b_j'')''\|_{L^\infty} \le C \|a^{(4)}\|_{L^\infty}. \tag{2.5.26}$$

Note that this implies that each function b_j is such that b_j^2 is $C^{3,1}$ and that N and C depend only on the dimension m.

Comment. Part of this theorem is a consequence of the classical proof of the Fefferman-Phong inequality ([44]) and of the more refined analysis of J.-M. Bony in [16] (see also [53] and [133]). However the control of the L^∞ norm of the quantities $(b_j' b_j'')'$, $(b_j b_j'')''$ is more difficult to achieve. Naturally the inequality (2.5.26) is a key element of our proof, since it is connected with the estimates (2.4.22).

Proof of the theorem. We define

$$\rho(x) = \left(|a(x)| + |a''(x)|^2 \right)^{1/4}, \quad \Omega = \{x, \rho(x) > 0\}, \tag{2.5.27}$$

assuming as we may $\|a^{(4)}\|_{L^\infty} \le 1$. Note that, since ρ is continuous, the set Ω is open. The metric $|dx|^2/\rho(x)^2$ is slowly varying in Ω (see Lemma 4.3.13): $\exists r_0 > 0, C_0 \ge 1$ such that

$$x \in \Omega, |y - x| \le r_0 \rho(x) \Longrightarrow y \in \Omega, C_0^{-1} \le \frac{\rho(x)}{\rho(y)} \le C_0. \tag{2.5.28}$$

The constants r_0, C_0 can be chosen as "universal" constants, thanks to the normalization on $a^{(4)}$ above. Moreover, using Lemma 4.3.11, the nonnegativity of a implies with $\gamma_j = 1$ for $j = 0, 2, 4$, $\gamma_1 = 3, \gamma_3 = 4$,

$$|a^{(j)}(x)| \le \gamma_j \rho(x)^{4-j}, \quad 1 \le j \le 4. \tag{2.5.29}$$

Applying Theorem 4.3.7, we get the following lemma.

---

[15]A $C^{3,1}$ function is a C^3 function whose third-order derivatives are Lipschitz-continuous.

Lemma 2.5.20. *Let a, ρ, Ω, r_0 be as above. There exists a positive number $r_0' \leq r_0$, such that for all $r \in]0, r_0']$, there exists a sequence $(x_\nu)_{\nu \in \mathbb{N}}$ of points in Ω and a positive number M_r, such that the following properties are satisfied. We define $U_\nu, U_\nu^*, U_\nu^{**}$ as the closed Euclidean balls with center x_ν and radius $r\rho_\nu, 2r\rho_\nu, 4r\rho_\nu$ with $\rho_\nu = \rho(x_\nu)$. There exist two families of non-negative smooth functions on $\mathbb{R}^m$, $(\varphi_\nu)_{\nu \in \mathbb{N}}$, $(\psi_\nu)_{\nu \in \mathbb{N}}$ such that*

$$\sum_\nu \varphi_\nu^2(x) = 1_\Omega(x), \; \operatorname{supp} \varphi_\nu \subset U_\nu, \quad \psi_\nu \equiv 1 \; on \; U_\nu^*, \; \operatorname{supp} \psi_\nu \subset U_\nu^{**} \subset \Omega.$$

Moreover, for all $l \in \mathbb{N}$, we have

$$\sup_{x \in \Omega, \nu \in \mathbb{N}} \|\varphi_\nu^{(l)}(x)\| \rho_\nu^l + \sup_{x \in \Omega, \nu \in \mathbb{N}} \|\psi_\nu^{(l)}(x)\| \rho_\nu^l < \infty.$$

The overlap of the balls U_ν^{**} is bounded, i.e.,

$$\bigcap_{\nu \in \mathcal{N}} U_\nu^{**} \neq \emptyset \quad \Longrightarrow \quad \#\mathcal{N} \leq M_r.$$

Moreover, $\rho(x) \sim \rho_\nu$ all over U_ν^{**} (i.e., the ratios $\rho(x)/\rho_\nu$ are bounded above and below by a fixed constant, provided that $x \in U_\nu^{**}$).

Since a is vanishing on Ω^c, we obtain

$$a(x) = \sum_{\nu \in \mathbb{N}} a(x) \varphi_\nu^2(x). \tag{2.5.30}$$

Definition 2.5.21. Let a, ρ, Ω be as above. Let θ be a positive number $\leq \theta_0$, where θ_0 is a fixed constant satisfying the requirements of Lemma 4.3.17. A point $x \in \Omega$ is said to be

(i) θ-elliptic whenever $a(x) \geq \theta \rho(x)^4$,

(ii) θ-non-degenerate whenever $a(x) < \theta \rho(x)^4$: we have then $\|a''(x)\|^2 \geq \rho(x)^4/2$.

We go on now with the proof of Theorem 2.5.19. We choose a positive number θ satisfying the condition in Definition 2.5.21. We choose a positive number $r \leq r_0'$ as defined in Lemma 2.5.20 and we consider a sequence (x_ν) as in that lemma. We assume also that $4r \leq \theta/8$, so that Lemma 2.5.20 can be applied on the ball U_ν^{**}.

Let us first consider the "elliptic" indices ν such that x_ν is θ-elliptic. According to Lemma 4.3.15, for $x \in U_\nu^{**}$, we have $a(x) \sim \rho_\nu^4$, so that with

$$b_\nu(x) = a(x)^{1/2} \psi_\nu(x), \quad b_\nu^2 = a\psi_\nu^2, \quad \varphi_\nu^2 b_\nu^2 = a\varphi_\nu^2 \tag{2.5.31}$$

and on $\operatorname{supp} \varphi_\nu$ (where $\psi_\nu \equiv 1$),

$$\begin{cases} b_\nu' = 2^{-1} a^{-1/2} a', \\ b_\nu'' = -2^{-2} a^{-3/2} a'^2 + 2^{-1} a^{-1/2} a'', \\ b_\nu''' = 3 \times 2^{-3} a^{-5/2} a'^3 - \frac{3}{4} a^{-3/2} a' a'' + 2^{-1} a^{-1/2} a''', \\ b_\nu^{(4)} = -\frac{15}{16} a^{-7/2} a'^4 + \frac{9}{4} a^{-5/2} a'^2 a'' - \frac{3}{4} a^{-3/2} a''^2 - a^{-3/2} a' a''' + \frac{1}{2} a^{-1/2} a^{(4)}, \end{cases}$$

yielding

$$
\begin{cases}
|b'_\nu| \le 2^{-1}a^{-1/2}|a'| \lesssim a^{-1/2}\rho^3 \lesssim \rho, \\
|b''_\nu| \lesssim a^{-3/2}\rho^6 + a^{-1/2}\rho^2 \lesssim 1, \\
|b'''_\nu| \lesssim a^{-5/2}\rho^9 + a^{-3/2}\rho^3\rho^2 + a^{-1/2}\rho \lesssim \rho^{-1}, \\
|b^{(4)}_\nu| \lesssim a^{-7/2}\rho^{12} + a^{-5/2}\rho^6\rho^2 + a^{-3/2}\rho^4 + a^{-3/2}\rho^3\rho + a^{-1/2} \lesssim \rho^{-2}.
\end{cases}
$$

Note in particular that

$$
|b_\nu b^{(4)}_\nu| + |b^{(1)}_\nu b^{(3)}_\nu| + |b^{(2)}_\nu b^{(2)}_\nu| \le C(\theta). \tag{2.5.32}
$$

The whole difficulty is concentrated on the next case.

The non-degenerate indices ν are those for which x_ν is θ-non-degenerate. Since $4r \le \theta/8 \le \theta^{1/2}$, we can apply Remark 4.3.18 on the product[16]

$$
Q_\nu = [-\theta^{1/4}\rho_\nu + x_{\nu 1}, \theta^{1/4}\rho_\nu + x_{\nu 1}] \times B_{\mathbb{R}^{m-1}}(x'_\nu, \theta^{1/2}\rho_\nu),
$$

where $x_\nu = (x_{\nu 1}, x'_\nu) \in \mathbb{R} \times \mathbb{R}^{m-1}$. There exists

$$
\alpha : B_{\mathbb{R}^{m-1}}(x'_\nu, \theta^{1/2}\rho_\nu) \to [x_{\nu 1} - \theta^{1/4}\rho_\nu, x_{\nu 1} + \theta^{1/4}\rho_\nu]
$$

such that

$$
\partial_1 a(\alpha(x'), x') = 0 \tag{2.5.33}
$$

and $\partial_1^2 a(x) \ge \rho_\nu^2/2$ for $|x - x_\nu| \le R_0\rho_\nu$ where $R_0 = 10^{-2}$ according to Lemma 4.3.16. We have on Q_ν,

$$
a(x) = a(x_1, x')
$$
$$
= \int_0^1 (1-t)\partial_1^2 a\big(\alpha(x') + t(x_1 - \alpha(x')), x'\big)dt\big(x_1 - \alpha(x')\big)^2 + a(\alpha(x'), x'). \tag{2.5.34}
$$

According to Remark 4.3.18, we recall that we have for $|x' - x'_\nu| \le \theta^{1/2}\rho_\nu$,

$$
\begin{cases}
|\alpha(x') - x_{\nu 1}| \le \theta^{1/4}\rho_\nu, \\
|\alpha'(x')| \quad\;\; \le 2\rho_\nu^{-2}\rho(\alpha(x'), x')^2 \le 2C_0^2 = C_1, \\
|\alpha''(x')| \quad\; \le 2\rho_\nu^{-2}\big(4^2C_0^4 + 4^2C_0^2 + 12\big)\rho(\alpha(x'), x') \le C_2\rho_\nu^{-1}, \\
|\alpha'''(x')| \quad \le C_3\rho_\nu^{-2},
\end{cases} \tag{2.5.35}
$$

with universal constants C_j. Let us now compute the derivatives of the function

$$
B' = B_{\mathbb{R}^{m-1}}(x'_\nu, \theta^{1/2}\rho_\nu) \ni x' \mapsto a(\alpha(x'), x') = c(x'). \tag{2.5.36}
$$

---

[16]Naturally the choice of the linear coordinates depends on the index ν, according to Remark 4.3.18. Note also that $U^{**}_\nu \subset Q_\nu \subset B(x_\nu, R_0\rho_\nu)$ since $4r \le \theta^{1/2} \le \theta^{1/4} \le R_0$, according to the previous requirements on r and θ and also to the condition on θ in Lemma 4.3.17.

We have, denoting by ∂_2 the partial derivative with respect to x',

$$c' = \alpha'\partial_1 a + \partial_2 a = \partial_2 a \qquad \text{(here we use the identity } \partial_1 a(\alpha(x'), x') \equiv 0\text{)},$$

$$c'' = \alpha'\partial_1\partial_2 a + \partial_2^2 a,$$

$$c''' = \alpha''\partial_1\partial_2 a + \alpha'^2\partial_1^2\partial_2 a + 2\alpha'\partial_1\partial_2^2 a + \partial_2^3 a,$$

$$c'''' = \alpha'''\partial_1\partial_2 a + 3\alpha''\alpha'\partial_1^2\partial_2 a + 3\alpha''\partial_1\partial_2^2 a + \alpha'^3\partial_1^3\partial_2 a + 3\alpha'^2\partial_1^2\partial_2^2 a$$
$$+ 3\alpha'\partial_1\partial_2^3 a + \partial_2^4 a,$$

and we obtain

$$|c'| \lesssim \rho^3, \quad |c''| \lesssim \rho^2, \quad |c'''| \lesssim \rho^{-1}\rho^2 + \rho \sim \rho, \quad |c''''| \lesssim \rho^{-2}\rho^2 + \rho^{-1}\rho + 1 \sim 1,$$

so that

$$c \in C^{3,1}(B'), \quad |c^{(j)}| \lesssim \rho_\nu^{4-j}, 0 \le j \le 4. \tag{2.5.37}$$

Since $\partial_1^2 a \gtrsim \rho^2$ on Q_ν, we can define

$$R(x) = \omega(x)^{1/2}, \quad \omega(x) = \int_0^1 (1-t)\partial_1^2 a\big(\alpha(x') + t(x_1 - \alpha(x')), x'\big)dt. \tag{2.5.38}$$

Note also that the identity (on Q_ν), $a = R(x)^2(x_1 - \alpha)^2 + a(\alpha(x'), x')$ forces the function

$$B(x) = R(x)^2(x_1 - \alpha)^2$$

to be $C^{3,1}(Q_\nu)$ with a j-th derivative bounded above in absolute value by ρ_ν^{4-j} ($0 \le j \le 4$) since it is the case for a and c (this fact is not obvious since the function R is a priori only $C^{1,1}$). Defining on Q_ν,

$$b(x) = R(x)\big(x_1 - \alpha(x')\big), \tag{2.5.39}$$

we see that

$$a = b^2 + c, \quad |(b^2)^{(j)}| = |B^{(j)}| \lesssim \rho_\nu^{4-j}, \ 0 \le j \le 4. \tag{2.5.40}$$

As a consequence with $\beta = x_1 - \alpha(x')$, $b^2 = R^2\beta^2 = B \in C^{3,1}$,

$$R^2\beta^2 = \overbrace{B(\alpha(x'), x')}^{=0} + \overbrace{\int_0^1 \partial_1 B(\alpha(x') + \theta(x_1 - \alpha(x')), x')d\theta}^{\in C^{2,1}} \beta,$$

$$|\beta^{(j)}| \lesssim \rho^{1-j}, 0 \le j \le 3,$$

and since β vanishes on a hypersurface

$$\begin{cases} R^2\beta &= \int_0^1 \partial_1 B(\alpha(x') + \theta(x_1 - \alpha(x')), x')d\theta \in C^{2,1}, \\ |(R^2\beta)^{(j)}| \lesssim \rho_\nu^{3-j}, \ 0 \le j \le 3, \quad \text{(from (2.5.40))}. \end{cases} \tag{2.5.41}$$

Also we have $0 < R^2 = \omega \in C^{1,1}, \omega \sim \rho_\nu^2$ and from (2.5.38), (2.5.35),

$$|\omega^{(j)}| \lesssim \rho_\nu^{2-j}, 0 \le j \le 2, \tag{2.5.42}$$

entailing that with $R = \omega^{1/2}$,

$$|R' = \frac{1}{2}\omega^{-1/2}\omega'| \lesssim 1, \quad |R'' = -\frac{1}{4}\omega^{-3/2}\omega'^2 + \frac{1}{2}\omega^{-1/2}\omega''| \lesssim \rho_\nu^{-3}\rho_\nu^2 + \rho_\nu^{-1} \sim \rho_\nu^{-1}. \tag{2.5.43}$$

Using Leibniz' formula, we get

$$(R^2\beta)''' = (\omega\beta)''' = \omega'''\beta + 3\omega''\beta' + 3\omega'\beta'' + \omega\beta''',$$

which makes sense since ω''' is a distribution of order 1 and β is $C^{2,1}$ (see (2.5.35)). From (2.5.41), we know that $(\omega\beta)'''$ is L^∞, and since it is also the case of $\omega''\beta'$, $\omega'\beta''$, $\omega\beta'''$ from (2.5.42) and (2.5.35), we get that $\omega'''\beta$ belongs to L^∞ and

$$|\omega'''\beta| \lesssim 1. \tag{2.5.44}$$

On the other hand we have

$$\omega''' = 2(RR')'' = 2(R'^2 + RR'')' = 4R'R'' + 2(RR'')' = 6R'R'' + 2\underbrace{R}_{C^{1,1}}\underbrace{R'''}_{\substack{\text{distribution} \\ \text{of order 1}}}$$

entailing from (2.5.44), that $\beta(6R'R'' + 2RR''')$ is L^∞ and since it is the case of $\beta R'R''$ (from (2.5.35) and (2.5.43)), we get that $\beta RR'''$ is L^∞ and, using Section 4.2.2, we obtain

$$|\beta RR'''| \lesssim 1, \quad \text{i.e., for all multi-indices } \gamma \text{ with length 3, } |\beta R\partial_x^\gamma R| \lesssim 1. \tag{2.5.45}$$

With $b = R\beta$, we get $b'b'' = (R'\beta + R\beta')(R''\beta + 2R'\beta' + R\beta'')$ and to check that $(b'b'')'$ is in L^∞ with

$$|(b'b'')'| \lesssim 1, \tag{2.5.46}$$

it is enough (see (2.5.35), (2.5.43)) to check the derivatives of $R''\beta R'\beta$, $R''\beta R\beta'$ which are, up to bounded terms (see Section 4.3.2 in the appendix for the meaning of the products)

$$R'''\beta R'\beta = R'''\beta RR'\frac{\beta}{R}, \quad R'''\beta R\beta'$$

which are bounded according to (2.5.45)–(2.5.43)–(2.5.35). Note that b'' is bounded from (2.5.43) and (2.5.35). We want also to verify that $(bb'')''$ is bounded. We use that $(b^2)^{(4)}$ is bounded from (2.5.40) and since we have

$$\underbrace{(b^2)''''}_{\substack{\text{bounded} \\ (2.5.40)}} = 2(b' \otimes b' + bb'')'' = 2\underbrace{(b' \otimes b'' + b'' \otimes b')'}_{\substack{\text{bounded} \\ (2.5.46)}} + 2(bb'')'', \tag{2.5.47}$$

we obtain[17] the boundedness of $(bb'')''$.

Remark 2.5.22. Before going on, we should note that our functions b, c above are only defined on Q_ν where the identity $a(x) = b(x)^2 + c(x')$ holds. We can replace the function c above by

$$\tilde{c}(x') = c(x')\chi\big((x' - x'_\nu)\theta^{-1/2}\rho_\nu^{-1}\big)$$

where $\chi \in C_c^\infty(\mathbb{R}^{m-1})$ is supported in the unit ball and equal to 1 in the ball of radius $1/2$, so that $\tilde{c}$ is defined on $\mathbb{R}^{m-1}$ and the identity $a = b^2 + \tilde{c}$ holds on

$$x_\nu + \frac{1}{2}(Q_\nu - x_\nu) \supset U_\nu^* \supset \operatorname{supp}\varphi_\nu.$$

The bounds on the derivatives are unchanged as long as θ is fixed, which is the case.

Taking that remark into account, as well as the above estimates on the derivatives, we have finally, with E_2 standing for the non-degenerate indices,

$$a(x) = \sum_{\nu \in \mathbb{N}} b_\nu(x)^2 \varphi_\nu^2(x) + \sum_{\nu \in E_2} a_\nu(x')\varphi_\nu^2(x),$$

$$|b_\nu| \lesssim \rho_\nu^2, \ |b'_\nu| \lesssim \rho_\nu, \ |b''_\nu| \lesssim 1, \ |(b_\nu b''_\nu)''| + |(b'_\nu b''_\nu)'| \lesssim 1$$

$$|a_\nu| \lesssim \rho_\nu^4, \ |a'_\nu| \lesssim \rho_\nu^3, \ |a''_\nu| \lesssim \rho_\nu^2, \ |a'''_\nu| \lesssim \rho_\nu, \ |a''''_\nu| \lesssim 1,$$

$$a_\nu \text{ is defined on } \mathbb{R}^{m-1}.$$

Now, we consider the function $\mathbb{R}^{m-1} \ni t \mapsto A(t) = \rho_\nu^{-4} a_\nu(\rho_\nu t)$ and we have

$$|A_\nu| \lesssim 1, \ |A'_\nu| \lesssim 1, \ |A''_\nu| \lesssim 1, \ |A'''_\nu| \lesssim 1, \ |A''''_\nu| \lesssim 1.$$

Following the main argument in the proof by C. Fefferman and D.H. Phong, we can use an induction on the dimension m to get

$$A(t) = \sum_{1 \le j \le N_{m-1}} B_j^2(t), \quad B_j \in C^{1,1}, \quad \text{and } B_j'', (B_j' B_j'')', (B_j B_j'')'' \in L^\infty.$$

Incorporated in the induction hypothesis is that the bounds on B depend only on the bounds on $A^{(4)}$. We obtain

$$a(x) = \sum_{\nu \in \mathbb{N}} b_\nu(x)^2 \varphi_\nu^2(x) + \sum_{\nu \in E_2} \sum_{1 \le j \le N_{m-1}} \rho_\nu^4 B_{j,\nu}^2\Big(\frac{x'}{\rho_\nu}\Big)\varphi_\nu^2(x)$$

---

[17] The equality (2.5.47) is an equality between tensors $(0,4)$ and it might look somewhat pedantic to resort to such notation: the reader may check directly the implication

$$\left.\begin{array}{r} \forall\gamma, |\gamma| = 4, \partial_x^\gamma(b^2) \in L^\infty \\ \forall\gamma_j, 1 \le j \le 3, |\gamma_1| = 1 = |\gamma_2|, |\gamma_3| = 2, \partial_x^{\gamma_1}(\partial_x^{\gamma_2} b \partial_x^{\gamma_3} b) \in L^\infty \end{array}\right\}$$

$$\Longrightarrow \forall\gamma_3, \gamma_4, |\gamma_3| = 2 = |\gamma_4|, \ \partial_x^{\gamma_3}(b\partial_x^{\gamma_4} b) \in L^\infty.$$

i.e.,

$$a(x) = \sum_{1 \leq j \leq N_{m-1}+1} \sum_{\nu \in \mathbb{N}} b_{\nu,j}(x)^2 \varphi_\nu^2(x).$$

One needs to pass to a finite sum, which is quite standard since the overlap of the support of the functions φ_ν is bounded; this last argument is given in the appendix 4.3.6. The proof of Theorem 2.5.19 is complete. $\qquad\square$

The proof, Step 3: application of the Wick calculus

Let a be a non-negative function defined on $\mathbb{R}^{2n}$ such that $a^{(4)}$ belongs to $\mathcal{A}$ (defined in Proposition 2.5.6). Applying Lemma 2.5.17 and the L^2-boundedness of the operators with Weyl symbol in $\mathcal{A}$, we see that it suffices to prove that the operator with Wick symbol $a - \frac{1}{8\pi} \operatorname{trace} a''$ is semi-bounded from below. Since $\mathcal{A} \subset L^\infty(\mathbb{R}^{2n})$, it is enough to prove the following lemma.

Lemma 2.5.23. *Let a be a non-negative function defined on $\mathbb{R}^{2n}$ such that $a^{(4)}$ belongs to $L^\infty(\mathbb{R}^{2n})$. Theorem 2.5.19 provides a decomposition $a = \sum_{1 \leq j \leq N} b_j^2$ along with the estimates (2.5.26). Then we have*

$$\left(a - \frac{1}{8\pi} \operatorname{trace} a''\right)^{Wick} = \sum_{1 \leq j \leq N} \left[\left(b_j - \frac{1}{8\pi} \operatorname{trace} b_j''\right)^{Wick}\right]^2 + R$$

where R is an L^2-bounded operator such that $\|R\|_{\mathcal{L}(L^2(\mathbb{R}^n))} \leq C\|a^{(4)}\|_{L^\infty(\mathbb{R}^{2n})}$, C depending only on the dimension n.

Proof. We have

$$a - \frac{1}{8\pi} \operatorname{trace} a'' = a - \frac{\Delta a}{8\pi} = \sum_{1 \leq j \leq N} b_j^2 - \frac{1}{4\pi}|\nabla b_j|^2 - \frac{1}{4\pi} b_j \Delta b_j. \qquad (2.5.48)$$

Then using Lemma 2.4.8, we get

$$b_j^{Wick} b_j^{Wick} = \left(b_j^2 - \frac{1}{4\pi}|\nabla b_j|^2\right)^{Wick} + S_j, \qquad (2.5.49)$$

with

$$\|S_j\|_{\mathcal{L}(L^2(\mathbb{R}^n))} \leq C_1 \left(\|b_j''\|_{L^\infty}^2 + \|(b_j'' b_j')'\|_{L^\infty} + \|(b_j b_j'')''\|_{L^\infty}\right) \leq C_2 \|a^{(4)}\|_{L^\infty(\mathbb{R}^{2n})},$$

where C_1, C_2 depend only on the dimension. Moreover, we have, from Lemma 2.4.6,

$$\operatorname{Re}\left(b_j^{Wick}(\Delta b_j)^{Wick}\right) = \left(b_j \Delta b_j - \frac{1}{4\pi}\nabla \cdot (\nabla b_j \Delta b_j) + \frac{1}{4\pi}(\Delta b_j)^2\right)^{Wick} + R_j,$$

$$(2.5.50)$$

with $\|R_j\|_{\mathcal{L}(L^2(\mathbb{R}^n))} \leq C_3\|b_j''\|_{L^\infty(\mathbb{R}^{2n})}^2 \leq C_4\|a^{(4)}\|_{L^\infty(\mathbb{R}^{2n})}$. As a consequence, from (2.5.49), (2.5.50), we get

$$\left(b_j - \frac{1}{8\pi}\operatorname{trace} b_j''\right)^{\text{Wick}}\left(b_j - \frac{1}{8\pi}\operatorname{trace} b_j''\right)^{\text{Wick}}$$

$$= \left(b_j^2 - \frac{1}{4\pi}|\nabla b_j|^2 - \frac{1}{4\pi}b_j\Delta b_j\right)^{\text{Wick}} + \frac{1}{16\pi^2}\left(\nabla\cdot(\nabla b_j\Delta b_j)\right)^{\text{Wick}} - \frac{1}{16\pi^2}\left((\Delta b_j)^2\right)^{\text{Wick}}$$

$$+ S_j - \frac{1}{4\pi}R_j + \frac{1}{64\pi^2}(\Delta b_j)^{\text{Wick}}(\Delta b_j)^{\text{Wick}}, \qquad (2.5.51)$$

so that from (2.4.5), (2.5.26) and the estimates above for R_j, S_j, we obtain from (2.5.48) that

$$\sum_{1\leq j\leq N}\left(b_j - \frac{1}{8\pi}\operatorname{trace} b_j''\right)^{\text{Wick}}\left(b_j - \frac{1}{8\pi}\operatorname{trace} b_j''\right)^{\text{Wick}} = \left(a - \frac{1}{8\pi}\operatorname{trace} a''\right)^{\text{Wick}} + S$$

with $\|S\|_{\mathcal{L}(L^2(\mathbb{R}^n))} \leq C_5\|a^{(4)}\|_{L^\infty(\mathbb{R}^{2n})}$, with C_5 depending only on the dimension. This is the result of the lemma, completing as well the proof of Theorem 2.5.10. $\square$

Proof of Corollary 2.5.11.

Let us begin with the statement (iv) in this corollary. Let us define

$$A(x,\xi) = h^{-2}a(xh^{1/2},\xi h^{-1/2},h). \qquad (2.5.52)$$

The function A satisfies

$$(\partial_\xi^\alpha\partial_x^\beta A)(x,\xi) = h^{-2-\frac{|\alpha|}{2}+\frac{|\beta|}{2}}(\partial_1^\beta\partial_2^\alpha a)(xh^{1/2},\xi h^{-1/2},h)$$

$$= h^{\frac{|\alpha|+|\beta|-4}{2}}(\partial_1^\beta\partial_2^\alpha a)(xh^{1/2},\xi h^{-1/2},h)h^{-|\alpha|}$$

so that for $|\alpha|+|\beta|=4$, we have $(\partial_\xi^\alpha\partial_x^\beta A)(x,\xi) = (\partial_1^\beta\partial_2^\alpha a)(xh^{1/2},\xi h^{-1/2},h)h^{-|\alpha|}$. We have supposed that for $|\alpha|+|\beta|=4$, the functions

$$(x,\xi)\mapsto(\partial_1^\beta\partial_2^\alpha a)(xh^{1/2},\xi h^{-1/2},h)h^{-|\alpha|}$$

belong to $\mathcal{A}$ with a norm bounded above independently by ν_0. As a result the function $A^{(4)}(x,\xi)$ belongs to $\mathcal{A}$ with a norm bounded above by ν_0. Since $A(x,\xi)\geq 0$, Theorem 2.5.10 implies that $A^w + C_n\nu_0 \geq 0$, i.e.,

$$\left(a(xh^{1/2},\xi h^{-1/2},h)\right)^w + C_n\nu_0 h^2 \geq 0$$

and since there is a unitary mapping U_h such that

$$U_h^* a(x,\xi,h)^w U_h = \left(a(xh^{1/2},\xi h^{-1/2},h)\right)^w,$$

we obtain

$$a(x,\xi,h)^w + C_n\nu_0 h^2 \geq 0. \qquad (2.5.53)$$

To get that $\operatorname{Re} a(x, D, h) + Ch^2 \geq 0$, one[18] should note that the symbols $A^{(4)}$ defined above belong to $\mathcal{A}$, which implies that it is also the case for $J^{-1/2}A^{(4)}$ and $J^{1/2}\overline{A^{(4)}}$. Now we have

$$2\operatorname{Re} a(x, D, h) = 2\operatorname{Re}(J^{-1/2}a)^w = (J^{-1/2}a + J^{1/2}\bar{a})^w,$$

so that rescaling[19] the symbol $J^{-1/2}a + J^{1/2}\bar{a}$, we find $J^{-1/2}A + J^{1/2}\bar{A}$. Since we have

$$J^{-1/2}A = e^{-i\pi D_x \cdot D_\xi}A = A - i\pi D_x \cdot D_\xi A - \int_0^1 (1-\theta)e^{-i\pi\theta D_x \cdot D_\xi}d\theta\pi^2(D_x \cdot D_\xi)^2 A,$$

and that A is real-valued, we get

$$\operatorname{Re}(J^{-1/2}A) = A - \int_0^1 (1-\theta)e^{-i\pi\theta D_x \cdot D_\xi}d\theta\pi^2 \underbrace{(D_x \cdot D_\xi)^2 A}_{\in \mathcal{A}}.$$

Now we have from the previous identity, since $\mathcal{A}$ is stable by the group J^t (Theorem 1.1 in [130]), with a uniform constant for t in a compact set,

$$2\operatorname{Re} A(x, D) = \left(2\operatorname{Re}(J^{-1/2}A)\right)^w \in 2A^w + \mathcal{A}^w.$$

We can then apply the result (2.5.53) and the L^2 boundedness of $\mathcal{A}^w$ to conclude. The proof of (iv) in Corollary 2.5.11 is complete.

Let us show that (iv) implies (iii). We define $b(x, \xi, h) = a(x, h\xi)$, which is non-negative; it is enough to check the functions

$$(x, \xi) \mapsto (\partial_1^\beta \partial_2^\alpha b)(xh^{1/2}, \xi h^{-1/2}, h)h^{-|\alpha|},$$

for $|\alpha| + |\beta| = 4$. We have in fact

$$(\partial_1^\beta \partial_2^\alpha b)(xh^{1/2}, \xi h^{-1/2}, h)h^{-|\alpha|} = (\partial_1^\beta \partial_2^\alpha a)(xh^{1/2}, \xi h^{1/2}).$$

Now, from Lemma 4.8.1 in our appendix, for $h \in (0, 1]$, the functions

$$(x, \xi) \mapsto (\partial_1^\beta \partial_2^\alpha a)(xh^{1/2}, \xi h^{1/2})$$

belong to $\mathcal{A}$ with a bounded norm since we have supposed that $a^{(4)} \in \mathcal{A}$. We can then apply the already proven result (iv) in the corollary to get

$$a(x, \xi h)^w + Ch^2\|a^{(4)}\|_{\mathcal{A}} \geq 0, \quad \operatorname{Re} a(x, hD) + Ch^2\|a^{(4)}\|_{\mathcal{A}} \geq 0.$$

---

[18]With the group J^t defined in Lemma 4.1.2, the formula linking the Weyl quantization with the ordinary quantization is $a(x, D) = (J^{-1/2}a)^w$.

[19]We define

$$B(x, \xi) = h^{-2}(J^{-1/2}a)(xh^{1/2}, \xi h^{-1/2}) + h^{-2}(J^{1/2}a)(xh^{1/2}, \xi h^{-1/2})$$
$$= (J^{-1/2}A)(x, \xi) + (J^{1/2}A)(x, \xi).$$

Let us show that (iv) implies (ii). We assume that $a(x, \xi, h)$ is a non-negative function satisfying the assumptions of (ii). According to the already proven (iv), we need only to check, for $|\alpha'| + |\beta'| = 4$, the norm in $\mathcal{A}$ of

$$(x, \xi) \mapsto (\partial_1^{\beta'} \partial_2^{\alpha'} a)(xh^{1/2}, \xi h^{-1/2}, h) h^{-|\alpha'|} = c_{\alpha' \beta'}(x, \xi).$$

Because of the second inclusion in (2.5.15), it is enough to find an L^∞ bound on the $2n + 1$ first derivatives of that function; we have, for $|\alpha''| + |\beta''| \le 2n + 1$,

$$(\partial_\xi^{\alpha''} \partial_x^{\beta''} c_{\alpha' \beta'})(x, \xi) = (\partial_1^{\beta' + \beta''} \partial_2^{\alpha' + \alpha''} a)(xh^{1/2}, \xi h^{-1/2}, h) h^{-|\alpha'|} h^{\frac{-|\alpha''| + |\beta''|}{2}}$$
$$(2.5.54)$$

and from the assumption in (ii), we get, since $4 \le |\alpha' + \alpha''| + |\beta' + \beta''| \le 2n + 5$,

$$|(\partial_1^{\beta' + \beta''} \partial_2^{\alpha' + \alpha''} a)(xh^{1/2}, \xi h^{-1/2}, h)| \le C_{\alpha' + \alpha'', \beta' + \beta''} h^{|\alpha'| + |\alpha''|}, \qquad (2.5.55)$$

so that (2.5.54)–(2.5.55) imply

$$|(\partial_\xi^{\alpha''} \partial_x^{\beta''} c_{\alpha' \beta'})(x, \xi)| \le C_{\alpha' + \alpha'', \beta' + \beta''} h^{|\alpha'| + |\alpha''|} h^{-|\alpha'|} h^{\frac{-|\alpha''| + |\beta''|}{2}}$$
$$= C_{\alpha' + \alpha'', \beta' + \beta''} h^{\frac{|\alpha''| + |\beta''|}{2}} \le C_{\alpha' + \alpha'', \beta' + \beta''}.$$

yielding the sought bound. The proof of (ii) is complete.

Proof of (i) *in Corollary* 2.5.11. Using a Littlewood-Paley decomposition, we have

$$1 = \sum_{\nu \ge 0} \varphi_\nu^2(\xi), \quad \varphi_\nu \in C_c^\infty(\mathbb{R}^n),$$

for $\nu \ge 1$, $\operatorname{supp} \varphi_\nu \subset \{2^{\nu-1} \le |\xi| \le 2^{\nu+1}\}$, $\sup_{\nu, \xi} |\partial_\xi^\alpha \varphi_\nu(\xi)| 2^{\nu|\alpha|} < \infty$.

We introduce also some smooth non-negative compactly supported functions $\psi_\nu(\xi)$, satisfying the same uniform estimates as φ_ν and supported in $2^{\nu-3} \le |\xi| \le 2^{\nu+3}$ for $\nu \ge 1$, identically 1 on $2^{\nu-2} \le |\xi| \le 2^{\nu+2}$ (in particular on the support of φ_ν). We consider a non-negative symbol a satisfying (2.5.10) for $4 \le |\alpha| + |\beta| \le 2n + 5$. We write

$$a = \sum_{\nu \ge 0} \varphi_\nu^2 a = \sum_{\nu \ge 0} (\psi_\nu \sharp \varphi_\nu^2 a \sharp \psi_\nu + r_\nu). \qquad (2.5.56)$$

The proof relies on the following

Claim 2.5.24. *The operator with Weyl symbol* $\sum_\nu r_\nu$ *is bounded on* $L^2(\mathbb{R}^n)$.

As a matter of fact, if this claim is proven, we are left with the operator $\sum_\nu \psi_\nu^w (\varphi_\nu^2 a)^w \psi_\nu^w$ and we can apply the already proven result (ii) in this corollary to get that with a uniform C,

$$\sum_\nu \psi_\nu^w (\varphi_\nu^2 a)^w \psi_\nu^w = \sum_\nu \psi_\nu^w \underbrace{\left((\varphi_\nu^2 a)^w + C\right)}_{\ge 0} \psi_\nu^w - C \underbrace{\left(\sum_\nu \psi_\nu^2\right)^w}_{L^2 \text{ bounded}}$$

and so this operator is semi-bounded from below as well as a^w. Let us prove the claim. The following composition formula is proven in Theorem 4.6.1:

$$(a_1 \sharp a_2 \sharp a_3)(X) = 2^{2n} \iint_{\mathbb{R}^{2n} \times \mathbb{R}^{2n}} a_1(Y_1) a_2(Y_2) a_3(X - Y_1 + Y_2)$$
$$\times e^{-4i\pi[X-Y_1, X-Y_2]} dY_1 dY_2. \quad (2.5.57)$$

Applying this to $\psi_\nu \sharp a_\nu \sharp \psi_\nu$ with $a_\nu = \varphi_\nu^2 a$, we get

$$r_\nu(x, \xi) = -2^n \iiint_0^1 (1-\theta) e^{-4i\pi y\eta} \psi_\nu(\xi + \eta) \psi_\nu(\xi - \eta) (\partial_x^2 a_\nu)(x + \theta y, \xi) y^2 \, dy \, d\eta \, d\theta$$

$$= \frac{2^n}{16\pi^2} \iiint_0^1 (1-\theta) \partial_\eta^2 (e^{-4i\pi y\eta}) \psi_\nu(\xi + \eta) \psi_\nu(\xi - \eta) (\partial_x^2 a_\nu)(x + \theta y, \xi) \, dy \, d\eta \, d\theta$$

$$= \frac{2^n}{16\pi^2} \iiint_0^1 (1-\theta) e^{-4i\pi y\eta} \partial_\eta^2 (\psi_\nu(\xi + \eta) \psi_\nu(\xi - \eta)) (\partial_x^2 a_\nu)(x + \theta y, \xi) \, dy \, d\eta \, d\theta.$$

From this formula we see that r_ν is supported where $2^{\nu-1} \leq |\xi| \leq 2^{\nu+1}$ since it is the case for a_ν ($\nu \geq 1$); since the overlap of the rings where $|\xi| \sim 2^\nu$ is bounded, it is enough to check some bounds on the derivatives of r_ν to get similar bounds on the $\sum_\nu r_\nu$. Moreover in the integrand, if the function $\psi_\nu(\xi + \eta)$ is differentiated, we get

$$2^{\nu+2} \leq |\xi + \eta| \leq 2^{\nu+3} \text{ or } 2^{\nu-3} \leq |\xi + \eta| \leq 2^{\nu-2}.$$

As a result, in the first case, we have $|\eta| \geq |\xi + \eta| - |\xi| \geq 2^{\nu+2} - 2^{\nu+1} = 2^{\nu+1}$, whereas in the second case $|\eta| \geq |\xi| - |\xi + \eta| \geq 2^{\nu-1} - 2^{\nu-2} = 2^{\nu-2}$, which implies that we always have $|\eta| \geq 2^{\nu-2}$. Since we have also $|\eta| \leq |\xi + \eta| + |\xi| \leq 2^{\nu+3} + 2^{\nu+1}$, we obtain (note that the case when the other function $\psi(\xi - \eta)$ is differentiated is similar) on the integrand

$$2^{\nu-2} \leq |\eta| \leq 2^{\nu+4}. \quad (2.5.58)$$

We write now

$$\frac{1}{\alpha!} (\partial_\xi^\alpha \partial_x^\beta r_\nu)(x, \xi) = \sum_{\alpha' + \alpha'' = \alpha} \frac{2^n}{\alpha'! \alpha''! 16\pi^2}$$

$$\times \iiint_0^1 (1-\theta) e^{-4i\pi y\eta} \partial_\xi^{\alpha'} \partial_\eta^2 (\psi_\nu(\xi + \eta) \psi_\nu(\xi - \eta)) (\partial_\xi^{\alpha''} \partial_x^\beta \partial_x^2 a_\nu)(x + \theta y, \xi) \, dy \, d\eta \, d\theta$$

and since the integral above is, for N, k even integers, $N > n$,

$$\iiint_0^1 (1-\theta) e^{-4i\pi y\eta} (1 + 4|\eta|^2)^{-k/2}$$

$$\times (1 + 4|y|^2)^{-N/2} (1 + D_\eta^2)^{N/2} \partial_\xi^{\alpha'} \partial_\eta^2 (\psi_\nu(\xi + \eta) \psi_\nu(\xi - \eta))$$

$$\times (1 + D_y^2)^{k/2} ((\partial_\xi^{\alpha''} \partial_x^\beta \partial_x^2 a_\nu)(x + \theta y, \xi)) \, dy \, d\eta \, d\theta,$$

we get, for $|\alpha| + |\beta| + k \leq 2n + 3$,

$$|(\partial_\xi^\alpha \partial_x^\beta r_\nu)(x, \xi)| \leq C_{\alpha\beta N}(2^\nu)^{-k-|\alpha'|-2+2-|\alpha''|+n} = C_{\alpha\beta N}(2^\nu)^{-|\alpha|+n-k}.$$

For α, β given such that $\max(|\alpha|, |\beta|) \leq n+1$, we choose $k = n-|\alpha|$ or $k = n-|\alpha|+1$ so that k is even, and we get, uniformly in ν, $|(\partial_\xi^\alpha \partial_x^\beta r_\nu)(x, \xi)| \lesssim 1$; note that then we have indeed

$$|\alpha| + |\beta| + k \leq |\beta| + n + 1 \leq 2n + 2 \leq 2n + 3.$$

Eventually, from (a mild version of) Theorem 1.2 in [22] we get Claim 2.5.24: we have proven that for $\max(|\alpha|, |\beta|) \leq n+1$, $\partial_\xi^\alpha \partial_x^\beta r$ is bounded. The proof of (2.5.12) is complete, under the assumptions of the corollary.

Proof of (1.1.2). To obtain also the result for the ordinary quantization is not a direct consequence of the previous result, because of our limitation on the regularity of a. So we have to revisit our argument above, replacing at each step the Weyl quantization by the standard quantization. It is a bit tedious, but unavoidable. We write

$$a = \sum_{\nu \geq 0} \varphi_\nu^2 a = \sum_{\nu \geq 0} (\psi_\nu \circ \varphi_\nu^2 a \circ \psi_\nu + s_\nu). \tag{2.5.59}$$

The proof relies on the following

Claim 2.5.25. *The operator with standard symbol $\sum_\nu s_\nu$ is bounded on $L^2(\mathbb{R}^n)$.*

As a matter of fact, if this claim is proven, we are left with the operator

$$\operatorname{Re} \sum_\nu \operatorname{op}(\psi_\nu) \operatorname{op}(\varphi_\nu^2 a) \operatorname{op}(\psi_\nu)$$

and we can apply the already proven result (ii) in this corollary to get that with a uniform C,

$$\sum_\nu \operatorname{op}(\psi_\nu) \operatorname{Re} \operatorname{op}(\varphi_\nu^2 a) \operatorname{op}(\psi_\nu)$$

$$= \sum_\nu \operatorname{op}(\psi_\nu) \underbrace{\left(\operatorname{Re} \operatorname{op}(\varphi_\nu^2 a) + C\right)}_{\geq 0} \operatorname{op}(\psi_\nu) - C \underbrace{\operatorname{op}(\sum_\nu \psi_\nu^2)}_{L^2 \text{bounded}},$$

and so this operator is semi-bounded from below as well as $\operatorname{Re} \operatorname{op}(a)$. Let us prove the claim. Recalling the ordinary composition formula, we have

$$(a \circ b)(x, \xi) = \iint_{\mathbb{R}^n \times \mathbb{R}^n} e^{-2i\pi y\eta} a(x, \xi + \eta) b(y + x, \xi) dy d\eta. \tag{2.5.60}$$

Applying this to $\psi_\nu \circ a_\nu \circ \psi_\nu$ with $a_\nu = \varphi_\nu^2 a$, we get $\psi_\nu \circ a_\nu \circ \psi_\nu = \psi_\nu \circ a_\nu \psi_\nu = \psi_\nu \circ a_\nu$ and

$$(\psi_\nu \circ a_\nu \circ \psi_\nu)(x,\xi) = \iint e^{-2i\pi y\eta} \psi_\nu(\xi+\eta) a_\nu(y+x,\xi) dy d\eta$$

$$= \iint e^{-2i\pi y\eta} \psi_\nu(\xi+\eta) \Big(a_\nu(x,\xi) + \int_0^1 (1-\theta)\partial_x^2 a_\nu(x+\theta y,\xi) y^2 d\theta \Big) dy d\eta$$

$$= (a_\nu \psi_\nu)(x,\xi) - s_\nu(x,\xi),$$

with

$$s_\nu(x,\xi) = -\iiint_0^1 (1-\theta) e^{-2i\pi y\eta} \psi_\nu(\xi+\eta)(\partial_x^2 a_\nu)(x+\theta y,\xi) y^2 dy d\eta d\theta.$$

That formula is so similar to the defining formula of r_ν above that we can resume the discussion and use (a mild version of) Theorem 1.1 in [22] to get the claim 2.5.25: The proof of (2.5.11) is complete, under the assumptions of the corollary.

2.5.4   Analytic functional calculus

Theorem 2.5.26. *Let g be an admissible metric on $\mathbb{R}^{2n}$ (Definition 2.2.15). There exists some positive integers κ_0, κ_1, a positive increasing function α defined on $\mathbb{N}$, depending only on the structure constants of g, as given by (2.2.16) and (2.2.3) such that the following property is true. Let $\nu \geq 2$ be an integer and $a \in S(1,g)$ (see Definition 2.2.6). Then $a^{\sharp\nu}$ belongs to $S(1,g)$ and for all $X, T \in \mathbb{R}^{2n}, k \in \mathbb{N}$,*

$$|(a^{\sharp\nu})^{(k)}(X) T^k| \leq g_X(T)^{k/2} \Big(\alpha(k) \|a\|_{S(1,g)}^{(\kappa_0 k + \kappa_1)} \Big)^\nu,$$

which implies

$$\|a^{\sharp\nu}\|_{S(1,g)}^{(k)} \leq \Big(\alpha(k) \|a\|_{S(1,g)}^{(\kappa_0 k + \kappa_1)} \Big)^\nu, \tag{2.5.61}$$

where the semi-norms in $S(1,g)$ are defined in (2.2.7).

N.B. The fact that $a^{\sharp\nu}$ belongs to $S(1,g)$ is clear from the symbolic calculus, as given by Theorem 2.3.7. The point of this lemma is that the k-th semi-norm of $a^{\sharp\nu}$ is bounded above by the ν-th power of the $(\kappa_0 k + \kappa_1)$-th semi-norm of a, a shift which *does not depend* on ν.

Proof. We use the partition of unity of Proposition 2.2.9 to get

$$(a^{\sharp\nu})(X) = \int_{\mathcal{P}^\nu} (a_{Y_1} \sharp \ldots \sharp a_{Y_\nu})(X) |g_{Y_1}|^{1/2} \ldots |g_{Y_\nu}|^{1/2} dY_1 \ldots dY_\nu.$$

We note that with $g_j = g_{Y_j}, U_j = U_{Y_j,r}$, from the first inequality in (2.2.29), the assumption (4.6.12) is satisfied with $\delta_{j,j+1} = \delta_r(Y_j, Y_{j+1})$ where the function δ_r

is given by (2.2.27) and N_0, C_0 depend only on the "structure constants" of the metric g (as given by (2.2.16) and (2.2.3)): this is

$$1 + \frac{g_{Y_{j+1}}(T)}{g_{Y_j}(T)} + \frac{g_{Y_j}(S)}{g_{Y_{j+1}}(S)} \leq C_0 \delta_r(Y_j, Y_{j+1})^{N_0}. \tag{2.5.62}$$

We use now the Corollaries 4.6.6, 4.6.7 to obtain for all $N, M \in \mathbb{N}$ to be chosen later[20],

$$|(a^{\sharp \nu})^{(k)}(X)T^k| \leq \kappa_n^{\nu/2} 2^{\nu N} \nu^k \omega(n, C_0, N_0, M)^{\nu/2}$$
$$\times \int_{\mathcal{P}^\nu} \max_{k_1 + \cdots + k_\nu = k} \left(\prod_{1 \leq j \leq \nu} \|a_{Y_j}\|_{g_{Y_j}, U_{Y_j}, r}^{(k_j + N + M + (2n+1)(2N_0+2))} \right) \left(\max_{1 \leq j \leq \nu} g_{Y_j}(T) \right)^{k/2}$$
$$\times \left(1 + \nu^{-1} \max_{1 \leq j \leq \nu} (g_{Y_1}^\sigma \wedge \cdots \wedge g_{Y_\nu}^\sigma)(X - U_{Y_j, r}) \right)^{-N/4}$$
$$\times \prod_{1 \leq j < \nu} \delta_r(Y_j, Y_{j+1})^{-M/8} |g_{Y_1}|^{1/2} \ldots |g_{Y_\nu}|^{1/2} dY_1 \ldots dY_\nu.$$

We have from (4.4.37)

$$1 + \nu^{-1} \max_{1 \leq j \leq \nu} (g_{Y_1}^\sigma \wedge \cdots \wedge g_{Y_\nu}^\sigma)(X - U_{Y_j, r})$$
$$\geq 1 + \nu^{-1}(g_{Y_1}^\sigma \wedge \cdots \wedge g_{Y_\nu}^\sigma)(X - U_{Y_\nu, r})$$
$$\geq \left(1 + g_{Y_\nu}^\sigma(X - U_{Y_\nu, r}) \right) \prod_{1 \leq j < \nu} \left(1 + C_0 \delta_r(Y_j, Y_{j+1})^{N_0} \right)^{-1}$$
$$\geq \left(1 + g_{Y_\nu}^\sigma(X - U_{Y_\nu, r}) \right) (2C_0)^{1-\nu} \prod_{1 \leq j < \nu} \delta_r(Y_j, Y_{j+1})^{-N_0}.$$

We note also that, with the partition of unity (φ_Y) given in Proposition 2.2.9,

$$\|a_Y\|_{g_Y, U_{Y, r}}^{(l)} = \|\varphi_Y a\|_{g_Y, U_{Y, r}}^{(l)} \leq C(l) \|a\|_{S(1,g)}^{(l)}$$

where the semi-norms of the symbol a are given by (2.2.7) and $C(l)$ depends only on the structure constants of the metric g. Thanks to the first inequality in (2.2.29), the ratios $g_{Y_j}(T)/g_{Y_{j+1}}(T)$ are bounded above by $C_3 \delta_r(Y_j, Y_{j+1})^{N_0}$. As a consequence, we have

---

[20]We have to keep in mind that the constants n, C_0, N_0 are "fixed". We want to track the dependence of the estimates with respect to ν and k. Incidentally, we have assumed, as we may, that N_0 is an integer.

$$|(a^{\sharp\nu})^{(k)}(X)T^k|$$
$$\leq \kappa_n^{\nu/2} 2^{\nu N} \nu^k \omega(n, C_0, N_0, M)^{\nu/2} C\big(k + N + M + (2n+1)(2N_0+2)\big)^{\nu}$$
$$\times \Big(\max_{k_1+\cdots+k_\nu=k} \prod_{1\leq j\leq \nu} \|a\|_{S(1,g)}^{(k_j+N+M+(2n+1)(2N_0+2))} \Big)$$
$$\times \int_{\mathcal{P}^\nu} g_{Y_\nu}(T)^{k/2} C_3^{(\nu-1)k/2} \big(1 + g_{Y_\nu}^\sigma(X - U_{Y_\nu,r})\big)^{-N/4} (2C_0)^{(\nu-1)N/4}$$
$$\times \prod_{1\leq j<\nu} \delta_r(Y_j, Y_{j+1})^{-\frac{M}{8}+\frac{N_0 N}{4}+\frac{N_0 k}{2}} |g_{Y_1}|^{1/2} \dots |g_{Y_\nu}|^{1/2} dY_1 \dots dY_\nu.$$

We note now that from the first inequality in (2.2.29),

$$\frac{g_{Y_\nu}(T)}{g_X(T)} \leq C_3 \delta_r(X, Y_\nu)^{N_0} \leq C_3 \big(1 + 2g_{Y_\nu}^\sigma(X - U_{Y_\nu,r})\big)^{N_0}$$

so that

$$|(a^{\sharp\nu})^{(k)}(X)T^k|$$
$$\leq \kappa_n^{\nu/2} 2^{\nu N} \nu^k \omega(n, C_0, N_0, M)^{\nu/2} C\big(k + N + M + (2n+1)(2N_0+2)\big)^{\nu}$$
$$\times \Big(\max_{k_1+\cdots+k_\nu=k} \prod_{1\leq j\leq \nu} \|a\|_{S(1,g)}^{(k_j+N+M+(2n+1)(2N_0+2))} \Big)$$
$$\times g_X(T)^{k/2} \int_{\mathcal{P}^\nu} (2C_3)^{k/2} C_3^{(\nu-1)k/2} \big(1 + g_{Y_\nu}^\sigma(X - U_{Y_\nu,r})\big)^{-\frac{N}{4}+\frac{N_0 k}{2}} (2C_0)^{(\nu-1)N/4}$$
$$\times \prod_{1\leq j<\nu} \delta_r(Y_j, Y_{j+1})^{-\frac{M}{8}+\frac{N_0 N}{4}+\frac{N_0 k}{2}} |g_{Y_1}|^{1/2} \dots |g_{Y_\nu}|^{1/2} dY_1 \dots dY_\nu. \qquad (2.5.63)$$

We know also from (2.2.27) and the second inequality in (4.4.26) that

$$\big(1 + g_{Y_\nu}^\sigma(X - U_{Y_\nu,r})\big) \geq \frac{1}{2} \delta_r(X, Y_\nu).$$

This implies that the integral above is, up to some constant terms that we shall make precise,

$$\int_{\mathcal{P}^\nu} \prod_{1\leq j<\nu} \delta_r(Y_j, Y_{j+1})^{-\frac{M}{8}+\frac{N_0 N}{4}+\frac{N_0 k}{2}} \delta_r(X, Y_\nu)^{-\frac{N}{4}+\frac{N_0 k}{2}}$$
$$\times |g_{Y_1}|^{1/2} \dots |g_{Y_\nu}|^{1/2} dY_1 \dots dY_\nu. \qquad (2.5.64)$$

We take a look at the second inequality in (2.2.29), and for a given k, we can choose N so that $\frac{N}{4} - \frac{N_0 k}{2} \geq N_1$, then choose M so that $\frac{M}{8} - \frac{N_0 N}{4} - \frac{N_0 k}{2} \geq N_1$ in such a way that the integral is bounded above by a fixed constant. At this point, we note that N_1, N_0 depend only on the structure constants of g and that we can choose, assuming as we may that $N_1 \in \mathbb{N}$,

$$N = 4N_1 + 2N_0 k, \ M = 8N_1 + 2N_0 N + 4N_0 k, \qquad (2.5.65)$$

so that

$$N = 4N_1 + 2N_0 k, \quad M = 8N_1 + 8N_0 N_1 + (4N_0 + 4N_0^2)k, \qquad (2.5.66)$$

giving an affine dependence of N, M on k (N_0, N_1 are fixed constants depending only on the structure constants of g). On the other hand, we have in (2.5.63) a constant

$$B = \kappa_n^{\nu/2} 2^{\nu N} \nu^k \omega(n, C_0, N_0, M)^{\nu/2} C(k + N + M + (2n+1)(2N_0+2))^\nu$$
$$\times (2C_3)^{k/2} C_3^{(\nu-1)k/2} (2C_0)^{(\nu-1)N/4}, \quad (2.5.67)$$

where C_0, C_3 are fixed depending only on the structure constants of g. The constant B can be estimated from above by

$$\beta(k)^\nu \nu^k, \quad \text{with } \beta(k) \text{ depending on } k, n \text{ and on the structure constants of } g.$$

Taking into account our choice (2.5.65) of N, M, we see that (2.5.64) is bounded above by C_4^ν, where C_4 is a fixed constant depending only on the structure constants of g. We have proven that

$$|(a^{\sharp\nu})^{(k)}(X)T^k| \le g_X(T)^{k/2}$$
$$\times \left(\max_{k_1 + \cdots + k_\nu = k} \prod_{1 \le j \le \nu} \|a\|_{S(1,g)}^{(k_j + N + M + (2n+1)(2N_0+2))} \right) \nu^k \beta(k)^\nu C_4^\nu$$

where N, M are given by (2.5.66) and depend in an affine way on k and on the structure constants of g. The constant $\beta(k)$ depends only on k, n and on the structure constants of the metric g. This gives readily (2.5.61), completing the proof of the lemma. $\qquad\square$

Corollary 2.5.27. *Let g be an admissible metric on $\mathbb{R}^{2n}$ (Definition 2.2.15), $a \in S(1,g)$, and f be an entire function of one complex variable. Then the operator $f(a^w)$ has a symbol in $S(1,g)$.*

Proof. The operator $f(a^w) = \left(\sum_{\nu \ge 0} c_\nu a^{\sharp\nu} \right)^w = b^w$ and from the estimates (2.5.61) of Theorem 2.5.26, b is a smooth function and

$$|b^{(k)}(X)T^k| \le \sum_{\nu \ge 0} |c_\nu| \left(\alpha(k) \|a\|_{S(1,g)}^{(\kappa_0 k + \kappa_1)} \right)^\nu g_X(T)^{k/2} \le C_{k,a} g_X(T)^{k/2},$$

since the radius of convergence of the series $\sum_{\nu \ge 0} c_\nu \rho^\nu$ is infinite. $\qquad\square$

2.6 Sobolev spaces attached to a pseudo-differential calculus

2.6.1 Introduction

The Weyl quantization associates an operator a^w to a function a defined on the phase space $\mathbb{R}^{2n}$. When the phase space is equipped with an admissible metric g

and M is an admissible weight for g (see Definition 2.2.15), we have obtained an algebra of operators, composition formulas, symbolic calculus and $L^2(\mathbb{R}^n)$ boundedness of the operators with symbols in $S(1,g)$. In the standard case of the metric $|dx|^2 + \langle\xi\rangle^{-2}|d\xi|^2$ which is such that

$$S(\langle\xi\rangle^m, |dx|^2 + \langle\xi\rangle^{-2}|d\xi|^2) = S_{1,0}^m \quad (m \in \mathbb{R}, \text{cf. Section 1.1.3}),$$

the natural Sobolev spaces (indexed by $s \in \mathbb{R}$) are the standard

$$H^s(\mathbb{R}^n) = \{u \in \mathscr{S}'(\mathbb{R}^n), \langle D\rangle^s u \in L^2(\mathbb{R}^n)\},$$

with the obvious Hilbertian structure; in that case, the operators of order m (i.e., of weight $\langle\xi\rangle^m$) are continuous from $H^{s+m}(\mathbb{R}^n)$ to $H^s(\mathbb{R}^n)$. We wish to generalize this to the case of a general admissible metric on $\mathbb{R}^{2n}$ and construct some Hilbert spaces $H(M,g)$ such that, for an operator a^w with $a \in S(M_1, g)$ (M_1 is another admissible weight for g), a^w will send continuously $H(M,g)$ into $H(MM_1^{-1}, g)$. The case of the classes $S_{1,0}^m$ is oversimplified by the fact that the Fourier transform provides the explicit isomorphism $\langle D\rangle^s$ of $H^s(\mathbb{R}^n)$ with $L^2(\mathbb{R}^n)$, which is the Fourier multiplier $\langle\xi\rangle^s$. With the metric corresponding to the classes (1.1.48),

$$g = (1 + |x|^2 + |\xi|^2)^{-1}(|dx|^2 + |d\xi|^2)$$

which is proven admissible in Lemma 2.2.18, although it is easy to define the natural Sobolev spaces when $m \in \mathbb{N}/2$,

$$\mathcal{H}((1 + |x|^2 + |\xi|^2)^m, g) = \{u \in \mathscr{S}'(\mathbb{R}^n), x^\alpha D^\beta u \in L^2(\mathbb{R}^n), |\alpha| + |\beta| \le 2m\},$$

a direct definition for all real m requires some techniques of interpolation and duality[21]. As a result, even in that rather simple case, the construction of the Sobolev spaces requires some effort and displaying an explicit pseudo-differential isomorphism of $L^2(\mathbb{R}^n)$ with $\mathcal{H}((1 + |x|^2 + |\xi|^2)^m, g)$ is not completely obvious. We shall mainly follow the presentation of J.-M. Bony and J.-Y. Chemin in [19], which generalized the work of A. Unterberger in [145], who dealt with symplectic metrics (see also [26], [88]).

2.6.2   Definition of the Sobolev spaces

In this section we consider an admissible metric g on $\mathbb{R}^{2n}$, a g-admissible weight m, $(\varphi_Y)_{Y \in \mathbb{R}^{2n}}$ a partition of unity given by Theorem 2.2.7 and $(\psi_Y)_{Y \in \mathbb{R}^{2n}}$, $(\theta_Y)_{Y \in \mathbb{R}^{2n}}$ the uniformly confined families of symbols given by Theorem 2.3.15, i.e., verifying

$$\varphi_Y = \psi_Y \sharp \theta_Y, \quad \int \varphi_Y |g_Y|^{1/2} dY = 1. \tag{2.6.1}$$

---

[21]Another way would be to use the spectral decomposition of the harmonic oscillator and Mehler's formula (see, e.g., [145] and [76]) to replace the Fourier transform, but it would be very specific of that example.

Definition 2.6.1. The Sobolev space $H(m, g)$ is the space of $u \in \mathscr{S}'(\mathbb{R}^n)$ such that

$$\int_{\mathbb{R}^{2n}} m(Y)^2 \|\theta_Y^w u\|_{L^2(\mathbb{R}^n)}^2 |g_Y|^{1/2} dY < \infty. \tag{2.6.2}$$

This is a pre-Hilbert space equipped with the dot product

$$\langle u, v \rangle_{H(m,g)} = \int_{\mathbb{R}^{2n}} m(Y)^2 \langle \theta_Y^w u, \theta_Y^w v \rangle_{L^2(\mathbb{R}^n)} |g_Y|^{1/2} dY. \tag{2.6.3}$$

We shall prove below (Corollary 2.6.16) that $H(m, g)$ is actually a Hilbert space but some preliminary work has to be done before obtaining this.

Remark 2.6.2. According to Lemma 1.1.12, for $u \in \mathscr{S}'(\mathbb{R}^n)$, we have $\theta_Y^w u \in \mathscr{S}(\mathbb{R}^n)$ so that the definition makes sense; the dot product (2.6.3) is indeed sesquilinear Hermitian and positive-definite, since $\theta_Y^w u = 0$ for all $Y \in \mathbb{R}^{2n}$ implies $\varphi_Y^w u = 0$ and the partition property yields $u = 0$.

We need first to prove that the space $H(m, g)$ does not depend on the choice of the partition and of the families (ψ_Y), (θ_Y).

Proposition 2.6.3. *A distribution $u \in \mathscr{S}'(\mathbb{R}^n)$ belongs to $H(m, g)$ if and only if for all uniformly confined families of symbols $(\phi_Y)_{Y \in \mathbb{R}^{2n}}$, we have*

$$\int_{\mathbb{R}^{2n}} m(Y)^2 \|\phi_Y^w u\|_{L^2(\mathbb{R}^n)}^2 |g_Y|^{1/2} dY < \infty.$$

Proof. The sufficiency follows from the definition of $H(m, g)$. Let us prove that the condition is necessary; if $(\phi_Y)_{Y \in \mathbb{R}^{2n}}$ is a uniformly confined family of symbols and $u \in H(m, g)$, we have (with $L^2(\mathbb{R}^n)$ scalar products)

$$\Omega(u) = \int m(Y)^2 \langle \phi_Y^w u, \phi_Y^w u \rangle |g_Y|^{1/2} dY$$

$$= \iiint m(Y)^2 \langle \phi_Y^w \psi_S^w \theta_S^w u, \phi_Y^w \psi_T^w \theta_T^w u \rangle |g_Y|^{1/2} |g_T|^{1/2} |g_S|^{1/2} dY \, dS \, dT$$

$$= \iiint m(Y)^2 \langle \bar{\psi}_T^w \bar{\phi}_Y^w \phi_Y^w \psi_S^w \theta_S^w u, \theta_T^w u \rangle |g_Y|^{1/2} |g_T|^{1/2} |g_S|^{1/2} dY \, dS \, dT.$$

According to the estimate (2.5.1) and to the biconfinement estimates (Corollary 2.3.3), we have for all N, $\|\psi_T^w \bar{\phi}_Y^w \phi_Y^w \psi_S^w\|_{\mathcal{L}(L^2(\mathbb{R}^n))} \leq C_N \delta_r(T, Y)^{-N} \delta_r(Y, S)^{-N}$ so that,

$$\Omega(u) \leq C_N \iiint \frac{m(Y)^2}{m(S)m(T)} \delta_r(T, Y)^{-N} \delta_r(Y, S)^{-N} |g_Y|^{1/2} dY$$

$$\times m(S) \|\theta_S^w u\|_{L^2} m(T) \|\theta_T^w u\|_{L^2} |g_T|^{1/2} |g_S|^{1/2} dS \, dT.$$

Thanks to Schur's property (see Remark 4.7.3), we need only to check, using (2.2.31),

$$\sup_{T} \iint \delta_r(T,Y)^{-N+N_0} \delta_r(Y,S)^{-N+N_0} |g_Y|^{1/2} |g_S|^{1/2} dY\, dS < \infty,$$

which follows from (2.2.29) if $N \geq N_1 + N_0$. The proof of the proposition is complete. $\square$

Theorem 2.6.4. *Let g be an admissible metric on $\mathbb{R}^{2n}$, let m_1, m_2 be g-admissible weights and let $a \in S(m_2 m_1^{-1}, g)$. Then a^w is continuous from $H(m_2, g)$ into $H(m_1, g)$.*

Proof. We have for $u \in H(m_2, g)$,

$$\|a^w u\|^2_{H(m_1,g)} = \int m_2(Y)^2 \|m_2(Y)^{-1} m_1(Y) \theta_Y^w a^w u\|^2_{L^2} |g_Y|^{1/2} dY,$$

and the family $(m_2(Y)^{-1} m_1(Y) \theta_Y \sharp a)_{Y \in \mathbb{R}^{2n}}$ is a uniformly confined family of symbols from Proposition 2.3.16. Proposition 2.6.3 gives the result. $\square$

Theorem 2.6.5. *Let g be an admissible metric on $\mathbb{R}^{2n}$. The space $H(1, g)$ is equal to $L^2(\mathbb{R}^n)$.*

Proof. The symbol $a = \int_{\mathbb{R}^{2n}} \bar{\theta}_Y \sharp \theta_Y |g_Y|^{1/2} dY$ belongs to $S(1, g)$ and a^w is thus bounded on $L^2(\mathbb{R}^n)$ (cf. Theorem 2.5.1). As a result for $u \in L^2(\mathbb{R}^n)$, we have

$$\|u\|^2_{H(1,g)} = \langle a^w u, u \rangle_{L^2(\mathbb{R}^n)} \leq C\|u\|^2_{L^2(\mathbb{R}^n)}.$$

Conversely, if $u \in H(1, g)$, we have

$$\|u\|^2_{L^2(\mathbb{R}^n)} = \iint \langle \psi_Y^w \theta_Y^w u, \psi_Z^w \theta_Z^w u \rangle_{L^2} |g_Y|^{1/2} |g_Z|^{1/2} dY\, dZ$$

$$\leq \iint \|\bar{\psi}_Z^w \psi_Y^w\|_{\mathcal{L}(L^2)} \|\theta_Y^w u\|_{L^2} \|\theta_Z^w u\|_{L^2} |g_Y|^{1/2} |g_Z|^{1/2} dY\, dZ,$$

and since the kernel $(Y, Z) \mapsto \|\overline{\psi_Z}^w \psi_Y^w\|_{\mathcal{L}(L^2)}$ satisfies the Schur property as in the proof of the theorem above, we get $\|u\|^2_{L^2(\mathbb{R}^n)} \leq C\|u\|^2_{H(1,g)}$, proving the result. $\square$

2.6.3   Characterization of pseudo-differential operators

We start with the result of R. Beals ([7]), characterizing the set of operators with a symbol in $S_{0,0}^0$ (the smooth bounded functions on $\mathbb{R}^{2n}$ as well as all their derivatives, see (2.2.2)). We follow essentially the presentation of [15].

The constant metric

For A a linear operator from $\mathscr{S}(\mathbb{R}^n)$ to $\mathscr{S}'(\mathbb{R}^n))$, L a linear form on $\mathbb{R}^{2n}$, we define $\operatorname{ad} L^w \cdot A = L^w A - A L^w$, which makes sense as an operator from $\mathscr{S}(\mathbb{R}^n)$ to $\mathscr{S}'(\mathbb{R}^n)$.

Theorem 2.6.6. *Let g be a constant metric on $\mathbb{R}^{2n}$ such that $g \leq g^\sigma$ (see (4.4.18)) and let A be a linear operator from $\mathscr{S}(\mathbb{R}^n)$ to $\mathscr{S}'(\mathbb{R}^n)$. The operator A is equal to a^w with $a \in S(1,g)$ (see Definition 2.2.6) if and only if all the iterated commutators*

$$\operatorname{ad} L_1^w \ldots \operatorname{ad} L_\nu^w \cdot A \in \mathcal{L}(L^2(\mathbb{R}^n)), \tag{2.6.4}$$

for linear forms L_j. Moreover, for all $l \in \mathbb{N}$, there exists $k(l) \in \mathbb{N}, C_l > 0$ such that, with the notation (2.2.7), we have

$$\|a\|_{S(1,g)}^{(l)} \leq C_l \sup_{\substack{\nu \leq k(l) \\ g(T_j) \leq 1, L_j(X) = [T_j, X]}} \| \operatorname{ad} L_1^w \ldots \operatorname{ad} L_\nu^w \cdot A \|_{\mathcal{L}(L^2(\mathbb{R}^n))}. \tag{2.6.5}$$

We shall use later on, for a linear operator A from $\mathscr{S}(\mathbb{R}^n)$ to $\mathscr{S}'(\mathbb{R}^n)$, the notation

$$\|A\|_{\operatorname{op}(g)}^{(k)} = \sup_{\substack{\nu \leq k \\ g(T_j) \leq 1, L_j(X) = [T_j, X]}} \| \operatorname{ad} L_1^w \ldots \operatorname{ad} L_\nu^w \cdot A \|_{\mathcal{L}(L^2(\mathbb{R}^n))}. \tag{2.6.6}$$

Proof. The conditions are obviously necessary. To prove their sufficiency, we start with a lemma.

Lemma 2.6.7. *Let $\mathcal{B}$ be a Banach space continuously contained in $\mathscr{S}'(\mathbb{R}^m)$ and translation invariant: for $a \in \mathcal{B}, t \in \mathbb{R}^m, a(\cdot - t) \in \mathcal{B}$ with the same norm as a. There exists C, N such that, for all $a \in \mathcal{B}$ such that $\partial^\alpha a \in \mathcal{B}$ for $|\alpha| \leq N$, a belongs to $L^\infty(\mathbb{R}^m)$ and*

$$\|a\|_{L^\infty(\mathbb{R}^m)} \leq C \sum_{|\alpha| \leq N} \|\partial^\alpha a\|_{\mathcal{B}}.$$

Proof of the lemma. Since the unit ball of $\mathcal{B}$ is a bounded subset of $\mathscr{S}'$, there exists a semi-norm p_0 of $\mathscr{S}$ such that for $a \in \mathcal{B}, \varphi \in \mathscr{S}$,

$$|\langle a, \varphi \rangle_{\mathscr{S}', \mathscr{S}}| \leq \|a\|_{\mathcal{B}} \, p_0(\varphi).$$

On the other hand, with a semi-norm p_1 on $\mathscr{S}$, we have for $\chi \in \mathscr{S}$,

$$(\chi(D)a)(x) = \langle a(\cdot - x), \hat{\chi} \rangle \implies \|\chi(D)a)\|_{L^\infty} \leq \|a\|_{\mathcal{B}} \, p_0(\hat{\chi}) = \|a\|_{\mathcal{B}} \, p_1(\chi).$$

Let $\psi \in \mathscr{S}$ such that $\psi \equiv 1$ on the unit ball. For $\epsilon \in (0,1]$ the function $\xi \mapsto \rho_{\epsilon,N}(\xi) = \big(\psi(\epsilon\xi) - \psi(\xi)\big)|\xi|^{-N}$ belongs to $\mathscr{S}$ (note that it vanishes for $|\xi| \leq 1$)

and since

$$\xi^\alpha(\partial_\xi^\beta \rho_{\epsilon,N})(\xi) = \sum_{\beta'+\beta''=\beta} \frac{\beta!}{\beta'!\beta''!} \xi^\alpha \partial_\xi^{\beta'}(|\xi|^{-N})\Big((\partial_\xi^{\beta''}\psi)(\epsilon\xi)\epsilon^{|\beta''|} - (\partial_\xi^{\beta''}\psi)(\xi)\Big)$$

$$= \mathbf{1}\{|\xi| \geq 1\} \sum_{\substack{\beta'+\beta''=\beta \\ |\beta''|\geq 1}} O(|\xi|^{-N-|\beta'|-|\beta''|+|\alpha|}) \underbrace{(\partial_\xi^{\beta''}\psi)(\epsilon\xi)(\epsilon|\xi|)^{|\beta''|}}_{\text{bounded}}$$

$$+ \mathbf{1}\{|\xi| \geq 1\} \sum_{\substack{\beta'+\beta''=\beta \\ |\beta''|\geq 1}} O(|\xi|^{-N-|\beta'|+|\alpha|})(\partial_\xi^{\beta''}\psi)(\xi)$$

$$+ \mathbf{1}\{|\xi| \geq 1\}\xi^\alpha \partial_\xi^\beta(|\xi|^{-N})\big(\psi(\epsilon\xi) - \psi(\xi)\big),$$

we find that for $N \geq |\alpha|$, $\|\xi^\alpha \partial_\xi^\beta \rho_{\epsilon,N}\|_{L^\infty} \leq C_{\psi,\alpha,\beta}$. As a consequence, for N large enough (chosen as an even integer), $p_1(\rho_{\epsilon,N})$ is bounded above by a constant independent of ϵ. We have thus, with C independent of ϵ,

$$\|\psi(\epsilon D)a\|_{L^\infty} \leq \|\psi(D)a\|_{L^\infty} + \|\rho_{\epsilon,N}(D)|D|^N a\|_{L^\infty} \leq C(\|a\|_{\mathcal{B}} + \||D|^N a\|_{\mathcal{B}}) = C_1,$$

and letting ϵ tends to 0 provides the result of the lemma, since the weak limit of $\psi(\epsilon D)a$ in $\mathscr{S}'(\mathbb{R}^m)$ is a and the above estimate gives the bound

$$|\langle a, \varphi \rangle_{\mathscr{S}',\mathscr{S}}| = \lim_{\epsilon \to 0} |\langle \psi(\epsilon D)a, \varphi \rangle| \leq C_1 \|\varphi\|_{L^1(\mathbb{R}^m)},$$

completing the proof. $\square$

We go on with the proof of the theorem. We consider the Banach space $\mathcal{B}$ of $b \in \mathscr{S}'(\mathbb{R}^{2n})$ such that $b^w \in \mathcal{L}(L^2(\mathbb{R}^n))$, equipped with the[22] norm $\|b\|_{\mathcal{B}} = \|b^w\|_{\mathcal{L}(L^2(\mathbb{R}^n))}$. The invariance by translation of $\mathcal{B}$ is an immediate consequence of (2.1.14) for the phase translation. We need now to prove the embedding in $\mathscr{S}'(\mathbb{R}^{2n})$. For $b \in \mathcal{B}, \Phi \in \mathscr{S}(\mathbb{R}^{2n})$, we have

$$\langle b, \Phi \rangle_{\mathscr{S}',\mathscr{S}} = \int e^{2i\pi x\xi} b(x,\xi) \langle x \rangle^{-\frac{n+1}{2}} \langle \xi \rangle^{-\frac{n+1}{2}} \Theta(\eta,y) e^{2i\pi(\eta x + y\xi)} dx d\xi dy d\eta$$

$$= \int e^{2i\pi x\xi} b(x,\xi) \underbrace{\langle x \rangle^{-\frac{n+1}{2}} e^{2i\pi\eta x}}_{=\overline{u_\eta(x)}} \underbrace{\langle \xi \rangle^{-\frac{n+1}{2}} e^{2i\pi y\xi}}_{=\widehat{v_y}(\xi)} dx d\xi \Theta(\eta,y) dy d\eta,$$

with $\Theta(\eta,y) = \iint e^{-2i\pi(\eta x' + y\xi')} \langle x' \rangle^{\frac{n+1}{2}} \langle \xi' \rangle^{\frac{n+1}{2}} \Phi(x',\xi') e^{-2i\pi x'\xi'} dx' d\xi'$. As a result, we obtain

$$|\langle b, \Phi \rangle| \leq \iint |\Theta(\eta,y)||\langle b(x,D)v_y, u_\eta \rangle| d\eta dy \leq \iint |\Theta(\eta,y)| d\eta dy \|b(x,D)\|_{\mathcal{L}(L^2)}$$

$$\leq p(\Phi)\|b(x,D)\|_{\mathcal{L}(L^2)},$$

---

[22]It is obviously a norm from the estimate (2.6.7) Moreover the space $\mathcal{B}$ is isomorphic to $\mathcal{L}(L^2(\mathbb{R}^n))$: from the kernel theorem, any linear continuous operator from $\mathscr{S}(\mathbb{R}^n)$ to $\mathscr{S}'(\mathbb{R}^n)$ has a distribution kernel in $\mathscr{S}'(\mathbb{R}^{2n})$, and it is the case in particular of $B \in \mathcal{L}(L^2(\mathbb{R}^n))$. As a consequence, the Weyl symbol b of B, linked to the kernel κ_b by the formula (2.1.22) belongs to $\mathscr{S}'(\mathbb{R}^{2n})$ as well.

where p is a semi-norm on $\mathscr{S}(\mathbb{R}^{2n})$. To get the result with b^w, we use the notation
(4.1.14) and, noting that the group J^t is an isomorphism of $\mathscr{S}(\mathbb{R}^{2n})$, $\langle b, \Phi \rangle = \langle J^{1/2}b, J^{-1/2}\Phi \rangle$ so that

$$|\langle b, \Phi \rangle| \leq p(J^{-1/2}\Phi)\|b^w\|_{\mathcal{L}(L^2)} = p_0(\Phi)\|b^w\|_{\mathcal{L}(L^2)}, \qquad (2.6.7)$$

where p_0 is a semi-norm of $\mathscr{S}(\mathbb{R}^{2n})$. Using Lemma 4.4.25, we can find some linear
symplectic coordinates y, η such that $g = \sum_{1 \leq j \leq n} h_j(|dy_j|^2 + |d\eta_j|^2)$, and the
condition $g \leq g^\sigma$ means $(0 <)h_j \leq 1$, so that the vectors $\partial_{y_j}, \partial_{\eta_j}$ have a g-norm
≤ 1. Since the assumptions of Lemma 2.6.7 are fulfilled, for a linear operator
$A = a^w \in \mathcal{L}(L^2(\mathbb{R}^n))$, satisfying (2.6.4), we have $a \in \mathcal{B}$ and

$$\|a\|_{L^\infty(\mathbb{R}^{2n})} \leq C \sum_{|\alpha|+|\beta| \leq N} \|(\partial_y^\alpha \partial_\eta^\beta a)^w\|_{\mathcal{L}(L^2(\mathbb{R}^n))}$$

$$\leq C' \sup_{\substack{\nu \leq N \\ g(T_j) \leq 1, L_j(X)=[T_j,X]}} \| \operatorname{ad} L_1^w \dots \operatorname{ad} L_\nu^w a^w\|_{\mathcal{L}(L^2(\mathbb{R}^n))},$$

which is (2.6.5) for $l = 0, k = N$. To get the estimates for the derivatives, we con-
sider for linear forms $(L_j = [T_j, \cdot], g(T_j) \leq 1)_{1 \leq j \leq l}$, the operator $\operatorname{ad} L_1^w \dots \operatorname{ad} L_l^w a^w$,
which belongs to $\mathcal{L}(L^2(\mathbb{R}^n))$ by the hypothesis (2.6.4), with symbol $\partial_{T_1} \dots \partial_{T_l} a$,
thus belonging to $\mathcal{B}$. We can apply the previous inequality to that symbol and
obtain (2.6.5) with $k(l) = l + N$, completing the proof of the theorem. $\square$

Remark 2.6.8. Let $\rho \in C_c^\infty(\mathbb{R}^{2n})$ with integral 1, supported in the unit ball for the
standard Euclidean metric Γ_0, $\epsilon > 0$ and $\rho_\epsilon(X) = \epsilon^{-2n}\rho(X/\epsilon)$. For $a \in \mathscr{S}'(\mathbb{R}^{2n})$,
the smooth function $a * \rho_\epsilon$ satisfies for any $X \in \mathbb{R}^{2n}$, $(a * \rho_\epsilon)(X) = \langle a(X - \cdot), \rho_\epsilon(\cdot) \rangle$
and thus for N an even integer, using (2.6.7) (and (2.1.14) for a phase translation),

$$|(a * \rho_\epsilon)(X)| = |\langle \langle D_Y \rangle^N (a(X - Y)), \langle D_Y \rangle^{-N}(\rho_\epsilon(Y)) \rangle|$$

$$\leq p_0(\langle D \rangle^{-N}\rho_\epsilon)\|(\langle D \rangle^N a)^w\|_{\mathcal{L}(L^2)}.$$

Since the semi-norm p_0 can be chosen as $p_0(\Phi) = C\|\langle X \rangle^{N_0}\langle D \rangle^{N_0}\Phi\|_{L^1(\mathbb{R}^{2n})}$, we
get

$$p_0(\langle D \rangle^{-N_0}\rho_\epsilon) = C\|\langle X \rangle^{N_0}\rho_\epsilon\|_{L^1(\mathbb{R}^{2n})} \leq C_0 \quad \text{if } \epsilon \leq 1,$$

implying that $\|a\|_{L^\infty(\mathbb{R}^{2n})} \leq C_0 \|a^w\|_{\operatorname{op}(\Gamma_0)}^{(N_0)}$. We get also

$$\|a\|_{S(1,\Gamma_0)}^{(k)} \leq C_k \|a^w\|_{\operatorname{op}(\Gamma_0)}^{(N_0+k)},$$

and the formula (2.1.14) implies as well

$$\|a\|_{S(1,g_0)}^{(k)} \leq C_k \|a^w\|_{\operatorname{op}(g_0)}^{(N_0+k)}, \qquad (2.6.8)$$

where g_0 is a positive-definite quadratic form such that $g_0 \leq g_0^\sigma$, which provides
another proof of Theorem 2.6.6. However Lemma 2.6.7 is of independent interest
and useful for the proof displayed above.

Lemma 2.6.9. *For all $k \in \mathbb{N}$, for all $\tau \in [0, 1[$, there exists $C_{k,\tau} > 0, l_{k,\tau} \in \mathbb{N}$, such that, for all $a \in C^\infty(\mathbb{R}^{2n})$, for all Euclidean norms g_0 on $\mathbb{R}^{2n}$ such that $g_0 \leq g_0^\sigma$,*

$$\|(a^{(k)})^w\|_{\mathcal{L}(L^2(\mathbb{R}^n))} \leq C_{k,\tau}\|a^w\|_{\mathcal{L}(L^2)}^\tau \|(a^{(l_{k,\tau})})^w\|_{\mathcal{L}(L^2)}^{1-\tau}, \tag{2.6.9}$$

where $\|(a^{(k)})^w\|_{\mathcal{L}(L^2(\mathbb{R}^n))} = \sup_{T \in \mathbb{R}^{2n}, g_0(T)=1} \|(a^{(k)}(\cdot)T^k)^w\|_{\mathcal{L}(L^2(\mathbb{R}^n))}.$ \quad (2.6.10)

N.B. It is of key importance to notice that the constants $C_{k,\tau}, l_{k,\tau}$ depend only on k and τ and *not* on g_0 or a.

Proof. For $X, T \in \mathbb{R}^{2n}$, we have

$$a(X + T) = a(X) + a'(X)T + \int_0^1 (1 - s)^2 a''(X + sT)ds\, T^2$$

and thus, with $T = \lambda e_j$ ($\lambda > 0$, e_j is the j-th vector of an orthonormal basis of $\mathbb{R}^{2n}$ for g_0)

$$\forall \lambda > 0, \quad \lambda\|(\partial_j a)^w\|_{\mathcal{L}(L^2)} \leq 2\|a^w\|_{\mathcal{L}(L^2)} + \frac{\lambda^2}{2}\|(\partial_j^2 a)^w\|_{\mathcal{L}(L^2)}$$

so that $\|(\partial_j a)^w\|_{\mathcal{L}(L^2)} \leq 2\|a^w\|_{\mathcal{L}(L^2)}^{1/2}\|(\partial_j^2 a)^w\|_{\mathcal{L}(L^2)}^{1/2}$ and consequently

$$\|(\partial_j \partial_k a)^w\|_{\mathcal{L}(L^2)} \leq 2^{3/2}\|a^w\|_{\mathcal{L}(L^2)}^{1/4}\|(\partial_k^2 a)^w\|_{\mathcal{L}(L^2)}^{1/4}\|(\partial_j^2 \partial_k a)^w\|_{\mathcal{L}(L^2)}^{1/2}.$$

Going on with that interpolation argument, using also (4.2.2), we get the result of the lemma. $\qquad\square$

Localized commutators

Theorem 2.6.10. *Let A be a linear mapping from $\mathscr{S}(\mathbb{R}^n)$ into $\mathscr{S}'(\mathbb{R}^n)$, g, m be admissible metric and weight on $\mathbb{R}^{2n}$. There exists $a \in S(m, g)$ such that $A = a^w$ if and only if, for all uniformly confined families of symbols $(\psi_Y)_{Y \in \mathbb{R}^{2n}}$ (see Definition 2.3.14), for all linear forms L_j given by $L_j(X) = [X, T_j]$, with $g_Y(T_j) \leq 1$, for all $k \in \mathbb{N}$,*

$$\sup_{Y \in \mathbb{R}^{2n}} m(Y)^{-1}\|\operatorname{ad} L_1^w \ldots \operatorname{ad} L_k^w \cdot \psi_Y^w A\|_{\mathcal{L}(L^2(\mathbb{R}^n))} < +\infty, \tag{2.6.11}$$

i.e., with the notation (2.6.6), $\sup_{Y \in \mathbb{R}^{2n}} m(Y)^{-1}\|\psi_Y^w A\|_{op(g_Y)}^{(k)} < +\infty$.

Proof. The condition is easily seen to be necessary: if $A = a^w$ with $a \in S(m, g)$, we have with the partition of unity of Theorem 2.2.7, $L(X) = [X, T], g_Y(T) \leq 1$, using (2.1.12), with $a_Z = \varphi_Z a$,

$$\psi_Y^w A = \int_{\mathbb{R}^{2n}} \psi_Y^w a_Z^w |g_Z|^{1/2} dZ,$$

$$2i\pi \operatorname{ad} L^w \cdot \psi_Y^w A = \int_{\mathbb{R}^{2n}} \left((\psi_Y' \cdot T)^w a_Z^w + \psi_Y^w (a_Z' \cdot T)^w\right)|g_Z|^{1/2} dZ,$$

and using the estimates in the proof of Theorem 2.3.7 as well as (2.5.1), we get (2.6.11) for $k = 1$ and by iteration for all k. Let us prove the sufficiency. First of all, with $(\varphi_Y)_{Y \in \mathbb{R}^{2n}}$ standing for a partition of unity, as given by Theorem 2.2.7, from Theorem 2.3.15, we can find two uniformly confined families of symbols $(\theta_Y)_{Y \in \mathbb{R}^{2n}}, (\psi_Y)_{Y \in \mathbb{R}^{2n}}$ such that $\varphi_Y = \psi_Y \sharp \theta_Y$. From Theorem 2.6.6 applied to the constant metric g_Y, we get that $m(Y)^{-1} \theta_Y^w A = c_Y^w$, where $c_Y \in S(1, g_Y)$ uniformly (i.e., with semi-norms bounded above independently of Y) and thus

$$m(Y)^{-1} \varphi_Y^w A = m(Y)^{-1} \psi_Y^w \theta_Y^w A = \psi_Y^w c_Y^w,$$

which gives

$$A = \int_{\mathbb{R}^{2n}} \varphi_Y^w A |g_Y|^{1/2} dY = \int_{\mathbb{R}^{2n}} m(Y)(\psi_Y \sharp c_Y)^w |g_Y|^{1/2} dY = a^w,$$

with $a \in S(m, g)$ since the family $(\psi_Y \sharp c_Y)_{Y \in \mathbb{R}^{2n}}$ is a uniformly confined family of symbols from Lemma 2.3.6. $\qquad\square$

Remark 2.6.11. The previous proof implies that

$$\forall k, \exists C, \exists l, \quad \|a\|_{S(m,g)}^{(k)} \leq C \|a^w\|_{op(m,g)}^{(l)}, \tag{2.6.12}$$

where $\|a\|_{S(m,g)}^{(k)}$ is given by (2.2.7) and with

$$\|a^w\|_{op(m,g)}^{(l)} = \sup_{\substack{Y \in \mathbb{R}^{2n}, p \leq l \\ g_Y(L_j) \leq 1}} m(Y)^{-1} \| \operatorname{ad} L_1^w \dots \operatorname{ad} L_p^w \cdot \theta_Y^w A \|_{\mathscr{L}(L^2(\mathbb{R}^n))}, \tag{2.6.13}$$

where the linear form L, given by $LX = [T, X]$ is identified with the vector T and $g_Y(L)$ is defined as $g_Y(T)$. We have also, using the notation (2.6.6),

$$\|a^w\|_{op(m,g)}^{(l)} = \sup_{Y \in \mathbb{R}^{2n}} m(Y)^{-1} \|\theta_Y^w a^w\|_{op(g_Y)}^{(l)}. \tag{2.6.14}$$

Theorem 2.6.12. *Let g, m be admissible metric and weight on $\mathbb{R}^{2n}$ and let A be a linear operator from $\mathscr{S}(\mathbb{R}^n)$ into $\mathscr{S}'(\mathbb{R}^n)$. There exists $a \in S(m, g)$ such that $A = a^w$ if and only if the semi-norms $\|A\|_{op(m,g)}^{(k)}$ defined in (2.6.14) are finite and*

$$\forall k, \exists C, \exists l, \quad \|a\|_{S(m,g)}^{(k)} \leq C \|A\|_{op(m,g)}^{(l)}, \tag{2.6.15}$$

$$\forall k, \forall \tau \in [0, 1[, \exists C, \exists l, \quad \|a\|_{S(m,g)}^{(k)} \leq C \left(\|A\|_{op(m,g)}^{(0)}\right)^\tau \left(\|A\|_{op(m,g)}^{(l)}\right)^{1-\tau}. \tag{2.6.16}$$

Proof. The equivalence is proven in Theorem 2.6.10 and the first inequality in the above remark. The second inequality follows from the first and from the interpolation result of Lemma 2.6.9. $\qquad\square$

2.6.4 One-parameter group of elliptic operators

Lemma 2.6.13. *Let g be an admissible metric on $\mathbb{R}^{2n}$ and m be a g-admissible weight. Then $1 + |\operatorname{Log} m|$ is also a g-admissible weight. Assuming $m \in S(m, g)$, we have*[23]

$$\forall k \geq 1, \ \exists C_k \quad |(\operatorname{Log} m)^{(k)}(X)T^k| \leq C_k g_X(T)^{k/2}, \tag{2.6.17}$$

and $\operatorname{Log} m \in S(1 + |\operatorname{Log} m|, g)$.

Proof. There exists $r > 0, c > 0$ such that whenever $g_X(Y - X) \leq r^2$, we have $m(X)m(Y)^{-1} \in [e^{-c}, e^c]$, so that $|\operatorname{Log} m(X) - \operatorname{Log} m(Y)| \leq c$, implying

$$\left(\frac{1 + |\operatorname{Log} m(Y)|}{1 + |\operatorname{Log} m(X)|}\right)^{\pm 1} \leq 1 + c,$$

proving the slow variation. Moreover, there exists $C > 1, N \geq 0$ such that

$$\frac{1 + |\operatorname{Log} m(Y)|}{1 + |\operatorname{Log} m(X)|} \leq 1 + \frac{|\operatorname{Log} m(Y) - \operatorname{Log} m(X)|}{1 + |\operatorname{Log} m(X)|}$$

$$\leq 1 + \frac{\operatorname{Log} C + N \operatorname{Log}\left(1 + (g_X^\sigma \wedge g_Y^\sigma)(Y - X)\right)}{1 + |\operatorname{Log} m(X)|},$$

giving the first result of the lemma. We have for $k \geq 1$ (see (4.3.1)),

$$\frac{1}{k!}(\operatorname{Log} m)^{(k)}T^k = \sum_{k_1 + \cdots + k_r = k} (-1)^{r-1} r^{-1} m^{-r} \frac{1}{k_1!} m^{(k_1)} T^{k_1} \cdots \frac{1}{k_r!} m^{(k_r)} T^{k_r},$$

and $|m^{(k_j)} T^{k_j}| \leq C m(X) g_X(T)^{k_j/2}$ which implies (2.6.17). $\square$

Lemma 2.6.14. *Let g be an admissible metric on $\mathbb{R}^{2n}$, m be a g-admissible weight such that $m \in S(m, g)$. If $(\psi_Y)_{Y \in \mathbb{R}^{2n}}$ is a uniformly confined family of symbols, the family $\left(\psi_Y(\cdot) \sharp \operatorname{Log}(m(\cdot)m(Y)^{-1})\right)_{Y \in \mathbb{R}^{2n}}$ is also uniformly confined.*

Proof. We have with the partition of unity of Theorem 2.2.7,

$$b_Y = \psi_Y \sharp \operatorname{Log}\left(m(\cdot)m(Y)^{-1}\right) = \int \psi_Y \sharp \varphi_Z \operatorname{Log}\left(m(\cdot)m(Y)^{-1}\right) |g_Z|^{1/2} dZ,$$

and $|\varphi_Z(X) \operatorname{Log}\left(m(X)m(Y)^{-1}\right)| \leq C\varphi_Z(X)\left(1 + \operatorname{Log}(\delta_r(Z, Y))\right)$, whereas the X-derivatives of $\operatorname{Log}\left(m(X)m(Y)^{-1}\right)$ satisfy the estimates of $S(1, g)$ (cf. (2.6.17)). We get that $\left(\delta_r(Y, Z)^{-1} \varphi_Z \operatorname{Log}\left(m(\cdot)m(Y)^{-1}\right)\right)_{Z \in \mathbb{R}^{2n}}$ is a uniformly confined family of symbols and the confinement estimates yield

$$|b_Y^{(k)}(X)T^k| \leq C_{N,N'} \int \delta_r(Z, Y) \delta_r(Z, Y)^{-N}\left(1 + g_Y^\sigma(X - U_{Y,r})\right)^{-N'} |g_Z|^{1/2} dZ,$$

which gives the result, if $N \geq N_1 + 1$, with N_1 given by (2.2.29). $\square$

---

[23] According to Remark 2.2.8, if m is a g-admissible weight, we may assume $m \in S(m, g)$.

Theorem 2.6.15. *Let g be an admissible metric on $\mathbb{R}^{2n}$, M, m be g-admissible weights such that $M \in S(M, g), m \in S(m, g)$ and let $a \in S(m, g), \alpha \in S(1, g)$. There exists a unique smooth application $\mathbb{R} \times \mathbb{R}^{2n} \ni (t, X) \mapsto a_t(X) \in \mathbb{C}$ such that $a_t \in S(mM^t, g)$ for all t and*

$$\frac{\partial a_t(X)}{\partial t} = ((\alpha + \operatorname{Log} M)\sharp a_t)(X), \quad a_0 = a. \tag{2.6.18}$$

When $m \equiv a_0 \equiv 1$, the unique solution of (2.6.18) satisfies as well $\frac{\partial a_t(X)}{\partial t} = (a_t\sharp(\alpha + \operatorname{Log} M))(X)$ and we have

$$a_t \sharp a_s = a_{t+s}, \quad a_t \in S(M^t, g). \tag{2.6.19}$$

First step of the proof. We suppose first the existence of a solution a_t of (2.6.18). We have

$$a_t(X) = a_0(X) + \int_0^t (\underbrace{(\alpha + \operatorname{Log} M)}_{S(\langle \operatorname{Log} M \rangle, g)} \sharp \underbrace{a_s}_{S(mM^s, g)})(X)ds$$

and thus $a_t(X) = a_0(X) + \int_0^t b(s, X)$, with $b(s) \in S(m\langle \operatorname{Log} M \rangle M^s, g)$, proving that the mapping $t \mapsto a_t$ is C^1 from $[-\tau, \tau]$ for any positive τ into the Fréchet space

$$S(m\langle \operatorname{Log} M \rangle (M + M^{-1})^\tau, g).$$

We define, using (2.6.1),

$$c_{t,Y} = m(Y)^{-1}M(Y)^{-t}\theta_Y \sharp a_t. \tag{2.6.20}$$

The symbol $c_{t,Y}(X)$ is C^1 in t and such that

$$\frac{\partial c_{t,Y}}{\partial t} = m(Y)^{-1}M(Y)^{-t}\theta_Y\sharp(\alpha + \operatorname{Log} M)\sharp a_t - \operatorname{Log} M(Y)c_{t,Y}$$

and verifying, with

$$\beta_Y(X) = \alpha(X) + \operatorname{Log}\frac{M(X)}{M(Y)}, \tag{2.6.21}$$

the equation $\frac{\partial c_{t,Y}}{\partial t} = m(Y)^{-1}M(Y)^{-t}\int \theta_Y\sharp\beta_Y\sharp\psi_Z\sharp\theta_Z\sharp a_t|g_Z|^{1/2}dZ$, i.e.,

$$\frac{\partial c_{t,Y}}{\partial t} = \int \frac{m(Z)M(Z)^t}{m(Y)M(Y)^t}\theta_Y\sharp\beta_Y\sharp\psi_Z\sharp c_{t,Z}|g_Z|^{1/2}dZ.$$

We have thus

$$\frac{\partial c_{t,Y}}{\partial t} = \int K_{t,Y,Z}\sharp c_{t,Z}|g_Z|^{1/2}dZ, \quad c_{0,Y} = m(Y)^{-1}\theta_Y\sharp a, \tag{2.6.22}$$

$$\text{with } K_{t,Y,Z} = \frac{m(Z)M(Z)^t}{m(Y)M(Y)^t}\theta_Y\sharp\beta_Y\sharp\psi_Z. \tag{2.6.23}$$

We shall also consider the analogue

$$\frac{\partial \tilde{c}_{t,Y}}{\partial t} = \int \tilde{K}_{t,Y,Z} \sharp \tilde{c}_{t,Z} |g_Z|^{1/2} dZ, \quad \tilde{c}_{0,Y} = m(Y)^{-1} \theta_Y \sharp a, \tag{2.6.24}$$

$$\text{with } \tilde{K}_{t,Y,Z} = \frac{m(Z)M(Z)^t}{m(Y)M(Y)^t} \theta_Y \sharp \beta_Z \sharp \psi_Z. \tag{2.6.25}$$

Next step: resolution of the equation (2.6.22). For each $k \in \mathbb{N}$, we consider the vector space E_k defined as follows: E_0 is the Banach space $L^\infty(\mathbb{R}^{2n}, \mathcal{L}(L^2(\mathbb{R}^n)))$, E_1 is the subspace of E_0 such that

$$\|A\|_{E_1} = \|A\|_{E_0} + \sup_{\substack{Y, g_Y(T)=1 \\ L(X)=[X,T]}} \|[L^w, A_Y]\|_{\mathcal{L}(L^2(\mathbb{R}^n))} < \infty.$$

More generally, for $k \geq 1$, the space E_k is defined, using the notation (2.6.6), as the subspace of E_0 such that

$$\|A\|_{E_k} = \|A_Y\|_{\mathrm{op}(g_Y)}^{(k)} < \infty.$$

If (A_j) is a Cauchy sequence in E_1, it converges in E_0 (with limit A) and for $g_Y(T) = 1$, $L_T(X) = [X,T]$, the sequence $([L_T^w, A_{j,Y}])$ converges in

$$L^\infty(\mathbb{G}_1, \mathcal{L}(L^2(\mathbb{R}^n))), \quad \mathbb{G}_1 = \{(Y,T) \in \mathbb{R}^{2n} \times \mathbb{R}^{2n}, g_Y(T) \leq 1\},$$

with limit B. Now we have also as operators from $\mathscr{S}(\mathbb{R}^n)$ to $\mathscr{S}'(\mathbb{R}^n)$, the equality $[L_T^w, A_Y] = B_{Y,T}$; this proves that $A \in E_1$ and that the sequence (A_j) converges in E_1 to A since

$$\sup_{Y, g_Y(T)=1} \|[L_T^w, A_{j,Y} - A_Y]\|_{\mathcal{L}(L^2(\mathbb{R}^n))}$$

$$= \sup_{Y, g_Y(T)=1} \|[L_T^w, A_{j,Y}] - B_{Y,T}\|_{L^\infty(\mathbb{G}_1, \mathcal{L}(L^2(\mathbb{R}^n)))} \underset{j \to +\infty}{\longrightarrow} 0. \tag{2.6.26}$$

The same type of argument works for $k \geq 2$, replacing the bundle $\mathbb{G}_1$ by the suitable $\mathbb{G}_k$. We define the operator $\mathcal{K}_t$ by $(\mathcal{K}_t A)_Y = \int K_{t,Y,Z}^w A_Z |g_Z|^{1/2} dZ$, and we want to prove that it is a bounded operator on E_k, uniformly for t bounded. Let $Y, T \in \mathbb{R}^{2n}$ such that $g_Y(T) = 1$, L the linear form $LX = [X,T]$. We check

$$\|(\operatorname{ad} L)^k (\mathcal{K}_t A)_Y\|_{\mathcal{L}(L^2(\mathbb{R}^n))}$$

$$\leq C \max_{k_1+k_2=k} \int \|(\operatorname{ad} L)^{k_1} K_{t,Y,Z}^w\|_{\mathcal{L}(L^2)} \|(\operatorname{ad} L)^{k_2} A_Z\|_{\mathcal{L}(L^2)} |g_Z|^{1/2} dZ$$

$$\leq C \max_{k_1+k_2=k} \int \|(\operatorname{ad} L)^{k_1} K_{t,Y,Z}^w\|_{\mathcal{L}(L^2)} \|A\|_{E_k} g_Z(T)^{k_2/2} g_Y(T)^{-k_2/2} |g_Z|^{1/2} dZ.$$

According to Lemma 2.6.14, the family $(\theta_Y \sharp \beta_Y)_{Y \in \mathbb{R}^{2n}}$ is uniformly confined and the biconfinement Theorem 2.3.2 implies that the families

$$(\delta_r(Y, Z)^N \theta_Y \sharp \beta_Y \sharp \psi_Z)_{Y \in \mathbb{R}^{2n}}$$

are uniformly confined (and this uniformly with respect to Z), so that, using (2.5.1) and (2.2.29) for N large enough,

$$\sup_{Y, |t| \leq T} \|(\operatorname{ad} L)^k (\mathcal{K}_t A)_Y\|_{\mathcal{L}(L^2)}$$

$$\leq C_N \max_{l \leq k} \sup_{Y, |t| \leq \tau} \int \delta_r(Y, Z)^{-N} \frac{m(Z) M(Z)^t}{m(Y) M(Y)^t} \|A\|_{E_k} \frac{g_Z(T)^{l/2}}{g_Y(T)^{l/2}} |g_Z|^{1/2} dZ < \infty.$$

Handling the same estimate with $(\operatorname{ad} L)^k$ replaced by $(\operatorname{ad} L_1)^{k_1} \ldots (\operatorname{ad} L_r)^{k_r}$ is strictly analogous to the previous argument. We have proven that the operator $\mathcal{K}_t$ is bounded on the Banach space E_k, with a norm bounded for t bounded. As a result, since $m(Y)^{-1} \theta_Y^w a_0^w$ belongs to all E_k, the ODE

$$\frac{dC}{dt} = \mathcal{K}C, \quad C(0) = m(Y)^{-1} \theta_Y^w a_0^w, \tag{2.6.27}$$

has a unique solution on $\mathbb{R}$ belonging to all E_k. The same holds for the equation (2.6.24). The characterization of Theorem 2.6.10 implies that $C_Y(t) = c_{t,Y}^w$ with $c_{t,Y} \in S(1, g_Y)$ uniformly in Y and for t bounded.

Third step: proof of the first statement of the theorem. If a_t is a solution of (2.6.18), we have proven that $C(t, Y) = c_{t,Y}^w$ given by (2.6.20) satisfies (2.6.27) and this implies the uniqueness of the solution of (2.6.18), since

$$a_t^w = \int \psi_Y^w \theta_Y^w a_t^w |g_Y|^{1/2} dY = \int m(Y) M(Y)^t \, \psi_Y^w c_{t,Y}^w |g_Y|^{1/2} dY.$$

To prove the existence of a solution of (2.6.18), we consider the solution $\tilde{c}_{t,Y}$ of (2.6.24), and we define

$$a_t = \int m(Y) M(Y)^t \, \psi_Y \sharp \tilde{c}_{t,Y} |g_Y|^{1/2} dY.$$

Thanks to Theorem 2.6.10 and to Proposition 2.3.16, the family $(\psi_Y \sharp \tilde{c}_{t,Y})_{Y \in \mathbb{R}^{2n}}$ is a uniformly confined family of symbols and a_t belongs to $S(mM^t, g)$. On the other hand, we have

$$\frac{\partial a_t}{\partial t} = \int m(Y) M(Y)^t \operatorname{Log} M(Y) \, \psi_Y \sharp \tilde{c}_{t,Y} |g_Y|^{1/2} dY$$

$$+ \iint m(Y) M(Y)^t \psi_Y \sharp \tilde{K}_{t,Y,Z} \sharp \tilde{c}_{t,Z} |g_Z|^{1/2} |g_Y|^{1/2} dY dZ$$

$$= \int m(Z)M(Z)^t \operatorname{Log} M(Z)\ \psi_Z \sharp \tilde{c}_{t,Z} |g_Z|^{1/2} dZ$$

$$+ \iint m(Z)M(Z)^t \psi_Y \sharp \theta_Y \sharp \beta_Z \sharp \psi_Z \sharp \tilde{c}_{t,Z} |g_Z|^{1/2} |g_Y|^{1/2} dY\, dZ$$

$$= \int m(Z)M(Z)^t (\alpha + \operatorname{Log} M) \sharp \psi_Z \sharp \tilde{c}_{t,Z} |g_Z|^{1/2} dZ$$

$$= (\alpha + \operatorname{Log} M) \sharp a_t,$$

along with $a_{t=0} = \int m(Y)\psi_Y \sharp m(Y)^{-1}\theta_Y \sharp a_0 |g_Y|^{1/2} dY = a_0$, providing a solution of (2.6.18). We may consider the unique solution a_t of (2.6.18) for $a_0 \equiv 1$. If moreover $m \equiv 1$, for s given in $\mathbb{R}$, the mappings $t \mapsto a_{t+s}, t \mapsto a_t \sharp a_s$ are both solutions of (2.6.18) with value a_s at $t = 0$ and the already proven uniqueness property (with $M = m^s$) implies

$$a_{t+s} = a_t \sharp a_s.$$

In particular, we have for all t, $a_t \sharp a_{-t} = 1$ and, taking the derivative of that identity, we get $\alpha \sharp a_t \sharp a_{-t} - a_t \sharp \alpha \sharp a_{-t}$, that is $\alpha = a_t \sharp \alpha \sharp a_{-t}$, implying the commutation for $\sharp$ of a_t with α, completing the proof of the theorem. $\square$

Corollary 2.6.16. *Let g be an admissible metric and m be an admissible g-weight. There exists $a \in S(m, g), b \in S(m^{-1}, g)$ such that $a \sharp b = b \sharp a = 1$ and for any g-admissible weight M, the operator a^w is an isomorphism from $H(M, g)$ onto $H(Mm^{-1}, g)$ with inverse b^w. The space $H(m, g)$ is a Hilbert space with the dot product (2.6.3). An equivalent Hilbertian structure on $H(m, g)$ is given by the dot product $\langle a^w u, a^w v \rangle_{L^2(\mathbb{R}^n)}$.*

Proof. We may assume that $m \in S(m, g)$ and solve the equation (2.6.19) with $M = m, \alpha = 0$: we get $a_1 \sharp a_{-1} = 1 = a_{-1} \sharp a_1$, $a_{\pm 1} \in S(m^{\pm 1}, g)$. Moreover, for a g-admissible weight M, thanks to Theorem 2.6.4, the operator a_1^w (resp. a_{-1}^w) sends continuously $H(M, g)$ into $H(Mm^{-1}, g)$ (resp. $H(Mm^{-1}, g)$ into $H(M, g)$) and since $a_1^w a_{-1}^w = \operatorname{Id}_{H(Mm^{-1}, g)}, a_{-1}^w a_1^w = \operatorname{Id}_{H(M, g)}$, we get the result of the theorem. We already know that $H(m, g)$ is a pre-Hilbertian space with the dot product (2.6.3); the completeness of that space follows from the completeness of $L^2(\mathbb{R}^n)$ and the above isomorphisms. Moreover $H(m, g) \ni u \mapsto \|a_1^w u\|_{L^2(\mathbb{R}^n)}$ is a Hilbertian norm on $H(m, g)$, which is equivalent to the norm given by (2.6.3) since a_1^w is an isomorphism from $H(m, g)$ onto $H(1, g) = L^2(\mathbb{R}^n)$. $\square$

Corollary 2.6.17. *Let g be an admissible metric and m be an admissible g-weight.*

(1) *Let $u \in \mathscr{S}'(\mathbb{R}^n)$; the following properties are equivalent,*

    (i) *$u \in H(m, g)$,*

    (ii) *For all $a \in S(m, g)$, $a^w u \in L^2(\mathbb{R}^n)$,*

    (iii) *There exist $c \in S(m^{-1}, g), v \in L^2(\mathbb{R}^n)$, such that $u = c^w v$.*

(2) *The space $\mathscr{S}(\mathbb{R}^n)$ is dense in $H(m,g)$.*

(3) *The dual space of $H(m,g)$ is canonically identified with $H(m^{-1},g)$.*

Proof. We prove (1): thanks to Theorem 2.6.4, we know (i)$\Longrightarrow$(ii). Corollary 2.6.16 provides an isomorphism a^w with $a \in S(m,g)$ of $H(m,g)$ with $L^2(\mathbb{R}^n)$, with inverse b^w with $b \in S(m^{-1},g)$; as a result if (ii) holds, we have $u = b^w a^w u$ with $a^w u \in L^2, b \in S(m^{-1},g)$, that is (iii). If (iii) is satisfied and a^w is the isomorphism from $H(m,g)$ onto L^2 (with inverse b^w, $a \in S(m,g), b \in S(m^{-1},g)$) of Corollary 2.6.16, we have

$$u = c^w a^w \underbrace{b^w v}_{\in H(m,g)} = (\underbrace{c \sharp a}_{\in S(1,g)})^w (b^w v),$$

so that Theorem 2.6.4 implies $u \in H(m,g)$, completing the proof of (1). Since $\mathscr{S}(\mathbb{R}^n)$ is dense in $L^2(\mathbb{R}^n)$, the isomorphism b^w sends $L^2(\mathbb{R}^n)$ onto $H(m,g)$ and since we have $\mathscr{S}(\mathbb{R}^n) \supset b^w\big(\mathscr{S}(\mathbb{R}^n)\big)$, the latter is dense in $H(m,g)$, implying (2). To prove (3), using an isomorphism a^w between $H(m,g)$ and $L^2(\mathbb{R}^n)$ with $a \in S(m,g)$, with an inverse b^w between $H(m^{-1},g)$ and $L^2(\mathbb{R}^n)$ with $b \in S(m^{-1},g)$, we consider the sesquilinear mapping

$$\begin{array}{rcl}
H(m,g) \times H(m^{-1},g) & \longrightarrow & \mathbb{C} \\
(u,v) & \mapsto & \langle a^w u, b^w v\rangle_{L^2(\mathbb{R}^n)},
\end{array} \qquad (2.6.28)$$

and we equip the space $H(m,g)$ with the norm $\|a^w u\|_{L^2}$, and $H(m^{-1},g)$ with the norm $\|b^w v\|_{L^2}$. The anti-linear mapping $\Phi : H(m^{-1},g) \longrightarrow \big(H(m,g)\big)^*$ given by

$$\langle \Phi(v), u\rangle_{H(m,g)^*, H(m,g)} = \langle a^w u, b^w v\rangle_{L^2(\mathbb{R}^n)},$$

is such that

$$\|\Phi(v)\|_{H(m,g)^*} = \sup_{\|a^w u\|_{L^2}=1} |\langle a^w u, b^w v\rangle_{L^2(\mathbb{R}^n)}| = \|b^w v\|_{L^2} = \|v\|_{H(m^{-1},g)},$$

since $a^w\big(\mathscr{S}(\mathbb{R}^n)\big) = \mathscr{S}(\mathbb{R}^n)$ (if $u \in \mathscr{S}$, $u = a^w b^w u$) so that Φ is isometric; it is also onto since the Riesz representation theorem implies that for $\xi_0 \in H(m,g)^*$, we can find $w_0 \in H(m,g)$ so that

$$\forall u \in H(m,g), \quad \xi(u) = \langle u, w_0\rangle_{H(m,g)} = \langle a^w u, a^w w_0\rangle_{L^2} = \langle a^w u, b^w a^w a^w w_0\rangle_{L^2},$$

i.e., $\xi = \Phi(v)$ with $v = a^w a^w w_0$ which belongs to $H(m^{-1},g)$. The open mapping theorem implies that Φ is an isomorphism between $H(m^{-1},g)$ and $H(m,g)^*$ and that the pairing (2.6.28) provides the identification of these spaces. $\qquad\square$

Lemma 2.6.18. *Let g_1, g_2 be two admissible metrics on $\mathbb{R}^{2n}$ and let m be an admissible g_j-weight such that $m \in S(m,g_j), j = 1,2$. Then $H(m,g_1) = H(m,g_2)$.*

Proof. If $a_{j,t} \in S(m^t, g_j)$ verify (2.6.19) with

$$\frac{\partial a_{j,t}(X)}{\partial t} = (\text{Log } m \sharp a_{j,t})(X) = (a_{j,t} \sharp \text{Log } m)(X), \quad a_{j,0}(X) \equiv 1,$$

the operators $a_{j,t}^w$ are continuous endomorphisms of the Fréchet space $\mathscr{S}(\mathbb{R}^n)$ and for all $u \in \mathscr{S}(\mathbb{R}^n)$, the mappings $t \mapsto a_{j,t}^w u \in \mathscr{S}(\mathbb{R}^n)$ are of class C^1. For $u \in \mathscr{S}(\mathbb{R}^n)$, we calculate

$$\partial_t \big(a_{2,-t}^w a_{1,t}^w u\big) = -(a_{2,-t} \sharp \text{Log } m)^w a_{1,t}^w u + a_{2,-t}^w (\text{Log } m \sharp a_{1,t})^w u = 0, \quad a_{2,0}^w a_{1,0}^w u = u,$$

so that $a_{2,-t} \sharp a_{1,t} = 1$ and $a_{2,t} = a_{2,t} \sharp a_{1,-t} \sharp a_{1,t} = a_{1,t}$, so that the operators $a_{2,t}^w, a_{1,t}^w$ defining for $t = 1$ the isomorphisms between $H(m, g_j)$ and L^2 are the same, giving the result of the lemma. $\square$

2.6.5   An additional hypothesis for the Wiener lemma: the geodesic temperance

We have seen so far that the pseudo-differential calculus can be developed in great generality using the rather simple notion of admissibility of the metric g, as given by Definition 2.2.15. We have seen in the previous sections that a large part of the theory of the Sobolev spaces attached to a pseudo-differential calculus given by an admissible metric can also be developed under this sole assumption (see also [19] which we have followed) and that the analytic functional calculus is possible under only the assumption of admissibility. However, we would like to prove a Wiener-type lemma (see, e.g., our Remark 2.5.9), as explained below. Let us consider a pseudo-differential operator A with a symbol in the class $S(1, g)$ (A is L^2 bounded if g is admissible) such that A is invertible as an operator in $L^2(\mathbb{R}^n)$: is it true that A^{-1} is a pseudo-differential operator with symbol in $S(1, g)$? A positive answer to that question seems to require an additional assumption, the so-called geodesic temperance. Although no example of an admissible metric which is not geodesically temperate seems to be available, we shall study in this section a version of the geodesic temperance which will allow us to prove a Wiener lemma and which is also invariant by conformal change.

Definition 2.6.19. Let g be an admissible metric on $\mathbb{R}^{2n}$ (see Definition 2.2.15) and let $g^\natural$ be the intermediate symplectic metric, as given by Definition 2.2.19. Let d be the geodesic distance on $\mathbb{R}^{2n}$ for the metric $g^\natural$. We shall say that the metric g is geodesically temperate if there exists $C > 0, N \geq 0$ such that, for all $X, Y, T \in \mathbb{R}^{2n}$,

$$\frac{g_X(T)}{g_Y(T)} \leq C\big(1 + d(X, Y)\big)^N. \tag{2.6.29}$$

Lemma 2.6.20. *Let g be an admissible metric on $\mathbb{R}^{2n}$ (Definition 2.2.15) so that g is geodesically temperate (Definition 2.6.19); then there exist $C, N > 0$ such that*

$$C^{-1}\big(1 + g_X^\natural(X - Y)\big)^{1/N} \leq 1 + d(X, Y) \leq C\big(1 + g_X^\natural(X - Y)\big)^N, \tag{2.6.30}$$

where d is the geodesic distance for $g^\natural$.

Proof. We have for $X_0, X_1 \in \mathbb{R}^{2n}$, using the admissibility of $g^\natural$,

$$
\begin{aligned}
d(X_0, X_1) &\leq \int_0^1 \left(g^\natural_{X_0 + t(X_1 - X_0)}(X_1 - X_0) \right)^{1/2} dt \\
&\leq \int_0^1 C'\left(1 + g^\natural_{X_0}(t(X_1 - X_0))\right)^{N'} dt\, g^\natural_{X_0}(X_1 - X_0)^{1/2} \\
&\leq C'\left(1 + g^\natural_{X_0}(X_1 - X_0)\right)^{N'} g^\natural_{X_0}(X_1 - X_0)^{1/2},
\end{aligned}
$$

providing the last inequality of (2.6.30). On the other hand, for a $g^\natural$-geodesic curve between X_0 and X_1, we have from (2.6.29)

$$
\begin{aligned}
d(X_0, X_1) &= \int_0^1 g^\natural_{X(t)}(\dot X(t))^{1/2} dt \\
&\geq \int_0^1 g^\natural_{X_0}(\dot X(t))^{1/2} C^{-1}\left(1 + \max_{t \in [0,1]} d(X_0, X(t))\right)^{-N} dt \\
&\geq C^{-1} g^\natural_{X_0}(X_1 - X_0)^{1/2}\left(1 + d(X_0, X_1)\right)^{-N},
\end{aligned}
$$

since $\int_0^1 g^\natural_{X_0}(\dot X(t))^{1/2} dt \geq g^\natural_{X_0}$-distance from X_0 to X_1 and

$$
\max_{t \in [0,1]} d(X_0, X(t)) \leq d(X_0, X_1)
$$

since $t \mapsto X(t)$ is a geodesic curve for the Riemannian metric $g^\natural$. We obtain thus the first inequality in (2.6.30), completing the proof of the lemma. $\square$

Remark 2.6.21. With the notation and assumptions of the previous lemma, there exists $C, N > 0$ such that, for all $X, Y \in \mathbb{R}^{2n}$,

$$
\frac{1}{C}\left(1 + (g^\natural_X \wedge g^\natural_Y)(X - Y)\right)^{1/N} \leq 1 + d(X, Y) \leq C\left(1 + (g^\natural_X \wedge g^\natural_Y)(X - Y)\right)^N. \tag{2.6.31}
$$

The left inequality above follows from the third inequality in (4.4.26) and the first inequality in (2.6.30). Moreover, we have from (4.4.27) (with T depending on X, Y),

$$
(g^\natural_X \wedge g^\natural_Y)(X - Y) = 2g^\natural_X(X - T) + 2g^\natural_Y(T - Y)
$$

and using the triangle inequality along with the right inequality in (2.6.30), we get

$$
\begin{aligned}
1 + d(X, Y) &\leq 1 + d(X, T) + d(Y, T) \leq C\left(1 + g^\natural_X(X - T)\right)^N + C\left(1 + g^\natural_Y(Y - T)\right)^N \\
&\leq C\left(2 + g^\natural_X(X - T) + g^\natural_Y(T - Y)\right)^N \leq C\left(2 + \frac{1}{2}(g^\natural_X \wedge g^\natural_Y)(X - Y)\right)^N,
\end{aligned}
$$

which gives (2.6.31) (with the same N as in (2.6.30)). Defining for $r > 0, X, Y \in \mathbb{R}^{2n}$,

$$\delta_r^\natural(X,Y) = 1 + (g_X^\natural \wedge g_Y^\natural)(U_{X,r}^\natural - U_{Y,r}^\natural), \tag{2.6.32}$$

$$\text{with } U_{X,r}^\natural = \{X' \in \mathbb{R}^{2n}, g_X^\natural(X'-X) \leq r^2\}, \tag{2.6.33}$$

we see that $\delta_r^\natural$ is the main distance function[24] for the admissible metric $g^\natural$ (see (2.2.27) and the fact that $g^\natural$ is symplectic, according to (4.4.29)). We have

$$\delta_r^\natural(X,Y) = 1 + (g_X^\natural \wedge g_Y^\natural)(X'-Y'), \quad X' \in U_{X,r}^\natural, \ Y' \in U_{Y,r}^\natural,$$

so that the ratio $\delta_r^\natural(X,Y)/1 + (g_{X'}^\natural \wedge g_{Y'}^\natural)(X'-Y')$ is bounded above and below by some fixed positive constants, implying from (2.6.31) that

$$\begin{aligned} d(X,Y) &\leq d(X,X') + d(X',Y') + d(Y',Y) \\ &\leq d(X,X') + C\big(1 + (g_{X'}^\natural \wedge g_{Y'}^\natural)(X'-Y')\big)^N + d(Y',Y) \\ &\leq d(X,X') + C'\underbrace{\big(1 + (g_X^\natural \wedge g_Y^\natural)(X'-Y')\big)^N}_{\delta_r^\natural(X,Y)^N} + d(Y',Y), \end{aligned}$$

and since $g^\natural$ is slowly varying, $X' \in U_{X,r}^\natural, Y' \in U_{Y,r}^\natural$, we get that

$$\exists C > 0, \exists N \geq 0, \exists r > 0, \quad \forall X, Y \in \mathbb{R}^{2n}, \quad 1 + d(X,Y) \leq C\delta_r^\natural(X,Y)^N. \tag{2.6.34}$$

Lemma 2.6.22. *Let g be an admissible and geodesically temperate metric on $\mathbb{R}^{2n}$ (Definitions 2.2.15, 2.6.19) and let G be a slowly varying metric on $\mathbb{R}^{2n}$ such that $G = \mu g$ is conformal to g, with $\mu \leq \lambda_g$ (defined in (2.2.13)). Then G is admissible and geodesically temperate on $\mathbb{R}^{2n}$.*

Proof. The fact that G is admissible follows from Lemma 2.2.21. Since G is conformal to g, Definition 2.2.19 implies that $G^\natural = g^\natural$; moreover, from Proposition 2.2.20, the ratio G_X/G_Y is bounded above by

$$C\big(1 + (g_X^\natural \wedge g_Y^\natural)(X-Y)\big)^N$$

and Lemma 2.6.20 entails that this quantity is bounded above by a constant multiplied by a power of $(1+d(X,Y))$, proving the second assertion of the lemma. $\qquad\square$

Lemma 2.6.23. *The metrics $\langle\xi\rangle^{2\delta}|dx|^2 + \langle\xi\rangle^{-2\rho}|d\xi|^2$, (with $0 \leq \delta \leq \rho \leq 1$, $\delta < 1$), $|dx|^2 + h^2|d\xi|^2$ (with $0 < h \leq 1$), $(1 + |x|^2 + |\xi|^2)^{-1}|dx|^2 + |d\xi|^2$ are admissible and geodesically temperate. Any slowly varying metric G (see Definition 2.2.1) such that $G \leq G^\sigma$ (i.e., (2.2.12) is satisfied) and G is conformal to one of the metrics above is admissible and geodesically temperate.*

---

[24]Note that $\delta_r^\natural(X,Y)$ is larger than $\Delta_r(X,Y)$ as defined by (2.2.32), since in the latter formula $U_{X,r}$ is the g_X-ball with center X, and radius r whereas here in (2.6.32), we are considering the smaller $g_X^\natural$-ball (see (2.2.22)).

Proof. The admissibility follows from Lemma 2.2.18; the geodesic temperance for the last two metrics follows from the previous Lemma 2.6.22 since they are both conformal to a constant metric (note that all the constants are uniform with respect to $h \in (0, 1]$). We see that the first metrics of the lemma are conformal to

$$|dx|^2 \langle \xi \rangle^{\delta + \rho} + |d\xi|^2 \langle \xi \rangle^{-\rho - \delta}$$

and, using again Lemma 2.6.22, it is enough to prove that the latter is geodesically temperate. Since $(\rho + \delta)/2 \in [0, 1)$, this amounts to proving that

$$\frac{|\eta|}{1 + |\xi|} \leq C\big(1 + d_0(\xi, \eta)\big)^N \tag{2.6.35}$$

where d_0 is the geodesic distance for $|d\xi|^2 \langle \xi \rangle^{-2\theta}$ on $\mathbb{R}^n$, where $\theta \in [0, 1)$. We have for a C^1 curve $t \mapsto \xi(t)$ in $\mathbb{R}^n$,

$$(1 - \theta)^{-1} \frac{d}{dt}(\langle \xi(t) \rangle^{1-\theta}) = \langle \xi(t) \rangle^{-\theta - 1} \dot{\xi}(t) \cdot \xi(t)$$

and consequently

$$\int_0^1 |\dot{\xi}(t)| \langle \xi(t) \rangle^{-\theta} dt \geq (1 - \theta)^{-1} \left| \int_0^1 \frac{d}{dt}(\langle \xi(t) \rangle^{1-\theta}) dt \right| = \frac{\big| \langle \xi(1) \rangle^{1-\theta} - \langle \xi(0) \rangle^{1-\theta} \big|}{(1 - \theta)},$$

which proves $\big| \langle \xi \rangle^{1-\theta} - \langle \eta \rangle^{1-\theta} \big| \leq (1 - \theta) d_0(\xi, \eta)$, implying for $\langle \xi \rangle^{1-\theta} \leq \frac{1}{2} \langle \eta \rangle^{1-\theta}$ that $\langle \eta \rangle^{1-\theta} \leq 2(1 - \theta) d_0(\xi, \eta)$, which gives (2.6.35) in that case, whereas (2.6.35) is obvious in the other case $\langle \xi \rangle^{1-\theta} > \frac{1}{2} \langle \eta \rangle^{1-\theta}$. The last statement of the lemma follows from Lemma 2.6.22. The proof of the lemma is complete. $\square$

Remark 2.6.24. The main point of the previous lemma is the stability of the admissible and geodesically temperate metrics by conformal change preserving the uncertainty principle and the slowness. We shall apply this property to some Calderón-Zygmund metrics which are precisely conformal to a constant metric or to the metric defining the $S_{1,0}$ class.

Lemma 2.6.25. *Let g be an admissible metric on $\mathbb{R}^{2n}$. Then $\forall N_0 \geq 0, \exists C_0 > 0, \exists k_0 \in \mathbb{N}, \forall N_1 \geq 0, \exists C_1, \exists k_1 \in \mathbb{N}, \forall \nu \in \mathbb{N}^*, \forall J \subset \{0, \ldots, \nu - 1\}, \forall c_0, \ldots, c_\nu \in \mathscr{S}(\mathbb{R}^{2n}), \forall Y_0, \ldots, Y_\nu \in \mathbb{R}^{2n}$ and we have*

$$\|c_0^w \cdots c_\nu^w\|_{\mathscr{L}(L^2)} \leq C_0^{\nu - \text{card } J} C_1^{1 + \text{card } J} \|c_0\|_{g_{Y_0}, U_{Y_0}, r}^{(k_1)} \Big(\max_{j \notin K, j \neq 0} \|c_j\|_{g_{Y_j}, U_{Y_j}, r}^{(k_0)} \Big)^{\nu - \text{card } J}$$

$$\times \Big(\max_{j \in K, j \neq 0} \|c_j\|_{g_{Y_j}, U_{Y_j}, r}^{(k_1)} \Big)^{\text{card } J} \prod_{0 \leq j < \nu} \delta_r(Y_j, Y_{j+1})^{-N_0} \prod_{j \in J} \delta_r(Y_j, Y_{j+1})^{-N_1},$$

with $K = J \cup \big((1 + J) \cap [0, \nu]\big) = \{j \in \{0, \ldots, \nu\} \text{ such that } j \in J \text{ or } j - 1 \in J\}$.

Proof. We have

$$\|c_0^w \ldots c_\nu^w\|_{\mathscr{L}(L^2)} \leq \|c_0^w c_1^w\|_{\mathscr{L}(L^2)} \|c_2^w c_3^w\|_{\mathscr{L}(L^2)} \cdots ,$$
$$\|c_0^w \ldots c_\nu^w\|_{\mathscr{L}(L^2)} \leq \|c_0^w\|_{\mathscr{L}(L^2)} \|c_1^w c_2^w\|_{\mathscr{L}(L^2)} \|c_3^w c_4^w\|_{\mathscr{L}(L^2)} \cdots ,$$

and thus

$$\|c_0^w \ldots c_\nu^w\|_{\mathscr{L}(L^2)} \leq \|c_0^w\|_{\mathscr{L}(L^2)}^{1/2} \|c_\nu^w\|_{\mathscr{L}(L^2)}^{1/2} \prod_{1 \leq j < \nu} \|c_j^w c_{j+1}^w\|_{\mathscr{L}(L^2)}^{1/2}. \qquad (2.6.36)$$

Using (2.5.1), the biconfinement estimates of Corollary 2.3.3 and the first estimates in (2.2.29), we get

$$\|c_0^w\|_{\mathscr{L}(L^2)} \leq C(n) \|c_0\|_{g_{Y_0}, U_{Y_0}, r}^{(2n+1)}, \quad \|c_\nu^w\|_{\mathscr{L}(L^2)} \leq C(n) \|c_\nu\|_{g_{Y_\nu}, U_{Y_\nu}, r}^{(2n+1)},$$

for $j \notin J$, $\quad \|c_j^w c_{j+1}^w\|_{\mathscr{L}(L^2)} \leq C_0 \delta_r(Y_j, Y_{j+1})^{-2N_0} \|c_j\|_{g_{Y_j}, U_{Y_j}, r}^{(k_0)} \|c_{j+1}\|_{g_{Y_{j+1}}, U_{Y_{j+1}}, r}^{(k_0)},$

and for $j \in J$,

$$\|c_j^w c_{j+1}^w\|_{\mathscr{L}(L^2)} \leq C_1 \delta_r(Y_j, Y_{j+1})^{-2N_0 - 2N_1} \|c_j\|_{g_{Y_j}, U_{Y_j}, r}^{(k_1)} \|c_{j+1}\|_{g_{Y_{j+1}}, U_{Y_{j+1}}, r}^{(k_1)},$$

with C_0, k_0 depending only on N_0 and C_1, k_1 on $N_0 + N_1$. We may assume $C_1 \geq C_0, k_1 \geq k_0$. Plugging the above estimates in (2.6.36) yields the result of the lemma. $\qquad \square$

Lemma 2.6.26. *Let g be an admissible and geodesically temperate metric such that, for all $X \in \mathbb{R}^{2n}$, $g_X = g_X^\sigma$ (implying $g \equiv g^\flat \equiv g^\sigma$). Then there exist some constants C_0, k_0 depending only on the structure constants[25] of g, and for all $k \in \mathbb{N}$, there exist some constants C_1, k_1, N_1, such that for $a \in S(1, g)$, $\nu \in \mathbb{N}$, with the notation of (2.6.13), (2.2.7),*

$$\|(a^w)^\nu\|_{op(1,g)}^{(k)} \leq C_1 (\nu + 1)^{N_1} \left(C_0 \|a\|_{S(1,g)}^{(k_0)} \right)^{\nu - k} \left(C_1 \|a\|_{S(1,g)}^{(k_1)} \right)^k.$$

Proof. For $Y_0 \in \mathbb{R}^{2n}$, we consider k linear forms on $\mathbb{R}^{2n}$, $L_j(X) = [T_j, X]$, $j = 1, \ldots, k$ such that $g_{Y_0}(T_j) = 1$. We have, using the notation of (2.6.13), with $a = \int a \varphi_Y \, dY$, $a_Y = a \varphi_Y$,

$$\omega_k = \| \operatorname{ad} L_1^w \ldots \operatorname{ad} L_k^w \cdot \theta_{Y_0}^w (a^w)^\nu \|_{\mathscr{L}(L^2)}$$
$$\leq \int \| \operatorname{ad} L_1^w \ldots \operatorname{ad} L_k^w \cdot \theta_{Y_0}^w a_{Y_1}^w \ldots a_{Y_\nu}^w \|_{\mathscr{L}(L^2)} \, dY_1 \ldots dY_\nu.$$

Since $\operatorname{ad} C \cdot AB = [C, A]B + A[C, B]$, we get, with the convention $\prod_{l \in \emptyset} A_l = \operatorname{Id}$ that

$$\operatorname{ad} C \cdot A_1 \ldots A_\mu = \sum_{1 \leq j \leq \mu} \left(\prod_{1 \leq l < j} A_l \right) [C, A_j] \left(\prod_{j < l \leq \mu} A_l \right),$$

---

[25] We mean here the constants occurring in (2.2.3), (2.2.16).

so that ω_k is bounded above by a sum of $(\nu+1)^k$ terms $\tilde{\omega}_k$, with

$$\tilde{\omega}_k = \int \|b_0^w \ldots b_\nu^w\|_{\mathscr{L}(L^2)} dY_1 \ldots dY_\nu, \quad b_j = (\prod_{\alpha \in E_j} \partial_{T_\alpha}) a_{Y_j},$$

with the convention $a_{Y_0} = \theta_{Y_0}$ and the $(E_j)_{0 \le j \le \nu}$ are (possibly empty) disjoint subsets of $\{1, \ldots, k\}$. In particular, defining $J = \{j \in [0, \nu], E_j \ne \emptyset\}$, we have

$$k \ge \sum_{j \in J} \operatorname{card} E_j \ge \sum_{j \in J} 1 = \operatorname{card} J.$$

We define for $j \notin J$, $c_j = b_j (= a_{Y_j})$ and for $j \in J$, $c_j = b_j(\prod_{\alpha \in E_j} g_{Y_j}(T_\alpha)^{-1/2})$. Since $g = g^\natural$ is geodesically temperate and $g_{Y_0}(T_\alpha) = 1$, we have from the triangle inequality for d (and Hölder's inequality)

$$g_{Y_j}(T_\alpha) \le C(1 + d(Y_0, Y_j))^N \le C\nu^{N-1} \sum_{0 \le l < j} (1 + d(Y_l, Y_{l+1}))^N.$$

We reach the estimate

$$\|b_0^w \ldots b_\nu^w\|_{\mathscr{L}(L^2)} \le C_2 \nu^{M_0} \|c_0^w \ldots c_\nu^w\|_{\mathscr{L}(L^2)} \prod_{0 \le j < \nu} \delta_r(Y_j, Y_{j+1})^{m_j M_0},$$

where the $m_j \in \mathbb{N}$ and such that $\sum_{0 \le j < \nu} m_j \le k$ and M_0 depends only on g. We define $\tilde{J} = \{j \in [0, \nu[, m_j > 0\}$ and we have

$$k \ge \sum_{0 \le j < \nu} m_j \ge \sum_{0 \le j < \nu} 1 = \operatorname{card} \tilde{J}.$$

We apply now Lemma 2.6.25 to get with $\tilde{K} = \tilde{J} \cup (1 + \tilde{J})$,

$$\tilde{\omega}_k \le C_0^{\nu - \operatorname{card} \tilde{J}} C_1^{1 + \operatorname{card} \tilde{J}} C_2 \nu^{M_0}$$

$$\times \|c_0\|_{g_{Y_0}, U_{Y_0}, r}^{(k_1)} \Big(\max_{j \notin \tilde{K}, j \ne 0} \|c_j\|_{g_{Y_j}, U_{Y_j}, r}^{(k_0)} \Big)^{\nu - \operatorname{card} \tilde{J}} \Big(\max_{j \in \tilde{K}, j \ne 0} \|c_j\|_{g_{Y_j}, U_{Y_j}, r}^{(k_1)} \Big)^{\operatorname{card} \tilde{J}}$$

$$\times \int \prod_{0 \le j < \nu} \delta_r(Y_j, Y_{j+1})^{-N_0} \prod_{j \in \tilde{J}} \delta_r(Y_j, Y_{j+1})^{-N_1}$$

$$\times \prod_{j \in \tilde{J}} \delta_r(Y_j, Y_{j+1})^{k M_0} dY_1 \ldots dY_\nu.$$

The semi-norms of the c_j for $j \ge 1$ depend only on the semi-norms of a in $S(1, g)$, with

$$\begin{cases} \|c_j\|_{g_{Y_j}, U_{Y_j}, r}^{(l)} \le \|a_{Y_j}\|_{g_{Y_j}, U_{Y_j}, r}^{(l+k)} & \text{if } j \in \tilde{K}, \\ \|c_j\|_{g_{Y_j}, U_{Y_j}, r}^{(l)} \le \|a_{Y_j}\|_{g_{Y_j}, U_{Y_j}, r}^{(l)} & \text{if } j \notin \tilde{K}, \end{cases}$$

whereas the semi-norms of c_0 are those of θ_{Y_0} and depend only on the structure constants of g. We choose N_0 large enough (depending only on the structure constants of g) to ensure $\sup_Y \int \delta_r(Y, Z)^{-N_0} dZ < +\infty$. For each k, we choose $N_1(k) = k M_0$ and, integrating successively with respect to $Y_\nu, Y_{\nu-1}, \ldots, Y_1$, we get with $k_1' = k_1 + k$, card $\tilde{J} = \tau k$ with $\tau \in [0, 1]$,

$$
\begin{aligned}
\tilde{\omega}_k &\leq \left(C_0 \|a\|_{S(1,g)}^{(k_0)}\right)^{\nu - \text{card } \tilde{J}} \left(C_1 \|a\|_{S(1,g)}^{(k_1')}\right)^{\text{card } \tilde{J}} C_3 \nu^{M_0} \\
&= \left(C_0 \|a\|_{S(1,g)}^{(k_0)}\right)^{\nu - k + k - \tau k} \left(C_1 \|a\|_{S(1,g)}^{(k_1')}\right)^{\tau k} C_3 \nu^{M_0}
\end{aligned}
$$

and using $k_0 \leq k_1, C_0 \leq C_1$

$$
\leq \left(C_0 \|a\|_{S(1,g)}^{(k_0)}\right)^{\nu - k} \left(C_1 \|a\|_{S(1,g)}^{(k_1')}\right)^{k - \tau k} \left(C_1 \|a\|_{S(1,g)}^{(k_1')}\right)^{\tau k} C_3 \nu^{M_0},
$$

providing the result of the lemma. $\qquad \Box$

Theorem 2.6.27. *Let g be an admissible and geodesically temperate metric on $\mathbb{R}^{2n}$ and let $a \in S(1, g)$ such that a^w is invertible in $\mathscr{L}(L^2(\mathbb{R}^n))$. Then there exists $b \in S(1, g)$ such that $b^w = (a^w)^{-1}$.*

Proof. (1) *We assume first that $g \equiv g^\natural \equiv g^\sigma$. Let us consider $r \in S(1, g)$ such that $\|r^w\|_{\mathscr{L}(L^2)} < 1$. We have $(\mathrm{Id} - r^w)^{-1} = \sum_{\nu \geq 0} (r^w)^\nu$ and for $\tau \in]0, 1[$ (to be chosen later), with $R = r^w$, using Theorem 2.6.12, we get for $k \in \mathbb{N}$,*

$$
\|R^\nu\|_{\mathrm{op}(1,g)}^{(k)} \leq C_{k,\tau} \left(\|R^\nu\|_{\mathrm{op}(1,g)}^{(0)}\right)^\tau \left(\|R^\nu\|_{\mathrm{op}(1,g)}^{(l_{k,\tau})}\right)^{1-\tau}. \tag{2.6.37}
$$

Using Lemma 2.6.26, we have

$$
\|R^\nu\|_{\mathrm{op}(1,g)}^{(l_{k,\tau})} \leq C_{k,\tau}' (\nu + 1)^{N_{k,\tau}} \left(C_0 \|r\|_{S(1,g)}^{(k_0)}\right)^{\nu - l_{k,\tau}} \left(C_1 \|r\|_{S(1,g)}^{(k_1(k,\tau))}\right)^{l_{k,\tau}}, \tag{2.6.38}
$$

and noticing that $\|R^\nu\|_{\mathrm{op}(1,g)}^{(0)} = \sup_{Y \in \mathbb{R}^{2n}} \|\theta_Y^w R^\nu\|_{\mathscr{L}(L^2)} \leq C_2 \|R\|_{\mathscr{L}(L^2)}^\nu$, we get, using (2.6.37), (2.6.38),

$$
\|R^\nu\|_{\mathrm{op}(1,g)}^{(k)}
$$

$$
\leq C_{k,\tau} \left(C_2 \|R\|_{\mathscr{L}(L^2)}^\nu\right)^\tau \left(C_{k,\tau}' (\nu+1)^{N_{k,\tau}} \left(C_0 \|r\|_{S(1,g)}^{(k_0)}\right)^{\nu - l_{k,\tau}} \left(C_1 \|r\|_{S(1,g)}^{(k_1(k,\tau))}\right)^{l_{k,\tau}}\right)^{1-\tau}
$$

$$
\leq \tilde{C}_{k,\tau} (\nu+1)^{N_{k,\tau}} C_0^\nu \|R\|_{\mathscr{L}(L^2)}^{\nu\tau} \left(\|r\|_{S(1,g)}^{(k_0)}\right)^{(\nu - l_{k,\tau})(1-\tau)} \left(\|r\|_{S(1,g)}^{(k_1(k,\tau))}\right)^{(1-\tau)l_{k,\tau}}.
$$

Choosing now $1 - \tau$ small enough, depending only on C_0, k_0 (in particular independent of k), so that

$$
\|R\|_{\mathscr{L}(L^2)}^\tau \left(C_0 \|r\|_{S(1,g)}^{(k_0)}\right)^{1-\tau} < 1,
$$

we obtain for $\nu \geq l_{k,\tau}$

$$
\|R^\nu\|_{\mathrm{op}(1,g)}^{(k)} \leq C_{k,\tau}'' (\nu+1)^{N_{k,\tau}} \left(\|R\|_{\mathscr{L}(L^2)}^\tau \left(C_0 \|r\|_{S(1,g)}^{(k_0)}\right)^{1-\tau}\right)^\nu
$$

which implies the convergence of $\sum_{\nu \geq 0} r^{\sharp \nu}$ in $S(1, g)$. Now if $a \in S(1, g)$ satisfies the invertibility assumption of the theorem, we get that the selfadjoint operator $(\bar{a} \sharp a)^w = (a^w)^* a^w$ is positive invertible and thus, with $0 < c \leq C$,

$$\forall u \in L^2, \quad c\|u\|_{L^2}^2 \leq \langle (a^w)^* a^w u, u \rangle \leq C\|u\|_{L^2}^2,$$

so that $C^{-1}(a^w)^* a^w = \mathrm{Id} - b^w$, $\|b^w\|_{\mathscr{L}(L^2)} < 1$. From the previous discussion, we get that $\left((a^w)^* a^w\right)^{-1} = d_1^w$ with $d_1 \in S(1, g)$, and also $\left(a^w (a^w)^*\right)^{-1} = d_2^w$ with $d_2 \in S(1, g)$, so that

$$\underbrace{d_1 \sharp \bar{a}}_{=\alpha_1} \sharp a = 1 = a \sharp \underbrace{\bar{a} \sharp d_2}_{=\alpha_2} \implies \alpha_1 = \alpha_1 \sharp a \sharp \alpha_2 = \alpha_2,$$

and a^w has in $\mathscr{L}(L^2)$ the inverse α_1^w with $\alpha_1 \in S(1, g)$.

(2) *We assume now only that g is admissible and geodesically temperate.* Then g satisfies (2.6.29), (2.6.31), (2.2.22), and $g^{\flat}$ is indeed admissible (see Proposition 2.2.20). Let $a \in S(1, g)$ with a^w invertible in $\mathscr{L}(L^2)$; since $S(1, g) \subset S(1, g^{\flat})$, we get from the first part of the proof that there exists $b \in S(1, g^{\flat})$ so that $a \sharp b = b \sharp a = 1$. For $S, T \in \mathbb{R}^{2n}$ with $g_S(T) = 1$, we set $M_{S,T}(X) = g_X(T)^{1/2}$; the functions $M_{S,T}$ are a family of weight functions with uniform structure constants for $g^{\flat}$ (and for g), thanks to (2.6.29) and (2.6.31): we have

$$\frac{M_{S,T}(X)^2}{M_{S,T}(Y)^2} = \frac{g_X(T)}{g_Y(T)} \leq C(1 + (g_X^{\flat} \wedge g_Y^{\flat})(X - Y))^N,$$

and if $g_X^{\flat}(Y - X) \leq r^2$, (2.2.22) implies $g_X(Y - X) \leq r^2$ so that with r chosen as in (2.2.3), we get $C^{-1} \leq M_{S,T}(X)/M_{S,T}(Y) \leq C$. We note also that $\partial_T a$ belongs to $S(M_{S,T}, g) \subset S(M_{S,T}, g^{\flat})$ with uniform semi-norms with respect to S, T. Using $0 = \partial_T b \sharp a + b \sharp \partial_T a$, we get

$$\partial_T b = -b \sharp \partial_T a \sharp b, \tag{2.6.39}$$

which implies, thanks to Theorem 2.3.7, that $\|\partial_T b\|_{S(M_{S,T}, g^{\flat})}^{(k)} \leq C_k$ with C_k independent of S, T. In particular, for $k = 0$, we get for $g_S(T) = 1$ that $|(\partial_T b)(S)| \leq C_0$, which implies $\|b\|_{S(1, g)}^{(1)} < \infty$. Moreover, from (2.6.39), we get with $T_1, T_2 \in \mathbb{R}^{2n}$,

$$\begin{aligned}
\partial_{T_1} \partial_{T_2} b &= -\partial_{T_1} b \sharp \partial_{T_2} a \sharp b - b \sharp \partial_{T_1} \partial_{T_2} a \sharp b - b \sharp \partial_{T_2} a \sharp \partial_{T_1} b \\
&= b \sharp \partial_{T_1} a \sharp b \sharp \partial_{T_2} a \sharp b - b \sharp \partial_{T_1} \partial_{T_2} a \sharp b + b \sharp \partial_{T_2} a \sharp b \sharp \partial_{T_1} a \sharp b,
\end{aligned} \tag{2.6.40}$$

so that with $g_S(T_1) = g_S(T_2) = 1$, using $\partial_{T_j} a \in S(M_{S,T_j}, g^{\flat})$ and $\partial_{T_2} \partial_{T_1} a \in S(M_{S,T_1} M_{S,T_2}, g^{\flat})$ we get

$$\|\partial_{T_1} \partial_{T_2} b\|_{S(M_{S,T_1} M_{S,T_2}, g^{\flat})}^{(k)} \leq C_k'$$

with C'_k independent of S, T_1, T_2. In particular, for $k = 0$, this gives

$$|(\partial_{T_1}\partial_{T_2}b)(S)| \leq C'_0, \quad \text{which implies} \quad \|b\|^{(2)}_{S(1,g)} < \infty.$$

We conclude with an induction, based upon the extension of (2.6.39), (2.6.40) to higher-order derivatives. $\qquad\qquad\square$

Corollary 2.6.28. *Let g, m be admissible metric and weight on $\mathbb{R}^{2n}$ with g geodesically temperate and let $a \in S(m, g)$. The existence of $b \in S(m^{-1}, g)$ such that $b\sharp a = a\sharp b = 1$ is equivalent to the invertibility of a^w as an operator from $H(m_1, g)$ onto $H(m_1 m^{-1}, g)$ for some g-admissible weight function m_1.*

Proof. We use Corollary 2.6.16 to find $c_1 \in S(m_1 m^{-1}, g), c_2 \in S(mm_1^{-1}, g), d_1 \in S(m_1, g), d_2 \in S(m_1^{-1}, g)$ with

$$c_1 \sharp c_2 = c_2 \sharp c_1 = d_1 \sharp d_2 = d_2 \sharp d_1 = 1.$$

If a^w is invertible from $H(m_1, g)$ onto $H(m_1 m^{-1}, g)$, we have that $d_2^w a^w c_1^w$ is invertible in $\mathcal{L}(L^2)$. Since $d_2 \sharp a \sharp c_1 \in S(1, g)$ we use the previous theorem, and get that there exists $d \in S(1, g)$ such that

$$d_2 \sharp a \sharp c_1 \sharp d = d \sharp d_2 \sharp a \sharp c_1 = 1 \Longrightarrow c_2 = d \sharp d_2 \sharp a, \ d_1 = a \sharp c_1 \sharp d$$

so that $c_1 \sharp d \sharp d_2 \sharp a = 1 = a \sharp c_1 \sharp d \sharp d_2$ and the symbol $c_1 \sharp d \sharp d_2 \in S(m^{-1}, g)$ is the inverse of a. Conversely, if for some $b \in S(m^{-1}, g)$, $a \sharp b = b \sharp a = 1$, then the operators $b^w a^w, a^w b^w$ are the identities of $H(m_1, g), H(m_1 m^{-1}, g)$, so that a^w is invertible from $H(m_1, g)$ onto $H(m_1 m^{-1}, g)$ with inverse b^w. This completes the proof of the corollary. $\qquad\square$

Theorem 2.6.29. *Let g, m be admissible metric and weight on $\mathbb{R}^{2n}$ with g geodesically temperate. Let $(\varphi_Y)_{Y \in \mathbb{R}^{2n}}$ be a uniformly confined family which is a partition of unity (see Theorem 2.2.7). Then there exists a uniformly confined family of symbols $(\psi_Y)_{Y \in \mathbb{R}^{2n}}$ such that*

$$\int (\psi_Y \sharp \varphi_Y) |g_Y|^{1/2} = 1.$$

Moreover, $H(m, g) = \{u \in \mathscr{S}'(\mathbb{R}^n), \int m(Y)^2 \|\varphi_Y^w u\|^2_{L^2(\mathbb{R}^n)} |g_Y|^{1/2} < \infty\}$.

Proof. Using Lemma 2.5.3 and Theorem 2.6.27, we get that the operator $a^w = A = \int (\varphi_Y^w)^2 |g_Y|^{1/2} dY$ has an inverse b^w with $b \in S(1, g)$. The family $(\psi_Y = b \sharp \varphi_Y)_{Y \in \mathbb{R}^{2n}}$ is uniformly confined, thanks to Proposition 2.3.16. As a result, with the definition given in (2.6.2), thanks to Proposition 2.6.3, we can choose on $H(m, g)$ the norm

$$\|u\|^2_{H(m,g)} = \int \|\varphi_Y^w u\|^2_{L^2} m(Y)^2 |g_Y|^{1/2} dY$$

since $1 = \int (\overbrace{b \sharp \varphi_Y}^{\psi_Y} \sharp \overbrace{\varphi_Y}^{\theta_Y}) |g_Y|^{1/2} dY$. This completes the proof. $\qquad\square$

Chapter 3

Estimates for Non-Selfadjoint Operators

3.1 Introduction

3.1.1 Examples

Principal type with non-negative (resp. non-positive) imaginary part

We have seen in the first chapter that it is very easy to prove a priori estimates for pseudo-differential operators of real principal type (see, e.g., the proof of Theorem 1.2.38), and also for principal-type operators whose imaginary part is either always non-negative or always non-positive (Theorem 1.2.39). Thanks to Remark 1.2.36, it boils down to proving an $L^2 - L^2$ injectivity estimate $\|Lu\|_0 \gtrsim \|u\|_0$ for

$$L = D_t + a(t,x,\xi)^w + ib(t,x,\xi)^w, \quad \text{where } a, b \in S^1_{1,0}, b \geq 0 \text{ (or } b \leq 0) \quad (3.1.1)$$

depending continuously on $t \in \mathbb{R}$. Note that Lemma 4.3.21 in the appendix provides an even more precise form of such an estimate.[1] At any rate, proving local solvability or an $L^2 - L^2$ injectivity estimate for that class of examples does not require much: for instance in the case $b \geq 0$, just conjugate the operator L by $e^{\lambda t}$ so that

$$L_\lambda = e^{2\pi\lambda t} L e^{-2\pi\lambda t} = D_t + a(t,x,\xi)^w + i\big(b(t,x,\xi)^w + \lambda\big),$$

choose λ large enough so that, thanks to Gårding's inequality (Theorem 1.1.26), $b^w + \lambda \geq 0$ as an operator, and apply Lemma 4.3.21. Note in particular that for

---

[1]Note also that this lemma and the multiplier method described in the second part of the proof of Theorem 1.2.39 fall short of proving the propagation-of-singularities result of Theorem 1.2.23, which requires a subtler estimate that should follow the flow of the bicharacteristics of the real part.

an evolution equation ($t \in \mathbb{R}$ is the time-variable)

$$\underbrace{\frac{d}{dt} + i \operatorname{Im} Q(t)}_{\substack{\text{propagation,} \\ \text{e.g., real vector field} \\ \text{with null divergence}}} + \underbrace{\operatorname{Re} Q(t)}_{\text{diffusion}}, \tag{3.1.2}$$

where $\operatorname{Re} Q(t)$ is a pseudo-differential operator with real-valued Weyl symbol, the inequality (4.3.47) shows that the forward Cauchy problem is well-posed whenever $\operatorname{Re} Q(t)$ is bounded from below, e.g., is non-negative. Most evolution equations or systems of mathematical physics are essentially of the type (3.1.2) with a non-negative $\operatorname{Re} Q(t)$ together with very significant complications coming from rough coefficients or nonlinearities.

Creation and annihilation operators

With $t \in \mathbb{R}, h > 0$,

$$C_+ = hD_t + it = \frac{1}{2i\pi}\Big(h\frac{d}{dt} - 2\pi t\Big), \quad \text{is the creation operator,} \tag{3.1.3}$$

$$C_- = hD_t - it = C_+^* = \frac{1}{2i\pi}\Big(h\frac{d}{dt} + 2\pi t\Big), \quad \text{is the annihilation operator} \tag{3.1.4}$$

and the identities, say for $v \in \mathscr{S}(\mathbb{R})$,

$$\|C_+ v\|_0^2 = \|hD_t v\|_0^2 + \|tv\|_0^2 + \frac{h}{2\pi}\|v\|^2 = \|C_- v\|_0^2 + \frac{h}{\pi}\|v\|^2,$$

imply

$$\pi^{1/2}\|C_+ v\|_0 \geq h^{1/2}\|v\|_0, \quad C_-\big(e^{-\frac{\pi t^2}{h}}\big) = 0, \tag{3.1.5}$$

so that the operator C_+ is strongly injective with loss of $1/2$ derivative (in a semi-classical scale) and that the operator C_- has a non-trivial one-dimensional kernel generated by $e^{-\pi t^2/h}$. Note also the identity

$$C_+ C_- = C_-^* C_- = h^2 D_t^2 + t^2 - \frac{h}{2\pi},$$

linking the harmonic oscillator to the creation and annihilation operators; this implies in particular that the ground-state of

$$H_0 = \pi(h^2 D_t^2 + t^2)/h \text{ is } e^{-\pi t^2/h} h^{-1/4} 2^{1/4} = \phi_{0,h}(t) \text{ at the energy } 1/2$$

and that using $[C_-, C_+] = h/\pi$, we get for $k \in \mathbb{N}$,

$$H_0 C_+^k \phi_{0,h} = (k + \frac{1}{2})C_+^k \phi_{0,h}, \tag{3.1.6}$$

providing the spectral decomposition[2] of the one-dimensional harmonic oscillator. The multi-dimensional creation and annihilation operators, with $(x_1, \ldots, x_n) \in \mathbb{R}^n$

$$C_{+,j} = hD_{x_j} + ix_j, \quad C_{-,j} = hD_{x_j} - ix_j \tag{3.1.7}$$

satisfy similar estimates and verify also that

$$H_0 = \frac{\pi}{h}(h^2|D_x|^2 + |x|^2) = \frac{n}{2} + \sum_{1 \leq j \leq n} C_{-,j}^* C_{-,j}. \tag{3.1.8}$$

Moreover $\phi_{0,h}(x) = e^{-\pi|x|^2/h}h^{-n/4}2^{-n/4}$ is the ground-state of H_0 at the energy $n/2$,

$$[C_{-,j}, C_{+,k}] = \frac{h}{\pi}\delta_{j,k}, \quad [C_{+,j}, C_{+,k}] = 0, \quad [C_{-,j}, C_{-,k}] = 0,$$

$$H_0 C_+^\alpha \phi_{0,h} = (|\alpha| + \frac{n}{2})C_+^\alpha \phi_{0,h}, \quad \alpha \in \mathbb{N}^n, \quad C_+^\alpha = C_{+,1}^{\alpha_1} \ldots C_{+,n}^{\alpha_n}.$$

Our expanding-the-square method used above relies on the general identity for operators J, K on some Hilbert space,

$$J = J^*, K^* = -K, \quad 2\operatorname{Re}\langle Ju, Ku \rangle = \langle [J, K]u, u \rangle. \tag{3.1.9}$$

Creation and annihilation operators, continued

The one-dimensional operators

$$C_+^{[k]} = hD_t + it^{2k+1}, \quad C_-^{[k]} = hD_t - it^{2k+1} \tag{3.1.10}$$

are similar to creation and annihilation operators for $k \in \mathbb{N}$. In particular, we have with $c_k > 0$, for $v \in \mathscr{S}(\mathbb{R})$,

$$\|C_+^{[k]}v\|_0 \geq c_k h^{\frac{2k+1}{2k+2}}\|v\|_0, \quad \ker C_-^{[k]} = \mathbb{C}e^{-\frac{\pi t^{2k+2}}{h(k+1)}}. \tag{3.1.11}$$

The second assertion is obvious whereas the first deserves a proof. With a linear change of coordinate $t \mapsto th^{1/(2k+2)}$, we see that $C_+^{[k]}$ is unitarily equivalent to $(D_t + it^{2k+1})h^{(2k+1)/(2k+2)}$, so it is enough to prove the estimate for $h = 1$. For $v \in \mathscr{S}(\mathbb{R})$, we know that $\dot{v} - 2\pi t^{2k+1}v = 2i\pi C_+^{[k]}v$ so that

$$v(t) = \begin{cases} 2i\pi \int_{+\infty}^t e^{-\frac{\pi(s^{2k+2}-t^{2k+2})}{k+1}}(C_+^{[k]}v)(s)ds & \text{for } t \geq 0, \\ 2i\pi \int_{-\infty}^t e^{-\frac{\pi(s^{2k+2}-t^{2k+2})}{k+1}}(C_+^{[k]}v)(s)ds & \text{for } t \leq 0, \end{cases}$$

---

[2]The space $S = span\{C_+^k\phi_{0,h}\}_{k \in \mathbb{N}}$ equals $L^2(\mathbb{R})$ since it contains $\{t^k\phi_{0,h}(t)\}_{k \in \mathbb{N}}$ and thus for all $z \in \mathbb{C}$, the function $t \mapsto e^{-\pi t^2 h^{-1}}e^{zt}$. If $u \in L^2(\mathbb{R})$ is orthogonal to S, the convolution $u * e^{-\pi h^{-1}t^2}$ vanishes identically, and thus, taking Fourier transforms, we see that $u = 0$.

and since for $t \geq 0$ we have

$$\int H(t)H(s-t)e^{-\frac{\pi(s^{2k+2}-t^{2k+2})}{k+1}}ds$$

$$= \frac{1}{2k+2}\int_0^{+\infty}e^{-\frac{\pi\sigma}{k+1}}(\sigma+t^{2k+2})^{-(\frac{2k+1}{2k+2})}d\sigma$$

$$\leq \frac{1}{2k+2}\int_0^{+\infty}e^{-\frac{\pi\sigma}{k+1}}\sigma^{-(\frac{2k+1}{2k+2})}d\sigma = \alpha_k < +\infty,$$

and also

$$\sup_s\int H(t)H(s-t)e^{-\frac{\pi(s^{2k+2}-t^{2k+2})}{k+1}}dt$$

$$= \frac{1}{2k+2}\int_0^{s^{2k+2}}e^{-\frac{\pi\sigma}{k+1}}(s^{2k+2}-\sigma)^{-(\frac{2k+1}{2k+2})}d\sigma$$

$$\leq \frac{1}{2k+2}\int_0^{+\infty}e^{-\frac{\pi\sigma}{k+1}}d\sigma + \frac{1}{2k+2}\int_{\max(0,s^{2k+2}-1)}^{s^{2k+2}}e^{-\frac{\pi\sigma}{k+1}}(s^{2k+2}-\sigma)^{-(\frac{2k+1}{2k+2})}d\sigma$$

$$\leq \frac{1}{2\pi}+e^{-\frac{\pi\max(0,s^{2k+2}-1)}{k+1}}\leq \frac{1}{2\pi}+1$$

along with analogous estimates for $t \leq 0$, Schur's Lemma (see, e.g., Remark 4.7.3) gives $\|v\|_0 \leq C_k\|C_+^{[k]}v\|_0$, which proves (3.1.11).

The operator $C_+^{[k]}$ with (dense) domain

$$D_k = \{u \in L^2(\mathbb{R}), \partial_t u \in L^2(\mathbb{R}), t^{2k+1}u \in L^2(\mathbb{R})\} = \{u \in L^2(\mathbb{R}), C_+^{[k]}u \in L^2(\mathbb{R})\}$$

is injective and has a closed image (thanks to (3.1.11)) of codimension 1: it is a Fredholm operator with index -1. The operator $C_-^{[k]}$ is the adjoint of $C_+^{[k]}$ and is onto with a one-dimensional kernel: it is a Fredholm operator with index $+1$.

Cauchy-type operators

For $k \in \mathbb{N}$, we define

$$C_0^{[k]} = hD_t + t^{2k}\sqrt{-1}. \tag{3.1.12}$$

There exists $c_k > 0$, such that for $v \in \mathscr{S}(\mathbb{R})$,

$$\|C_0^{[k]}v\|_0 \geq c_k h^{\frac{2k}{2k+1}}\|v\|_0. \tag{3.1.13}$$

As above, the linear change of variable $t \mapsto th^{1/(2k+1)}$ shows that $C_0^{[k]}$ is unitarily equivalent to $h^{2k/(2k+1)}(D_t + \sqrt{-1}t^{2k})$ so that it suffices to prove (3.1.13) for $h = 1$. Although a direct resolution of the ODE, such as for proving (3.1.11), would provide the answer, we shall prove a more general lemma, implying both (3.1.12) and (3.1.11).

Lemma 3.1.1. *Let $\phi \in C^0(\mathbb{R}; \mathbb{R})$ such that*

$$\phi(t) > 0, s > t \implies \phi(s) \geq 0. \tag{3.1.14}$$

Then for all $v \in W^{1,1}(\mathbb{R})$ with $\phi v \in L^1(\mathbb{R})$, we have

$$\sup_{t \in \mathbb{R}} |v(t)| \leq \int_{\mathbb{R}} |\frac{dv}{dt} - \phi v| dt, \tag{3.1.15}$$

and if v is compactly supported, $\mathrm{diameter}(\mathrm{supp}\, v) \leq \delta$, $v \in H^1(\mathbb{R}), \phi v \in L^2(\mathbb{R})$,

$$\|v\|_{L^2(\mathbb{R})} \leq \delta \|\frac{dv}{dt} - \phi v\|_{L^2(\mathbb{R})}. \tag{3.1.16}$$

Moreover defining for $\lambda > 0$, $m(\lambda) = |\{t \in \mathbb{R}, |\phi(t)| \leq \lambda^{-1}\}|$ and assuming that $\kappa(\phi) = \inf_{\lambda > 0}(m(\lambda) + \lambda) < +\infty$, we have for $v \in H^1(\mathbb{R})$ with $\phi v \in L^2(\mathbb{R})$,

$$\|v\|_{L^2(\mathbb{R})} \leq 2\|\frac{dv}{dt} - \phi v\|_{L^2(\mathbb{R})} \kappa(\phi). \tag{3.1.17}$$

Comment. This lemma implies the estimates (3.1.13) and (3.1.11): first of all the hypothesis (3.1.14) holds for $t^{2k+1}, \pm t^{2k}$ (violated for $-t^{2k+1}$). Moreover for $\phi = h^{-1} t^l$, we have

$$\kappa(\phi) \leq |\{t \in \mathbb{R}, h^{-1}|t|^l \leq h^{-\frac{1}{l+1}}\}| + h^{\frac{1}{l+1}} \leq 2h^{\frac{1}{l+1}}$$

and thus $\|v\| \leq 4h^{\frac{1}{l+1}} \|\frac{d}{idt} v + ih^{-1} t^l v\|$, so that $h^{\frac{l}{l+1}} \|v\| \leq 4\|h\frac{d}{idt} v + it^l v\|$ and for l even the same estimate for $h\frac{d}{idt} v - it^l v$. Also the reader may have noticed that the estimates (3.1.15) and (3.1.17) hold true without any condition on the support of v; on the other hand $\kappa(0) = +\infty$ and although the estimate (3.1.16) holds for $\phi \equiv 0$, no better estimate is true in that simple case.

Proof of the lemma. We define $T = \inf\{t \in \mathbb{R}, \phi(t) > 0\}$ ($T = \pm\infty$ if $\pm\phi \leq 0$). The condition (3.1.14) ensures that

$$t > T \implies \exists t' \in (T, t) \text{ with } \phi(t') > 0 \implies \phi(t) \geq 0, \tag{3.1.18}$$

$$t < T \implies \phi(t) \leq 0. \tag{3.1.19}$$

For $v \in C_c^1(\mathbb{R})$, we have with $\dot{v} - \phi v = f$, and $t \geq T$,

$$v(t) = \int_{+\infty}^t f(s) e^{\int_s^t \phi(\sigma) d\sigma} ds = -\int_t^{+\infty} f(s) e^{-\int_t^s \phi(\sigma) d\sigma} ds$$

and since $\phi \geq 0$ on $[T, +\infty)$, we get for $t \geq T$, $|v(t)| \leq \int_t^{+\infty} |f(s)| ds$ and similarly for $t \leq T$, $|v(t)| \leq \int_{-\infty}^t |f(s)| ds$, so that (3.1.15) follows as well as its immediate consequence (3.1.16). For future reference we give another proof of (3.1.15) which uses a more flexible energy method. We calculate with $L = \frac{d}{idt} + i\phi$ and $v \in \mathscr{S}(\mathbb{R})$

$$\text{for } t'' \geq T, \ 2\operatorname{Re}\langle Lv, iH(t-t'')v\rangle = |v(t'')|^2 + 2\int_{t''}^{+\infty} |\phi(t)||v(t)|^2 dt,$$

$$\text{for } t' \leq T, \ 2\operatorname{Re}\langle Lv, -iH(t'-t)v\rangle = |v(t')|^2 + 2\int_{-\infty}^{t'} |\phi(t)||v(t)|^2 dt,$$

and we get

$$\sup_{t\in\mathbb{R}} |v(t)|^2 + 2\int_{\mathbb{R}} |\phi(t)||v(t)|^2 dt \leq 2\int_{\mathbb{R}} |(Lv)(t)||v(t)| dt, \tag{3.1.20}$$

proving (3.1.15) (with a constant 2), which implies also

$$\int_{\mathbb{R}} |\phi(t)||v(t)|^2 dt \leq \|Lv\|_{L^2}\|v\|_{L^2}. \tag{3.1.21}$$

Now, we have also with $\lambda > 0$,

$$\int_{\mathbb{R}} |v(t)|^2 dt \leq \int_{\lambda|\phi(t)|\leq 1} |v(t)|^2 dt + \int_{\lambda|\phi(t)|>1} \lambda|\phi(t)||v(t)|^2 dt$$
$$\leq |\{t\in \operatorname{supp} v, |\phi(t)| \leq 1/\lambda\}| \sup |v(t)|^2 + \lambda\|Lv\|_{L^2}\|v\|_{L^2}$$
$$\leq 2\|Lv\|_{L^2}\|v\|_{L^2}\Big(|\{t\in \operatorname{supp} v, |\phi(t)| \leq 1/\lambda\}| + \lambda/2\Big), \tag{3.1.22}$$

which gives (3.1.17). $\square$

The last estimate is of particular interest when the function ϕ has a polynomial behaviour, in the sense of the following lemma.

Lemma 3.1.2. *Let $k \in \mathbb{N}^*, \delta > 0$ and $C > 0$ be given. Let I be an interval of $\mathbb{R}$ and $q : I \to \mathbb{R}$ be a C^k function such that*

$$\inf_{t\in I} |\partial_t^k q| \geq \delta. \tag{3.1.23}$$

Then for all $h > 0$, the set

$$\{t\in I, |q(t)| \leq Ch^k\} \subset_{1\leq l\leq k} \cup J_l \tag{3.1.24}$$

where J_l is an interval with length $h(\alpha_k C\delta^{-1})^{1/k}, \alpha_k = 2^{2k}k!$. As a consequence, the Lebesgue measure of $\{t\in I, |q(t)| \leq Ch^k\}$ is smaller than

$$hC^{1/k}\delta^{-1/k}4k(k!)^{1/k} \leq hC^{1/k}\delta^{-1/k}4k^2.$$

Proof. Let $k \in \mathbb{N}^*$, $h > 0$ and set $E_k(h, C, q) = \{t \in I, |q(t)| \leq Ch^k\}$. Let us first assume $k = 1$. Assume that $t, t_0 \in E_1(h, C, q)$; then the mean value theorem and (3.1.23) imply $2Ch \geq |q(t) - q(t_0)| \geq \delta|t - t_0|$ so that

$$E_1(h, C, q) \cap \{t, |t - t_0| > h2C\delta^{-1}\} = \emptyset :$$

otherwise we would have $2Ch > \delta h2C/\delta$. As a result, for any $t_0, t \in E_1(h, C, q)$, we have $|t - t_0| \leq h2C\delta^{-1}$. Either $E_1(h, C, q)$ is empty or it is not empty and then included in an interval with length $\leq h4C\delta^{-1}$.

Let us assume now that $k \geq 2$. If $E_k(h, C, q) = \emptyset$, then (3.1.24) holds true. We assume that there exists $t_0 \in E_k(h, C, q)$ and we write for $t \in I$,

$$q(t) = q(t_0) + \underbrace{\int_0^1 q'(t_0 + \theta(t - t_0))d\theta(t - t_0)}_{Q(t)}. \tag{3.1.25}$$

Then if $t \in E_k(h, C, q)$, we have $2Ch^k \geq |Q(t)(t - t_0)|$. Now for a given $\omega > 0$, either $|t - t_0| \leq \omega h/2$ and $t \in [t_0 - \omega h/2, t_0 + \omega h/2]$, or $|t - t_0| > \omega h/2$ and from the previous inequality, we infer $|Q(t)| \leq \omega^{-1}4Ch^{k-1}$, i.e., we get that

$$E_k(h, C, q) \subset [t_0 - \omega h/2, t_0 + \omega h/2] \cup E_{k-1}(h, \omega^{-1}4C, Q). \tag{3.1.26}$$

But the function Q satisfies the assumptions of the lemma with $k - 1, \delta/k$ instead of k, δ: in fact for $t \in I$,

$$Q^{(k-1)}(t) = \int_0^1 q^{(k)}(t_0 + \theta(t - t_0))\theta^{k-1}d\theta$$

and if $q^{(k)}(t) \geq \delta$ on I, we get $Q^{(k-1)}(t) \geq \delta/k$. By induction on k and using (3.1.26), we get that

$$E_k(h, C, q) \subset [t_0 - \omega h/2, t_0 + \omega h/2] \cup_{1 \leq l \leq k-1} J_l, \quad |J_l| \leq h(4C\omega^{-1}k\delta^{-1}\alpha_{k-1})^{1/(k-1)}. \tag{3.1.27}$$

We choose now ω so that $\omega = (4C\omega^{-1}k\delta^{-1}\alpha_{k-1})^{1/(k-1)}$, i.e., $\omega^k = 4C\delta^{-1}k\alpha_{k-1}$, that is $\omega = (C\delta^{-1}4k\alpha_{k-1})^{1/k}$, yielding the result if $\alpha_k = 4k\alpha_{k-1}$, i.e., with $\alpha_1 = 2$,

$$\alpha_k = (4k)(4(k-1))\dots(4 \times 2)\alpha_1 = 4^{k-1}k!2^2 = 2^{2k}k!.$$

The proof of the lemma is complete. $\qquad\square$

A consequence of Lemma 3.1.2 and of the estimate (3.1.22) is that for $q :$ $\mathbb{R} \to \mathbb{R}$ satisfying (3.1.23),(3.1.14) and $h > 0$,

$$\|v\|_{L^2(\mathbb{R})} \leq 2\|\dot{v} - h^{-1}q(t)v\|_{L^2(\mathbb{R})} \left(\frac{h^{\frac{1}{k+1}}}{2} + |\{t \in \mathbb{R}, h^{-1}|q(t)| \leq h^{-\frac{1}{k+1}}\}| \right)$$

$$|\{t \in \mathbb{R}, h^{-1}|q(t)| \leq h^{-\frac{1}{k+1}}\}| = |\{t \in \mathbb{R}, |q(t)| \leq h^{\frac{k}{k+1}}\}| \leq 4k^2 h^{\frac{1}{k+1}}\delta^{-\frac{1}{k}},$$

so that
$$h^{\frac{k}{k+1}}\|v\|_{L^2(\mathbb{R})} \leq \|h\dot{v} - q(t)v\|_{L^2(\mathbb{R})}(1 + 8k^2\delta^{-1/k}). \qquad (3.1.28)$$

On the other hand (3.1.21) implies as well

$$\int h^{-1}|q(t)||v(t)|^2 dt \leq \|\dot{v} - h^{-1}q\|_{L^2(\mathbb{R})}\|v\|_{L^2(\mathbb{R})}$$

so that we have proven the following result.

Lemma 3.1.3. *Let* $q \in C^\infty(\mathbb{R};\mathbb{R})$ *such that (3.1.14) and (3.1.23) (for* $I = \mathbb{R}$ *and some* $k \in \mathbb{N}^*$*) hold. Then for all* $h > 0$ *and all* $v \in C_c^\infty(\mathbb{R})$ *we have*

$$h^{\frac{k}{k+1}}\|v\|_{L^2(\mathbb{R})}^2 + \int |q(t)||v(t)|^2 dt \leq \|h\dot{v} - q(t)v\|_{L^2(\mathbb{R})}\|v\|_{L^2(\mathbb{R})}(2 + 8k^2\delta^{-1/k}). \qquad (3.1.29)$$

Mizohata-Nirenberg-Treves operators

We shall first consider the two-dimensional operators, in fact vector fields in $\mathbb{R}^2$,

$$M_k = D_{x_1} + ix_1^k D_{x_2}. \qquad (3.1.30)$$

Using a Fourier transformation with respect to the x_2 variable, and the estimates (3.1.13), it is easy to see that M_{2k}, M_{2k}^* are locally solvable and satisfy even a *subelliptic estimate*

$$C\|M_{2k}v\|_0 \geq \|v\|_{\frac{1}{2k+1}}, \quad C\|M_{2k}^*v\|_0 \geq \|v\|_{\frac{1}{2k+1}}. \qquad (3.1.31)$$

It is not difficult either to see that M_{2k+1}, M_{2k+1}^* are not locally solvable: let us for instance consider with $\nu \in \mathscr{S}(\mathbb{R})$, supported in $[1, +\infty[$, the function

$$v(x_1, x_2) = \int e^{2i\pi x_2\xi_2} e^{-\pi x_1^2\xi_2}\nu(\xi_2)d\xi_2 \qquad (3.1.32)$$

which belongs to $\mathscr{S}(\mathbb{R}^2)$ since it is the case of its Fourier transform with respect to x_2 and we have $M_1^*v = 0$. The Fourier transform of v is

$$\omega(\xi) = \nu(\xi_2)\xi_2^{-1/2}e^{-\pi\xi_1^2\xi_2^{-1}}. \qquad (3.1.33)$$

If M_1 were locally solvable at 0, thanks to Lemma 1.2.28, we would find a neighborhood V_0 of 0, N_0, C_0 such that for all $\chi \in C_c^\infty(V_0)$ and all $\nu \in \mathscr{S}(\mathbb{R})$ with supp $\nu \subset [1, +\infty)$,

$$C_0\|\nu M_1^*(\chi)\|_{N_0} = C_0\|M_1^*(\chi v)\|_{N_0} \geq \|\chi v\|_{-N_0} = \|\langle D\rangle^{-N_0}(\chi v)\|_{L^2}. \qquad (3.1.34)$$

We have, for $\lambda \geq 1$, choosing $\nu(\xi_2) = \chi_1(\xi_2\lambda^{-3})$ with χ_1 given $C_c^\infty(\mathbb{R})$ supported in $[1, 2]$, equal to 1 on $[5/4, 3/2]$,

$$1 \geq |x_1| \geq \lambda^{-1} \Longrightarrow |v^{(\alpha)}(x)| \leq C_\alpha e^{-\pi\lambda}\lambda^3\lambda^{3|\alpha|},$$

$$|x_1| < \lambda^{-1}, \ |x_2| \geq \lambda^{-2} \Longrightarrow |v^{(\alpha)}(x)| \leq C_{N,\alpha}(|x_2| + \lambda^{-2})^{-N}\lambda^{-3N+3+3|\alpha|}.$$

Let us now choose χ such that $\chi(x) = \chi_0(x_1\lambda)\chi_0(x_2\lambda^2)$ where χ_0 is given in $C_c^\infty(\mathbb{R})$ supported in $[-2, 2]$, equal to 1 on $[-1, 1]$. We have thus, assuming as we may that $N_0 \in \mathbb{N}$, using that on support of $\nabla\chi$, either $|x_1| \geq \lambda^{-1}$ or $|x_1| < \lambda^{-1}, |x_2| \geq \lambda^{-2}$,

$$
\|vM_1^*(\chi)\|_{N_0} \leq C_{N_0} \max_{|\alpha|+|\beta|=N_0} \|v^{(\alpha)}\nabla\chi^{(\beta)}\|_0
$$

$$
\leq C_{N_0} \max_{|\alpha|+|\beta|=N_0} (e^{-\pi\lambda}\lambda^{3+3|\alpha|+2|\beta|+2-3}, C'_{N,\alpha}\lambda^{-3N+3+3|\alpha|+2N-3})
$$

$$
\leq C_N\lambda^{-N}, \tag{3.1.35}
$$

meaning that the lhs of (3.1.34) is rapidly decreasing with respect to λ. On the other hand, from (3.1.32), we get

$$
(\chi v)(x) = \chi_0(\lambda x_1)\chi_0(\lambda^2 x_2)\int e^{2i\pi\lambda^3 x_2\xi_2}e^{-\pi x_1^2\lambda^3\xi_2}\chi_1(\xi_2)d\xi_2\lambda^3,
$$

so that

$$
\lambda^{-3}(\chi v)(x_1\lambda^{-3/2}, x_2\lambda^{-3}) = \chi_0(\lambda^{-1/2}x_1)\chi_0(\lambda^{-1}x_2)\int e^{2i\pi x_2\xi_2}e^{-\pi x_1^2\xi_2}\chi_1(\xi_2)d\xi_2
$$

implying that in $\mathscr{S}(\mathbb{R}^2)$,

$$
\lim_{\lambda\to+\infty} \underbrace{\lambda^{-3}(\chi v)(x_1\lambda^{-3/2}, x_2\lambda^{-3})}_{f_\lambda(x_1,x_2)} = \int e^{2i\pi x_2\xi_2}e^{-\pi x_1^2\xi_2}\chi_1(\xi_2)d\xi_2 = w_0(x).
$$

The Fourier transform of w_0 is $\omega_0(\xi) = \chi_1(\xi_2)\xi_2^{-1/2}e^{-\pi\xi_1^2\xi_2^{-1}}$ which is not the zero function. We have

$$
\lambda^{-6}\|\chi v\|_{-N_0}^2 = \iint (1 + \lambda^3\xi_1^2 + \lambda^6\xi_2^2)^{-N_0}|\lambda^{-3}\widehat{\chi v}(\lambda^{3/2}\xi_1, \lambda^3\xi_2)|^2 d\xi_1 d\xi_2\lambda^{9/2}
$$

$$
= \iint (1 + \lambda^3\xi_1^2 + \lambda^6\xi_2^2)^{-N_0}|\widehat{f_\lambda}(\xi)|^2 d\xi_1 d\xi_2\lambda^{-9/2}
$$

$$
\geq \lambda^{-(6N_0+\frac{9}{2})}\iint (1 + \xi_1^2 + \xi_2^2)^{-N_0}|\widehat{f_\lambda}(\xi)|^2 d\xi_1 d\xi_2
$$

and thus $\liminf_{\lambda\to+\infty} \lambda^{6N_0-6+\frac{9}{2}}\|\chi v\|_{-N_0}^2 = c_0 > 0$ whereas the inequality (3.1.34) and (3.1.35) imply that for all N, $\lim_{\lambda\to+\infty} \lambda^N\|\chi v\|_{-N_0}^2 = 0$, proving by reductio ad absurdum that M_1 is not solvable. The proof of non-local-solvability of M_{2k+1}, M_{2k+1}^* is similar.

It is also interesting to take a look at the operators[3]

$$
N_k = D_{x_1} + ix_1^k|D_{x_2}|, \quad k \in \mathbb{N}. \tag{3.1.36}
$$

---

[3] Or to the operators with homogeneous symbol $\xi_1 + ix_1^k|\xi|$, quantized by $(\xi_1 + ix_1^k|\xi|)\omega(\xi)$, where ω is as in (1.2.19).

The operators N_{2k}, N_{2k}^* are both locally solvable and even subelliptic with loss of $2k/(2k+1)$ derivatives (same argument as for (3.1.31)). Something more interesting occurs for odd k: the operator N_{2k+1} is not locally solvable (essentially same construction as for M_1) and is subelliptic with loss of $(2k+1)/(2k+2)$ derivatives (use the properties of $C_+^{[k]}$ in (3.1.10) or Lemma 3.1.1). The operator N_{2k+1}^* is locally solvable (from the estimates satisfied by N_{2k+1}).

The Hans Lewy operator

In 1957, Hans Lewy [102] found a nonsolvable complex-valued non-vanishing vector field

$$\mathcal{L}_0 = \partial_{x_1} + i\partial_{x_2} + i(x_1 + ix_2)\partial_{x_3}. \tag{3.1.37}$$

The symbol of $\mathcal{L}_0$ is (up to a non-zero constant)

$$\ell_0(x,\xi) = \xi_1 + i\xi_2 + i(x_1 + ix_2)\xi_3 = (\xi_1 - x_2\xi_3) + i(\xi_2 + x_1\xi_3)$$

so that $\{\xi_1 - x_2\xi_3, \xi_2 + x_1\xi_3\} = 2\xi_3$ and in particular for every $x \in \mathbb{R}^3$,

$$\ell_0(x_1, x_2, x_3, x_2, -x_1, 1) = 0, \quad \{\operatorname{Re}\ell_0, \operatorname{Im}\ell_0\}(x_1, x_2, x_3, x_2, -x_1, 1) = 2 > 0.$$

In that sense $\mathcal{L}_0$ is very close to the operator M_1 (at $\xi_1 = 0 = x_1, \xi_2 = 1$) and a construction of a quasi-mode, i.e., of an approximate solution of $\mathcal{L}_0^* v = 0$ can be performed in terms quite similar to the construction of a quasi-mode for M_1^*. The operator $\mathcal{L}_0$ is non-locally-solvable at every point of $\mathbb{R}^3$, even taking local solvability in the very weak sense of Definition 1.2.25.

The complex harmonic oscillator

For a change we consider now an operator with complex symbol which is not of principal type: we define on the real line the operator

$$Q_\theta = \pi(D_x^2 + e^{2i\theta}x^2), \quad |\theta| < \pi/2, \tag{3.1.38}$$

and using the spectral decomposition of the standard harmonic oscillator Q_0 (see (3.1.6)), we obtain that the spectrum of Q_θ is discrete with eigenvalues $e^{i\theta}(\frac{1}{2} + \mathbb{N})$; in particular the spectrum of $\pi(D_x^2 + ix^2)$ is $e^{i\pi/4}(\frac{1}{2} + \mathbb{N})$. Of course the operator Q_θ is non-selfadjoint for $0 < |\theta| < \pi/2$ and for $z \notin \sigma(Q_\theta)$, although the resolvent $(z - Q_\theta)^{-1}$ is a bounded operator on $L^2(\mathbb{R})$, the sets in the complex plane, so-called pseudo-spectrum of index ε,

$$\sigma_\varepsilon(Q_\theta) = \{z \in \sigma(Q_\theta)^c, \|(z - Q_\theta)^{-1}\| > \varepsilon^{-1}\}$$

may be very far from the spectrum. To simplify matters, let us only consider $L = D_x^2 + ix^2$. The numerical range of this operator is by definition the closure of $\{\xi^2 + ix^2\}_{(x,\xi) \in \mathbb{R}^2} = \mathbb{R}_+^2$ and if $a + ib = z \notin \mathbb{R}_+^2$, we can prove $\|(z - L)^{-1}\| \leq \frac{1}{|z - \mathbb{R}_+^2|}$.

Looking at the boundary of the numerical range, we take $z = \lambda > 0$ and we examine $L - \lambda = D_x^2 - \lambda + ix^2$. The article [120] by K. Pravda-Starov provides a complete and self-contained study of the complex harmonic oscillator and his proposition on page 754 gives the following lemma.

Lemma 3.1.4. *There exists a constant $c_0 > 0$ such that for all $\lambda \geq 1$ and all $u \in \mathscr{S}(\mathbb{R})$, $\|D_x^2 u + ix^2 u - \lambda u\|_{L^2(\mathbb{R})} \geq c_0 \lambda^{1/3} \|u\|_{L^2(\mathbb{R})}$.*

There is now a huge literature on pseudo-spectrum and it is very interesting to note that numerical analysts are playing a very important rôle in promoting its importance for the analysis of non-selfadjoint operators; the book by L.N. Trefethen and M. Embree [138] provides a good overview of the subject. Also the work of E.B. Davies (see, e.g., the book [30] and the article [31]) introduced the subject to spectral theorists whereas the article by N. Dencker, J. Sjöstrand, M. Zworski [36] made explicitly the link with subellipticity theory. The article [122] by K. Pravda-Starov provides a complete overview of pseudo-spectrum for differential operators with quadratic symbols.

3.1.2 First-bracket analysis

For a non-selfadjoint pseudo-differential operator, the analysis of the first bracket at characteristic points is already providing a lot of information.

Theorem 3.1.5. *Let Ω be an open subset of $\mathbb{R}^n$, $m \in \mathbb{R}$, $P \in \Psi_{ps}^m(\Omega)$ with positively-homogeneous principal symbol p_m so that (1.2.18) holds.*

(1) *Assume that there exists $(x_0, \xi_0) \in \dot{T}^*(\Omega)$, such that*

$$p_m(x_0, \xi_0) = 0 \text{ and } \{\operatorname{Re} p_m, \operatorname{Im} p_m\}(x_0, \xi_0) > 0. \tag{3.1.39}$$

Then the operator P is not locally solvable at x_0 (see Definition 1.2.25). Moreover the operator P is subelliptic with loss of $1/2$ derivative at (x_0, ξ_0), i.e., satisfies (see Definition 1.2.21) for all $s \in \mathbb{R}, u \in \mathscr{D}'(\Omega)$,

$$Pu \in H^s_{(x_0, \xi_0)} \Longrightarrow u \in H^{s+m-\frac{1}{2}}_{(x_0, \xi_0)}. \tag{3.1.40}$$

(2) *Assume that for all $(x, \xi) \in \dot{T}^*(\Omega)$,*

$$p_m(x, \xi) = 0 \Longrightarrow \{\operatorname{Re} p_m, \operatorname{Im} p_m\}(x, \xi) < 0. \tag{3.1.41}$$

Then the operator P is locally solvable at every point of Ω and the operator P^* is subelliptic with loss of $1/2$ derivatives on Ω.

Remark 3.1.6. We may define microlocal solvability at a point (x_0, ξ_0) in the cotangent bundle: there exists $N_0 \in \mathbb{N}$ such that for all $f \in H^{N_0}_{loc}(\Omega)$, there exists

$u \in \mathscr{D}'(\Omega)$ with $(x_0, \xi_0) \notin WF(Pu - f)$. Now we could change the formulation of the theorem above by saying that at a characteristic point (x_0, ξ_0),

$$\{\operatorname{Re} p_m, \operatorname{Im} p_m\}(x_0, \xi_0) > 0 \Longrightarrow P \text{ subelliptic } \frac{1}{2} \text{ and nonsolvable at } (x_0, \xi_0),$$

$$\{\operatorname{Re} p_m, \operatorname{Im} p_m\}(x_0, \xi_0) < 0 \Longrightarrow P^* \text{ subelliptic } \frac{1}{2} \text{ and } P \text{ solvable at } (x_0, \xi_0).$$

Remark 3.1.7. As a result, local solvability at x_0 implies that $\{\operatorname{Re} p_m, \operatorname{Im} p_m\} \leq 0$ on the characteristic set, and in particular if P is a differential operator, that Poisson bracket is a polynomial with odd degree $2m - 1$ in the momentum variable, whereas $(x, \xi) \in \operatorname{char} P$ is equivalent to $(x, -\xi) \in \operatorname{char} P$ so that a necessary condition for local solvability of differential operators is that $\operatorname{char} P \subset \{\{\operatorname{Re} p_m, \operatorname{Im} p_m\} = 0\}$, implying in particular the nonsolvability at every point of the Hans Lewy operator (3.1.37), thanks to the calculations of Poisson bracket there.

Theorem 3.1.5 follows from its semi-classical version. We start with the positive first bracket.

Theorem 3.1.8. *Let* $p \in C_b^\infty(\mathbb{R}^{2n})$, $(x_0, \xi_0) \in \mathbb{R}^{2n}$ *such that* $p(x_0, \xi_0) = 0$ *and* $\{\operatorname{Re} p, \operatorname{Im} p\}(x_0, \xi_0) > 0$. *Then there exists* $\varphi \in C_c^\infty(\mathbb{R}^{2n})$ *with* $\varphi(x_0, \xi_0) \neq 0$, $C > 0$, *such that* $\forall u \in L^2(\mathbb{R}^n)$,

$$\|\varphi(x, h\xi)^w u\|_{L^2} h^{1/2} \leq C \|p(x, h\xi)^w u\|_{L^2} + Ch\|u\|_{L^2}.$$

Proof. We define $a(x, \xi, h) = p(x, h\xi)$ and we calculate

$$\|a^w u\|^2 = \|(\operatorname{Re} a)^w u\|^2 + \|(\operatorname{Im} a)^w u\|^2 + \langle [(\operatorname{Re} a)^w, i(\operatorname{Im} a)^w] u, u \rangle,$$

so that the symbol c of $[(\operatorname{Re} a)^w, i(\operatorname{Im} a)^w]$ is

$$c = \frac{h}{2\pi} \{\operatorname{Re} p, \operatorname{Im} p\}(x, h\xi) + h^2 r_0(x, \xi, h), \quad r_0 \in S(1, |dx|^2 + h^2 |d\xi|^2) = S_{scl}^0.$$

Now taking $\chi_0 \in C_c^\infty(\mathbb{R}^{2n}; [0, 1])$ supported in the Euclidean ball $B(0, 2)$, equal to 1 on $B(0, 1)$, the symbol $(x, \xi) \mapsto \chi_0(x - x_0, h\xi - \xi_0) = \chi(x, \xi, h)$ belongs to S_{scl}^0 and is such that

$$c \geq \alpha_0 h \chi^2 + h^2 r_0 \chi^2 + (1 - \chi^2) c, \quad \text{with some positive } \alpha_0,$$

and Gårding's inequality (Theorem 1.1.39) implies

$$c^w \geq \alpha_0 h \chi^w \chi^w - C_0 h^2 + ((1 - \chi^2) c)^w$$

so that

$$C_0 h^2 \|u\|^2 + \|a^w u\|^2 \geq \alpha_0 h \|\chi^w u\|^2 + \langle ((1 - \chi^2) c)^w u, u \rangle$$

and if $\tilde{\chi}(x, \xi, h) = \chi_0(2(x - x_0), 2(h\xi - \xi_0))$ (note that $\tilde{\chi}$ belongs to S_{scl}^0 and is supported in $\{\chi = 1\}$), we get

$$C_0 h^2 \|\tilde{\chi}^w u\|^2 + \|a^w \tilde{\chi}^w u\|^2 \geq \alpha_0 h \|\tilde{\chi}^w u\|^2 + O(h^\infty)\|u\|^2$$

and since $[a^w, \tilde{\chi}^w]$ has a symbol in $h S_{scl}^0$, we get that

$$O(h^2)\|u\|^2 + \|\tilde{\chi}^w a^w u\|^2 \geq \alpha_0 h \|\tilde{\chi}^w u\|^2,$$

giving the result. □

We examine now the negative bracket case and we construct a *quasi-mode*.

Theorem 3.1.9. *Let* $p \in C_b^\infty(\mathbb{R}^{2n}), (x_0, \xi_0) \in \mathbb{R}^{2n}$ *such that* $p(x_0, \xi_0) = 0$ *and* $\{\operatorname{Re} p, \operatorname{Im} p\}(x_0, \xi_0) < 0$. *Then there exist* $(u_h)_{0 < h \leq 1}$ *unit vectors in* $L^2(\mathbb{R}^n)$ *such that*

$$\|p(x, h\xi)^w u_h\|_{L^2} = O(h^\infty).$$

Moreover $WF_{scl}(u_h) = \{(x_0, \xi_0)\}$, *i.e., for all* $(x_1, \xi_1) \neq (x_0, \xi_0)$, *there exists* $a \in C_c^\infty(\mathbb{R}^{2n}), a(x_1, \xi_1) \neq 0$, *with* $\|a(x, h\xi)^w u_h\|_{L^2} = O(h^\infty)$.

Proof. Using a symplectic reduction (see Theorem 21.3.3 in [73]), we can find symplectic coordinates centered at (x_0, ξ_0) so that $e(x, \xi)p(x, \xi) = \eta_1 - i y_1$. Quantizing the symplectic transformation with a semi-classical Fourier integral operator (see, e.g., Section 4.7.3), we are reduced to construction proving nonsolvability of M_1 defined in (3.1.30). □

The conclusion for this section is clear and was already completely clarified in 1966, after the works of H. Lewy ([102]), L. Hörmander ([63], [64]), L. Nirenberg and F. Treves ([112]): when the first bracket is positive at a characteristic point, we have nonsolvability and $1/2$ subellipticity and when it is negative we have (microlocal) solvability. The semi-classical versions are analogous: when the first bracket is positive at a characteristic point, we have a subelliptic estimate with loss of $1/2$ derivative, and when it is negative, we can construct a quasi-mode (approximate null solution). Of course the next question is: what happens if that first bracket vanishes at a characteristic point? The examples (3.1.30), (3.1.36), (3.1.10), (3.1.4), (3.1.12), show that different behaviours are possible, depending in these cases on the way the imaginary part changes sign along the bicharacteristic flow of the real part: for $\xi_1 + i x_1^{2k+1}$ (some kind of degenerate creation operator) an $L^2 - L^2$ injectivity estimate holds ($x_1 \mapsto x_1^{2k+1}$ changes sign from $-$ to $+$), whereas for $\xi_1 - i x_1^{2k+1}$ (degenerate annihilation operator) no $L^2 - L^2$ injectivity estimate holds, the kernel is non-trivial ($x_1 \mapsto -x_1^{2k+1}$ changes sign from $+$ to $-$). Moreover an $L^2 - L^2$ injectivity estimate holds for $\xi_1 \pm i x_1^{2k}$ ($x_1 \mapsto \pm x_1^{2k}$ does not change sign). In the next section, we provide some more examples, hopefully shedding some light on the change-of-sign condition obviously concerned with proving a priori estimates for non-selfadjoint pseudo-differential operators.

3.1.3 Heuristics on condition (Ψ)

The previous section with examples, and in particular Lemma 3.1.1 show that the change-of-sign condition (3.1.14) implies that

$$\forall \lambda \geq 0, \forall v \in C_c^\infty(\mathbb{R}), \quad \|v\|_{L^\infty(\mathbb{R})} \leq \|\dot{v} - \lambda\phi v\|_{L^1(\mathbb{R})}. \tag{3.1.42}$$

Conversely the following lemma proves that (3.1.14) follows from (3.1.42).

Lemma 3.1.10. *Let $\phi \in C^\infty(\mathbb{R}, \mathbb{R})$ such that there exists $t_1 < t_2$ with $\phi(t_1) > 0 > \phi(t_2)$. Then*

$$\inf_{\substack{\lambda > 0 \\ v \in C_c^\infty(\mathbb{R}), \|v\|_{L^\infty(\mathbb{R})}=1}} \|\dot{v} - \lambda\phi v\|_{L^1(\mathbb{R})} = 0.$$

Proof. When ϕ changes sign from $+$ to $-$ at a finite-order at some point t_0, we have near t_0, $\phi(t) = e(t)(t - t_0)^{2k+1}$ with $e(t_0) < 0$ and we consider with $\chi_0 \in C_{[-2,2]}^\infty(\mathbb{R}; [0,1])$, equal to 1 on $[-1,1]$, $r > 0$ small enough,

$$v(t) = \chi_0((t - t_0)/r) \exp \int_{t_0}^t \lambda\phi(s)ds, \quad |v(t)| \leq 1 = v(t_0) = \|v\|_{L^\infty(\mathbb{R})}.$$

We have $\dot{v} - \lambda\phi v = r^{-1}\chi_0'((t - t_0)/r) \exp \int_{t_0}^t \lambda\phi(s)ds$ so that, with $c_0 > 0$,

$$\|\dot{v} - \lambda\phi v\|_{L^1(\mathbb{R})} = \int |\chi_0'(\theta)|e^{\lambda \int_{t_0}^{t_0+\theta r} \phi(s)ds}d\theta, \quad \int_{t_0}^{t_0+\theta r} \phi(s)ds \leq -c_0\theta^{2k+2},$$

so that $\lim_{\lambda \to +\infty} \|\dot{v} - \lambda\phi v\|_{L^1(\mathbb{R})} = 0$.

The general case: ϕ is positive on (t_1, t_1'), negative on (t_2', t_2) and vanishes at t_1', t_2' with $t_1 < t_1' \leq t_2' < t_2$. We consider

$$v(t) = \chi(t) \exp \int_{t_1'}^t \lambda\phi(s)ds, \quad \text{with } \chi = 1 \text{ on } [t_1', t_2'], \text{ supp}\chi \subset (t_1, t_2).$$

We have (assuming also χ valued in $[0,1]$), $\|v\|_{L^\infty} \geq |v(t_1')| = 1$ and

$$\dot{v} - \lambda\phi v = \chi'(t) \exp \int_{t_1'}^t \lambda\phi(s)ds = \begin{cases} \chi'(t) \exp \int_{t_1'}^t \lambda\phi(s)ds & \text{if } t \leq t_1', \\ 0 & \text{if } t_1' \leq t \leq t_2', \\ \chi'(t) \exp \int_{t_2'}^t \lambda\phi(s)ds & \text{if } t_2' \leq t, \end{cases}$$

and as before we get $\lim_{\lambda \to +\infty} \|\dot{v} - \lambda\phi v\|_{L^1(\mathbb{R})} = 0$. The proof of the lemma is complete. $\qquad\square$

As a consequence, we see that proving (3.1.42) is equivalent to the analytical condition (3.1.14). This means that proving an a priori estimate for

$$D_t + iq(t, x, \xi) = -i(\frac{d}{dt} - q),$$

where q is a real-valued function depending on some parameters x, ξ, is equivalent to the analytical condition

$t \mapsto q(t, x, \xi)$ does not change sign from $+$ to $-$ when t is increasing.

Since proving an priori estimate for $D_t + iq(t, x, \xi)$ is akin to proving some solvability properties for $D_t - iq(t, x, \xi)$, which has the "symbol" $\tau - iq(t, x, \xi)$, the analytical condition linked to local solvability of $D_t - iq(t, x, \xi)$ should be $t \mapsto q(t, x, \xi)$ does not change sign from $+$ to $-$ while t is increasing. Taking a complex-valued symbol $p = p_1 + ip_2$, with p_j real-valued (and $dp_1 \neq 0$), we see that the analytical condition for local solvability of p^w should be

p_2 *does not change sign from* $-$ *to* $+$ *along the oriented flow of* H_{p_1}.

This geometrical condition on the principal symbol $p_1 + ip_2$ of a principal-type operator will be called condition (Ψ); it is obviously invariant by change of coordinates but it is also important to give a version which is invariant by multiplication by a complex-valued non-vanishing factor. We discuss these matters in the next section, and give the proper references in the literature. On the other hand, with a complex-valued symbol $p = p_1 + ip_2$, with p_j real-valued (and $dp_1 \neq 0$), p^w should satisfy some a priori estimate whenever the complex conjugate $\bar{p}$ satisfies condition (Ψ), i.e.,

p_2 *does not change sign from* $+$ *to* $-$ *along the oriented flow of* H_{p_1}.

The latter condition will be called condition $(\overline{\Psi})$ and means simply that $\bar{p}$ satisfies condition (Ψ). In some sense, condition $(\overline{\Psi})$ is an injectivity condition, since an operator P with a symbol satisfying condition $(\overline{\Psi})$ should be such that

$$\|Pu\| \gtrsim \|u\|,$$

whereas condition (Ψ) is a surjectivity condition: an operator P with a symbol satisfying condition (Ψ) should be such that

$$\|P^*u\| \gtrsim \|u\|,$$

thus be solvable.

The following picture gives an explanation for the necessity of condition (Ψ) for local solvability, i.e., the necessity of condition $(\overline{\Psi})$ for local solvability of P^*.

We consider a complex-valued symbol p with $d\,\mathrm{Re}\,p \neq 0$ and we draw a bicharacteristic curve of $\mathrm{Re}\,p$, with the orientation given by the small arrows. On the set $\{\mathrm{Im}\,p \geq 0\}$ the propagation of singularities is forward; we keep the original orientation and the top large arrow has the same direction as the small one. On the set $\{\mathrm{Im}\,p \leq 0\}$ the propagation of singularities is backward; we reverse the original orientation and the bottom large arrow has the opposite direction to the

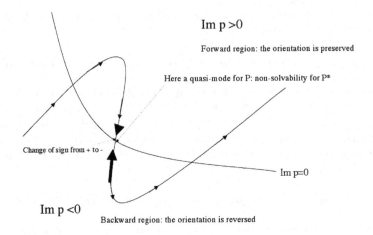

Figure 3.1: Violation of condition $(\overline{\Psi})$

small one. Since condition $(\overline{\Psi})$ is violated, we have a change of sign of $\operatorname{Im} p$ from $+$ to $-$, resulting in a collision of the thick arrows, triggering a quasi-mode for P, i.e., a non-zero approximate solution of $Pu = 0$, and non-local solvability of P^*. On the other hand, looking at the other changes of sign (e.g., top left of the picture, from $-$ to $+$), we see that changing the orientation in the backward region does not create a collision.

Remark 3.1.11. For a principal-type operator, using the Malgrange-Weierstrass preparation theorem (see Proposition 4.5.1), we shall see that the principal symbol can be assumed of the form

$$\xi_1 + a(x_1, x', \xi') + ib(x_1, x', \xi'), \quad a, b \quad \text{real-valued.}$$

and Remark 1.2.36 can be used to reduce local solvability questions to prove an a priori estimate for the evolution operator, slightly changing notation, $D_t + a(t, x, D) + ib(t, x, D)$, at least when the loss of derivatives is less than 1 (when the loss is > 1, as in the general case under condition (Ψ), we devote the whole Section 3.7.3 to the delicate transfer from semi-classical to local estimates). This type of estimate can also be reduced to a semi-classical estimate for

$$hD_t + a(t, x, hD) + ib(t, x, hD), \quad \text{with } a, b \in C_b^\infty, \text{ real-valued,}$$

and using a semi-classical Fourier integral operator (see Section 4.7.3), the problem can be further reduced to deal with the evolution equation

$$hD_t + iq(t, x, h\xi)^w, \quad \text{with } q \in C_b^\infty, \text{ real-valued.}$$

3.2 The geometry of condition (Ψ)

3.2.1 Definitions and examples

Definition 3.2.1. Let $\mathcal{M}$ be a C^∞ symplectic manifold[4], $p : \mathcal{M} \to \mathbb{C}$ be a smooth complex-valued function. The function p is said to satisfy the condition (Ψ) if for all $m \in \mathcal{M}$ such that $p(m) = 0$, there exists an open neighborhood U of m such that, either for $z = 1$ or for $z = i$, $H_{\mathrm{Re}\, zp} \neq 0$ on U and for γ a bicharacteristic curve[5] of $\mathrm{Re}\, zp$,

$$\forall s, t \in \mathbb{R} \quad \text{such that} \quad \mathrm{Im}\, zp(\gamma(t)) < 0, s > t \Longrightarrow \mathrm{Im}\, zp(\gamma(s)) \leq 0. \qquad (3.2.1)$$

The function p is said to satisfy the condition ($\bar{\Psi}$) on $\mathcal{M}$ if the complex conjugate $\bar{p}$ satisfies the condition (Ψ).

N.B. Another formulation is to say that p satisfies condition (Ψ) whenever $\mathrm{Im}\, p$ does not change sign from $-$ to $+$ while moving on the positive direction on a bicharacteristic of $\mathrm{Re}\, p$ or that $\mathrm{Re}\, p$ does not change sign from $+$ to $-$ while moving on the positive direction on a bicharacteristic of $\mathrm{Im}\, p$.

Let us give a couple of examples: we consider on the phase space $\mathbb{R}^n_x \times \mathbb{R}^n_\xi$, equipped with its standard symplectic structure (4.5.5) the function

$$\mathbb{R} \times \mathbb{R}^{n-1} \times \mathbb{R} \times \mathbb{R}^{n-1} \ni (x_1, x', \xi_1, \xi') \mapsto p(x, \xi) \quad \text{with}$$
$$p(x, \xi) = \xi_1 + ia(x_1, x', \xi')b(x_1, x', \xi'), \quad \text{with } a \geq 0 \text{ and } \partial_{x_1} b \leq 0. \qquad (3.2.2)$$

The bicharacteristic curve of $\mathrm{Re}\, p$ are $\gamma(t) = (x_1 + t, x', 0, \xi')$ and the imaginary part of $p(\gamma(t))$ is $\beta(t) = a(x_1 + t, x', \xi')b(x_1 + t, x', \xi')$. Since $a \geq 0$ and $t \mapsto b(x_1 + t, x', \xi')$ is non-increasing, the function β cannot change sign from $-$ to $+$ while t increases and (3.2.1) holds for $z = 1$. Note that the function given by the formula (3.2.2) with $a \geq 0$ and $\partial_{x_1} b \geq 0$ satisfies condition ($\bar{\Psi}$). On the other hand, for $k \in \mathbb{N}$, the function

$$\xi_1 + ix_1^{2k+1} \qquad (3.2.3)$$

violates condition (Ψ), but from the previous discussion satisfies condition ($\bar{\Psi}$). We can prove as well that

$$\xi_1 - ix_1^{2k+1} \qquad (3.2.4)$$

satisfies condition (Ψ) and violates condition ($\bar{\Psi}$)(see also (3.1.10)). Note also the examples

$$\xi_1 \pm ix_1^{2k} \qquad (3.2.5)$$

---

[4] A short reminder on symplectic geometry is given in Section 4.5.
[5] A bicharacteristic curve of a (C^2, real-valued) function r is an integral curve of the Hamiltonian vector field H_r where $r = 0$. Along such a curve, one has $r(\gamma(t)) = r(\gamma(0))(= 0)$ since $\frac{d}{dt}(r(\gamma(t))) = H_r(r)(\gamma(t)) = \{r, r\} = 0$.

which satisfy both condition (Ψ) and condition $(\overline{\Psi})$(see (3.1.12)). It may be worth noticing that (3.2.2) produces more general examples than the three previous two-dimensional ones: in particular, with $k, l, m \in \mathbb{N}$, the functions

$$p_{\pm}(x, \xi) = \xi_1 \pm ix_1^{2l}(\xi_2 + x_1^{2k+1}x_2^{2m})$$

are such that p_+ satisfies condition $(\overline{\Psi})$, while p_- satisfies condition (Ψ).

Theorem 3.2.2. *, Let $\mathcal{M}$ be a C^∞ symplectic manifold, $p \in C^\infty(\mathcal{M}; \mathbb{C})$ be a smooth complex-valued function satisfying condition (Ψ) and $e \in C^\infty(\mathcal{M}; \mathbb{C}^*)$ be a non-vanishing complex-valued smooth function. Then the function ep satisfies condition (Ψ).*

References for the proof. This result of invariance of condition (Ψ) by multiplication by an "elliptic" factor was proven by L. Nirenberg and F. Treves in [113] and another proof can be found in the four-volume book of L. Hörmander (Theorem 26.4.12 in [74]). A key lemma for that theorem is a result on flow-invariant sets proven in the papers by J.-M. Bony [13] and H. Brézis [24] (see Lemma 26.4.11 in [74]).   □

This is a difficult theorem but the first-bracket analysis offers a very simple special case: assume that $\{\operatorname{Re} p, \operatorname{Im} p\} < 0$ at $p = 0$; then p satisfies the condition (Ψ), since $H_{\operatorname{Re} p} \neq 0$ and $\operatorname{Im} p$ has a simple zero, with a negative derivative along the bicharacteristic of $\operatorname{Re} p$. With a non-vanishing factor e, we calculate at $ep = 0$, i.e., at $p = 0$ (with $p = p_1 + ip_2, e = e_1 + ie_2$, and p_j, e_j real-valued),

$$\{e_1 p_1 - e_2 p_2, e_2 p_1 + e_1 p_2\} = (e_1^2 + e_2^2)\{p_1, p_2\} < 0.$$

Theorem 3.2.3. *Let $\mathcal{M}$ be a C^∞ symplectic manifold, $p \in C^\infty(\mathcal{M}, \mathbb{C})$ be a smooth complex-valued function satisfying condition (Ψ), $m \in \mathcal{M}$ such that $p(m) = 0$. Then there exists an open neighborhood U of m, a function $e \in C^\infty(U; \mathbb{C}^*)$ and a symplectomorphism χ of a neighborhood V of 0 in the symplectic $(\mathbb{R}_x^n \times \mathbb{R}_\xi^n, \sum d\xi_j \wedge dx_j)$ onto U such that*

$$((ep) \circ \chi)(x, \xi) = \xi_1 + if(x, \xi'), \quad f \text{ real-valued, independent of } \xi_1, \qquad (3.2.6)$$
$$f(x_1, x', \xi') < 0, h > 0 \Longrightarrow f(x_1 + h, x', \xi') \leq 0. \qquad (3.2.7)$$

References for the proof. This result follows from Theorem 3.2.2, the Malgrange-Weierstrass theorem and the Darboux theorem. A proof is given in Proposition 26.4.13 of [74].   □

A typical example of a function satisfying condition (Ψ) was already given with (3.2.2). Let us consider the following smooth function f, defined on $\mathbb{R}^3$ by

$$f(x_1, x_2, \xi_2) = e^{-x_1^{-2}}\left(\xi_2 H(-x_1) + (\xi_2^3 + x_2^2)H(x_1)\right), \quad H = \mathbf{1}_{\mathbb{R}_+}.$$

Note that $f(0, x_2, \xi_2) = 0$ and for $h > 0$, we have

$$x_1 < 0, f(x_1, x_2, \xi_2) > 0 \Longrightarrow \quad \xi_2 > 0 \quad \Longrightarrow f(x_1 + h, x_2, \xi_2) \geq 0,$$
$$x_1 > 0, f(x_1, x_2, \xi_2) > 0 \Longrightarrow \xi_2^3 + x_2^2 > 0 \Longrightarrow f(x_1 + h, x_2, \xi_2) \geq 0,$$

so that $\xi_1 + if(x_1, x_2, \xi_2)$ actually satisfies condition $(\overline{\Psi})$. However, the function f is not a priori given as a product ab with $a, b \in C^\infty$ and $a \geq 0, b'_{x_1} \geq 0$.

3.2.2 Condition (P)

Definition 3.2.4. Let $\mathcal{M}$ be a C^∞ symplectic manifold, $p : \mathcal{M} \to \mathbb{C}$ be a smooth complex-valued function. The function p is said to satisfy condition (P) if p and $\bar{p}$ both satisfy condition (Ψ).

N.B. Equivalently, the function p satisfies condition (P) if for all $m \in \mathcal{M}$ such that $p(m) = 0$, there exists an open neighborhood U of m such that, either for $z = 1$ or for $z = i$, $H_{\mathrm{Re}\,zp} \neq 0$ on U and for γ a bicharacteristic curve of $\mathrm{Re}\,zp$,

$$\forall s, t \in \mathbb{R} \quad \mathrm{Im}\big(zp(\gamma(t))\big)\,\mathrm{Im}\big(zp(\gamma(s))\big) \geq 0. \tag{3.2.8}$$

In other words, the imaginary part of $\mathrm{Im}\,zp$ does not change sign along the bicharacteristic curves of $\mathrm{Re}\,zp$. Using Theorem 3.2.2, it means that if $d\,\mathrm{Re}\,p \neq 0, p = 0$ at some point, then $\mathrm{Im}\,p$ does not change sign along the bicharacteristic curve of $\mathrm{Re}\,p$.

Remark 3.2.5. In Remark 3.1.7, we have already seen that differential operators have a particular behaviour. Let $p : \dot{T}^*(M) \longrightarrow \mathbb{C}$ be a smooth complex-valued function satisfying condition (Ψ), such that $p(x, -\xi) = (-1)^m p(x, \xi)$. Then p satisfies condition (P). In fact, if $d\,\mathrm{Re}\,p \neq 0$, $p = 0$ at (x_0, ξ_0) and $\mathrm{Im}\,p$ were to change sign from $+$ to $-$ along the bicharacteristic curve of $\mathrm{Re}\,p$ through (x_0, ξ_0), then with

$$\overbrace{\qquad\qquad}^{\gamma(t, x_0, \xi_0)}$$
$$\dot{x}(t, x_0, \xi_0) = (\partial_\xi\,\mathrm{Re}\,p)\big(x(t, x_0, \xi_0), \xi(t, x_0, \xi_0)\big),$$
$$\dot{\xi}(t, x_0, \xi_0) = -(\partial_x\,\mathrm{Re}\,p)\big(x(t, x_0, \xi_0), \xi(t, x_0, \xi_0)\big)$$

we could find $t < s$ with $\mathrm{Im}\,p(\gamma(t, x_0, \xi_0)) > 0 > \mathrm{Im}\,p(\gamma(s, x_0, \xi_0))$. Since we have

$$\dot{x}(t, x_0, \xi_0) = (\partial_\xi(-1)^{m-1}\,\mathrm{Re}\,p)\big(x(t, x_0, \xi_0), -\xi(t, x_0, \xi_0)\big),$$
$$-\dot{\xi}(t, x_0, \xi_0) = -(\partial_x(-1)^{m-1}\,\mathrm{Re}\,p)\big(x(t, x_0, \xi_0), -\xi(t, x_0, \xi_0)\big)$$

we get that $t \mapsto \tilde{\gamma}(t, x_0, \xi_0) = \big(x(t, x_0, \xi_0), -\xi(t, x_0, \xi_0)\big)$ is the integral curve of the Hamiltonian field of $(-1)^{m-1}\,\mathrm{Re}\,p(x, \xi)$ through $(x_0, -\xi_0)$, and

$$\mathrm{Im}\big((-1)^{m-1}p(\tilde{\gamma}(t))\big) = -\,\mathrm{Im}\big(p(\gamma(t))\big) < 0 < -\,\mathrm{Im}\big(p(\gamma(s))\big) = \mathrm{Im}\big((-1)^{m-1}p(\tilde{\gamma}(s))\big),$$

so that $(-1)^{m-1}p$ violates condition (Ψ). If m is odd it contradicts the fact that p satisfies condition (Ψ). If m is even, this implies that $-p$ violates condition (Ψ), which is equally impossible since condition (Ψ) is (trivially) invariant by multiplication by a real non-zero factor.

Following (3.2.2), we can give some examples of functions satisfying condition (P). In particular it is the case of

$$p(x,\xi) = \xi_1 + ia(x_1, x', \xi')b(x', \xi'), \quad \text{with } a \geq 0. \tag{3.2.9}$$

In fact, the product $a(t, x', \xi')b(x', \xi')a(s, x', \xi')b(x', \xi') \geq 0$ so that condition (P) holds and that example is also interesting to show that condition (P) does not require a specific sign of the imaginary part but simply rules out change of sign along the integral curves of $H_{\mathrm{Re}\, p}$. The most typical example is the degenerate Cauchy-Riemann operators (see also (3.1.12)),

$$D_{x_1} + ia(x_1, x_2)D_{x_2}, \quad \text{with } a \geq 0, \tag{3.2.10}$$

which is a particular case of (3.2.9). The imaginary part of the symbol is $a(x_1, x_2)\xi_2$ and although the function $t \mapsto a(t, x_2)\xi_2$ does not change sign, it will take in general positive and negative values and thus is not in the category (3.1.1). In Lemma 2.1 of [114], Nirenberg and Treves proved that an analytic function satisfying condition (P) can be written via a factorization as in (3.2.9).

However, it was pointed out by these authors that this factorization form does not hold in the C^∞ category, as shown by the following example. Let q be the function defined on $\mathbb{R}^3$ by

$$q(t, x, \xi) = \begin{cases} (\xi - te^{-1/x})^2 & \text{if } x > 0, \\ \xi^2 & \text{if } x = 0, \\ \xi(\xi - e^{1/x}) & \text{if } x < 0. \end{cases} \tag{3.2.11}$$

The function q is C^∞ since

$$q(t, x, \xi) = \xi^2 - 2t\xi H(x)e^{-1/x} + t^2 H(x)e^{-2/x} - \xi H(-x)e^{1/x}.$$

For every fixed (x, ξ), the function $t \mapsto q(t, x, \xi)$ does not change sign since $q(t, x, \xi)q(s, x, \xi) \geq 0$. Nevertheless it is not possible to find some C^∞ functions a, b such that a is non-negative and b independent of t such that $q = ab$: if it were so, we would have

$$\forall x < 0, \quad \xi(\xi - e^{1/x}) = b(x, \xi)a(t, x, \xi) \Longrightarrow b(x, 0) = b(x, e^{1/x}) = 0,$$

since for each $x < 0$ the function $\xi \mapsto \xi(\xi - e^{1/x})$ has simple zeroes $0, e^{1/x}$. As a consequence, for $x < 0$, we have $b(x, 0) = 0$ and

$$0 = e^{-1/x}b(x, e^{1/x}) = e^{-1/x}b(x, 0) + (\partial_\xi b)(x, 0) + O(e^{1/x}) = (\partial_\xi b)(x, 0) + O(e^{1/x})$$

and this implies as well $\partial_\xi b = 0$ at $(0, 0)$. If $x > 0$, we must have $b(x, \xi) \neq 0$, otherwise

$$\exists x_0 > 0, \exists \xi_0, \forall t, \quad 0 = (\xi_0 - te^{-1/x_0})^2 \Longrightarrow \xi_0 = te^{-1/x_0} \quad \text{which is absurd.}$$

Hence for $x > 0$ we must have $a(t, x, te^{-1/x}) = 0$. When $x \to 0_+$, it follows that $a(t, 0, 0) = 0$. But then we obtain for all t that $q(t, 0, 0) = a(t, 0, 0)b(0, 0) = 0$, and $(\partial_\xi q)(t, 0, 0) = a(t, 0, 0)\partial_\xi b(0, 0) + b(0, 0)\partial_\xi a(t, 0, 0) = 0$ as well as

$$\partial_\xi^2 q(t, 0, 0) = (\partial_\xi^2 a)(t, 0, 0)b(0, 0) + 2\partial_\xi a(t, 0, 0)\partial_\xi b(0, 0) + a(t, 0, 0)\partial_\xi^2 b(0, 0) = 0.$$

But this is a contradiction since we have $q(t, 0, \xi) = \xi^2$ and thus $\partial_\xi^2 q(t, 0, 0) = 2$.

3.2.3 Condition (Ψ) for semi-classical families of functions

In this section, we shall consider that the phase space $\mathbb{R}^{2n}$ is equipped with a *symplectic quadratic form* Γ (Γ is a positive-definite quadratic form such that $\Gamma = \Gamma^\sigma$, see (4.4.18)). According to Lemma 4.4.25 it is possible to find some linear symplectic coordinates (x, ξ) in $\mathbb{R}^{2n}$ such that

$$\Gamma(x, \xi) = |(x, \xi)|^2 = \sum_{1 \le j \le n} x_j^2 + \xi_j^2.$$

The running point of our Euclidean symplectic $\mathbb{R}^{2n}$ will be usually denoted by X or by an upper-case letter such as Y, Z. The open Γ-ball with center X and radius r will be denoted by $B(X, r)$.

Definition 3.2.6. Let $q(t, X, \Lambda)$ be a smooth real-valued function defined on $\Xi = \mathbb{R} \times \mathbb{R}^{2n} \times [1, +\infty)$, vanishing for $|t| \ge 1$ and satisfying

$$\forall k \in \mathbb{N}, \sup_{\Xi} \|\partial_X^k q\|_\Gamma \Lambda^{-1+\frac{k}{2}} = \gamma_k < +\infty, \text{ i.e., } q(t, \cdot) \in S(\Lambda, \Lambda^{-1}\Gamma), \quad (3.2.12)$$

$$s > t \quad \text{and} \quad q(t, X, \Lambda) > 0 \implies q(s, X, \Lambda) \ge 0. \quad (3.2.13)$$

Note[6] that the condition (3.2.13) expresses that, for all $\Lambda \ge 1$, the functions $\tau - iq(t, x, \xi, \Lambda)$, defined on the symplectic $\mathbb{R}_t \times \mathbb{R}_x^n \times \mathbb{R}_\tau \times \mathbb{R}_\xi^n$, with its canonical symplectic form $d\tau \wedge dt + \sum d\xi_j \wedge dx_j$, satisfy condition ($\Psi$), i.e., that $\tau + iq(t, x, \xi, \Lambda)$ satisfies condition ($\overline{\Psi}$). In this section, the Euclidean norm $\Gamma(X)^{1/2}$ is fixed and the norms of the vectors and of the multilinear forms are taken with respect to that norm. We shall write everywhere $|\cdot|$ instead of $\|\cdot\|_\Gamma$. Furthermore, we shall say that C is a "fixed" constant if it depends only on a finite number of γ_k above and on the dimension n. We shall always omit the dependence of q with respect to the large parameter Λ and write $q(t, X)$ instead of $q(t, X, \Lambda)$. We define now for $t \in \mathbb{R}$,

$$\mathbb{X}_+(t) = \cup_{s \le t}\{X \in \mathbb{R}^{2n}, q(s, X) > 0\}, \quad (3.2.14)$$

$$\mathbb{X}_-(t) = \cup_{s \ge t}\{X \in \mathbb{R}^{2n}, q(s, X) < 0\}, \quad (3.2.15)$$

$$\mathbb{X}_0(t) = \mathbb{X}_-(t)^c \cap \mathbb{X}_+(t)^c. \quad (3.2.16)$$

---

[6]We shall say that a family of functions $(a_\Lambda)_{\Lambda \ge 1}$ belongs to $S(\Lambda, \Lambda^{-1}\Gamma)$ whenever $a_\Lambda \in S(\Lambda, \Lambda^{-1}\Gamma)$ with semi-norms uniformly bounded with respect to the parameter $\Lambda \ge 1$. Naturally that uniformity with respect to the large parameter is of crucial importance.

Lemma 3.2.7. *Let q be as in Definition 3.2.6. The sets $\mathbb{X}_+(t), \mathbb{X}_-(t)$ are disjoint open subsets of $\mathbb{R}^{2n}$. The sets $\mathbb{X}_0(t), \mathbb{X}_0(t) \cup \mathbb{X}_\pm(t)$ are closed. The three sets $\mathbb{X}_0(t), \mathbb{X}_\pm(t)$ are two-by-two disjoint with union $\mathbb{R}^{2n}$. Moreover*

$$\overline{\mathbb{X}_\pm(t)} \subset \mathbb{X}_0(t) \cup \mathbb{X}_\pm(t)$$

and when t increases, $\mathbb{X}_+(t)$ increases and $\mathbb{X}_-(t)$ decreases.

Proof. If $X \in \mathbb{X}_+(t)$, i.e., is such that $q(s,X) > 0$ for some $s \leq t$, the condition (3.2.13) implies $q(t',X) \geq 0$ for all $t' \geq t$, i.e., $X \notin \mathbb{X}_-(t)$, proving $\mathbb{X}_+(t) \subset \mathbb{X}_-(t)^c$. Thus we have $\mathbb{R}^{2n} = \mathbb{X}_0(t) \cup \mathbb{X}_+(t) \cup \mathbb{X}_-(t)$ as a disjoint union and this implies that $\mathbb{X}_0(t), \mathbb{X}_0(t) \cup \mathbb{X}_\pm(t)$ are closed since their complements are open. We have also $\overline{\mathbb{X}_\pm(t)} \subset \mathbb{X}_0(t) \cup \mathbb{X}_\pm(t)$ since $\mathbb{X}_0(t) \cup \mathbb{X}_\pm(t)$ are closed. The very definitions (3.2.14), (3.2.15) imply the last statement. $\square$

Lemma 3.2.8. *Let (E,d) be a metric space, $A \subset E$ and $\kappa > 0$ be given. We define $\Psi_{A,\kappa}(x) = \kappa$ if $A = \emptyset$ and if $A \neq \emptyset$, we define $\Psi_{A,\kappa}(x) = \min(d(x,A), \kappa)$. The function $\Psi_{A,\kappa}$ is valued in $[0, \kappa]$, Lipschitz-continuous with a Lipschitz constant ≤ 1. Moreover, the following implication holds: $A_1 \subset A_2 \subset E \Longrightarrow \Psi_{A_1,\kappa} \geq \Psi_{A_2,\kappa}$.*

Proof. The Lipschitz continuity assertion is obvious since $x \mapsto d(x,A)$ is Lipschitz-continuous with Lipschitz constant 1. The monotonicity property is trivially inherited from the distance function. $\square$

Lemma 3.2.9. *Let $\kappa > 0$ be given and the functions Ψ be defined by the previous lemma. For each $X \in \mathbb{R}^{2n}$, the function $t \mapsto \Psi_{\mathbb{X}_+(t),\kappa}(X)$ is decreasing and for each $t \in \mathbb{R}$, the function $X \mapsto \Psi_{\mathbb{X}_+(t),\kappa}(X)$ is supported in $\mathbb{X}_+(t)^c = \mathbb{X}_-(t) \cup \mathbb{X}_0(t)$. For each $X \in \mathbb{R}^{2n}$, the function $t \mapsto \Psi_{\mathbb{X}_-(t),\kappa}(X)$ is increasing and for each $t \in \mathbb{R}$, the function $X \mapsto \Psi_{\mathbb{X}_-(t),\kappa}(X)$ is supported in $\mathbb{X}_-(t)^c = \mathbb{X}_+(t) \cup \mathbb{X}_0(t)$. As a consequence the function $X \mapsto \Psi_{\mathbb{X}_+(t),\kappa}(X)\Psi_{\mathbb{X}_-(t),\kappa}(X)$ is supported in $\mathbb{X}_0(t)$.*

Proof. The monotonicity in t follows from the fact that $\mathbb{X}_+(t)$(resp. $\mathbb{X}_-(t)$) is increasing (resp. decreasing) with respect to t and from Lemma 3.2.8. Moreover, if X belongs to the open set $\mathbb{X}_\pm(t)$, one has $\Psi_{\mathbb{X}_\pm(t),\kappa}(X) = 0$, implying the support property. $\square$

Lemma 3.2.10. *For $\kappa > 0, t \in \mathbb{R}, X \in \mathbb{R}^{2n}$, we define[7]*

$$\sigma(t, X, \kappa) = \Psi_{\mathbb{X}_-(t),\kappa}(X) - \Psi_{\mathbb{X}_+(t),\kappa}(X). \tag{3.2.17}$$

The function $t \mapsto \sigma(t, X, \kappa)$ is increasing and valued in $[-\kappa, \kappa]$, the function $X \mapsto \sigma(t, X, \kappa)$ is Lipschitz-continuous with Lipschitz constant less than 2; we have

$$\sigma(t, X, \kappa) = \begin{cases} \min(|X - \mathbb{X}_-(t)|, \kappa) & \text{if } X \in \mathbb{X}_+(t), \\ -\min(|X - \mathbb{X}_+(t)|, \kappa) & \text{if } X \in \mathbb{X}_-(t). \end{cases}$$

---

[7]When the distances of X to both $\mathbb{X}_\pm(t)$ are less than κ, we have $\sigma(t, X, \kappa) = |X - \mathbb{X}_-(t)| - |X - \mathbb{X}_+(t)|$.

We have $\{X \in \mathbb{R}^{2n}, \sigma(t, X, \kappa) = 0\} \subset \mathbb{X}_0(t) \subset \{X \in \mathbb{R}^{2n}, q(t, X) = 0\}$, *and*

$$\{X \in \mathbb{R}^{2n}, \pm q(t, X) > 0\} \subset \mathbb{X}_{\pm}(t) \subset \{X \in \mathbb{R}^{2n}, \pm\sigma(t, X, \kappa) > 0\}$$
$$\subset \{X \in \mathbb{R}^{2n}, \pm\sigma(t, X, \kappa) \geq 0\} \subset \{X \in \mathbb{R}^{2n}, \pm q(t, X) \geq 0\}. \quad (3.2.18)$$

Proof. Everything follows from the previous lemmas, except for the first, fourth and sixth inclusions. Note that if $X \in \mathbb{X}_+(t)$, $\sigma(t, X, \kappa) = \min(|X - \mathbb{X}_-(t)|, \kappa)$ is positive: otherwise it vanishes and

$$X \in \mathbb{X}_+(t) \cap \overline{\mathbb{X}_-(t)} \subset \mathbb{X}_+(t) \cap \left(\mathbb{X}_-(t) \cup \mathbb{X}_0(t)\right) = \emptyset.$$

As a consequence, we get the penultimate inclusions

$$\mathbb{X}_+(t) \subset \{X \in \mathbb{R}^{2n}, \sigma(t, X, \kappa) > 0\}$$

and similarly $\mathbb{X}_-(t) \subset \{X \in \mathbb{R}^{2n}, \sigma(t, X, \kappa) < 0\}$, so that

$$\{X \in \mathbb{R}^{2n}, \sigma(t, X, \kappa) = 0\} \subset \mathbb{X}_+(t)^c \cap \mathbb{X}_-(t)^c = \mathbb{X}_0(t),$$

giving the first inclusion. The last inclusion follows from the already established

$$\{X \in \mathbb{R}^{2n}, q(t, X)) < 0\} \subset \mathbb{X}_-(t) \subset \{X \in \mathbb{R}^{2n}, \sigma(t, X, \kappa) < 0\}. \qquad \square$$

Definition 3.2.11. Let $q(t, X)$ be as above. We define

$$\delta_0(t, X) = \sigma(t, X, \Lambda^{1/2}) \qquad (3.2.19)$$

and we notice that from the previous lemmas, $t \mapsto \delta_0(t, X)$ is increasing, valued in $[-\Lambda^{1/2}, \Lambda^{1/2}]$, satisfying

$$|\delta_0(t, X) - \delta_0(t, Y)| \leq 2|X - Y| \qquad (3.2.20)$$
$$\text{and such that } \{X \in \mathbb{R}^{2n}, \delta_0(t, X) = 0\} \subset \{X \in \mathbb{R}^{2n}, q(t, X) = 0\}, \qquad (3.2.21)$$
$$\{X \in \mathbb{R}^{2n}, \pm q(t, X) > 0\} \subset \{X, \pm\delta_0(t, X) > 0\} \subset \{X, \pm q(t, X) \geq 0\}. \qquad (3.2.22)$$

Lemma 3.2.12. *Let f be a symbol in $S(\Lambda^m, \Lambda^{-1}\Gamma)$ where m is a positive real number. We define*

$$\lambda(X) = 1 + \max_{\substack{0 \leq j < 2m \\ j \in \mathbb{N}}} \left(\|f^{(j)}(X)\|_{\Gamma}^{\frac{2}{2m-j}}\right). \qquad (3.2.23)$$

Then $f \in S(\lambda^m, \lambda^{-1}\Gamma)$ and the mapping from $S(\Lambda^m, \Lambda^{-1}\Gamma)$ to $S(\lambda^m, \lambda^{-1}\Gamma)$ is continuous. Moreover, with $\gamma = \max_{\substack{0 \leq j < 2m \\ j \in \mathbb{N}}} \gamma_j^{\frac{2}{2m-j}}$, where the γ_j are the seminorms of f, we have for all $X \in \mathbb{R}^{2n}$,

$$1 \leq \lambda(X) \leq 1 + \gamma\Lambda. \qquad (3.2.24)$$

The metric $\lambda^{-1}\Gamma$ is admissible (Definition 2.2.15), with structure constants depending only on γ. It will be called the m-proper metric of f. The function λ above is a weight for the metric $\lambda^{-1}\Gamma$ and will be called the m-proper weight of f.

Proof. The proof of this lemma is given in the appendix, Section 4.9.2. □

Comment. Note also that this result provides a Calderón-Zygmund decomposition of the function f, namely a cutting procedure for this function, which depends on the function f itself: a typical "box" around the point X is the Euclidean ball with center X and radius $\lambda(X)^{1/2}$. The powers occurring in (3.2.23) follow from a simple dimensional analysis due to Taylor's formula. In fact, if $f : \mathbb{R}^N \to \mathbb{R}$ has the dimension of (length)^{2m}, the quantities $f^{(j)}(X)$ have the dimension L^{2m-j}, and thus

$$\|f^{(j)}(X)\|^{\frac{1}{2m-j}}$$

has the dimension of a length. It is then natural to compare the quantities,

$$\|f^{(j)}(X)\|^{\frac{1}{2m-j}} \quad \text{for } 0 \le j < 2m, \text{ all with the dimension of a length.}$$

The largest will provide the radius $\lambda(X)^{1/2}$. A peculiarity of the phase space decomposition is that the stopping procedure must ensure that the uncertainty principle should be satisfied, i.e., λ should remain larger than 1.

   The first occurrence of these ideas in pseudo-differential analysis is due to R. Beals and C. Fefferman in [8] with the calculations detailed in Section 3.5.1 (see in particular the formula (3.5.5), some kind of proper class for the family $\{q(t,\cdot)\}_{t\in\mathbb{R}}$). However their presentation of the Calderón-Zygmund decomposition was based on some sort of dynamical procedure, in which the phase space is first cut into large boxes, then some of the large boxes are kept if one of the quantities $(\|f^{(j)}(X)\|^{1/(2m-j)})_{0\le j<2m}$ is the largest all over the box, the other large boxes are cut into 2^N "generation 1" boxes; the latter are kept whenever for some $j \in [0, 2m[, \|f^{(j)}(X)\|^{1/(2m-j)}$ is the largest all over the box, then each generation 1 box which did not satisfy the above criterion is cut into 2^N generation 2 boxes... and so on. That very suggestive but complicated presentation was by-passed by L. Hörmander's metrics, and the previous lemma encapsulates all the details of the Calderón-Zygmund lemma by the simple introduction of the metric $\lambda^{-1}\Gamma$, quite easily proven admissible. It turns out that the metrics are much easier to handle than the procedure described above and that the flexibility of this tool was important in subsequent developments of pseudo-differential analysis.

   On the other hand, it is not clear if some kind of metric can be used for the Calderón-Zygmund decomposition occurring in the proof of the atomic decomposition of the Hardy space, where the classical proof uses a dynamical cutting and stopping procedure.

Lemma 3.2.13. *Let $q(t, X)$, $\delta_0(t, X)$ be as above. We define, with $\langle s \rangle = (1 + s^2)^{1/2}$,*

$$\mu(t, X) = \langle \delta_0(t, X) \rangle^2 + |\Lambda^{1/2} q'_X(t, X)| + |\Lambda^{1/2} q''_{XX}(t, X)|^2. \tag{3.2.25}$$

The metric $\mu^{-1}(t, \cdot)\Gamma$ is slowly varying with structure constants depending only on a finite number of semi-norms of q in $S(\Lambda, \Lambda^{-1}\Gamma)$. Moreover, there exists $C > 0$, *depending only on a finite number of semi-norms of q, such that*

$$\mu(t, X) \leq C\Lambda, \quad \frac{\mu(t, X)}{\mu(t, Y)} \leq C(1 + |X - Y|^2), \qquad (3.2.26)$$

and we have

$$\Lambda^{1/2} q(t, X) \in S(\mu(t, X)^{3/2}, \mu^{-1}(t, \cdot)\Gamma), \qquad (3.2.27)$$

so that the semi-norms depend only on the semi-norms of q in $S(\Lambda, \Lambda^{-1}\Gamma)$.

Proof. We notice first that

$$1 + \max(|\Lambda^{1/2} q'_X(t, X)|, |\Lambda^{1/2} q''_{XX}(t, X)|^2)$$

is the 1-proper weight of the vector-valued symbol $\Lambda^{1/2} q'_X(t, \cdot)$. Using Lemma 4.9.2, we get that $\mu^{-1}\Gamma$ is slowly varying, and Lemma 4.9.1 provides the second part of (3.2.26). From Definition 3.2.11 and (3.2.12), we obtain that

$$\mu(t, X) \leq C\Lambda + \langle \delta_0(t, X) \rangle^2 \leq C'\Lambda \quad \text{and} \quad \Lambda^{1/2} q'_X(t, \cdot) \in S(\mu(t, X), \mu^{-1}(t, \cdot)\Gamma).$$

We are left with the proof of $|\Lambda^{1/2} q(t, X)| \leq C\mu^{3/2}(t, X)$. Let us consider $\widetilde{\mu}(t, X)$ the 3/2-proper weight of $\Lambda^{1/2} q(t, X)$:

$$\widetilde{\mu}(t, X) = 1 + \max_{j=0,1,2} |\Lambda^{1/2} q^{(j)}(t, X)|^{\frac{2}{3-j}},$$

where all the derivatives are taken with respect to X; if the maximum is realized for $j \in \{1, 2\}$, we get from Lemma 3.2.12 and (3.2.27) that

$$|\Lambda^{1/2} q(t, X)| \leq \widetilde{\mu}(t, X)^{3/2} = \left(1 + \max_{j=1,2} |\Lambda^{1/2} q^{(j)}(t, X)|^{\frac{2}{3-j}}\right)^{\frac{3}{2}}$$

$$\leq \left(1 + \max_{j=1,2} \mu^{(\frac{3}{2} - \frac{j}{2})(\frac{2}{3-j})}\right)^{\frac{3}{2}} \leq 2\mu(t, X)^{3/2}, \qquad (3.2.28)$$

which is the result that we had to prove. We have eventually to deal with the case where the maximum in the definition of $\widetilde{\mu}$ is realized for $j = 0$; note that if $\widetilde{\mu}(t, X) \leq C_0$, we obtain

$$|\Lambda^{1/2} q(t, X)| \leq \widetilde{\mu}(t, X)^{3/2} \leq C_0^{3/2} \leq C_0^{3/2} \mu(t, X)^{3/2},$$

so we may also assume $\widetilde{\mu}(t, X) > C_0$. If $C_0 > 1$, we have

$$C_0 < \widetilde{\mu}(t, X) = 1 + (\Lambda^{1/2} |q(t, X)|)^{\frac{2}{3}}$$

entailing $(1 - C_0^{-1})\widetilde{\mu}(t, X) \leq |\Lambda^{1/2} q(t, X)|^{\frac{2}{3}} \leq \widetilde{\mu}(t, X)$. Now if $h \in \mathbb{R}^{2n}$ is such that $|h| \leq r\widetilde{\mu}(t, X)^{1/2}$, we get from the slow variation of the metric $\widetilde{\mu}^{-1}\Gamma$, that the ratio $\widetilde{\mu}(t, X + h)/\widetilde{\mu}(t, X)$ is bounded above and below, provided r is small enough. Using now that $\Lambda^{1/2} q(t, \cdot) \in S(\widetilde{\mu}^{3/2}(t, \cdot), \widetilde{\mu}^{-1}(t, \cdot)\Gamma)$, we get by Taylor's formula

$$\Lambda^{1/2} q(t, X + h) = \Lambda^{1/2} q(t, X) + \Lambda^{1/2} q'(t, X)h + \frac{1}{2}\Lambda^{1/2} q''(t, X)h^2 + O(\gamma_3 |h|^3/6),$$

so that

$$\Lambda^{1/2}|q(t, X + h)| \geq \Lambda^{1/2}|q(t, X)| - \widetilde{\mu}(t, X)|h| - \frac{1}{2}|h|^2\widetilde{\mu}(t, X)^{1/2} - \gamma_3|h|^3/6$$

$$\geq \Lambda^{1/2}|q(t, X)| - \widetilde{\mu}(t, X)^{3/2}\underbrace{\left(r + \frac{r^2}{2} + \gamma_3\frac{r^3}{6}\right)}_{=\epsilon(r)}.$$

This gives $\Lambda^{1/2}|q(t, X + h)| \geq \Lambda^{1/2}|q(t, X)| - \epsilon(r)\widetilde{\mu}(t, X)^{3/2}$, $\lim_{r\to 0}\epsilon(r) = 0$, so that, for r, C_0^{-1} small enough,

$$|\Lambda^{1/2}q(t, X + h)| \geq \left((1 - C_0^{-1})^{3/2} - \epsilon(r)\right)\widetilde{\mu}(t, X)^{3/2} \geq \frac{1}{2}\widetilde{\mu}(t, X)^{3/2}.$$

As a consequence, the Γ-ball $B\left(X, r\widetilde{\mu}(t, X)^{1/2}\right)$ is included in $\mathbb{X}_+(t)$ or in $\mathbb{X}_-(t)$ and thus, in the first case (the second case is similar), we have

$$|X - \mathbb{X}_+(t)| = 0, \quad |X - \mathbb{X}_-(t)| \geq r\widetilde{\mu}(t, X)^{1/2},$$

otherwise $|X - \mathbb{X}_-(t)| < r\widetilde{\mu}(t, X)^{1/2}$ and

$$\emptyset \neq B\left(X, r\widetilde{\mu}(t, X)^{1/2}\right) \cap \mathbb{X}_-(t) \subset \mathbb{X}_+(t) \cap \mathbb{X}_-(t) = \emptyset,$$

implying that, with a fixed $r_0 > 0$,

$$\delta_0(t, X) \geq \min(\Lambda^{1/2}, r\widetilde{\mu}(t, X)^{1/2}) \geq r_0\widetilde{\mu}(t, X)^{1/2} \geq r_0|\Lambda^{1/2}q(t, X)|^{1/3},$$

so that, in both cases, $|\Lambda^{1/2}q(t, X)| \leq r_0^{-3}|\delta_0(t, X)|^3 \leq r_0^{-3}\mu(t, X)^{3/2}$. $\square$

Lemma 3.2.14. *Let* $q(t, X), \delta_0(t, X), \mu(t, X)$ *be as above. We define,*

$$\nu(t, X) = \langle\delta_0(t, X)\rangle^2 + |\Lambda^{1/2}q'_X(t, X)\mu(t, X)^{-1/2}|^2. \qquad (3.2.29)$$

The metric $\nu^{-1}(t, \cdot)\Gamma$ *is slowly varying with structure constants depending only on a finite number of semi-norms of* q *in* $S(\Lambda, \Lambda^{-1}\Gamma)$. *There exists* $C > 0$, *depending only on a finite number of semi-norms of* q, *such that*

$$\nu(t, X) \leq 2\mu(t, X) \leq C\Lambda, \quad \frac{\nu(t, X)}{\nu(t, Y)} \leq C(1 + |X - Y|^2), \qquad (3.2.30)$$

and we have

$$\Lambda^{1/2}q(t, X) \in S(\mu(t, X)^{1/2}\nu(t, X), \nu(t, \cdot)^{-1}\Gamma), \qquad (3.2.31)$$

so that the semi-norms of this symbol depend only on those of q *in* $S(\Lambda, \Lambda^{-1}\Gamma)$. *Moreover the function* $\mu(t, X)$ *is a weight for the metric* $\nu(t, \cdot)^{-1}\Gamma$.

Proof. Let us check the two first inequalities in (3.2.30). From

$$|\Lambda^{1/2}q'| \leq \mu(t, X) \leq C\Lambda,$$

established in the previous lemma, we get

$$\nu(t, X) \leq \langle \delta_0(t, X) \rangle^2 + \mu(t, X) \leq 2\mu(t, X) \leq 2C\Lambda.$$

We introduce now the weight $\mu_*(t, X) = \tilde{\mu}(t, X)$ as in Remark 2.2.8 so that the ratios $\mu_*(t, X)/\mu(t, X)$ are bounded above and below by some constants depending only on a finite number of semi-norms of q. That weight $\mu_*(t, X)$ belongs to $S(\mu, \mu^{-1}\Gamma) = S(\mu_*, \mu_*^{-1}\Gamma)$. We notice first that

$$|\Lambda^{1/2}(q\mu_*^{-1/2})'|^2 \leq 2|\Lambda^{1/2}q'\mu_*^{-1/2}|^2 + C_1|\Lambda^{1/2}q\mu^{-1}|^2$$

$$\leq C_2|\Lambda^{1/2}q'\mu^{-1/2}|^2 + C_1|\Lambda^{1/2}q\mu^{-1/2}|\overbrace{|\Lambda^{1/2}q\mu^{-3/2}|}^{\lesssim 1}$$

$$\leq C_2|\Lambda^{1/2}q'\mu^{-1/2}|^2 + C_3|\Lambda^{1/2}q\mu^{-1/2}|.$$

Since we have also[8]

$$|\Lambda^{1/2}q'\mu^{-1/2}| \sim |\Lambda^{1/2}q'\mu_*^{-1/2}| \lesssim |\Lambda^{1/2}(q\mu_*^{-1/2})'| + |\Lambda^{1/2}q\mu_*^{-1}|$$

$$\lesssim |\Lambda^{1/2}(q\mu_*^{-1/2})'| + \underbrace{|\Lambda^{1/2}q\mu_*^{-3/2}|^{1/2}}_{\lesssim 1}|\Lambda^{1/2}q\mu_*^{-1/2}|^{1/2}$$

we get that

$$\tilde{\nu}(t, X) = 1 + \max\left(|\Lambda^{1/2}q'_X(t, X)\mu(t, X)^{-1/2}|^2, |\Lambda^{1/2}q(t, X)\mu(t, X)^{-1/2}|\right) \quad (3.2.32)$$

is equivalent to the 1-proper weight of the symbol $\Lambda^{1/2}q(t, X)\mu_*(t, X)^{-1/2}$ in $S(\mu, \mu^{-1}\Gamma)$. As a consequence, from Lemma 4.9.2, we get that $(\tilde{\nu} + \langle \delta_0 \rangle^2)^{-1}\Gamma$ is slowly varying.

We need only to prove that

$$|\Lambda^{1/2}q(t, X))\mu(t, X)^{-1/2}| \leq C\nu(t, X). \quad (3.2.33)$$

In fact, from (3.2.33), we shall obtain

$$\nu(t, X) \leq \tilde{\nu}(t, X) + \langle \delta_0(t, X) \rangle^2 \leq (C + 1)\nu(t, X)$$

so that the metrics $(\tilde{\nu} + \langle \delta_0 \rangle^2)^{-1}\Gamma$ and $\nu^{-1}\Gamma$ are equivalent and thus both slowly varying (that property will also give the last inequality in (3.2.30) from Lemma 4.9.1). Moreover, from Lemma 3.2.12, we have

$$\Lambda^{1/2}q(t, X)\mu_*(t, X)^{-1/2} \in S(\tilde{\nu}, \tilde{\nu}^{-1}\Gamma),$$

---

[8]Below, the inequality $a \lesssim b$ means that $a \leq Cb$ where C is a constant depending only on a finite number of semi-norms of q. The equivalence $a \sim b$ stands for $a \lesssim b$ and $b \lesssim a$.

so that

$$\Lambda^{1/2}(q\mu_*^{-1/2})^{(k)}$$

$$\lesssim \begin{cases} \nu^{1-k/2} & \text{for } k \leq 2, \text{ since } \Lambda^{1/2}q\mu_*^{-1/2} \in S(\tilde{\nu}; , \tilde{\nu}^{-1}\Gamma) \text{ and } \tilde{\nu} \lesssim \nu, \\ \mu^{1-k/2} \lesssim \nu^{1-k/2} & \text{for } k \geq 2, \text{ since } \Lambda^{1/2}q\mu_*^{-1/2} \in S(\mu; \mu^{-1}\Gamma) \text{ and } \nu \lesssim \mu, \end{cases}$$

which implies that

$$\Lambda^{1/2}q\mu_*^{-1/2} \in S(\nu, \nu^{-1}\Gamma).$$

Moreover, we have $\mu_*^{1/2} \in S(\mu_*^{1/2}, \nu^{-1}\Gamma)$ since, using $\nu \lesssim \mu$, we get

$$|(\mu_*^{1/2})^{(k)}| \lesssim \mu^{\frac{1-k}{2}} \lesssim \mu^{\frac{1}{2}}\nu^{-k/2},$$

entailing $\Lambda^{1/2}q \in S(\mu^{1/2}\nu, \nu^{-1}\Gamma)$, i.e., (3.2.31). On the other hand, μ is slowly varying for $\nu^{-1}\Gamma$, since

$$|X - Y| \ll \nu(t, X)^{1/2}(\lesssim \mu(t, X)^{1/2}) \quad \text{implies } |X - Y| \ll \mu(t, X)^{1/2}$$

and thus $\mu(t, X) \sim \mu(t, Y)$, which proves along with (3.2.26) that μ is a weight for $\nu^{-1}\Gamma$.

Let us now check (3.2.33). This inequality is obvious if

$$|\Lambda^{1/2}q\mu^{-1/2}| \leq |\Lambda^{1/2}q'\mu^{-1/2}|^2.$$

Note that if $\tilde{\nu}(t, X) \leq C_0$, we obtain $|\Lambda^{1/2}q\mu^{-1/2}| \leq C_0 \leq C_0\nu$ so we may also assume $\tilde{\nu}(t, X) > C_0$. If $C_0 > 1$, we have $C_0 < \tilde{\nu}(t, X) = 1 + (\Lambda^{1/2}|q|\mu^{-1/2})$ entailing

$$(1 - C_0^{-1})\tilde{\nu}(t, X) \leq |\Lambda^{1/2}q\mu^{-1/2}| \leq \tilde{\nu}(t, X).$$

Now if $h \in \mathbb{R}^{2n}$ is such that $|h| \leq r\tilde{\nu}(t, X)^{1/2}$, we get from the slow variation of the metric $\tilde{\nu}^{-1}\Gamma$, that the ratio $\tilde{\nu}(t, X + h)/\tilde{\nu}(t, X)$ is bounded above and below, provided r is small enough. Using now that $\Lambda^{1/2}q\mu_*^{-1/2} \in S(\tilde{\nu}, \tilde{\nu}^{-1}\Gamma)$, we get by Taylor's formula

$$\Lambda^{1/2}q(t, X + h)\mu_*^{-1/2}(t, X + h) = \Lambda^{1/2}q(t, X)\mu_*^{-1/2}(t, X) + \epsilon(r)\tilde{\nu}(t, X),$$

with $\lim_{r \to 0} \epsilon(r) = 0$, so that, for r, C_0^{-1} small enough,

$$|\Lambda^{1/2}q(t, X + h)\mu_*^{-1/2}(t, X + h)| \geq ((1 - C_0^{-1}) - \epsilon(r))\tilde{\nu}(t, X) \geq \frac{1}{2}\tilde{\nu}(t, X).$$

As a consequence, the Γ-ball $B(X, r\tilde{\nu}(t, X)^{1/2})$ is included in $\mathbb{X}_+(t)$ or in $\mathbb{X}_-(t)$ and thus, in the first case (the second case is similar) $|X - \mathbb{X}_+(t)| = 0$, $|X - \mathbb{X}_-(t)| \geq r\tilde{\nu}(t, X)^{1/2}$, implying that, with a fixed $r_0 > 0$,

$$\delta_0(t, X) \geq \min(\Lambda^{1/2}, r\tilde{\nu}(t, X)^{1/2}) \geq r_0\tilde{\nu}(t, X)^{1/2} \geq r_0|\Lambda^{1/2}q(t, X)\mu(t, X)^{-1/2}|^{1/2},$$

so that, in both cases, $|\Lambda^{1/2}q(t, X)\mu(t, X)^{-1/2}| \leq C_0|\delta_0(t, X)|^2 \leq C_0\nu(t, X)$. The proof of the lemma is complete. $\qquad \square$

We wish now to discuss the normal forms attached to the metric $\nu^{-1}(t, \cdot)\Gamma$ for the symbol $q(t, \cdot)$. In the sequel of this section, we consider that t is fixed.

Definition 3.2.15. Let $0 < r_1 \leq 1/2$ be given. With ν defined in (3.2.29), we shall say that

(1) Y is a non-negative (resp. non-positive) point at level t if $\delta_0(t, Y) \geq r_1\nu(t, Y)^{1/2}$, (resp. $\delta_0(t, Y) \leq -r_1\nu(t, Y)^{1/2}$).

(2) Y is a gradient point at level t if $|\Lambda^{1/2}q'_Y(t, Y)\mu(t, Y)^{-1/2}|^2 \geq \nu(t, Y)/4$ and $\delta_0(t, Y)^2 < r_1^2\nu(t, Y)$.

(3) Y is a negligible point in the remaining cases $|\Lambda^{1/2}q'_Y(t, Y)\mu(t, Y)^{-1/2}|^2 < \nu(t, Y)/4$ and $\delta_0(t, Y)^2 < r_1^2\nu(t, Y)$. Note that this implies $\nu(t, Y) \leq 1 + r_1^2\nu(t, Y) + \nu(t, Y)/4 \leq 1 + \nu(t, Y)/2$ and thus $\nu(t, Y) \leq 2$.

Note that if Y is a non-negative point, from (3.2.20) we get, for $T \in \mathbb{R}^{2n}$, $|T| \leq 1, 0 \leq r \leq r_1/4$,

$$\delta_0\big(t, Y + r\nu^{1/2}(t, Y)T\big) \geq \delta_0(t, Y) - 2r\nu^{1/2}(t, Y) \geq \frac{r_1}{2}\nu^{1/2}(t, Y),$$

and from (3.2.22), this implies that $q(t, X) \geq 0$ on the ball $B(Y, r\nu^{1/2}(t, Y))$. Similarly if Y is a non-positive point, $q(t, X) \leq 0$ on the ball $B(Y, r\nu^{1/2}(t, Y))$. Moreover if Y is a gradient point, we have $|\delta_0(t, Y)| < r_1\nu(t, Y)^{1/2}$ so that, if $Y \in \mathbb{X}_+(t)$, we have $\min(|Y - \mathbb{X}_-(t)|, \Lambda^{1/2}) < r_1\nu(t, Y)^{1/2}$ and if r_1 is small enough, since $\nu \lesssim \Lambda$, we get that $|Y - \mathbb{X}_-(t)| < r_1\nu(t, Y)^{1/2}$ which implies that there exists $Z_1 \in \mathbb{X}_-(t)$ such that $|Y - Z_1| < r_1\nu(t, Y)^{1/2}$. On the segment $[Y, Z_1]$, the Lipschitz-continuous function is such that $\delta_0(t, Y) > 0$ ($Y \in \mathbb{X}_+(t)$ cf. Lemma 3.2.10) and $\delta_0(t, Z_1) < 0$ ($Z_1 \in \mathbb{X}_-(t)$); as a result, there exists a point Z (on that segment) such that $\delta_0(t, Z) = 0$ and thus $q(t, Z) = 0$. Naturally the discussion for a gradient point Y in $\mathbb{X}_-(t)$ is analogous. If the gradient point Y belongs to $\mathbb{X}_0(t)$, we get right away $q(t, Y) = 0$, also from Lemma 3.2.10. The function

$$f(T) = \Lambda^{1/2}q\big(t, Y + r_1\nu^{1/2}(t, Y)T\big)\mu(t, Y)^{-1/2}\nu(t, Y)^{-1} \tag{3.2.34}$$

satisfies for r_1 small enough with respect to the semi-norms of q and c_0, C_0, C_1, C_2 fixed positive constants, $|T| \leq 1$, from (3.2.31),

$$|f(T)| \leq |S - T|C_0r_1 \leq C_1r_1^2, \quad |f'(T)| \geq r_1c_0, \quad |f''(T)| \leq C_2r_1^2.$$

The standard analysis (see our appendix, Section 4.9.5) of the Beals-Fefferman metric [8] shows that, on $B(Y, r_1\nu^{1/2}(t, Y))$,

$$q(t, X) = \Lambda^{-1/2}\mu^{1/2}(t, Y)\nu^{1/2}(t, Y)e(t, X)\beta(t, X), \tag{3.2.35}$$

$$1 \leq e \in S(1, \nu(t, Y)^{-1}\Gamma), \ \beta \in S(\nu(t, Y)^{1/2}, \nu(t, Y)^{-1}\Gamma), \tag{3.2.36}$$

$$\beta(t, X) = \nu(t, Y)^{1/2}(X_1 + \alpha(t, X')), \alpha \in S(\nu(t, Y)^{1/2}, \nu(t, Y)^{-1}\Gamma). \tag{3.2.37}$$

Lemma 3.2.16. *Let $q(t, X)$ be a smooth function satisfying (3.2.12), (3.2.13) and let $t \in [-1, 1]$ be given. The metric g_t on $\mathbb{R}^{2n}$ is defined as $\nu(t, X)^{-1}\Gamma$ where ν is defined in (3.2.29). There exists $r_0 > 0$, depending only on a finite number of semi-norms of q in (3.2.12) such that, for any $r \in]0, r_0]$, there exists a sequence of points (X_k) in $\mathbb{R}^{2n}$, and sequences of functions $(\chi_k), (\psi_k)$ satisfying the properties in Theorem 4.3.7 such that there exists a partition of $\mathbb{N}$,*

$$\mathbb{N} = E_+ \cup E_- \cup E_0 \cup E_{00}$$

so that, according to Definition 3.2.15, $k \in E_+$ means that X_k is a non-negative point, ($k \in E_-$:X_k non-positive point; $k \in E_0$:X_k gradient point, $k \in E_{00}$:X_k negligible point).

Proof. This lemma is an immediate consequence of Definition 3.2.15, of Theorem 4.3.7 and of Lemma 3.2.14, asserting that the metric g_t is admissible.   □

3.2.4   Some lemmas on C^3 functions

We prove in this section a key result on the second derivative f''_{XX} of a real-valued smooth function $f(t, X)$ such that $\tau - if(t, x, \xi)$ satisfies condition (Ψ). The following claim gives a good qualitative version of what is needed for our estimates; although we shall not use this result, and the reader may skip the proof to proceed directly to the more technical Lemma 3.2.19; the very simple formulation and proof of this claim is a good warm-up for the sequel.

Lemma 3.2.17 (Dencker's lemma). *Let f_1, f_2 be two real-valued twice differentiable functions defined on an open set Ω of $\mathbb{R}^N$ and such that $f_1^{-1}(\mathbb{R}_+^*) \subset f_2^{-1}(\mathbb{R}_+)$ (i.e., $f_1(x) > 0 \Longrightarrow f_2(x) \geq 0$). If for some $\omega \in \Omega$, the conditions $f_1(\omega) = f_2(\omega) = 0$, $df_1(\omega) \neq 0, df_2(\omega) = 0$ are satisfied, we have $f_2''(\omega) \geq 0$ (as a quadratic form).*

Proof. Using the obvious invariance by change of coordinates of the statement, we may assume $f_1(x) \equiv x_1$ and $\omega = 0$. The assumption is then for $x = (x_1, x') \in \mathbb{R} \times \mathbb{R}^{N-1}$ in a neighborhood of the origin

$$f_2(0) = 0, df_2(0) = 0, \quad x_1 > 0 \Longrightarrow f_2(x_1, x') \geq 0.$$

Using the second-order Taylor-Young formula for f_2, we get $f_2(x) = \frac{1}{2}\langle f_2''(0)x, x\rangle + \epsilon(x)|x|^2$, $\lim_{x \to 0} \epsilon(x) = 0$, and thus for $T = (T_1, T'), |T| = 1, \rho \neq 0$ small enough, the implication $T_1 > 0 \Longrightarrow \langle f_2''(0)T, T\rangle + 2\epsilon(\rho T) \geq 0$. Consequently we have $\{S, \langle f_2''(0)S, S\rangle \geq 0\} \supset \{S, S_1 > 0\}$ and since the larger set is closed and stable by the symmetry with respect to the origin, we get that it contains also $\{S, S_1 \leq 0\}$, which is the result $f_2''(0) \geq 0$.   □

Remark 3.2.18. This claim has the following consequence: take three functions f_1, f_2, f_3, twice differentiable on Ω, such that, for $1 \leq j \leq k \leq 3$, $f_j(x) > 0 \Longrightarrow f_k(x) \geq 0$. Assume that, at some point ω we have $f_1(\omega) = f_2(\omega) = f_3(\omega) = 0$, $df_1(\omega) \neq 0, df_3(\omega) \neq 0, df_2(\omega) = 0$. Then one has $f_2''(\omega) = 0$. Lemma 3.2.17 gives $f_2''(\omega) \geq 0$ and it can be applied to the couple $(-f_3, -f_2)$ to get $-f_2''(\omega) \geq 0$.

Notation. The open Euclidean ball of $\mathbb{R}^N$ with center 0 and radius r will be denoted by B_r. For a k-multilinear symmetric form A on $\mathbb{R}^N$, we shall note $\|A\| = \max_{|T|=1} |AT^k|$ which is easily seen to be equivalent to the norm

$$\max_{|T_1|=\cdots=|T_k|=1} |A(T_1,\ldots,T_k)|$$

since the symmetrized $T_1 \otimes \cdots \otimes T_k$ can be written a sum of kth powers (see the appendix, Section 4.2.2).

Lemma 3.2.19. *Let $R_0 > 0$ and f_1, f_2 be real-valued functions defined in $\bar{B}_{R_0}$. We assume that f_1 is C^2, f_2 is C^3 and for $x \in \bar{B}_{R_0}$,*

$$f_1(x) > 0 \Longrightarrow f_2(x) \geq 0. \tag{3.2.38}$$

We define the non-negative numbers ρ_1, ρ_2, by

$$\rho_1 = \max\left(|f_1(0)|^{\frac{1}{2}}, |f_1'(0)|\right), \quad \rho_2 = \max\left(|f_2(0)|^{\frac{1}{3}}, |f_2'(0)|^{\frac{1}{2}}, |f_2''(0)|\right), \tag{3.2.39}$$

and we assume that, with a positive C_0,

$$0 < \rho_1, \quad \rho_2 \leq C_0\rho_1 \leq R_0. \tag{3.2.40}$$

We define the non-negative numbers C_1, C_2, C_3, by

$$C_1 = 1 + C_0\|f_1''\|_{L^\infty(\bar{B}_{R_0})}, \quad C_2 = 4 + \frac{1}{3}\|f_2'''\|_{L^\infty(\bar{B}_{R_0})}, \quad C_3 = C_2 + 4\pi C_1. \tag{3.2.41}$$

Assume that for some $\kappa_2 \in [0,1]$, with $\kappa_2 C_1 \leq 1/4$,

$$\rho_1 = |f_1'(0)| > 0, \tag{3.2.42}$$

$$\max\left(|f_2(0)|^{1/3}, |f_2'(0)|^{1/2}\right) \leq \kappa_2|f_2''(0)|, \tag{3.2.43}$$

$$B(0, \kappa_2^2\rho_2) \cap \{x \in \bar{B}_{R_0}, f_1(x) \geq 0\} \neq \emptyset. \tag{3.2.44}$$

Then we have

$$|f_2''(0)_-| \leq C_3\kappa_2\rho_2, \tag{3.2.45}$$

where $f_2''(0)_-$ stands for the negative part of the quadratic form $f_2''(0)$. Note that, whenever (3.2.44) is violated, we get $B(0, \kappa_2^2\rho_2) \subset \{x \in \bar{B}_{R_0}, f_1(x) < 0\}$ (note that $\kappa_2^2\rho_2 \leq \rho_2 \leq R_0$) and thus

$$\text{distance}(0, \{x \in \bar{B}_{R_0}, f_1(x) \geq 0\}) \geq \kappa_2^2\rho_2. \tag{3.2.46}$$

Proof. We may assume that for $x = (x_1, x') \in \mathbb{R} \times \mathbb{R}^{N-1}$, $\rho_1 = |f_1'(0)| = \frac{\partial f_1}{\partial x_1}(0,0)$, $\frac{\partial f_1}{\partial x'}(0,0) = 0$, so that

$$f_1(x) \geq f_1(0) + \rho_1 x_1 - \frac{1}{2}\|f_1''\|_\infty|x|^2. \tag{3.2.47}$$

Moreover, from (3.2.44), we know that there exists $z \in B(0, \kappa_2^2 \rho_2)$ such that $f_1(z) \geq 0$. As a consequence, we have $0 \leq f_1(z) \leq f_1(0) + \rho_1 z_1 + \frac{1}{2} \| f_1'' \|_\infty \kappa_2^4 \rho_2^2$ and thus

$$f_1(x) \geq \rho_1 x_1 - \rho_1 \kappa_2^2 \rho_2 - \frac{1}{2} \| f_1'' \|_\infty (|x|^2 + \kappa_2^4 \rho_2^2). \tag{3.2.48}$$

On the other hand, we have

$$f_2(x) \leq f_2(0) + f_2'(0)x + \frac{1}{2} f_2''(0)x^2 + \frac{1}{6} \| f_2''' \|_\infty |x|^3$$

$$\leq \kappa_2^3 \rho_2^3 + \kappa_2^2 \rho_2^2 |x| + \frac{1}{6} \| f_2''' \|_\infty |x|^3 + \frac{1}{2} f_2''(0)x^2$$

and the implications, for $|x| \leq R_0$,

$$\rho_1 x_1 > \rho_1 \kappa_2^2 \rho_2 + \frac{1}{2} \| f_1'' \|_\infty (|x|^2 + \kappa_2^4 \rho_2^2) \Longrightarrow f_1(x) > 0 \Longrightarrow f_2(x) \geq 0$$

$$\Longrightarrow -\frac{1}{2} f_2''(0)x^2 \leq \kappa_2^3 \rho_2^3 + \kappa_2^2 \rho_2^2 |x| + \frac{1}{6} \| f_2''' \|_\infty |x|^3. \tag{3.2.49}$$

Let us take $x = \kappa_2 \rho_2 y$ with $|y| = 1$ (note that $|x| = \kappa_2 \rho_2 \leq R_0$); the property (3.2.49) gives, using $\rho_2 / \rho_1 \leq C_0$,

$$y_1 > \kappa_2 (1 + \| f_1'' \|_\infty C_0) \Longrightarrow -f_2''(0)y^2 \leq \kappa_2 \rho_2 (4 + \frac{1}{3} \| f_2''' \|_\infty),$$

so that

$$\{y \in \mathbb{S}^{N-1}, -f_2''(0)y^2 \leq \kappa_2 \rho_2 (4 + \frac{1}{3} \| f_2''' \|_\infty)\} \supset \{y \in \mathbb{S}^{N-1}, y_1 > \kappa_2 (1 + \| f_1'' \|_\infty C_0)\}$$

and since the larger set is closed and stable by symmetry with respect to the origin, we get, with

$$C_1 = 1 + \| f_1'' \|_\infty C_0, \quad C_2 = 4 + \frac{1}{3} \| f_2''' \|_\infty,$$

the implication

$$y \in \mathbb{S}^{N-1}, |y_1| \geq \kappa_2 C_1 \Longrightarrow -f_2''(0)y^2 \leq \kappa_2 \rho_2 C_2. \tag{3.2.50}$$

Let us now take $y \in \mathbb{S}^{N-1}$, such that $|y_1| < \kappa_2 C_1 (\leq 1/4)$. We may assume $y = y_1 \vec{e_1} \oplus y_2 \vec{e_2}$, with $\vec{e_1}, \vec{e_2}$ orthogonal unit vectors and $y_2 = (1 - y_1^2)^{1/2}$. We consider the following rotation in the $(\vec{e_1}, \vec{e_2})$ plane with $\epsilon_0 = \kappa_2 C_1 \leq 1/4$,

$$R = \begin{pmatrix} \cos(2\pi\epsilon_0) & \sin(2\pi\epsilon_0) \\ -\sin(2\pi\epsilon_0) & \cos(2\pi\epsilon_0) \end{pmatrix}, \quad \text{so that } |(Ry)_1| = |y_1 \cos(2\pi\epsilon_0) + y_2 \sin(2\pi\epsilon_0)|,$$

and since $\epsilon_0 \leq 1/4$,

$$|(Ry)_1| \geq -|y_1| + (1 - y_1^2)^{1/2} 4\epsilon_0 \geq \epsilon_0 (\sqrt{15} - 1) > \epsilon_0 = \kappa_2 C_1.$$

Moreover the rotation R satisfies $\|R - \mathrm{Id}\| \leq 2\pi\epsilon_0 = 2\pi\kappa_2 C_1$. We have, using (3.2.50) and $|(Ry)_1| \geq \kappa_2 C_1, |y| = 1$,

$$
\begin{aligned}
-f_2''(0)y^2 &= -f_2''(0)(Ry)^2 - \langle f_2''(0)(y - Ry), y + Ry \rangle \\
&\leq -f_2''(0)(Ry)^2 + |f_2''(0)||y - Ry||y + Ry| \\
&\leq \kappa_2\rho_2 C_2 + 2\rho_2|y - Ry| \leq \kappa_2\rho_2 C_2 + 2\rho_2 2\pi\kappa_2 C_1.
\end{aligned}
$$

Eventually, for all $y \in \mathbb{S}^{N-1}$, we have

$$
-f_2''(0)y^2 \leq \kappa_2\rho_2(C_2 + 4\pi C_1) = C_3\kappa_2\rho_2. \tag{3.2.51}
$$

Considering now the quadratic form $Q = f_2''(0)$ and its canonical decomposition $Q = Q_+ - Q_-$, we have, for all $y \in \mathbb{R}^N$, $\langle Q_- y, y \rangle \leq \kappa_2\rho_2 C_3|y|^2 + \langle Q_+ y, y \rangle$. Using now the canonical orthogonal projections $E_\pm$ on the positive (resp. negative) eigenspaces, we write $y = E_+ y \oplus E_- y$ and we get that

$$
\begin{aligned}
\langle Q_- y, y \rangle = \langle Q_- E_- y, E_- y \rangle &\leq C_3\kappa_2\rho_2|E_- y|^2 + \langle Q_+ E_- y, E_- y \rangle \\
&= C_3\kappa_2\rho_2|E_- y|^2 \leq C_3\kappa_2\rho_2|y|^2, \tag{3.2.52}
\end{aligned}
$$

yielding (3.2.45). The proof of Lemma 3.2.19 is complete. $\qquad\square$

Lemma 3.2.20. *Let f_1, f_2, f_3 be real-valued functions defined in $\bar{B}_{R_0}$. We assume that f_1, f_3 are C^2, f_2 is C^3 and for $x \in \bar{B}_{R_0}, 1 \leq j \leq k \leq 3$,*

$$
f_j(x) > 0 \implies f_k(x) \geq 0. \tag{3.2.53}
$$

We define the non-negative numbers ρ_1, ρ_2, ρ_3 by

$$
\begin{aligned}
&\rho_1 = \max\left(|f_1(0)|^{\frac{1}{2}}, |f_1'(0)|\right), & \rho_2 = \max\left(|f_2(0)|^{\frac{1}{3}}, |f_2'(0)|^{\frac{1}{2}}, |f_2''(0)|\right), \\
&\rho_3 = \max\left(|f_3(0)|^{\frac{1}{2}}, |f_3'(0)|\right) &
\end{aligned} \tag{3.2.54}
$$

and we assume that, with a positive C_0,

$$
0 < \rho_1, \rho_3 \quad \text{and} \quad \rho_2 \leq C_0 \min(\rho_1, \rho_3) \leq C_0 \max(\rho_1, \rho_3) \leq R_0. \tag{3.2.55}
$$

We define the non-negative numbers C_1, C_2, C_3, by

$$
C_1 = 1 + C_0 \max(\|f_1''\|_{L^\infty(\bar{B}_{R_0})}, \|f_3''\|_{L^\infty(\bar{B}_{R_0})}), \tag{3.2.56}
$$

$$
C_2 = 4 + \frac{1}{3}\|f_2'''\|_{L^\infty(\bar{B}_{R_0})}, \quad C_3 = C_2 + 4\pi C_1. \tag{3.2.57}
$$

Assume that for some $\kappa_1, \kappa_3 \in [0, 1]$, and $0 < \kappa_2 C_3 \leq 1/2$,

$$
|f_1(0)|^{1/2} \leq \kappa_1|f_1'(0)|, \quad |f_3(0)|^{1/2} \leq \kappa_3|f_3'(0)|, \tag{3.2.58}
$$

$$
B(0, \kappa_2^2\rho_2) \cap \{x \in \bar{B}_{R_0}, f_1(x) \geq 0\} \neq \emptyset, \tag{3.2.59}
$$

$$
B(0, \kappa_2^2\rho_2) \cap \{x \in \bar{B}_{R_0}, f_3(x) \leq 0\} \neq \emptyset. \tag{3.2.60}
$$

Then we have

$$\max\left(|f_2(0)|^{1/3}, |f_2'(0)|^{1/2}\right) \leq \rho_2 \leq \kappa_2^{-1} \max\left(|f_2(0)|^{1/3}, |f_2'(0)|^{1/2}\right). \quad (3.2.61)$$

Note that, whenever (3.2.59) or (3.2.60) is violated, we get

$$B(0, \kappa_2^2 \rho_2) \subset \{x \in \bar{B}_{R_0}, f_1(x) < 0\} \text{ or } B(0, \kappa_2^2 \rho_2) \subset \{x \in \bar{B}_{R_0}, f_3(x) > 0\}$$

and thus

$$\text{dist}\left(0, \{x \in \bar{B}_{R_0}, f_1(x) \geq 0\}\right) \geq \kappa_2^2 \rho_2 \text{ or } \text{dist}\left(0, \{x \in \bar{B}_{R_0}, f_3(x) \leq 0\}\right) \geq \kappa_2^2 \rho_2.$$
$$(3.2.62)$$

Proof. This follows almost immediately from the previous lemma and it is analogous to the remark following Lemma 3.2.17: assuming that we have

$$\max\left(|f_2(0)|^{1/3}, |f_2'(0)|^{1/2}\right) \leq \kappa_2 |f_2''(0)| \quad (3.2.63)$$

will yield $|f_2''(0)| \leq C_3 \kappa_2 \rho_2$ by applying Lemma 3.2.19 (note that $\kappa_2 C_1 \leq \kappa_2 \frac{C_3}{4\pi} \leq \frac{1}{8\pi} < 1/4$) to the couples (f_1, f_2) and $(-f_3, -f_2)$; consequently, if (3.2.63) is satisfied, we get

$$\max\left(|f_2(0)|^{1/3}, |f_2'(0)|^{1/2}\right) \leq \rho_2 \leq \max\left(|f_2(0)|^{1/3}, |f_2'(0)|^{1/2}, C_3 \kappa_2 \rho_2\right)$$

and since $C_3 \kappa_2 < 1$, it yields

$$\max\left(|f_2(0)|^{1/3}, |f_2'(0)|^{1/2}\right) = \rho_2, \quad (3.2.64)$$

which implies (3.2.61). Let us now suppose that (3.2.63) does not hold, and that we have $\kappa_2 |f_2''(0)| < \max\left(|f_2(0)|^{1/3}, |f_2'(0)|^{1/2}\right)$. This implies (3.2.61):

$$\max\left(|f_2(0)|^{1/3}, |f_2'(0)|^{1/2}\right) \leq \rho_2 \leq \kappa_2^{-1} \max\left(|f_2(0)|^{1/3}, |f_2'(0)|^{1/2}\right).$$

The proof of the lemma is complete. $\qquad \square$

Remark 3.2.21. We shall apply this lemma to a "fixed" κ_2, depending only on the constant C_3 such as $\kappa_2 = 1/(2C_3)$.

3.2.5   Inequalities for symbols

In this section, we apply the results of the previous section to obtain various inequalities on symbols linked to our symbol q introduced in (3.2.12). Our main result is the following theorem.

Theorem 3.2.22. *Let q be a symbol satisfying (3.2.12), (3.2.13) and δ_0, μ, ν as defined above in (3.2.19), (3.2.25) and (3.2.29). For the real numbers t', t, t'', and $X \in \mathbb{R}^{2n}$, we define*

$$N(t', t'', X) = \frac{\langle \delta_0(t', X) \rangle}{\nu(t', X)^{1/2}} + \frac{\langle \delta_0(t'', X) \rangle}{\nu(t'', X)^{1/2}}, \quad (3.2.65)$$

$$R(t, X) = \Lambda^{-1/2} \mu(t, X)^{1/2} \nu(t, X)^{-1/2} \langle \delta_0(t, X). \quad (3.2.66)$$

Then there exists a constant $C_0 \geq 1$, depending only on a finite number of semi-norms of q in (3.2.12), such that, for $t' \leq t \leq t''$, we have

$$C_0^{-1} R(t,X) \leq N(t',t'',X) + \frac{\delta_0(t'',X) - \delta_0(t,X)}{\nu(t'',X)^{1/2}} + \frac{\delta_0(t,X) - \delta_0(t',X)}{\nu(t',X)^{1/2}}. \quad (3.2.67)$$

Proof. We are given $X \in \mathbb{R}^{2n}$ and $t' \leq t \leq t''$ real numbers.
First reductions. First of all, we may assume that, for some positive (small) κ to be chosen later, we have

$$\langle \delta_0(t',X) \rangle \leq \kappa \nu(t',X)^{1/2} \quad \text{and} \quad \langle \delta_0(t'',X) \rangle \leq \kappa \nu(t'',X)^{1/2}. \quad (3.2.68)$$

In fact, otherwise, we have $N(t',t'',X) > \kappa$ and since from (3.2.26), we have $\mu(t,X) \leq C\Lambda$ where C depends only on a finite number of semi-norms of q, we get from (3.2.66), (3.2.29)

$$R(t,X) \leq C^{1/2} \nu(t,X)^{-1/2} \langle \delta_0(t,X) \rangle \leq C^{1/2} \leq C^{1/2} \kappa^{-1} N(t',t'',X),$$

so that we shall only need

$$C_0 \geq C^{1/2} \kappa^{-1} \quad (3.2.69)$$

to obtain (3.2.67). Also, we may assume that, with the same positive (small) κ,

$$\nu(t,X) \leq \kappa^2 \nu(t',X) \quad \text{and} \quad \nu(t,X) \leq \kappa^2 \nu(t'',X). \quad (3.2.70)$$

Otherwise, we would have for instance $\nu(t,X) > \kappa^2 \nu(t',X)$ and since $t \geq t'$,

$$R(t,X) \leq \Lambda^{-1/2} \mu(t,X)^{1/2} \kappa^{-1} \frac{\langle \delta_0(t,X) \rangle}{\nu(t',X)^{1/2}}$$

$$\leq C^{1/2} \kappa^{-1} \left(\frac{\langle \delta_0(t',X) \rangle + \overbrace{|\delta_0(t,X) - \delta_0(t',X)|}^{\geq 0}}{\nu(t',X)^{1/2}} \right)$$

$$\leq C^{1/2} \kappa^{-1} N(t',t'',X) + C^{1/2} \kappa^{-1} \frac{\delta_0(t,X) - \delta_0(t',X)}{\nu(t',X)^{1/2}},$$

which implies (3.2.67) provided that (3.2.69) holds. Finally, we may also assume that

$$\nu(t,X) \leq \kappa^2 \mu(t,X), \quad (3.2.71)$$

otherwise we would have, using that $\delta_0(t',X) \leq \delta_0(t,X) \leq \delta_0(t'',X)$ and the convexity of $s \mapsto \sqrt{1+s^2} = \langle s \rangle$,

$$R(t,X) \leq \kappa^{-1} \frac{\langle \delta_0(t,X) \rangle}{\Lambda^{1/2}} \leq \kappa^{-1} \frac{\langle \delta_0(t',X) \rangle}{\Lambda^{1/2}} + \kappa^{-1} \frac{\langle \delta_0(t'',X) \rangle}{\Lambda^{1/2}}$$

and this implies, using $\nu(t',X), \nu(t'',X) \leq C\Lambda$ (see (3.2.30)),

$$R(t,X) \leq C^{1/2} \kappa^{-1} \frac{\langle \delta_0(t',X) \rangle}{\nu(t',X)^{1/2}} + C^{1/2} \kappa^{-1} \frac{\langle \delta_0(t'',X) \rangle}{\nu(t'',X)^{1/2}},$$

which gives (3.2.67) provided that (3.2.69) holds. On the other hand, we may assume that

$$\max\left(\langle \delta_0(t,X)\rangle, \kappa^{1/2}|\Lambda^{1/2}q'(t,X)|^{1/2}\right) \leq 2\kappa\mu(t,X)^{1/2}. \qquad (3.2.72)$$

Otherwise, we would have either

$$\mu(t,X)^{1/2} \leq \frac{1}{2}\kappa^{-1}\langle \delta_0(t,X)\rangle \leq \frac{1}{2}\kappa^{-1}\nu(t,X)^{1/2} \underbrace{\leq}_{\text{from (3.2.71)}} \frac{1}{2}\mu(t,X)^{1/2}$$

which is impossible, or we would have

$$\mu(t,X)^{1/2} \leq \frac{1}{2}\kappa^{-1/2}|\Lambda^{1/2}q'(t,X)|^{1/2} \overbrace{\leq}^{\text{from (3.2.29)}} \frac{1}{2}\kappa^{-1/2}\nu(t,X)^{1/4}\mu(t,X)^{1/4}$$

$$\underbrace{\leq}_{\text{from (3.2.71)}} \frac{1}{2}\mu(t,X)^{1/2}, \quad \text{(which is also impossible)}.$$

The estimate (3.2.72) implies that, for $\kappa < 1/16$,

$$\Lambda|q''(t,X)|^2 \underbrace{\leq}_{(3.2.25)} \mu(t,X) \underbrace{\leq}_{(3.2.25)} \langle \delta_0(t,X)\rangle^2 + |\Lambda^{1/2}q'(t,X)| + \Lambda|q''(t,X)|^2$$

$$\underbrace{\leq}_{(3.2.72)} (4\kappa^2 + 4\kappa)\mu(t,X) + \Lambda|q''(t,X)|^2,$$

and thus

$$\Lambda|q''(t,X)|^2 \leq \mu(t,X) \leq \frac{1}{1-8\kappa}\Lambda|q''(t,X)|^2 \leq 2\Lambda|q''(t,X)|^2. \qquad (3.2.73)$$

This implies that

$$R(t,X) \leq \Lambda^{-1/2}2^{1/2}\Lambda^{1/2}|q''(t,X)|\frac{\langle \delta_0(t,X)\rangle}{\left(\langle \delta_0(t,X)\rangle^2 + \Lambda|q'(t,X)|^2\mu(t,X)^{-1}\right)^{1/2}}$$

$$\leq 2^{1/2}|q''(t,X)|. \qquad (3.2.74)$$

Rescaling the symbols. We sum-up our situation, changing the notation so that $X = 0, t' = t_1, t = t_2, t'' = t_3, \nu_1 = \nu(t',0), \nu_2 = \nu(t,0), \nu_3 = \nu(t'',0), \delta_j = \delta_0(t_j,0), \mu_j = \mu(t_j,0)$. The following conditions are satisfied:

$$\left.\begin{array}{c} \langle \delta_1\rangle \leq \kappa\nu_1^{1/2}, \qquad \langle \delta_3\rangle \leq \kappa\nu_3^{1/2}, \\[2mm] \nu_2 \leq \kappa^2\nu_1, \qquad \nu_2 \leq \kappa^2\nu_3, \qquad \nu_2 \leq \kappa^2\mu_2, \\[2mm] R(t_2,0) \leq 2|q''(t_2,0)|\dfrac{\langle \delta_2\rangle}{\langle \delta_2\rangle + |q'(t_2,0)|/|q''(t_2,0)|} \leq 2|q''(t_2,0)|, \\[2mm] \Lambda|q''(t_2,0)|^2 \leq \mu_2 \leq 2\Lambda|q''(t_2,0)|^2, \\[2mm] \kappa < 1/16, \qquad C_0 \geq \kappa^{-1}C^{1/2}, \end{array}\right\}$$

$$(3.2.75)$$

where $\kappa > 0$ is to be chosen later and C depends only on a finite number of semi-norms of q. We define now the smooth functions f_1, f_2 defined on $\mathbb{R}^{2n}$ by

$$f_1(Y) = q(t_1, Y)\Lambda^{1/2}\mu_1^{-1/2}, \quad f_2(Y) = \nu_1^{1/2}q(t_2, Y), \tag{3.2.76}$$

and we note (see (3.2.12)-(3.2.27)) that $\|f_1''\|_{L^\infty}$ and $\|f_2'''\|_{L^\infty}$ are bounded above by semi-norms of q; moreover the assumption (3.2.38) holds for that couple of functions, from (3.2.13).

Lemma 3.2.23. *We define*

$$\mu_{12}^{1/2} = \max\big(\langle\delta_2\rangle, |\nu_1^{1/2}q'(t_2, 0)|^{1/2}, |\nu_1^{1/2}q''(t_2, 0)|\big). \tag{3.2.77}$$

If $\max\big(\langle\delta_2\rangle, \kappa^{1/2}|\nu_1^{1/2}q'(t_2, 0)|^{1/2}\big) > 2\kappa\mu_{12}^{1/2}$, *then* (3.2.67) *is satisfied provided* $C_0 \geq 3/\kappa$.

Proof. We have either $|\nu_1^{1/2}q''(t_2, 0)| \leq \mu_{12}^{1/2} \leq \frac{1}{2}\kappa^{-1}\langle\delta_2\rangle$ implying

$$|q''(t_2, 0)| \leq \frac{1}{2\kappa}\frac{\langle\delta_2\rangle}{\nu_1^{1/2}} \leq \frac{1}{2\kappa}\frac{\langle\delta_1\rangle}{\nu_1^{1/2}} + \frac{1}{2\kappa}\frac{\delta_2 - \delta_1}{\nu_1^{1/2}}$$

which gives (3.2.67) (using $R(t_2, 0) \leq 2|q''(t_2, 0)|$ in (3.2.75)), provided $C_0 \geq 1/\kappa$, or we have

$$|\nu_1^{1/2}q''(t_2, 0)| \leq \mu_{12}^{1/2} < \frac{1}{2}\kappa^{-1/2}|\nu_1^{1/2}q'(t_2, 0)|^{1/2},$$

implying $\dfrac{|q''(t_2, 0)|^2}{|q'(t_2, 0)|} \leq \dfrac{1}{4\kappa\nu_1^{1/2}}$ so that (using $R(t_2, 0) \leq 2|q''(t_2, 0)|^2\langle\delta_2\rangle/|q'(t_2, 0)|$ in (3.2.75)), we get $R(t_2, 0) \leq \frac{1}{2\kappa}\frac{\langle\delta_2\rangle}{\nu_1^{1/2}}$, which gives similarly (3.2.67), provided $C_0 \geq 1/(2\kappa)$. $\qquad\square$

A consequence of this lemma is that we may assume

$$\max\big(\langle\delta_2\rangle, \kappa^{1/2}|\nu_1^{1/2}q'(t_2, 0)|^{1/2}\big)$$
$$\leq 2\kappa\mu_{12}^{1/2} = 2\kappa\max\big(\langle\delta_2\rangle, |\nu_1^{1/2}q'(t_2, 0)|^{1/2}, |\nu_1^{1/2}q''(t_2, 0)|\big),$$

and since $\kappa < 1/4$, we get

$$\mu_{12}^{1/2} = |\nu_1^{1/2}q''(t_2, 0)|, \quad \max\big(\langle\delta_2\rangle, \kappa^{1/2}|\nu_1^{1/2}q'(t_2, 0)|^{1/2}\big) \leq 2\kappa|\nu_1^{1/2}q''(t_2, 0)|. \tag{3.2.78}$$

Lemma 3.2.24. *The functions* f_1, f_2 *defined in* (3.2.76) *satisfy the assumptions* (3.2.38), (3.2.39), (3.2.40), (3.2.41), (3.2.42), (3.2.43) *in Lemma 3.2.19.*

Proof. We have already checked (3.2.38). We know from Lemma 3.2.14 that, with a constant C depending only on a finite number of semi-norms of q (see (3.2.31)),

$$|f_1(0) = q(t_1,0)\Lambda^{1/2}\mu_1^{-1/2}|^{1/2} \leq C\nu_1^{1/2},$$

but we may assume here that $C \leq 1/2$: if we had $|f_1(0)| > \nu_1^{1/2}/2$, the function f_1 would be positive (resp.negative) on $B(0, r_0\nu_1^{1/2})$, with some fixed $r_0 > 0$ and consequently we would have $|\delta_1| \geq r_0\nu_1^{1/2}$. But we know that $\langle \delta_1 \rangle \leq \kappa\nu_1^{1/2}$, so we can choose a priori κ small enough so that $|\delta_1| \geq r_0\nu_1^{1/2}$ does not occur. From (3.2.75), we have $\langle \delta_1 \rangle \leq \kappa\nu_1^{1/2}$, the latter implying $f_1'(0) \neq 0$ from (3.2.29) since $\kappa^2 < 3/4$ and more precisely

$$\rho_1 = |f_1'(0)| \geq (1-\kappa^2)^{1/2}\nu_1^{1/2} \geq \nu_1^{1/2}/2. \tag{3.2.79}$$

Moreover we have, from (3.2.31) and $\nu_2 \leq \kappa^2\nu_1$ in (3.2.75),

$$\max(|\nu_1^{1/2}q'(t_2,0)|^{1/2}, |\nu_1^{1/2}q''(t_2,0)|) \leq \mu_{12}^{1/2} \leq C_1\nu_1^{1/2},$$

with a constant C_1 depending only on a finite number of semi-norms of q and thus

$$\max(|f_2'(0)|^{1/2}, |f_2''(0)|) \leq 2C_1\rho_1. \tag{3.2.80}$$

Moreover, we have from Lemma 3.2.14, $\Lambda^{1/2}|q(t_2,0)|\mu_2^{-1/2} \leq C_2\nu_2$, so that with constants C_2, C_3 depending only on a finite number of semi-norms of q, using (3.2.72), we get

$$|f_2(0)| \leq \nu_1^{1/2}C_2\nu_2\Lambda^{-1/2}\mu_2^{1/2} \leq \nu_1^{1/2}C_3\nu_2 \leq C_3\kappa^2\nu_1^{3/2}.$$

That property and (3.2.80), (3.2.79) give (3.2.40) with $R_0 = C\rho_1$, where C depends only on a finite number of semi-norms of q. We have already seen that the constants occurring in (3.2.41) are bounded above by semi-norms of q and that (3.2.42) holds. Let us now check (3.2.43). We already know that, from (3.2.78),

$$|f_2'(0)|^{1/2} = |\nu_1^{1/2}q'(t_2,0)|^{1/2} \leq 2\kappa^{1/2}|\nu_1^{1/2}q''(t_2,0)| = 2\kappa^{1/2}|f_2''(0)|. \tag{3.2.81}$$

If we have $|\nu_1^{1/2}q(t_2,0)| \geq \kappa^{1/2}\mu_{12}^{3/2}$, then for $|h| \leq \kappa^{1/3}\mu_{12}^{1/2}$, we get, using $\nu_1 \lesssim \Lambda$ and Taylor's formula along with (3.2.77), (3.2.78),

$$|\nu_1^{1/2}q(t_2,h)| \geq \kappa^{1/2}\mu_{12}^{3/2} - 4\kappa^{4/3}\mu_{12}^{3/2} - \frac{1}{2}\kappa^{2/3}\mu_{12}^{3/2} - C\nu_1^{1/2}\Lambda^{-1/2}\kappa\mu_{12}^{3/2}$$

$$= \mu_{12}^{3/2}(\kappa^{1/2} - 4\kappa^{4/3} - \frac{\kappa^{2/3}}{2} - C'\kappa) \geq \mu_{12}^{3/2}\kappa^{1/2}/2 > 0,$$

provided κ is small enough with respect to a constant depending only on a finite number of semi-norms of q; that inequality implies that the ball $B(0, \kappa^{1/3}\mu_{12}^{1/2})$ is included in $\mathbb{X}_+(t_2)$ or in $\mathbb{X}_-(t_2)$ implying that $|\delta_0(t_2, 0) = \delta_2| \geq \kappa^{1/3}\mu_{12}^{1/2}$ which is incompatible with (3.2.78), provided $\kappa < 2^{-3/2}$, since (3.2.78) implies $|\delta_2| \leq 2\kappa\mu_{12}^{1/2}$. Eventually, we get

$$|f_2(0)|^{1/3} = |\nu_1^{1/2}q(t_2, 0)|^{1/3} \leq \kappa^{1/6}\mu_{12}^{1/2} = \kappa^{1/6}|f_2''(0)| \qquad (3.2.82)$$

and with (3.2.82) we obtain (3.2.43) with

$$\kappa_2 = \kappa^{1/6}. \qquad (3.2.83)$$

The proof of Lemma 3.2.24 is complete. $\square$

End of the proof of Theorem 3.2.22. To apply Lemma 3.2.19, we have to suppose (3.2.44). In that case we get

$$\nu_1^{1/2}|q''(t_2, 0)_-| = |f_2''(0)_-| \leq C\kappa_2\rho_2 = C\kappa^{1/6}\nu_1^{1/2}|q''(t_2, 0)|,$$
$$\text{i.e.,} \quad |q''(t_2, 0)_-| \leq C\kappa^{1/6}|q''(t_2, 0)|. \qquad (3.2.84)$$

If (3.2.44) is not satisfied, we obtain, according to (3.2.46), (3.2.83) and $\mu_{12} = \nu_1^{1/2}|q''(t_2, 0)|$,

$$\delta_0(t_1, 0) = \delta_1 \leq -\kappa^{1/3}\nu_1^{1/2}|q''(t_2, 0)|,$$

which gives $\frac{1}{3}R(t_2, 0) \leq |q''(t_2, 0)| \leq \kappa^{-1/3}\frac{|\delta_1|}{\nu_1^{1/2}}$ and (3.2.67) provided $C_0 \geq 3\kappa^{-1/3}$. If we introduce now the smooth functions F_1, F_2 defined on $\mathbb{R}^{2n}$ by

$$F_1(Y) = -q(t_3, Y)\Lambda^{1/2}\mu_3^{-1/2}, \quad F_2(Y) = -\nu_3^{1/2}q(t_2, Y), \qquad (3.2.85)$$

starting over our discussion, we see that (3.2.67) is satisfied, provided

$$\kappa \leq \kappa_0 \quad \text{and} \quad C_0 \geq \gamma_0\kappa^{-1}, \qquad (3.2.86)$$

where κ_0, γ_0 are positive constants depending only on the semi-norms of q, except in the case where we have (3.2.84) and

$$|q''(t_2, 0)_+| \leq C\kappa^{1/6}|q''(t_2, 0)|. \qquad (3.2.87)$$

Naturally, since $|q''(t_2, 0)| = |q''(t_2, 0)_+| + |q''(t_2, 0)_-|$, the estimates (3.2.84), (3.2.87) cannot be both true for a κ small enough with respect to a constant depending on a finite number of semi-norms of q and a non-vanishing $q''(t_2, 0)$ (that vanishing is prevented by the penultimate line in (3.2.75)). The proof of Theorem 3.2.22 is complete. $\square$

Remark 3.2.25. The reader may find our proof quite tedious, but referring him to the simpler remark following Lemma 3.2.17, we hope that he can find there some motivation to read the details of our argument, which is the rather natural quantitative statement following from that remark. On the other hand, Theorem

3.2.22 is analogous to one of the key arguments provided by N. Dencker in [35] in which he proves, using our notation in the theorem,

$$R(t, X) \lesssim N(t', t'', X) + \delta_0(t'', X) - \delta_0(t', X) \qquad (3.2.88)$$

which is weaker than our (3.2.67). In particular, R (and N) looks like a symbol of order 0 (weight 1) whereas the right-hand side of (3.2.88) contains the difference $\delta_0(t'', X) - \delta_0(t', X)$, which looks like a symbol of order $1/2$. Our theorem gives a stronger and in some sense more homogeneous version of N. Dencker's result, which will lead to improvements in the remainder's estimates. Also, we note the (inhomogeneous) estimate

$$\Lambda^{-1/2} \mu(t, X)^{1/2} \nu(t, X)^{-1/2} \lesssim N(t', t'', X),$$

which is in fact a consequence of our proof, but is not enough to handle the remainder's estimate below in our proof, and which will not be used: in fact (3.2.67) implies

$$\Lambda^{-1/2} \mu^{1/2} \nu^{-1/2} = R \langle \delta_0 \rangle^{-1}$$
$$\lesssim \frac{N(t', t'', X)}{\langle \delta_0(t, X) \rangle} + \frac{\delta_0(t'', X) - \delta_0(t, X)}{\nu(t'', X)^{1/2} \langle \delta_0(t, X) \rangle} + \frac{\delta_0(t, X) - \delta_0(t', X)}{\nu(t', X)^{1/2} \langle \delta_0(t, X) \rangle}$$
$$\lesssim \frac{N(t', t'', X)}{\langle \delta_0(t, X) \rangle} + \frac{1}{\nu(t'', X)^{1/2}} + \frac{1}{\nu(t', X)^{1/2}} \lesssim N(t', t'', X).$$

3.2.6   Quasi-convexity

A differentiable function ψ of one variable is said to be quasi-convex on $\mathbb{R}$ if $\dot{\psi}(t)$ does not change sign from $+$ to $-$ for increasing t (see [75]). In particular, a differentiable convex function is such that $\dot{\psi}(t)$ is increasing and is thus quasi-convex.

Definition 3.2.26. Let $\sigma_1 : \mathbb{R} \to \mathbb{R}$ be an increasing function, $C_1 > 0$ and let $\rho_1 : \mathbb{R} \to \mathbb{R}_+$. We shall say that ρ_1 is quasi-convex with respect to (C_1, σ_1) if for $t_1, t_2, t_3 \in \mathbb{R}$,

$$t_1 \leq t_2 \leq t_3 \implies \rho_1(t_2) \leq C_1 \max\big(\rho_1(t_1), \rho_1(t_3)\big) + \sigma_1(t_3) - \sigma_1(t_1). \qquad (3.2.89)$$

When σ_1 is a constant function and $C_1 = 1$, this is the definition of quasi-convexity.

Lemma 3.2.27. *Let $\sigma_1 : \mathbb{R} \to \mathbb{R}$ be an increasing function and let $\omega : \mathbb{R} \to \mathbb{R}_+$. We define*

$$\rho_1(t) = \inf_{t' \leq t \leq t''} \Big(\omega(t') + \omega(t'') + \sigma_1(t'') - \sigma_1(t') \Big). \qquad (3.2.90)$$

Then the function ρ_1 is quasi-convex with respect to $(2, \sigma_1)$.

Proof. We consider $t_1 \leq t_2 \leq t_3$ three real numbers. We have

$$\rho_1(t_2) = \inf_{t' \leq t_2 \leq t''} \left(\omega(t') + \omega(t'') + \sigma_1(t'') - \sigma_1(t') \right)$$

$$\leq \inf_{t' \leq t_1, t_3 \leq t''} \left(\omega(t') + \omega(t'') + \sigma_1(t'') - \sigma_1(t_3) + \sigma_1(t_1) - \sigma_1(t') \right) + \sigma_1(t_3) - \sigma_1(t_1)$$

$$\leq \inf_{\substack{t' \leq t_1 \leq t''_1, \\ t'_3 \leq t_3 \leq t''}} \left(\omega(t') + \omega(t''_1) + \omega(t'_3) + \omega(t'') + \sigma_1(t'') - \sigma_1(t'_3) + \sigma_1(t''_1) - \sigma_1(t') \right)$$

$$+ \sigma_1(t_3) - \sigma_1(t_1)$$

$$= \rho_1(t_1) + \rho_1(t_3) + \sigma_1(t_3) - \sigma_1(t_1) \leq 2\max(\rho_1(t_1), \rho_1(t_3)) + \sigma_1(t_3) - \sigma_1(t_1). \qquad \square$$

The following lemma is due to L. Hörmander [78].

Lemma 3.2.28. *Let* $\sigma_1 : \mathbb{R} \to \mathbb{R}$ *be an increasing function and let* $\omega : \mathbb{R} \to \mathbb{R}_+$. *Let* $T > 0$ *be given. We consider the function* ρ_1 *as defined in Lemma 3.2.27 and we define*

$$\Theta_T(t) = \sup_{-T \leq s \leq t} \left\{ \sigma_1(s) - \sigma_1(t) + \frac{1}{2T} \int_s^t \rho_1(r) dr - \rho_1(s) \right\}. \qquad (3.2.91)$$

Then we have

$$2T \partial_t(\Theta_T + \sigma_1) \geq \rho_1, \quad \text{and for } |t| \leq T, \quad |\Theta_T(t)| \leq \rho_1(t). \qquad (3.2.92)$$

Proof. We have $\Theta_T(t) \geq -\rho_1(t)$, and

$$\Theta_T(t) + \sigma_1(t) = \underbrace{\sup_{-T \leq s \leq t} \left\{ \sigma_1(s) + \frac{1}{2T} \int_s^0 \rho_1(r) dr - \rho_1(s) \right\}}_{\text{increasing with } t} + \frac{1}{2T} \int_0^t \rho_1(r) dr,$$

so that $\partial_t(\Theta_T + \sigma_1) \geq \frac{1}{2T} \rho_1$. Moreover, from the proof of Lemma 3.2.27, we obtain for $s \leq r \leq t$ that $\rho_1(r) \leq \rho_1(s) + \rho_1(t) + \sigma_1(t) - \sigma_1(s)$ and thus

$$\frac{1}{2T} \int_s^t \rho_1(r) dr \leq \frac{1}{t-s} \int_s^t \rho_1(r) dr \leq \rho_1(s) + \rho_1(t) + \sigma_1(t) - \sigma_1(s)$$

which gives $\Theta_T(t) \leq \rho_1(t)$, ending the proof of the lemma. $\qquad \square$

Definition 3.2.29. For $T > 0, X \in \mathbb{R}^{2n}, |t| \leq T$, we define

$$\left. \begin{aligned} \omega(t, X) &= \frac{\langle \delta_0(t, X) \rangle}{\nu(t, X)^{1/2}}, \\ \sigma_1(t, X) &= \delta_0(t, X), \\ \eta(t, X) &= \int_{-T}^t \delta_0(s, X) \Lambda^{-1/2} ds + 2T, \end{aligned} \right\} \qquad (3.2.93)$$

where δ_0, ν are defined in (3.2.19), (3.2.29). For $T > 0$, $(t, X) \in \mathbb{R} \times \mathbb{R}^{2n}$, we define $\Theta(t, X)$ by the formula (3.2.91)

$$\Theta(t, X) = \sup_{-T \leq s \leq t} \left\{ \sigma_1(s, X) - \sigma_1(t, X) + \frac{1}{2T} \int_s^t \rho_1(r, X) dr - \rho_1(s, X) \right\},$$
(3.2.94)

where ρ_1 is defined by (3.2.90). We define also

$$m(t, X) = \delta_0(t, X) + \Theta(t, X) + T^{-1} \delta_0(t, X) \eta(t, X).$$
(3.2.95)

Theorem 3.2.30. *With the notation above for Θ, ρ_1, m, with R and C_0 defined in Theorem 3.2.22, we have for $T > 0$, $|t| \leq T$, $X \in \mathbb{R}^{2n}$, $\Lambda \geq 1$,*

$$|\Theta(t, X)| \leq \rho_1(t, X) \leq 2 \frac{\langle \delta_0(t, X) \rangle}{\nu(t, X)^{1/2}}, \quad |\sigma_1(t, X)| = |\delta_0(t, X)|,$$
(3.2.96)

$$C_0^{-1} R(t, X) \leq \rho_1(t, X) \leq 2T \frac{\partial}{\partial t} \Big(\Theta(t, X) + \sigma_1(t, X) \Big),$$
(3.2.97)

$$0 \leq \eta(t, X) \leq 4T, \quad \frac{d}{dt}(\delta_0 \eta) \geq \delta_0^2 \Lambda^{-1/2}, \quad |\eta'_X(t, X)| \leq 4T\Lambda^{-1/2},$$
(3.2.98)

$$T \frac{d}{dt} m \geq \frac{1}{2} \rho_1 + \delta_0^2 \Lambda^{-1/2} \geq \frac{1}{2C_0} R + \delta_0^2 \Lambda^{-1/2} \geq \frac{1}{2^{3/2} C_0} \langle \delta_0 \rangle^2 \Lambda^{-1/2}.$$
(3.2.99)

Proof. It follows immediately from the previous results: the first estimate in (3.2.96) is (3.2.92), whereas the second is due to $\rho_1 \leq 2\omega$, which follows from (3.2.90). The equality in (3.2.96) follows from Definition 3.2.29. The first inequality in (3.2.97) is a consequence of (3.2.90) and (3.2.67) and the second is (3.2.92). The first two inequalities in (3.2.98) are a consequence of $|\delta_0(t, X)| \leq \Lambda^{1/2}$ which follows from Definition 3.2.11. The third inequality reads

$$\frac{d}{dt}(\delta_0 \eta) = \dot{\delta}_0 \eta + \delta_0 \dot{\eta} \geq \delta_0 \dot{\eta} = \delta_0^2 \Lambda^{-1/2},$$

and the fourth inequality in (3.2.98) follows from (3.2.20). Let us check finally (3.2.99): since $m = \delta_0 + \Theta + T^{-1} \delta_0 \eta$, (3.2.92) and the already proven (3.2.98) imply $T \frac{d}{dt} m \geq \frac{1}{2} \rho_1 + \delta_0^2 \Lambda^{-1/2}$ and (3.2.97)(proven) gives

$$\frac{1}{2} \rho_1 + \delta_0^2 \Lambda^{-1/2} \geq \frac{1}{2C_0} R + \delta_0^2 \Lambda^{-1/2} = \frac{1}{2C_0} \Lambda^{-1/2} \mu^{1/2} \nu^{-1/2} \langle \delta_0 \rangle + \delta_0^2 \Lambda^{-1/2}$$

$$\underset{\text{from (3.2.30)}}{\geq} \frac{1}{2C_0} \Lambda^{-1/2} (2^{-1/2} \langle \delta_0 \rangle + \delta_0^2) \geq \frac{1}{2^{3/2} C_0} \Lambda^{-1/2} \langle \delta_0 \rangle^2,$$

completing the proof of Theorem 3.2.30. $\square$

3.3 The necessity of condition (Ψ)

In the present section we shall review the classical result due to R. Moyer in two dimensions ([108]) and to L. Hörmander ([70], see also Section 26.4 in [74]) in the general case, that condition (Ψ) is indeed necessary for local solvability. That necessity result was settled before by L. Nirenberg and F. Treves for operators having symbols whose imaginary part vanishes at a finite order along the Hamiltonian flow of the real part ([113]).

Theorem 3.3.1. *Let Ω be an open subset of $\mathbb{R}^n$, $x_0 \in \Omega$ and P be a principal-type properly supported pseudo-differential operator on Ω with principal symbol p_m. If P is locally solvable at x_0 (Definition 1.2.25), then there is a neighborhood of x_0 on which p_m satisfies condition (Ψ)(Definition 3.2.1).*

This result is a consequence of Corollary 26.4.8 in [74] and there is no need for reproducing its proof here. Let us also note that the book [74] contains a more general result involving semi-global solvability.

Definition 3.3.2. Let Ω be an open subset of $\mathbb{R}^n$, K be a compact subset of Ω and P be a properly supported pseudo-differential operator on Ω . P is said to be solvable at K if there exist an open neighborhood V of K and a space $\mathcal{F}$ of finite codimension in $C^\infty(\Omega)$, such that

$$\forall f \in \mathcal{F}, \quad \exists u \in \mathscr{D}'(\Omega) \quad \text{with } Pu = f \text{ in } V.$$

The following result follows from Theorem 26.4.2 in [74].

Theorem 3.3.3. *Let Ω be an open subset of $\mathbb{R}^n$, K be a compact subset of Ω and P be a properly supported pseudo-differential operator on Ω. The following conditions are equivalent.*

(i) *P is solvable at K.*

(ii) *There exists $N \in \mathbb{N}$, V an open neighborhood of K such that*

$$\forall f \in H^N_{loc}(\Omega), \quad \exists u \in \mathscr{D}'(\Omega) \quad \text{with } Pu - f \in C^\infty(V).$$

Definition 3.3.4. Let Ω be an open subset of $\mathbb{R}^n$, $\mathcal{K}$ be a compactly based cone in $\dot{T}^*(\Omega)$ and P be a properly supported pseudo-differential operator on Ω. P is said to be solvable at $\mathcal{K}$ if there exists $N \in \mathbb{N}$, such that

$$\forall f \in H^N_{loc}(\Omega), \quad \exists u \in \mathscr{D}'(\Omega) \quad \text{with } WF(Pu - f) \cap \mathcal{K} = \emptyset.$$

A consequence of that definition and of the previous theorem is that solvability at a compact set $K \subset \Omega$ is equivalent to solvability at $K \times \mathbb{S}^{n-1}$.

We shall give here a short discussion of the steps of the proof in a semi-classical framework, as it was already done with Theorem 3.1.9, containing a construction of a quasi-mode when the first bracket is negative at a characteristic point. Let us recall at this point that a function p satisfies condition $(\overline{\Psi})$ when $\bar{p}$ satisfies condition (Ψ). Thus, disproving local solvability for an operator with symbol violating condition (Ψ) amounts to disproving an a priori estimate for the adjoint operator, whose symbol will then violate condition $(\overline{\Psi})$.

We consider then a symbol $p \in S^0_{scl}$, i.e., a C^∞ function on $\mathbb{R}^{2n}$ depending also on a parameter $h \in (0, 1]$ such that

$$\forall \alpha, \beta \in \mathbb{N}^n \times \mathbb{N}^n, \qquad \sup_{(x,\xi,h) \in \mathbb{R}^n \times \mathbb{R}^n \times (0,1]} |(\partial_x^\alpha \partial_\xi^\beta p)(x, \xi, h)| h^{-|\beta|} < +\infty. \qquad (3.3.1)$$

We shall assume also that there exists a sequence $(p_j)_{j\geq 0} \in C^\infty_b(\mathbb{R}^{2n})$ such that

$$p(x, \xi, h) \sim \sum_{j \geq 0} h^j p_j(x, h\xi) : \forall N, p(x, \xi, h) - \sum_{0 \leq j < N} h^j p_j(x, h\xi) \in h^N S^0_{scl}. \qquad (3.3.2)$$

We consider the bicharacteristic curve γ of the real part of p_0 starting at $(x, \xi) \in \mathbb{R}^{2n}$, i.e.,

$$\dot{\gamma}(t, x, \xi) = H_{\mathrm{Re}\, p_0}(\gamma(t, x, \xi)), \quad \gamma(0, x, \xi) = (x, \xi). \qquad (3.3.3)$$

We assume that there exists $(x_0, \xi_0) \in \mathbb{R}^{2n}$ such that $p_0(x_0, \xi_0) = 0$ and that in any neighborhood V of (x_0, ξ_0), there exists $(x, \xi) \in V, t < s \in \mathbb{R}$ such that $\mathrm{Re}\, p_0(x, \xi) = 0$ and $\gamma([t, s], x, \xi) \subset V$,

$$\mathrm{Im}\, p_0(\gamma(t, x, \xi)) > 0 > \mathrm{Im}\, p_0(\gamma(s, x, \xi)). \qquad (3.3.4)$$

Theorem 3.3.5. *Let $p \in S^0_{scl}$ be as above with p_0 satisfying (3.3.4) near (x_0, ξ_0). Then for all neigborhoods V of (x_0, ξ_0), there exist $(u_h)_{h \in (0,1]}$ unit vectors in $L^2(\mathbb{R}^n)$ such that*

$$\|p(x, \xi, h)^w u_h\|_{L^2} = O(h^\infty),$$

with $WF_{scl}(u_h) \subset V$, i.e., for all $(x_1, \xi_1) \notin V$, there exists $a \in C^\infty_c(\mathbb{R}^{2n})$, $a(x_1, \xi_1) \neq 0$, with $\|a(x, h\xi)^w u_h\|_{L^2} = O(h^\infty)$.

Using the normal form given in Proposition 4.5.1 and Theorem 4.7.8 on quantization of semi-classical canonical transformations, we are reduced to proving the result for $p(x, \xi, h) = h\xi_1 + if(x_1, x', h\xi')$. Changing the notation, we need to construct some approximate null solutions for the evolution equation

$$D_t + ih^{-1}q(t, x, h\xi)^w$$

provided $q \in C^\infty_b(\mathbb{R} \times \mathbb{R}^{2n})$ is real-valued, $q(0, 0, 0) = 0$ and for all $\varepsilon > 0$, there exist $-\varepsilon < t < s < \varepsilon$, $(x, \xi) \in B((0, 0), \varepsilon)$ with $q(t, x, \xi) > 0 > q(s, x, \xi)$.

The construction is similar to the one in Theorem 3.1.9 or in Lemma 3.1.10 only in one particular case: $t \mapsto q(t, x, \xi)$ has a zero of odd order $2k+1$ of constant multiplicity with respect to the parameters (x, ξ). In that case there is a normal form given by Theorem 21.3.5 in [73], which reduces the study to the ODE

$$D_t - ih^{-1}t^{2k+1}.$$

Even with a finite order vanishing at $(x, \xi) = (0,0)$, the construction of a quasi-mode is more complicated. Just to give an example with a third order 0, we may have

$$q(t, x, \xi) = -(t - \xi)(t + x)^2, \quad \text{or} \quad q(t, x, \xi) = -t^3 + x.$$

In the first case the quasi-mode should be located near $t = \xi$ and in the second case near $t = x^{1/3}$ where the change of sign from $+$ to $-$ occurs. In fact, in the third order 0 case, anything of the following type may occur:

$$q(t, x, \xi) = -t^3 + a_2(x, \xi)t^2 + a_1(x, \xi)t + a_0(x, \xi), \quad a_j(0, 0) = 0.$$

Naturally, leaving the finite order of vanishing increases dramatically the complexity. The general construction of an approximate null solution is based upon a complex WKB method where one looks for a solution of type

$$u_h \sim e^{ih^{-1}\phi} \sum_{j \geq 0} h^j a_j$$

where the phase function ϕ has a non-negative imaginary part and should satisfy approximately some eiconal equation $\partial_t \phi - iq(t, x, \partial_x \phi) = 0$. The reader will find in the proof of Lemma 26.4.14 of [74] the details of the construction of that phase function, the key step in the proof.

3.4 Estimates with loss of $k/k+1$ derivative

3.4.1 Introduction

We have seen in the previous section that condition (Ψ) is necessary for local solvability. Starting an investigation of the sufficiency of that condition, we take p as the (complex-valued) principal symbol of a principal-type pseudo-differential operator P satisfying condition (Ψ). The principal symbol

$$\bar{p} = \ell_1 + i\ell_2, \quad \ell_1, \ell_2 \text{ real-valued,} \tag{3.4.1}$$

of the adjoint operator P^* satisfies condition ($\overline{\Psi}$). We have seen already in Theorem 3.1.5 that the first-bracket analysis is complete: since condition ($\overline{\Psi}$) holds for $\bar{p}$, non-vanishing of the Poisson bracket at a characteristic point implies $\{\ell_1, \ell_2\} > 0$ and this entails subellipticity with loss of $1/2$ derivatives for P^* and local solvability for P. The first class of cases that we want to investigate is linked to some sort of finite-type assumption, related to subellipticity. First we assume that one of the iterated bracket of the real and imaginary part is not vanishing: we shall say that the geometry is of finite type.

We consider $\ell = \ell_1 + i\ell_2$ a complex-valued function in C_b^∞. We shall assume that condition ($\overline{\Psi}$) holds for ℓ in a neighborhood V_0 of a point $(x_0, \xi_0) \in \mathbb{R}^{2n}$.

Moreover, we define

$$\ell_{12} = \{\ell_1, \ell_2\} = -\ell_{21}, \quad H_j = H_{\ell_j}, \tag{3.4.2}$$

$$\ell_{112} = H_1^2(\ell_2), \ell_{212} = -\ell_{221}, \ell_{221} = H_2^2(\ell_1), \ell_{121} = -\ell_{112}, \tag{3.4.3}$$

$$\text{for } j_k \in \{1,2\}, \quad \ell_{j_1,\dots,j_{l+1}} = H_{j_1} \dots H_{j_l}(\ell_{j_{l+1}}), \quad |(j_1,\dots,j_l)| = l. \tag{3.4.4}$$

We assume that one of these iterated brackets is non-zero and we define the integer k so that

$$\text{for all } |J| \le k, \quad \ell_J(x_0, \xi_0) = 0, \text{ and}$$
$$\text{there exists } |J| \text{ with } |J| = k+1 \text{ such that } \quad \ell_J(x_0, \xi_0) \neq 0. \tag{3.4.5}$$

Remark 3.4.1. Note that

$$|\ell| \neq 0 \text{ means } k = 0, \text{ that is ellipticity}, \tag{3.4.6}$$

$$\ell = 0, \{\ell_1, \ell_2\} > 0, \text{ means } k = 1, \tag{3.4.7}$$

$$\ell = \{\ell_1, \ell_2\} = 0, H_{\ell_1}^2(\ell_2) \neq 0 \text{ or } H_{\ell_2}^2(\ell_1) \neq 0, \text{ means } k = 2. \tag{3.4.8}$$

We have already seen some of the simplest ODE-like models (3.1.10), (3.1.12) such as

$$\xi_1 + ix_1^k, \ k \in 2\mathbb{N} + 1, \qquad \xi_1 \pm ix_1^k, \ k \in 2\mathbb{N}.$$

The following simple-looking example is interesting: we consider

$$\ell(s, V) = \xi_1 + ix_1^s(\xi_2 + V(x_1, x_2)), \quad s \in 2\mathbb{N}, \ \partial_{x_1} V \ge 0, \tag{3.4.9}$$

where $\partial_{x_1} V$ is a non-zero polynomial. Let us check for instance that for (3.4.9) with $V = x_1 x_2^2$, we have indeed $k = 9$ at $(0,0)$: for $\ell_1 + i\ell_2 = \xi_1 + ix_1^2(\xi_2 + x_1 x_2^2)$, we have

$$\frac{1}{6} H_1^3(\ell_2) = x_2^2, \quad \frac{1}{2} H_1^2(\ell_2) = \xi_2 + 3x_1 x_2^2 \quad \text{so that}$$

$$\frac{1}{24}\{H_1^2(\ell_2), H_1^3(\ell_2)\} = x_2, \quad \{H_1^2(\ell_2), \{H_1^2(\ell_2), H_1^3(\ell_2)\}\} \neq 0$$

i.e.,

$$0 \neq H_{H_1^2(\ell_2)}^2(H_1^3(\ell_2)) = H_{\{\ell_1, H_1(\ell_2)\}}^2(H_1^3(\ell_2))$$

and since $H_{\{\ell_1, H_1(\ell_2)\}} = [H_1, [H_1, H_2]]$ we get

$$[H_1, [H_1, H_2]]^2 H_1^3(\ell_2) \neq 0$$

which forces $H^I(\ell_2) \neq 0$ with some $|I| = 9$. We leave it for the reader to check that $\ell_J = 0$ at $(0,0)$ for $|J| \le 8$.

3.4.2 The main result on subellipticity

Subellipticity is defined as follows.

Definition 3.4.2. Let Ω be an open subset of $\mathbb{R}^n$, $m \in \mathbb{R}$, $P \in \Psi_{ps}^m(\Omega)$ and $(x_0, \xi_0) \in \dot{T}^*(\Omega)$. The operator P is subelliptic at (x_0, ξ_0) with loss of μ derivative if for all $s \in \mathbb{R}$,

$$u \in \mathscr{D}'(\Omega), \ Pu \in H_{(x_0, \xi_0)}^s \implies u \in H_{(x_0, \xi_0)}^{s+m-\mu}. \tag{3.4.10}$$

The main result on subelliptic equations is proven in Chapter 27 of L. Hörmander's book [74].

Theorem 3.4.3. Let $\Omega, m, P, (x_0, \xi_0)$ be as above and let p_m be the principal symbol of P. For $k \in \mathbb{N}$, the operator P is subelliptic at (x_0, ξ_0) with loss of $\frac{k}{k+1}$ derivative if and only if the following conditions are satisfied.

(1) There exists a neighborhood V of (x_0, ξ_0) such that condition $(\overline{\Psi})$ holds for p_m on V.

(2) There exists $z \in \mathbb{C}, \mathbb{N} \ni j \leq k$ such that $H_{\mathrm{Re}(zp_m)}^j(\mathrm{Im}(zp_m))(x_0, \xi_0) \neq 0$.

We shall not reproduce here the proof of that very difficult theorem[9] but we intend to point out that subellipticity with loss of $k/(k+1)$ derivatives at (x_0, ξ_0) is equivalent to the existence of a classical symbol a of order 0, non-characteristic at (x_0, ξ_0) and of a constant C such that, for all $u \in C_c^\infty(V_0)$,

$$\|a^w u\|_{m-\frac{k}{k+1}} \leq C(\|Pu\|_0 + \|u\|_{m-1}), \tag{3.4.11}$$

where V_0 is a neighborhood of x_0, thus implying microlocal solvability at (x_0, ξ_0) for the adjoint operator P^*. It means that condition (Ψ) reinforced with a finite type assumption such as (2) in Theorem 3.4.3 does imply local solvability.

3.4.3 Simplifications under a more stringent condition on the symbol

As it is pointed out in by F. Treves in [140] and by L. Hörmander in Section 27.3 of [74], the proof of (3.4.11) is considerably simplified when the principal symbol satisfies condition (P)(see Definition 3.2.4) along with the finite-type assumption (2) in Theorem 3.4.3. We want to display here a short proof in the same spirit, using a coherent states method, which essentially reduces the problem to an ODE,

---

[9]In L. Hörmander's contribution to the book *Fields medallists' lectures* [4], one can read the following. *"For the scalar case, Egorov [41] found necessary and sufficient conditions for subellipticity with loss of δ derivatives; the proof of sufficiency was completed in [68]. A slight modification of the presentation is given in [74], but it is still very complicated technically. Another approach which also covers systems operating on scalars has been given by Nourrigat [116] (see also the book [60] by Helffer and Nourrigat), but it is also far from simple so the study of subelliptic operators may not yet be in a final form"*.

via the Wick calculus exposed here in Section 2.4.1. When the geometry is finite-type, one can get a microlocal reduction to a model

$$hD_t + iq(t, x, h\xi)^w, \quad q \geq 0, \partial_t^k q \neq 0, \text{ with } k \text{ even.}$$

The plan of the proof is quite clear: first we prove an estimate for an ODE with parameters, such as $h\frac{d}{dt} - q(t, x, h\xi)$, then using a non-negative quantization for the symbol q, we show that the ODE estimates can be transferred to the semi-classical level. We go slightly beyond condition (P) with the following result (here and thereafter we use Remark 3.1.11 to get a reduction to a semi-classical estimate).

Theorem 3.4.4. *Let n be an integer and $q(t, x, \xi)$ be a real-valued symbol in C_b^∞: q is defined on $\mathbb{R} \times \mathbb{R}^n \times \mathbb{R}^n$, smooth with respect to t, x, ξ and such that, for all multi-indices α, β, $\sup |(\partial_x^\alpha \partial_\xi^\beta q)(t, x, \xi)| < +\infty$. Assume moreover that $\tau + iq$ satisfies condition $(\overline{\Psi})$ and*

$$q(t, x, \xi) = 0 \implies d_{x,\xi} q(t, x, \xi) = 0, \tag{3.4.12}$$

$$for some k \in \mathbb{N}, \quad \inf |\partial_t^k q(t, x, \xi)| > 0. \tag{3.4.13}$$

Then there exists some positive constants C, h_0, such that, for $h \in (0, h_0]$, for any $u(t, x) \in C_c^1(\mathbb{R}, L^2(\mathbb{R}^n))$,

$$C \|hD_t u + iq(t, x, h\xi)^w u\|_{L^2(\mathbb{R}^{n+1})} \geq h^{\frac{k}{k+1}} \|u\|_{L^2(\mathbb{R}^{n+1})}. \tag{3.4.14}$$

Let us note right now that the condition (3.4.12) is satisfied by non-negative (and non-positive) functions. However that condition may be satisfied by some functions which may change sign such as $q = ta(t, x, \xi)$, $a \geq 0$. In fact, if $a = 0$ we have $da = 0$, so that $dq = 0$; at $t = 0$, we have $d_{x,\xi} q = 0$.

Going back to our operator $hD_t + iq(t, x, h\xi)^w$, we shall first replace it by the unitary equivalent $hD_t + iq(t, h^{1/2}x, h^{1/2}\xi)^w$ acting on $C_c^1(\mathbb{R}_t; L^2(\mathbb{R}^n))$. Next, one can check, using Proposition 2.4.3, that

$$q(t, h^{1/2}x, h^{1/2}\xi)^w = q(t, h^{1/2}x, h^{1/2}\xi)^{\text{Wick}} + O(h), \quad \text{in } \mathcal{L}(L^2(\mathbb{R}^n)).$$

Also, defining $\Phi(t) = Wu(t) \in L^2(\mathbb{R}^{2n})$ (see (2.4.2)), we are reduced to proving an estimate for

$$P = hD_t + i\pi_H q(t, h^{1/2}X)\pi_H,$$

where the Toeplitz orthogonal projection $\pi_H = WW^*$. Now we see that for $\Phi \in C_c^1(\mathbb{R}_t; H)$, with $K = H^\perp$, $L^2(\mathbb{R}^{2n})$ norms,

$$\|(hD_t + iq(t, h^{1/2}X))\Phi\|^2 = \|P\Phi\|^2 + \|\pi_K q(t, h^{1/2}X)\Phi\|^2$$

$$= \|P\Phi\|^2 + \|[\pi_H, q(t, h^{1/2}\cdot)]\Phi\|^2.$$

Handling the linear ODE given by $\mathcal{P} = hD_t + iq(t, h^{1/2}X)$ is a simple matter using the estimate (3.1.29) in Lemma 3.1.3 and we obtain

$$C \|\mathcal{P}\Phi\| \|\Phi\| \geq \langle |q(t, h^{1/2}X)|\Phi, \Phi\rangle + h^{k/k+1} \|\Phi\|^2.$$

The nasty term $\|[\pi_H, q(t, h^{1/2} \cdot)]\Phi\|^2$ can be estimated from above by

$$h\|\nabla q(t, h^{1/2} \cdot)\Phi\|^2 + O(h^2)\|\Phi\|^2,$$

since the kernel of $[\pi_H, q(t, h^{1/2} \cdot)]$ is, with the notation (2.4.4),

$$\Pi(X, Y)\big(q(t, h^{1/2}Y) - q(t, h^{1/2}X)\big) = \Pi(X, Y)h^{1/2}\nabla q(t, h^{1/2}Y)(Y - X)$$
$$- \Pi(X, Y)\int_0^1 (1 - \theta)h\nabla^2 q(t, h^{1/2}(Y + \theta(X - Y)))d\theta(X - Y)^2.$$

Moreover, thanks to Lemma 4.3.10 and to the assumption (3.4.12), we have

$$\langle |\nabla q(t, h^{1/2}X)|^2\Phi, \Phi\rangle \leq 2\langle |q(t, h^{1/2}X)|\Phi, \Phi\rangle \sup|\nabla^2 q(t, \cdot)|,$$

yielding

$$\langle |q(t, h^{1/2} \cdot)|\Phi, \Phi\rangle + h^{k/k+1}\|\Phi\|^2$$
$$\leq C_1\|P\Phi\|\|\Phi\| + C_1\|[\pi_H, q(t, h^{1/2} \cdot)]\Phi\|\|\Phi\|$$
$$\leq C_1\|P\Phi\|\|\Phi\| + \underbrace{C_2\langle |q(t, h^{1/2} \cdot)|\Phi, \Phi\rangle^{1/2}h^{1/2}\|\Phi\| + C_3h\|\Phi\|^2}_{\text{can be absorbed in the lhs, first line above.}},$$

completing the proof of the theorem.

3.5 Estimates with loss of one derivative

3.5.1 Local solvability under condition (P)

Statement of the result

Theorem 3.5.1. *Let Ω be an open subset of $\mathbb{R}^n$, $m \in \mathbb{R}$, $x_0 \in \Omega$ and $P \in \Psi_{ps}^m(\Omega)$ a principal-type pseudo-differential operator with a principal symbol satisfying condition (P) (Definition 3.2.4) in a neighborhood of x_0. Then P is locally solvable at x_0 with loss of one derivative.*

A most important consequence of that theorem is the fact that for *differential* operators, condition (Ψ) is equivalent to local solvability since from Remark 3.2.5 condition (P) is equivalent to condition (Ψ) for differential operators, whereas the necessity is given by Theorem 3.3.1. The previous result was proven in 1970 by L. Nirenberg and F. Treves in [114] under an analyticity assumption for the principal symbol, or more precisely under some factorization hypothesis which holds under an analyticity assumption and is not true in the C^∞ category (see here the example (3.2.11)). That analyticity assumption was removed in 1973 by R. Beals and C. Fefferman in [8], who invented a new calculus of pseudo-differential operators, based upon a Calderón-Zygmund decomposition of a family of symbols in the phase space. The decomposition of the symbol is such that the calculus is tailored on the symbol under scope. We give the details of the construction below, using the metrics devices introduced by L. Hörmander and exposed here in Chapter 2. The main step to proving the previous result is the following semi-classical theorem (see Remark 3.1.11).

Theorem 3.5.2. *Let $q : \mathbb{R} \times \mathbb{R}^n \times \mathbb{R}^n \times (0,1] \mapsto \mathbb{R}$ be a function such that*

$$\forall \alpha, \beta \in \mathbb{N}^n, \quad C_{\alpha\beta} = \sup_{\substack{(t,x,\xi,h) \in \\ \mathbb{R} \times \mathbb{R}^n \times \mathbb{R}^n \times (0,1]}} |(\partial_x^\alpha \partial_\xi^\beta q)(t,x,\xi,h)| h^{-|\beta|} < +\infty, \quad (3.5.1)$$

$$\forall (t,s,x,\xi,h) \in \mathbb{R} \times \mathbb{R} \times \mathbb{R}^n \times \mathbb{R}^n \times (0,1], \quad q(t,x,\xi,h)q(s,x,\xi,h) \geq 0. \quad (3.5.2)$$

Then there exist positive constants T, C, h_0 depending only on the $(C_{\alpha\beta})$ such that for all $v \in \mathscr{S}(\mathbb{R}^{n+1})$ with $v(t,x) = 0$ for $|t| \geq T$, $0 < h \leq h_0$,

$$hT^{-1}\|v\|_{L^2(\mathbb{R}^{n+1})} \leq C\|hD_t v + iq(t,x,\xi,h)^w v\|_{L^2(\mathbb{R}^{n+1})}. \quad (3.5.3)$$

First reductions

The operators $q(t,x,\xi,h)^w$ and $q(t,h^{1/2}x, h^{-1/2}\xi, h)^w$ are unitarily equivalent and, defining for $(t,X) \in \mathbb{R} \times \mathbb{R}^{2n}$, $Q(t,X) = h^{-1}q(t,h^{1/2}x, h^{-1/2}\xi, h)$, we may thus assume, omitting the h dependence of Q that for all $k \in \mathbb{N}$, with $\Lambda = h^{-1}$,

$$|(\partial_X^k Q)(t,X)| \leq C_k \Lambda^{1-\frac{k}{2}}, \quad (3.5.4)$$

where the C_k depend only on the $(C_{\alpha\beta})$ above. It means that $Q \in S(\Lambda, \Lambda^{-1}\Gamma)$, where Γ is the standard Euclidean norm on $\mathbb{R}^{2n}$. Following Lemma 3.2.12, we define, for $X \in \mathbb{R}^{2n}$,

$$\lambda(t,X) = 1 + \max\left(|Q(t,X)|, |\partial_X Q(t,X)|^2\right), \quad \lambda(X) = \sup_{t \in \mathbb{R}} \lambda(t,X). \quad (3.5.5)$$

According to Lemma 3.2.12, we have $1 \leq \lambda(t,X) \leq 1 + \gamma\Lambda$ where γ is a semi-norm of Q in $S(\Lambda, \Lambda^{-1}\Gamma)$, depending thus only on the (C_k) of (3.5.4) which implies

$$1 \leq \lambda(X) \leq 1 + \gamma\Lambda. \quad (3.5.6)$$

We know also from the same lemma that the metrics $\lambda(t,X)^{-1}\Gamma$ are slowly varying with structure constants depending only on γ. In particular we know that there exists $r_0 > 0, C_0 > 0$ such that

$$|X - Y| \leq r_0 \lambda(t,X)^{1/2} \implies C_0^{-1} \leq \frac{\lambda(t,X)}{\lambda(t,Y)} \leq C_0.$$

As a result, $|X - Y| \leq \frac{r_0}{2}\lambda(X)^{1/2}$ implies $|X - Y| \leq r_0 \lambda(t,X)^{1/2}$, provided $\lambda(X)/4 \leq \lambda(t,X)$ for some t and this gives

$$\frac{1}{4}\lambda(X) \leq \lambda(t,X) \leq C_0 \lambda(t,Y) \leq C_0 \lambda(Y)$$

and the slow variation of $\lambda(X)^{-1}\Gamma$ according to Remark 2.2.2. Since $\lambda(X) \geq 1$, Lemma 2.2.21 implies that $\lambda(X)^{-1}\Gamma$ is indeed admissible. Moreover, for all $t \in \mathbb{R}$, $Q(t,\cdot) \in S(\lambda(X), \lambda(X)^{-1}\Gamma)$ uniformly with respect to t: this is a consequence of Lemma 3.2.12 for the derivatives of order ≤ 2 and (3.5.6) along with (3.5.4) give for $k > 2$, $|(\partial_X^k Q)(t,X)| \leq C_k \Lambda^{1-\frac{k}{2}} \leq C_k \lambda(X)^{1-\frac{k}{2}}(1+\gamma)^{(k-2)/2}$. We have proven the following result.

Lemma 3.5.3. *With Γ standing for the Euclidean norm on $\mathbb{R}^{2n}$, $\lambda(X)$ defined by (3.5.5), the metric $\lambda(X)^{-1}\Gamma$ is admissible with structure constants depending only on the (C_k) in (3.5.4) and for all $t \in \mathbb{R}$, $Q(t,\cdot) \in S(\lambda(X), \lambda(X)^{-1}\Gamma)$ with for all $k \in \mathbb{N}$, $\gamma_k = \sup_{(t,X,\Lambda)\in\mathbb{R}^{1+2n}\times[1,+\infty)} |(\partial_X^k Q)(t,X)|\lambda(X)^{\frac{k}{2}-1} < +\infty$ depends only on the (C_l). Moreover, we have*

$$\text{for } j = 0,1, \quad |(\partial_X^j Q)(t,X)| \leq \lambda(X)^{1-\frac{j}{2}}. \tag{3.5.7}$$

Classification of points

That metric is so well-tailored to the symbol Q that we can classify the points in $\mathbb{R}^{2n}$ according to the respective size of the quantities

$$\sup_{t\in\mathbb{R}} |Q(t,X)|, \quad \sup_{t\in\mathbb{R}} |Q'_X(t,X)|^2, \quad 1.$$

Let $0 < r < 1$ be given. A point X will be called an r-elliptic point whenever

$$r\lambda(X) < \sup_{t\in\mathbb{R}} |Q(t,X)|. \tag{3.5.8}$$

In that case, we find $t_0 \in \mathbb{R}$ such that $r\lambda(X) < |Q(t_0,X)| \leq \lambda(X)$ and with $S \in \mathbb{R}^{2n}, |S| = 1, \rho \geq 0$,

$$Q(t_0, X + \rho\lambda(X)^{1/2}S) = Q(t_0,X) + Q'_X(t_0,X)S\rho\lambda(X)^{1/2} + O(1)\rho^2\lambda(X)$$

and thus using (3.5.7),

$$|Q(t_0, X + \rho\lambda(X)^{1/2}S)| \geq r\lambda(X) - \rho\lambda(X) - \frac{1}{2}\gamma_2\rho^2\lambda(X) \geq \frac{r}{2}\lambda(X)$$

if

$$r/2 > \rho + \rho^2\gamma_2/2. \tag{3.5.9}$$

As a consequence, there exists $\rho_r > 0$ (depending only on r and γ_2) such that, for $Y \in U_{X,\rho_r}$ (the $\lambda(X)^{-1}\Gamma$ ball with center X and radius ρ_r, see the notation (2.2.8)),

$$|Q(t_0,Y)| \geq \frac{r}{2}\lambda(X).$$

Assuming that $Q(t_0,X) > 0$ (resp. < 0), we get that $Q(t_0,Y) > 0$ (resp. < 0) on U_{X,ρ_r} and the condition (3.5.2) shows that, for all $t \in \mathbb{R}$, we have $Q(t,Y) \geq 0$ on

U_{X,ρ_r} (resp. $Q(t, Y) \leq 0$ on U_{X,ρ_r}). We say that $X \in \mathbb{R}^{2n}$ is an r-non-degenerate point if

$$\sup_{t \in \mathbb{R}} |Q(t, X)| \leq r\lambda(X) \leq r^{1/2}\lambda(X) < \sup_{t \in \mathbb{R}} |Q'_X(t, X)|^2. \qquad (3.5.10)$$

In particular, an r-non-degenerate point is not an r-elliptic point and the same discussion as above gives that there exists $t_0 \in \mathbb{R}$ such that, for all $Y \in U_{X,\rho}$,

$$|Q(t_0, Y)| \leq 2r\lambda(X), \quad |Q'_X(t_0, Y)|^2 \geq \frac{r^{1/2}}{2}\lambda(X),$$

provided

$$4\rho^4\gamma_2^4 \leq r \quad \text{and} \quad \rho + \gamma_2\rho^2/2 \leq r. \qquad (3.5.11)$$

For r given in $(0, 1)$, it is possible to find $\rho_r > 0$ depending only on r and γ_2, satisfying the latter conditions as well as (3.5.9). Defining now, for $S \in \mathbb{R}^{2n}$,

$$F(S) = Q(t_0, X + \lambda(X)^{1/2}S)\lambda(X)^{-1}$$

we get that $|F(0)| \leq r$, $|F'(0)|^2 \geq r^{1/2}/2$, $\|F''\|_{L^\infty} \leq \gamma_2$ and thus

$$|(1+\gamma_2)^{-1}F'(0)|^2 \geq (1+\gamma_2)^{-1}2^{-1}r^{-1/2}|(1+\gamma_2)^{-1}F(0)|, \quad \|(1+\gamma_2)^{-1}F''\|_{L^\infty} \leq 1$$

so that we can apply Lemma 4.9.14 to the function $S \mapsto (1+\gamma_2)^{-1}F(S)$, provided

$$r < (1+\gamma_2)^{-2}2^{-14}. \qquad (3.5.12)$$

As a result, using also Remark 4.9.15, we find that for $|S| \leq r^{1/4}2^{-1/2}2^{-5}(1+\gamma_2)^{-1}$,

$$F(S) = e(S)(S_1 + \alpha(S')), \quad e \geq r^{1/4}2^{-1/2}2^{-5}(1+\gamma_2)^{-1},$$

with $e \in C_b^\infty(\mathbb{R}^{2n}), \alpha \in C_b^\infty(\mathbb{R}^{2n-1})$, with semi-norms controlled by those of F. Finally we choose $r \in (0, 1)$ such that (3.5.12) holds, then we choose $\rho > 0$ such that (3.5.11) and (3.5.9) are verified. We obtain readily that for $Y \in U_{X,\rho}$

$$Q(t_0, Y) = e_0(Y)\overbrace{\left(Y_1 - X_1 + \beta(Y' - X')\right)}^{b(Y)}\lambda(X)^{1/2}, \quad e_0(Y) \geq c_0 > 0, \qquad (3.5.13)$$

$$|e_0^{(k)}(Y)| + |\beta^{(k)}(Y')|\lambda(X)^{-1/2} \leq C_k'\lambda(X)^{-k/2}, \qquad (3.5.14)$$

where c_0, C_k' depend only on the (C_k) of (3.5.4). Now if $U_{X,\rho} \ni Y \mapsto Q(t_0, Y)$ does not vanish, e.g., stays positive, we know from condition (3.5.2) that for all $t \in \mathbb{R}$, $Q(t, Y)$ stays non-negative on $U_{X,\rho}$. On the other hand, if X is an r-non-degenerate point and if $Y \mapsto Q(t_0, Y)$ vanishes on $U_{X,\rho}$, it does on the smooth hypersurface $\equiv b(Y) = 0$ (with $db(Y) \neq 0$) and we have, for $Y \in U_{X,\rho}$,

$$\pm Q(t_0, Y) > 0 \iff \pm b(Y) > 0.$$

As a consequence, the condition (3.5.2) implies that

$$\text{for all } t \in \mathbb{R}, \quad \pm b(Y) > 0 \Longrightarrow \pm Q(t, Y) \geq 0,$$

and this gives that

$$Q(t, Y) = a(t, Y)b(Y), \quad 0 \leq a \in S(1, \lambda^{-1}\Gamma), \quad b \in S(\lambda, \lambda^{-1}\Gamma) \text{ on } U_{X,\rho}, \quad (3.5.15)$$

since from the discussion above we have that $b(Y) = Y_1 - B(Y')$ and thus since $Q(t, B(Y'), Y') \equiv 0$, Taylor expansion from the point $(B(Y'), Y')$ gives the factorization (3.5.15).

If X is neither r-elliptic, nor r-non-degenerate, we shall call it negligible and it satisfies $\sup_{t \in \mathbb{R}} |Q(t, X)| \leq r\lambda(X)$, $\sup_{t \in \mathbb{R}} |Q'_X(t, X)|^2 \leq r^{1/2}\lambda(X)$ so that if $r \leq 1/4$ as we may also assume, we get $\lambda(X) \leq 1 + r^{1/2}\lambda(X)$ and thus

$$\lambda(X) \leq 2. \tag{3.5.16}$$

Localization

We consider now $v \in C_c^\infty(\mathbb{R}_t; L^2(\mathbb{R}_x^n)), \operatorname{supp} v \subset [-T, T]$ and $(\varphi_Z(x, \xi))_{Z \in \mathbb{R}^{2n}}$ a partition of unity in $\mathbb{R}^{2n}$ related to the admissible metric $\lambda^{-1}\Gamma$. According to Lemma 2.5.3, we have

$$\|v(t)\|_{L^2(\mathbb{R}^n)}^2 \sim \int_{\mathbb{R}^{2n}} \|\varphi_Z^w v(t)\|_{L^2(\mathbb{R}^n)}^2 \underbrace{\lambda(Z)^{-n} dZ}_{\tilde{d}Z}.$$

Theorem 2.2.7 allows us to choose the support of φ_Y in $U_{Y,r}$ with $r > 0$ as small as we wish, so that we can assume that r is small enough for the properties of the previous section to be satisfied. Defining

$$\mathcal{L} = D_t + iQ(t, X)^w, \quad Q(t) = Q(t, X)^w, \tag{3.5.17}$$

we claim that it is enough to prove

$$\|\varphi_Z^w v(t)\|_{L^2 \mathbb{R}^n)} \leq C_0 \int_{\mathbb{R}} \|\mathcal{L}\varphi_Z^w v(t)\|_{L^2(\mathbb{R}^n)} dt + C_0' \int_{\mathbb{R}} \|\psi_Z^w v(t)\|_{L^2(\mathbb{R}^n)} dt, \quad (3.5.18)$$

where $(\psi_Z)_{Z \in \mathbb{R}^{2n}}$ is a uniformly confined family of symbols. In fact, assuming first $C_0' = 0$, the latter inequality along with Lemma 2.5.3 implies that

$$\|v(t)\|_{L^2(\mathbb{R}^n)}^2 \leq C_1 \int_{\mathbb{R}^{2n}} \|\varphi_Z^w v(t)\|_{L^2(\mathbb{R}^n)}^2 \tilde{d}Z$$

$$\leq C_1 C_0^2 \int_{\mathbb{R}^{2n}} \left(\int_{\mathbb{R}} \|\mathcal{L}\varphi_Z^w v(t)\|_{L^2(\mathbb{R}^n)} dt \right)^2 \tilde{d}Z$$

$$\leq C_1 C_0^2 \int_{\mathbb{R}^{2n}} \left(\int_{\mathbb{R}} (\|[Q(t), \varphi_Z^w]v(t)\|_{L^2(\mathbb{R}^n)} + \|\varphi_Z^w \mathcal{L}v(t)\|_{L^2(\mathbb{R}^n)}) dt \right)^2 \tilde{d}Z$$

$$\leq C_1 C_0^2 \int_{\mathbb{R}^{2n}} \left(\int_{\mathbb{R}} \| [Q(t), \varphi_Z^w] v(t) \|_{L^2(\mathbb{R}^n)}^2 dt + \int_{\mathbb{R}} \| \varphi_Z^w \mathcal{L} v(t) \|_{L^2(\mathbb{R}^n)}^2 dt \right) 2T \widetilde{d}Z$$

$$\leq C_2 T \int_{\mathbb{R}} \| v(t) \|_{L^2(\mathbb{R}^n)}^2 dt + C_2 T \int_{\mathbb{R}} \| \mathcal{L} v(t) \|_{L^2(\mathbb{R}^n)}^2 dt,$$

since $[Q(t), \varphi_Z^w]$ is L^2 bounded from the calculus with the metric $\lambda^{-1}\Gamma$ (note that $Q(t, \cdot) \in S(\lambda, \lambda^{-1}\Gamma), \varphi_Z \in S(1, \lambda^{-1}\Gamma))$ but also

$$\int_{\mathbb{R}^{2n}} [Q(t), \varphi_Z^w]^* [Q(t), \varphi_Z^w] \widetilde{d}Z$$

is L^2 bounded from Cotlar's lemma and the confinement estimates: the family $\big(Q(t, \cdot) \sharp \varphi_Z - \varphi_Z \sharp Q(t, \cdot)\big)_{Z \in \mathbb{R}^{2n}}$ is a uniformly confined family of symbols for the metric $\lambda^{-1}\Gamma$ (see Definition 2.3.14 and Proposition 2.3.16). As a consequence we obtain from (3.5.18), taking now into account the last term in the factor of C_0',

$$\big(1 - 2(C_2 T^2 + C_3 C_0'^2 T)\big) \sup_{t \in \mathbb{R}} \| v(t) \|_{L^2(\mathbb{R}^n)}^2 \leq C_2 T \| \mathcal{L} v \|_{L^2(\mathbb{R}^{n+1})}^2,$$

which implies (3.5.3) for T small enough. We note also that

$$(Q(t, \cdot) \sharp \varphi_Z)(X) = Q(t, X) \varphi_Z(X) + r_Z(X), \quad r_Z \in S(1, \lambda(Z)^{-1}\Gamma),$$

but also $\int \bar{r}_Z^w r_Z^w \widetilde{d}Z$ is L^2 bounded since the family $(r_Z)_{Z \in \mathbb{R}^{2n}}$ is a uniformly confined family of symbols. Note also that, if for each $Z \in \mathbb{R}^{2n}$, $X \in \operatorname{supp} \varphi_Z$,

$$Q(t, X) = Q(t, X, Z), \quad Q(t, \cdot, Z) \in S(\lambda(Z), \lambda(Z)^{-1}\Gamma),$$

then

$$Q(t, \cdot, Z) \sharp \varphi_Z = Q(t, \cdot) \sharp \varphi_Z + s_Z,$$

where $(s_Z)_{Z \in \mathbb{R}^{2n}}$ is a uniformly confined family of symbols. Eventually, it will be enough to prove

$$\| v(t) \|_{L^2(\mathbb{R}^n)} \leq C \int_{\mathbb{R}} \| D_t v + i Q(t, \cdot, Z)^w v \|_{L^2(\mathbb{R}^n)} dt + C \int_{\mathbb{R}} \| v(t) \|_{L^2(\mathbb{R}^n)}. \quad (3.5.19)$$

A priori estimates

We shall of course use the classification of points above: when Z is an r-elliptic point, we know that $Q(t, X, Z)$ is non-negative (or non-positive) for all t, X. Lemma 4.3.21 along with Gårding's inequality (Theorem 2.5.4) provides the easy answer (see also the short discussion with the example (3.1.1)). When Z is a negligible point, we know that the symbol $X \mapsto Q(t, X, Z)$ is bounded as well as all its derivatives since $\lambda(Z) \leq 2$ and thus Lemma 4.3.21 is also enough to handle that case. If Z is an r-non-degenerate point, we have

$$Q(t, X, Z) = a(t, X)b(X), \quad a(t, X) \geq 0,$$

$$a(t, X), \ b(X)\lambda(Z)^{-1} \in S(1, \lambda(Z)^{-1}\Gamma),$$

$$\lambda(Z)^{-1/2}b(X) = X_1 + \beta(X_2, \ldots, X_{2n}), \quad \beta \in S(\lambda(Z)^{1/2}, \lambda(Z)^{-1}\Gamma).$$

To simplify notation, we consider Z as fixed and note $\lambda = \lambda(Z)$. Using the "sharp Egorov principle" as given in Theorem 4.7.8, we are reduced to proving for $v \in C_c^\infty(\mathbb{R}; L^2(\mathbb{R}^n))$,

$$\sup_{t \in \mathbb{R}} \|v(t)\|_{L^2(\mathbb{R}^n)} \leq C \int_{\mathbb{R}} \|D_t v + i\lambda^{1/2}(X_1 a_0(t, X))^w v\|_{L^2(\mathbb{R}^n)} dt,$$

with a non-negative smooth a_0 defined on $\mathbb{R}_t \times \mathbb{R}_X^{2n}$ such that $|\partial_X^k a_0| \leq \gamma_k \lambda^{-k/2}$, $\operatorname{supp} a_0 \subset \{X \in \mathbb{R}^{2n}, |X| \leq r\lambda^{1/2}\}$ and $\lambda \geq 1$. We shall use now the metric

$$g = \frac{|dX_1|^2}{(\lambda^{-1/2} + |X_1|)^2} + \sum_{2 \leq j \leq 2n} \frac{|dX_j|^2}{\lambda}. \tag{3.5.20}$$

The metric g satisfies the uncertainty principle (see (2.2.12)) and its Planck function is

$$\mu(X) = \min(\lambda, (1 + \lambda^{1/2}|X_1|)) \geq 1. \tag{3.5.21}$$

Moreover g is slowly varying since $|X_1 - Y_1| \leq \frac{\lambda^{-1/2} + |X_1|}{2}$ implies

$$\frac{\lambda^{-1/2} + |X_1|}{\lambda^{-1/2} + |Y_1|} \leq 1 + \frac{|X_1 - Y_1|}{\lambda^{-1/2} + |Y_1|} \leq 1 + \frac{1}{2}\frac{\lambda^{-1/2} + |X_1|}{\lambda^{-1/2} + |Y_1|} \implies \frac{\lambda^{-1/2} + |X_1|}{\lambda^{-1/2} + |Y_1|} \leq 2$$

and the first inequality also gives the temperance. We have for $\mathcal{Y} \in C^\infty(\mathbb{R}; [0, 1])$, $\operatorname{supp} \mathcal{Y} \subset [1/2, +\infty)$, $\mathcal{Y} = 1$ on $[1, +\infty)$, nondecreasing,

$$\lambda^{1/2} X_1 a_0(t, X) \in S(\mu, g), \ \mathcal{Y}(\lambda^{1/2} X_1) \in S(1, g)$$

$$\implies [\mathcal{Y}(\lambda^{1/2} X_1)^w, \lambda^{1/2} X_1 a_0(t, X)^w] \in \operatorname{op}(S(1, g)),$$

since for $k \geq 1$, $\mathcal{Y}^{(k)}$ is supported in $[1/2, 1]$ so that

$$|\mathcal{Y}^{(k)}(\lambda^{1/2} X_1)|\lambda^{k/2} \leq \|\mathcal{Y}^{(k)}\|_{L^\infty}|X_1|^{-k/2}\mathbf{1}\{1/2 \leq \lambda^{1/2}|X_1| \leq 1\},$$

and $|\lambda^{1/2} X_1 a_0(t, X)| \leq C\mu(X)$, $\lambda^{-1}|dX|^2 \leq g$ on the support of a_0. From the properties of $\mathcal{Y}$ above, we have with $\mathcal{Y}_\lambda(X_1) = \mathcal{Y}(\lambda^{1/2} X_1)$,

$$1 = \mathcal{Y}_\lambda^2 + \breve{\mathcal{Y}}_\lambda^2 + \mathcal{I}_\lambda^2, \quad \mathcal{I}_\lambda(X_1) = \mathcal{I}(\lambda^{1/2} X_1), \quad \mathcal{I} \in C_c^\infty(\mathbb{R}).$$

We have $\|v(t)\|_{L^2(\mathbb{R}^n)}^2 = \|\mathcal{Y}_\lambda^w v(t)\|_{L^2(\mathbb{R}^n)}^2 + \|\breve{\mathcal{Y}}_\lambda^w v(t)_{L^2(\mathbb{R}^n)}\|^2 + \|\mathcal{I}_\lambda^w v(t)\|_{L^2(\mathbb{R}^n)}^2$, and

$$\mathcal{Y}_\lambda^w \left(D_t + i\lambda^{1/2}(X_1 a_0(t, X))^w\right) = \left(D_t + i\lambda^{1/2}(X_1 a_0(t, X))^w\right)\mathcal{Y}_\lambda^w + r_0^w, \ r_0 \in S(1, g),$$

so that with $\widetilde{\mathcal{Y}} \in C^\infty(\mathbb{R}; [0,1])$, $\mathrm{supp}\, \widetilde{\mathcal{Y}} \subset [1/4, +\infty)$, $\widetilde{\mathcal{Y}} = 1$ on $[1/2, +\infty)$,

$$\mathcal{Y}_\lambda^w\left(D_t + i\lambda^{1/2}(X_1 a_0(t,X))^w\right) = \left(D_t + i\lambda^{1/2}(\widetilde{\mathcal{Y}}_\lambda(X_1)X_1 a_0(t,X))^w\right)\mathcal{Y}_\lambda^w + s_0^w,$$

with $s_0 \in S(1,g)$. The symbol $\lambda^{1/2}\widetilde{\mathcal{Y}}_\lambda(X_1)X_1 a_0(t,X)$ is non-negative and belongs to $S(\mu, g)$. We can then apply again Lemma 4.3.21 along with Gårding's inequality to get the a priori estimate,

$$\|\mathcal{Y}_\lambda^w v(t)\|$$
$$\le C \int_\mathbb{R} \|\mathcal{Y}_\lambda^w(D_t v + i\lambda^{1/2}(X_1 a_0(t,X))^w v(t))\|_{L^2(\mathbb{R}^n)} dt + C \int_\mathbb{R} \|v(t)\|_{L^2(\mathbb{R}^n)} dt$$

as well as (using the nonpositivity of $\lambda^{1/2}\check{\mathcal{Y}}_\lambda(-X_1)X_1 a_0(t,X)$),

$$\|\check{\mathcal{Y}}_\lambda^{\ w} v(t)\|$$
$$\le C \int_\mathbb{R} \|\check{\mathcal{Y}}_\lambda^{\ w}(D_t v + i\lambda^{1/2}(X_1 a_0(t,X))^w v(t))\|_{L^2(\mathbb{R}^n)} dt + C \int_\mathbb{R} \|v(t)\|_{L^2(\mathbb{R}^n)} dt.$$

These estimates and the similar one with $\mathcal{I}_\lambda$, even easier to get since

$$\mathcal{I}_\lambda(X_1)\lambda^{1/2}X_1 a_0(t,X_1) \in S(1,g),$$

give the sought estimate (3.5.19). The proof of Theorem 3.5.2 is complete.

Remark 3.5.4. We have avoided using a Hilbertian lemma such as the Nirenberg-Treves estimate (Lemma 3.1 of [114], Theorem 26.8.1 in [74]) and we chose a further microlocalization with the metric g given by (3.5.20). Although the Hilbertian lemma referred to above is of great interest, it does not use the complete information at hand: nonetheless we know that the imaginary part can be factorized as $a_0(t,X)b_1(X)$ with a_0 non-negative of order 0 (i.e., in $S(1, \lambda^{-1}\Gamma)$) and b_1 of order 1 (i.e., in $S(\lambda, \lambda^{-1}\Gamma)$), but we know also that $db \ne 0$ at $b = 0$, so that the change-of-sign takes place on a smooth hypersurface. A semi-classical canonical transformation can thus map b to a coordinate and we are left after the quantization of that canonical transformation with a degenerate Cauchy-Riemann operator

$$D_t + iD_{x_1} a_0(t,x,D_x), \qquad a_0 \ge 0, \quad \text{order } 0.$$

Also, we were motivated by the fact that this procedure highlights an interesting example of second microlocalization with respect to the hypersurface $b = 0$ (see [20], [96]).

3.5.2   The two-dimensional case, the oblique derivative problem

Let P be a properly supported pseudodiffferential operator of principal type with a homogeneous principal symbol satisfying condition (Ψ). The homogeneity in two dimensions leads to a great simplification since the principal symbol of P^* can be reduced to

$$\tau + iq(t,x)\xi, \quad (t,x,\tau,\xi) \in \mathbb{R}^4, \quad \text{near } \tau = 0, \xi = 1, \qquad (3.5.22)$$

$$q(t,x) > 0, s > t \Longrightarrow q(s,x) \geq 0. \qquad (3.5.23)$$

The oblique derivative problem is a classical non-elliptic boundary value problem: consider for instance an open subset Ω of $\mathbb{R}^n$ with smooth boundary $\partial\Omega$ and X a real non-vanishing smooth vector field. The oblique derivative problem is given by

$$\begin{cases} \Delta u = 0 & \text{in } \Omega, \\ Xu = f & \text{on } \partial\Omega. \end{cases} \qquad (3.5.24)$$

We write $X = \alpha\frac{\partial}{\partial\nu} + T$, where α is a smooth real-valued function, $\frac{\partial}{\partial\nu}$ is the interior unit normal and T is a real vector field tangential to the boundary. If w is the restriction of u on the boundary, then $u = Gw$ is the Poisson integral of w and the oblique derivative problem reduces to $XGw = f$, which is a pseudodiffferential equation on $\partial\Omega$. The fact that X is non-vanishing is equivalent to the principal-type assumption for XG and the ellipticity is equivalent to α non-vanishing, as for the Neumann problem ($\alpha = 1, T = 0$). Now if $\alpha(m_0) = 0$ then $T(m_0) \neq 0$, the problem is no longer elliptic, but still principal type.

After the works of Y.V. Egorov and V.A. Kondrat'ev [42], A. Melin and J. Sjöstrand [106] gave a construction of a right parametrix for the oblique-derivative problem with a transversality assumption: with the present notation, they assumed that the regions $\{\pm\alpha > 0\}$ are separated by a smooth manifold to which T is transverse and points into $\{\alpha \leq 0\}$. This was a very strong geometrical assumption which can be naturally weakened down to condition (Ψ): The condition (Ψ) for the operator XG can be expressed as follows: if Φ_T^t is the flow of the vector field T,

$$\alpha(m) < 0, t > 0 \Longrightarrow \alpha(\Phi_T^t(m)) \leq 0. \qquad (3.5.25)$$

It turns out that the adjoint operator of XG can be reduced locally near a point where α vanishes to

$$D_t + i\beta(t,x)\omega(t,x,\xi), \quad 0 \leq \omega \in S_{1,0}^1, \quad \beta \in C^\infty, \qquad (3.5.26)$$

$$\beta(t,x) > 0, s > t \Longrightarrow \beta(s,x) \geq 0. \qquad (3.5.27)$$

The local solvability of 2D pseudo-differential operators satisfying condition (Ψ) and the local solvability of the oblique-derivative problem under condition (Ψ) were proven by the author in [89], [91].

Theorem 3.5.5. *Let Ω be an open subset of $\mathbb{R}^2$, $m \in \mathbb{R}$, $x_0 \in \Omega$ and $P \in \Psi_{ps}^m(\Omega)$ a pseudo-differential operator of principal type with a principal symbol satisfying condition (Ψ) in a neighborhood of x_0. Then P is locally solvable at x_0 with loss of one derivative.*

Theorem 3.5.6. *Let Ω be an open subset of $\mathbb{R}^n$ with a smooth boundary $\partial\Omega$ and P a second-order uniformly elliptic operator with real-principal symbol. Let T be a smooth real vector field on $\partial\Omega$ and $\alpha \in C^\infty(\partial\Omega, ; \mathbb{R})$ such that (3.5.25) holds. Then with $\frac{\partial}{\partial\nu}$ the unit interior normal, $X = \alpha\frac{\partial}{\partial\nu} + T$, G the Dirichlet kernel (PGw = 0 on Ω, $(Gw)|_{\partial\Omega} = w$), the pseudo-differential operator XG is locally solvable at every point on $\partial\Omega$ with loss of one derivative.*

Proof of both theorems. The oblique derivative problem and the 2D problem can be reduced to the study of the semi-classical operator,

$$Q(t,x,\xi) = \alpha(t,x)\Omega(t,x,\xi), \quad \Omega \geq 0, \quad |\partial_X^k\Omega| \leq C_k\Lambda^{1-\frac{k}{2}}, \quad |\partial_x^k\alpha| \leq C_k\Lambda^{-\frac{k}{2}}$$

with $\Lambda = h^{-1}$ and the condition

$$\alpha(t,x) > 0, s > t \Longrightarrow \alpha(s,x) \geq 0. \tag{3.5.28}$$

We define for $T > 0$ given,

$$\theta(x) = \sup\{t \in [-T,T], \alpha(t,x) < 0\}, \tag{3.5.29}$$

with $\theta(x) = -T$ if $\alpha(t,x) \geq 0$ for all $t \in [-T,T]$. For $t \in [-T,T]$, we have

$$(t - \theta(x))\alpha(t,x) \geq 0$$

since it is true when $\theta(x) = -T$ and if $-T < \theta(x) < t \leq T$ we have, $\alpha(t,x) \geq 0$, whereas for $-T \leq t < \theta(x)$, there exists $s \in (t, \theta(x)]$ with $\alpha(s,x) < 0$ and (3.5.28) gives $\alpha(t,x) \leq 0$. We calculate with $\sigma(t,x) = \text{sign}(t - \theta(x))$ (the operator $\sigma(t)$ of multiplication by the L^∞ function $\sigma(t,x)$ is L^2 bounded), using (4) in Proposition 2.4.3 since $t \mapsto \sigma(t,x)$ is non-decreasing, and replacing $Q(t)$ by $\alpha\Omega^w$(an L^2 bounded round-off),

$$2\,\text{Re}\langle D_t v + iQ(t)v, i\sigma(t)v + i\frac{1}{2}H(t - t_0)v\rangle$$

$$\geq 2\,\text{Re}\langle|\alpha|\Omega^w v, v\rangle + \frac{1}{2}\|v(t_0)\|^2 + \text{Re}\langle H(t - t_0)\alpha\Omega^w v, v\rangle. \tag{3.5.30}$$

We check

$$2\,\text{Re}\,|\alpha|\Omega^w$$

$$= 2\,\text{Re}(\alpha_+^{1/2}[\alpha_+^{1/2}, \Omega^w]) + 2\,\text{Re}(\alpha_-^{1/2}[\alpha_-^{1/2}, \Omega^w]) + 2\alpha_+^{1/2}\Omega^w\alpha_+^{1/2} + 2\alpha_-^{1/2}\Omega^w\alpha_-^{1/2}$$

$$= [\alpha_+^{1/2}, [\alpha_+^{1/2}, \Omega^w]] + [\alpha_-^{1/2}, [\alpha_-^{1/2}, \Omega^w]]$$

$$+ 2\alpha_+^{1/2}(\Omega^w + \gamma_0)\alpha_+^{1/2} + 2\alpha_-^{1/2}(\Omega^w + \gamma_0)\alpha_-^{1/2} - 2\gamma_0|\alpha|,$$

so that with $\nu_0 = \frac{1}{2}H(t - t_0)$

$$2\,\text{Re}\,|\alpha|\Omega^w + H(t - t_0)\,\text{Re}\,\alpha\Omega^w$$

$$= [\alpha_+^{1/2}, [\alpha_+^{1/2}, \Omega^w]](1 + \nu_0) + [\alpha_-^{1/2}, [\alpha_-^{1/2}, \Omega^w]](1 - \nu_0)$$

$$+ 2\alpha_+^{1/2}(\Omega^w + \gamma_0 + \nu_0)\alpha_+^{1/2} + 2\alpha_-^{1/2}(\Omega^w + \gamma_0 - \nu_0)\alpha_-^{1/2} - 2\gamma_0|\alpha|.$$

Using Gårding's inequality, we can choose γ_0 depending only on the semi-norms of the non-negative Ω such that $\Omega^w + \gamma_0 - \nu_0 \geq 0$. To handle the double brackets, we shall use the already mentioned following lemma.

Lemma 3.5.7 (Nirenberg-Treves estimate, Lemma 26.8.2 in [74]). *Let A, B be bounded operators on a Hilbert space with B selfadjoint. Then, with operator norms, we have*

$$\| [B_+^{1/2}, [B_+^{1/2}, A]] \| \leq \frac{10}{3} \|A\|^{1/4} \|[B, A]\|^{1/2} \|[B, [B, A]]\|^{1/4}. \tag{3.5.31}$$

That lemma and the symbolic calculus in $S(\Lambda^m, \Lambda^{-1}\Gamma)$ imply that

$$[\alpha_\pm^{1/2}, [\alpha_\pm^{1/2}, \Omega^w]]$$

are L^2-bounded. We get thus that

$$2\operatorname{Re}\langle D_t v + iQ(t)v, i\sigma(t)v + i\frac{1}{2}H(t-t_0)v \rangle \geq \frac{1}{2}\sup_t \|v(t)\|_{L^2(\mathbb{R}^n)}^2 - C\int_{\mathbb{R}} \|v(t)\|_{L^2(\mathbb{R}^n)}^2 dt,$$

entailing the result

$$C_1 \int_{\mathbb{R}} \|D_t v + iQ(t)v\|_{L^2(\mathbb{R}^n)} dt \geq \sup_t \|v(t)\|_{L^2(\mathbb{R}^n)},$$

for $v \in C_{[-T,T]}^1(\mathbb{R}; L^2(\mathbb{R}^n))$ and T small enough. $\qquad\square$

Both results are obtained with the optimal loss of one derivative and were generalized by Hörmander in Section 8, *Lagrangean change of sign* of [77] (we use here the slightly modified version of [78]).

Theorem 3.5.8. *Let $q : \mathbb{R} \times \mathbb{R}^n \times \mathbb{R}^n \times (0,1] \to \mathbb{R}$ be a function satisfying (3.5.1) and such that*

$$q(t, x, \xi, h) > 0, s > t \Longrightarrow q(s, x, \eta, h) \geq 0. \tag{3.5.32}$$

Then there exist positive constants T, C, h_0 depending only on the $(C_{\alpha\beta})$ such that for all $v \in \mathscr{S}(\mathbb{R}^{n+1})$ with $v(t, x) = 0$ for $|t| \geq T$, $0 < h \leq h_0$,

$$hT^{-1}\|v\|_{L^2(\mathbb{R}^{n+1})} \leq C\|hD_t v + iq(t, x, \xi, h)^w v\|_{L^2(\mathbb{R}^{n+1})}. \tag{3.5.33}$$

The assumption (3.5.32) is of course stronger than condition ($\overline{\Psi}$) for $D_t + iq^w$, but, as noted above, is fulfilled for operators satisfying condition ($\overline{\Psi}$) in two dimensions as well as for the oblique derivative problem.

3.5.3   Transversal sign changes

Solvability results

In this section, we prove a version of the following two theorems.

Theorem 3.5.9. *Let* $q(t, x, \xi) \in C^0\big([-1, 1], C^\infty(\mathbb{R}^n_x \times \mathbb{R}^n_\xi)\big)$ *such that*

$$q(t, x, \xi) > 0, s > t \Longrightarrow q(s, x, \xi) \geq 0, \qquad (3.5.34)$$

for $s, t \in [-1, 1], (x, \xi) \in \mathbb{R}^n \times \mathbb{R}^n$ *and such that for all multi-indices* α, β,

$$\sup_{|t|\leq 1, \ (x,\xi)\in\mathbb{R}^{2n}} |(\partial_x^\alpha \partial_\xi^\beta q)(t, x, \xi)|(1 + |\xi|)^{-1+|\beta|} = \gamma_{\alpha\beta}(q) < +\infty. \qquad (3.5.35)$$

We assume also that there exists a constant D_0 *such that, for* $|\xi| \geq 1$,

$$|\xi|^{-1}\left|\frac{\partial q}{\partial x}(t, x, \xi)\right|^2 + |\xi|\left|\frac{\partial q}{\partial \xi}(t, x, \xi)\right|^2 \leq D_0 \frac{\partial q}{\partial t}(t, x, \xi), \text{ when } q(t, x, \xi) = 0. \ (3.5.36)$$

Then, there exist positive constants ρ, C *depending only on* n *and on a finite number of* $\gamma_{\alpha\beta}(q)$ *in* (3.5.35) *such that, for* $v(t, x) \in C_c^\infty\big((-\rho, \rho), \mathscr{S}(\mathbb{R}^n_x)\big)$,

$$\sup_t \|v(t)\|_{L^2(\mathbb{R}^n)} \leq C \int \|D_t v + iq(t, x, \xi)^w v\|_{L^2(\mathbb{R}^n)} dt. \qquad (3.5.37)$$

Theorem 3.5.10. *Let* Ω *be an open subset of* $\mathbb{R}^n$ *and let* $P \in \Psi_{ps}^m(\Omega)$ *of principal type with a principal symbol* p_m *satisfying condition* (Ψ). *We assume that there exists a constant* D *such that, for* (x, ξ) *in the cosphere bundle* $S^*(\Omega)$,

$$p(x, \xi) = 0 \text{ and } (dp \wedge d\bar{p})(x, \xi) \neq 0$$
$$\Longrightarrow \|(dp \wedge d\bar{p})(x, \xi)\|^2 \leq D|\{\bar{p}, p\}(x, \xi)|. \qquad (3.5.38)$$

Then, the operator P *is locally solvable at any point of* Ω *with loss of one derivative.*

Condition (3.5.38) is concerned only with the non-singular zero set

$$\mathcal{Z} = \{(x, \xi) \in S^*(\Omega), p = 0, dp \wedge d\bar{p} \neq 0\}, \qquad (3.5.39)$$

and in particular, if $\mathcal{Z}$ is empty, we get that condition (Ψ) implies local solvability with loss of one derivative. Note also that (3.5.38) is invariant by multiplication by an elliptic factor, and thus is easily shown to be equivalent to (3.5.36) for $\tau - iq$ in these coordinates. Since condition (Ψ) for p implies

$$i\{\bar{p}, p\} \geq 0 \quad \text{at} \quad p = 0,$$

one could replace the last term in the right-hand side of (3.5.38) by $\text{Im}\{p, \bar{p}\}D$.

Remark 3.5.11. The existence of a Lipschitz-continuous function, homogeneous of degree 0, $\theta(x, \xi)$, so that

$$(t - \theta(x, \xi)) \, q(t, x, \xi) \geq 0 \tag{3.5.40}$$

implies (3.5.36): in fact differentiating the identity $0 = q(\theta(x, \xi), x, \xi)$ we get

$$0 = q'_x(\theta(x, \xi), x, \xi) + q'_t(\theta(x, \xi), x, \xi)\theta'_x = q'_\xi(\theta(x, \xi), x, \xi) + q'_t(\theta(x, \xi), x, \xi)\theta'_\xi$$

and thus

$$|q'_x(\theta(x, \xi), x, \xi)|^2 |\xi|^{-1} + |q'_\xi(\theta(x, \xi), x, \xi)|^2 |\xi| \leq 2|q'_t(\theta(x, \xi), x, \xi)|^2 \|\theta'\|^2_{L^\infty(S^*(\Omega))} |\xi|^{-1}$$

and this gives (3.5.36) at $t = \theta(x, \xi)$ since q'_t is bounded on $S^*(\Omega)$. On the open set $\{t \neq \theta(x, \xi)\}$, the function q is of constant sign and thus its gradient vanishes at $q = 0$.

The theorems above are proven in a paper of the author ([93]) with a rather complicated method. We have chosen to display here a simpler proof leading to a weaker statement where some limitations occur on the size of the constants D_0, D in (3.5.36), (3.5.38). Nevertheless, the proof given here captures the main ideas of [93], and relies on the Wick calculus.

Energy estimates

Let $q_\Lambda(t, X, \Lambda)$ be a smooth function on $\mathbb{R}_t \times \mathbb{R}^{2n}_X$, defined for $\Lambda \geq 1$, supported in $\mathbf{B} = \{|t| \leq 1\} \times \{|X| \leq \Lambda^{1/2}\}$, so that, for each k,

$$\sup_{\substack{t \in \mathbb{R}, X \in \mathbb{R}^{2n} \\ \Lambda \geq 1}} |D^k_X q_\Lambda(t, X)| \Lambda^{-1 + \frac{k}{2}} < \infty. \tag{3.5.41}$$

We omit below the subscript Λ on q_Λ. We assume that (3.5.34) holds. We define

$$\theta(X) = \begin{cases} \inf\{t \in (-1, +1), q(t, X) > 0\}, & \text{if this set is not empty,} \\ +1, & \text{otherwise.} \end{cases} \tag{3.5.42}$$

The function θ is bounded measurable (actually, θ is upper semi-continuous). We set

$$s(t, X) = \begin{cases} +1, & \text{if } t > \theta(X), \\ 0, & \text{if } t = \theta(X), \\ -1, & \text{if } t < \theta(X). \end{cases} \tag{3.5.43}$$

The function s is bounded measurable and (3.5.34) implies that

$$s(t, X) q(t, X) = |q(t, X)|. \tag{3.5.44}$$

Moreover, the distribution derivative $\dfrac{\partial s}{\partial t}$ is a positive measure satisfying

$$\langle \frac{\partial s}{\partial t}, \Psi(t, X) \rangle_{S'(\mathbb{R}^{2n+1}), S(\mathbb{R}^{2n+1})} = 2 \int_{\mathbb{R}^{2n}} \Psi(\theta(X), X) dX. \tag{3.5.45}$$

We need to examine a couple of properties for the distribution

$$T = \sum_{1 \leq j \leq 2n} \frac{\partial q}{\partial X_j} \frac{\partial s}{\partial X_j} = \frac{\partial q}{\partial X} \cdot \frac{\partial s}{\partial X} \quad . \tag{3.5.46}$$

First of all, we claim that

$$\text{supp } T \subset \{(t, X) \in \mathbf{B}, \quad q(t, X) = 0\}. \tag{3.5.47}$$

In fact, from (3.5.44), the restriction of s to the open set

$$\{q(t, X) > 0\} \ (\text{resp.}\{q(t, X) < 0\}) \text{ is } 1 \ (\text{resp.}-1).$$

Thus the support of $\partial s / \partial X_j$ is included in $\{q(t, X) = 0\}$. Since the restriction of q to the open set $\mathbf{B}^c$ is zero, (3.5.47) is proved. Let $\chi_0 : \mathbb{R} \to [0, 1]$ be a smooth function, equal to 1 on $[-1, 1]$, vanishing outside $(-2, 2)$, and set

$$T_0 = \chi_0(|q'_X|^2)T, \quad \omega = 1 - \chi_0, \quad T_1 = \omega(|q'_X|^2)T. \tag{3.5.48}$$

We have from (3.5.47), (3.5.48),

$$\text{supp } T_1 \subset \{(t, X) \in \mathbf{B}, \ q(t, X) = 0 \text{ and } |q'_X(t, X)| \geq 1\} = \mathbf{K}. \tag{3.5.49}$$

We are able to give an explicit expression for T_1. The open set

$$\Omega = \{q'_X(t, X) \neq 0\} \cap \{|q(t, X)| < 1\} \tag{3.5.50}$$

is a neighborhood of the compact $\mathbf{K}$. From (3.5.44) and the fact that the Lebesgue measure of $\Omega \cap \{q(t, X) = 0\}$ is zero, the restriction $s_{|\Omega}$ of s to Ω is the L^∞ function $q/|q|$. This proves that

$$T_{1|\Omega} = \omega(|q'_X|^2)q'_X \cdot \frac{\partial}{\partial X}\left[\frac{q}{|q|}\right] = 2\delta(q)|q'_X|^2\omega(|q'_X|^2). \tag{3.5.51}$$

Since Ω is a neighborhood of the support of T_1, (3.5.51) determines completely T_1.

Lemma 3.5.12. *Let q and s be as above . Using Definition 2.4.1, we set*

$$Q(t) = \int_{\mathbb{R}^{2n}} q(t, X)\Sigma_X dX = q(t, \cdot)^{\text{Wick}}, \quad J(t) = s(t, \cdot)^{\text{Wick}}. \tag{3.5.52}$$

Let $u(t, x) \in \mathcal{S}(\mathbb{R} \times \mathbb{R}^n)$, set $u(t)(x) = u(t, x)$, and for $(t, X) \in \mathbb{R} \times \mathbb{R}^{2n}$,

$$\Phi(t, X) = [Wu(t)](X) = \langle u(t), \tau_X \varphi \rangle_{L^2(\mathbb{R}^n)}. \tag{3.5.53}$$

The function Φ belongs to $\mathcal{S}(\mathbb{R} \times \mathbb{R}^{2n})$ and, with ω defined in (3.5.48), Ω in (3.5.50), $\Psi \in C_c^\infty(\Omega, [0, 1])$, $\Psi \equiv 1$ on a neighborhood of $\mathbf{K}$ (see (3.5.49)), we have

$$\text{Re}\langle D_t u, iJ(t)u(t) \rangle_{L^2(\mathbb{R}^{n+1})} = \frac{1}{2\pi} \int_{\mathbb{R}^{2n}} |\Phi(\theta(X), X)|^2 dX, \tag{3.5.54}$$

$$\text{Re}\langle Q(t)u(t), J(t)u(t) \rangle_{L^2(\mathbb{R}^{n+1})} \geq \iint_{\mathbb{R}_t \times \mathbb{R}_X^{2n}} |q(t, X)||\Phi(t, X)|^2 dt dX$$

$$- \frac{1}{2\pi}\langle \delta(q)|q'_X|^2\omega(|q'_X|^2), \Psi(t, X)|\Phi(t, X)|^2 \rangle_{\mathscr{D}'(\Omega), \mathscr{D}(\Omega)} - C\|u\|^2_{L^2(\mathbb{R}^{n+1})}, \tag{3.5.55}$$

where C is a constant depending only on the dimension and the semi-norms of q.

Proof. Let us first notice that from Definition 2.4.1 and (3.5.45), the left-hand side of (3.5.54) is

$$-\frac{1}{4\pi}\iint \frac{\partial}{\partial t}\left[\langle \Sigma_X u(t), u(t)\rangle_{L^2(\mathbb{R}^n)}\right] s(t,X)\, dtdX = \frac{1}{2\pi}\int_{\mathbb{R}^{2n}} |\Phi(\theta(X),X)|^2 dX.$$

We use (3.5.54) and Lemma 2.4.6 to write, with $L^2(\mathbb{R}^{n+1}) = L^2(\mathbb{R}_t, L^2(\mathbb{R}^n))$ dot products,

$$\mathrm{Re}\langle Q(t)u(t), J(t)u(t)\rangle = \langle \mathrm{Re}[J(t)Q(t)]u(t), u(t)\rangle$$

$$= \langle \left[|q(t,\cdot)| - \frac{1}{4\pi}\frac{\partial q}{\partial X}(t,\cdot)\cdot\frac{\partial s}{\partial X}(t,\cdot)\right]^{\mathrm{Wick}} u(t), u(t)\rangle + \langle S(t)\, u(t), u(t)\rangle,$$

where $\|S(t)\|_{\mathcal{L}(L^2(\mathbb{R}^n))} \leq c_n\gamma_2(q)$. We get then the following inequality, using (3.5.46), (3.5.48), (3.5.51) and (3.5.53), with Ψ as in the lemma,

$$\mathrm{Re}\langle Q(t)u(t),J(t)u(t)\rangle \geq \iint_{\mathbb{R}_t\times\mathbb{R}_X^{2n}} |q(t,X)||\Phi(t,X)|^2 dX$$

$$-\frac{1}{2\pi}\langle \delta(q)|q'_X|^2\omega(|q'_X|^2), \Psi(t,X)|\Phi(t,X)|^2\rangle_{\mathscr{D}'(\Omega),\mathscr{D}(\Omega)}$$

$$-\frac{1}{4\pi}\langle \chi_0(|q'_X|^2)\frac{\partial q}{\partial X}(t,X)\cdot\frac{\partial s}{\partial X}(t,X), |\Phi(t,X)|^2\rangle_{\mathscr{S}'(\mathbb{R}^{2n+1}),\mathscr{S}(\mathbb{R}^{2n+1})}$$

$$- c_n\gamma_2(q)\|u\|^2_{L^2(\mathbb{R}^{n+1})}.$$

To obtain (3.5.55), we need only to check the duality bracket with χ_0. This term is

$$\frac{1}{4\pi}\iint s(t,X)\frac{\partial}{\partial X}\cdot\left[\chi_0(|q'_X|^2)\frac{\partial q}{\partial X}(t,X)|\Phi(t,X)|^2\right]dtdX$$

$$= \frac{1}{4\pi}\iint s(t,X)\frac{\partial}{\partial X}\cdot\left[\chi_0(|q'_X|^2)\frac{\partial q}{\partial X}(t,X)\right]|\Phi(t,X)|^2 dtdX$$

$$+ \frac{1}{4\pi}\iint s(t,X)\,\chi_0(|q'_X|^2)\frac{\partial q}{\partial X}(t,X)\cdot\frac{\partial}{\partial X}\left[\langle \Sigma_X u(t), u(t)\rangle_{L^2(\mathbb{R}^n)}\right] dtdX.$$

$$(3.5.56)$$

We calculate

$$\frac{\partial}{\partial X}\cdot\left[\chi_0(|q'_X|^2)\frac{\partial q}{\partial X}(t,X)\right] = \chi_0'(|q'_X|^2)2q''_{XX}(q'_X,q'_X) + \chi_0(|q'_X|^2)\,\mathrm{Tr}q''_{XX}. \quad (3.5.57)$$

From (3.5.41) and the fact that the support of χ_0 is bounded by 2, we get that (3.5.57) is bounded by a semi-norm of q. This proves that the absolute value of the first term in the right-hand side of (3.5.56) is bounded above by the product of a semi-norm of q with $\|u\|^2_{L^2(\mathbb{R}^{n+1})}$. We claim that, from Cotlar's lemma (Lemma 4.7.1) and (2.4.10), we have

$$\|\int_{\mathbb{R}^{2n}} \alpha(Y)\frac{\partial}{\partial Y_j}(\Sigma_Y)\,dY\|_{\mathcal{L}(L^2(\mathbb{R}^n))} \le \|\alpha\|_{L^\infty(\mathbb{R}^{2n})}d_n, \qquad (3.5.58)$$

where d_n depends only on the dimension : in fact, the Weyl symbol of

$$\frac{\partial}{\partial Y_j}(\Sigma_Y)\frac{\partial}{\partial Z_j}(\Sigma_Z) = \frac{\partial^2}{\partial Y_j\partial Z_j}\Sigma_Y\Sigma_Z \text{ is } q_{YZ}(X) = p_{YZ}(X)L_j(Y-X,Z-X),$$

where L_j is a polynomial of degree 2. Now, we have

$$|q_{YZ}(X)| \le 16\pi 2^{n/2}\sqrt{|p_{YZ}(X)|} \le 16\pi 2^n e^{-\frac{\pi}{4}|Y-Z|^2}e^{-\pi|X-\frac{Y+Z}{2}|^2}, \qquad (3.5.59)$$

so that the $\mathcal{L}(L^2(\mathbb{R}^n))$ norm of $\frac{\partial}{\partial Y_j}(\Sigma_Y)\frac{\partial}{\partial Z_j}(\Sigma_Z)$ is bounded above by the $L^1(\mathbb{R}^{2n})$ norm of its symbol q_{YZ}, which is estimated by $16\pi 2^n\ e^{-\frac{\pi}{4}|Y-Z|^2}$ from (3.5.59). Cotlar's lemma implies then (3.5.58). We note that

$$s(t,X)\chi_0(|q'_X|^2)\frac{\partial q}{\partial X}(t,X)$$

is bounded by 2, so that (3.5.58) implies that the absolute value of the second term in the right-hand side of (3.5.56) is bounded above by $\pi^{-1}nd_n\|u\|^2_{L^2(\mathbb{R}^{n+1})}$. This concludes the proof of Lemma 3.5.12. $\qquad\square$

Theorem 3.5.13. *Let q, Q, J, u be as in Lemma 3.5.12. We assume that*

$$q(t,X) = 0 \quad \text{and} \quad |q'_X(t,X)|^2 \ge 1 \Longrightarrow \quad |q'_X(t,X)|^2 \le q'_t(t,X). \qquad (3.5.60)$$

Then, assuming supp $u \subset \{|t| \le T\}$ for a positive T, the following estimate holds (with $L^2(\mathbb{R}^{n+1})$ dot products and norms):

$$\mathrm{Re}\langle D_t u + iQ(t)u, iJ(t)u + i\frac{t}{T}u\rangle \ge (4\pi T)^{-1}\|u\|^2 - C\|u\|^2, \qquad (3.5.61)$$

where C is the constant given in (3.5.55). Thus, for $0 < T \le \frac{1}{36\pi C}$,

$$\|D_t u + iQ(t)u\| \ge (9\pi T)^{-1}\|u\|. \qquad (3.5.62)$$

Proof. Let (t_0, X_0) be a point in $\mathbf{K}$ (see (3.5.49)). From (3.5.60), $q'_t(t_0, X_0) > 0$, so that the implicit function theorem and (3.5.42) give that, in an open neighborhood of (t_0, X_0),

$$q(t,X) = e(t,X)(t - \theta(X)) \text{ with } e > 0 \text{ and } e, \theta \in C^\infty.$$

This implies that, on this neighborhood,

$$\delta(t - \theta(X)) = \delta(q)q'_t(t,X). \qquad (3.5.63)$$

Eventually, (3.5.63) makes sense and is satisfied in an open neighborhood $\tilde{\Omega}$ of $\mathbf{K}$. Thus, setting $\Omega_0 = \Omega \cap \tilde{\Omega}$, where Ω is defined in (3.5.50), we obtain, with ω defined in (3.5.48) and $\Psi \in C_c^\infty(\Omega_0, [0,1])$, $\Psi \equiv 1$ in a neighborhood of $\mathbf{K}$, Φ given by (3.5.53),

$$\langle \delta(q) q'_t(t,X), \Psi(t,X) |\Phi(t,X)|^2 \rangle_{\mathcal{D}'(\Omega_0), \mathcal{D}(\Omega_0)} \leq \int_{\mathbb{R}^{2n}} |\Phi(\theta(X), X)|^2 dX. \quad (3.5.64)$$

Moreover, from the assumption (3.5.60) and (3.5.64), we have

$$\frac{1}{2\pi} \langle \delta(q) |q'_X|^2 \omega(|q'_X|^2), \Psi(t,X) |\Phi(t,X)|^2 \rangle_{\mathcal{D}'(\Omega_0), \mathcal{D}(\Omega_0)}$$
$$\leq \frac{1}{2\pi} \langle \delta(q) q'_t, \omega(|q'_X|^2) \Psi(t,X) |\Phi(t,X)|^2 \rangle_{\mathcal{D}'(\Omega_0), \mathcal{D}(\Omega_0)}$$
$$\leq \frac{1}{2\pi} \int_{\mathbb{R}^{2n}} |\Phi(\theta(X), X)|^2 dX. \quad (3.5.65)$$

Inequalities (3.5.64)-(3.5.65) and Lemma 3.5.12 imply that

$$\mathrm{Re}\langle D_t u + iQ(t)u, iJ(t)u + i\frac{t}{T}u \rangle$$
$$\geq ((4\pi T)^{-1} - C)\|u\|^2 + \iint_{\mathbb{R}_t \times \mathbb{R}_X^{2n}} \left[|q(t,X)| + \frac{t}{T} q(t,X) \right] |\Phi(t,X)|^2 dX dt. \quad (3.5.66)$$

Since u is supported in $|t| \leq T$, so is Φ. The inequality (3.5.66) implies (3.5.61). The estimate (3.5.62) follows from (3.5.61), $|t| \leq T, \|J(t)\| \leq 1$. The proof of Theorem 3.5.13 is complete. $\qquad\qquad\square$

3.5.4 Semi-global solvability under condition (P)

We have seen that condition (P) implies local solvability with loss of one derivative (Section 3.5.1), so that for P of order m satisfying condition (P) near a point x_0, $f \in H_{loc}^s$, the equation $Pu = f$ has a solution $u \in H_{loc}^{s+m-1}$ near x_0. However the neighborhood on which the equation holds depends a priori on s and may shrink when s increases; as a consequence, that result does not imply that for $f \in C^\infty$, there exists a C^∞ solution u. That problem was solved by L. Hörmander in [67] (see also Section 26.11 in [74]) where a semi-global existence theory was also added. Definition 3.3.2 introduced a notion of solvability at a compact set K, that we shall call semi-global solvability. We shall give here a short description of some of these results, referring the reader to Chapter 26 of [74] for the proofs and details.

When p is a smooth complex-valued function defined on a symplectic manifold, we shall say that an integral curve of $H_{\mathrm{Re}(ep)}$ where e does not vanish is a semi-bicharacteristic of p. Let $\mathcal{K}$ be a subset of a symplectic manifold $\mathcal{M}$ and

$p : \mathcal{M} \to \mathbb{C}$ a smooth function. We shall say that $\mathcal{K}$ is non-trapping for p when for all m such that $p(m) = 0$, there exists a semi-bicharacteristic $\gamma(t)$ of p such that $\gamma(0) = m$ and there exist $t_- < 0 < t_+$ with $\gamma(t_\pm) \notin \mathcal{K}$ or $p(\gamma(t_\pm)) \neq 0$.

Theorem 3.5.14. *Let Ω be an open subset of $\mathbb{R}^n$ and $P \in \Psi_{ps}^m(\Omega)$ a pseudo-differential operator of principal type with a principal symbol p_m satisfying condition (P) on Ω. Let K be a compact subset of Ω such that $\dot{T}^*(K)$ is non-trapping for p_m. Then*

$$\mathcal{N}_K = \{u \in \mathscr{E}_K', P^*u = 0\} \tag{3.5.67}$$

is a finite-dimensional subspace of C_K^∞ orthogonal to $P\mathscr{D}'(\Omega)$. For every $f \in H_{loc}^s(\Omega)$, orthogonal to $\mathcal{N}_K$ and every $t < s + m - 1$, one can find $u \in H_{loc}^t(\Omega)$ satisfying the equation $Pu = f$ near K (if $s = +\infty$, one can take $t = +\infty$).

We can note the strict inequality $t < s + m - 1$ for this semi-global solvability result, a contrast with the real principal type case and also with the local solvability result under condition (P). The proof of this result involves a propagation-of-singularities theorem where various cases are examined. The most difficult case is exposed in Section 26.10 of [74], *The Singularities on One-Dimensional Bicharacteristics*; to handle the propagation result, a localized form of the estimates (3.5.3) is required (Proposition 26.10.4 in [74]).

3.6 Condition (Ψ) does not imply solvability with loss of one derivative

3.6.1 Introduction

So far in this chapter, we have seen that condition (Ψ) is necessary for local solvability, and various cases in which it is also sufficient. In particular, condition (Ψ) along with a finite-type assumption ensures local solvability (see Section 3.4), and also condition (Ψ) ensures local solvability of differential operators and more generally of operators satisfying condition (P) (Section 3.5.1). We have shown also that in two dimensions and for the oblique derivative problem (Section 3.5.2) condition (Ψ) also implies local solvability and the case of transversal sign changes was also investigated in Section 3.5.3. In all previous cases, condition (Ψ) implies local solvability with loss of one derivative (or less in the finite-type case). In this section, we show that condition (Ψ) does *not* imply solvability with loss of one derivative for all pseudo-differential equations of principal type.

We begin with an informal introduction to the ideas governing our counterexample. We want to discuss the solvability of an evolution operator

$$\frac{d}{dt} + Q(t) \tag{3.6.1}$$

where each $Q(t)$ is a selfadjoint unbounded operator on a Hilbert space $\mathcal{H}$. This is equivalent to discussing a priori estimates for

$$\frac{d}{dt} - Q(t) = i\big(D_t + iQ(t)\big). \tag{3.6.2}$$

It is quite clear that the above problem is far too general, and so we wish to start our discussion with the simplest non-trivial example: instead of dealing with infinite-dimensional Hilbert space, let us take $\mathcal{H} = \mathbb{R}^2$, so that $Q(t)$ is a 2×2 symmetric matrix, defining a (bounded!) operator on $\mathbb{R}^2$, allowed to depend on large parameters. Since it could still be complicated, let us assume

$$Q(t) = H(-t)Q_1 + H(t)Q_2, \tag{3.6.3}$$

where H is the Heaviside function (characteristic function of $\mathbb{R}_+$), Q_1 and Q_2 are 2×2 symmetric matrices. There is of course no difficulty solving the equation

$$\frac{dv}{dt} + Q(t)v = f.$$

However, if we want to get uniform estimates with respect to the size of the coefficients of $Q(t)$, we have to choose carefully our solutions, even in finite dimension. When $Q_1 = Q_2$, a good fundamental solution is given by

$$H(t)E_1^+ \exp -tQ_1 - H(-t)E_1^- \exp -tQ_1$$

where E_1^+, E_1^- are the spectral projections corresponding to the half-axes. If we go back to (3.6.3) with $Q_1 \neq Q_2$, there is a trivial case in which the operator $\frac{d}{dt} + Q(t)$ is uniformly solvable: the monotone-increasing situation $Q_1 \leq Q_2$ yielding the estimate

$$\|D_t u + iQ(t)u\|_{L^2} \geq \|D_t u\|_{L^2}, \quad \text{where } L^2 = L^2(\mathbb{R}, \mathcal{H}). \tag{3.6.4}$$

We eventually come to our first point: is it true that solvability for $(d/dt) + Q(t)$ implies the same property for $(d/dt) + \alpha(t)Q(t)$, where α is a non-negative scalar function? This question is naturally linked to condition (Ψ), since whenever $q(t, x, \xi)$ satisfies (3.2.13) so does $a(t, x, \xi)q(t, x, \xi)$ for a non-negative symbol a. We are thus quite naturally led to discuss the uniform solvability of $(d/dt) + \alpha(t)Q(t)$, with

$$Q(t) = H(-t)Q_1 + H(t - \theta)Q_2,$$
$$Q_1 \leq Q_2, \ 2 \times 2 \text{ symmetric matrices, } \theta > 0. \tag{3.6.5}$$

The most remarkable fact about the pair of matrices $Q_1 \leq Q_2$ is the *"drift"*: the best way to understand it is to look at Figure 3.2 (E_j^+ stands also for its range). The condition $Q_1 \leq Q_2$ implies that the cones $\{\omega, \langle Q_2\omega, \omega \rangle < 0\}$ and $\{\omega, \langle Q_1\omega, \omega \rangle > 0\}$ are disjoint but does not prevent E_1^+ and E_2^- from getting very close. Let us define the drift $d(Q_1, Q_2)$ of the pair Q_1, Q_2 as the absolute value of

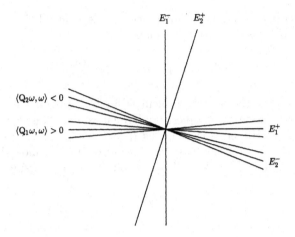

Figure 3.2: Position of eigenspaces and cones.

the cotangent of the angle between E_1^+ and E_2^- so that

$d(Q_1, Q_2) = 0$ when $E_2^- \subset E_1^-$ and $E_1^+ \subset E_2^+$,

$d(Q_1, Q_2) = +\infty$ when $E_1^+ \cap E_2^- \neq \{0\}$,

$d(Q_1, Q_2)$ unbounded when the distance between the spheres of E_1^+ and E_2^- is 0.

It is easy to see that a bounded drift is equivalent to the invertibility of the non-negative operator $E_2^+ + E_1^-$ and this provides a good definition in infinite dimension. If we consider for instance the pair of symmetric matrices

$$Q_{1,\nu} = \begin{pmatrix} \nu^2 & 0 \\ 0 & -\nu^2 \end{pmatrix} \leq e^{-i\alpha_\nu} \begin{pmatrix} \nu^3 & 0 \\ 0 & -\nu \end{pmatrix} e^{i\alpha_\nu} = Q_{2,\nu} \qquad (3.6.6)$$

where ν is a large positive parameter, $e^{i\alpha_\nu}$ the rotation of angle α_ν, with $\cos^2 \alpha_\nu = 2/\nu$, the drift goes to infinity with ν since

$$\begin{array}{c} \text{the square of the distance between the unit spheres} \\ \text{of } E_1^+ \text{ and } E_2^- \text{ is equivalent to } 2/\nu. \end{array} \qquad (3.6.7)$$

We now claim non-uniform solvability of the operator $d/dt + Q(t)$, with $Q(t)$ given by (3.6.5), Q_1, Q_2 by (3.6.6). We begin with ω_1 and ω_2 unit vectors respectively in E_1^+ and E_2^-,

$$u(t) = \begin{cases} e^{tQ_1}\omega_1 & \text{on } t < 0, \\ \omega_1 + \frac{t}{\theta}(\omega_2 - \omega_1) & \text{on } 0 \leq t \leq \theta, \\ e^{(t-\theta)Q_2}\omega_2 & \text{on } t > \theta. \end{cases} \qquad (3.6.8)$$

We now compute

$$\left\| \frac{du}{dt} - Q(t)u \right\|^2_{L^2} = \int_0^\theta \frac{|\omega_2 - \omega_1|^2}{\theta^2} dt = \theta^{-1}|\omega_2 - \omega_1|^2, \qquad (3.6.9)$$

where $|\cdot|$ stands for the norm on $\mathcal{H}$. On the other hand, since ω_1, ω_2 are unit vectors, we get

$$\|u\|^2_{L^2} \geq \int_0^\theta \left|\omega_1 + \frac{t}{\theta}(\omega_2 - \omega_1)\right|^2 dt \geq \frac{\theta}{2} - \theta|\omega_2 - \omega_1|^2 \geq \frac{\theta}{4} \qquad (3.6.10)$$

if $|\omega_2 - \omega_1|^2 \leq 1/4$, an easily satisfied requirement because of (3.6.7). Consequently, using (3.6.9), (3.6.10), we get

$$\|u\|^{-2}_{L^2} \left\| \frac{du}{dt} - Q(t)u \right\|^2_{L^2} \leq \theta^{-2}|\omega_2 - \omega_1|^2 4. \qquad (3.6.11)$$

Since ω_1 and ω_2 can be chosen arbitrarily close and independently of the size of the "hole" θ, we easily get a nonsolvable operator on $\ell^2(\mathbb{N})$ by taking direct sums. Note that (3.6.11) can be satisfied by a compactly supported u since the eigenvalues corresponding to ω_1 and ω_2 go to infinity with ν in such a way that there is no difficulty in multiplying u by a cut-off function.

What we have done so far is to get an "abstract" nonsolvable operator obtained by change of time-scale from a monotone-increasing situation; the basic device for the construction was the unbounded drift of $Q_1 \leq Q_2$. Since we are interested in pseudo-differential operators, the next question is obviously: Is an unbounded drift possible for $Q_1 \leq Q_2$, both of them pseudo-differential? We shall see that the answer is yes, leading to our counterexample. It is quite interesting to note that operators satisfying condition (P) do not drift (in particular differential operators satisfying condition (Ψ), equivalent to (P) in the differential case), as shown by the Beals-Fefferman reduction: After an inhomogeneous microlocalization and canonical transformation, their procedure leads to an evolution operator

$$\frac{d}{dt} + Q(t), \quad \text{with } Q(t) = Q(t, x, D_x) \text{ and } Q(t, x, \xi) = \xi_1 a(t, x, \xi),$$

where a is a non-negative symbol of order 0 (in a nonhomogeneous class). Then a Nirenberg-Treves commutator argument gives way to an estimate after multiplication by the sign of ξ_1.

Quite noticeable too is the fact that 2-dimensional pseudo-differential operators do not drift, since the sign function is a monotone matrix function on operators whose symbols are defined on a Lagrangean manifold; the last remark led the author to a proof of local solvability in two dimensions [89] and for oblique-derivative-type operators [91].

Our example below shows that subelliptic operators can drift, but in a bounded way. We shall see that condition (Ψ) prevents the drift from becoming infinite, but allows unbounded drifting.

We first study the very simple case

$$Q_1 = D_{x_1} = \frac{1}{2i\pi}\frac{\partial}{\partial x_1} \le Q_2 = D_{x_1} + \Lambda x_1^2 = e^{-2i\pi\frac{\Lambda x_1^3}{3}} D_{x_1} e^{2i\pi\frac{\Lambda x_1^3}{3}}, \qquad (3.6.12)$$

where Λ is a large positive parameter. Consider ω_1 a unit vector in E_1^+, i.e.,

$$\omega_1(x) = \int \kappa_1(\xi)e^{2i\pi x\xi}d\xi, \quad 1 = \|\kappa_1\|_{L^2}, \text{ supp } \kappa_1 \subset \mathbb{R}^+, \qquad (3.6.13)$$

and ω_2 a unit vector in E_2^-, i.e.,

$$\omega_2(x)e^{2i\pi\frac{\Lambda x_1^3}{3}} = \int \kappa_2(-\xi)e^{2i\pi x\xi}d\xi, \quad 1 = \|\kappa_2\|_{L^2}, \text{ supp } \kappa_2 \subset \mathbb{R}^+. \qquad (3.6.14)$$

A convenient way of estimating the drift of the pair (Q_1, Q_2) is to get an upper bound smaller than 1 for $|\langle\omega_1, \omega_2\rangle_{L^2}|$: This quantity is 0 if the pair is not drifting and is 1 if the drift is infinite. Now

$$\langle\omega_1, \omega_2\rangle = \iiint \kappa_1(\xi)e^{2i\pi x(\xi+\eta)}\overline{\kappa_2(\eta)}e^{2i\pi\frac{\Lambda x^3}{3}}dx\,d\eta\,d\xi$$

and thus

$$\langle\omega_1, \omega_2\rangle = \iint \kappa_1(\Lambda^{1/3}\xi)\Lambda^{1/6}\overline{\kappa_2(\Lambda^{1/3}\eta)}\Lambda^{1/6}A(\xi+\eta)d\eta\,d\xi, \qquad (3.6.15)$$

where

$$A(\xi) = \int e^{2i\pi\frac{x^3}{3}}e^{2i\pi x\xi}dx \quad \text{is the Airy function.} \qquad (3.6.16)$$

Consequently,

$$\sup_{\substack{\omega_1\in E_1^+, \omega_2\in E_2^- \\ \|\omega_1\|=\|\omega_2\|=1}} |\langle\omega_1, \omega_2\rangle| > 0 \qquad (3.6.17)$$

since for $\kappa_1(\xi) = \overline{\kappa_2(\xi)} = \kappa(\xi\Lambda^{-1/3})\Lambda^{-1/6}$ with a non-negative κ, supported in the interval $[1, 2]$, $1 = \|\kappa\|_{L^2}$, (3.6.15) gives

$$\langle\omega_1, \omega_2\rangle = \iint \kappa(\xi)\kappa(\eta)A(\xi+\eta)d\eta\,d\xi = \int A(\xi)(\kappa * \kappa)(\xi)d\xi, \qquad (3.6.18)$$

and the last term is a positive constant, independent of Λ (the Airy function given by (3.6.16) is positive on $\mathbb{R}^+$). Figure 3.3 will be useful for the understanding of these inequalities. This picture in the (x_1, ξ_1) symplectic plane shows that even though ω_1 and ω_2 are "living" in two faraway strips, one of which is curved, their dot product could be large. If we add one dimension to get a homogeneous version, Λ would be $|\xi_1| + |\xi_2|$, so that the above localizations in the phase space appear as two different second microlocalizations with respect to the hypersurfaces

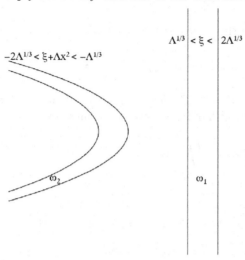

$\Lambda^{1/3} < \xi < 2\Lambda^{1/3}$

$-2\Lambda^{1/3} < \xi + \Lambda x^2 < -\Lambda^{1/3}$

ω_2

ω_1

Figure 3.3: Localization in the phase space of ω_1, ω_2.

$\{\xi_1 = 0\}$ on the one hand and $\{\xi_1 + |\xi|x_1^2 = 0\}$ on the other hand (see [14], [20]). These second microlocalizations are somehow incompatible so that the long range interaction between two faraway boxes corresponding to two different calculi could be large, as shown by the equality (3.6.18).

However, the pair given by (3.6.12) has a bounded drift, i.e., the quantity (3.6.17) is bounded above by a number strictly smaller than 1. This implies the solvability of

$$\frac{d}{dt} + \alpha(t, x, D_x)\left[H(-t)D_{x_1} + H(t)\left(D_{x_1} + x_1^2\sqrt{D_{x_1}^2 + D_{x_2}^2}\right)\right], \qquad (3.6.19)$$

where $\alpha(t, x, \xi)$ is a non-negative symbol of order 0, flat at $t = 0$. Since we are not going to use that result, we leave its proof to the reader with a hint: Compute the real parts of

$$\langle D_t u + iQ(t)u, iH(t-T)E_2^+ u\rangle,$$
$$\langle D_t u + iQ(t)u, -iH(T-t)H(t)E_2^- u\rangle, \qquad \text{for non-negative } T,$$
$$\langle D_t u + iQ(t)u, -iH(T-t)E_1^- u\rangle,$$
$$\langle D_t u + iQ(t)u, iH(t-T)H(-t)E_1^+ u\rangle, \qquad \text{for non-positive } T,$$

and use the bounded drift, meaning $E_2^+ + E_1^-$ invertible, and the Nirenberg-Treves commutator argument (see, e.g., Lemma 26.8.2 in [74]).

We now restart our discussion on pseudo-differential operators and study the following case, which turns out to be the generic one, using the microlocalization procedure of [8] and the Egorov principle of [45]:

$$Q_1 = D_{x_1} = \frac{1}{2i\pi}\frac{\partial}{\partial x_1} \le Q_2 = D_{x_1} + V(x_1) = e^{-2i\pi\phi(x_1)}D_{x_1}e^{2i\pi\phi(x_1)} \quad (3.6.20)$$

with a non-negative $V = \phi'$. To follow the lines of the computations starting at (3.6.12), the problem at hand is to estimate from above

$$\langle\omega_1,\omega_2\rangle = \iint \kappa_1(\xi)e^{2i\pi x(\xi+\eta)}\overline{\kappa_2(\eta)}e^{2i\pi\phi(x)}dxd\eta d\xi, \quad (3.6.21)$$

with

$$1 = \|\kappa_1\|_{L^2} = \|\kappa_2\|_{L^2}, \quad \operatorname{supp}\kappa_1 \subset \mathbb{R}_+, \quad \operatorname{supp}\kappa_2 \subset \mathbb{R}_+. \quad (3.6.22)$$

This means estimating from above the $\mathcal{L}(L^2)$ norm of the product

$$\Pi = H(-D_x)e^{2i\pi\phi}H(D_x),$$

where $H(D_x)$ is the Fourier multiplier by the Heaviside function H. If ϕ is $\frac{1}{2}H(x)$, then

$$i\Pi = iH(-D_x)\big(-H(x) + H(-x)\big)H(D_x) = F\Omega F, \quad (3.6.23)$$

where F is the Fourier transform and Ω the Hardy operator whose kernel is $H(\xi)H(\eta)/\pi(\xi+\eta)$ (the norm of Ω is obviously ≤ 1 from (3.6.23)). It is not difficult to see that the norm of the Hardy operator is exactly 1, as shown in Section 4.1.5. As a consequence, we get an unbounded drift for the pair $(D_{x_1}; D_{x_1}+\frac{1}{2}\delta(x_1))$, at least in a formal way; we will approximate the Dirac mass by a sequence of smooth functions $\frac{1}{2}\nu W(\nu x_1)$, where W is non-negative with integral 1. In order to get a symbol, we shall perform this approximation at the frequencies equivalent to 2^ν. Moreover, we shall choose carefully the size and regularization of the "hole" θ depending on this frequency.

Theorem 3.6.1. *There exists a C^∞ function $Q : \mathbb{R}^3 \times \mathbb{R}^2 \to \mathbb{R}$ such that*

$$Q \in S^1_{1,0}, \quad i.e., \quad \sup_{(x,\xi')\in\mathbb{R}^3\times\mathbb{R}^2}|(\partial_x^\alpha\partial_{\xi'}^\beta Q)(x,\xi')|(1+|\xi'|)^{|\beta|-1} < +\infty,$$

$$Q - \sum_{0\le j<N} q_j \in S^{1-N}_{1,0}, \quad q_j \in S^{1-j}_{1,0}, \ q_j \ \textit{positively-homogeneous of degree } 1 - j,$$

such that $q_0(x_1,x_2,x_3,\xi_1,\xi_2) > 0, y_3 > x_3 \implies q_0(x_1,x_2,y_3,\xi_1,\xi_2) \ge 0$, and there exists a sequence $(u_k)_{k\ge1}$ in $C^\infty_c(\mathbb{R}^3)$ with $\operatorname{supp}u_k \subset \{x \in \mathbb{R}^3, |x| \le 1/k\}$ such that

$$\lim_{k\to+\infty}\left\|\frac{\partial u_k}{\partial x_3} - Q(x,D')u_k\right\|_{L^2(\mathbb{R}^3)} = 0, \quad \|u_k\|_{L^2(\mathbb{R}^3)} = 1. \quad (3.6.24)$$

3.6.2 Construction of a counterexample

We follow the presentation of [77].

The operator

Let $\alpha \in C^\infty(\mathbb{R}; \mathbb{R}_+)$ vanishing on $[-1/2, 1/2]$ and $W \in C_c^\infty(\mathbb{R}; \mathbb{R}_+)$. We define for $t, x \in \mathbb{R}$, H the Heaviside function,

$$Q(t) = \alpha(t)\big(D_x + H(t)W(x)\big) = \alpha(t)\big(H(\tfrac{1}{2} - t)Q_1 + H(t - \tfrac{1}{2})Q_2\big), \quad (3.6.25)$$

$$Q_1 = D_x, \quad Q_2 = D_x + W(x). \quad (3.6.26)$$

We set $a(t) = \int_0^t \alpha(s)ds$ and we consider a function

$$\sigma \in C^\infty(\mathbb{R}; [0, 1]), \sigma' \geq 0, \ \sigma(t) = 0 \text{ on } t \leq -1/2, \quad \sigma(t) = 1 \text{ on } t \geq 1/2. \quad (3.6.27)$$

We have $\int \sigma'(t)^2 dt \geq (\int_{|t| \leq 1/2} \sigma'(t)dt)^2 = 1$ and we can choose σ such that

$$\int \sigma'(t)^2 dt \leq 3/2. \quad (3.6.28)$$

We define for ω_1, ω_2 in $L^2(\mathbb{R})$ with norm 1,

$$u(t) = \begin{cases} e^{a(t)Q_1}\omega_1 & \text{for } t \leq -1/2, \\ (1 - \sigma(t))\omega_1 + \sigma(t)\omega_2 & \text{for } |t| < 1/2, \\ e^{a(t)Q_2}\omega_2 & \text{for } t \geq 1/2, \end{cases} \quad (3.6.29)$$

and we have $\dot{u} - Q(t)u(t) = \sigma'(t)(\omega_2 - \omega_1)$ so that

$$\int_\mathbb{R} \|\dot{u} - Q(t)u(t)\|^2 dt = \|\omega_2 - \omega_1\|^2 \int \sigma'(t)^2 dt \leq 3(1 - \text{Re}\langle\omega_1, \omega_2\rangle). \quad (3.6.30)$$

On the other hand, we have

$$\int_{-1/2}^{1/2} \|u(t)\|^2 dt = 1 - \int_{-1/2}^{1/2} 2\sigma(t)\big(1 - \sigma(t)\big)\big(1 - \text{Re}\langle\omega_1, \omega_2\rangle\big)dt$$

$$\geq 1 - \frac{1}{2}(1 - \text{Re}\langle\omega_1, \omega_2\rangle) = \frac{1}{2}(1 + \text{Re}\langle\omega_1, \omega_2\rangle). \quad (3.6.31)$$

The point of the sequel of the construction, following the previous introductory Section 3.6.1, is to find $\omega_1 \in E_1^+, \omega_2 \in E_2^-$, where E_1^+ is the subspace corresponding to the positive spectrum of Q_1, E_2^- is the subspace corresponding to the negative spectrum of Q_2, with ω_2, ω_1 very close.

Choice of the function

Let ω_1 be an $L^2(\mathbb{R})$ function with norm 1 and Fourier transform κ_1 supported in $\mathbb{R}_+$. Noting that

$$Q_2 = D_x + W(x) = e^{-2i\pi\Phi(x)}D_x e^{2i\pi\Phi(x)}, \quad \Phi(x) = \int_{-\infty}^x W(y)dy,$$

we take $\omega_2 = e^{-2i\pi\Phi(x)}\widehat{\kappa_2}(x)$ where $\kappa_2 \in L^2(\mathbb{R})$, $\|\kappa_2\| = 1$, supp $\kappa_2 \subset \mathbb{R}_+$. We have for any $c \in \mathbb{C}$, using $\langle \omega_1, \widehat{\kappa_2} \rangle = \langle \kappa_1, \check{\kappa}_2 \rangle = 0$,

$$\langle \omega_1, \omega_2 \rangle = \int \omega_1(x) e^{2i\pi\Phi(x)}\overline{\widehat{\kappa_2}(x)}dx = \int \omega_1(x)\big(e^{2i\pi\Phi(x)} - c\big)\overline{\widehat{\kappa_2}(x)}dx, \quad (3.6.32)$$

and thus, since ω_1, κ_2 have L^2-norm 1, we get

$$|\langle \omega_1, \omega_2 \rangle| \leq \inf_{c \in \mathbb{C}} \|e^{2i\pi\Phi} - c\|_{L^\infty(\mathbb{R})}. \quad (3.6.33)$$

Note that the right-hand side is < 1 if the range of $e^{2i\pi\Phi}$ does not cover a half circle, that is if $\int_\mathbb{R} W(y)dy < 1/2$. On the other hand, when $\int_\mathbb{R} W(y)dy = 1/2$, we can actually find ω_1, ω_2 as above so that $\langle \omega_1, \omega_2 \rangle$ is arbitrarily close to 1.

Lemma 3.6.2. *Let*

$$W \in C_c^\infty([-1, 1]; \mathbb{R}_+), \int W(y)dy = 1/2, \quad \Phi(x) = \int_{-\infty}^x W(y)dy, \quad (3.6.34)$$

and set, for $\varepsilon > 0, x \in \mathbb{R}$,

$$U_{\varepsilon,\pm}(x) = \pm e^{-(\frac{1}{2}+\varepsilon)\operatorname{Log}(x\pm i)} = \pm(x \pm i)^{-\frac{1}{2}-\varepsilon} \quad (3.6.35)$$

with the logarithm given by (4.1.8) (the argument is in $(-\pi, \pi)$). Then it follows that

$$\big|\|U_{\varepsilon,\pm}\|_{L^2}^2 - \varepsilon^{-1}\big| \leq 3, \quad \big|\langle U_{\varepsilon,+}, e^{-2i\pi\Phi}U_{\varepsilon,-}\rangle - \varepsilon^{-1}\big| \leq \pi + 4, \quad (3.6.36)$$

$$\|U_{\varepsilon,\pm}^{(k)}\|_{L^2}^2 \leq \|U_{\varepsilon,\pm}\|_{L^2}^2 \prod_{0 \leq j < k} \Big(\frac{1}{2} + \varepsilon + j\Big)^2, \quad (3.6.37)$$

$$for \; X > 0, \quad \int_{|x| \geq X} |U_{\varepsilon,\pm}(x)|^2 dx \leq 2^{\frac{1}{2}+\varepsilon} X^{-2\varepsilon}\|U_{\varepsilon,\pm}\|_{L^2}^2. \quad (3.6.38)$$

Proof. Taylor's formula implies $\big|(x^2+1)^{-\frac{1}{2}-\varepsilon} - (x^2)^{-\frac{1}{2}-\varepsilon}\big| \leq (\frac{1}{2}+\varepsilon)(x^2)^{-\frac{3}{2}-\varepsilon}$ and since (we omit the subscript ε of U in the proof)

$$\|U_\pm\|_{L^2}^2 = 2\int_0^1 (1+x^2)^{-\frac{1}{2}-\varepsilon}dx + 2\int_1^{+\infty} x^{-1-2\varepsilon}dx$$

$$+ 2\int_1^{+\infty}\big((x^2+1)^{-\frac{1}{2}-\varepsilon} - (x^2)^{-\frac{1}{2}-\varepsilon}\big)dx,$$

we get

$$\big|\|U_\pm\|_{L^2}^2 - \varepsilon^{-1}\big| \leq 2\int_0^1 (1+x^2)^{-\frac{1}{2}-\varepsilon}dx + (1+2\varepsilon)\int_1^{+\infty} x^{-3-2\varepsilon}dx \leq 2 + \frac{1+2\varepsilon}{2+2\varepsilon} \leq 3,$$

which is the first estimate in (3.6.36). We check now

$$\langle U_+, e^{-2i\pi\Phi}U_-\rangle = -\int_{|x|\leq1}(x+i)^{-1-2\varepsilon}e^{2i\pi\Phi(x)}dx - \int_{|x|>1}(x+0i)^{-1-2\varepsilon}e^{2i\pi\Phi(x)}dx$$
$$-\int_{|x|>1}\left((x+i)^{-1-2\varepsilon} - (x+0i)^{-1-2\varepsilon}\right)e^{2i\pi\Phi(x)}dx. \qquad (3.6.39)$$

We have for $|x| \geq 1$,

$$(x+0i)^{-1-2\varepsilon}e^{2i\pi\Phi(x)} = e^{-(1+2\varepsilon)(\ln|x|+i\pi H(-x))}e^{i\pi H(x)} = -|x|^{-1-2\varepsilon}e^{-2i\pi\varepsilon H(-x)}$$

and this implies

$$-\int_{|x|>1}(x+0i)^{-1-2\varepsilon}e^{2i\pi\Phi(x)}dx = \int_1^{+\infty}x^{-1-2\varepsilon}dx + \int_{-\infty}^{-1}|x|^{-1-2\varepsilon}dx\,e^{-2i\pi\varepsilon}$$
$$= \frac{1}{2\varepsilon}(1 + e^{-2i\pi\varepsilon}) = \frac{1}{\varepsilon} + \frac{e^{-2i\pi\varepsilon} - 1}{2\varepsilon}. \qquad (3.6.40)$$

Moreover we have $|\int_{|x|\leq1}(x+i)^{-1-2\varepsilon}e^{2i\pi\Phi(x)}dx| \leq 2$, and

$$\left|\int_{|x|>1}\left((x+i)^{-1-2\varepsilon} - (x+0i)^{-1-2\varepsilon}\right)e^{2i\pi\Phi(x)}dx\right|$$
$$\leq (1+2\varepsilon)\int_{|x|>1}|x|^{-2-2\varepsilon}dx = 2. \qquad (3.6.41)$$

Collecting the information (3.6.39), (3.6.40), (3.6.41), we obtain

$$|\langle U_+, e^{-2i\pi\Phi}U_-\rangle - \varepsilon^{-1}| \leq \pi + 2 + 2 = \pi + 4,$$

which is the second inequality in (3.6.36). The estimate (3.6.37) follows immediately from the expression

$$U_{\pm}^{(k)}(x) = U_{\pm}(x)(x\pm i)^{-k}(-1)^k\prod_{0\leq j<k}\left(\frac{1}{2} + \varepsilon + j\right)$$

and for $X > 0$, we have

$$\int_{|x|\geq X}|U_{\pm}(x)|^2dx \leq \int_{|x|\geq X}|x|^{-1-2\varepsilon}dx = \varepsilon^{-1}X^{-2\varepsilon},$$

and $\|U_{\pm}\|_{L^2}^2 \geq 2^{-\frac{1}{2}-\varepsilon}\int_{|x|\geq1}|x|^{-1-2\varepsilon}dx = \varepsilon^{-1}2^{-\frac{1}{2}-\varepsilon}$, so that (3.6.38) holds, completing the proof of the lemma. □

We set, for $\varepsilon > 0, \delta > 0$,

$$U_{\pm}^{\varepsilon,\delta}(x) = \pm e^{\pm2i\pi\delta x}(x\pm i)^{-\frac{1}{2}-\varepsilon} = U_{\varepsilon,\pm}(x)e^{\pm2i\pi\delta x}, \qquad (3.6.42)$$

and we have from (4.1.50) that $\operatorname{supp}\widehat{U_{\varepsilon,\pm}} \subset \mathbb{R}_\pm$ and thus

$$\operatorname{supp}\widehat{U_\pm^{\varepsilon,\delta}} \subset \{\xi, \pm\xi \geq \delta\}. \tag{3.6.43}$$

Naturally the first estimate in (3.6.36) is unchanged if $U_{\varepsilon,\pm}$ is replaced by $U_\pm^{\varepsilon,\delta}$. We note also that for $0 < \varepsilon < \frac{1}{17\pi}$,

$$\left| \int (e^{4i\pi\delta x} - 1)U_{\varepsilon,+}(x)^2 e^{2i\pi\Phi(x)} dx \right| = \left| \int (e^{4i\pi\delta x} - 1)(x+i)^{-1-2\varepsilon} e^{2i\pi\Phi(x)} dx \right|$$

$$\leq \int_{|x|\leq 1/\delta} 4\pi\delta|x||x|^{-1-2\varepsilon} dx + \int_{|x|>1/\delta} 2|x|^{-1-2\varepsilon} dx \leq 8\pi\frac{\delta^{2\varepsilon}}{1-2\varepsilon} + 4\frac{\delta^{2\varepsilon}}{2\varepsilon} < 5/2,$$

if $0 < \delta \leq \varepsilon^{1/2\varepsilon}$ since $8\pi\varepsilon(1-2\varepsilon)^{-1} < 1/2$. As a consequence, we have

$$\big|\|U_\pm^{\varepsilon,\delta}\|_{L^2}^2 - \varepsilon^{-1}\big| \leq 3, \quad \big|\langle U_+^{\varepsilon,\delta}, e^{-2i\pi\Phi}U_-^{\varepsilon,\delta}\rangle - \varepsilon^{-1}\big| \leq \pi + 4 + \frac{5}{2} < 10, \tag{3.6.44}$$

$$\text{provided}\quad 0 < \varepsilon < \frac{1}{17\pi} \quad \text{and}\quad 0 < \delta \leq \varepsilon^{1/2\varepsilon}. \tag{3.6.45}$$

From now on, we assume that condition (3.6.45) is satisfied by the positive parameters ε, δ. We get in particular from the first inequality in (3.6.44) that

$$\left.\begin{aligned} \frac{9}{10} &\leq 1 - 3\varepsilon \leq \varepsilon\|U_\pm^{\varepsilon,\delta}\|^2 \leq 1 + 3\varepsilon \leq \frac{11}{10},\\[2mm] \left|1 - \frac{1}{\varepsilon\|U_+^{\varepsilon,\delta}\|\|U_-^{\varepsilon,\delta}\|}\right| &\leq \frac{6\varepsilon}{1-9\varepsilon^2} \leq 7\varepsilon. \end{aligned}\right\} \tag{3.6.46}$$

Lemma 3.6.3. *Let W and Φ as in Lemma 3.6.2 and set $V_\pm^{\varepsilon,\delta} = U_\pm^{\varepsilon,\delta}/\|U_\pm^{\varepsilon,\delta}\|$ where $U_\pm^{\varepsilon,\delta}$ is defined by (3.6.42) and ε, δ satisfy (3.6.45). Then we have*

$$\|V_\pm^{\varepsilon,\delta}\| = 1, \quad \operatorname{supp}\widehat{V_\pm^{\varepsilon,\delta}} \subset \{\pm\xi \geq \delta\}, \quad \|D^k V_\pm^{\varepsilon,\delta}\| \leq C_k, \tag{3.6.47}$$

$$|\langle V_+^{\varepsilon,\delta}, e^{-2i\pi\Phi}V_-^{\varepsilon,\delta}\rangle - 1| \leq 20\varepsilon, \tag{3.6.48}$$

$$\forall X > 0, \quad \int_{|x|>X} |V_\pm^{\varepsilon,\delta}(x)|^2 dx \leq 2X^{-2\varepsilon}. \tag{3.6.49}$$

Proof. From the first estimate in (3.6.36), (3.6.42) and (3.6.45), we get $U_\pm^{\varepsilon,\delta} \neq 0$ and $V_\pm^{\varepsilon,\delta}$ makes sense with norm 1, whereas the spectrum condition is (3.6.43) and the derivatives estimates follow from (3.6.37) and $|\delta| \leq 1$, giving (3.6.47). From the second inequality in (3.6.44), we get

$$\left|\langle V_+^{\varepsilon,\delta}, e^{-2i\pi\Phi}V_-^{\varepsilon,\delta}\rangle - \frac{1}{\varepsilon\|U_+^{\varepsilon,\delta}\|\|U_-^{\varepsilon,\delta}\|}\right| \leq \frac{10}{\|U_+^{\varepsilon,\delta}\|\|U_-^{\varepsilon,\delta}\|},$$

and thus from (3.6.46), we get

$$|\langle V_+^{\varepsilon,\delta}, e^{-2i\pi\Phi}V_-^{\varepsilon,\delta}\rangle - 1| \leq |1 - \frac{1}{\varepsilon\|U_+^{\varepsilon,\delta}\|\|U_-^{\varepsilon,\delta}\|}| + \frac{10}{\|U_+^{\varepsilon,\delta}\|\|U_-^{\varepsilon,\delta}\|} \leq 7\varepsilon + \frac{100}{9}\varepsilon \leq 20\varepsilon,$$

yielding (3.6.48). Finally (3.6.38) implies (3.6.49) since $\varepsilon \leq 1/2$. $\qquad\square$

A construction with parameters

N.B. In the sequel as above in this section, the norm $\|\cdot\|$ without subscript will stand for the $L^2(\mathbb{R})$-norm, unless otherwise specified.

We consider now

$$\chi \in C_c^\infty((-1,1);[0,1]), \quad \chi = 1 \text{ on } [-1/2,1/2], \quad \|\chi'\|_{L^\infty} \leq 3, \tag{3.6.50}$$

$$\alpha = 1 - \chi, \quad a(t) = \int_0^t \alpha(s)ds, \quad \text{so that } \alpha \in C^\infty(\mathbb{R};[0,1]), \tag{3.6.51}$$

$$\alpha(t) = 1 \text{ if } |t| \geq 1, \quad \alpha(t) = 0 \text{ if } |t| \leq 1/2, \quad \|\alpha'\|_{L^\infty} \leq 3, \tag{3.6.52}$$

$$\frac{a(t)}{t} \geq \frac{1}{2} \text{ if } |t| \geq 2, \tag{3.6.53}$$

since for $t \geq 2$ (resp.≤ -2), we have $a(t) = \int_0^1 \alpha(s)ds + (t-1) \geq \frac{t}{2}$ (resp. $a(t) = \int_0^{-1} \alpha(s)ds + (t+1) \leq \frac{t}{2}$). Let μ be a positive (large) parameter and define

$$u(t,x) = \begin{cases} e^{\mu a(t)D_x}V_+^{\varepsilon,\delta}, & \text{if } t \leq -1/2, \\ (1-\sigma(t))V_+^{\varepsilon,\delta} + \sigma(t)e^{-2i\pi\Phi}V_-^{\varepsilon,\delta}, & \text{if } |t| \leq 1/2, \\ e^{\mu a(t)(D_x+W)}\left(e^{-2i\pi\Phi}V_-^{\varepsilon,\delta}\right) = e^{-2i\pi\Phi}e^{\mu a(t)D_x}V_-^{\varepsilon,\delta}, & \text{if } t \geq 1/2, \end{cases} \tag{3.6.54}$$

where σ satisfies (3.6.27), $W, \Phi, V_\pm^{\varepsilon,\delta}$ are given in Lemma 3.6.3, and ε, δ satisfy (3.6.45). From (3.6.30), (3.6.31) and (3.6.48), it follows that, with $\mu > 0$ and $\mu Q(t) = \mu\alpha(t)(D_x + H(t)W(x))$,

$$\int_\mathbb{R} \|(\frac{d}{dt} - \mu Q(t))u(t)\|^2 dt \leq 60\varepsilon, \quad \int_{-1/2}^{1/2} \|u(t)\|^2 dt \geq 1 - 10\varepsilon. \tag{3.6.55}$$

Although the previous inequalities look essentially sufficient for our counterexample, we need also to make sure that they will still hold for a t-compactly supported u; for this reason we multiply u by a cut-off function. With χ satisfying (3.6.50), we get

$$(\frac{d}{dt} - \mu Q(t))(\chi(t/4)u(t)) = \underbrace{\frac{1}{4}\chi'(\frac{t}{4})\,u(t)}_{\substack{\text{supported in} \\ 2 \leq |t| \leq 4}} + \chi(t/4)\underbrace{(\frac{d}{dt} - \mu Q(t))u(t)}_{\substack{\text{supported in} \\ |t| \leq 1/2}},$$

and we note that if $|t| \geq 2$, from (3.6.53) and the inclusion in (3.6.47), we obtain $\|u(t)\| \leq e^{-\mu\delta|t|/2} \leq e^{-\mu\delta}$, so that from (3.6.55) and (3.6.50),

$$\int_{\mathbb{R}} \|(\frac{d}{dt} - \mu Q(t))(\chi(\frac{t}{4})u(t))\|^2 dt \leq \frac{9}{4}e^{-2\mu\delta} + 60\varepsilon. \qquad (3.6.56)$$

We need also to multiply by a cut-off function in the x-variable and for this purpose we set for (large) $X > 0$ to be chosen later, $\chi_x(x) = \chi(x/X)$. When $|t| \leq 1/2$, we have by (3.6.49), (3.6.50),

$$\|\chi_x u(t)\|^2 \geq \|u(t)\|^2 - \int_{|x|>X/2} |(1 - \sigma(t))V_+^{\varepsilon,\delta}(x) + \sigma(t)e^{-2i\pi\Phi}V_-^{\varepsilon,\delta}(x)|^2 dx$$

$$\geq \|u(t)\|^2 - 2^{1+2\varepsilon}X^{-2\varepsilon} \qquad (3.6.57)$$

and this implies from (3.6.55)

$$\int_{-1/2}^{1/2} \|\chi_x u(t)\|^2 dt \geq 1 - 10\varepsilon - 4X^{-2\varepsilon}. \qquad (3.6.58)$$

We compute also

$$(\frac{d}{dt} - \mu Q(t))(\chi_x\chi(\frac{t}{4})u(t)) = \chi_x(\frac{d}{dt} - \mu Q(t))(\chi(\frac{t}{4})u(t))$$

$$- \mu\alpha(t)\chi(\frac{t}{4})u(t)\frac{1}{2i\pi}X^{-1}\chi'(x/X),$$

and we have

$$\int_{\mathbb{R}} \|\mu\alpha(t)\chi(\frac{t}{4})u(t)\frac{1}{2i\pi}X^{-1}\chi'(\cdot/X)\|^2 dt$$

$$\leq \mu^2 X^{-2}\frac{1}{4\pi^2}\int_{|t|\leq 4} \|\chi'\|_{L^\infty}^2 \|u(t)\|^2 dt \leq \mu^2 X^{-2}\frac{72}{4\pi^2} \leq 2\mu^2 X^{-2}, \qquad (3.6.59)$$

so that from (3.6.56), (3.6.58),

$$\int_{\mathbb{R}} \|(\frac{d}{dt} - \mu Q(t))(\chi_x\chi(\frac{t}{4})u(t))\|^2 dt \leq \frac{9}{2}e^{-2\mu\delta} + 120\varepsilon + 4\mu^2 X^{-2}, \qquad (3.6.60)$$

$$\int_{|t|\leq 1/2} \|\chi_x\chi(t/4)u(t)\|^2 dt \geq 1 - 10\varepsilon - 4X^{-2\varepsilon}. \qquad (3.6.61)$$

To apply the inequalities above, involving several parameters, we have to make sure that $\varepsilon, X^{-2\varepsilon}, e^{-\mu\delta}, \mu^2 X^{-2}$ are all small. We shall choose later on the proper relationships between all these parameters, and since we want the support to shrink to the origin, we devise a rescaling.

Rescaling

With u defined by (3.6.54), $\theta > 0$ (small), $\nu > 0$ (large) we define v on $\mathbb{R}^2$ by

$$v(s,y) = \nu^{1/2}\theta^{-1/2}u(\frac{s}{\theta}, \nu y)\chi(\frac{s}{4\theta})\chi(\frac{\nu y}{X}), \tag{3.6.62}$$

so that $\operatorname{supp} v \subset [-4\theta, 4\theta] \times [-\nu^{-1}X, \nu^{-1}X]$. With

$$q(t,x,\xi) = \alpha(t)(\xi + H(t)W(x)) \tag{3.6.63}$$

the symbol of $Q(t)$ as defined by (3.6.25), we have (e.g., using (2.1.14), but a direct verification is easier),

$$(\mathscr{U}v)(t,x) = \theta^{1/2}\nu^{-1/2}v(\theta t, \nu^{-1}x), \quad \mathscr{U} \text{ unitary on } L^2(\mathbb{R}^2),$$
$$\mathscr{U}^*(\partial_t - \mu q(t,x,\xi)^w)\mathscr{U} = \theta\partial_s - \mu q(\theta^{-1}s, \nu y, \nu^{-1}\eta)^w,$$

and thus

$$\iint_{\mathbb{R}^2} |\theta\partial_s v - \mu q(\theta^{-1}t, \nu y, \nu^{-1}\eta)^w v|^2 dsdy$$
$$= \iint_{\mathbb{R}^2} |(\partial_t - \mu q(t,x,\xi)^w)(\chi_x \chi(t/4)u(t))|^2 dtdx.$$

As a result, we have, with v given by (3.6.62), q by (3.6.63),

$$\widetilde{Q}(s) = \mu\theta^{-1}q(\theta^{-1}s, \nu y, \nu^{-1}\eta)^w = \mu\theta^{-1}\alpha(\theta^{-1}s)\nu^{-1}(D_y + H(s)\nu W(\nu y)), \tag{3.6.64}$$

$$\|\partial_s v - \widetilde{Q}(s)v\|^2_{L^2(\mathbb{R}^2)} \leq \theta^{-2}(\frac{9}{2}e^{-2\mu\delta} + 120\varepsilon + 4\mu^2 X^{-2}), \tag{3.6.65}$$

$$\iint_{|s|\leq\theta/2} |v(s,y)|^2 dsdy \geq 1 - 10\varepsilon - 4X^{-2\varepsilon}. \tag{3.6.66}$$

We shall need that the rhs of (3.6.65), $\varepsilon, X^{-2\varepsilon}$ tend to zero to ensure that v is almost in the kernel of $\partial_s - \widetilde{Q}(s)$, but also that $\theta, X\nu^{-1}$ tend to zero to make sure that the support of v shrinks to the origin. Last but not least, we need also a control of the size of the derivatives of the coefficient of D_y in (3.6.64), i.e., $\mu\theta^{-k}\nu^{-1}$ is bounded for every fixed non-negative integer k. We choose for ν a large positive integer,

$$\varepsilon = (\operatorname{Log}\nu)^{-1/2}, \quad \delta = \varepsilon^{\frac{1}{2\varepsilon}}, \quad X = \nu^{2/3}, \quad \theta = (\operatorname{Log}\nu)^{-1/8}, \quad \mu = \nu^{1/3}. \tag{3.6.67}$$

We have

$$\operatorname{Log}\delta = -\frac{\operatorname{Log}\operatorname{Log}\nu}{4\varepsilon} = -\frac{1}{4}(\operatorname{Log}\nu)^{1/2}\operatorname{Log}\operatorname{Log}\nu, \tag{3.6.68}$$

$$\varepsilon\theta^{-2} = (\operatorname{Log}\nu)^{-1/4}, \quad (\mu^2 X^{-2}\theta^{-2}) = \nu^{-2/3}(\operatorname{Log}\nu)^{1/4}, \tag{3.6.69}$$

$$e^{-2\mu\delta}\theta^{-2} = \frac{(\operatorname{Log}\nu)^{1/4}}{e^{2\mu\delta}} = \frac{(\operatorname{Log}\nu)^{1/4}}{\exp(2\nu^{1/3}(\operatorname{Log}\nu)^{-\frac{(\operatorname{Log}\nu)^{1/2}}{4}})}, \tag{3.6.70}$$

so that for ν large enough we have

$$\|\partial_s v - \widetilde{Q}(s)v\|_{L^2(\mathbb{R}^2)} \leq 11(\mathrm{Log}\,\nu)^{-1/8}. \tag{3.6.71}$$

Using (3.6.66) we get for ν large enough,

$$1 - 11(\mathrm{Log}\,\nu)^{-1/2} \leq 1 - 10(\mathrm{Log}\,\nu)^{-1/2} - \frac{4}{\nu^{\frac{4}{3(\mathrm{Log}\,\nu)^{1/2}}}} \leq \|v\|^2_{L^2(\mathbb{R}^2)},$$

which implies for ν large enough

$$1 - 11(\mathrm{Log}\,\nu)^{-1/2} \leq \|v\|_{L^2(\mathbb{R}^2)}, \tag{3.6.72}$$

and we have also from (3.6.62), (3.6.50),

$$\mathrm{supp}\,v \subset \{(s,y), |s| \leq 4(\mathrm{Log}\,\nu)^{-1/8}, |y| \leq \nu^{-1/3}\}. \tag{3.6.73}$$

Finally, for all $k \in \mathbb{N}$, we have $\mu\nu^{-1}\theta^{-k} = \nu^{-2/3}(\mathrm{Log}\,\nu)^{k/8} = O_k(\nu^{-1/2})$.

Proposition 3.6.4. *For each $\nu \in \mathbb{N}^*$, there exist $a_\nu \in C^\infty(\mathbb{R};\mathbb{R}), b_\nu \in C^\infty(\mathbb{R}^2;\mathbb{R})$ such that, with $q_\nu \in C^\infty(\mathbb{R}^3,\mathbb{R})$ defined by $q_\nu(t,x,\xi) = a_\nu(t)\xi + b_\nu(t,x)$, we have $q_\nu(t,x,\xi) > 0, s \geq t \Longrightarrow q_\nu(s,x,\xi) \geq 0$ and*

$$\forall(k,l,m) \in \mathbb{N}^3, \sup_{\substack{(t,x)\in\mathbb{R}^2 \\ \nu\geq 1}} \left(|(\partial_t^k a_\nu)(t)| + |(\partial_t^l \partial_x^m b_\nu)(t,x)|\nu^{-1-m}\right)\nu^{1/2} < +\infty. \tag{3.6.74}$$

For each $\nu \in \mathbb{N}^*$, there exists $u_\nu \in C_c^\infty(\mathbb{R}^2)$ with $\mathrm{supp}\,u_\nu \subset B(0,r_\nu)$ and such that $\lim_{\nu\to+\infty} r_\nu = 0$, with

$$\sup_{\nu\geq 1}\|\partial_t u_\nu - q_\nu(t,x,D_x)u_\nu\|_{L^2(\mathbb{R}^2)}(\mathrm{Log}\,\nu)^{1/8} < +\infty, \quad \|u_\nu\|_{L^2(\mathbb{R}^2)} = 1, \tag{3.6.75}$$

$$\forall k \in \mathbb{N}, \quad \sup_{\nu\geq 1}\|\partial_x^k u_\nu\|_{L^2(\mathbb{R}^2)}\nu^{-k} < +\infty. \tag{3.6.76}$$

Proof. We take, according to (3.6.64), (3.6.67), (3.6.51), (3.6.34),

$$q_\nu(t,x,\xi) = \mu_\nu\theta_\nu^{-1}\alpha(\theta_\nu^{-1}t)\nu^{-1}\big(\xi + H(t)\nu W(\nu x)\big), \tag{3.6.77}$$

$$a_\nu(t) = \mu_\nu\theta_\nu^{-1}\alpha(\theta_\nu^{-1}t)\nu^{-1}, \quad b_\nu(t,x) = \mu_\nu\theta_\nu^{-1}\alpha(\theta_\nu^{-1}t)H(t)W(\nu x), \tag{3.6.78}$$

$$u_\nu = v/\|v\|_{L^2(\mathbb{R}^2)}, \quad v \text{ is given in (3.6.62).} \tag{3.6.79}$$

We divide the inequality (3.6.71) by $\|v\|$ and using (3.6.72), we get all the results, except for the very last inequality. Since $u_\nu = v/\|v\|_{L^2(\mathbb{R}^2)}$ and v is given by (3.6.62), where χ is a "fixed" function, $X \geq 1$, we examine the L^2 norm of the x-derivatives of u, and since Φ is also fixed, it amounts to checking the $L^2(\mathbb{R}^2_{t,\xi})$-norm of

$$\xi^k V_\pm^{\varepsilon,\delta}(\xi)\mathbf{1}_{[-2,2]}(t) \text{ and, using (3.6.53), } \xi^k V_\pm^{\varepsilon,\delta}(\xi)e^{-\frac{1}{2}\mu|t||\xi|}\mathbf{1}\{|t| \geq 2\}.$$

The first is taken care of by (3.6.47) and for the second function, the integral with respect to t is bounded above by $\mu^{-1}\delta^{-1}\int|\xi|^{2k}|V_{\pm}^{\varepsilon,\delta}(\xi)|^2d\xi$ to which we can also apply (3.6.47) since

$$\mu\delta = \nu^{1/3}(\mathrm{Log}\,\nu)^{-\frac{1}{4}}(\mathrm{Log}\,\nu)^{1/2} \geq 1, \quad \text{for } \nu \text{ large enough.}$$

The proof of the lemma is complete.   □

Remark 3.6.5. Note that the function $s \mapsto H(s)\alpha(s)$ is C^∞ and $H(s)\alpha(s) = H(s)\alpha(s)\alpha(2s)$. We have also

$$b_\nu(t,x) = a_\nu(t)\alpha(2t\theta_\nu^{-1})H(t)\nu W(\nu x) \tag{3.6.80}$$

so that

$$q_\nu(t,x,\xi) = a_\nu(t)\big(\xi + \underbrace{\alpha(2t\theta_\nu^{-1})H(t)\nu W(\nu x)}_{\beta_\nu(t,x)}\big). \tag{3.6.81}$$

From (3.6.78) we have the slight improvement of (3.6.74)

$$\forall k \in \mathbb{N}, \quad \sup_{\substack{(t,x)\in\mathbb{R}^2 \\ \nu\geq 1}} |(\partial_t^k a_\nu)(t)|\nu^{\frac{7}{12}} < +\infty, \tag{3.6.82}$$

and also

$$\forall (k,l,m) \in \mathbb{N}^3, \quad \sup_{\substack{(t,x)\in\mathbb{R}^2 \\ \nu\geq 1}} |(\partial_t^l\partial_x^m \beta_\nu)(t,x)|\nu^{-1-m}\nu^{-\frac{1}{12}} < +\infty. \tag{3.6.83}$$

$\mathbf{S}_{1,0}^1$ counterexample

The operator q_ν^w in the proposition above is $a_\nu(t)D_x + b_\nu(t,x)$ and the derivatives of a_ν are bounded as well as those of $2^{-\nu}b_\nu$, so that we can see the symbol q_ν as a first-order semi-classical symbol $2^\nu\big(a_\nu(t)2^{-\nu}\xi_1 + b_\nu(t,x_1)2^{-\nu}\big)$ with small parameter $2^{-\nu}$; to realize this as a three-dimensional operator, we shall consider 2^ν as $|\xi_2|$, following a standard path, with

$$a_\nu(t)\xi_1 + b(t,x_1)\xi_2 2^{-\nu} = \xi_2 2^{-\nu}q_\nu(t,x_1,2^\nu\xi_1/\xi_2). \tag{3.6.84}$$

We shall of course have to multiply that symbol by a suitable cut-off function localizing near $\xi_2 \sim 2^\nu$ and also to localize in the x_2-variable, say by multiplying by $\chi((x_2 - 2\rho_j)/\rho_j)$, which is supported in $(\rho_j, 3\rho_j)$; these are disjoint sets if $\rho_j = 3^{-j}$, but the derivatives of the cut-off function are powers of 3^j which should remain small with respect to ν. The balance of the parameters is made precise in the following statement.

Lemma 3.6.6. *Let f be defined on $\mathbb{R}_t \times \mathbb{R}_x^2 \times \mathbb{R}_\xi^2$ by*

$$f(t,x,\xi) = \sum_{j\geq 1} \chi\big((x_2 - 2\rho_j)/\rho_j\big)\Big(a_{\nu_j}(t)\xi_1 + b_{\nu_j}(t,x_1)\xi_2 2^{-\nu_j}\Big), \qquad (3.6.85)$$

where χ is given by (3.6.50), a_ν, b_ν by Proposition 3.6.4, $\rho_j = 3^{-j}, \nu_j = 2^{j^2}$. The function f is a real-valued linear function of ξ with $C^\infty(\mathbb{R}^3_{t,x_1,x_2})$ coefficients all of whose derivatives are bounded. If $\xi_2 f(t,x,\xi) > 0$ and $s \geq t$, then we have $\xi_2 f(s,x,\xi) \geq 0$.

Proof. The supports of the smooth functions $x_2 \mapsto \chi\big((x_2 - 2\rho_j)/\rho_j\big)$ are disjoint and from (3.6.74), for $k, m, l \in \mathbb{N}$,

$$\sup_{j\geq 1}|\chi^{(m)}\big((x_2 - 2\rho_j)/\rho_j\big)\rho_j^{-m}\partial_t^k a_{\nu_j}(t)| \leq C_{k,m}\sup_{j\geq 1}(\nu_j^{-1/2}\rho_j^{-m}) \leq C_{k,m}2^{2m^2},$$

$$\sup_{j\geq 1}|\chi^{(m)}\big((x_2 - 2\rho_j)/\rho_j\big)\rho_j^{-m}(\partial_t^k\partial_{x_1}^l b_{\nu_j})(t)2^{-\nu_j}| \leq C_{k,m,l}\sup_{j\geq 1}\nu_j^{l+\frac{1}{2}}\rho_j^{-m}2^{-\nu_j}$$

$$\leq C_{k,m,l}\sup_{j\geq 1} 2^{j^2(l+\frac{1}{2})}3^{jm}2^{-2^{j^2}} = C'_{k,m,l},$$

so that the first statements on f are valid. Since the supports of $\chi((x_2 - 2\rho_j)/\rho_j)$ are disjoint, the last property follows from the same one on q_ν proven in Proposition 3.6.4 and from (3.6.84). $\qquad\square$

Remark 3.6.7. Using the expressions (3.6.81), the estimates (3.6.82), (3.6.83) and introducing a function $\widetilde{\chi} \in C_c^\infty((-1,1);[0,1])$ equal to 1 on the support of χ, we get

$$f(t,x,\xi) = \sum_{j\geq 1} \chi((x_2 - 2\rho_j)/\rho_j)a_{\nu_j}(t)\Big(\xi_1 + \beta_{\nu_j}(t,x_1)\xi_2 2^{-\nu_j}\widetilde{\chi}((x_2 - 2\rho_j)/\rho_j)\Big),$$

and thus, since the supports of $x_2 \mapsto \widetilde{\chi}((x_2 - 2\rho_j)/\rho_j)$ are disjoint and $\chi\widetilde{\chi} = 1$, we get

$$f(t,x,\xi)$$
$$= \Big(\sum_{j\geq 1}\chi((x_2 - 2\rho_j)/\rho_j)a_{\nu_j}(t)\Big)\Big(\xi_1 + \xi_2\sum_{j\geq 1}\widetilde{\chi}((x_2 - 2\rho_j)/\rho_j)\beta_{\nu_j}(t,x_1)2^{-\nu_j}\Big)$$

$$= g_0(t,x_2)(\xi_1 + \xi_2 g_1(t,x_1,x_2)) \qquad (3.6.86)$$
$$\text{with}\quad g_0 \in C_b^\infty(\mathbb{R}^2;\mathbb{R}_+), \quad g_1 \in C_b^\infty(\mathbb{R}^3;\mathbb{R}), \quad \partial_t g_1 \geq 0. \qquad (3.6.87)$$

We shall consider below a Fourier multiplier $c(D_x)$ such that

$$c(\xi) = \chi(\xi_1/\xi_2)H(\xi_2)H(\xi_2^2 - \xi_1^2)\big(1 - \chi(|\xi|^2)\big), \qquad (3.6.88)$$

where χ is given by (3.6.50). The symbol c belongs to $S_{1,0}^0$ and is homogeneous with degree 0 for $|\xi| \geq 1$ (see Definition 1.2.9) since $\xi_2 \geq |\xi_1|, |\xi|^2 \geq 1/2 \Longrightarrow \xi_2 \geq 1/2$,

$$\text{so that}\quad c(\xi) = \begin{cases} \chi(\xi_1/\xi_2)(1 - \chi(|\xi|^2)) & \text{if } \xi_2 \geq 1/2, \\ 0 & \text{if } \xi_2 < 1/2. \end{cases}$$

which is C^∞ since it is zero for $\xi_2 \leq \frac{1}{2} + \frac{\varepsilon_0}{4}$, assuming $\operatorname{supp}\chi \subset [-1+\varepsilon_0, 1-\varepsilon_0]$ with $1/2 > \varepsilon_0 > 0$:

$$\frac{1}{2} \leq \xi_2 \leq \frac{1}{2} + \frac{\varepsilon_0}{4},\ |\xi|^2 \geq \frac{1}{2},\ |\xi_1| \leq \xi_2(1 - \varepsilon_0)$$

$$\Longrightarrow \frac{1}{2} \leq \xi_2^2(1 + (1-\varepsilon_0)^2) \leq (\frac{1}{2} + \frac{\varepsilon_0}{4})^2(2 - 2\varepsilon_0 + \varepsilon_0^2)$$

$$\Longrightarrow 1 \leq (1 + \frac{\varepsilon_0}{2})^2(1 - \varepsilon_0 + \frac{\varepsilon_0^2}{2}) = 1 - \frac{\varepsilon_0^2}{4} + \frac{\varepsilon_0^3}{4} + \frac{\varepsilon_0^4}{8} < 1,$$

which is impossible so that for $\frac{1}{2} \leq \xi_2 \leq \frac{1}{2} + \frac{\varepsilon_0}{4}$, either $|\xi|^2 < \frac{1}{2}$ or $|\xi_1| > \xi_2(1-\varepsilon_0)$ and in both cases $c(\xi) = 0$. The operators

$$c(D_x)f(t, x, D_x),\ c(D_x)f(t, x, \xi)^w,\ f(t, x, D_x)c(D_x),\ f(t, x, \xi)^w c(D_x),$$

have the principal symbol $c(\xi)f(t, x, \xi)$, i.e., can be written as $\big(c(\xi)f(t, x, \xi)\big)^w +$ an operator with symbol in $S_{1,0}^0$ which is an asymptotic sum of symbols homogeneous of degree $l,\ l \in \mathbb{Z}_-$. Since $\xi_2 \geq 0$ on the support of c, the product cf is real-valued, belongs to $S_{1,0}^1$, is homogeneous of degree 1 for $|\xi| \geq 1$ and satisfies

$$c(\xi)f(t, x, \xi) > 0, s \geq t \Longrightarrow c(\xi)f(s, x, \xi) > 0. \tag{3.6.89}$$

Theorem 3.6.8. *Let f be defined in (3.6.85) and c be given by (3.6.88). The pseudo-differential operator*

$$c(D_x)f(t, x, D_x)$$

has the homogeneous principal symbol $c(\xi)f(t, x, \xi)$ and its symbol is an asymptotic sum of homogeneous symbols of degree $l \in 1 + \mathbb{Z}_-$ There exists a sequence $(U_j)_{j\geq 1}$ of functions in $C_c^\infty(\mathbb{R}^3)$ such that

$$\lim_j \|\partial_t U_j - c(D_x)f(t, x, D_x)U_j\|_{L^2(\mathbb{R}^3)} = 0, \tag{3.6.90}$$

with $\ \|U_j\|_{L^2(\mathbb{R}^3)} = 1,\ \ \operatorname{supp} U_j \subset B(0, R_j),\ \ R_j > 0,\ \lim_j R_j = 0.$ (3.6.91)

Note that this theorem implies Theorem 3.6.1.

Comment. Choosing $\chi_1 \in C_c^\infty((-1, 1); [0, 1])$ equal to 1 on the support of χ and defining

$$a(t, x_2) = \sum_{j\geq 1} \chi((x_2 - 2\rho_j)/\rho_j)\mu_{\nu_j}\theta_{\nu_j}^{-1}\alpha(t/\theta_{\nu_j})\nu_j^{-1}, \tag{3.6.92}$$

$$\Omega(x_1, x_2) = \sum_{j\geq 1} \chi_1((x_2 - 2\rho_j)/\rho_j)\nu_j W(\nu_j x_1)2^{-\nu_j}, \tag{3.6.93}$$

we get, using that the supports of $\chi_1\big((x_2 - 2\rho_j)/\rho_j\big)$ are disjoint and Lemma 3.6.6, that

$$f(t, x, \xi) = a(t, x_2)\big(\xi_1 + H(t)\Omega(x_1, x_2)\xi_2\big), \tag{3.6.94}$$

with

$$a \in C_b^\infty(\mathbb{R}^2), 0 \leq \Omega \in C_c^\infty(\mathbb{R}^2),\ 0 \leq a(t, x_2)H(t)\Omega(x_1, x_2) \in C_b^\infty(\mathbb{R}^3).$$

As a consequence, for $t < 0$, we have $f(t, x, \xi)\xi_1 \geq 0$ and for $t > 0$, we get $f(t, x, \xi)(\xi_1 + \Omega(x)\xi_2) \geq 0$, so that Condition (P) holds on both sets where $\pm t > 0$. For a given (x, ξ) with $\xi_2 > 0$, a sign change of $t \mapsto f(t, x, \xi)$ occurs when $\xi_1 < 0 < \xi_1 + \Omega(x)\xi_2$ and t should cross the origin. Still with $\xi_2 > 0$, if $\xi_1 \geq 0$, the function $t \mapsto f(t, x, \xi)$ remains non-negative and if $\xi_1 + \Omega(x)\xi_2 \leq 0$, $t \mapsto f(t, x, \xi)$ remains non-positive.

Proof of Theorem 3.6.8. The statements on $c(D_x)f(t, x, \xi)^w$ are proven in Lemma 3.6.6. Choose a function $\psi \in C_c^\infty((-1/2, 1/2))$ (so that $\chi\psi = \psi$) with $\|\psi\|_{L^2(\mathbb{R})} = 1$ and set

$$U_j(t, x_1, x_2) = u_{\nu_j}(t, x_1)\psi_j(x_2), \quad \psi_j(x_2) = \psi\big(2(x_2 - 2\rho_j)/\rho_j\big)2^{1/2}\rho_j^{-1/2}e^{2i\pi 2^{\nu_j}x_2} \tag{3.6.95}$$

where u_ν is given in Proposition 3.6.4, and ρ_j, ν_j in Lemma 3.6.6. We have $\|U_j\|_{L^2(\mathbb{R}^3)} = 1$,

$$\mathrm{supp}\, U_j \subset \{(t, x_1, x_2) \in \mathbb{R}^3, t^2 + x_1^2 \leq r_{\nu_j}^2, \frac{7\rho_j}{4} \leq x_2 \leq \frac{9\rho_j}{4}\}, \tag{3.6.96}$$

so that (3.6.91) holds. Since $f(t, x, D_x)$ is a differential operator (in fact a vector field as proven in Lemma 3.6.6), and the support of $x_2 \mapsto \chi((x_2 - 2\rho_j)/\rho_j)$ is included in $(\rho_j, 3\rho_j) = (3^{-j}, 3^{1-j})$ and $[\frac{7\rho_j}{4}, \frac{9\rho_j}{4}] \subset (3^{-j}, 3^{1-j})$ we have, since the $(3^{-j}, 3^{1-j})$ are disjoint, with a_ν, b_ν given in (3.6.78),

$$\big(f(t, x, D_x)U_j\big)(t, x) = \chi((x_2 - 2\rho_j)/\rho_j)a_{\nu_j}(t)D_{x_1}U_j$$
$$+ \chi((x_2 - 2\rho_j)/\rho_j)b_{\nu_j}(t, x_1)2^{-\nu_j}D_{x_2}U_j. \tag{3.6.97}$$

From (3.6.75), we have

$$\partial_t U_j - f(t, x, D_x)U_j$$
$$= \psi_j\partial_t u_{\nu_j} - \chi((x_2 - 2\rho_j)/\rho_j)\psi_j a_{\nu_j}D_1 u_{\nu_j} - \chi((x_2 - 2\rho_j)/\rho_j)b_{\nu_j}2^{-\nu_j}D_{x_2}(u_{\nu_j}\psi_j)$$

$$= \underbrace{\chi((x_2 - 2\rho_j)/\rho_j)\psi_j(x_2)}_{\text{bounded in } L^2(\mathbb{R}_{x_2})} \overbrace{\big(\partial_t u_{\nu_j} - q_{\nu_j}(t, x_1, D_{x_1})u_{\nu_j}\big)(t, x_1)}^{\to 0 \text{ in } L^2(\mathbb{R}^2_{t, x_1})}$$
$$+ \underbrace{u_{\nu_j}(t, x_1)b_{\nu_j}(t, x_1)}_{\substack{\text{with } L^2(\mathbb{R}^2_{t,x_1}) \text{ norm} \leq C\nu_j^{1/2} \\ \text{from (3.6.75), (3.6.74)}}} \chi((x_2 - 2\rho_j)/\rho_j)(1 - 2^{-\nu_j}D_{x_2})\psi_j.$$

Since

$$\chi((x_2 - 2\rho_j)/\rho_j)(1 - 2^{-\nu_j}D_{x_2})\psi_j$$

$$= \psi\big(2(x_2 - 2\rho_j)/\rho_j\big)2^{1/2}\rho_j^{-1/2}e^{2i\pi 2^{\nu_j}x_2}$$

$$- 2^{-\nu_j}\psi\big(2(x_2 - 2\rho_j)/\rho_j\big)2^{1/2}\rho_j^{-1/2}e^{2i\pi 2^{\nu_j}x_2}2^{\nu_j}$$

$$- 2^{-\nu_j}\frac{2}{\rho_j}\psi'\big(2(x_2 - 2\rho_j)/\rho_j\big)2^{1/2}\rho_j^{-1/2}e^{2i\pi 2^{\nu_j}x_2}\frac{1}{2i\pi}$$

$$= -2^{-\nu_j}\frac{2}{\rho_j}\psi'\big(2(x_2 - 2\rho_j)/\rho_j\big)2^{1/2}\rho_j^{-1/2}e^{2i\pi 2^{\nu_j}x_2}\frac{1}{2i\pi},$$

we get $\lim_j \|\partial_t U_j - f(t,x,D_x)U_j\|_{L^2(\mathbb{R}^3)} = 0$ from the factor $2^{-\nu_j}$. Since $c(D_x)$ is of order 0, thus L^2 bounded, we have also

$$\lim_j \|\partial_t c(D_x)U_j - c(D_x)f(t,x,D_x)U_j\|_{L^2(\mathbb{R}^3)} = 0,$$

so that we are left with checking the term $(c(D_x) - 1)\partial_t U_j$ to verify (3.6.90). We know that $c(\xi) - 1$ does vanish if $\xi_2 \geq \max(2|\xi_1|, 1/2)$ and $|\xi| \geq 1$. By (3.6.75) and (3.6.95), we have

$$\partial_t U_j = \psi_j(x_2)(\partial_t u_{\nu_j})(t,x_1) = \overbrace{\psi_j(x_2)}^{\text{bounded in } L^2(\mathbb{R})} \overbrace{\big(\partial_t u_{\nu_j} - q_\nu(t,x_1,D_{x_1})u_{\nu_j}\big)}^{\to 0 \text{ in } L^2(\mathbb{R}^2)}$$

$$+ \psi_j(x_2)\big(a_{\nu_j}(t)D_{x_1}u_{\nu_j} + b_{\nu_j}(t,x_1)u_{\nu_j}\big),$$

so that we are reduced to studying

$$G_j(t,x) = \big(c(D_x) - 1\big)\Big(\psi_j(x_2)\underbrace{\big(a_{\nu_j}(t)D_{x_1}u_{\nu_j} + b_{\nu_j}(t,x_1)u_{\nu_j}\big)}_{g_j(t,x_1)}\Big).$$

We have since $c(\xi) = 1$ for $\xi_2 \geq 1 + 2|\xi_1|$,

$$\|G_j(t,\cdot)\|_{L^2(\mathbb{R}^2)}^2 \leq \iint |c(\xi) - 1|^2 |\hat{\psi}(\tfrac{\rho_j}{2}(\xi_2 - 2^{\nu_j}))|^2 \frac{\rho_j}{2}d\xi_2|\widehat{g_j}(t,\xi_1)|^2 d\xi_1$$

$$\leq \iint_{\xi_2 \leq 1 + 2|\xi_1|} |\hat{\psi}(\tfrac{\rho_j}{2}(\xi_2 - 2^{\nu_j}))|^2 \frac{\rho_j}{2}d\xi_2|\widehat{g_j}(t,\xi_1)|^2 d\xi_1$$

$$\leq \iint_{s \leq \frac{\rho}{2}(1+2|\xi_1|-2^{\nu_j})} |\hat{\psi}(s)|^2 ds|\widehat{g_j}(t,\xi_1)|^2 d\xi_1.$$

When $|\xi_1| \leq 2^{\nu_j - 2}$ in the integrand, (for $j \geq 2$ we have $\nu_j \geq 16$) we have $s \leq -\rho_j 2^{\nu_j - 3}$ and we can estimate the integral with respect to s by any negative power of $\rho_j 2^{\nu_j - 3}$. When $|\xi_1| \geq 2^{\nu_j - 2}$, we estimate the integral with respect to s

by $\|\hat{\psi}\|^2_{L^2(\mathbb{R})} = 1$. We have thus

$$\|G_j(t,\cdot)\|^2_{L^2(\mathbb{R}^2)}$$

$$\leq C_{N,\psi}(\rho_j 2^{\nu_j-3})^{-N}\int_{|\xi_1|\leq 2^{\nu_j-2}}|\widehat{g_j}(t,\xi_1)|^2 d\xi_1 + \int_{|\xi_1|\geq 2^{\nu_j-2}}|\widehat{g_j}(t,\xi_1)|^2 d\xi_1$$

$$\leq C_{2,\psi}(\rho_j 2^{\nu_j-3})^{-2}\int_{|\xi_1|\leq 2^{\nu_j-2}}|\widehat{g_j}(t,\xi_1)|^2 d\xi_1 + \int |\xi_1|^2 2^{4-2\nu_j}|\widehat{g_j}(t,\xi_1)|^2 d\xi_1$$

$$\leq 2^{-2\nu_j}(\rho_j^{-2}2^6 C_{2,\psi}\|g_j\|^2_{L^2(\mathbb{R}^2)} + 2^4\|D_{x_1}g_j\|^2_{L^2(\mathbb{R}^2)}).$$

Thanks to the estimates (3.6.74) and (3.6.76), the term

$$\rho_j^{-2}2^6 C_{2,\psi}\|g_j\|^2_{L^2(\mathbb{R}^2)} + 2^4\|D_{x_1}g_j\|^2_{L^2(\mathbb{R}^2)}$$

is increasing at most polynomially in ν_j, implying that

$$\lim_j \|G_j\|_{L^2(\mathbb{R}^3)} = 0,$$

since using the r_ν defined in Proposition 3.6.4, we have, with $\lim_j r_{\nu_j} = 0$,

$$\|G_j\|^2_{L^2(\mathbb{R}^3)} = \int_{|t|\leq r_{\nu_j}}\|G_j(t,\cdot)\|^2_{L^2(\mathbb{R}^2)}dt \leq 2r_{\nu_j}\sup_{|t|\leq r_{\nu_j}}\|G_j(t,\cdot)\|^2_{L^2(\mathbb{R}^2)}$$

$$\leq C_0 r_{\nu_j}\nu_j^{N_0}2^{-2\nu_j}.$$

The proof of Theorem 3.6.8 is complete. $\square$

3.6.3   More on the structure of the counterexample

In this section, we provide some more information on the counterexample, and in particular we show that it could be modified to yield a counterexample with some homogeneous principal symbol with respect to the momentum variable in $\mathbb{R}^3$. Going back to (3.6.85), we define

$$f_j(t,x_1,x_2,\xi_1,\xi_2) = \chi((x_2 - 2\rho_j)/\rho_j)(a_{\nu_j}(t)\xi_1 + b_{\nu_j}(t,x_1)\xi_2 2^{-\nu_j}),$$

and we see that,

for $m_1 \geq 1$, $\dfrac{\partial^{m_1}f_j}{\partial x_1^{m_1}} = \chi((x_2 - 2\rho_j)/\rho_j)\dfrac{\partial^{m_1}b_{\nu_j}}{\partial x_1^{m_1}}(t,x_1)\xi_2 2^{-\nu_j},$

for $m_2 \geq 1$, $\dfrac{\partial^{m_2}f_j}{\partial x_2^{m_2}} = \chi^{(m_2)}((x_2 - 2\rho_j)/\rho_j)\rho_j^{-m_2}(a_{\nu_j}(t)\xi_1 + b_{\nu_j}(t,x_1)\xi_2 2^{-\nu_j}),$

for $m_1 \geq 1, m_2 \geq 1$, $\dfrac{\partial^{m_1+m_2}f_j}{\partial x_1^{m_1}\partial x_2^{m_2}} = \chi^{(m_2)}((x_2 - 2\rho_j)/\rho_j)\rho_j^{-m_2}\dfrac{\partial^{m_1}b_{\nu_j}}{\partial x_1^{m_1}}(t,x_1)\xi_2 2^{-\nu_j}.$

As a consequence, we get from (3.6.74)

$$\text{for } m_1 \geq 1, \quad \left|\frac{\partial^{m_1} f_j}{\partial x_1^{m_1}}\right| \leq |\xi_2| \nu_j^{\frac{1}{2}+m_1} 2^{-\nu_j}, \tag{3.6.98}$$

$$\text{for } m_2 \geq 1, \quad \left|\frac{\partial^{m_2} f_j}{\partial x_2^{m_2}}\right| \leq \|\chi^{(m_2)}\|_{L^\infty} \left(|\xi_1| \nu_j^{-1/2} \rho_j^{-m_2} + |\xi_2| 2^{-\nu_j} \nu_j^{1/2} \rho_j^{-m_2}\right), \tag{3.6.99}$$

$$\text{for } m_1 \geq 1, m_2 \geq 1, \quad \left|\frac{\partial^{m_1+m_2} f_j}{\partial x_1^{m_1} \partial x_2^{m_2}}\right| \leq |\xi_2| \|\chi^{(m_2)}\|_{L^\infty} \rho_j^{-m_2} \nu_j^{\frac{1}{2}+m_1} 2^{-\nu_j}, \tag{3.6.100}$$

and also repeating the calculations with the derivatives with respect to t and using again the estimates (3.6.74), we obtain, since

$$\lim_j \nu_j^{-1/2} \rho_j^{-m_2} = \lim_j 2^{-j^2/2} 3^{jm_2} = 0, \quad \lim_j 2^{-\nu_j} \nu_j^{m_1+1} \rho_j^{-m_2} = \lim_j 2^{-2^{j^2}-1} 3^{jm_2} = 0,$$

that

$$\lim_j \left(\sup_{t,x,\xi} |(\partial_t^l \partial_{x_1}^{m_1} \partial_{x_2}^{m_2} \partial_\xi^\beta f_j)(t,x,\xi)| |\xi|^{-1+|\beta|}|\right) = 0. \tag{3.6.101}$$

With $\sigma_1(\xi,\tau)$ homogeneous of degree 0, supported in

$$\mathcal{N}_{\kappa_1} = \{(\xi,\tau) \in \mathbb{R}^2 \times \mathbb{R}, |\xi| \leq \kappa_1|\tau|\}, \quad \text{with } \kappa_1 > 0, \tag{3.6.102}$$

denoting by σ_1 the Fourier multiplier $\sigma_1(\xi,\tau)$, we have

$$\sigma_1 D_t U_j = \sigma_1 \big(D_t U_j + ic(D_x)f(t,x,D_x))U_j\big) - i\sigma_1 c(D_x)f(t,x,D_x)U_j,$$

and thus, with $L^2(\mathbb{R}^3)$ norms,

$$\|\sigma_1 D_t U_j\| \leq \|D_t U_j + ic(D_x)f(t,x,D_x)U_j\| + \|\sigma_1 c(D_x)f(t,x,D_x)U_j\|. \tag{3.6.103}$$

On the other hand, we have from (3.6.97)

$$
\begin{aligned}
\big(\sigma_1 c(D_x)&f(t,x,D_x)U_j\big)(t,x)\\
&= c(D_x)\big[\sigma_1, \chi((x_2-2\rho_j)/\rho_j)a_{\nu_j}(t)\big]D_{x_1}U_j\\
&\quad + c(D_x)\big[\sigma_1, \chi((x_2-2\rho_j)/\rho_j)b_{\nu_j}(t,x_1)2^{-\nu_j}\big]D_{x_2}U_j\\
&\quad + c(D_x)\chi((x_2-2\rho_j)/\rho_j)a_{\nu_j}(t)D_{x_1}\sigma_1 U_j\\
&\quad + c(D_x)\chi((x_2-2\rho_j)/\rho_j)b_{\nu_j}(t,x_1)2^{-\nu_j}D_{x_2}\sigma_1 U_j.
\end{aligned}
$$

With $\widetilde{\sigma_1}$ also homogeneous of degree 0 supported in $\mathcal{N}_{\kappa_2}$ and equal to 1 near the support of σ_1, we have

$$D_{x_1}\sigma_1 U_j = D_{x_1}\langle D_x\rangle^{-1}\widetilde{\sigma_1}\langle D_x\rangle D_t^{-1}\sigma_1 D_t U_j$$

which implies, noting from (3.6.78) that

$$\sup_{t,x_2} |\chi((x_2-2\rho_j)/\rho_j)a_{\nu_j}(t)| \leq \nu_j^{-1/2} \leq 1,$$

$$\|D_{x_1}\sigma_1 U_j\| \le \|\sigma_1 D_t U_j\|\kappa, \quad \text{with } \kappa = \sup_{(\tau,\xi)\in\text{supp }\widetilde{\sigma}_1}(1+|\xi|)|\tau|^{-1}, \qquad (3.6.104)$$

and we get similarly from (3.6.78)

$$\|D_{x_2}\sigma_1 U_j\| \le \|\sigma_1 D_t U_j\|\kappa \sup |W|. \qquad (3.6.105)$$

We thus obtain that

$$\|\sigma_1 c(D_x)f(t,x,D_x)U_j\| \le$$
$$\|[\sigma_1,\chi((x_2-2\rho_j)/\rho_j)a_{\nu_j}(t)]D_{x_1}U_j + [\sigma_1,\chi((x_2-2\rho_j)/\rho_j)b_{\nu_j}(t,x_1)2^{-\nu_j}]D_{x_2}U_j\|$$
$$+ (1+\sup|W|)\kappa\|\sigma_1 D_t U_j\|$$
$$= \|[\sigma_1, f_j(t,x,D_x)]U_j\| + (1+\sup|W|)\kappa\|\sigma_1 D_t U_j\|. \qquad (3.6.106)$$

From (3.6.103), (3.6.90), we obtain with $\lim_j \varepsilon_j = 0$,

$$\|\sigma_1 D_t U_j\| \le \varepsilon_j + \|[\sigma_1, f_j(t,x,D_x)]U_j\| + (1+\sup|W|)\kappa\|\sigma_1 D_t U_j\|. \qquad (3.6.107)$$

Since κ depends only on $\widetilde{\sigma}_1$, we can choose it such that

$$(1+\sup|W|)\kappa \le 1/2.$$

Now the operator $[\sigma_1, f_j(t,x,D_x)]$ has a symbol in $S^0_{1,0}(\mathbb{R}^3 \times \mathbb{R}^3)$ (see Lemma 4.9.16) and thus is bounded on $L^2(\mathbb{R}^3)$; moreover from (3.6.101), the semi-norms of this symbol tend to zero when j goes to infinity and we get from (3.6.107) and $\|U_j\|_{L^2(\mathbb{R}^3)} = 1$ that $\lim_j \|\sigma_1 D_t U_j\| = 0$. Since we have from (3.6.90) that $\lim_j \|(1-\sigma_1)(D_t U_j + ic(D_x)f(t,x,D_x)U_j)\| = 0$, we obtain that

$$\lim_j \|D_t U_j + (1-\sigma_1)c(D_x)f(t,x,D_x)U_j\| = 0. \qquad (3.6.108)$$

The operator $(1-\sigma_1)c(D_x)f(t,x,D_x)$ has a symbol in $S^1_{1,0}(\mathbb{R}^3 \times \mathbb{R}^3)$, since $1-\sigma_1$ is supported in $\{(\xi,\tau)\in\mathbb{R}^2\times\mathbb{R}, |\tau|\le C|\xi|\}$ and we may also choose $\sigma_1(\xi,\tau)$ valued in $[0,1]$, so that the operator $D_t + i(1-\sigma_1)c(D_x)f(t,x,D_x)$ satisfies condition $(\overline{\Psi})$. We have proven the following extension of Theorem 3.6.1.

Theorem 3.6.9. *There exists a C^∞ function $F : \mathbb{R}^3 \times \mathbb{R}^3 \to \mathbb{R}$ such that*

$$F \in S^1_{1,0}, \quad i.e., \quad \sup_{(x,\xi)\in\mathbb{R}^3\times\mathbb{R}^3}|(\partial_x^\alpha\partial_\xi^\beta F)(x,\xi)|(1+|\xi|)^{|\beta|-1} < +\infty,$$

$$F - \sum_{0\le j<N} F_j \in S^{1-N}_{1,0}, \quad F_j \in S^{1-j}_{1,0}, \quad F_j \text{ positively-homogeneous of degree } 1-j,$$

such that

$$F_0(x_1,x_2,x_3,\xi_1,\xi_2,\xi_3) > 0, y_3 > x_3 \Longrightarrow F_0(x_1,x_2,y_3,\xi_1,\xi_2,\xi_3) \ge 0,$$

and there exists a sequence $(u_k)_{k\ge 1}$ in $C_c^\infty(\mathbb{R}^3)$ with $\text{supp } u_k \subset \{x \in \mathbb{R}^3, |x| \le 1/k\}$ such that

$$\lim_{k\to+\infty}\left\|\frac{\partial u_k}{\partial x_3} - F(x,D)u_k\right\|_{L^2(\mathbb{R}^3)} = 0, \quad \|u_k\|_{L^2(\mathbb{R}^3)} = 1. \qquad (3.6.109)$$

Moreover, according to Remark 3.6.7, the principal symbol F_0 can be chosen of the form

$$F_0(x,\xi) = \xi_3 + ig_0(x_3, x_2)\big(\xi_1 + \xi_2 g_1(x_3, x_1, x_2)\big)\sigma_0(\xi_1, \xi_2, \xi_3), \qquad (3.6.110)$$

$$\text{with } g_0 \in C_b^\infty(\mathbb{R}^2; \mathbb{R}_+), \quad g_1 \in C_b^\infty(\mathbb{R}^3; \mathbb{R}), \quad \partial_{x_3} g_1 \geq 0, \qquad (3.6.111)$$

$$0 \leq \sigma_0 \in S_{1,0}^0, \text{ homogeneous of degree } 0, \text{ supported in } \{\xi_2 > 0\}. \qquad (3.6.112)$$

Remark 3.6.10. Let Ω be an open subset of $\mathbb{R}^n$, $x_0 \in \Omega$, and $P \in \Psi_{ps}^1(\Omega)$ a properly supported pseudo-differential operator. We want to compare the two following properties. We shall say that P is L^2-locally solvable at x_0 if

$$\exists V \in \mathcal{V}_{x_0}, \ \forall f \in L_{loc}^2(\Omega), \ \exists u \in L_{loc}^2(\Omega) \quad \text{with } Pu = f \text{ in } V. \qquad (3.6.113)$$

According to Lemmas 1.2.29, 1.2.30, (3.6.113) is equivalent to (1.2.39) for $s = 0, m = \mu = 1$. Another apparently weaker statement is

$$\forall f \in L_{loc}^2(\Omega), \ \exists u \in L_{loc}^2(\Omega), \exists V \in \mathcal{V}_{x_0} \quad \text{with } Pu = f \text{ in } V. \qquad (3.6.114)$$

Let us show that (3.6.114) implies (3.6.113). First of all, if L_0 is a compact neighborhood of x_0, since P is properly supported, there exists a compact subset K_0 of Ω such that $\text{supp}\, v \subset K_0^c \Longrightarrow \text{supp}\, Pv \subset L_0^c$. As a consequence, taking $\phi_0 \in C_c^\infty(\Omega)$ equal to 1 on a neighborhood of K_0 gives

$$Pu = P(\phi_0 u) + P\big(\underbrace{(1-\phi_0)u}_{\substack{\text{supported in} \\ K_0^c}}\big) \Longrightarrow (Pu)_{|\text{ int } L_0} = \big(P(\phi_0 u)\big)_{|\text{ int } L_0}.$$

In particular, if (3.6.114) holds, we may assume that $u \in L_{comp}^2(\Omega)$ with $\text{supp}\, u \subset \text{supp}\, \phi_0$. We consider now for $N \in \mathbb{N}^*$,

$$\mathcal{F}_N = \{f \in L_{loc}^2(\Omega), \exists u \in L_{\text{supp}\, \phi_0}^2(\Omega), \ Pu = f \text{ in } B(x_0, N^{-1}), \|u\|_{L^2} \leq N\}$$

and the union of the $\mathcal{F}_N$ is $L_{loc}^2(\Omega)$. Moreover each set $\mathcal{F}_N$ is convex, symmetric and closed and thus it follows from Baire's theorem that 0 must be an interior point of one of the $\mathcal{F}_N$, i.e., there exists $N_1 \in \mathbb{N}^*, K_1 \Subset \Omega, R_1 > 0$ such that

$$f \in L_{loc}^2(\Omega), \quad \|f\|_{L^2(K_1)} \leq R_1$$
$$\Longrightarrow \exists u \in L_{\text{supp}\, \phi_0}^2(\Omega), \ Pu = f \text{ in } B(x_0, N_1^{-1}), \quad (3.6.115)$$

which implies (3.6.113). As a result, when (1.2.39) for $s = 0, m = \mu = 1$ is violated,

$$\exists f_0 \in L_{loc}^2(\Omega), \ \forall u \in L_{loc}^2(\Omega), \forall V \in \mathcal{V}_{x_0}, \quad Pu \neq f_0 \text{ on } V. \qquad (3.6.116)$$

Remark 3.6.11. In particular the operator $P = \partial_{x_3} + F(x, D)$ with F given by Theorem 3.6.9 is a principal type first-order pseudo-differential operator in $\mathbb{R}^3$, satisfying condition (Ψ), that we may also assume properly supported[10]. From the previous theorem and remark, there exists $f \in L^2_{loc}(\mathbb{R}^3)$ such that for all V open neighborhoods of 0 in $\mathbb{R}^3$, for all $u \in L^2_{loc}(\mathbb{R}^3)$, $Pu \neq f$ in V. A similar argument gives also that the set of $f \in L^2_{loc}(\mathbb{R}^3)$ such that $Pu = f$ on a neighborhood of 0 for some $u \in L^2_{loc}(\mathbb{R}^3)$ is of first category: this set is the union of the $\mathcal{F}_N$ and if it were second category, we would obtain (3.6.115) and thus (3.6.113), that is local L^2-solvability at 0, which does not hold.

3.7 Condition (Ψ) does imply solvability with loss of 3/2 derivatives

3.7.1 Introduction

Statement of the result of that section

Let P be a properly supported principal-type pseudo-differential operator in a C^∞ manifold $\mathcal{M}$, with principal (complex-valued) symbol p. The symbol p is assumed to be a C^∞ homogeneous[11] function of degree m on $\dot{T}^*(\mathcal{M})$, the cotangent bundle minus the zero section. The principal type assumption that we shall use here is that

$$(x, \xi) \in \dot{T}^*(\mathcal{M}), \quad p(x, \xi) = 0 \Longrightarrow \partial_\xi p(x, \xi) \neq 0. \qquad (3.7.1)$$

Also, the operator P will be assumed of polyhomogeneous type, which means that its total symbol is equivalent to

$$p + \sum_{j \geq 1} p_{m-j},$$

where p_k is a smooth homogeneous function of degree k on $\dot{T}^*(\mathcal{M})$.

Theorem 3.7.1. *Let P be as above, such that its principal symbol p satisfies condition (Ψ) on $\dot{T}^*(\mathcal{M})$ (see Definition 3.2.1). Let s be a real number. Then, for all $x \in \mathcal{M}$, there exists a neighborhood V such that*

$$\text{for all } f \in H^s_{loc}, \text{ there exists } u \in H^{s+m-\frac{3}{2}}_{loc} \text{ such that } \quad Pu = f \quad \text{in } V.$$

Proof. The proof of this theorem is given in Section 3.7.2. $\square$

Note that our loss of derivatives is equal to 3/2. Section 3.6 proves that solvability with loss of one derivative does *not* follow from condition (Ψ), so we have to content ourselves with a loss strictly greater than 1. However, the number

---

[10]Modifying P by an operator R of order -1 (see Proposition 1.2.5) does not change the property (3.6.109) since (see (4.3.9)) $\|Ru_k\|_{L^2(\mathbb{R}^3)} \leq C\|u_k\|_{H^{-1}} \leq C'\text{diameter}(\text{supp } u_k)\|u_k\|_{L^2} \to 0$ when $k \to +\infty$.

[11]Here and in the sequel, "homogeneous" will always mean positively-homogeneous.

3/2 is not likely to play any significant rôle and one should probably expect a loss of $1+\epsilon$ derivatives under condition (Ψ). In fact, for the counterexamples given in Section 3.6 (see also [92], [77]), it seems (but it has not been proven) that there is only a "logarithmic" loss, i.e., the solution u should satisfy $u \in \log \langle D_x \rangle (H^{s+m-1})$.

Nevertheless, the methods used in the present article are strictly limited to providing a 3/2 loss. We refer the reader to our Subsection 3.7.2 for an argument involving a Hilbertian lemma on a simplified model. This is of course in sharp contrast with operators satisfying condition (P) such as differential operators satisfying condition (Ψ). Let us recall that condition (P) is simply ruling out any change of sign of Im(zp) along the oriented Hamiltonian flow of Re(zp). Under condition (P) ([8]) or under condition (Ψ) in two dimensions ([89]), local solvability occurs with a loss of one derivative, the "optimal" loss, and in fact the same as for $\partial/\partial x_1$ (see Section 3.5). One should also note that the semi-global existence theorems of [67] (see also Theorem 26.11.2 in [74]) involve a loss of $1+\epsilon$ derivatives. However in that case there is no known counterexample which would ensure that this loss is unavoidable.

Remark 3.7.2. Theorem 3.7.1 will be proved by a multiplier method, involving the computation of $\langle Pu, Mu \rangle$ with a suitably chosen operator M. It is interesting to notice that, the greater is the loss of derivatives, the more regular should be the multiplier in the energy method. As a matter of fact, the Nirenberg-Treves multiplier of [114] is not even a pseudo-differential operator in the $S^0_{1/2,1/2}$ class, since it could be as singular as the operator sign D_{x_1}; this does not create any difficulty, since the loss of derivatives is only 1. On the other hand, in [35], [95], where estimates with loss of 2 derivatives are handled, the regularity of the multiplier is much better than $S^0_{1/2,1/2}$, since we need to consider it as an operator of order 0 in an asymptotic class defined by an admissible metric on the phase space.

Remark 3.7.3. For microdifferential operators acting on microfunctions, the sufficiency of condition (Ψ) was proven by J.-M. Trépreau [139](see also [75]), so the present theorems are concerned only with the C^∞ category.

3.7.2 Energy estimates

Some a priori estimates and loss of derivatives

In this section, we prove that, at least when a factorization occurs, it is possible to prove an energy estimate with loss of 3/2 derivatives (the loss is always counted with respect to the elliptic case). The factorization assumption is a very strong one, and is far from being satisfied under condition (Ψ). However, the fact that a rather straightforward argument does exist under that hypothesis is also an indication that a natural multiplier can be found. When condition (P) holds for $D_t + iq(t, x, D_x)$, the multiplier in the energy method is essentially a quantization of sign $q(t, x, \xi)$, which can be very singular. A way of reformulating condition $(\overline{\Psi})$ for $D_t + iq(t, x, D_x)$ is to say that a properly defined sign $q(t, x, \xi)$ is actually non-decreasing: nevertheless no quantization of that rough function seems at hand to handle the estimates, but since we have to give up the hope of proving an estimate with loss of one derivative (see Section 3.6, [92]), we may accept some smoother multiplier, expressed like the signed distance to the change-of-sign set (see (3.2.19)). That program has been initially carried out by N. Dencker in [35] with loss of two derivatives, a loss which was cut off to 3/2 later in the paper [98]. The details of the argument are quite involved and exposed below ; in this section, we study a far simpler model, which might serve as a motivation for the reader.

Let us study the model-case

$$L = D_t + iA_0 B_1, \quad A_0 \in \mathrm{op}(S^0), B_1 \in \mathrm{op}(S^1),$$

with $S^j = S_{scl}^j = S(\Lambda^j, \Lambda^{-1}|dX|^2)$ (see page 67) and real-valued Weyl symbols such that $A_0 \geq c_0 \Lambda^{-1}, \dot{B}_1 \geq 0$. We compute, using the notation

$$\|u\| = \left(\int |u(t)|^2 dt \right)^{1/2}, \quad |v| = \|v\|_{\mathbb{H}}, \quad \mathbb{H} = L^2(\mathbb{R}^n), \quad |u|_\infty = \sup_{t \in \mathbb{R}} |u(t)|,$$

$$2 \operatorname{Re}\langle Lu, iB_1 u \rangle = \langle \dot{B}_1(t)u(t), u(t) \rangle + 2 \operatorname{Re}\langle A_0 B_1 u, B_1 u \rangle \geq 2c_0 \Lambda^{-1} \|B_1 u\|^2.$$

As a consequence, for $\operatorname{supp} u \subset [-T, T]$,

$$2 \operatorname{Re}\langle Lu, iB_1 u \rangle + 2 \operatorname{Re}\langle Lu, iH(t - T_0)u \rangle$$

$$\geq c_0 \Lambda^{-1} \|B_1 u\|^2 + |u|_\infty^2 + \|A_0^{1/2} B_1 u\|^2 + 2 \operatorname{Re}\langle A_0^{1/2} B_1 u, iH_{T_0} A_0^{1/2} u \rangle$$

$$\geq c_0 \Lambda^{-1} \|B_1 u\|^2 + |u|_\infty^2 (1 - \sup_{|t| \leq T} \|A_0(t)\| T)$$

$$\text{(for T small enough)} \quad \geq c_0 \Lambda^{-1} \|B_1 u\|^2 + \frac{1}{2} |u|_\infty^2, \quad (3.7.2)$$

so that $c_0^{-1} \Lambda \|Lu\|^2 + c_0 \Lambda^{-1} \|B_1 u\|^2 + 2\|Lu\|\|u\| \geq c_0 \Lambda^{-1} \|B_1 u\|^2 + \frac{1}{2}|u|_\infty^2$ and thus

$$(c_0^{-1} \Lambda + 1)\|Lu\|^2 + T|u|_\infty^2 \geq \frac{1}{2}|u|_\infty^2$$

entailing for $T \leq 1/4$, $(c_0^{-1}\Lambda + 1)\|Lu\|^2 \geq \frac{1}{4}|u|_\infty^2$, which gives $\|Lu\| \gg \Lambda^{-1/2}\|u\|$, an estimate with loss of 3/2 derivatives.

The next question is obviously: how do we manage to get the estimate $A_0 \geq \Lambda^{-1}$? Assuming $A_0 \geq -C\Lambda^{-1}$, we can always consider instead $A_0 + (C+1)\Lambda^{-1} \geq \Lambda^{-1}$; now this modifies the operator L and although our estimate is too weak to absorb a zeroth order perturbation, it is enough to check that the energy method is stable by zeroth order perturbation. We consider then

$$D_t + iA_0 B_1 + S + iR, \quad A_0 \geq \Lambda^{-1}, \quad S, R \in \mathrm{op}(S^0).$$

Inspecting the method above, we see that S will not produce any trouble, since we shall commute it with B_1, producing an operator of order 0. The terms produced by R are more delicate to handle: we shall have to deal with

$$2\langle Ru, B_1 u \rangle + 2\langle Ru, H_{T_0} u \rangle.$$

The second term is L^2 bounded and can be absorbed. There is no simple way to absorb the first term, which is of size $\|B_1u\|\|u\|$, way too large with respect to the terms that we dominate. However we can consider the L^2-bounded invertible operator $U(t)$ (which is in $\mathrm{op}(S^0)$ and selfadjoint) such that $U(0) = \mathrm{Id}$ and $\dot{U}(t) = -U(t)R(t)$ so that

$$
\begin{aligned}
L &= D_t + iR + iA_0B_1 + S = U(t)^{-1}D_tU(t) + iA_0B_1 + S \\
&= U(t)^{-1}\big(D_t + iA_0B_1 + S\big)U(t) - U(t)^{-1}\big[iA_0B_1 + S, U(t)\big] \\
&= U(t)^{-1}\Big(D_t + iA_0B_1 + S + \big[U(t), iA_0B_1 + S\big]U(t)^{-1}\Big)U(t). \quad (3.7.3)
\end{aligned}
$$

Now the term $\big[U(t), iA_0B_1\big]U(t)^{-1}$ has a real-valued principal symbol in S^0 and amounts to a modification of S, up to unimportant terms of order -1. The term $\big[U(t), S\big]U(t)^{-1}$ is of order -1 and can be absorbed. We have proven the following lemma.

Lemma 3.7.4. *Let $\Lambda \geq 1$ be given. We consider the metric $G = |dx|^2 + \Lambda^{-2}|d\xi|^2$ on $\mathbb{R}^n \times \mathbb{R}^n$. Let $a_0(t, x, \xi)$ be in $S(1, G)$ such that $a_0(t, x, \xi) \geq 0$. Let $b_1(t, x, \xi)$ be real-valued and in $S(\Lambda, G)$ such that*

$$
\big(b_1(t, x, \xi) - b(s, x, \xi)\big)(t - s) \geq 0.
$$

Let $r(t, x, \xi)$ be a complex-valued symbol in $S(1, G)$. Assuming that a_0, b_1, r_0 are continuous functions, there exists a constant $C > 0$ depending only on the semi-norms of the symbols a_0, b_1, r_0, such that, for all $u \in C_c^1([-T, T], L^2(\mathbb{R}^n))$ with $CT \leq 1$,

$$
C\|Lu\|_{L^2(\mathbb{R}^{n+1})} \geq \Lambda^{-1/2}T^{-1}\|u\|_{L^2(\mathbb{R}^{n+1})}.
$$

Preliminaries

Definition 3.7.5. Let $T > 0$ be given. With m defined in (3.2.95), we define for $|t| \leq T$,

$$
M(t) = m(t, X)^{\mathrm{Wick}}, \quad (3.7.4)
$$

where the Wick quantization is given by Definition 2.4.1.

Lemma 3.7.6. *With $T > 0$ and M given above, we have with ρ_1 given in (3.2.90), R defined in (3.2.66), for $|t| \leq T$, $\Lambda \geq 1$,*

$$
\begin{aligned}
\frac{d}{dt}M(t) &\geq \frac{1}{2T}\rho_1(t, X)^{\mathrm{Wick}} + \frac{1}{T}(\delta_0^2)^{\mathrm{Wick}}\Lambda^{-1/2} \\
&\geq \frac{1}{2C_0T}R^{\mathrm{Wick}} + T^{-1}(\delta_0^2)^{\mathrm{Wick}}\Lambda^{-1/2} \\
&\geq \frac{1}{2^{3/2}C_0T}(\langle\delta_0\rangle^2)^{\mathrm{Wick}}\Lambda^{-1/2}. \quad (3.7.5)
\end{aligned}
$$

$$|\Theta(t,X)| \le \rho_1(t,X) \le 2\frac{\langle\delta_0(t,X)\rangle}{\nu(t,X)^{1/2}}, \tag{3.7.6}$$

$$T^{-1}|\delta_0(t,X)\eta(t,X)| \le 4|\delta_0(t,X)|, \tag{3.7.7}$$

$$T^{-1}|\delta'_{0X}(t,X)\eta(t,X)| + T^{-1}|\delta_{0X}(t,X)\eta'_X(t,X)| \le 12. \tag{3.7.8}$$

Proof. The derivative in (3.7.5) is taken in the distribution sense, i.e., the first inequality in (3.7.5) means that (2.4.8) is satisfied with

$$a(t,X) = m(t,X) - \frac{1}{2T}\int_{-T}^{t}\rho_1(s,X)ds - \frac{1}{T}\Lambda^{-1/2}\int_{-T}^{t}\delta_0(s,X)^2 ds.$$

It follows in fact from (3.2.99). The other inequalities in (3.7.5) follow directly from (3.2.99) and the fact that the Wick quantization is positive (see (2.4.6)). The inequality (3.7.6) is (3.2.96) and (3.7.7) follows from (3.2.98) whereas (3.7.8) is a consequence of (3.2.20) (3.2.98) and Definition 3.2.11. $\qquad\square$

Lemma 3.7.7. *Using the definitions above and the notation (2.4.7), we have*

$$\Theta(t,\cdot)*\exp-2\pi\Gamma \in S(\langle\delta_0(t,\cdot)\rangle\nu(t,\cdot)^{-1/2},\Gamma), \tag{3.7.9}$$

$$\delta_0(t,\cdot)*\exp-2\pi\Gamma \in S(\langle\delta_0(t,\cdot)\rangle,\Gamma), \tag{3.7.10}$$

$$\delta'_{0X}(t,\cdot)*\exp-2\pi\Gamma \in S(1,\Gamma), \tag{3.7.11}$$

$$T^{-1}\eta(t,\cdot)*\exp-2\pi\Gamma \in S(1,\Gamma), \tag{3.7.12}$$

$$T^{-1}\eta(t,\cdot)'_X*\exp-2\pi\Gamma \in S(\Lambda^{-1/2},\Gamma), \tag{3.7.13}$$

with semi-norms independent of $T \le 1$ and of t for $|t| \le T$. According to Definition 2.2.15, the function $X \mapsto \langle\delta_0(t,X)\rangle$ is a Γ-weight.

Proof. The last statement follows from (3.2.20). The inequalities ensuring (3.7.9)–(3.7.13) are then immediate consequences of Lemmas 3.7.6 and 4.9.3. $\qquad\square$

Stationary estimates for the model cases

Let $T > 0$ be given and $Q(t) = q(t)^w$ given by (3.2.12), (3.2.13). We define $M(t)$ according to (3.7.4). We consider

$$\mathrm{Re}\big(Q(t)M(t)\big) = \frac{1}{2}Q(t)M(t) + \frac{1}{2}M(t)Q(t) = P(t). \tag{3.7.14}$$

We have, omitting now the variable t fixed throughout all this subsection,

$$P = \mathrm{Re}\left[q^w\big(\delta_0(1+T^{-1}\eta)\big)^{\mathrm{Wick}} + q'\Theta^{\mathrm{Wick}}\right]. \tag{3.7.15}$$

[1] *Let us assume first that $q = \Lambda^{-1/2}\mu^{1/2}\nu^{1/2}\beta e_0$ with $\beta \in S(\nu^{1/2},\nu^{-1}\Gamma), 1 \le e_0 \in S(1,\nu^{-1}\Gamma)$ and $\delta_0 = \beta$. Moreover, we assume $0 \le T^{-1}\eta \le 4, T^{-1}|\eta'| \le 4\Lambda^{-1/2}$,*

$|\Theta| \le C\langle\delta_0\rangle\nu^{-1/2}$. *Here* Λ, μ, ν *are assumed to be positive constants such that* $\Lambda \ge \mu \ge \nu \ge 1$.

Then using Lemma 4.9.5 with

$$a_1 = \beta e_0, \quad m_1 = \langle\beta\rangle, \quad a_2 = (1 + T^{-1}\eta)e_0^{-1}, m_2 = \nu^{-1/2},$$

we get, with obvious notation,

$$\left(\delta_0 e_0\right)^{\text{Wick}}\left(e_0^{-1}(1 + T^{-1}\eta)\right)^{\text{Wick}} = \left(\delta_0(1 + T^{-1}\eta)\right)^{\text{Wick}} + S(\langle\delta_0\rangle\nu^{-1/2}, \Gamma)^w$$

and as a consequence from Proposition 2.4.3, we obtain, with

$$\beta_0 = \beta e_0, \quad \eta_0 = e_0^{-1}(1 + T^{-1}\eta), \tag{3.7.16}$$

the identity

$$\left(\beta_0^{\ w} + S(\nu^{-1/2}, \nu^{-1}\Gamma)^w\right)\eta_0^{\text{Wick}} = \left(\delta_0(1 + T^{-1}\eta)\right)^{\text{Wick}} + S(\langle\delta_0\rangle\nu^{-1/2}, \Gamma)^w,$$

entailing $\left(\delta_0(1 + T^{-1}\eta)\right)^{\text{Wick}} = \beta_0^{\ w}\eta_0^{\text{Wick}} + S(\langle\delta_0\rangle\nu^{-1/2}, \Gamma)^w$. As a result, we have

$$QM = \Lambda^{-1/2}\mu^{1/2}\nu^{1/2}\beta_0^w\beta_0^w\eta_0^{\text{Wick}} + \beta_0^w S(\overbrace{\Lambda^{-1/2}\mu^{1/2}\nu^{1/2}\langle\delta_0\rangle\nu^{-1/2}}^{=\Lambda^{-1/2}\mu^{1/2}\langle\delta_0\rangle}, \Gamma)^w$$
$$+ \beta_0^w S(\underbrace{\Lambda^{-1/2}\mu^{1/2}\nu^{1/2}\langle\delta_0\rangle\nu^{-1/2}}_{=\Lambda^{-1/2}\mu^{1/2}\langle\delta_0\rangle}, \Gamma)^w.$$

This implies that, with $\gamma_0 = 1/\sup e_0 > 0$, (so that $1 \le e_0 \le \gamma_0^{-1}$), we have

$$2\operatorname{Re} QM = 2\Lambda^{-1/2}\mu^{1/2}\nu^{1/2}\beta_0^w\eta_0^{\text{Wick}}\beta_0^w + 2\operatorname{Re}\beta_0^w \Lambda^{-1/2}\mu^{1/2}\nu^{1/2}\overbrace{\left[\beta_0^w, \eta_0^{\text{Wick}}\right]}^{\in S(\nu^{-1/2}, \Gamma)^w.}$$
$$+ \operatorname{Re}\beta_0^w S(\Lambda^{-1/2}\mu^{1/2}\langle\delta_0\rangle, \Gamma)^w$$

and thus

$$2\operatorname{Re} QM = 2\Lambda^{-1/2}\mu^{1/2}\nu^{1/2}\beta_0^w\eta_0^{\text{Wick}}\beta_0^w + \operatorname{Re}\beta_0^w S(\Lambda^{-1/2}\mu^{1/2}\langle\delta_0\rangle, \Gamma)^w$$
$$\underbrace{\ge}_{\substack{\text{since } \eta_0 \ge e_0^{-1} \\ \text{from } \eta \ge 0 \text{ in } (3.2.98)}} 2\Lambda^{-1/2}\mu^{1/2}\nu^{1/2}\beta_0^w\gamma_0\beta_0^w + \beta_0^w b_0^w + \bar{b}_0^w\beta_0^w, \tag{3.7.17}$$

with $b_0 \in S(\Lambda^{-1/2}\mu^{1/2}\langle\delta_0\rangle, \Gamma)$. With the notation $\lambda = \Lambda^{-1/2}\mu^{1/2}\nu^{1/2}\gamma_0$, we use the identity,

$$\Lambda^{1/2}\mu^{1/2}\nu^{1/2}\beta_0^w\gamma_0\beta_0^w + \beta_0^w b_0^w + \bar{b}_0^w\beta_0^w$$
$$= \left(\lambda^{1/2}\beta_0^w + \lambda^{-1/2}\bar{b}_0^w\right)\left(\lambda^{1/2}\beta_0^w + \lambda^{-1/2}b_0^w\right) - \lambda^{-1}\bar{b}_0^w b_0^w,$$

so that from (3.7.17), we obtain

with b_1 real-valued in $S(\underbrace{\Lambda^{1/2}\mu^{-1/2}\nu^{-1/2}\Lambda^{-1}\mu\langle\delta_0\rangle^2}_{\Lambda^{-1/2}\mu^{1/2}\nu^{-1/2}\langle\delta_0\rangle^2},\Gamma)$,

the inequality

$$2\operatorname{Re}QM + b_1^w \geq \Lambda^{-1/2}\mu^{1/2}\nu^{1/2}\gamma_0\beta_0^w\beta_0^w. \qquad (3.7.18)$$

Using now (4.9.3), we get, with a "fixed" constant C, that

$$b_1^w \leq C\Lambda^{-1/2}\mu^{1/2}\nu^{-1/2}(1+\beta^2)^{\text{Wick}}$$
$$= C\Lambda^{-1/2}\mu^{1/2}\nu^{-1/2}\operatorname{Id} + C\Lambda^{-1/2}\mu^{1/2}\nu^{-1/2}(\beta_0^2 e_0^{-2})^{\text{Wick}}$$
$$\leq C\Lambda^{-1/2}\mu^{1/2}\nu^{-1/2}\operatorname{Id} + C\Lambda^{-1/2}\mu^{1/2}\nu^{-1/2}(\beta_0^2)^{\text{Wick}},$$

and since, from Proposition 2.4.3(2), we have

$$(\beta_0^2)^{\text{Wick}} = (\beta_0^2)^w + S(1,\nu^{-1}\Gamma)^w = \beta_0^w\beta_0^w + S(1,\nu^{-1}\Gamma)^w,$$

the inequality (3.7.17) implies

$$2\operatorname{Re}QM + C\Lambda^{-1/2}\mu^{1/2}\nu^{-1/2}\operatorname{Id} + C\Lambda^{-1/2}\mu^{1/2}\nu^{-1/2}\beta_0{}^w\beta_0{}^w$$
$$+ S(\Lambda^{-1/2}\mu^{1/2}\nu^{-1/2},\nu^{-1}\Gamma)^w$$
$$\geq 2\operatorname{Re}QM + b_1^w \geq \Lambda^{-1/2}\mu^{1/2}\nu^{1/2}\gamma_0\beta_0^w\beta_0^w,$$

so that

$$\operatorname{Re}QM + S(\Lambda^{-1/2}\mu^{1/2}\nu^{-1/2},\Gamma)^w$$
$$\geq \beta_0^w\beta_0^w(\Lambda^{-1/2}\mu^{1/2}\nu^{1/2}\gamma_0 - C'\Lambda^{-1/2}\mu^{1/2}\nu^{-1/2}). \qquad (3.7.19)$$

The rhs of (3.7.19) is non-negative provided $\nu \geq C'\gamma_0^{-1}$ and since $C'\gamma_0^{-1}$ is a fixed constant, we may first suppose that this condition is satisfied; if it is not the case, we would have that ν is bounded above by a fixed constant and since $\nu \geq 1$, that would imply $q \in S(\Lambda^{-1/2}\mu^{1/2},\Gamma)$ and $P \in S(\Lambda^{-1/2}\mu^{1/2},\Gamma)^w$. In both cases, we get

$$\operatorname{Re}QM + S(\Lambda^{-1/2}\mu^{1/2}\nu^{-1/2},\Gamma)^w \geq 0. \qquad (3.7.20)$$

[2] *Let us assume now that* $q \geq 0$, $q \in S(\Lambda^{-1/2}\mu^{1/2}\nu,\nu^{-1}\Gamma)$, $\gamma_0\nu^{1/2} \leq \delta_0 \leq \gamma_0^{-1}\nu^{1/2}$ *with a positive fixed constant* γ_0. *Moreover, we assume* $0 \leq T^{-1}\eta \leq 4, T^{-1}|\eta'| \leq 4\Lambda^{-1/2}$, $|\Theta(X)| \leq C$, Θ *real-valued. Here* Λ,μ,ν *are assumed to be positive constants such that* $\Lambda \geq \mu \geq \nu \geq 1$.

We start over our discussion from the identity (3.7.15):

$$P = \operatorname{Re}\left[q^w\left(\delta_0(1+T^{-1}\eta)+\Theta\right)^{\text{Wick}}\right]. \qquad (3.7.21)$$

We define

$$a_0 = \delta_0(1 + T^{-1}\eta) \tag{3.7.22}$$

and we note that $\gamma_0\nu^{1/2} \leq a_0 \leq 5\gamma_0^{-1}\nu^{1/2}$.

Remark 3.7.8. We may assume that $\nu^{1/2} \geq 2C/\gamma_0$ which implies $C \leq \frac{1}{2}\gamma_0\nu^{1/2}$ so that

$$\frac{1}{2}\gamma_0\nu^{1/2} \leq a_0 + \Theta \leq (5\gamma_0^{-1} + C\gamma_0/2)\nu^{1/2}. \tag{3.7.23}$$

In fact if $\nu^{1/2} < 2C/\gamma_0$ we have $\left(\delta_0(1 + T^{-1}\eta) + \Theta\right)^{\text{Wick}} \in S(1,\Gamma)^w$, $\Lambda^{1/2}\mu^{-1/2}q \in S(1,\Gamma)$ and $P \in S(\Lambda^{-1/2}\mu^{1/2},\Gamma)^w$ so that (3.7.20) holds also in that case.

We have the identity

$$q^w\left(\delta_0(1 + T^{-1}\eta)\right)^{\text{Wick}} = q^w a_0^{\text{Wick}} \quad \text{with} \quad \begin{cases} \gamma_0\nu^{1/2} \leq a_0 \leq 5\gamma_0^{-1}\nu^{1/2}, \\ |a_0'| \leq 10 + \gamma_0^{-1}4\frac{|\delta_0|}{\Lambda^{1/2}} \leq 10 + 4\gamma_0^{-1}. \end{cases}$$
$$\tag{3.7.24}$$

The Weyl symbol of $(a_0 + \Theta)^{\text{Wick}}$, which is

$$a = (a_0 + \Theta) * 2^n \exp -2\pi\Gamma, \tag{3.7.25}$$

belongs to $S_1(\nu^{1/2}, \nu^{-1}\Gamma)$ (see Definition 4.9.10): this follows from Lemma 4.9.12 and (3.7.24) for $a_0 * \exp -2\pi\Gamma$ and is obvious for $\Theta * 2^n \exp -2\pi\Gamma$ which belongs to $S(1,\Gamma)$. Moreover the estimates (3.7.23) imply that the symbol a satisfies

$$\frac{1}{2}\gamma_0\nu^{1/2} \leq a(X) = \int (a_0 + \Theta)(X + Y)2^n e^{-2\pi|Y|^2} dY \leq (5\gamma_0^{-1} + C\gamma_0/2)\nu^{1/2}.$$
$$\tag{3.7.26}$$

As a result, the symbol $b = a^{1/2}$ belongs to $S_1(\nu^{1/4}, \nu^{-1}\Gamma)$ and $1/b$ belongs to $S_1(\nu^{-1/4}, \nu^{-1}\Gamma)$: we have

$$2^{-1/2}\gamma_0^{1/2}\nu^{1/4} \leq |b| \leq (5\gamma_0^{-1} + C\gamma_0/2)^{1/2}\nu^{1/4}$$

and moreover $a' = a_0' * 2^n \exp -2\pi\Gamma + \Theta * 2^n(\exp -2\pi\Gamma)'$, so that, using

$$|a'| \leq 10 + 4\gamma_0^{-1} + C\|2^n(\exp -2\pi\Gamma)'\|_{L^1(\mathbb{R}^{2n})} = C_1,$$

we get $2|b'| = |a'(X)|a(X)^{-1/2} \leq 2^{1/2}\gamma_0^{-1/2}\nu^{-1/4}C_1$, and the derivatives of $a^{1/2}$ of order $k \geq 2$ are a sum of terms of type

$$a^{\frac{1}{2}-m}a^{(k_1)}\dots a^{(k_m)}, \quad \text{with } k_1 + \dots + k_m = k, \text{ all } k_j \geq 1,$$

which can be estimated by $C\nu^{\frac{1}{4}-\frac{m}{2}} \leq C\nu^{-\frac{1}{4}}$ since $m \geq 1$. Similarly we obtain that $b^{-1} \in S_1(\nu^{-1/4}, \nu^{-1}\Gamma)$. From Lemma 4.9.11, we have

$$b^w b^w = a^w + S(\nu^{-1/2}, \Gamma)^w = (a_0 + \Theta)^{\text{Wick}} + S(\nu^{-1/2}, \Gamma)^w,$$

which means $\quad (a_0 + \Theta)^{\text{Wick}} = b^w b^w + r_0^w, \quad r_0 \in S(\nu^{-1/2}, \Gamma), \text{ real-valued.}$

Using that $1/b$ belongs to $S_1(\nu^{-1/4}, \nu^{-1}\Gamma)$, we write, using again Lemma 4.9.11,

$$\left(b + \frac{1}{2}b^{-1}r_0\right)^w \left(b + \frac{1}{2}b^{-1}r_0\right)^w = b^w b^w + r_0^w + S(\nu^{-1/4}\nu^{-1/2}\nu^{-1/4}, \Gamma)^w,$$

which gives,

$$(a_0 + \Theta)^{\text{Wick}} = \left(b + \frac{1}{2}b^{-1}r_0\right)^w \left(b + \frac{1}{2}b^{-1}r_0\right)^w + S(\nu^{-1}, \Gamma)^w. \qquad (3.7.27)$$

Note that $b_0 = b + \frac{1}{2}b^{-1}r_0$ belongs to $S_1(\nu^{1/4}, \nu^{-1}\Gamma)$ since it is true for b and $b^{-1}r_0 \in S(\nu^{-3/4}, \Gamma)$: we get then

$$2\operatorname{Re}\big(q^w(a_0+\Theta)^{\text{Wick}}\big) = 2b_0^w q^w b_0^w + \big[\ \overbrace{[q^w, b_0^w]}^{S(\Lambda^{-1/2}\mu^{1/2}\nu^{1/4}\nu^{-1/4}, \Gamma)}\underset{S(\Lambda^{-1/2}\mu^{1/2}\nu^{1/2}\nu^{-1/4}, \Gamma)}{}, b_0^w\big] + \operatorname{Re}(q^w S(\nu^{-1}, \Gamma)^w)$$

so that

$$P = b_0^w q^w b_0^w + S(\Lambda^{-1/2}\mu^{1/2}, \Gamma)^w. \qquad (3.7.28)$$

Using now the Fefferman-Phong inequality ([44], Theorem 18.6.8 in [73]) for the non-negative symbol q, we get

$$b_0^w q^w b_0^w = b_0^w(q^w + C\Lambda^{-1/2}\mu^{1/2}\nu^{-1})b_0^w + S(\Lambda^{-1/2}\mu^{1/2}\nu^{-1/2}, \Gamma)^w$$
$$\geq S(\Lambda^{-1/2}\mu^{1/2}\nu^{-1/2}, \Gamma)^w,$$

so that, from (3.7.28) we get eventually

$$\operatorname{Re}(QM) + S(\Lambda^{-1/2}\mu^{1/2}, \Gamma)^w \geq 0. \qquad (3.7.29)$$

Stationary estimates

Let $T > 0$ be given and $Q(t) = q(t)^w$ given by (3.2.12), (3.2.13). We define $M(t)$ according to (3.7.4). We consider

$$\operatorname{Re}\big(Q(t)M(t)\big) = \frac{1}{2}Q(t)M(t) + \frac{1}{2}M(t)Q(t) = P(t). \qquad (3.7.30)$$

We have, omitting now the variable t fixed throughout all the present subsection,

$$P = \operatorname{Re}\left[q^w\big(\delta_0(1 + T^{-1}\eta)\big)^{\text{Wick}} + q^w\Theta^{\text{Wick}}\right]. \qquad (3.7.31)$$

Lemma 3.7.9. *Let p be the Weyl symbol of P defined in (3.7.31) and $\widetilde{\Theta} = \Theta *$ $2^n \exp -2\pi\Gamma$, where Θ is defined in (3.2.94) (and satisfies (3.2.96)). Then we have*

$$p(t, X) \equiv p_0(t, X) = q(t, X)\big(\delta_0(1 + T^{-1}\eta) * 2^n \exp -2\pi\Gamma\big) + q(t, X)\widetilde{\Theta}(t, X),$$
$$(3.7.32)$$

modulo $S(\Lambda^{-1/2}\mu^{1/2}\nu^{-1/2}\langle\delta_0\rangle, \Gamma)$.

Proof. Using the results of Section 3.2.3, we know that the symbol $X \mapsto q(t,X)$ belongs to the class

$$S(\Lambda^{-1/2}\mu(t,X)^{1/2}\nu(t,X),\nu(t,X)^{-1}\Gamma)$$

as shown in Lemma 3.2.14. In fact from (3.2.31) we know that the symbol q belongs to $S(\Lambda^{-1/2}\mu^{1/2}\nu,\nu^{-1}\Gamma)$, and from (3.7.6) and Lemma 4.9.3, we obtain, using Theorem 2.3.8 (see also Theorem 18.5.5 in [73]),

$$q\sharp\widetilde{\Theta} = q\widetilde{\Theta} + \frac{1}{4i\pi}\left\{q,\widetilde{\Theta}\right\} + S(\Lambda^{-1/2}\mu^{1/2}\nu^{-1/2}\langle\delta_0\rangle,\Gamma).$$

This implies that $\mathrm{Re}\,(q\sharp\widetilde{\Theta}) \in q\widetilde{\Theta} + S(\Lambda^{-1/2}\mu^{1/2}\nu^{-1/2}\langle\delta_0\rangle,\Gamma)$. On the other hand, we know that

$$\mathrm{Re}\left(q\sharp\,\overbrace{\left[\delta_0(1+T^{-1}\eta) * \exp-2\pi\Gamma\right]}^{\omega}\right)$$
$$= q\omega + \sum_{|\alpha|=|\beta|=2} c_{\alpha\beta}q^{(\alpha)}\omega^{(\beta)} + S(\Lambda^{-1/2}\mu^{1/2}\nu^{-1},\Gamma)$$

so that it is enough to concentrate our attention on the "products" $q''\omega''$. We have

$$\left(\delta_0(1+T^{-1}\eta)\right)'' * \exp-2\pi\Gamma \in S(1,\Gamma)$$

and since $q'' \in S(\Lambda^{-1/2}\mu^{1/2},\nu^{-1}\Gamma)$, we get a remainder in $S(\Lambda^{-1/2}\mu^{1/2},\Gamma)$, which is fine as long as $\langle\delta_0\rangle \geq c\nu^{1/2}$. However when $\langle\delta_0\rangle \leq c\nu^{1/2}$, we know that, for a good choice of the fixed positive constant c, the function δ_0 satisfies the estimates of $S(\nu^{1/2},\nu^{-1}\Gamma)$, since it is the Γ-distance function to the set of (regular) zeroes of the function q so that

$$q''\delta_0'' \in S(\Lambda^{-1/2}\mu^{1/2}\nu^{-1/2},\nu^{-1}\Gamma)$$

which is what we are looking for. However, we are left with

$$q''(\delta_0\eta * \exp-2\pi\Gamma)''T^{-1}.$$

Since we have

$$(\delta_0\eta)'' = \delta_0''\eta + 2\delta_0'\eta' + \delta_0\eta''$$

and

$$|\delta_0''\eta + 2\delta_0'\eta'| \leq CT(\nu^{-1/2} + \Lambda^{-1/2}),$$

we have only to deal with the term

$$\delta_0\eta'' * \exp-2\pi\Gamma = \int \delta_0(Y)\eta''(Y)\exp-2\pi\Gamma(X-Y)dY$$

$$= - \int \underbrace{\delta_0'(Y)\eta'(Y)}_{\lesssim T\Lambda^{-1/2}} \exp -2\pi\Gamma(X-Y)dY$$

$$- \int \underbrace{\delta_0(Y)\eta'(Y)}_{\lesssim T\Lambda^{-1/2}\langle\delta_0\rangle} 4\pi(X-Y)\exp -2\pi\Gamma(X-Y)dY.$$

For future reference we summarize part of the previous discussion by the following result.

Lemma 3.7.10. *With the notation above, we have*

$$\left|\Big(\delta_0(1+T^{-1}\eta)*\exp -2\pi\Gamma\Big)\right| \le C\langle\delta_0\rangle, \quad \left|\Big(\delta_0(1+T^{-1}\eta)*\exp -2\pi\Gamma\Big)'\right| \le C,$$

$$\left|\Big(\delta_0(1+T^{-1}\eta)*\exp -2\pi\Gamma\Big)''\right| \le C\langle\delta_0\rangle\nu^{-1/2}.$$

Proof of Lemma 3.7.10. Restarting the discussion, we have already seen that the result is true whenever $\langle\delta_0\rangle \gtrsim \nu^{1/2}$. Moreover when $\langle\delta_0\rangle \ll \nu^{1/2}$, we have seen that $|\delta_0''| \lesssim \nu^{-1/2}$ and $T^{-1}|\eta| \lesssim 1$; moreover we have already checked $|\eta'| \lesssim T\Lambda^{-1/2}$ and $T^{-1}|\delta_0'\eta'| \lesssim \Lambda^{-1/2} \lesssim \nu^{-1/2}$ as well as $|\delta_0\eta''*\exp -2\pi\Gamma| \lesssim \Lambda^{-1/2}\langle\delta_0\rangle \lesssim \langle\delta_0\rangle\nu^{-1/2}$, completing the proof of Lemma 3.7.10. $\qquad\square$

Eventually, using Lemma 4.9.3, we get that the first integral above is in $S(T\Lambda^{-1/2}, \Gamma)$ whereas the second belongs to $S(T\Lambda^{-1/2}\langle\delta_0\rangle, \Gamma)$. Finally, it means that, up to terms in $S(\Lambda^{-1/2}\mu^{1/2}\nu^{-1/2}\langle\delta_0\rangle, \Gamma)$, the operator $P(t)$ has a Weyl symbol equal to the rhs of (3.7.32). The proof of Lemma 3.7.9 is complete. $\qquad\square$

We shall use a partition of unity $1 = \sum_k \chi_k^2$ related to the metric $\nu(t, X)^{-1}\Gamma$ and a sequence (ψ_k) as in Theorem 4.3.7. We have, omitting the variable t, with p_0 defined in (3.7.32),

$$p_0(X) = \sum_k \chi_k(X)^2 q(X) \int \delta_0(Y)\big(1+T^{-1}\eta(Y)\big)2^n \exp -2\pi\Gamma(X-Y)dY$$

$$+ \sum_k \chi_k(X)^2 q(X) \int \Theta(Y)2^n \exp -2\pi\Gamma(X-Y)dY.$$

Using Lemma 4.9.6, we obtain, assuming $\delta_0 = \delta_{0k}, \Theta = \Theta_k, q = q_k$ on U_k

$$p_0 = \sum_k \chi_k^2 q_k\big(\delta_{0k}(1+T^{-1}\eta)*2^n \exp -2\pi\Gamma\big) + \sum_k \chi_k^2 q_k\big(\Theta_k*2^n \exp -2\pi\Gamma\big)$$

$$+ S(\Lambda^{-1/2}\mu^{1/2}\nu^{-\infty}, \Gamma). \tag{3.7.33}$$

Lemma 3.7.11. *With* $\widetilde{\Theta}_k = \Theta_k * 2^n \exp -2\pi\Gamma$, $d_k = \delta_{0k}(1+T^{-1}\eta)*2^n \exp -2\pi\Gamma$ *and* q_k, χ_k *defined above, we have*

$$\sum_k \chi_k \sharp q_k d_k \sharp \chi_k + \sum_k \chi_k \sharp q_k \widetilde{\Theta}_k \sharp \chi_k = p_0 + S(\Lambda^{-1/2}\mu^{1/2}\nu^{-1/2}\langle\delta_0\rangle, \Gamma). \tag{3.7.34}$$

Proof. We already know that $|d_k| \lesssim \langle \delta_0 \rangle$, $|d_k'| \lesssim 1$, $|d_k''| \lesssim \langle \delta_0 \rangle \nu^{-1/2}$, so that

$$|(q_k d_k)''| = q_k'' d_k + 2q_k' d_k' + q_k d_k'' \lesssim \Lambda^{-1/2} \mu^{1/2} (\langle \delta_0 \rangle + \nu^{1/2} + \nu^{1/2} \langle \delta_0 \rangle)$$
$$\lesssim \Lambda^{-1/2} \mu^{1/2} \nu^{1/2} \langle \delta_0 \rangle. \tag{3.7.35}$$

As a consequence, we get

$$\sum_k \chi_k \sharp q_k d_k \sharp \chi_k$$

$$= \sum_k \Big(\chi_k q_k d_k + \frac{1}{4i\pi} \{\chi_k, q_k d_k\} + S(\nu^{-1}(\Lambda^{-1/2} \mu^{1/2} \langle \delta_0 \rangle \nu^{1/2}), \Gamma) \Big) \sharp \chi_k$$

$$= \sum_k \Big(\chi_k q_k d_k + \frac{1}{4i\pi} \{\chi_k, q_k d_k\} \Big) \sharp \chi_k + \sum_k S(\Lambda^{-1/2} \mu^{1/2} \langle \delta_0 \rangle \nu^{-1/2}, \Gamma) \sharp \chi_k$$

$$= \sum_k \Big(\chi_k q_k d_k + \frac{1}{4i\pi} \{\chi_k, q_k d_k\} \Big) \chi_k + \frac{1}{4i\pi} \sum_k \Big\{ \chi_k q_k d_k + \frac{1}{4i\pi} \{\chi_k, q_k d_k\}, \chi_k \Big\}$$

$$+ S(\Lambda^{-1/2} \mu^{1/2} \langle \delta_0 \rangle \nu^{-1/2}, \Gamma)$$

since

$$|(\chi_k q_k d_k)'' \chi_k''| \lesssim \Lambda^{-1/2} \mu^{1/2} (\langle \delta_0 \rangle + \nu^{1/2} + \nu \langle \delta_0 \rangle \nu^{-1/2}) \nu^{-1} \lesssim \langle \delta_0 \rangle \nu^{-1/2} \Lambda^{-1/2} \mu^{1/2}.$$

Using now that $\chi_k \sharp q_k d_k \sharp \chi_k$ is real-valued, we obtain

$$\sum_k \chi_k \sharp q_k d_k \sharp \chi_k$$

$$= \sum_k \chi_k^2 q_k d_k - \frac{1}{16\pi^2} \sum_k \{\{\chi_k, q_k d_k\}, \chi_k\} + S(\Lambda^{-1/2} \mu^{1/2} \langle \delta_0 \rangle \nu^{-1/2}, \Gamma). \tag{3.7.36}$$

We note now that, using (3.7.35), we have

$$\{\{\chi_k, q_k d_k\}, \chi_k\} = -H_{\chi_k}^2(q_k d_k) \in S(\Lambda^{-1/2} \mu^{1/2} \langle \delta_0 \rangle \nu^{1/2} \nu^{-1}, \Gamma). \tag{3.7.37}$$

We examine now the term

$$\chi_k \sharp q_k \widetilde{\Theta}_k \sharp \chi_k = (\chi_k q_k \widetilde{\Theta}_k) \sharp \chi_k + \frac{1}{4i\pi} \Big\{ \chi_k, q_k \widetilde{\Theta}_k \Big\} \sharp \chi_k$$
$$+ S(\nu^{-1} \Lambda^{-1/2} \mu^{1/2} \nu \langle \delta_0 \rangle \nu^{-1/2}, \Gamma) \sharp \chi_k.$$

We have

$$\mathrm{Re}(\chi_k q_k \widetilde{\Theta}_k \sharp \chi_k) \in \chi_k^2 q_k \widetilde{\Theta}_k + S(\nu^{-1} \Lambda^{-1/2} \mu^{1/2} \nu \langle \delta_0 \rangle \nu^{-1/2}, \Gamma),$$
$$\frac{1}{4i\pi} \Big\{ \chi_k, q_k \widetilde{\Theta}_k \Big\} \in i\mathbb{R} + S(\nu^{-1/2} \Lambda^{-1/2} \mu^{1/2} \nu \langle \delta_0 \rangle \nu^{-1/2}, \Gamma).$$

Since $\chi_k \sharp q_k \widetilde{\Theta}_k \sharp \chi_k$ is real-valued, we get

$$\sum_k \chi_k \sharp q_k \widetilde{\Theta}_k \sharp \chi_k = \sum_k \chi_k^2 q_k \widetilde{\Theta}_k + S(\Lambda^{-1/2}\mu^{1/2}\langle \delta_0 \rangle \nu^{-1/2}, \Gamma). \tag{3.7.38}$$

Collecting the information (3.7.33),(3.7.36),(3.7.37) and (3.7.38), we obtain (3.7.34) and the lemma. $\square$

From this lemma and Lemma 3.7.9 we obtain that

$$\mathrm{Re}\big(Q(t)M(t)\big) = \sum_k \chi_k^w \big(q_k d_k + q_k \widetilde{\Theta}_k\big)^w \chi_k^w + S(\Lambda^{-1/2}\mu^{1/2}\langle \delta_0 \rangle \nu^{-1/2}, \Gamma)^w. \tag{3.7.39}$$

Moreover the same arguments as above in Lemma 3.7.9 give also that

$$\mathrm{Re}(q_k^w d_k^w + q_k^w \widetilde{\Theta}_k^w) = (q_k d_k + q_k \widetilde{\Theta}_k)^w + S(\Lambda^{-1/2}\mu^{1/2}\langle \delta_0 \rangle \nu^{-1/2}, \Gamma)^w. \tag{3.7.40}$$

Proposition 3.7.12. *Let $T > 0$ be given and $Q(t) = q(t)^w$ given by (3.2.12), (3.2.13). We define $M(t)$ according to (3.7.4). Then, with a partition of unity $1 = \sum_k \chi_k^2$ related to the metric $\nu(t, X)^{-1}\Gamma$ we have*

$$\mathrm{Re}\,(Q(t)M(t)) = \sum_k \chi_k^w \,\mathrm{Re}\big(q_k^w d_k^w + q_k^w \widetilde{\Theta}_k^w\big)\chi_k^w + S(\Lambda^{-1/2}\mu^{1/2}\langle \delta_0 \rangle \nu^{-1/2}, \Gamma)^w$$

$$\tag{3.7.41}$$

$$and \quad \mathrm{Re}\,(Q(t)M(t)) + S(\Lambda^{-1/2}\mu^{1/2}\langle \delta_0 \rangle \nu^{-1/2}, \Gamma)^w \geq 0. \tag{3.7.42}$$

Proof. The equality (3.7.41) follows from (3.7.39), (3.7.40). According to Lemma 3.2.16, we have to deal with four subsets of indices, $E_\pm, E_0, E_{00}$. The classification in Definition 3.2.15 shows that **[1]** on page 254 takes care of the cases E_0 and shows that, from (3.7.20),

$$\text{for } k \in E_0, \quad \mathrm{Re}\big(q_k^w d_k^w + q_k^w \widetilde{\Theta}_k^w\big) + S(\Lambda^{-1/2}\mu^{1/2}\nu^{-1/2}, \Gamma)^w \geq 0. \tag{3.7.43}$$

Furthermore, the estimate (3.7.29) on page 256 shows that

$$\text{for } k \in E_\pm, \quad \mathrm{Re}\big(q_k^w d_k^w + q_k^w \widetilde{\Theta}_k^w\big) + S(\Lambda^{-1/2}\mu^{1/2}\nu^{-1/2}\langle \delta_0 \rangle, \Gamma)^w \geq 0. \tag{3.7.44}$$

Moreover if $k \in E_{00}$, the weight ν is bounded above and

$$q_k^w d_k^w + q_k^w \widetilde{\Theta}_k^w \in S(\Lambda^{-1/2}\mu^{1/2}\nu^{-1/2}, \Gamma)^w. \tag{3.7.45}$$

The equality (3.7.41) and (3.7.43)–(3.7.44)–(3.7.45) give (3.7.42). $\square$

The multiplier method

Theorem 3.7.13. *Let $T > 0$ be given and $Q(t) = q(t)^w$ given by (3.2.12)–(3.2.13). We define $M(t)$ according to (3.7.4). There exist $T_0 > 0$ and $c_0 > 0$ depending only on a finite number of γ_k in (3.2.12) such that, for $0 < T \leq T_0$, with $D(t,X) = \langle \delta_0(t,X) \rangle$, (D is Lipschitz-continuous with Lipschitz constant 2, as δ_0 in (3.2.20) and thus a Γ-weight),*

$$\frac{d}{dt}M(t) + 2\operatorname{Re}\left(Q(t)M(t)\right) \geq T^{-1}(D^2)^{Wick}\Lambda^{-1/2}c_0. \qquad (3.7.46)$$

Moreover we have with m defined in (3.2.95), $\widetilde{m}(t,\cdot) = m(t,\cdot) * 2^n \exp -2\pi\Gamma$,

$$M(t) = m(t,X)^{Wick} = \widetilde{m}(t,X)^w, \text{ with } \widetilde{m} \in S_1(D, D^{-2}\Gamma) + S(1,\Gamma). \qquad (3.7.47)$$
$$m(t,X) = a(t,X) + b(t,X), \ |a/D| + |a'_X| + |b| \text{ bounded, } \dot{m} \geq 0, \qquad (3.7.48)$$
$$a = \delta_0(1 + T^{-1}\eta), \quad b = \widetilde{\Theta}. \qquad (3.7.49)$$

Proof. From the estimate (3.7.5), we get, with a positive fixed constant C_0,

$$\frac{d}{dt}M(t) \geq \frac{1}{2C_0 T}(\Lambda^{-1/2}\mu^{1/2}\nu^{-1/2}\langle\delta_0\rangle)^{Wick} + T^{-1}(\delta_0^2)^{Wick}\Lambda^{-1/2},$$

and from (3.7.6) and Lemma 4.9.4 we know that, with a fixed (non-negative) constant C_1,

$$2\operatorname{Re}\left(Q(t)M(t)\right) + C_1(\Lambda^{-1/2}\mu^{1/2}\nu^{-1/2}\langle\delta_0\rangle)^{Wick} \geq 0.$$

As a result we get, if $4C_1 C_0 T \leq 1$ (we shall choose $T_0 = \frac{1}{4C_0(C_1+1)}$),

$$\frac{d}{dt}M(t) + 2\operatorname{Re}\left(Q(t)M(t)\right) \geq \frac{1}{4C_0 T}(\Lambda^{-1/2}\mu^{1/2}\nu^{-1/2}\langle\delta_0\rangle)^{Wick} + T^{-1}(\delta_0^2)^{Wick}\Lambda^{-1/2}.$$

Using (3.2.30)($\mu \geq \nu/2$), this gives

$$\frac{d}{dt}M(t) + 2\operatorname{Re}\left(Q(t)M(t)\right) \geq T^{-1}\Lambda^{-1/2}(1 + \delta_0^2)^{Wick}\left(\frac{1}{2^{5/2}C_0 + 1}\right),$$

which is the sought result. □

3.7.3 From semi-classical to local estimates

From semi-classical to inhomogeneous estimates

Let us consider a smooth real-valued function f defined on $\mathbb{R} \times \mathbb{R}^n \times \mathbb{R}^n$, satisfiying (3.2.13) and such that, for all multi-indices α, β,

$$\sup_{\substack{t \in \mathbb{R} \\ (x,\xi) \in \mathbb{R}^{2n}}} |(\partial_x^\alpha \partial_\xi^\beta f)(t,x,\xi)|(1 + |\xi|)^{-1+|\beta|} = C_{\alpha\beta} < \infty. \qquad (3.7.50)$$

Using a Littlewood-Paley decomposition, we have

$$f(t, x, \xi) = \sum_{j \in \mathbb{N}} f(t, x, \xi) \varphi_j(\xi)^2, \quad \text{supp } \varphi_0 \text{ compact},$$

for $j \geq 1$, $\text{supp } \varphi_j \subset \{\xi \in \mathbb{R}^n, 2^{j-1} \leq |\xi| \leq 2^{j+1}\}$, $\sup_{j, \xi} |\partial_\xi^\alpha \varphi_j(\xi)| 2^{j|\alpha|} < \infty$.

We introduce also some smooth non-negative compactly supported functions $\psi_j(\xi)$, satisfying the same estimates as φ_j and supported in $2^{j-2} \leq |\xi| \leq 2^{j+2}$ for $j \geq 1$, identically 1 on the support of φ_j. For each $j \in \mathbb{N}$, we define the symbol

$$q_j(t, x, \xi) = f(t, x, \xi) \psi_j(\xi) \tag{3.7.51}$$

and we remark that (3.2.13) is satisfied for q_j and the following estimates hold:

$$|(\partial_x^\alpha \partial_\xi^\beta q_j)| \leq C'_{\alpha\beta} \Lambda_j^{1-|\beta|}, \quad \text{with } \Lambda_j = 2^j.$$

Note that the semi-norms of q_j can be estimated from above independently of j. We can reformulate this by saying that

$$q_j \in S(\Lambda_j, \Lambda_j^{-1} \Gamma_j), \quad \text{with } \Gamma_j(t, \tau) = |t|^2 \Lambda_j + |\tau|^2 \Lambda_j^{-1} \quad \text{(note } \Gamma_j = \Gamma_j^\sigma). \tag{3.7.52}$$

Lemma 3.7.14. *There exists $T_0 > 0, c_0 > 0$, depending only on a finite number of semi-norms of f such that, for each $j \in \mathbb{N}$, we can find D_j a Γ_j-uniformly Lipschitz-continuous function with Lipschitz constant 2, valued in $[1, \sqrt{2\Lambda_j}]$, a_j, b_j real-valued such that*

$$\sup_{\substack{j \in \mathbb{N}, |t| \leq T_0 \\ X \in \mathbb{R}^{2n}}} \left(\left| \frac{a_j(t, X)}{D_j(t, X)} \right| + \|\nabla_X a_j(t, X)\|_{\Gamma_j} + |b_j(t, X)| \right) < \infty. \tag{3.7.53}$$

With $m_j = a_j + b_j$, $\widetilde{m}_j(t, \cdot) = m_j(t, \cdot) * 2^n \exp{-2\pi\Gamma_j}$, $Q_j(t) = q_j(t)^w$, we have

$$M_j(t) = m_j(t, X)^{Wick(\Gamma_j)} = \widetilde{m}_j(t, X)^w,$$

$$\text{with } \widetilde{m}_j \in S_1(D_j, D_j^{-2}\Gamma_j) + S(1, \Gamma_j), \tag{3.7.54}$$

(the $\text{Wick}(\Gamma_j)$ quantization is defined in Definition 4.9.7) the estimate

$$\frac{d}{dt} M_j(t) + 2\,\text{Re}\,(Q_j(t) M_j(t)) \geq T^{-1}(D_j^2)^{Wick(\Gamma_j)} \Lambda_j^{-1/2} c_0. \tag{3.7.55}$$

Proof. It is a straightforward consequence of Definition 4.9.7 and of Theorem 3.7.13: let us check this. Considering the linear symplectic mapping

$$L : (t, \tau) \mapsto (\Lambda_j^{-1/2} t, \Lambda_j^{1/2} \tau),$$

we see that the symbols $q_j \circ L$ belong uniformly to $S(\Lambda_j, \Lambda_j^{-1}\Gamma_0)$. Applying Theorem 3.7.13 to $q_j \circ L$, we find D, a Γ_0–uniformly Lipschitz-continuous function ≥ 1, a, b real-valued such that

$$\sup_{\substack{j \in \mathbb{N}, |t| \leq T_0 \\ X \in \mathbb{R}^{2n}}} \left(\left| \frac{a(t, X)}{D(t, X)} \right| + \|\nabla_X a(t, X)\|_{\Gamma_0} + |b(t, X)| \right) < \infty, \tag{3.7.56}$$

and so that, with

$$m = a + b, \quad \widetilde{m}(t, \cdot) = m(t, \cdot) * 2^n \exp -2\pi\Gamma_0, \quad Q(t) = (q_j(t) \circ L)^w,$$

we have

$$M(t) = m(t, X)^{\text{Wick}} = \widetilde{m}(t, X)^w, \quad \text{with } \widetilde{m} \in S_1(D, D^{-2}\Gamma_0) + S(1, \Gamma_0), \tag{3.7.57}$$

$$\frac{d}{dt} M(t) + 2\,\text{Re}\,\big(Q(t)M(t)\big) \geq T^{-1}\big(D^2\big)^{\text{Wick}(\Gamma_0)} \Lambda_j^{-1/2} c_0. \tag{3.7.58}$$

Now we define the real-valued functions $a_j = a \circ L^{-1}, b_j = b \circ L^{-1}, D_j = D \circ L^{-1}$ and we have, since $\Gamma_0(S) = \Gamma_j(LS)$,

$$\left| \frac{a_j(t, X)}{D_j(t, X)} \right| + \|\nabla_X a_j(t, X)\|_{\Gamma_j} + |b_j(t, X)|$$

$$= \left| \frac{a(t, L^{-1}X)}{D(t, L^{-1}X)} \right| + \sup_{T \in \mathbb{R}^{2n}} \frac{|a_j'(t, X) \cdot T|}{\Gamma_j(T)^{1/2}} + |b(t, L^{-1}X)|$$

$$= \left| \frac{a(t, L^{-1}X)}{D(t, L^{-1}X)} \right| + \sup_{T \in \mathbb{R}^{2n}} \frac{|a'(t, X) \cdot L^{-1}T|}{\Gamma_j(T)^{1/2}} + |b(t, L^{-1}X)|$$

$$= \left| \frac{a(t, L^{-1}X)}{D(t, L^{-1}X)} \right| + \|a'(t, X)\|_{\Gamma_0} + |b(t, L^{-1}X)|,$$

so that (3.7.56) implies (3.7.53). Considering now $m_j = a_j + b_j$ and for a metaplectic U in the fiber of the symplectic L (see Definition 4.9.7), we have

$$M_j(t) = m_j(t, X)^{\text{Wick}(\Gamma_j)} = U(m_j \circ L)^{\text{Wick}(\Gamma_0)} U^*. \tag{3.7.59}$$

Thus we obtain

$$\frac{d}{dt} M_j(t) + 2\,\text{Re}\,\big(Q_j(t)M_j(t)\big)$$

$$\text{from (3.7.59)} = U \frac{d}{dt}(m_j \circ L)^{\text{Wick}(\Gamma_0)} U^*$$

$$+ 2\,\text{Re}\,\Big(UU^* q_j(t)^w U(m_j \circ L)^{\text{Wick}(\Gamma_0)} U^*\Big)$$

$$= U\Big[\frac{d}{dt}(m_j \circ L)^{\text{Wick}(\Gamma_0)}$$

$$+ 2\,\text{Re}\,\Big(U^* q_j(t)^w U(m_j \circ L)^{\text{Wick}(\Gamma_0)}\Big) \Big] U^*$$

$$\text{using } {}^{m=m_j \circ L}_{(q \circ L)^w = U^* q^w U} \Big] \quad = U \Big[\frac{d}{dt}(m)^{\text{Wick}(\Gamma_0)} + 2 \operatorname{Re} \Big((q_j \circ L)^w (m)^{\text{Wick}(\Gamma_0)} \Big) \Big] U^*$$

$$\text{from (3.7.58)} \quad \geq U \Big[T^{-1}(D^2)^{\text{Wick}(\Gamma_0)} \Lambda_j^{-1/2} c_0 \Big] U^*$$

$$\text{from (4.9.8)} \quad = T^{-1} U U^* \big(D^2 \circ L^{-1} \big)^{\text{Wick}(\Gamma_j)} U U^* \Lambda_j^{-1/2} c_0$$

$$= T^{-1} \big(D_j^2 \big)^{\text{Wick}(\Gamma_j)} \Lambda_j^{-1/2} c_0,$$

which is (3.7.55), completing the proof of the lemma. $\square$

We define now, with φ_j given after (3.7.50), M_j in (3.7.54)

$$\mathcal{M}(t) = \sum_{j \in \mathbb{N}} \varphi_j^w \Lambda_j^{-1/2} M_j(t) \varphi_j^w. \tag{3.7.60}$$

Lemma 3.7.15. *With M_j defined in (3.7.54) and φ_j, ψ_j as above,*

$$\sum_j \varphi_j^w M_j(t) \big((1 - \psi_j) f(t) \big)^w \varphi_j^w \in S(\langle \xi \rangle^{-\infty}, \langle \xi \rangle |dx|^2 + \langle \xi \rangle^{-1} |d\xi|^2)^w, \tag{3.7.61}$$

$$\sum_j \varphi_j^w M_j(t) \varphi_j^w \big((1 - \psi_j) f(t) \big)^w \in S(\langle \xi \rangle^{-\infty}, \langle \xi \rangle |dx|^2 + \langle \xi \rangle^{-1} |d\xi|^2)^w. \tag{3.7.62}$$

Proof. Since $\psi_j \equiv 1$ on the support of φ_j, we get that, uniformly with respect to the index j,

$$\big((1 - \psi_j) f(t) \big)^w \varphi_j^w \in S(\langle \xi \rangle^{-\infty}, |dx|^2 + \langle \xi \rangle^{-2} |d\xi|^2)^w. \tag{3.7.63}$$

Since $\widetilde{m_j} \in S(\Lambda_j^{1/2}, \Lambda_j |dx|^2 + \Lambda_j^{-1} |d\xi|^2)$, we get that

$$\psi_j \widetilde{m_j} \in S(\langle \xi \rangle^{1/2}, \langle \xi \rangle |dx|^2 + \langle \xi \rangle^{-1} |d\xi|^2),$$

and consequently $\varphi_j \sharp \psi_j \widetilde{m_j} \in S(\langle \xi \rangle^{1/2}, \langle \xi \rangle |dx|^2 + \langle \xi \rangle^{-1} |d\xi|^2)$ so that

$$\varphi_j \sharp \psi_j \widetilde{m_j} \sharp (1 - \psi_j) f(t) \sharp \varphi_j \in S(\langle \xi \rangle^{-\infty}, \langle \xi \rangle |dx|^2 + \langle \xi \rangle^{-1} |d\xi|^2)$$
$$\subset S(\langle \xi \rangle^{-\infty}, |dx|^2 + |d\xi|^2). \tag{3.7.64}$$

Moreover we have

$$\varphi_j \sharp (1 - \psi_j) \widetilde{m_j} \in S(\Lambda_j^{-\infty}, \Lambda_j |dx|^2 + \Lambda_j^{-1} |d\xi|^2) \subset S(\Lambda_j^{-\infty}, |dx|^2 + |d\xi|^2)$$

so that (3.7.63) implies

$$\varphi_j \sharp (1 - \psi_j) \widetilde{m_j} \sharp (1 - \psi_j) f(t) \sharp \varphi_j \in S(\langle \xi \rangle^{-\infty}, |dx|^2 + |d\xi|^2)$$
$$\subset S(\langle \xi \rangle^{-\infty}, \langle \xi \rangle |dx|^2 + \langle \xi \rangle^{-1} |d\xi|^2). \tag{3.7.65}$$

As a consequence, from (3.7.64) and (3.7.65) we get, uniformly in j, that

$$\varphi_j \sharp \widetilde{m_j} \sharp (1 - \psi_j) f(t) \sharp \varphi_j \in S(\langle \xi \rangle^{-\infty}, \langle \xi \rangle |dx|^2 + \langle \xi \rangle^{-1} |d\xi|^2). \tag{3.7.66}$$

Since φ_j, ψ_j depend only on the variable ξ, the support condition implies $\varphi_j^w \psi_j^w = \varphi_j^w$ and we obtain that from (3.7.66)

$$\sum_j \varphi_j \sharp \widetilde{m_j} \sharp (1 - \psi_j) f(t) \sharp \varphi_j$$

$$= \sum_j \psi_j \sharp \varphi_j \sharp \widetilde{m_j} \sharp (1 - \psi_j) f(t) \sharp \varphi_j \sharp \psi_j \in S(\langle \xi \rangle^{-\infty}, \langle \xi \rangle |dx|^2 + \langle \xi \rangle^{-1} |d\xi|^2),$$

completing the proof of (3.7.61). The proof of (3.7.62) follows almost in the same way: we get as in (3.7.66) that

$$\varphi_j \sharp \widetilde{m_j} \sharp \varphi_j \sharp (1 - \psi_j) f(t) \in S(\langle \xi \rangle^{-\infty}, \langle \xi \rangle |dx|^2 + \langle \xi \rangle^{-1} |d\xi|^2).$$

Now with $\Phi_j = \varphi_j \sharp (1 - \psi_j) f(t)$, we have $\Phi_j \in S(\langle \xi \rangle^{-\infty}, |dx|^2 + \langle \xi \rangle^{-2} |d\xi|^2)$ and from the formula (4.9.14) we have also

$$|(\partial_x^\alpha \partial_\xi^\beta \Phi_j)(x, \xi)| \le C_{\alpha \beta N} 2^{jn} (1 + |\xi - \operatorname{supp} \varphi_j|)^{-N} (1 + |\xi|),$$

so that

$$|(\partial_x^\alpha \partial_\xi^\beta \Phi_j)(x, \xi)| \le \begin{cases} C_{\alpha \beta N} 2^{jn} 2^{-j(N-1)} & \text{if } |\xi| \ge 2^{j+2}, \\ C_{\alpha \beta N} 2^{jn} 2^{-jN} & \text{if } 2^{j-2} < |\xi| < 2^{j+2}, \\ C_{\alpha \beta N} 2^{jn} 2^{-j(N-1)} & \text{if } |\xi| \le 2^{j-2}, \end{cases}$$

implying that $\sum_j \varphi_j \sharp \widetilde{m_j} \sharp \Phi_j$ belongs to $S(\langle \xi \rangle^{-\infty}, |dx|^2 + \langle \xi \rangle^{-2} |d\xi|^2)$. $\qquad \square$

Lemma 3.7.16. *With* $F(t) = f(t, x, \xi)^w$, $\mathcal{M}$ *defined in* (3.7.60)), M_j *in* (3.7.54),

$$\frac{d}{dt} \mathcal{M}(t) + 2 \operatorname{Re}(\mathcal{M}(t) F(t)) = \sum_j \Lambda_j^{-1/2} \varphi_j^w \left(\dot{M}_j(t) + 2 \operatorname{Re} \left(M_j(t) (\psi_j f(t))^w \right) \right) \varphi_j^w$$

$$+ \sum_j 2 \operatorname{Re} \left(\varphi_j^w M_j(t) [\varphi_j^w, (\psi_j f(t))^w] \Lambda_j^{-1/2} \right) + S(\langle \xi \rangle^{-\infty}, \langle \xi \rangle |dx|^2 + \langle \xi \rangle^{-1} |d\xi|^2)^w.$$

Proof. We have

$$\frac{d}{dt} \mathcal{M}(t) + 2 \operatorname{Re}(\mathcal{M}(t) F(t))$$

$$= \sum_j \varphi_j^w \dot{M}_j(t) \Lambda_j^{-1/2} \varphi_j^w + 2 \operatorname{Re} \left(\varphi_j^w M_j(t) \Lambda_j^{-1/2} \varphi_j^w F(t) \right)$$

$$= \sum_j \varphi_j^w \dot{M}_j(t) \Lambda_j^{-1/2} \varphi_j^w + 2 \operatorname{Re} \left(\varphi_j^w \Lambda_j^{-1/2} M_j(t) F(t) \varphi_j^w \right)$$

$$+ 2 \operatorname{Re} \left(\varphi_j^w \Lambda_j^{-1/2} M_j(t) [\varphi_j^w, F(t)] \right). \tag{3.7.67}$$

On the other hand, we have

$$2\operatorname{Re}\!\left(\varphi_j^w \Lambda_j^{-1/2} M_j(t) F(t)\varphi_j^w\right) = 2\operatorname{Re}\!\left(\varphi_j^w \Lambda_j^{-1/2} M_j(t)\big(\psi_j f(t)\big)^w \varphi_j^w\right)$$
$$+\, 2\operatorname{Re}\!\left(\varphi_j^w \Lambda_j^{-1/2} M_j(t)\big((1-\psi_j)f(t)\big)^w \varphi_j^w\right)$$

and since we have also

$$2\operatorname{Re}\!\left(\varphi_j^w \Lambda_j^{-1/2} M_j(t)[\varphi_j^w, F(t)]\right) = 2\operatorname{Re}\!\left(\varphi_j^w \Lambda_j^{-1/2} M_j(t)[\varphi_j^w, \big(\psi_j f(t)\big)^w]\right)$$
$$+\, 2\operatorname{Re}\!\left(\varphi_j^w \Lambda_j^{-1/2} M_j(t)[\varphi_j^w, \big((1-\psi_j)f(t)\big)^w]\right),$$

we get the result of the lemma from Lemma 3.7.15 and (3.7.67). $\square$

Lemma 3.7.17. *With the above notation, the operator*

$$\sum_j \operatorname{Re}\!\left(\varphi_j^w \Lambda_j^{-1/2} M_j(t)[\varphi_j^w, (\psi_j f(t))^w]\right) \qquad (3.7.68)$$

has a symbol in $S(\langle\xi\rangle^{-1}, \langle\xi\rangle|dx|^2 + \langle\xi\rangle^{-1}|d\xi|^2)$.

Proof. The Weyl symbol of the bracket $[\varphi_j^w, (\psi_j f(t))^w]$ is

$$\frac{1}{2i\pi}\{\varphi_j, \psi_j f(t)\} + r_j,\, r_j \in S(\Lambda_j^{-1}, \Lambda_j^{-1}\Gamma_j)$$

where (r_j) is a confined sequence in $S(\langle\xi\rangle^{-1}, |dx|^2 + \langle\xi\rangle^{-2}|d\xi|^2)$. As a consequence, we have

$$\sum_j \varphi_j^w \Lambda_j^{-1/2} M_j(t) r_j^w \in S(\langle\xi\rangle^{-1}, \langle\xi\rangle|dx|^2 + \langle\xi\rangle^{-1}|d\xi|^2)^w.$$

With $\Psi_j = -\frac{1}{2\pi}\{\varphi_j, \psi_j f(t)\}$ (real-valued $\in S(1, \Lambda_j^{-1}\Gamma_j)$), we are left with

$$\sum_j \Lambda_j^{-1/2} \operatorname{Re}(\varphi_j \sharp \widetilde{m}_j(t)\sharp i\Psi_j)$$

which belongs to $S(\langle\xi\rangle^{-1}, \langle\xi\rangle|dx|^2 + \langle\xi\rangle^{-1}|d\xi|^2)$. $\square$

Definition 3.7.18. The symplectic metric Υ on $\mathbb{R}^{2n}$ is defined as

$$\Upsilon_\xi = \langle\xi\rangle|dx|^2 + \langle\xi\rangle^{-1}|d\xi|^2. \qquad (3.7.69)$$

With D_j given in Lemma 3.7.14, we define

$$d(t, x, \xi) = \sum_j \varphi_j(\xi)^2 D_j(t, x, \xi). \qquad (3.7.70)$$

Lemma 3.7.19. *The function $d(t, \cdot)$ is uniformly Lipschitz-continuous for the metric Υ in the strongest sense, namely, there exists a positive fixed constant C such that*

$$C^{-1}|d(t, x, \xi) - d(t, y, \eta)|$$
$$\leq \min\big(\langle\xi\rangle^{1/2}, \langle\eta\rangle^{1/2}\big)|x - y| + \frac{|\xi - \eta|}{\max\big(\langle\xi\rangle^{1/2}, \langle\eta\rangle^{1/2}\big)}. \quad (3.7.71)$$

Moreover it satisfies $d(t, x, \xi) \in [1, 2\langle\xi\rangle^{1/2}]$. It is thus a weight for that metric Υ.

Proof. Since the φ_j are non-negative with $\sum_j \varphi_j^2 = 1$, we get from Lemma 3.7.14 that

$$1 = \sum_j \varphi_j^2 \leq \sum_j \varphi_j^2 D_j = d \leq \sum_j \varphi_j(\xi)^2 \Lambda_j^{1/2} 2^{1/2} \leq \sum_j \varphi_j(\xi)^2 \langle\xi\rangle^{1/2} 2 = \langle\xi\rangle^{1/2} 2.$$

Also, we have

$$d(t, x, \xi) - d(t, y, \eta)$$
$$= \sum_j \varphi_j(\xi)^2 \big(D_j(t, x, \xi) - D_j(t, y, \eta)\big) + \sum_j D_j(t, y, \eta)\big(\varphi_j(\xi)^2 - \varphi_j(\eta)^2\big),$$

so that, with $X = (x, \xi), Y = (y, \eta)$, Γ_j given in (3.7.52),

$$|d(t, x, \xi) - d(t, y, \eta)|$$
$$\leq \sum_j \varphi_j(\xi)^2 2\Gamma_j(X - Y)^{1/2} + \sum_{\substack{j, \\ \varphi_j(\xi) \neq 0 \text{ or } \varphi_j(\eta) \neq 0}} 2^{1/2} 2^{j/2} |\xi - \eta| 2^{-j} C$$
$$\lesssim \sum_j \varphi_j(\xi)^2 4\big(\langle\xi\rangle^{1/2}|x - y| + \langle\xi\rangle^{-1/2}|\xi - \eta|\big) + |\xi - \eta| \sum_{\substack{j, \\ \varphi_j(\xi) \neq 0 \text{ or } \varphi_j(\eta) \neq 0}} 2^{-j/2}$$
$$\lesssim \langle\xi\rangle^{1/2}|x - y| + \langle\xi\rangle^{-1/2}|\xi - \eta| + |\xi - \eta|\big(\langle\xi\rangle^{-1/2} + \langle\eta\rangle^{-1/2}\big).$$

We get thus, if $\langle\xi\rangle \sim \langle\eta\rangle$,

$$|d(t, x, \xi) - d(t, y, \eta)| \lesssim \langle\xi\rangle^{1/2}|x - y| + \langle\xi\rangle^{-1/2}|\xi - \eta|. \quad (3.7.72)$$

If $2^{j_0} \sim \langle\xi\rangle \ll \langle\eta\rangle \sim 2^{k_0}$, we have

$$|d(t, x, \xi) - d(t, y, \eta)| \leq \sum_{j, \varphi_j(\xi) \neq 0} \varphi_j(\xi)^2 2^{(j+1)/2} + \sum_{j, \varphi_j(\eta) \neq 0} \varphi_j(\eta)^2 2^{(j+1)/2}$$
$$\lesssim 2^{j_0/2} + 2^{k_0/2} \sim 2^{k_0/2} \sim |\eta - \xi| 2^{-k_0/2} \sim \langle\eta\rangle^{-1/2}|\eta - \xi|. \quad (3.7.73)$$

Eventually, (3.7.73)–(3.7.72) give (3.7.71), completing the proof of the lemma. $\quad\square$

Note also that $\langle \xi \rangle$ is a Υ-weight and is even such that

$$|\langle\xi\rangle^{1/2} - \langle\eta\rangle^{1/2}| \leq \frac{|\xi - \eta|}{\langle\xi\rangle^{1/2} + \langle\eta\rangle^{1/2}}. \tag{3.7.74}$$

Lemma 3.7.20. *With $F(t) = f(t, x, \xi)^w$, $\mathcal{M}$ defined in (3.7.60), M_j in (3.7.54), the positive constant c_0 defined in Lemma 3.7.14,*

$$\frac{d}{dt}\mathcal{M}(t) + 2\operatorname{Re}(\mathcal{M}(t)F(t)) \geq c_0 T^{-1} \sum_j \varphi_j^w \big(\Lambda_j^{-1} D_j^2\big)^{Wick(\Gamma_j)} \varphi_j^w + S(\langle\xi\rangle^{-1}, \Upsilon)^w. \tag{3.7.75}$$

The operator $\mathcal{M}(t)$ has a Weyl symbol in the class $S_1(\langle\xi\rangle^{-1/2}d, d^{-2}\Upsilon)$. Moreover the selfadjoint operator $\mathcal{M}(t)$ satisfies, with a fixed constant C,

$$\mathcal{M}(t)\mathcal{M}(t) \leq C^2 \sum_j \varphi_j^w \big(\Lambda_j^{-1} D_j^2\big)^{Wick(\Gamma_j)} \varphi_j^w. \tag{3.7.76}$$

Proof. The estimate (3.7.75) is a consequence of Lemmas 3.7.16, 3.7.17 and 3.7.14. From (3.7.60), we get that

$$\mathcal{M}(t) \in \sum_j \varphi_j^w S_1(D_j \Lambda_j^{-1/2}, D_j^{-2}\Gamma_j)^w \varphi_j^w \subset S_1(d\langle\xi\rangle^{-1/2}, d^{-2}\Upsilon)^w.$$

From Lemma 3.7.14 and the finite overlap of the φ_j, we get

$$\|\mathcal{M}(t)u\|^2 \lesssim \sum_j \Lambda_j^{-1} \|\varphi_j^w M_j(t)\varphi_j^w u\|^2 = \sum_j \Lambda_j^{-1} \langle \varphi_j^w M_j \varphi_j^w u, \varphi_j^w M_j \varphi_j^w u \rangle$$

$$= \sum_j \Lambda_j^{-1} \langle \varphi_j^w u, \underbrace{M_j(\varphi_j^2)^w M_j}_{\in S(D_j^2, \Gamma_j)^w} \varphi_j^w u \rangle \underset{\text{from Lemma 4.9.4}}{\lesssim} \sum_j \langle \varphi_j^w u, \big(\Lambda_j^{-1} D_j^2\big)^{Wick(\Gamma_j)} \varphi_j^w u \rangle,$$

which is (3.7.76). $\square$

Lemma 3.7.21. *Let a be a symbol in $S(\langle\xi\rangle^{-1}, \Upsilon)$. Then, with constants C_1, C_2 depending on a finite number of semi-norms of a, we have*

$$|\langle a^w u, u\rangle| \leq C_1 \|u\|_{H^{-1/2}}^2 \leq C_2 \sum_j \langle \big(\Lambda_j^{-1} D_j^2\big)^{Wick(\Gamma_j)} \varphi_j^w u, \varphi_j^w u \rangle.$$

Proof. We have, since $D_j \geq 1$ and the Wick quantizations are non-negative

$$\sum_j \langle \big(\Lambda_j^{-1} D_j^2\big)^{Wick(\Gamma_j)} \varphi_j^w u, \varphi_j^w u \rangle \geq \sum_j \langle \big(\Lambda_j^{-1}\big)^{Wick(\Gamma_j)} \varphi_j^w u, \varphi_j^w u \rangle$$

$$= \langle (\sum_j \Lambda_j^{-1}\varphi_j^2)^w u, u \rangle \sim \|u\|_{H^{-1/2}}^2,$$

where $H^{-1/2}$ is the standard Sobolev space of index $-1/2$. Now, it is a classical result that

$$\langle a^w u, u \rangle = \langle \underbrace{((\langle \xi \rangle^{1/2})^w a^w ((\langle \xi \rangle^{1/2})^w}_{\in S(1,\Upsilon)^w \subset \mathcal{L}(L^2)}((\langle \xi \rangle^{-1/2})^w u, ((\langle \xi \rangle^{-1/2})^w u \rangle$$

which implies that $|\langle a^w u, u \rangle| \lesssim \|u\|^2_{H^{-1/2}}$.   □

Theorem 3.7.22. *Let $f(t,x,\xi)$ be a smooth real-valued function defined on $\mathbb{R} \times \mathbb{R}^n \times \mathbb{R}^n$, satisfiying (3.2.13) and (3.7.50). Let $f_0(t,x,\xi)$ be a smooth complex-valued function defined on $\mathbb{R} \times \mathbb{R}^n \times \mathbb{R}^n$, such that $\langle \xi \rangle f_0(t,x,\xi)$ satisfies (3.7.50). Then there exists $T_0 > 0, c_0 > 0$ depending on a finite number of semi-norms of f, f_0, such that, for all $T \le T_0$ and all $u \in C_c^\infty((-T,T); \mathscr{S}(\mathbb{R}^n))$,*

$$\|D_t u + if(t,x,\xi)^w u + f_0(t,x,\xi)^w u\|_{L^2(\mathbb{R}^{n+1})} \ge c_0 T^{-1} \left(\int \|u(t)\|^2_{H^{-1/2}(\mathbb{R}^n)} dt \right)^{1/2}.$$

Proof. (i) *We assume first that $f_0 \equiv 0$.* Using Lemmas 3.7.20 – 3.7.21, we get

$$2\,\mathrm{Re}\langle D_t u + if(t)^w u, i\mathcal{M}(t)u \rangle \ge (c_0 T^{-1} - C_2) \sum_j \langle \varphi_j^w (\Lambda_j^{-1} D_j^2)^{\mathrm{Wick}(\Gamma_j)} \varphi_j^w u, u \rangle,$$

(3.7.77)

and from the estimate (3.7.76), provided that

$$c_0/(2C_2) \ge T, \tag{3.7.78}$$

we get

$$2\|D_t u + if(t)^w u\|_{L^2(\mathbb{R}^{n+1})} \left[\sum_j \langle \varphi_j^w (\Lambda_j^{-1} D_j^2)^{\mathrm{Wick}(\Gamma_j)} \varphi_j^w u, u \rangle \right]^{1/2} C$$

$$\ge \frac{c_0}{2T} \sum_j \langle \varphi_j^w (\Lambda_j^{-1} D_j^2)^{\mathrm{Wick}(\Gamma_j)} \varphi_j^w u, u \rangle$$

so that, with fixed positive constants c_1, c_2, using again Lemma 3.7.21,

$$\|D_t u + if(t)^w u\|_{L^2(\mathbb{R}^{n+1})} \ge \frac{c_1}{T} \left[\sum_j \langle \varphi_j^w (\Lambda_j^{-1} D_j^2)^{\mathrm{Wick}(\Gamma_j)} \varphi_j^w u, u \rangle \right]^{1/2}$$

$$\ge \frac{c_2}{T} \left(\int \|u(t)\|^2_{H^{-1/2}(\mathbb{R}^n)} dt \right)^{1/2},$$

which is our result. Let us check now the case $f_0 \not\equiv 0$.

(ii) *Let us assume that $\mathrm{Im}(f_0) \in S(\langle \xi \rangle^{-1}, \langle \xi \rangle^{-1}\Upsilon)$.* Going back to the computation in (3.7.77), with (3.7.78) fulfilled, we have

$$2\,\mathrm{Re}\langle D_t u + if(t)^w + f_0(t)^w u, i\mathcal{M}(t)u \rangle \ge \frac{c_0}{2T} \sum_j \langle \varphi_j^w (\Lambda_j^{-1} D_j^2)^{\mathrm{Wick}(\Gamma_j)} \varphi_j^w u, u \rangle$$

$$+ 2\,\mathrm{Re}\langle \mathrm{Re}(f_0(t))^w u, i\mathcal{M}(t)u \rangle + 2\,\mathrm{Re}\langle \mathrm{Im}(f_0(t))^w u, \mathcal{M}(t)u \rangle.$$

From the identity $2\operatorname{Re}\langle\operatorname{Re}(f_0(t))^w u, i\mathcal{M}(t)u\rangle = \langle[\operatorname{Re}(f_0(t))^w, i\mathcal{M}(t)]u, u\rangle$ and the fact that, from Theorem 2.3.8 (see also Theorem 18.5.5 in [73]) we have

$$\big[\operatorname{Re}(f_0(t))^w, i\mathcal{M}(t)\big] \in S(\langle\xi\rangle^{-1/2}dd^{-1}\langle\xi\rangle^{-1/2}, \Upsilon)^w = S(\langle\xi\rangle^{-1}, \Upsilon)^w$$

we can use Lemma 3.7.21 to control this term by

$$C\sum_j \langle\varphi_j^w(\Lambda_j^{-1}D_j^2)^{\operatorname{Wick}(\Gamma_j)}\varphi_j^w u, u\rangle.$$

On the other hand, from our assumption on $\operatorname{Im} f_0$, we get that

$$\mathcal{M}(t)\operatorname{Im}(f_0(t))^w \in S(\langle\xi\rangle^{-1/2}d\langle\xi\rangle^{-1}, \Upsilon)^w \subset S(\langle\xi\rangle^{-1}, \Upsilon)^w,$$

which can be also controlled by $C\sum_j \langle\varphi_j^w(\Lambda_j^{-1}D_j^2)^{\operatorname{Wick}(\Gamma_j)}\varphi_j^w u, u\rangle$. Eventually, we obtain the result in that case too, for T small enough.

(iii) *We are left with the general case* $\operatorname{Im}(f_0) \in S(1, \langle\xi\rangle^{-1}\Upsilon)$; we note that, with

$$\omega_0(t, x, \xi) = \int_0^t \operatorname{Im} f_0(s, x, \xi)ds, \quad \text{(which belongs to } S(1, \langle\xi\rangle^{-1}\Upsilon)), \quad (3.7.79)$$

we have

$$D_t + if(t)^w + (\operatorname{Re} f_0(t))^w + i(\operatorname{Im} f_0(t))^w$$
$$= (e^{\omega_0(t)})^w D_t(e^{-\omega_0(t)})^w + if(t)^w + (\operatorname{Re} f_0(t))^w$$
$$= (e^{\omega_0(t)})^w\Big(D_t + if(t)^w + (\operatorname{Re} f_0(t))^w\Big)(e^{-\omega_0(t)})^w$$
$$+ \big(if(t) - e^{\omega_0(t)}\sharp if(t)\sharp e^{-\omega_0(t)}\big)^w + S(\langle\xi\rangle^{-1}, \langle\xi\rangle^{-1}\Upsilon)^w.$$

Noting that $e^{\pm\omega_0}$ belongs to $S(1, \langle\xi\rangle^{-1}\Upsilon)$, we compute

$$e^{\omega_0}\sharp if\sharp e^{-\omega_0} = \Big(e^{\omega_0}if + \frac{1}{4i\pi}\{e^{\omega_0}, if\}\Big)\sharp e^{-\omega_0} + S(\langle\xi\rangle^{-1}, \langle\xi\rangle^{-1}\Upsilon)$$
$$= if + \frac{1}{4i\pi}\{e^{\omega_0}if, e^{-\omega_0}\} + \frac{1}{4i\pi}\{e^{\omega_0}, if\}e^{-\omega_0} + S(\langle\xi\rangle^{-1}, \langle\xi\rangle^{-1}\Upsilon)$$
$$= if + \frac{1}{2\pi}\{\omega_0, f\} + S(\langle\xi\rangle^{-1}, \langle\xi\rangle^{-1}\Upsilon).$$

We obtain

$$L = D_t + if(t)^w + f_0(t)^w$$
$$= (e^{\omega_0(t)})^w\Big(D_t + if(t)^w + (\operatorname{Re} f_0(t) + \frac{1}{2\pi}\{f, \omega_0\})^w\Big)(e^{-\omega_0(t)})^w$$
$$+ S(\langle\xi\rangle^{-1}, \langle\xi\rangle^{-1}\Upsilon)^w, \qquad (3.7.80)$$

and analogously

$$L_0 = D_t + if(t)^w + (\operatorname{Re} f_0(t) + \frac{1}{2\pi}\{f,\omega_0\})^w + S(\langle\xi\rangle^{-1}, \langle\xi\rangle^{-1}\Upsilon)^w$$
$$= (e^{-\omega_0(t)})^w L(e^{\omega_0(t)})^w. \qquad (3.7.81)$$

Using now the fact that the symbol $\operatorname{Re} f_0(t) + \frac{1}{2\pi}\{f,\omega_0\}$ is real-valued and belongs to $S(1, \langle\xi\rangle^{-1}\Upsilon)$, we can use (ii) to prove the estimate in the theorem for the operator

$$L_0 = D_t + if(t)^w + (\operatorname{Re} f_0(t) + \frac{1}{2\pi}\{f,\omega_0\})^w + S(\langle\xi\rangle^{-1}, \langle\xi\rangle^{-1}\Upsilon)^w.$$

We note also that $e^{\omega_0}\#e^{-\omega_0} = 1 + t^2 S(\langle\xi\rangle^{-2}, \langle\xi\rangle^{-1}\Upsilon)$ so that, for $|t|$ small enough,

$$\left.\begin{array}{c} \text{the operators } (e^{\pm\omega_0})^w \text{ are invertible in } L^2(\mathbb{R}^n) \\ \text{and their inverses are pseudo-differential operators in } S(1, \langle\xi\rangle^{-1}\Upsilon)^w. \end{array}\right\}$$
$$(3.7.82)$$

From the previous identity and (ii), we get for $u \in C_c^\infty((-T,T), \mathscr{S}(\mathbb{R}^n))$

$$\int \left\| (e^{-\omega_0(t)})^w L(e^{\omega_0(t)})^w u(t) \right\|_{L^2(\mathbb{R}^n)}^2 dt \geq \frac{c_0^2}{T^2} \int \|u(t)\|_{H^{-1/2}(\mathbb{R}^n)}^2 \, dt.$$

Applying this to

$$u(t) = \left((e^{\omega_0(t)})^w\right)^{-1} v(t), \qquad (3.7.83)$$

we obtain

$$\int \left\| (e^{-\omega_0(t)})^w Lv(t) \right\|_{L^2(\mathbb{R}^n)}^2 dt \geq \frac{c_0^2}{T^2} \int \left\| \left((e^{\omega_0(t)})^w\right)^{-1} v(t) \right\|_{H^{-1/2}(\mathbb{R}^n)}^2 dt. \quad (3.7.84)$$

We have

$$\left\| \left((e^{\omega_0(t)})^w\right)^{-1} v(t) \right\|_{H^{-1/2}(\mathbb{R}^n)}^2$$
$$= \|(\langle\xi\rangle^{-1/2})^w \left((e^{\omega_0(t)})^w\right)^{-1} (\langle\xi\rangle^{1/2})^w (\langle\xi\rangle^{-1/2})^w v(t)\|_{L^2(\mathbb{R}^n)}^2. \quad (3.7.85)$$

Now the operator $(\langle\xi\rangle^{-1/2})^w \left((e^{\omega_0(t)})^w\right)^{-1} (\langle\xi\rangle^{1/2})^w$ is invertible with inverse

$$\Omega(t) = (\langle\xi\rangle^{-1/2})^w (e^{\omega_0(t)})^w (\langle\xi\rangle^{1/2})^w \qquad (3.7.86)$$

which is a bounded operator on $L^2(\mathbb{R}^n)$ so that

$$\|v\|_{L^2} = \|\Omega\Omega^{-1}v\|_{L^2} \leq \|\Omega\|_{\mathcal{L}(L^2)}\|\Omega^{-1}v\|_{L^2}. \qquad (3.7.87)$$

As a result, from the inequality (3.7.84), we get

$$\int \left\| (e^{-\omega_0(t)})^w Lv(t) \right\|_{L^2(\mathbb{R}^n)}^2 dt \geq \frac{c_0^2}{T^2} \int \left\| \Omega(t)^{-1}(\langle \xi \rangle^{-1/2})^w v(t) \right\|_{L^2(\mathbb{R}^n)}^2 dt$$

$$\geq \frac{c_0^2}{T^2} \int \left\| (\langle \xi \rangle^{-1/2})^w v(t) \right\|_{L^2(\mathbb{R}^n)}^2 \frac{1}{\|\Omega(t)\|^2} dt \geq \frac{c_1^2}{T^2} \int \|v(t)\|_{H^{-1/2}(\mathbb{R}^n)}^2 dt,$$

which is the result. The proof of Theorem 3.7.22 is complete. $\square$

Remark 3.7.23. Although Theorem 3.7.22 provides a solvability result with loss of $3/2$ derivatives for the evolution equation

$$\partial_t + f(t, x, \xi)^w + f_0(t, x, \xi)^w,$$

where f, f_0 satisfy the assumptions of this theorem, the statement does not, for two reasons, seem quite sufficient to handle operators with homogeneous symbols. The first one is that the reduction of homogeneous symbols in the cotangent bundle of a manifold will lead to a model operator like the one above, but only at the cost of some microlocalization in the cotangent bundle. We need thus to get a microlocal version of our estimates. The second reason is that the function $f(t, x, \xi)$ is not a classical symbol in the phase space $\mathbb{R}_t \times \mathbb{R}_x^n \times \mathbb{R}_\tau \times \mathbb{R}_\xi^n$ and we have to pay attention to the discrepancy between homogeneous localization in the phase space $\mathbb{R}^{2n+2}$ and localization in $\mathbb{R}^{2n}$ with parameter t. That difficulty should be taken seriously, since the loss of derivatives is strictly larger than 1; in fact, commuting a cutoff function with the operator will produce an error of order 0, larger than what is controlled by the estimate. In the next section, we prove a localized version of Theorem 3.7.22 which will be suitable for future use in the homogeneous framework.

From semi-classical to localized inhomogeneous estimates

We begin with a modified version of Lemma 3.7.20, involving a microlocalization in $\mathbb{R}^{2n}$.

Lemma 3.7.24. *Let $f(t, x, \xi)$ be real-valued satisfying (3.2.13) and (3.7.50); we shall note $F(t) = f(t, x, \xi)^w$. Let $\mathcal{M}$ be defined in (3.7.60). We define $c_1 = c_0/C^2$, where c_0 is given by Lemma 3.7.14 and C appears in (3.7.76). Let $\psi(x, \xi)$ be a real-valued symbol in $S(1, \langle \xi \rangle^{-1}\Upsilon)$. We have*

$$\frac{d}{dt}\left(\psi^w \mathcal{M}(t)\psi^w \right) + 2\operatorname{Re}\left(\psi^w \mathcal{M}(t)\psi^w F(t) \right)$$

$$\geq c_1 T^{-1}\psi^w \mathcal{M}(t)\mathcal{M}(t)\psi^w + S(\langle \xi \rangle^{-1}, \Upsilon)^w. \quad (3.7.88)$$

Proof. We compute, using (3.7.75) on the fourth line below,

$$\frac{d}{dt}\left(\psi^w \mathcal{M}(t)\psi^w\right) + 2\operatorname{Re}\left(\psi^w \mathcal{M}(t)\psi^w F(t)\right)$$

$$= \psi^w \dot{\mathcal{M}}(t)\psi^w + \psi^w \mathcal{M}(t)\psi^w F(t) + F(t)\psi^w \mathcal{M}(t)\psi^w$$

$$= \psi^w \left(\dot{\mathcal{M}}(t) + 2\operatorname{Re}\mathcal{M}(t)F(t)\right)\psi^w + \psi^w \mathcal{M}(t)\left[\psi^w, F(t)\right] + \left[F(t), \psi^w\right]\mathcal{M}(t)\psi^w$$

$$\geq c_1 T^{-1}\psi^w \mathcal{M}(t)\mathcal{M}(t)\psi^w + c_2 \psi^w S(\langle\xi\rangle^{-1}, \Upsilon)^w \psi^w$$

$$\quad + \psi^w \left[\mathcal{M}(t), \left[\psi^w, F(t)\right]\right] + \psi^w \left[\psi^w, F(t)\right]\mathcal{M}(t) - \left[\psi^w, F(t)\right]\mathcal{M}(t)\psi^w$$

$$= c_1 T^{-1}\psi^w \mathcal{M}(t)\mathcal{M}(t)\psi^w + c_2 \psi^w S(\langle\xi\rangle^{-1}, \Upsilon)^w \psi^w$$

$$\quad + \psi^w \left[\mathcal{M}(t), \left[\psi^w, F(t)\right]\right] + \left[\psi^w, \left[\psi^w, F(t)\right]\right]\mathcal{M}(t) + \left[\psi^w, F(t)\right]\left[\psi^w, \mathcal{M}(t)\right].$$

Next we analyze each term on the last line. We have

- $\psi^w \left[\mathcal{M}(t), \left[\psi^w, F(t)\right]\right] \in S(d\langle\xi\rangle^{-1/2}1d^{-1}\langle\xi\rangle^{-1/2}, \Upsilon)^w = S(\langle\xi\rangle^{-1}, \Upsilon)^w$ since

$$\psi^w, \ \left[\psi^w, F(t)\right] \in S(1, \langle\xi\rangle^{-1}\Upsilon)^w, \quad \mathcal{M}(t) \in S_1(d\langle\xi\rangle^{-1/2}, d^{-2}\Upsilon)^w,$$

- $\left[\psi^w, \left[\psi^w, F(t)\right]\right]\mathcal{M}(t) \in S(d\langle\xi\rangle^{-3/2}, \Upsilon)^w \subset S(\langle\xi\rangle^{-1}, \Upsilon)^w$ since $d \leq 2\langle\xi\rangle^{1/2}$ and

$$\left[\psi^w, \left[\psi^w, F(t)\right]\right] \in S(\langle\xi\rangle^{-1}, \langle\xi\rangle^{-1}\Upsilon)^w, \quad \mathcal{M}(t) \in S_1(d\langle\xi\rangle^{-1/2}, d^{-2}\Upsilon)^w,$$

- $\left[\psi^w, F(t)\right]\left[\psi^w, \mathcal{M}(t)\right] \in S(d\langle\xi\rangle^{-1/2}\langle\xi\rangle^{-1/2}d^{-1}, \Upsilon)^w = S(\langle\xi\rangle^{-1}, \Upsilon)^w$ since

$$\left[\psi^w, F(t)\right] \in S(1, \langle\xi\rangle^{-1}\Upsilon)^w, \quad \mathcal{M}(t) \in S_1(d\langle\xi\rangle^{-1/2}, d^{-2}\Upsilon)^w.$$

We have proven in particular that

$$\frac{d}{dt}\left(\psi^w \mathcal{M}(t)\psi^w\right) + 2\operatorname{Re}\left(\psi^w \mathcal{M}(t)\psi^w F(t)\right)$$

$$= \psi^w \left(\dot{\mathcal{M}}(t) + 2\operatorname{Re}\mathcal{M}(t)F(t)\right)\psi^w + S(\langle\xi\rangle^{-1}, \Upsilon)^w. \quad (3.7.89)$$

Also, we have

$$\frac{d}{dt}\left(\psi^w \mathcal{M}(t)\psi^w\right) + 2\operatorname{Re}\left(\psi^w \mathcal{M}(t)\psi^w F(t)\right) \geq c_1 T^{-1}\psi^w \mathcal{M}(t)\mathcal{M}(t)\psi^w + S(\langle\xi\rangle^{-1}, \Upsilon)^w,$$

which is (3.7.88). $\qquad\qquad\square$

Theorem 3.7.25. *Let $f(t, x, \xi)$ be a smooth real-valued function defined on $\mathbb{R} \times \mathbb{R}^n \times \mathbb{R}^n$, satisfying (3.2.13) and (3.7.50). Let $f_0(t, x, \xi)$ be a smooth complex-valued function defined on $\mathbb{R} \times \mathbb{R}^n \times \mathbb{R}^n$, such that $\langle \xi \rangle f_0(t, x, \xi)$ satisfies (3.7.50). We define*

$$L = D_t + if(t, x, \xi)^w + f_0(t, x, \xi)^w.$$

Let $\psi(x, \xi) \in S(1, \langle \xi \rangle^{-1}\Upsilon)$ be a real-valued symbol. Then there exists $T_0 > 0, c_0 > 0, C \geq 0$, depending on a finite number of semi-norms of f, f_0, ψ, such that, for all $T \leq T_0$, all $u \in C_c^\infty((-T, T); \mathscr{S}(\mathbb{R}^n))$, with ω_0 given by (3.7.79),

$$c_0 \left(\int \|\psi^w u(t)\|_{H^{-1/2}(\mathbb{R}^n)}^2 \, dt \right)^{1/2} \leq T \|\psi^w (e^{-\omega_0})^w Lu\|_{L^2(\mathbb{R}^{n+1})}$$

$$+ CT^{1/2} \left(\int \|u(t)\|_{H^{-1/2}(\mathbb{R}^n)}^2 \, dt \right)^{1/2} + C \left(\int \|u(t)\|_{H^{-3/2}(\mathbb{R}^n)}^2 \, dt \right)^{1/2}. \quad (3.7.90)$$

Proof. We compute, noting $F(t) = f(t, x, \xi)^w$,

$$2 \operatorname{Re}\langle Lu, i\psi^w \mathcal{M}(t)\psi^w u \rangle = \left\langle \left(\psi^w \dot{\mathcal{M}}(t)\psi^w + 2\operatorname{Re}(\psi^w \mathcal{M}(t)\psi^w F(t)) \right) u, u \right\rangle$$

$$+ \left\langle \left[(\operatorname{Re} f_0(t))^w, i\psi^w \mathcal{M}(t)\psi^w \right] u, u \right\rangle + 2 \operatorname{Re} \langle \psi^w \mathcal{M}(t)\psi^w \operatorname{Im} f_0(t)^w u, u \rangle.$$

(i) *Let us assume that* $\operatorname{Im}(f_0) \in S(\langle \xi \rangle^{-1}, \langle \xi \rangle^{-1}\Upsilon)$. *Then we get that*

$$\psi^w \mathcal{M}(t)\psi^w \operatorname{Im} f_0(t)^w \in S(d\langle \xi \rangle^{-1/2}\langle \xi \rangle^{-1}, \Upsilon)^w \subset S(\langle \xi \rangle^{-1}, \Upsilon)^w$$

and since

$$\left[(\operatorname{Re} f_0(t))^w, i\psi^w \mathcal{M}(t)\psi^w \right] \in S(d\langle \xi \rangle^{-1/2}\langle \xi \rangle^{-1/2}d^{-1}, \Upsilon)^w = S(\langle \xi \rangle^{-1}, \Upsilon)^w,$$

the inequality (3.7.75), the identity (3.7.89) and Lemmas 3.7.21, 3.7.24 show that

$$2 \operatorname{Re}\langle Lu, i\psi^w \mathcal{M}(t)\psi^w u \rangle = \left\langle \left(\psi^w \dot{\mathcal{M}}(t)\psi^w + 2\operatorname{Re}(\psi^w \mathcal{M}(t)\psi^w F(t)) \right) u, u \right\rangle$$

$$\geq \frac{c_1}{2} T^{-1} \int \|\mathcal{M}(t)\psi^w u(t)\|_{L^2(\mathbb{R}^n)}^2 \, dt + \frac{c_0}{2} T^{-1} \int \|\psi^w u(t)\|_{H^{-1/2}(\mathbb{R}^n)}^2 \, dt$$

$$- C \int \|u(t)\|_{H^{-1/2}(\mathbb{R}^n)}^2 \, dt.$$

As a consequence, we have

$$2T \int \|\psi^w Lu(t)\|_{L^2(\mathbb{R}^n)} \|\mathcal{M}(t)\psi^w u(t)\|_{L^2(\mathbb{R}^n)} \, dt + CT \int \|u(t)\|_{H^{-1/2}(\mathbb{R}^n)}^2 \, dt$$

$$\geq \frac{c_1}{2} \int \|\mathcal{M}(t)\psi^w u(t)\|_{L^2(\mathbb{R}^n)}^2 \, dt + \frac{c_0}{2} \int \|\psi^w u(t)\|_{H^{-1/2}(\mathbb{R}^n)}^2 \, dt,$$

so that, with $\alpha > 0$,

$$T \int \left(T\alpha^{-1} \|\psi^w Lu(t)\|^2_{L^2(\mathbb{R}^n)} + \alpha T^{-1} \|\mathcal{M}(t)\psi^w u(t)\|^2_{L^2(\mathbb{R}^n)} \right) dt$$

$$+ CT \int \|u(t)\|^2_{H^{-1/2}(\mathbb{R}^n)} dt$$

$$\geq \frac{c_1}{2} \int \|\mathcal{M}(t)\psi^w u(t)\|^2_{L^2(\mathbb{R}^n)} dt + \frac{c_0}{2} \int \|\psi^w u(t)\|^2_{H^{-1/2}(\mathbb{R}^n)} dt.$$

Choosing $\alpha \leq c_1/2$ yields the result

$$T^2 \alpha^{-1} \int \|\psi^w Lu(t)\|^2_{L^2(\mathbb{R}^n)} dt + CT \int \|u(t)\|^2_{H^{-1/2}(\mathbb{R}^n)} dt$$

$$\geq \frac{c_0}{2} \int \|\psi^w u(t)\|^2_{H^{-1/2}(\mathbb{R}^n)} dt,$$

which is a better estimate than the sought one.

(ii) *Let us deal now with the general case* $\text{Im}(f_0) \in S(1, \langle\xi\rangle^{-1}\Upsilon)$. Using Definitions (3.7.79), (3.7.81) and the property (3.7.80), we can use (i) above to get the estimate for L_0, so that with a fixed $c_2 > 0$,

$$T\|\psi^w L_0 u\|_{L^2(\mathbb{R}^{n+1})} + T^{1/2} \left(\int \|u(t)\|^2_{H^{-1/2}(\mathbb{R}^n)} dt \right)^{1/2}$$

$$\geq c_2 \left(\int \|\psi^w u(t)\|^2_{H^{-1/2}(\mathbb{R}^n)} dt \right)^{1/2}, \quad (3.7.91)$$

so that $T\|\psi^w (e^{-\omega_0})^w L(e^{\omega_0})^w u\|_{L^2(\mathbb{R}^{n+1})} + T^{1/2} \left(\int \|u(t)\|^2_{H^{-1/2}(\mathbb{R}^n)} dt \right)^{1/2}$

$$\geq c_2 \left(\int \|\psi^w u(t)\|^2_{H^{-1/2}(\mathbb{R}^n)} dt \right)^{1/2}. \quad (3.7.92)$$

Applying this to $u(t)$ given by (3.7.83), we obtain

$$T\|\psi^w (e^{-\omega_0})^w Lv\|_{L^2(\mathbb{R}^{n+1})} + T^{1/2} \left(\int \left\| \left((e^{\omega_0})^w \right)^{-1} v(t) \right\|^2_{H^{-1/2}(\mathbb{R}^n)} dt \right)^{1/2}$$

$$\geq c_2 \left(\int \left\| \psi^w \left((e^{\omega_0})^w \right)^{-1} v(t) \right\|^2_{H^{-1/2}(\mathbb{R}^n)} dt \right)^{1/2}. \quad (3.7.93)$$

Using that $((e^{\omega_0})^w)^{-1}$ is a pseudo-differential operator with symbol in $S(1, \langle\xi\rangle^{-1}\Upsilon)$, we obtain, using the notation (3.7.86),

$$T\|\psi^w(e^{-\omega_0})^w Lv\|_{L^2(\mathbb{R}^{n+1})} + CT^{1/2}\left(\int \|v(t)\|_{H^{-1/2}(\mathbb{R}^n)}^2 dt\right)^{1/2}$$

$$\geq c_2\left(\int \|\Omega(t)^{-1}(\langle\xi\rangle^{-1/2})^w\psi^w v(t)\|_{L^2(\mathbb{R}^n)}^2 dt\right)^{1/2}$$

$$- C_1\left(\int \|v(t)\|_{H^{-3/2}(\mathbb{R}^n)}^2 dt\right)^{1/2},\quad (3.7.94)$$

so that, using (3.7.87),

$$T\|\psi^w(e^{-\omega_0})^w Lv\|_{L^2(\mathbb{R}^{n+1})} + CT^{1/2}\left(\int \|v(t)\|_{H^{-1/2}(\mathbb{R}^n)}^2 dt\right)^{1/2}$$

$$+ C_1\left(\int \|v(t)\|_{H^{-3/2}(\mathbb{R}^n)}^2 dt\right)^{1/2}$$

$$\geq c_2\left(\int \left\|(\langle\xi\rangle^{-1/2})^w\psi^w v(t)\right\|_{L^2(\mathbb{R}^n)}^2 \frac{1}{\|\Omega(t)\|^2}dt\right)^{1/2}$$

$$\geq c_3\left(\int \left\|(\langle\xi\rangle^{-1/2})^w\psi^w v(t)\right\|_{L^2(\mathbb{R}^n)}^2 dt\right)^{1/2} = c_3\left(\int \|\psi^w v(t)\|_{H^{-1/2}(\mathbb{R}^n)}^2 dt\right)^{1/2},$$

$$(3.7.95)$$

which is the result. The proof of the theorem is complete. $\qquad\square$

From inhomogeneous localization to homogeneous localization

In this section, we are given a positive integer n, and we define $N = n+1$. The running point of $T^*(\mathbb{R}^N)$ will be denoted by (y, η). We are also given a point $(y_0; \eta_0) \in \mathbb{R}^N \times \mathbb{S}^{N-1}$ such that

$$Y_0 = (y_0; \eta_0) = (t_0, x_0; \tau_0, \xi_0) \in \mathbb{R} \times \mathbb{R}^n \times \mathbb{R} \times \mathbb{R}^n, \text{ with } \tau_0 = 0, \ \xi_0 \in \mathbb{S}^{n-1}, t_0 = 0.$$
$$(3.7.96)$$

We consider $F(t, x, \xi) = f(t, x, \xi) - if_0(t, x, \xi)$, with f, f_0 satisfying the assumptions of Theorem 3.7.25. Let $\psi_0(\xi)$ be a function supported in a conic-neighborhood of ξ_0 and $\chi_0(\tau, \xi)$ be a homogeneous localization near $\tau = 0$ as in Section 4.9.6 with some positive r_0. We consider also a classical first-order pseudo-differential operator R in $\mathbb{R}^N$ such that $Y_0 \notin WFR$. We consider the first-order operator

$$\mathcal{L} = D_t + i\big(F(t, x, \xi)\psi_0(\xi)\chi_0(\tau, \xi)\big)^w + R. \qquad (3.7.97)$$

We have

$$\mathcal{L} = D_t + i\big(F(t, x, \xi)\psi_0(\xi)\big)^w + \underbrace{i\Big(F(t, x, \xi)\psi_0(\xi)\big(\chi_0(\tau, \xi) - 1\big)\Big)^w}_{=F_1(t, x, \tau, \xi)^w} + R. \quad (3.7.98)$$

Let $\psi_1(\xi)$ be a function supported in a conic-neighborhood of ξ_0 and $\chi_1(\tau, \xi)$ be a homogeneous localization near $\tau = 0$ as in Section 4.9.6 with some positive $r_1 < r_0$ and such that

$$\operatorname{supp}\chi_1 \subset \{\chi_0 = 1\}, \quad \operatorname{supp}(\psi_1\chi_1) \subset \{\psi_0\chi_0 = 1\}, \tag{3.7.99}$$
$$[-T_1, T_1] \times K_1 \times \operatorname{supp}\psi_1\chi_1 \subset (WFR)^c, \tag{3.7.100}$$

where $T_1 > 0$ and K_1 is a compact neighborhood of x_0. Let $\psi(x, \xi)$ be a symbol satisfying the assumptions of Theorem 3.7.25 and let $\rho_1 \in C_c^\infty(\mathbb{R})$, such that

$$\operatorname{supp}\psi \subset K_1 \times \{\psi_1 = 1\}, \quad \operatorname{supp}\rho_1 \subset [-T_1, T_1]. \tag{3.7.101}$$

We can apply Theorem 3.7.25 to the operator $L = D_t + i\big(F(t, x, \xi)\psi_0(\xi)\big)^w$. We have, with $u \in \mathscr{S}(\mathbb{R}^N)$,

$$T_1 \|\psi^w(e^{-\omega_0})^w(\mathcal{L} - F_1 - R)\rho_1\chi_1^w u\|_{L^2(\mathbb{R}^{n+1})}$$
$$+ CT_1^{1/2}\left(\int \|\rho_1\chi_1^w u(t)\|_{H^{-1/2}(\mathbb{R}^n)}^2 \, dt\right)^{1/2} + C\left(\int \|\rho_1\chi_1^w u(t)\|_{H^{-3/2}(\mathbb{R}^n)}^2 \, dt\right)^{1/2}$$
$$\geq c_0 \left(\int \|\psi^w \rho_1\chi_1^w u(t)\|_{H^{-1/2}(\mathbb{R}^n)}^2 \, dt\right)^{1/2}.$$

We get then

$$T_1 \|\psi^w(e^{-\omega_0})^w\rho_1\chi_1^w \mathcal{L}u + \psi^w(e^{-\omega_0})^w\rho_1[\mathcal{L}, \chi_1^w]u\|_{L^2(\mathbb{R}^{n+1})}$$
$$+ T_1 \|\psi^w(e^{-\omega_0})^w[\mathcal{L}, \rho_1]\chi_1^w u\|_{L^2(\mathbb{R}^{n+1})}$$
$$+ T_1 \|\psi^w(e^{-\omega_0})^w F_1^w \rho_1\chi_1^w u\|_{L^2(\mathbb{R}^{n+1})} + T_1 \|\psi^w(e^{-\omega_0})^w R\rho_1\chi_1^w u\|_{L^2(\mathbb{R}^{n+1})}$$
$$+ CT_1^{1/2}\left\|(\langle\xi\rangle^{-1/2})^w\rho_1\chi_1^w u\right\|_{L^2(\mathbb{R}^{n+1})} + C\left\|(\langle\xi\rangle^{-3/2})^w\rho_1\chi_1^w u\right\|_{L^2(\mathbb{R}^{n+1})}$$
$$\geq c_0\left(\int \|\psi^w \rho_1\chi_1^w u\|_{H^{-1/2}(\mathbb{R}^n)}^2 \, dt\right)^{1/2}. \tag{3.7.102}$$

We assume now that $u \in \mathscr{S}(\mathbb{R}^N)$, $\operatorname{supp} u \subset \{(t, x), |t| \leq T_1/2\}$ and also that ρ_1 is 1 on $[-3T_1/4, 3T_1/4]$. We introduce two admissible[12] metrics on $\mathbb{R}^{2N}$,

$$G = |dt|^2 + |dx|^2 + \frac{|d\xi|^2 + |d\tau|^2}{1 + |\xi|^2 + \tau^2}$$
$$\leq g = |dt|^2 + |dx|^2 + \frac{|d\xi|^2}{1 + |\xi|^2} + \frac{|d\tau|^2}{1 + |\xi|^2 + \tau^2}. \tag{3.7.103}$$

(1) The operator $[\mathcal{L}, \chi_1^w]$ has a symbol in $S(1, G)$ which is essentially supported in the region where $|\tau| \sim |\xi|$.

---

[12]See Lemma 4.9.17.

(2) The quantity $[\mathcal{L}, \rho_1]\chi_1^w u = [\mathcal{L}, \rho_1]\chi_1^w \rho_2 u$ if $\rho_2(t)$ is 1 on $[-T_1/2, T_1/2]$ and supported in $[-3T_1/4, 3T_1/4]$ and thus the operator $[\mathcal{L}, \rho_1]\chi_1^w \rho_2$ has a symbol in $S((1 + |\xi| + |\tau|)^{-\infty}, G)$.

(3) The operator $F_1^w \rho_1 \chi_1^w$ is the composition of the symbol $F_1 \in S(\langle \xi \rangle, g)$ with the symbol in $\rho_1 \sharp \chi_1 \in S(1, G)$ and thus is a priori in $S(\langle \xi \rangle, g)$; however, looking at the expansion, and using (3.7.99), we see that it has a symbol in $S((1 + |\xi| + |\tau|)^{-\infty}, G)$: it is not completely obvious though and we refer the reader to Lemma 4.9.18 for a complete argument.

(4) The operator $\psi^w (e^{-\omega_0})^w R\rho_1 \chi_1^w$ is also the composition of an operator in $S(1, g)^w$ with an operator in $S(\langle \xi, \tau \rangle, G)^w$; however, using (3.7.99), (3.7.100), (3.7.101) and Section 4.9.7, we see that $\psi^w (e^{-\omega_0})^w R\rho_1 \chi_1^w$ has a symbol in $S((1 + |\xi| + |\tau|)^{-\infty}, G)$.

(5) The operator $(\langle \xi \rangle^s)^w \rho_1 \chi_1^w$ is also the sum of an operator in $S(\langle \tau, \xi \rangle^s, G)$ plus a symbol in $S((1 + |\xi| + |\tau|)^{-\infty}, G)$.

With R_1 of order $-\infty$ (weight $\langle \xi, \tau \rangle^{-\infty}$) for G, E_0 of order 0 (weight 1) for G, supported in

$$\{(t, x, \tau, \xi), |t| \leq T_1, x \in K_1, (\tau, \xi) \in \operatorname{supp} \nabla\chi_1, (x, \xi) \in \operatorname{supp} \psi\},$$

we can write now

$$T_1 \|\psi^w (e^{-\omega_0})^w \rho_1 \chi_1^w \mathcal{L}u + E_0 u\|_{L^2(\mathbb{R}^{n+1})} + T_1 \|R_1 u\|_{L^2(\mathbb{R}^{n+1})}$$

$$+ CT_1^{1/2} \|u\|_{H^{-1/2}(\mathbb{R}^{n+1})} + C\|u\|_{H^{-3/2}(\mathbb{R}^{n+1})}$$

$$\geq c_0 \left(\int \|\psi^w \rho_1 \chi_1^w u\|_{H^{-1/2}(\mathbb{R}^n)}^2 dt \right)^{1/2}. \qquad (3.7.104)$$

Theorem 3.7.26. *Let $\mathcal{L}$ be the pseudo-differential operator given by (3.7.97) and $Y_0 = (y_0, \eta_0)$ be given by (3.7.96). We assume that $\{Y_0\} \subset \Delta_0 \subset (WFR)^c$, where Δ_0 is a compact-conic-neighborhood of Y_0. Then, there exists two pseudo-differential operators Φ_0, Ψ_0 of order 0 (weight 1) for G, both essentially supported in Δ_0 with Φ_0 is elliptic at Y_0, and there exists $r > 0$ such that, for all $u \in \mathcal{S}(\mathbb{R}^N), \operatorname{supp} u \subset \{(t, x), |t| \leq r\}$,*

$$r\|\Psi_0 \mathcal{L}u\|_{L^2(\mathbb{R}^N)} + r^{1/2}\|u\|_{H^{-1/2}(\mathbb{R}^N)} + \|u\|_{H^{-3/2}(\mathbb{R}^N)} \geq \|\Phi_0 u\|_{H^{-1/2}(\mathbb{R}^N)}. \quad (3.7.105)$$

Proof. It is a direct consequence of (3.7.105) since, using the ellipticity of $\mathcal{L}$ in the support of the symbol of E_0, we get $E_0 = \mathcal{K}\mathcal{L} + R_2$, where $\mathcal{K}$ is a pseudo-differential operator of order 0 such that $WF\mathcal{K} \subset \Delta_0$ and R_2 is a pseudo-differential operator of order $-\infty$ for G. $\qquad \square$

Proof of the solvability result stated in Theorem 3.7.1

Let P be a first-order pseudo-differential operator with principal symbol p satisfying the assumptions of Theorem 3.7.1 and let (y_0, η_0) be a point in the cosphere bundle. If $p(y_0, \eta_0) \neq 0$, then there exists a pseudo-differential operator Φ_0 of order 0, elliptic at (y_0, η_0) such that

$$\|P^* u\|_0 + \|u\|_{-1} \geq \|\Phi_0 u\|_1. \tag{3.7.106}$$

In fact, the ellipticity assumption implies that there exist a pseudo-differential operator K of order -1 and a pseudo-differential operator R of order 0 such that

$$\mathrm{Id} = K P^* + R, \quad (y_0, \eta_0) \notin WFR.$$

As consequence, for Φ_0 of order 0 essentially supported close enough to (y_0, η_0), we get $\Phi_0 = \Phi_0 K P^* + \Phi_0 R$ with $\Phi_0 R$ of order $-\infty$, which gives (3.7.107).

Let us assume now that $p(y_0, \eta_0) = 0$. We know from the assumption (3.7.1) that $\partial_\eta p(y_0, \eta_0) \neq 0$ and we may suppose that $(\partial_\eta \operatorname{Re} p)(y_0, \eta_0) \neq 0$. Using the Malgrange-Weierstrass theorem, we can find a conic-neighborhood of (y_0, η_0) in which

$$p(y, \eta) = \big(\sigma + a(s, z, \zeta) + ib(s, z, \zeta)\big) e_0(y, \eta)$$

where a, b are real-valued positively-homogeneous of degree 1, e_0 is homogeneous of degree 0, elliptic near (y_0, η_0), $(s, z; \sigma, \zeta) \in \mathbb{R} \times \mathbb{R}^n \times \mathbb{R} \times \mathbb{R}^n$ a choice of symplectic coordinates in $T^*(\mathbb{R}^N)$ $(N = n + 1)$, with $y_0 = (0, 0), \eta_0 = (0, \ldots, 0, 1)$. Noting that the Poisson bracket

$$\{\sigma + a, s\} = 1$$

we see that there exists an homogeneous canonical transformation Ξ^{-1}, from a (conic) neighborhood of (y_0, η_0) to a conic-neighborhood of $(0; 0, \ldots 0, 1)$ in $\mathbb{R}^N \times \mathbb{R}^N$ such that

$$p \circ \Xi = \big(\tau + iq(t, x, \xi)\big)(e \circ \Xi).$$

Note in particular that, setting $\tau = \sigma + a, t = s$, *(which preserves the coordinate s)* yields

$$-\partial_\tau q = \{t, q\} = \{s, b\} \circ \chi = 0.$$

We see now that there exists some elliptic Fourier integral operators A, B and E a pseudo-differential operator of order 0, elliptic at (y_0, η_0) such that

$$AEP^* B = D_t + i(f(t, x, \xi)\chi_0(\tau, \xi))^w + R,$$

$$BA = \mathrm{Id} + S, (y_0, \eta_0) \in \Gamma_0(\text{conic-neighborhood of } {}_{(y_0, \eta_0)}) \subset (WFS)^c,$$

where f satisfies (3.2.13), R is a pseudo-differential operator of order 0, and χ_0 is a non-negative homogeneous localization near $\tau = 0$. Using the fact that the coordinate s is preserved by the canonical transformation, we can assume that A, B are local operators in the t variable, i.e., are such that

$$u \in C_c^\infty, \operatorname{supp} u \subset \{(t, x) \in \mathbb{R} \times \mathbb{R}^n, |t| \leq r\} \Longrightarrow \operatorname{supp} Bu \subset \{(s, z) \in \mathbb{R} \times \mathbb{R}^n, |s| \leq r\}.$$

Using the fact that the operator P is polyhomogeneous, one can iterate the use of the Malgrange-Weierstrass theorem to reduce our case to $AEP^*B = \mathcal{L}$ of the type given in (3.7.97). We can apply Theorem 3.7.26, giving the existence of a pseudo-differential operator Ψ_0 of order 0, elliptic at $\Xi^{-1}(y_0, \eta_0)$, essentially supported in $\Xi^{-1}(\Gamma_0)$ such that for all $u \in C_c^\infty(\mathbb{R}^N)$, supp $u \subset \{|t| \leq r\}$,

$$r\|\Psi_0 AEP^*Bu\|_0 + r^{1/2}\|u\|_{-1/2} + \|u\|_{-3/2} \geq \|\Phi_0 u\|_{-1/2}.$$

We may assume that A and B are properly supported and apply the previous inequality to $u = Av$, whose support in the s variable is unchanged. We get

$$r\|\Psi_0 AEP^*BAv\|_0 + r^{1/2}\|Av\|_{-1/2} + \|Av\|_{-3/2} \geq \|\Phi_0 Av\|_{-1/2},$$

so that

$$r\|\Psi_0 AEP^*v\|_0 + Cr^{1/2}\|v\|_{-1/2} + C_1\|v\|_{-3/2} \geq \|\Phi_0 Av\|_{-1/2} \geq C_2^{-1}\|B\Phi_0 Av\|_{-1/2},$$

which gives, for all $v \in C_c^\infty(\mathbb{R}^N)$, supp $v \subset \{y \in \mathbb{R}^N, |y - y_0| \leq r\}$,

$$r\|P^*v\|_0 + r^{1/2}\|v\|_{-1/2} + \|v\|_{-3/2} \geq \|\Phi v\|_{-1/2}, \tag{3.7.107}$$

where $\Phi = cB\Phi_0 A$ is a pseudo-differential operator of order 0, elliptic near (y_0, η_0). By compactness of the cosphere bundle, one gets, using (3.7.108) or (3.7.107),

$$\|v\|_{-1/2} \leq C \sum_{1 \leq \kappa \leq l} \|\Phi_{0\kappa} v\|_{-1/2} + C\|v\|_{-1} \leq C_1 r\|P^*v\|_0 + C_1 r^{1/2}\|v\|_{-1/2} + C_1\|v\|_{-1}, \tag{3.7.108}$$

which entails, by shrinking r, the existence of $r_0 > 0, C_0 > 0$, such that for $v \in C_c^\infty(\mathbb{R}^N)$, supp $v \subset \{y \in \mathbb{R}^N, |y - y_0| \leq r_0\} = B_{r_0}$,

$$\|v\|_{-1/2} \leq C_0\|P^*v\|_0. \tag{3.7.109}$$

Let s be a real number and P be an operator of order m, satisfying the assumptions of Theorem 3.7.1. Let E_σ be a properly supported operator with symbol $\langle \xi \rangle^\sigma$. Then the operator $E_{1-m-s}PE_s$ is of first order, satisfies condition (Ψ) and from the previous discussion, there exists $C_0 > 0, r_0 > 0$ such that

$$\|v\|_{-1/2} \leq C_0\|E_sP^*E_{1-m-s}v\|_0, \quad v \in C_c^\infty(\mathbb{R}^N), \text{ supp } v \subset B_{r_0}.$$

We get, with χ_r supported in B_r and $\chi_r = 1$ on $B_{r/2}$, with supp $u \subset B_{r_0/4}$,

$$\|\chi_{r_0} E_{m+s-1}\chi_{r_0/2}u\|_{-1/2} \leq C_0\|E_sP^*E_{1-m-s}\chi_{r_0}E_{m+s-1}\chi_{r_0/2}u\|_0$$
$$\leq C_0\|E_sP^*E_{1-m-s}\underbrace{[\chi_{r_0}, E_{m+s-1}]\chi_{r_0/2}}_{S-\infty}u\|_0 + C_0\|E_sP^*\underbrace{E_{1-m-s}E_{m+s-1}}_{=\text{Id}+S-\infty}\underbrace{\chi_{r_0/2}u}_{=u}\|_0$$
$$\leq C_0\|P^*u\|_s + \|Ru\|_0,$$

where R is of order $-\infty$. Since we have

$$\chi_{r_0} E_{m+s-1} \chi_{r_0/2} u = \underbrace{[\chi_{r_0}, E_{m+s-1}] \chi_{r_0/2}}_{S-\infty} u + E_{m+s-1} \underbrace{\chi_{r_0} \chi_{r_0/2} u}_{=u},$$

we get $\|u\|_{s+m-\frac{3}{2}} \le C_0 \|P^* u\|_s + C_1 \|u\|_{s+m-2}$ and, shrinking the support of u (see Lemma 1.2.34), we obtain the estimate

$$\|u\|_{s+m-\frac{3}{2}} \le C_2 \|P^* u\|_s, \tag{3.7.110}$$

for $u \in C_c^\infty$ with support in a neighborhood of y_0. This implies the local solvability of P, with the loss of derivatives claimed by Theorem 3.7.1, whose proof is now complete.

3.8 Concluding remarks

3.8.1 A (very) short historical account of solvability questions

In 1957, Hans Lewy [102] constructed a counterexample showing that very simple and natural differential equations can fail to have local solutions; his example is the complex vector field (3.1.37) and one can show that there exists some C^∞ function f such that the equation $\mathcal{L}_0 u = f$ has no distribution solution, even locally. A geometric interpretation and a generalization of this counterexample were given in 1960 by L. Hörmander in [63] and extended in [64] to pseudo-differential operators. In 1970, L. Nirenberg and F. Treves ([113], [114], [115]), after a study of complex vector fields in [112] (see also the S. Mizohata paper [107]), refined this condition on the principal symbol to the so-called condition (Ψ), and provided strong arguments suggesting that it should be equivalent to local solvability. As a matter of fact, the Nirenberg-Treves conjecture states the equivalence between local solvability of a principal-type pseudo-differential operator and the geometric property condition (Ψ) of its principal symbol.

The necessity of condition (Ψ) for local solvability of principal-type pseudo-differential equations was proved in two dimensions by R. Moyer in [108] and in general by L. Hörmander ([70]) in 1981.

The sufficiency of condition (Ψ) for local solvability of differential equations was proved by R. Beals and C. Fefferman ([8]) in 1973; they created a new type of pseudo-differential calculus, based on a Calderón-Zygmund decomposition, and were able to remove the analyticity assumption required by L. Nirenberg and F. Treves. For differential equations in any dimension ([8]) and for pseudo-differential equations in two dimensions (N. Lerner's [89], see also [91]), it was shown more precisely that (Ψ) implies local solvability with a loss of one derivative with respect to the elliptic case: for a differential operator P of order m (or a pseudo-differential operator in two dimensions), satisfying condition (Ψ), $f \in H_{\text{loc}}^s$, the equation $Pu = f$ has a solution $u \in H_{\text{loc}}^{s+m-1}$.

In 1994, it was proved by N. Lerner in [92] (see also L. Hörmander's survey article [77] and [97]) that condition (Ψ) does not imply local solvability with loss of one derivative for pseudo-differential equations, contradicting repeated claims by several authors. However in 1996, N. Dencker in [33], proved that these counterexamples were indeed locally solvable, but with a loss of two derivatives. In 2006, N. Dencker [35] proved that condition (Ψ) implies local solvability with loss of two derivatives, providing the final step in the proof of the Nirenberg-Treves conjecture. The 2004 paper [34] by N. Dencker provided a loss of $\frac{3}{2} + \varepsilon$ derivatives for all ε positive. Following the pattern of [35], N. Lerner [98] showed in 2006 that the loss can actually be limited to $3/2$ derivatives.

That short historical summary was only concerned with the C^∞ category, but for microdifferential operators acting on microfunctions, the necessity of condition (Ψ) for microlocal solvability was proven by M. Sato, T. Kawai and M. Kashiwara in [127]: when condition (Ψ) is violated for such an operator, it is microlocally equivalent to $D_{x_1} + ix_1^{2k+1}D_{x_2}$ at $(x = 0; \xi_1 = 0, \xi_2 = 1, 0)$, which is the model M_{2k+1} studied in Section 3.1.1. The sufficiency of condition (Ψ) in that framework was established by J.-M. Trépreau in [139] (see also [75]).

3.8.2 Open problems

Although the Nirenberg-Treves conjecture is proven, the situation for pseudo-differential equations in three or more dimensions is not completely satisfactory and several questions remain.

Question 1. *Let $P \in \Psi_{ps}^m(\Omega)$ be a principal type pseudo-differential operator on Ω whose principal symbol p satisfies condition (Ψ). What is the minimal loss μ of derivatives for the local solvability of P?*

When P is a differential operator (or p satisfies the stronger condition (P), see Theorem 3.5.1) or the dimension is 2 (Theorem 3.5.6), or in various other cases (Sections 3.5.2, 3.5.3), the loss is known to be 1, the smallest possible in general and the same as for $\partial/\partial x_1$. When the geometry is finite-type, i.e., when P^* is subelliptic, local solvability occurs with a loss < 1 (Theorem 3.4.3). We know that

$$1 < \mu \leq \frac{3}{2}, \tag{3.8.1}$$

with the counterexample of Section 3.6.1 and Theorem 3.6.1 implying the first inequality, whereas Theorem 3.7.1 implies the other one. Also the construction of Section 3.6.1 uses heavily the nonanalyticity of the symbol and this leads to the following question.

Question 2. *Let P be a principal type pseudo-differential operator whose principal symbol p satisfies condition (Ψ) and is real-analytic. Is P locally solvable with loss of one derivative?*

On the other hand, it remains very unnatural to deal with a principal-type pseudo-differential operator and to have solvability with a loss > 1. An interesting question would be as follows.

Question 3. *Is there a geometric condition, say* ($\Psi^\sharp$), *stronger than condition* (Ψ) *on p which is equivalent to local solvability with loss of one derivative?*

No hint of such condition is in sight, and the question might be irrelevant if the loss under condition (Ψ) is in fact $1+\varepsilon$ for all positive ε. It is puzzling however to note that even for the simple class of examples

$$L = D_t + i\big(a_0(t,x,\xi)b_1(t,x,\xi)\big)^w + r_0(t,x,\xi)^w, \qquad (3.8.2)$$

$$0 \le a_0 \in S^0_{1,0}, \quad \partial_t b_1 \ge 0, \quad b_1 \text{ real-valued} \in S^1_{1,0}, \quad r_0 \in S^0_{1,0},$$

the methods developed in Section 3.7 (see also the introductory discussion in Section 3.7.2) do not give a better result than $\|Lu\|_{H^0} \gtrsim \|u\|_{H^{-1/2}}$, that is an estimate with loss of $3/2$ derivatives. The simplification given by this factorization is enormous, however if $\partial_t b_1$ is not identically 0 (in particular condition (P) does not hold), but if a characteristic curve of the real part stays in $\operatorname{Im} p = 0$ (here $b_1(t, x_0, \xi_0) \equiv 0$ at some point (x_0, ξ_0)), no better estimate seems at hand for that class, except if some very particular geometric structure of the change-of-sign-set is available as in Theorem 3.5.8. A more technical question should certainly be answered before attacking any of the previous questions.

Question 4. *Let* $a \in C_c^\infty(\mathbb{R}^3; \mathbb{R}_+), b \in C_c^\infty(\mathbb{R}^3; \mathbb{R}), \partial_t b \ge 0, \mu \in]1, 3/2[$. *We define for* $h > 0$, $L_h = h\partial_t - a(t, x, hD)b(t, x, hD)$. *Does there exist* $r > 0$ *such that*

$$\forall h \in (0, r], \ \forall u \in C_c^\infty(\mathbb{R}^2), \ \operatorname{supp} u \subset B(r), \quad \|Lu\|_{L^2(\mathbb{R}^2)} \ge r h^\mu \|u\|_{L^2(\mathbb{R}^2)},$$

where $B(r)$ *is the ball in* $\mathbb{R}^2$ *of radius* r *with center* 0? *Even the case where* $b(t, x, \xi) = \xi + \beta(t, x)$ *(say for* $|x| + |\xi| \le 1$*), with* $\partial_t \beta \ge 0$, *does not seem so easy.*

Some questions have very satisfactory answers under condition (P) and they may also be raised under condition (Ψ).

Question 5. *Let* P *be as in Question 1,* $x_0 \in \Omega$. *Is it true that for every* $f \in C^\infty$, *there is some* $u \in C^\infty$ *such that* $Pu = f$ *in a neighborhood of* x_0.

Question 6. *Let* P *be as in Question 1,* K *a compact subset of* Ω. *Is there a semiglobal solvability result analogous to Theorem 3.5.14?*

The answer to these questions would certainly require as for condition (P) a detailed study of the propagation of singularities and very little seems to be known on this point in the general case.

3.8.3 Pseudo-spectrum and solvability

Some interesting relationships do exist between the analysis of the pseudo-spectrum for non-selfadjoint operators (see the short discussion on page 170 with the references there) and local solvability of pseudo-differential equations. In particular for a (possibly unbounded) operator P on $L^2(\mathbb{R}^n)$ and $\varepsilon > 0$, the ε-pseudo-spectrum $\sigma_\varepsilon(P)$ is defined as

$$\sigma_\varepsilon(P) = \sigma_0(P) \cup \{z \in \mathbb{C} \backslash \sigma_0(P), \|(P - z\operatorname{Id})^{-1}\| > \varepsilon^{-1}\}, \qquad (3.8.3)$$

where $\sigma_0(P)$ stands for the spectrum of P. That notion is not interesting for selfadjoint operators since the spectral theorem for normal operators gives in that case $\|(P - z\,\mathrm{Id})^{-1}\| = \mathrm{dist}\big(z, \sigma_0(P)\big)^{-1}$. This is a drastic contrast with the non-selfadjoint case, where

$$\frac{1}{\mathrm{dist}\big(z, \sigma_0(P)\big)} \leq \|(P - z\,\mathrm{Id})^{-1}\| \leq \frac{1}{\mathrm{dist}\big(z, \mathcal{N}_P\big)}, \quad \mathcal{N}_P = \overline{\{\langle Pu, u\rangle\}}_{\substack{u \in D(P), \\ \|u\|=1}},$$

and "generically" for a nonnormal operator, $\mathrm{dist}(z, \sigma_0(P))^{-1} \ll \|(P - z\,\mathrm{Id})^{-1}\|$ so that the resolvent may be very large far away from the spectrum ($\mathcal{N}_P$ is the numerical range). On the other hand the inclusion $\sigma_\varepsilon(P) \subset \{z \in \mathbb{C}, |z - \mathcal{N}_P| < \varepsilon\}$ is in general not very sharp in the sense that the larger set can indeed be much larger than $\sigma_\varepsilon(P)$. The description of the set $\sigma_\varepsilon(P)$ (or some semi-classical versions of it as in [123]) is closely linked to proving some a priori estimates for the non-selfadjoint $P - z$, for instance estimates of type

$$\|(P - z)v\| \geq \delta\|v\|, \quad \delta > 0, \quad v \in D(P), \quad z \notin \sigma_0(P) \implies z \notin \sigma_\delta(P).$$

The latter estimate is indeed some type of injectivity estimate for $P - z$ and when P is a pseudo-differential operator, that type of estimate is related to some solvability property of the adjoint operator $P^* - \bar{z}$. Conversely, when this injectivity estimate is violated, in the sense that there exists a sequence (v_n) of unit vectors in the domain of P with $\lim_n (P - z)v_n = 0$, v_n is close to a quasi-mode. A great deal of attention should be paid to the domains of P, P^* but there is more than an analogy between the determination of the pseudo-spectrum of P and solvability properties of $P^* - \bar{z}$. In the semi-classical setting in particular, the techniques developed in this chapter provide right away several results on the semi-classical pseudo-spectrum (see [123]). One of the important features for the understanding of pseudo-spectrum is certainly to investigate the properties of the commutator $[P^*, P]$ as we have done in Section 3.1.2. More brackets and global conditions analogous to condition (Ψ) may also be formulated and developed in this framework.

Chapter 4

Appendix

4.1 Some elements of Fourier analysis

4.1.1 Basics

Let $n \geq 1$ be an integer. The Schwartz space $\mathscr{S}(\mathbb{R}^n)$ is defined as the space of C^∞ functions u from $\mathbb{R}^n$ to $\mathbb{C}$ such that, for all multi-indices[1] $\alpha, \beta \in \mathbb{N}^n$,

$$\sup_{x \in \mathbb{R}^n} |x^\alpha \partial_x^\beta u(x)| < +\infty.$$

A simple example of such a function is $e^{-|x|^2}$, ($|x|$ is the Euclidean norm of x) and more generally if A is a symmetric positive-definite $n \times n$ matrix the function

$$v_A(x) = e^{-\pi \langle Ax, x \rangle}$$

belongs to the Schwartz class. For $u \in \mathscr{S}(\mathbb{R}^n)$, we define its Fourier transform $\hat{u}$ as

$$\hat{u}(\xi) = \int_{\mathbb{R}^n} e^{-2i\pi x \cdot \xi} u(x) dx. \tag{4.1.1}$$

It is an easy matter to check that the Fourier transform sends $\mathscr{S}(\mathbb{R}^n)$ into itself[2]. Moreover, for A as above, we have

$$\widehat{v_A}(\xi) = (\det A)^{-1/2} e^{-\pi \langle A^{-1}\xi, \xi \rangle}. \tag{4.1.2}$$

---

[1] Here we use the multi-index notation: for $\alpha = (\alpha_1, \ldots, \alpha_n) \in \mathbb{N}^n$ we define

$$x^\alpha = x_1^{\alpha_1} \ldots x_n^{\alpha_n}, \quad \partial_x^\alpha = \partial_{x_1}^{\alpha_1} \ldots \partial_{x_n}^{\alpha_n}, \quad |\alpha| = \sum_{1 \leq j \leq n} \alpha_j.$$

[2] Just notice that $\xi^\alpha \partial_\xi^\beta \hat{u}(\xi) = \int e^{-2i\pi x\xi} \partial_x^\alpha (x^\beta u)(x) dx (2i\pi)^{|\beta| - |\alpha|} (-1)^{|\beta|}.$

In fact, diagonalizing the symmetric matrix A, it is enough to prove the one-dimensional version of (4.1.2), i.e., to check

$$\int_{\mathbb{R}} e^{-2i\pi x\xi} e^{-\pi x^2}\, dx = \int_{\mathbb{R}} e^{-\pi(x+i\xi)^2}\, dx\, e^{-\pi\xi^2} = e^{-\pi\xi^2},$$

where the second equality can be obtained by taking the derivative with respect to ξ of $\int e^{-\pi(x+i\xi)^2}\, dx$. Using (4.1.2) we calculate for $u \in \mathscr{S}(\mathbb{R}^n)$ and $\epsilon > 0$, dealing with absolutely converging integrals,

$$
\begin{aligned}
u_\epsilon(x) &= \int e^{2i\pi x\xi}\hat{u}(\xi)e^{-\pi\epsilon^2|\xi|^2}\, d\xi \\
&= \iint e^{2i\pi x\xi}e^{-\pi\epsilon^2|\xi|^2}u(y)e^{-2i\pi y\xi}\, dy\, d\xi \\
&= \int u(y)e^{-\pi\epsilon^{-2}|x-y|^2}\epsilon^{-n}\, dy \\
&= \int \underbrace{\big(u(x+\epsilon y) - u(x)\big)}_{\text{with absolute value}\le\epsilon|y|\,\|u'\|_{L^\infty}}\ e^{-\pi|y|^2}\, dy + u(x).
\end{aligned}
$$

Taking the limit when ϵ goes to zero, we get the Fourier inversion formula

$$u(x) = \int e^{2i\pi x\xi}\hat{u}(\xi)\, d\xi. \tag{4.1.3}$$

We have thus proved that the Fourier transform is an isomorphism of the Schwartz class and provided the explicit inversion formula (4.1.3). We note also that using the notation

$$D_{x_j} = \frac{1}{2i\pi}\frac{\partial}{\partial x_j}, \quad D_x^\alpha = \prod_{j=1}^{n} D_{x_j}^{\alpha_j} \quad \text{with } \alpha = (\alpha_1, \dots, \alpha_n) \in \mathbb{N}^n, \tag{4.1.4}$$

we have, for $u \in \mathscr{S}(\mathbb{R}^n)$,

$$\widehat{D_x^\alpha u}(\xi) = \xi^\alpha \hat{u}(\xi), \qquad \widehat{x^\alpha u}(\xi) = (-1)^{|\alpha|}(D_\xi^\alpha \hat{u})(\xi). \tag{4.1.5}$$

The space of tempered distributions $\mathscr{S}'(\mathbb{R}^n)$ is the topological dual of the Fréchet space $\mathscr{S}(\mathbb{R}^n)$ and the Fourier transform can be extended to it. Let T be a tempered distribution ; the Fourier transform $\hat{T}$ of T is defined by the formula

$$\prec \hat{T}, \varphi \succ_{\mathscr{S}',\mathscr{S}} = \prec T, \hat{\varphi} \succ_{\mathscr{S}',\mathscr{S}}. \tag{4.1.6}$$

The linear form $\hat{T}$ is obviously a tempered distribution since the Fourier transform is continuous on $\mathscr{S}$ and moreover, the inversion formula (4.1.3) holds on $\mathscr{S}'$: using the notation

$$(C\varphi)(x) = (\check{\varphi})(x) = \varphi(-x), \quad \text{for } \varphi \in \mathscr{S}, \tag{4.1.7}$$

we define $\check{S}$ for $S \in \mathscr{S}'$ by $\prec \check{S}, \varphi \succ_{\mathscr{S}',\mathscr{S}} = \prec S, \check{\varphi} \succ_{\mathscr{S}',\mathscr{S}}$ and we obtain, for $T \in \mathscr{S}'$,

$$\prec \check{\hat{T}}, \varphi \succ_{\mathscr{S}',\mathscr{S}} = \prec \hat{T}, \check{\varphi} \succ_{\mathscr{S}',\mathscr{S}} = \prec \hat{T}, \hat{\hat{\varphi}} \succ_{\mathscr{S}',\mathscr{S}} = \prec T, \hat{\hat{\varphi}} \succ_{\mathscr{S}',\mathscr{S}} = \prec T, \varphi \succ_{\mathscr{S}',\mathscr{S}},$$

where the last equality is due to the fact that $\varphi \mapsto \check{\varphi}$ commutes with the Fourier transform and (4.1.3) means $\hat{\check{\varphi}} = \varphi$, a formula also proven true on $\mathscr{S}'$ by the previous line of equality. The formula (4.1.5) is true as well for $T \in \mathscr{S}'$ since, with $\varphi \in \mathscr{S}$ and $\varphi_\alpha(\xi) = \xi^\alpha \varphi(\xi)$, we have

$$\prec \widehat{D^\alpha T}, \varphi \succ_{\mathscr{S}',\mathscr{S}} = \prec T, (-1)^{|\alpha|} D^\alpha \hat{\varphi} \succ_{\mathscr{S}',\mathscr{S}} = \prec T, \widehat{\varphi_\alpha} \succ_{\mathscr{S}',\mathscr{S}} = \prec \hat{T}, \varphi_\alpha \succ_{\mathscr{S}',\mathscr{S}} .$$

The formula (4.1.1) can be used to define directly the Fourier transform of a function in $L^1(\mathbb{R}^n)$ and this gives an $L^\infty(\mathbb{R}^n)$ function which coincides with the Fourier transform: for a test function $\varphi \in \mathscr{S}(\mathbb{R}^n)$, and $u \in L^1(\mathbb{R}^n)$, we have by Definition (4.1.6) above and the Fubini theorem

$$\prec \hat{u}, \varphi \succ_{\mathscr{S}',\mathscr{S}} = \int u(x)\hat{\varphi}(x)dx = \iint u(x)\varphi(\xi)e^{-2i\pi x\cdot\xi}dxd\xi = \int \tilde{u}(\xi)\varphi(\xi)d\xi$$

with $\tilde{u}(\xi) = \int e^{-2i\pi x\cdot\xi}u(x)dx$ which is thus the Fourier transform of u. The Fourier transform can also be extended into a unitary operator of $L^2(\mathbb{R}^n)$: first for test functions $\varphi, \psi \in \mathscr{S}(\mathbb{R}^n)$, using the Fubini theorem and (4.1.3), we get

$$\langle \hat{\psi}, \hat{\varphi} \rangle_{L^2(\mathbb{R}^n)} = \int \hat{\psi}(\xi)\overline{\hat{\varphi}(\xi)}d\xi = \iint \hat{\psi}(\xi)e^{2i\pi x\cdot\xi}\overline{\varphi(x)}dxd\xi = \langle \psi, \varphi \rangle_{L^2(\mathbb{R}^n)}.$$

Next, the density of $\mathscr{S}$ in L^2 shows that there is a unique continuous extension F of the Fourier transform to L^2 and that extension is an isometric operator. Now let $u \in L^2$ and $\varphi \in \mathscr{S}$ and we check, with a sequence (u_k) of function of $\mathscr{S}$ converging in L^2 to u,

$$\prec Fu, \varphi \succ_{\mathscr{S}',\mathscr{S}} = \langle Fu, \bar{\varphi} \rangle_{L^2} = \lim_k \langle \widehat{u_k}, \bar{\varphi} \rangle_{L^2} = \lim_k \int u_k(x)\hat{\varphi}(x)dx$$

$$= \prec u, \hat{\varphi} \succ_{\mathscr{S}',\mathscr{S}} = \prec \hat{u}, \varphi \succ_{\mathscr{S}',\mathscr{S}}$$

so that $Fu = \hat{u}$. Moreover, for $u \in L^2$, we have proven that $Fu = \hat{u}$ belongs to L^2 and we have $F^2 u = \widehat{Fu} = \hat{\hat{u}} = \check{u}$ so that F is onto and satisfies $F^* F = \mathrm{Id}_{L^2} = FF^*$ and $F^* = CF = FC$ with $Cu = \check{u}$.

4.1.2   The logarithm of a non-singular symmetric matrix

The set $\mathbb{C}\backslash\mathbb{R}_-$ is star-shaped with respect to 1, so that we can define the principal determination of the logarithm for $z \in \mathbb{C}\backslash\mathbb{R}_-$ by the formula

$$\mathrm{Log}\, z = \oint_{[1,z]} \frac{d\zeta}{\zeta}. \tag{4.1.8}$$

The function Log is holomorphic on $\mathbb{C}\backslash\mathbb{R}_-$ and we have $\mathrm{Log}\,z = \ln z$ for $z \in \mathbb{R}_+^*$ and by analytic continuation $e^{\mathrm{Log}\,z} = z$ for $z \in \mathbb{C}\backslash\mathbb{R}_-$. We get also by analytic continuation, that $\mathrm{Log}\,e^z = z$ for $|\,\mathrm{Im}\,z| < \pi$.

Let Υ_+ be the set of symmetric non-singular $n \times n$ matrices with complex entries and non-negative real part. The set Υ_+ is star-shaped with respect to the Id: for $A \in \Upsilon_+$, the segment $[1, A] = \big((1 - t)\,\mathrm{Id} + tA\big)_{t\in[0,1]}$ is obviously made with symmetric matrices with non-negative real part which are invertible, since for $0 \leq t < 1$, $\mathrm{Re}\,\big((1 - t)\,\mathrm{Id} + tA\big) \geq (1 - t)\,\mathrm{Id} > 0$ and for $t = 1$, A is assumed to be invertible[3]. We can now define, for $A \in \Upsilon_+$,

$$\mathrm{Log}\,A = \int_0^1 (A - I)\big(I + t(A - I)\big)^{-1} dt. \qquad (4.1.9)$$

We note that A commutes with $(I + sA)$ (and thus with $\mathrm{Log}\,A$), so that, for $\theta > 0$,

$$\frac{d}{d\theta}\,\mathrm{Log}(A + \theta I) = \int_0^1 \big(I + t(A + \theta I - I)\big)^{-1} dt$$
$$- \int_0^1 (A + \theta I - I)t\big(I + t(A + \theta I - I)\big)^{-2} dt,$$

and since $\frac{d}{dt}\big\{\big(I + t(A + \theta I - I)\big)^{-1}\big\} = -\big(I + t(A + \theta I - I)\big)^{-2}(A + \theta I - I)$, we obtain by integration by parts $\frac{d}{d\theta}\,\mathrm{Log}(A + \theta I) = (A + \theta I)^{-1}$. As a result, we find that for $\theta > 0, A \in \Upsilon_+$, since all the matrices involved are commuting,

$$\frac{d}{d\theta}\left((A + \theta I)^{-1} e^{\mathrm{Log}(A+\theta I)}\right) = 0,$$

so that, using the limit $\theta \to +\infty$, we get that $\forall A \in \Upsilon_+, \forall \theta > 0,\ e^{\mathrm{Log}(A+\theta I)} = (A + \theta I)$, and by continuity

$$\forall A \in \Upsilon_+,\quad e^{\mathrm{Log}\,A} = A,\quad \text{which implies}\quad \det A = e^{\mathrm{trace}\,\mathrm{Log}\,A}. \qquad (4.1.10)$$

Using (4.1.10), we can define for $A \in \Upsilon_+$,

$$(\det A)^{-1/2} = e^{-\frac{1}{2}\,\mathrm{trace}\,\mathrm{Log}\,A} = |\det A|^{-1/2} e^{-\frac{i}{2}\,\mathrm{Im}(\mathrm{trace}\,\mathrm{Log}\,A)}. \qquad (4.1.11)$$

- When A is a positive-definite matrix, $\mathrm{Log}\,A$ is real-valued and $(\det A)^{-1/2} = |\det A|^{-1/2}$.

---

[3]If A is a $n \times n$ symmetric matrix with complex entries such that $\mathrm{Re}\,A$ is positive-definite, then A is invertible: if $AX = 0$, then,

$$0 = \langle AX, \bar{X}\rangle = \langle A\,\mathrm{Re}\,X, \mathrm{Re}\,X\rangle + \langle A\,\mathrm{Im}\,X, \mathrm{Im}\,X\rangle + \overbrace{\langle A\,\mathrm{Re}\,X, -i\,\mathrm{Im}\,X\rangle + \langle Ai\,\mathrm{Im}\,X, \mathrm{Re}\,X\rangle}^{=0\ \text{since}\ A\ \text{symmetric}}$$

and taking the real part gives $\langle\mathrm{Re}\,A\,\mathrm{Re}\,X, \mathrm{Re}\,X\rangle + \langle\mathrm{Re}\,A\,\mathrm{Im}\,X, \mathrm{Im}\,X\rangle = 0$, implying $X = 0$ from the positive-definiteness of $\mathrm{Re}\,A$.

- When $A = -iB$ where B is a real non-singular symmetric matrix, we note that $B = PDP^*$ with $P \in O(n)$ and D diagonal. We see directly on the formulas (4.1.9),(4.1.8) that

$$\text{Log}\, A = \text{Log}(-iB) = P(\text{Log}(-iD))P^*, \quad \text{trace}\,\text{Log}\, A = \text{trace}\,\text{Log}(-iD)$$

and thus, with (μ_j) the (real) eigenvalues of B, we have $\text{Im}\,(\text{trace}\,\text{Log}\, A) = \text{Im} \sum_{1 \leq j \leq n} \text{Log}(-i\mu_j)$, where the last Log is given by (4.1.8). Finally we get,

$$\text{Im}\,(\text{trace}\,\text{Log}\, A) = -\frac{\pi}{2} \sum_{1 \leq j \leq n} \text{sign}\,\mu_j = -\frac{\pi}{2} \text{sign}\, B$$

where $\text{sign}\, B$ is the signature of B. As a result, we have when $A = -iB$, B a real symmetric non-singular matrix

$$(\det A)^{-1/2} = |\det B|^{-1/2} e^{i\frac{\pi}{4} \text{sign}\, B}. \qquad (4.1.12)$$

4.1.3   Fourier transform of Gaussian functions

Proposition 4.1.1. *Let A be a symmetric non-singular $n \times n$ matrix with complex entries such that $\text{Re}\, A \geq 0$. We define the Gaussian function v_A on $\mathbb{R}^n$ by $v_A(x) = e^{-\pi \langle Ax, x \rangle}$. The Fourier transform of v_A is*

$$\widehat{v_A}(\xi) = (\det A)^{-1/2} e^{-\pi \langle A^{-1}\xi, \xi \rangle}, \qquad (4.1.13)$$

where $(\det A)^{-1/2}$ is defined according to the formula (4.1.11). In particular, when $A = -iB$ with a symmetric real non-singular matrix B, we get

$$\text{Fourier}(e^{i\pi \langle Bx, x \rangle})(\xi) = \widehat{v_{-iB}}(\xi) = |\det B|^{-1/2} e^{i\frac{\pi}{4} \text{sign}\, B} e^{-i\pi \langle B^{-1}\xi, \xi \rangle}.$$

Proof. Let us define Υ_+^* as the set of symmetric $n \times n$ complex matrices with a positive-definite real part (naturally these matrices are non-singular since $Ax = 0$ for $x \in \mathbb{C}^n$ implies $0 = \text{Re}\langle Ax, \bar{x} \rangle = \langle (\text{Re}\, A)x, \bar{x} \rangle$, so that $\Upsilon_+^* \subset \Upsilon_+$).

Let us assume first that $A \in \Upsilon_+^*$; then the function v_A is in the Schwartz class (and so is its Fourier transform). The set Υ_+^* is an open convex subset of $\mathbb{C}^{n(n+1)/2}$ and the function $\Upsilon_+^* \ni A \mapsto \widehat{v_A}(\xi)$ is holomorphic and given on $\Upsilon_+^* \cap \mathbb{R}^{n(n+1)/2}$ by (4.1.13). On the other hand the function $\Upsilon_+^* \ni A \mapsto e^{-\frac{1}{2} \text{trace}\,\text{Log}\, A} e^{-\pi \langle A^{-1}\xi, \xi \rangle}$ is also holomorphic and coincides with the previous one on $\mathbb{R}^{n(n+1)/2}$. By analytic continuation this proves (4.1.13) for $A \in \Upsilon_+^*$.

If $A \in \Upsilon_+$ and $\varphi \in \mathscr{S}(\mathbb{R}^n)$, we have $\langle \widehat{v_A}, \varphi \rangle_{\mathscr{S}', \mathscr{S}} = \int v_A(x) \hat{\varphi}(x) dx$ so that

$\Upsilon_+ \ni A \mapsto \langle \widehat{v_A}, \varphi \rangle$ is continuous and thus[4] using the previous result on Υ_+^*,

$$\langle \widehat{v_A}, \varphi \rangle = \lim_{\epsilon \to 0_+} \langle \widehat{v_{A+\epsilon I}}, \varphi \rangle = \lim_{\epsilon \to 0_+} \int e^{-\frac{1}{2} \operatorname{trace} \operatorname{Log}(A+\epsilon I)} e^{-\pi \langle (A+\epsilon I)^{-1} \xi, \xi \rangle} \varphi(\xi) d\xi$$

(by continuity of Log on Υ_+ and domin. cv.) $= \int e^{-\frac{1}{2} \operatorname{trace} \operatorname{Log} A} e^{-\pi \langle A^{-1} \xi, \xi \rangle} \varphi(\xi) d\xi,$

which is the sought result.   □

Lemma 4.1.2. *Let $n \geq 1$ be an integer and $t \in \mathbb{R}^*$. We define the operator*

$$J^t = \exp 2i\pi t D_x \cdot D_\xi \tag{4.1.14}$$

on $\mathscr{S}'(\mathbb{R}_x^n \times \mathbb{R}_\xi^n)$ by $(F J^t a)(\xi, x) = e^{2i\pi t \xi \cdot x} \hat{a}(\xi, x)$, where F stands here for the Fourier transform in $2n$ dimensions. The operator J^t sends also $\mathscr{S}(\mathbb{R}_x^n \times \mathbb{R}_\xi^n)$ into itself continuously, satisfies (for $s, t \in \mathbb{R}$) $J^{s+t} = J^s J^t$ and is given by

$$(J^t a)(x, \xi) = |t|^{-n} \iint e^{-2i\pi t^{-1} y \cdot \eta} a(x + y, \xi + \eta) dy d\eta. \tag{4.1.15}$$

We have

$$J^t a = e^{i\pi t \langle BD, D \rangle} a = |t|^{-n} e^{-i\pi t^{-1} \langle B \cdot, \cdot \rangle} * a, \tag{4.1.16}$$

with the $2n \times 2n$ matrix $B = \begin{pmatrix} 0 & I_n \\ I_n & 0 \end{pmatrix}$. The operator J^t sends continuously $C_b^\infty(\mathbb{R}^{2n})$ into itself.

Proof. We have indeed $(F J^t a)(\xi, x) = e^{2i\pi t \xi \cdot x} \hat{a}(\xi, x) = e^{i\pi t \langle B\Xi, \Xi \rangle} \hat{a}(\Xi)$. Note that B is a $2n \times 2n$ symmetric matrix with null signature, determinant $(-1)^n$ and that $B^{-1} = B$. According to Proposition 4.1.1, the inverse Fourier transform of $e^{i\pi t \langle B\Xi, \Xi \rangle}$ is $|t|^{-n} e^{-i\pi t^{-1} \langle BX, X \rangle}$ so that $J^t a = |t|^{-n} e^{-i\pi t^{-1} \langle B \cdot, \cdot \rangle} * a$. Since the Fourier multiplier $e^{i\pi t \langle B\Xi, \Xi \rangle}$ is smooth bounded with derivatives polynomially bounded, it defines a continuous operator from $\mathscr{S}(\mathbb{R}^{2n})$ into itself.

In the sequel of the proof, we take $t = 1$, which will simplify the notation without corrupting the arguments (see nevertheless Remark 4.1.4). Let us consider $a \in \mathscr{S}(\mathbb{R}^{2n})$: we have with $k \in 2\mathbb{N}$ and the polynomial on $\mathbb{R}^n$ defined by $P_k(y) = (1 + |y|^2)^{k/2}$,

$$(Ja)(x, \xi) = \iint e^{-2i\pi y \cdot \eta} P_k(y)^{-1} P_k(D_\eta) \Big(P_k(\eta)^{-1} (P_k(D_y) a)(x + y, \xi + \eta) \Big) dy d\eta,$$

so that, with $|T_{\alpha\beta}(\eta)| \leq P_k(\eta)^{-1}$ and constants $c_{\alpha\beta}$, we obtain

---

[4]Note that the mapping $A \mapsto A^{-1}$ is a homeomorphism of Υ_+: with $X = AY$,

$$\operatorname{Re}\langle A^{-1} X, \bar{X} \rangle = \operatorname{Re}\langle Y, \overline{AY} \rangle = \operatorname{Re}\langle AY, \bar{Y} \rangle \geq 0, \quad \text{for } A \in \Upsilon_+.$$

$$(Ja)(x,\xi) = \sum_{\substack{|\beta|\le k \\ |\alpha|\le k}} c_{\alpha\beta} \iint e^{-2i\pi y\cdot\eta} P_k(y)^{-1} T_{\alpha\beta}(\eta)(D_\xi^\alpha D_x^\beta a)(x+y,\xi+\eta)dyd\eta.$$

$$(4.1.17)$$

Let us denote by $\tilde{J}a$ the right-hand side of (4.1.17). We already know that $\tilde{J}a = Ja$ for $a \in \mathscr{S}(\mathbb{R}^{2n})$. We also note that, using an even integer $k > n$, the previous integral converges absolutely whenever $a \in C_b^\infty(\mathbb{R}^{2n})$; moreover we have

$$\|\tilde{J}a\|_{L^\infty} \le C_n \sup_{\substack{|\alpha|\le n+2 \\ |\beta|\le n+2}} \|D_\xi^\alpha D_x^\beta a\|_{L^\infty},$$

and since the derivations commute with J and $\tilde{J}$, we also get that

$$\|\partial^\gamma \tilde{J}a\|_{L^\infty} \le C_n \sup_{\substack{|\alpha|\le n+2 \\ |\beta|\le n+2}} \|D_\xi^\alpha D_x^\beta \partial^\gamma a\|_{L^\infty}. \qquad (4.1.18)$$

It implies that $\tilde{J}$ is continuous from $C_b^\infty(\mathbb{R}^{2n})$ to itself. Let us now consider $a \in C_b^\infty(\mathbb{R}^{2n} \times \mathbb{R}^m)$; we define the sequence (a_k) in $\mathscr{S}(\mathbb{R}^{2n})$ by

$$a_k(x,\xi) = e^{-(|x|^2+|\xi|^2)/k^2} a(x,\xi).$$

We have

$$\langle Ja, \Phi\rangle_{\mathscr{S}^*(\mathbb{R}^{2n}),\mathscr{S}(\mathbb{R}^{2n})}$$
$$= \iint a(x,\xi)\overline{(J^{-1}\Phi)}(x,\xi)dxd\xi = \lim_{k\to+\infty}\iint a_k(x,\xi)\overline{(J^{-1}\Phi)}(x,\xi)dxd\xi$$
$$= \lim_{k\to+\infty}\iint (Ja_k)(x,\xi)\bar{\Phi}(x,\xi)dxd\xi = \iint (\tilde{J}a)(x,\xi)\bar{\Phi}(x,\xi)dxd\xi,$$

so that we indeed have $\tilde{J}a = Ja$ and from (4.1.18) the continuity property of the lemma whose proof is now complete. □

Remark 4.1.3. For $a \in \mathscr{S}'(\mathbb{R}^{2n})$, the complex conjugate $\bar{a}$ is defined by

$$\langle \bar{a}, \Phi\rangle_{\mathscr{S}',\mathscr{S}} = \overline{\langle a, \bar{\Phi}\rangle_{\mathscr{S}',\mathscr{S}}}$$

and we can check that for $t \in \mathbb{R}$, $\overline{(J^t a)} = J^{-t}(\bar{a})$ an obvious formula for $a \in \mathscr{S}(\mathbb{R}^{2n})$ (from the expression of J^t in the previous lemma) which can be extended by duality to $\mathscr{S}'(\mathbb{R}^{2n})$.

Remark 4.1.4. It will be interesting later on to use the fact that J^t sends $C_b^\infty(\mathbb{R}^{2n})$ into itself "polynomially" with respect to t. The point here is to note that, for $a \in \mathscr{S}(\mathbb{R}^{2n})$,

$$(J^t a)(x,\xi) = \iint e^{-2i\pi y\cdot\eta} a(x+ty,\xi+\eta)dyd\eta.$$

It is then easy to reproduce the proof of Lemma 4.1.2 to obtain

$$\|\partial^\gamma J^t a\|_{L^\infty} \leq C_n (1 + |t|)^{n+2} \sup_{\substack{|\alpha| \leq n+2 \\ |\beta| \leq n+2}} \|D_\xi^\alpha D_x^\beta \partial^\gamma a\|_{L^\infty}. \tag{4.1.19}$$

Lemma 4.1.5. *Let $n \geq 1$ be an integer and $m, t \in \mathbb{R}$. The operator J^t sends continuously $S_{1,0}^m(\mathbb{R}^{2n})$ into itself and for all integers $N \geq 0$,*

$$(J^t a)(x, \xi) = \sum_{|\alpha| < N} \frac{t^{|\alpha|}}{\alpha!} (D_\xi^\alpha \partial_x^\alpha a)(x, \xi) + r_N(t)(x, \xi), \quad r_N(t) \in S_{1,0}^{m-N},$$

$$r_N(t)(x, \xi) = t^N \int_0^1 \frac{(1 - \theta)^{N-1}}{(N - 1)!} \left(J^{\theta t}(D_\xi \cdot \partial_x)^N a \right)(x, \xi) d\theta.$$

Proof. We apply Taylor's formula on $J^t = \exp 2i\pi t D_x \cdot D_\xi$ to get for operators on $\mathscr{S}'(\mathbb{R}^{2n})$,

$$J^t = \sum_{0 \leq k < N} \frac{t^k}{k!} (D_\xi \cdot \partial_x)^k + \int_0^1 \frac{(1 - \theta)^{N-1}}{(N - 1)!} J^{\theta t}(t D_\xi \cdot \partial_x)^N d\theta, \tag{4.1.20}$$

and since

$$\frac{1}{k!}(D_\xi \cdot \partial_x)^k = \sum_{\substack{\alpha_1 + \cdots + \alpha_n = k \\ \alpha_j \in \mathbb{N}}} \frac{(D_{\xi_1} \partial_{x_1})^{\alpha_1}}{\alpha_1!} \cdots \frac{(D_{\xi_n} \partial_{x_n})^{\alpha_n}}{\alpha_n!},$$

we obtain the above formulas for $a \in \mathscr{S}'(\mathbb{R}^{2n})$. On the other hand, we get from (1.1.13) that the term $D_\xi^\alpha \partial_x^\alpha a$ belongs to $S_{1,0}^{m-|\alpha|}$. It is thus enough that we show that J^t sends continuously $S_{1,0}^m$ into itself. For that purpose, we can use the formula (4.1.17) (and assume that $t = 1$) in the proof of Lemma 4.1.2; also the same reasoning as in the proof of this lemma shows that the right-hand side of (4.1.17) is meaningful for $a \in S_{1,0}^m$ if $k > n + |m|$ and is indeed the expression of Ja. We get, for all $k \in \mathbb{N}$,

$$|Ja(x, \xi)| \leq C_{k,n} \iint \langle y \rangle^{-k} \langle \eta \rangle^{-k} \langle \xi + \eta \rangle^m d\xi d\eta$$

so that Peetre's inequality (1.1.17) yields, for $k > n + |m|$, $|Ja(x, \xi)| \leq C'_{k,n} \langle \xi \rangle^m$. The estimates for the derivatives are obtained similarly since they commute with J. The terms involving integrals of J^t can be handled via Remark 4.1.4, which provides a polynomial control with respect to t. $\qquad\square$

Remark 4.1.6. The mapping $\mathscr{S}'(\mathbb{R}^{2n}) \ni a \mapsto a(x, D)$ is (obviously) linear and one-to-one: if $a(x, D) = 0$, choosing $v(x) = e^{-\pi|x - x_0|^2}$, $\hat{u}(\xi) = e^{-\pi|\xi - \xi_0|^2}$, we get that the convolution of the distribution $\tilde{a}(x, \xi) = a(x, \xi)e^{2i\pi x \cdot \xi}$ with the Gaussian function $e^{-\pi(|x|^2+|\xi|^2)}$ is zero, so that, taking the Fourier transform shows that the product of the same Gaussian function with $\hat{a}$ is zero, implying that $\tilde{a}$ and thus a is zero. It is a consequence of a version of the Schwartz kernel theorem that the same mapping $\mathscr{S}'(\mathbb{R}^{2n}) \ni a \mapsto a(x, D) \in$ continuous linear operators from $\mathscr{S}(\mathbb{R}^n)$ to $\mathscr{S}'(\mathbb{R}^n)$ is indeed onto. However the "onto" part of our statement is highly non-trivial and a version of this theorem can be found in Theorem 5.2.1 of [71].

4.1.4 Some standard examples of Fourier transform

Some examples

Let us consider the Heaviside function defined on $\mathbb{R}$ by $H(x) = 1$ for $x > 0$, $H(x) = 0$ for $x \leq 0$; it is obviously a tempered distribution, so that we can compute its Fourier transform. With the notation of this section, we have, with δ_0 the Dirac mass at 0, $\check{H}(x) = H(-x)$,

$$\hat{H} + \hat{\check{H}} = \hat{1} = \delta_0, \quad \hat{H} - \hat{\check{H}} = \widehat{\text{sign}}, \quad \frac{1}{i\pi} = \frac{1}{2i\pi}2\widehat{\delta_0}(\xi) = \widehat{D\,\text{sign}}(\xi) = \xi\widehat{\text{sign}}\xi$$

so that $\xi\big(\widehat{\text{sign}}\xi - \frac{1}{i\pi}\text{pv}(1/\xi)\big) = 0$ and $\widehat{\text{sign}}\xi - \frac{1}{i\pi}\text{pv}(1/\xi) = c\delta_0$ with $c = 0$ since the lhs is odd. We get

$$\widehat{\text{sign}}(\xi) = \frac{1}{i\pi}\text{pv}\,\frac{1}{\xi}, \quad \widehat{\text{pv}(\frac{1}{\pi x})} = -i\,\text{sign}\,\xi, \quad \hat{H} = \frac{\delta_0}{2} + \frac{1}{2i\pi}\text{pv}(\frac{1}{\xi}). \quad (4.1.21)$$

We have also for $a \leq b$ real and $\alpha > 0$,

$$\widehat{1_{[a,b]}}(\xi) = \frac{\sin\big(\pi(b-a)\xi\big)}{\pi\xi}e^{-i\pi(a+b)\xi}, \quad \widehat{\frac{1}{x^2+\alpha^2}} = \frac{\pi}{\alpha}e^{-2\pi\alpha|\xi|}. \quad (4.1.22)$$

Trivia on the Gamma function and a few formulas

The Gamma function is defined for $z \in \mathbb{C}$ with $\text{Re}\,z > 0$ by the formula

$$\Gamma(z) = \int_0^{+\infty} t^{z-1}e^{-t}dt. \quad (4.1.23)$$

A simple integration by parts yields

$$\Gamma(z+1) = z\Gamma(z) \quad (4.1.24)$$

and we take advantage of that to define, for $z \notin \mathbb{Z}_-$,

$$\Gamma(z) = \frac{\Gamma(z+k)}{\prod_{0 \leq j < k}(z+j)}, \quad \text{provided } k \in \mathbb{N} \text{ such that } k + \text{Re}\,z > 0, \quad (4.1.25)$$

which makes sense since, if $k_1, k_2 \in \mathbb{N}, z \in \mathbb{C}$ such that $k_j + \operatorname{Re} z > 0, j = 1, 2$, we have from (4.1.24),

$$\frac{\Gamma(z + k_1)}{\prod_{0 \leq j < k_1} (z + j)} = \frac{\Gamma(z + k_2)}{\prod_{0 \leq j < k_2} (z + j)}.$$

The function Γ is thus a meromorphic function on $\mathbb{C}$ with simple poles at $-\mathbb{N}$ with, for $k \in \mathbb{N}$,

$$\operatorname{res}(\Gamma, -k) = \frac{(-1)^k}{k!}. \tag{4.1.26}$$

We get also easily that for $n \in \mathbb{N}$, $\Gamma(n+1) = n!$ and $\Gamma(1/2) = \sqrt{\pi}$. The classical Weierstrass formula[5] is

$$\frac{1}{\Gamma(z)} = z e^{\gamma z} \prod_{k=1}^{+\infty} (1 + \frac{z}{k}) e^{-z/k} \tag{4.1.28}$$

where $\gamma = \lim_N \left[(\sum_{1 \leq k \leq N} \frac{1}{k}) - \ln N \right]$ is the Euler-Mascheroni constant; the function $1/\Gamma$ is thus entire and the meromorphic function Γ does not vanish. A straightforward consequence of Weierstrass' formula is Euler's reflection formula:

$$\frac{1}{\Gamma(1-z)\Gamma(z)} = \frac{\sin(\pi z)}{\pi}. \tag{4.1.29}$$

We have, for $a > -1, b > 0$,

$$\int_0^{+\infty} x^a e^{-bx} dx = b^{-a-1} \Gamma(1 + a). \tag{4.1.30}$$

Let us compute the Fourier transform of $x_+^a e^{-bx}$. We have, for $a, b \in \mathbb{C}, \operatorname{Re} a > -1, \operatorname{Re} b > 0$, by analytic continuation of (4.1.30), for $\xi \in \mathbb{R}$,

$$\int_0^{+\infty} x^a e^{-bx} e^{-2i\pi x \xi} dx = \int_0^{+\infty} x^a e^{-x(b + 2i\pi \xi)} dx = \Gamma(1+a)(b + 2i\pi \xi)^{-a-1}. \tag{4.1.31}$$

Some homogeneous distributions and their Fourier transform

We define for $a \in \mathbb{C}, \operatorname{Re} a > -1$, the distribution χ_+^a on $\mathbb{R}$ by $H(x) x^a / \Gamma(1 + a)$, i.e.,

$$\langle \chi_+^a, \varphi \rangle_{\mathscr{D}'(\mathbb{R}), \mathscr{D}(\mathbb{R})} = \Gamma(1 + a)^{-1} \int_0^{+\infty} \varphi(x) x^a dx \tag{4.1.32}$$

---

[5]That formula is easily deduced from the Gauss formula:

$$\text{for } z \in \mathbb{C} \backslash \mathbb{Z}_-, \quad \Gamma(z) = \lim_{n \to +\infty} \frac{n^z \, n!}{\prod_{0 \leq j \leq n} (z + j)}, \tag{4.1.27}$$

which is a consequence of $\Gamma(x) = \lim_{n \to +\infty} \int_0^n t^{x-1} (1 - \frac{t}{n})^n dt$ for $x > 0$.

and we note that

$$\text{for } \operatorname{Re} a > -1, k \in \mathbb{N}, \quad \chi_+^a = (\frac{d}{dx})^k (\chi_+^{a+k}) \tag{4.1.33}$$

since

$$(-1)^k \Gamma(1+a+k)^{-1} \int_0^{+\infty} \varphi^{(k)}(x) x^{a+k} dx$$

$$= \Gamma(1+a+k)^{-1} \int_0^{+\infty} \varphi(x) x^a dx \prod_{0 \le j < k} (a+k-j) = \Gamma(1+a)^{-1} \int_0^{+\infty} \varphi(x) x^a dx.$$

For $a \in \mathbb{C}, k_1, k_2 \in \mathbb{N}$ such that $k_j + \operatorname{Re} a > -1, j = 1, 2$, we have from (4.1.33) if $k_2 \ge k_1$, $(\chi_+^{a+k_1}) = (\frac{d}{dx})^{k_2-k_1} (\chi_+^{a+k_2})$ and thus $(\frac{d}{dx})^{k_1} (\chi_+^{a+k_1}) = (\frac{d}{dx})^{k_2} (\chi_+^{a+k_2})$ so that we may define

$$\chi_+^a = (\frac{d}{dx})^k (\chi_+^{a+k}), \quad k \in \mathbb{N}, \text{ such that } k + \operatorname{Re} a > -1. \tag{4.1.34}$$

As a consequence, for all $\varphi \in \mathscr{D}(\mathbb{R})$, the function $a \mapsto \langle \chi_+^a, \varphi \rangle$ is entire on $\mathbb{C}$. We note in particular that for $k \in \mathbb{N}^*$,

$$\chi_+^{-k} = (\frac{d}{dx})^k H = \delta_0^{(k-1)}. \tag{4.1.35}$$

The distribution χ_-^a is defined as the reflexion of χ_+^a, i.e.,

$$\langle \chi_-^a, \varphi \rangle = \langle \chi_+^a, \check{\varphi} \rangle, \quad \text{with } \check{\varphi}(x) = \varphi(-x). \tag{4.1.36}$$

On the other hand, for $a \in \mathbb{C}, \operatorname{Re} a > -1$, we have

$$(x + i0)^a = x_+^a + e^{i\pi a} H(-x)|x|^a = x_+^a + e^{i\pi a} x_-^a$$

and thus $(x + i0)^a = (\chi_+^a + e^{i\pi a} \chi_-^a) \Gamma(1 + a)$. The function $a \mapsto \Gamma(1 + a)$ is meromorphic on $\mathbb{C}$ with simple poles at $\mathbb{Z}_-^*$, but the entire (distribution-valued) function $\chi_+^a + e^{i\pi a} \chi_-^a$ vanishes there since for $k \in \mathbb{N}^*$, we have from (4.1.35)

$$\chi_+^{-k} = \delta_0^{(k-1)}, \quad \chi_-^{-k} = \delta_0^{(k-1)}(-1)^{k-1}. \tag{4.1.37}$$

As a result, for all $a \in \mathbb{C}$, the formulas

$$(x+i0)^a = (\chi_+^a + e^{i\pi a} \chi_-^a)\Gamma(1+a), \quad (x-i0)^a = (\chi_+^a + e^{-i\pi a} \chi_-^a)\Gamma(1+a), \tag{4.1.38}$$

define two distributions on $\mathbb{R}$, depending holomorphically on a. We have also

$$(x + i0)^{-1} = \frac{d}{dx}(\operatorname{Log}(x + i0)) = \frac{d}{dx}(\ln|x| + i\pi H(-x)) = \operatorname{pv}\frac{1}{x} - i\pi\delta_0, \tag{4.1.39}$$

and for $k \in \mathbb{N}^*$, we have

$$(x + i0)^{-k} = \frac{(-1)^{k-1}}{(k-1)!}(\frac{d}{dx})^k(\ln|x|) + i\pi\frac{(-1)^k}{(k-1)!}\delta_0^{(k-1)}, \qquad (4.1.40)$$

$$(x - i0)^{-k} = \frac{(-1)^{k-1}}{(k-1)!}(\frac{d}{dx})^k(\ln|x|) - i\pi\frac{(-1)^k}{(k-1)!}\delta_0^{(k-1)}. \qquad (4.1.41)$$

Lemma 4.1.7. *For any $a \in \mathbb{C}$ we have, using the previous notation,*

$$\widehat{\chi_+^a}(\xi) = e^{-\frac{i\pi(a+1)}{2}}(2\pi)^{-a-1}(\xi - i0)^{-a-1}, \qquad (4.1.42)$$

$$\widehat{\chi_-^a}(\xi) = e^{\frac{i\pi(a+1)}{2}}(2\pi)^{-a-1}(\xi + i0)^{-a-1}, \qquad (4.1.43)$$

$$\widehat{(x+i0)^a}(\xi) = e^{\frac{i\pi a}{2}}(2\pi)^{-a}\chi_+^{-a-1}, \qquad (4.1.44)$$

$$\widehat{(x-i0)^a}(\xi) = e^{\frac{-i\pi a}{2}}(2\pi)^{-a}\chi_-^{-a-1}. \qquad (4.1.45)$$

The distributions $\chi_\pm^a, (x \pm i0)^a$ are homogeneous distributions on $\mathbb{R}$ with degree a.

Proof. Assuming (4.1.42), since the reflexion commutes with the Fourier transformation we get (4.1.43), and also (4.1.45) as well as (4.1.44); note that the reflexion of $(x + i0)^a$ is $e^{i\pi a}(x - i0)^a$ since for $a > 0$,

$$\langle (x + i0)^a, \varphi(-x)\rangle = \int e^{a(\ln|x|+i\pi H(-x))}\varphi(-x)dx = \int e^{a(\ln|x|+i\pi H(x))}\varphi(x)dx$$

$$= \int e^{a(\ln|x|-i\pi H(-x))}\varphi(x)dx\, e^{i\pi a} = e^{i\pi a}\langle (x - i0)^a, \varphi(x)\rangle$$

and the identity extends by analytic continuation (the reflexion of $(x - i0)^a$ is $e^{-i\pi a}(x + i0)^a$). Let us prove (4.1.42): by analyticity, it is enough to get it for $-1 < a < 0$. We have

$$\langle x_+^a, \hat\varphi\rangle = \int_0^{+\infty} x^a\hat\varphi(x)dx = \lim_{\varepsilon \to 0_+}\iint H(x)x^a\varphi(\xi)e^{-2i\pi x(\xi-i\varepsilon)}dxd\xi$$

(here we use (4.1.31)) $= \displaystyle\lim_{\varepsilon \to 0_+}\int \varphi(\xi)\Gamma(a+1)(2\pi\varepsilon + 2i\pi\xi)^{-a-1}d\xi$

$$= \Gamma(a+1)(2\pi)^{-a-1}\lim_{\varepsilon \to 0_+}\int \varphi(\xi)\big(i(\xi - i\varepsilon)\big)^{-a-1}d\xi$$

and since for $\xi \in \mathbb{R}$,

$$\mathrm{Log}(+0 + i\xi) = \ln|\xi| + i\frac{\pi}{2}\,\mathrm{sign}\,\xi = \ln|\xi| - i\pi H(-\xi) + i\frac{\pi}{2} = \mathrm{Log}(\xi - i0) + i\frac{\pi}{2},$$

we obtain (4.1.42). The last statement follows from (4.1.34) since x_+^{a+k} is homogeneous of degree $a + k$ whenever $k + \mathrm{Re}\,a > -1$ and also from the fact that the Fourier transform of a homogeneous distribution on $\mathbb{R}^n$ with degree a is homogeneous with degree $-a - n$. $\qquad\square$

In particular we recover the formulas

$$\widehat{(x+i0)^{-1}} = -2i\pi H, \quad \widehat{(x-i0)^{-1}} = 2i\pi \check{H}, \tag{4.1.46}$$

for $k \in \mathbb{N}^*$, $\widehat{(x+i0)^{-k}} = i^{-k}(2\pi)^k \dfrac{\xi_+^{k-1}}{(k-1)!}, \quad \widehat{(x-i0)^{-k}} = i^k(2\pi)^k \dfrac{\xi_-^{k-1}}{(k-1)!},$ (4.1.47)

and for $a, b \in \mathbb{C}$, $\operatorname{Re} b > 0$, we have also from (4.1.31)

$$(\widehat{x_+^a e^{-2\pi bx}})(\xi) = (2\pi)^{-a-1} e^{\frac{-i\pi(a+1)}{2}} (\xi - ib)^{-a-1}, \tag{4.1.48}$$

$$(\widehat{x_-^a e^{2\pi bx}})(\xi) = (2\pi)^{-a-1} e^{\frac{i\pi(a+1)}{2}} (\xi + ib)^{-a-1}, \tag{4.1.49}$$

$$\widehat{(x+ib)^a}(\xi) = (2\pi)^{-a} e^{i\frac{\pi a}{2}} \chi_+^{-1-a} e^{-2\pi b\xi}, \tag{4.1.50}$$

$$\widehat{(x-ib)^a}(\xi) = (2\pi)^{-a} e^{-i\frac{\pi a}{2}} \chi_-^{-1-a} e^{2\pi b\xi}, \tag{4.1.51}$$

and the reader is invited to check that it is indeed consistent with Lemma 4.1.7.

4.1.5 The Hardy Operator

The Hilbert transform $\mathcal{H}$ is the selfadjoint operator defined on $u \in L^2(\mathbb{R})$ by (see the formulas (4.1.21))

$$\mathcal{H}u = \operatorname{sign}(D_x)u, \quad \text{i.e.,} \quad \widehat{\mathcal{H}u}(\xi) = \operatorname{sign}\xi \, \hat{u}(\xi), \quad \mathcal{H}u = i\operatorname{pv}\frac{1}{\pi x} * u. \tag{4.1.52}$$

It is obvious to see that $\mathcal{H}$ is bounded on $L^2(\mathbb{R})$ with $\mathcal{L}(L^2(\mathbb{R}))$ norm equal to 1. It is also a classical result of harmonic analysis that $\mathcal{H}$ is also bounded on $L^p(\mathbb{R})$ for $1 < p < \infty$. We are here interested in the Hardy operator Ω defined by

$$\Omega = -iH\mathcal{H}\check{H}C = H\frac{1}{i}\mathcal{H}CH, \tag{4.1.53}$$

where H is the operator of multiplication by the Heaviside function and C is defined on $L^2(\mathbb{R})$ by $(C\kappa)(x) = \kappa(-x)$. We note that

$$C^2 = \operatorname{Id}, \ C^* = C, \ \mathcal{H}C = -C\mathcal{H} \Longrightarrow \Omega^* = C\check{H}\mathcal{H}Hi = HiC\mathcal{H}H = H\frac{1}{i}\mathcal{H}CH = \Omega. \tag{4.1.54}$$

Let us compute the kernel of Ω: we have, writing (abusively) integrals instead of brackets of duality,

$$(\Omega u)(x) = -iH(x) \iint e^{2i\pi(x-y)\xi} \operatorname{sign}\xi \, H(-y)u(-y)dyd\xi$$

$$= H(x)\frac{1}{\pi} \int \operatorname{pv} \frac{1}{x+y} H(y)u(y)dy$$

so that the kernel of Ω is $\dfrac{H(x)H(y)}{\pi} \operatorname{pv}\frac{1}{x+y}$, that we may write $\dfrac{H(x)H(y)}{\pi(x+y)}$. It is obvious from (4.1.53) that the $\mathcal{L}(L^2(\mathbb{R}))$ norm of Ω is smaller than 1; the following lemma states that in fact that norm is 1.

Lemma 4.1.8. *Let Ω be the Hardy operator defined by (4.1.53). We define for $\varepsilon > 0$,*

$$\kappa_\varepsilon(x) = \Gamma(\varepsilon)^{-1/2} e^{-x/2} x^{\frac{\varepsilon-1}{2}} H(x), \tag{4.1.55}$$

where Γ stands for the Gamma function. The function κ_ε belongs to $L^2(\mathbb{R})$ with norm 1 and the selfadjoint operator Ω satisfies

$$1 > \langle \Omega \kappa_\varepsilon, \kappa_\varepsilon \rangle > 1 - \varepsilon, \quad \|\Omega\|_{\mathcal{L}(L^2(\mathbb{R}))} = 1. \tag{4.1.56}$$

Proof. Note first that Ω is selfadjoint from the expression of its kernel and the calculation (4.1.54). We have $\|\kappa_\varepsilon\|_{L^2} = 1$ and

$$
\begin{aligned}
\langle \Omega \kappa_\varepsilon, \kappa_\varepsilon \rangle &= \Gamma(\varepsilon)^{-1} \iint \frac{H(x)H(y)}{\pi(x+y)} e^{-\frac{x+y}{2}} (xy)^{\frac{\varepsilon-1}{2}} \, dx \, dy \quad (x = t+s, y = t-s) \\
&= \Gamma(\varepsilon)^{-1} \iint \frac{H(t-|s|)}{\pi t} e^{-t} (t^2 - s^2)^{\frac{\varepsilon-1}{2}} \, dt \, ds \quad (s = t \sin \theta) \\
&= \frac{2}{\pi} \int_0^{\pi/2} \cos^\varepsilon \theta \, d\theta, \quad \text{and since } \cos \theta \geq \frac{2}{\pi}(\frac{\pi}{2} - \theta) \text{ for } \theta \in [0, \frac{\pi}{2}], \text{ we get,} \\
&\geq (\frac{2}{\pi})^{1+\varepsilon} \int_0^{\pi/2} (\frac{\pi}{2} - \theta)^\varepsilon \, d\theta = \frac{1}{1+\varepsilon} > 1 - \varepsilon. \qquad \square
\end{aligned}
$$

4.2 Some remarks on algebra

4.2.1 On simultaneous diagonalization of quadratic forms

Lemma 4.2.1. *Let V be a finite-dimensional real (resp. complex) vector space, and q_0, q_1 be bilinear symmetric (resp. sesquilinear Hermitian) forms on V so that q_0 is positive-definite (i.e., $q_0(v,v) \geq 0$, $q_0(v,v) = 0 \iff v = 0$). Then there exists a basis $\mathcal{B} = (e_j)_{1 \leq j \leq d}$ of V and real scalars $(\lambda_j)_{1 \leq j \leq d}$ such that*

$$q_0(e_j, e_k) = \delta_{j,k}, \qquad q_1(e_j, e_k) = \lambda_j \delta_{j,k}. \tag{4.2.1}$$

N.B. This lemma means that in the Euclidean (resp. Hermitian) vector space V equipped with the scalar product $\langle v, w \rangle = q_0(v, w)$, the basis $\mathcal{B}$ is orthonormal and the quadratic form q_1 is diagonal in that basis.

Proof. We define on V the scalar product $\langle v, w \rangle = q_0(v, w)$ (linear in v, antilinear in w if V is complex), so that $(V, \langle \cdot, \cdot \rangle)$ is a finite-dimensional (possibly complex) Hilbert space. We define the endomorphism A of V by the identity $\langle Av, w \rangle = q_1(v, w)$ and we have

$$\langle v, Aw \rangle = \overline{\langle Aw, v \rangle} = \overline{q_1(w,v)} = q_1(v,w) = \langle Av, w \rangle$$

so that A is selfadjoint and thus there exists an orthonormal basis $\mathcal{B} = (e_j)_{1 \leq j \leq d}$ of eigenvectors, giving the result. $\qquad \square$

N.B. It is not possible to weaken the assumption on q_0 down to q_0 non-negative: in $\mathbb{R}^2_{x,y}$ it is not possible to diagonalize simultaneously the quadratic forms x^2 and xy, otherwise, we would find a basis of $\mathbb{R}^2$,

$$e_1 = \begin{pmatrix} x_1 \\ y_1 \end{pmatrix}, \quad e_2 = \begin{pmatrix} x_2 \\ y_2 \end{pmatrix}$$

with $x_1 x_2 = 0 = x_1 y_2 + x_2 y_1$, implying either $(x_1 = 0, y_1 \neq 0, x_2 = 0)$ or $(x_2 = 0, y_2 \neq 0, x_1 = 0)$ and in both cases $e_1 \wedge e_2 = 0$ which is impossible.

The assumption q_0 positive-definite is important and cannot be replaced by non-degenerate: in $\mathbb{R}^2_{x,y}$ it is not possible to diagonalize simultaneously the quadratic forms $x^2 - y^2$ and xy. Otherwise, we would find a basis of $\mathbb{R}^2$,

$$e_1 = \begin{pmatrix} x_1 \\ y_1 \end{pmatrix}, \quad e_2 = \begin{pmatrix} x_2 \\ y_2 \end{pmatrix}$$

with $x_1 x_2 - y_1 y_2 = 0 = x_1 y_2 + x_2 y_1$, implying e_1 orthogonal (for the standard dot product on $\mathbb{R}^2$) to $(x_2, -y_2)$ and to (y_2, x_2) so that the determinant $\begin{vmatrix} x_2 & -y_2 \\ y_2 & x_2 \end{vmatrix} = 0$ and $e_2 = 0$, which is impossible.

Remark 4.2.2. A slightly different point of view would be to consider Q_0, Q_1 two $n \times n$ symmetric (resp. hermitian) matrices with Q_0 positive-definite: there exists $R \in O(n)$ (resp. $U(n)$) such that, $R^* Q_0 R = D$, D diagonal with positive entries, so that

$$D^{-1/2} R^* Q_0 R D^{-1/2} = \mathrm{Id}, D^{-1/2} R^* Q_1 R D^{-1/2} = Q_2,$$

with Q_2 symmetric (resp. Hermitian). Then there exists $S \in O(n)$ (resp. $U(n)$) such that

$$S^* D^{-1/2} R^* Q_1 R D^{-1/2} S = S^* Q_2 S = D_2, \quad \text{with } D_2 \text{ diagonal and real,}$$

whereas $S^* D^{-1/2} R^* Q_0 R D^{-1/2} S = S^* S = \mathrm{Id}$, proving (more constructively) the simultaneous diagonalization of Q_0, Q_1.

4.2.2 Some remarks on commutative algebra

Let us consider a normed real vector space V and let $\mathcal{T}$ be the tensor algebra of V. Since for $T_j \in V$, the symmetrized products $T_1 \otimes \cdots \otimes T_k$ can be written as a linear combination of k-th powers, the norm of the k-linear symmetric form A given by

$$\|A\| = \sup_{\|T\|=1} |AT^k| \tag{4.2.2}$$

is equivalent to the natural norm

$$\|A\| = \sup_{\substack{\|T_j\|=1, \\ 1 \leq j \leq k}} |AT_1 \ldots T_k| \tag{4.2.3}$$

and we have the inequalities $\|A\| \le \|A\| \le \kappa_k \|A\|$ with a constant κ_k depending only on k. The best constant in general is $\kappa_k = k^k/k!$. In fact, in a commutative algebra on a field with characteristic 0, using the polarization formula, the products $T_1 \ldots T_k$ are linear combination of k-th powers since

$$T_1 T_2 \ldots T_k = \frac{1}{2^k k!} \sum_{\varepsilon_j = \pm 1} \varepsilon_1 \ldots \varepsilon_k (\varepsilon_1 T_1 + \cdots + \varepsilon_k T_k)^k. \tag{4.2.4}$$

To prove the previous formula, we note that for $\alpha \in \mathbb{N}^k$, $|\alpha| = k$, we have if $\alpha_1 = 0$,

$$\sum_{\substack{\varepsilon_j = \pm 1 \\ 1 \le j \le k}} \varepsilon_1^{1+\alpha_1} \ldots \varepsilon_k^{1+\alpha_k} = \sum_{\varepsilon_1 = \pm 1} \varepsilon_1 \Big(\sum_{\substack{\varepsilon_j = \pm 1 \\ 2 \le j \le k}} \varepsilon_2^{1+\alpha_2} \ldots \varepsilon_k^{1+\alpha_k} \Big) = 0,$$

proving that the coefficient of T^α in the rhs of (4.2.4) is 0 if $\alpha_j = 0$ for some j; as a consequence, for all j, $\alpha_j \ge 1$ and since $|\alpha| = k$, we have $\alpha_j = 1$ for all j, proving that the rhs of (4.2.4) is proportional to the lhs. We check the constant, using the same remark as above,

$$\sum_{\varepsilon_j = \pm 1} \varepsilon_1 \ldots \varepsilon_k (\varepsilon_1 + \cdots + \varepsilon_k)^k = \sum_{\substack{\varepsilon_j = \pm 1 \\ \alpha_j = 1}} \Big(\prod_{1 \le j \le k} \varepsilon_j^{1+\alpha_j} \Big) \frac{k!}{\alpha!} = 2^k k!,$$

which gives (4.2.4). Using the triangle inequality, we get $\|A\| \le \frac{1}{2^k k!} 2^k k^k \|A\|$, and thus $\kappa_k \le \frac{k^k}{k!}$. On the other hand, for $T_j \in \mathbb{R}^k$ and A defined by

$$A(T_1, \ldots, T_k) = \frac{1}{k!} \sum_{\sigma \in \mathcal{S}_k} T_{\sigma(1),1} \ldots T_{\sigma(k),k},$$

we have $A(e_1, \ldots, e_k) = 1/k!$ so that $\|A\| \ge \frac{1}{k!}$ and for $\theta \in \mathbb{R}^k$ (with the norm $\sum |\theta_j|$),

$$|A\theta^k| = |\theta_1 \ldots \theta_k| \le \Big(\frac{\sum |\theta_j|}{k} \Big)^k \implies \|A\| \le k^{-k},$$

so that $\kappa_k \ge \frac{\|A\|}{\|A\|} \ge \frac{k^k}{k!}$.

Lemma 4.2.3. *Let V be a Euclidean finite-dimensional vector space, and A a symmetric k-multilinear form. We have, with $\|A\|, \|A\|$ defined in (4.2.3), (4.2.2),*

$$\|A\| = \|A\|. \tag{4.2.5}$$

Proof. This is a consequence of the 1928 paper by O.D. Kellogg [83]. Let us give a short proof, following the note by T. Muramatu & S. Wakabayashi [109]. We note first that $\big[(x+y)^2 + (ix - iy)^2 \big]^l = 2^{2l} x^l y^l$ so that

$$\sum_{0 \le j \le l} C_l^j (x+y)^{2j} (x-y)^{2l-2j} (-1)^{l-j} = 2^{2l} x^l y^l, \tag{4.2.6}$$

and that identity holds for x, y in a real commutative algebra. The property (4.2.5) is true for $k = 2$ (and for $k = 1$) from the identity (4.2.6) for $k = 2$ since

$$|A(x, y)| = \frac{1}{4} |A(x + y, x + y) - A(x - y, x - y)|$$

$$\leq \|A\| \frac{1}{4} (\|x + y\|^2 + \|x - y\|^2) = \|A\| \frac{\|x\|^2 + \|y\|^2}{2} \leq \|A\| \|x\| \|y\|. \quad (4.2.7)$$

Since the inequality $\|A\| \leq \|A\|$ is trivial, we may assume $\|A\| = 1$ and we have to prove $\|A\| \geq 1$. We shall use an induction on k. From the induction hypothesis, for all $1 \leq l \leq k - 1$, we can find $x, y \in V, \|x\| = \|y\| = 1$, such that $1 = |A(x^l y^{k-l})|$. We set

$$\gamma = \max_{\substack{x,y \in V, \|x\| = \|y\| = 1, \\ 1 = |A(x^l y^{k-l})| \text{ for some } l}} |\langle x, y \rangle|, \quad (4.2.8)$$

and we consider x, y both with norm 1, such that $\gamma = \langle x, y \rangle, 1 = |A(x^l y^{k-l})|$. We may assume that $k - l \geq l$ (otherwise we exchange the rôle of x and y). We have, using (4.2.6),

$$1 = |A(x^l y^l y^{k-2l})| = 2^{-2l} | \sum_{0 \leq j \leq l} C_l^j (-1)^{l-j} A(x + y)^{2j} (x - y)^{2l-2j} y^{k-2l}|$$

$$\leq 2^{-2l} \sum_{0 \leq j \leq l-1} C_l^j \|x + y\|^{2j} \|x - y\|^{2l-2j} + 2^{-2l} |A(x + y)^{2l} y^{k-2l}|$$

$$= 2^{-2l} \sum_{0 \leq j \leq l} C_l^j (2 + 2\gamma)^j (2 - 2\gamma)^{l-j} - 2^{-2l} \|x + y\|^{2l} + 2^{-2l} |A(x + y)^{2l} y^{k-2l}|$$

$$= 1 - 2^{-2l} \|x + y\|^{2l} + 2^{-2l} |A(x + y)^{2l} y^{k-2l}|,$$

which implies $|A(x + y)^{2l} y^{k-2l}| \geq \|x + y\|^{2l}$. If $x + y = 0$, we have $1 = |A(x^k)|$ and $\|A\| \geq 1$. Otherwise we get $1 \geq |A \frac{(x+y)^{2l}}{\|x+y\|^{2l}} y^{k-2l}| \geq 1$ and thus, from (4.2.8), we get

$$\frac{\gamma + 1}{\sqrt{2\gamma + 2}} = \langle \frac{x + y}{\|x + y\|}, y \rangle \leq \gamma \implies \gamma^2 (\gamma + 1) 2 \geq (\gamma + 1)^2 \implies 2\gamma^2 - \gamma - 1 \geq 0,$$

and since $\gamma \in [0, 1]$, it means $\gamma = 1$ and $x = y$ ($x = -y$ was ruled out before); finally, we have $|A(x^k)| = 1$ and the result of the lemma. $\square$

4.3 Lemmas of classical analysis

4.3.1 On the Faà di Bruno formula

Let us first recall the multinomial formula

$$(\sum_{1 \leq j \leq N} a_j)^m = \sum_{m_1 + \cdots + m_N = m} \frac{m!}{\prod m_j!} a_j^{m_j}.$$

The Faà di Bruno formula[6] deals with the iterated derivative of a composition of functions. First of all, let us consider (smooth) functions of one real variable

$$U \xrightarrow{f} V \xrightarrow{g} W, \qquad U, V, W \text{ open sets of } \mathbb{R}.$$

With $g^{(r)}$ always evaluated at $f(x)$, we have

$$(g \circ f)' = g' f',$$
$$(g \circ f)'' = g'' f'^2 + g' f'',$$
$$(g \circ f)''' = g''' f'^3 + g'' 3 f'' f' + g' f''',$$
$$(g \circ f)^{(4)} = g^{(4)} (f')^4 + 6 g^{(3)} f'^2 f'' + g'' (4 f''' f' + 3 f''^2) + g' f^{(4)},$$

i.e.,

$$\frac{1}{4!}(g \circ f)^{(4)} = \frac{g^{(4)}}{4!} \left(\frac{f'}{1!}\right)^4 + 3\frac{g^{(3)}}{3!} \left(\frac{f''}{2!}\right)\left(\frac{f'}{1!}\right)^2 + \frac{g^{(2)}}{2!}\left[\left(\frac{f''}{2!}\right)^2 + 2\frac{f'''}{3!}f'\right] + \frac{g^{(1)}}{1!}\frac{f^{(4)}}{4!}.$$

More generally we have the remarkably simple

$$\frac{(g \circ f)^{(k)}}{k!} = \sum_{1 \le r \le k} \frac{g^{(r)} \circ f}{r!} \prod_{\substack{k_1 + \cdots + k_r = k \\ k_j \ge 1}} \frac{f^{(k_j)}}{k_j!}. \qquad (4.3.1)$$

- There is only one multi-index $(1,1,1,1) \in \mathbb{N}^{*4}$ such that $\sum_{1 \le j \le 4} k_j = 4$.

- There are three multi-indices $(1,1,2),(1,2,1),(2,1,1) \in \mathbb{N}^{*3}$ with $\sum_{1 \le j \le 3} k_j = 4$.

- There is one multi-index $(2,2) \in \mathbb{N}^{*2}$ with $\sum_{1 \le j \le 2} k_j = 4$ and two multi-indices $(1,3),(3,1)$ such that $\sum_{1 \le j \le 2} k_j = 4$.

- There is one index $4 \in \mathbb{N}^*$ with $\sum_{1 \le j \le 1} k_j = 4$.

Usually the formula is written in a different way with the more complicated

$$\frac{(g \circ f)^{(k)}}{k!} = \sum_{\substack{l_1 + 2l_2 + \cdots + kl_k = k \\ r = l_1 + \cdots + l_k}} \frac{g^{(r)} \circ f}{l_1! \ldots l_k!} \prod_{1 \le j \le k} \left(\frac{f^{(j)}}{j!}\right)^{l_j}. \qquad (4.3.2)$$

Let us show that the two formulas coincide. We start from (4.3.1)

$$\frac{(g \circ f)^{(k)}}{k!} = \sum_{1 \le r \le k} \frac{g^{(r)} \circ f}{r!} \prod_{\substack{k_1 + \cdots + k_r = k \\ k_j \ge 1}} \frac{f^{(k_j)}}{k_j!}.$$

---

[6]Francesco Faà di Bruno (1825–1888) was an italian mathematician and priest, born at Alessandria. He was beatified in 1988, probably the only mathematician to reach sainthood so far. The "Chevalier François Faà di Bruno, Capitaine honoraire d'État-Major dans l'armée Sarde", defended his thesis in 1856, in the Faculté des Sciences de Paris in front of the following jury: Cauchy (chair), Lamé and Delaunay.

If we consider a multi-index

$$(k_1, \dots, k_r) = (\underbrace{1, \dots, 1}_{l_1 \text{ times}}, \underbrace{2, \dots, 2}_{l_2 \text{ times}}, \dots, \underbrace{j, \dots, j}_{l_j \text{ times}}, \dots, \underbrace{k, \dots, k}_{l_k \text{ times}})$$

we get as a factor of $g^{(r)}/r!$ the term $\prod_{1 \leq j \leq k} \left(\frac{f^{(j)}}{j!} \right)^{l_j}$ with $l_1 + 2l_2 + \dots + kl_k = k$, $l_1 + \dots + l_k = r$ and since we can permute the $(k_1, \dots, k_r)$ above, we get indeed a factor $\frac{r!}{l_1! \dots l_k!}$ which gives (4.3.2). The proof above can easily be generalized to a multi-dimensional setting with

$$U \xrightarrow{f} V \xrightarrow{g} W, \qquad U, V, W \text{ open sets of } \mathbb{R}^m, \mathbb{R}^n, \mathbb{R}^p, \ f, g \text{ of class } C^k.$$

Since the derivatives are multilinear symmetric mappings, they are completely determined by their values on the "diagonal" $T \otimes \dots \otimes T$: the symmetrized products of $T_1 \otimes \dots \otimes T_k$, noted as $T_1 \dots T_k$, can be written as a linear combination of k-th powers (see (4.2.4)). For $T \in \mathcal{T}_x(U)$, we have

$$\frac{(g \circ f)^{(k)}}{k!} T^k = \sum_{1 \leq r \leq k} \frac{g^{(r)} \circ f}{r!} \prod_{\substack{k_1 + \dots + k_r = k \\ k_j \geq 1}} \frac{f^{(k_j)}}{k_j!} T^{k_j},$$

which is consistent with the fact that $f^{(k_j)}(x) T^{k_j}$ belongs to the tangent space $T_{f(x)}(V)$ of V at $f(x)$ and $\otimes_{1 \leq j \leq r} f^{(k_j)}(x) T^{k_j}$ is a tensor product in $T^{r,0}(T_{f(x)}(V))$ on which $g^{(r)}(f(x))$ acts to send it on $T_{g(f(x))}(W)$. More details on the history of that formula can be obtained in the article [81].

4.3.2 On Leibniz formulas

Let a be a function in L^1_{loc} of some open set Ω of $\mathbb{R}^m$ and let u be a locally Lipschitz-continuous function on Ω. Although a' may be a distribution of order 1 and u is not C^1, it is possible to define the product $T = a'u$ as follows (φ is a test function):

$$\langle T, \varphi \rangle = - \int a(u'\varphi + u\varphi') dx$$

so that T is a distribution of order 1 satisfying the identity

$$(au)' = T + au'.$$

As a matter of fact, we have

$$\langle (au)', \varphi \rangle = - \int au\varphi' dx = \langle T, \varphi \rangle + \int au'\varphi dx = \langle T + au', \varphi \rangle.$$

It means in particular that one can multiply the first-order distribution $\frac{d}{dx}(\ln |x|) = \text{pv}\frac{1}{x}$ by the Lipschitz-continuous function $|x|$ and get

$$(\text{pv}\frac{1}{x})|x| = \frac{d}{dx}\left((\ln |x|)|x| \right) - (\ln |x|) \, \text{sign} \, x = \text{sign} \, x$$

as it is easily verified. On the other hand it is not possible to multiply the first order distribution δ_0' by the Lipschitz-continuous function $|x|$.

4.3.3 On Sobolev norms

For $s \in \mathbb{R}$, the space $H^s(\mathbb{R}^n) = \{u \in \mathscr{S}'(\mathbb{R}^n), \langle\xi\rangle^s \hat{u}(\xi) \in L^2(\mathbb{R}^n)\}$ is a Hilbert space equipped with the norm $\|u\|_s = \|\langle D\rangle^s u\|_{L^2(\mathbb{R}^n)}$.

Lemma 4.3.1 (Logarithmic convexity of the Sobolev norm). *Let $s_0 \leq s_1$ be real numbers. Then for all $u \in H^{s_1}(\mathbb{R}^n)$ and $\theta \in [0, 1]$,*

$$\|u\|_{(1-\theta)s_0 + \theta s_1} \leq \|u\|_{s_0}^{1-\theta} \|u\|_{s_1}^\theta. \tag{4.3.3}$$

Proof. Using Hölder's inequality, we have

$$\|u\|_{(1-\theta)s_0+\theta s_1}^2 = \int \langle\xi\rangle^{2(1-\theta)s_0}|\hat{u}(\xi)|^{2(1-\theta)} \langle\xi\rangle^{2\theta s_1}|\hat{u}(\xi)|^{2\theta} d\xi$$

$$\leq \left(\int \langle\xi\rangle^{2s_0}|\hat{u}(\xi)|^2 d\xi\right)^{1-\theta} \left(\int \langle\xi\rangle^{2s_1}|\hat{u}(\xi)|^2 d\xi\right)^\theta. \qquad \square$$

N.B. A consequence of that inequality is that for $s_0 < s < s_1$ and $u \in H^{s_1}(\mathbb{R}^n)$, we have with $s = (1-\theta)s_0 + \theta s_1$ and $\varepsilon > 0$,

$$\|u\|_s \leq \varepsilon^{-\theta}\theta^\theta \|u\|_{s_0}^{1-\theta} \varepsilon^\theta \theta^{-\theta} \|u\|_{s_1}^\theta \leq \varepsilon\|u\|_{s_1} + \varepsilon^{-\frac{\theta}{1-\theta}}(1-\theta)\theta^{\frac{\theta}{1-\theta}}\|u\|_{s_0}. \tag{4.3.4}$$

Lemma 4.3.2. *Let $s > -n/2$ and K be a compact subset of $\mathbb{R}^n$. On the space $H_K^s = \{u \in H^s(\mathbb{R}^n), \operatorname{supp} u \subset K\}$, the H^s-norm is equivalent to the "homogeneous" norm*

$$\|u\|_s = \left(\int |\xi|^{2s}|\hat{u}(\xi)|^2 d\xi\right)^{1/2}. \tag{4.3.5}$$

Proof. Note first that, if $u \in \mathscr{E}_K'(\mathbb{R}^n)$, its Fourier transform is an entire function and thus $|\xi|^{2s}|\hat{u}(\xi)|^2$ is locally integrable. Moreover, if $s \geq 0$, we have $\|u\|_s \leq \|u\|_s$ and for $r > 0$, $\chi_K \in C_c^\infty(\mathbb{R}^n)$ equal to 1 on K,

$$\|u\|_s^2 = \int_{|\xi| \leq r} (1 + |\xi|^2)^s |\hat{u}(\xi)|^2 d\xi + \int (r^{-2} + 1)^s |\xi|^{2s}|\hat{u}(\xi)|^2 d\xi$$

$$\leq \int_{|\xi| \leq r} (1 + |\xi|^2)^s d\xi \|\chi_K\|_0^2 \|u\|_0^2 + (r^{-2} + 1)^s \|u\|_s^2.$$

If r is such that $\int_{|\xi| \leq r}(1 + |\xi|^2)^s d\xi \|\chi_K\|_0^2 \leq 1/2$, we get the result. If $0 > s > -n/2$, we have $\|u\|_s \leq \|u\|_s$, and

$$\|u\|_s^2 = \int_{|\xi| \leq 1} |\xi|^{2s}|\hat{u}(\xi)|^2 d\xi + 2^{-s}\int \langle\xi\rangle^{2s}|\hat{u}(\xi)|^2 d\xi$$

$$\leq \int_{|\xi| \leq 1} |\xi|^{2s}\|\chi_K(\cdot)e^{2i\pi\cdot\xi}\|_{-s}^2 d\xi\|u\|_s^2 + 2^{-s}\|u\|_s^2 = C(K,s)\|u\|_s^2. \qquad \square$$

Lemma 4.3.3. *Let $s_1 > s_0 > -n/2$. There exists $C(s_0, s_1) > 0$ such that for all $r > 0$ and all $u \in H^{s_1}(\mathbb{R}^n)$ with diameter(supp u) $\leq r$,*

$$\|u\|_{s_0} \leq C(s_0, s_1) r^{s_1 - s_0} \|u\|_{s_1}. \tag{4.3.6}$$

Proof. We may assume that u is supported in $B(0, r)$ and $r \leq 1$. We define $v(x) = u(rx)$ so that v is supported in the unit ball of $\mathbb{R}^n$ and we apply Lemma 4.3.2: since $\|v\|_s^2 = \int |\xi|^{2s} |\hat{u}(\xi/r)|^2 r^{-2n} d\xi = r^{2s-n} \|u\|_s^2$, we obtain (for $u \neq 0$)

$$C_{s_1}^{-1} C_{s_0}^{-1} \frac{\|u\|_{s_0}}{\|u\|_{s_1}} \leq \frac{\|u\|_{s_0}}{\|u\|_{s_1}} = r^{s_1 - s_0} \frac{\|v\|_{s_0}}{\|v\|_{s_1}} \leq r^{s_1 - s_0} C_{s_1} C_{s_0} \frac{\|v\|_{s_0}}{\|v\|_{s_1}} \leq r^{s_1 - s_0} C_{s_1} C_{s_0}.$$

$\square$

Remark 4.3.4. The previous lemma implies in particular that when $n \geq 3$, $u \in L^2(\mathbb{R}^n)$, supp $u \subset B(x_0, r)$, then $\|u\|_{-1} \leq Cr\|u\|_0$. Although the same estimate is not true for $n = 2$, we need a two-dimensional substitute inequality: we have more generally (for $0 < r \leq 1$),

$$\|u\|_{-n/2}^2 = \int_{|\xi| \leq 1/r} (1 + |\xi|^2)^{-n/2} |\hat{u}(\xi)|^2 d\xi + \int_{|\xi| \geq 1/r} \langle \xi \rangle^{-n} |\hat{u}(\xi)|^2 d\xi$$

$$\leq \int_{|\xi| \leq 1/r} (1 + |\xi|^2)^{-n/2} d\xi \|u\|_0^2 \|\mathbf{1}_{B(x_0, r)}\|_0^2 + r^n \|u\|_0^2$$

$$\leq \int_0^{1/r} (1 + \rho^2)^{-n/2} \rho^{n-1} d\rho \alpha_n r^n \|u\|_0^2 + r^n \|u\|_0^2$$

$$\leq r^n \|u\|_0^2 (\alpha_n' + \alpha_n'' \ln(1/r)),$$

which gives that for all $n \in \mathbb{N}^*$, there exists β_n, such that for all $u \in L^2(\mathbb{R}^n)$ and $r \in]0, 1]$,

$$\text{supp } u \subset B(x_0, r) \Longrightarrow \|u\|_{-n/2} \leq (1 + \ln(1/r))^{1/2} r^{n/2} \beta_n \|u\|_0. \tag{4.3.7}$$

Lemma 4.3.5. *Let $s_1 > s_0$ be real numbers, $s_1 > -n/2$. For all $r > 0$, there exists $\phi(s_0, s_1, r)$ such that for all $u \in H^{s_1}(\mathbb{R}^n)$ with diameter(supp u) $\leq r$,*

$$\|u\|_{s_0} \leq \phi(s_0, s_1, r) \|u\|_{s_1}, \quad \lim_{r \to 0} \phi(s_0, s_1, r) = 0, \tag{4.3.8}$$

and more precisely, for $r \in]0, 1]$,

$$\left. \begin{array}{ll} \text{If } -n/2 < s_0 < s_1, & \phi = Cr^{s_1 - s_0}, \\ \text{If } -n/2 = s_0 < s_1, & \phi = Cr^{s_1 - s_0} (1 + \ln(1/r))^{1/2}, \\ \text{If } s_0 < -n/2 < s_1, & \phi = Cr^{s_1 + \frac{n}{2}} (1 + \ln(1/r))^{1/2}. \end{array} \right\} \tag{4.3.9}$$

Proof. We may assume $r < 1/2$ and $s_0 \leq -n/2$, supp $u \subset B(0, r)$.

(1) *Let us first assume that $s_0 = -n/2$ and $s_1 \geq 0$. Then the inequalities* (4.3.7), (4.3.6) *give*

$$\|u\|_{-n/2} \leq \left(1 + \ln(1/r)\right)^{1/2} r^{n/2} \beta_n \|u\|_0 \leq \left(1 + \ln(1/r)\right)^{1/2} r^{\frac{n}{2} + s_1} \beta_n C(0, s_1) \|u\|_{s_1}.$$

(2) *If $s_0 = -n/2$ and $-n/2 < s_1 < 0$. We have, for $\chi_0 \in C_c^\infty(\mathbb{R}^n)$, equal to 1 on the unit ball,*

$$
\|u\|_{s_0}^2 \leq \int_{|\xi| \leq 1/r} \langle \xi \rangle^{2s_0} |\hat{u}(\xi)|^2 d\xi + r^{2s_1 - 2s_0} \int_{|\xi| \geq 1/r} \langle \xi \rangle^{2s_1} |\hat{u}(\xi)|^2 d\xi
$$

$$
\leq \underbrace{\int_{|\xi| \leq 1/r} \langle \xi \rangle^{2s_0} \|\chi_0(\cdot/r) e^{2i\pi \cdot \xi}\|_{-s_1}^2 d\xi}_{I(r)} \|u\|_{s_1}^2 + r^{2s_1 - 2s_0} \|u\|_{s_1}^2,
$$

and since $I(r) \leq C_{n,s_1} \displaystyle\int_0^{1/r} (1 + \rho^2)^{-n/2} \rho^{n-1} r^n (r^{-2|s_1|} + \rho^{2|s_1|}) d\rho$

$$\leq C'_{n,s_1} \left(r^{n - 2|s_1|} \ln(1/r) + r^{n - 2|s_1|} \right)$$

we get in that case $\|u\|_{s_0}^2 \leq C''_{n,s_1} \|u\|_{s_1}^2 r^{2s_1 - 2s_0} \ln(1/r)$

(3) *Let us assume now that $s_0 < -n/2 < s_1$. We have from the previous discussion*

$$\|u\|_{s_0} \leq \|u\|_{-n/2} \leq C r^{s_1 + \frac{n}{2}} \left(\ln(1/r) \right)^{1/2} \|u\|_{s_1}. \qquad \square$$

4.3.4   On partitions of unity

A locally finite partition of unity

Lemma 4.3.6. *Let Ω be an open subset of $\mathbb{R}^n$. There exists a sequence $(\varphi_j)_{j \geq 1}$ of functions in $C_c^\infty(\Omega; [0,1])$ such that, for all compact subsets $K \subset \Omega$, $\varphi_{j|K} = 0$ for all j but a finite number and $\forall x \in \Omega$, $1 = \sum_{j \geq 1} \varphi_j(x)$. Moreover if K is a compact subset of Ω, $\phi_K = \sum_{j, \operatorname{supp} \varphi_j \cap K \neq \emptyset} \varphi_j$ belongs to $C_c^\infty(\Omega; [0,1])$ and is equal to 1 on a neighborhood of K.*

Proof. Considering for $j \in \mathbb{N}^*$ the set $K_j = \{x \in \Omega, |x - \Omega^c| \geq \frac{1}{j}, |x| \leq j\}$, we see immediately that K_j is compact, $K_j \subset \operatorname{int} K_{j+1}, \cup_{j \geq 1} K_j = \Omega$. We take now $\psi_j \in C_c^\infty(\operatorname{int} K_{j+1}; [0,1]), \psi_j = 1$ on K_j. We define

$$\varphi_1 = \psi_1, \quad \varphi_2 = \psi_2(1 - \psi_1), \quad \cdots \quad, \varphi_j = \psi_j \prod_{1 \leq l < j} (1 - \psi_l)$$

and we note that $\varphi_j \in C_c^\infty(\operatorname{int} K_{j+1}; [0,1])$. If K is a compact subset of Ω (we have also $\Omega = \cup_{j \geq 1} \operatorname{int} K_{j+1}$), then $K \subset K_{j_0}$ for some j_0 and $\varphi_{j|K} = 0$ for $j > j_0$. Moreover, we have by induction on j that $1 - \sum_{1 \leq l \leq j} \varphi_l = \prod_{1 \leq l \leq j}(1 - \psi_l)$, which implies $\sum_{1 \leq l} \varphi_l = 1_\Omega$. Let K be a compact subset of Ω: there exists $\varepsilon_0 > 0$ such that $K + \varepsilon_0 \subset \Omega$ and $\phi_{K+\varepsilon_0}$ is 1 on $K + \varepsilon_0 = K + \varepsilon_0 B_1$, where B_1 is the closed unit Euclidean ball of $\mathbb{R}^n$. We have also

$$\phi_{K+\varepsilon_0} = \phi_K + \overbrace{\sum_{\substack{j,\operatorname{supp}\varphi_j\cap K=\emptyset \\ \operatorname{supp}\varphi_j\cap(K+\varepsilon_0)\neq\emptyset}} \varphi_j}^{\omega}$$

and the sum defining ω is finite, so that $\omega \in C_c^\infty(K^c)$. If $L = \operatorname{supp}\omega \subset K^c$, on $(K + \varepsilon_0) \cap L^c$ (this is a neighborhood of K), we have $\phi_{K+\varepsilon_0} = \phi_K$, which gives that $\phi_K(x) = 1$ for $x \in (K + \varepsilon_0) \cap L^c$. $\qquad\square$

An example of a slowly varying metric

Let Ω be a proper open subset of $\mathbb{R}^n$. We define $\rho(x) = \operatorname{dist}(x, \Omega^c)$ and we see that ρ is positive on Ω. Moreover, if $x \in \Omega$ and $y \in \mathbb{R}^n$ are such that

$$|y - x| \le \frac{1}{2}\rho(x)$$

then y belongs to Ω (otherwise $y \in \Omega^c$ and $0 < \rho(x) = \operatorname{dist}(x, \Omega^c) \le |x - y| \le \rho(x)/2$, which is impossible) and the ratio $\rho(y)/\rho(x)$ belongs to the interval $[1/2, 2]$. In fact, the function ρ is Lipschitz-continuous with a Lipschitz constant smaller than 1. As a consequence, we get that

$$\rho(y) \le \rho(x) + |y - x| \le \frac{3}{2}\rho(x), \qquad \rho(x) \le \rho(y) + |y - x| \le \rho(y) + \frac{1}{2}\rho(x)$$

which gives $\rho(y)/\rho(x) \in [1/2, 3/2]$. We shall say that the metric $|dx|^2\rho(x)^{-2}$ is slowly varying on Ω.

A discrete partition of unity

The following result provides a discrete version of Theorem 2.2.7.

Theorem 4.3.7. *Let g be a slowly varying metric on $\mathbb{R}^m$ (cf. Definition 2.2.1). Then there exists $r_0 > 0$ such that, for all $r \in (0, r_0]$, there exists a sequence $(X_k)_{k\in\mathbb{N}}$ of points in the phase space $\mathbb{R}^{2n}$ and a positive number N_r such that the following properties are satisfied. We define U_k, U_k^*, U_k^{**} as the $g_k = g_{X_k}$ closed balls with center X_k and radius $r, 2r, 4r$. There exist two families of non-negative smooth functions on $\mathbb{R}^{2n}$, $(\chi_k)_{k\in\mathbb{N}}, (\psi_k)_{k\in\mathbb{N}}$ such that*

$$\sum_k \chi_k(X) = 1, \ \operatorname{supp}\chi_k \subset U_k, \quad \psi_k \equiv 1 \ \ on \ U_k^*, \ \operatorname{supp}\psi_k \subset U_k^{**}.$$

Moreover, $\chi_k, \psi_k \in S(1, g_k)$ with semi-norms bounded independently of k. The overlap of the balls U_k^{**} is bounded, i.e.,

$$\bigcap_{k\in\mathcal{N}} U_k^{**} \neq \emptyset \quad \Longrightarrow \quad \#\mathcal{N} \le N_r.$$

Moreover, $g_X \sim g_k$ all over U_k^{**} (i.e., the ratios $g_X(T)/g_k(T)$ are bounded above and below by a fixed constant, provided that $X \in U_k^{**}$).

4.3.5   On non-negative functions

A classical lemma

Lemma 4.3.8. *Let $f : \mathbb{R}^n \to \mathbb{R}_+$ be a C^2 non-negative function such that f'' is bounded. Then, for all $x \in \mathbb{R}^n$,*

$$|f'(x)|^2 \leq 2f(x)\|f''\|_{L^\infty(\mathbb{R}^n)}. \tag{4.3.10}$$

Here the norms of the (multi)linear forms are taken with respect to any fixed norm on $\mathbb{R}^n$.

Proof. Let x be given in $\mathbb{R}^n$ and T be a unit vector (for a given norm on $\mathbb{R}^n$). We have

$$0 \leq f(x + tT) = f(x) + tf'(x) \cdot T + t^2 \int_0^1 (1-\theta)f''(x + \theta tT)d\theta T^2$$

$$\leq f(x) + tf'(x) \cdot T + \frac{1}{2}t^2 \sup_{y \in \mathbb{R}^n} |f''(y)T^2|.$$

Consequently, this second-degree polynomial in the t variable has a non-positive discriminant, i.e., $|f'(x) \cdot T|^2 \leq 2f(x) \sup_{y \in \mathbb{R}^n} \|f''(y)\|$, which gives the result. $\square$

Remark 4.3.9. The constant 2 is optimal in the inequality (4.3.10) as shown by the one-variable function x^2. The same conclusion holds assuming only $f \in C^1$ with a (distribution) second derivative bounded: the same proof works, since the Taylor formula with integral remainder is true for a distribution. Moreover, a direct consequence of this lemma is that a non-negative $C^{1,1}$ function (a C^1 function with a Lipschitz-continuous second derivative) has a Lipschitz-continuous square root. A classical counterexample due to G. Glaeser [48] shows that this cannot be improved, even if the function F is C^∞ and flat at its zeroes: there exists a non-negative C^∞ function of one real variable such that $F^{-1}(\{0\}) = \{0\}$, F is flat at zero and $F^{1/2}$ is C^1 but not C^2. An example of such a function is

$$F(x) = e^{-x^{-2}} + e^{-x^{-1}}H(x)\sin^2(x^{-1}), \qquad H = \mathbf{1}_{\mathbb{R}_+}.$$

Lemma 4.3.10. *Let $f : \mathbb{R}^n \to \mathbb{R}$ be a $C^{1,1}$ function (i.e., $f \in C^1$, f' Lipschitz-continuous) such that f'' is bounded and $f(x) = 0 \implies f'(x) = 0$. Then, for all $x \in \mathbb{R}^n$,*

$$|f'(x)|^2 \leq 2|f(x)|\|f''\|_{L^\infty(\mathbb{R}^n)}. \tag{4.3.11}$$

Here the norms of the (multi)linear forms are taken with respect to any fixed norm on $\mathbb{R}^n$.

Proof. We define the open set $\Omega = \{x \in \mathbb{R}^n, f(x) \neq 0\}$ and consider $g(x) = |f(x)|$. The function g is C^1 on Ω and if $x_0 \notin \Omega, h \in \mathbb{R}^n$, we have

$$g(x_0 + h) - g(x_0) = |f(x_0) + f'(x_0)h + o(h)| = o(h),$$

so that $g'(x_0) = 0$. The function g is thus differentiable on $\mathbb{R}^n$ and

$$g' = sf' \text{ with } s(x) = f(x)/|f(x)| \text{ if } x \in \Omega \text{ and } s(x) = 0 \text{ if } x \notin \Omega.$$

If x_1, x_2 are such that $f(x_1)f(x_2) > 0$, we have $|g'(x_2) - g'(x_1)| = |f'(x_2) - f'(x_1)|$ and thus

$$|g'(x_2) - g'(x_1)| \leq \|f''\|_{L^\infty}|x_1 - x_2|. \tag{4.3.12}$$

If $s(x_1) = s(x_2) = 0$, the same is true since $g'(x_j) = 0, j = 1, 2$. If $s(x_1) = 0, s(x_2) = 1$, we have $g'(x_2) - g'(x_1) = f'(x_2) = f'(x_2) - f'(x_1)$ and thus (4.3.12) holds as well in that case and in the other remaining cases. The function g is non-negative, $C^{1,1}$ and $\|g''\|_{L^\infty} \leq \|f''\|_{L^\infty}$. We can regularize g by a non-negative mollifier $\rho_\varepsilon(x) = \varepsilon^{-n}\rho(x/\varepsilon), \rho \in C_c^\infty(\mathbb{R}^n; \mathbb{R}_+), \int \rho = 1$ and apply Lemma 4.3.8 to $g * \rho_\varepsilon$: we get

$$|(g' * \rho_\varepsilon)(x)|^2 \leq 2(g * \rho_\varepsilon)(x)\|g'' * \rho_\varepsilon\|_{L^\infty} \leq 2(g * \rho_\varepsilon)(x)\|f''\|_{L^\infty}.$$

Since g, g' are continuous functions and $g' = sf'$ with s as given above, we get $|f'(x)|^2 \leq 2|f(x)|\|f''\|_{L^\infty}$. □

On $C^{3,1}$ non-negative functions

Let a be a non-negative $C^{3,1}$ function defined on $\mathbb{R}^m$ such that $\|a^{(4)}\|_{L^\infty} \leq 1$; we recall the definition

$$\rho(x) = \left(|a(x)| + |a''(x)|^2\right)^{1/4}, \quad \Omega = \{x, \rho(x) > 0\}. \tag{4.3.13}$$

Lemma 4.3.11. *Let a, ρ, Ω be as above. For $0 \leq j \leq 4$, we have $\|a^{(j)}(x)\| \leq \gamma_j \rho(x)^{4-j}$, with $\gamma_0 = \gamma_2 = \gamma_4 = 1, \gamma_1 = 3, \gamma_3 = 4$.*

Proof. The inequalities for $j = 0, 2, 4$ are obvious. Let us write Taylor's formula,

$$a(x+h) = a(x) + a'(x)h + \frac{1}{2}a''(x)h^2 + \frac{1}{6}a^{(3)}(x)h^3 + \int_0^1 \frac{(1-\theta)^3}{3!}a^{(4)}(x+\theta h)d\theta h^4.$$

We get $a(x+h) - a(x) - \frac{1}{2}a''(x)h^2 - \frac{|h|^4}{24} \leq a'(x)h + \frac{1}{6}a^{(3)}(x)h^3$ and since $a(x+h) \geq 0$, we have

$$-a(x) - \frac{1}{2}a''(x)h^2 - \frac{|h|^4}{24} \leq a'(x)h + \frac{1}{6}a^{(3)}(x)h^3.$$

Since the rhs is odd in the variable h, we obtain

$$\left|a'(x)h + \frac{1}{6}a^{(3)}(x)h^3\right| \leq a(x) + \frac{1}{2}a''(x)h^2 + \frac{|h|^4}{24}. \tag{4.3.14}$$

Let us choose $h = \rho(x)sT$ where T is a unit vector and s is a real parameter. We have

$$|s\rho(x)a'(x)T + s^3\rho(x)^3\frac{1}{6}a^{(3)}(x)T^3| \le \rho(x)^4\Big(1 + \frac{1}{2}s^2 + \frac{s^4}{24}\Big). \qquad (4.3.15)$$

Remark 4.3.12. Let $\alpha, \beta, \gamma \in \mathbb{R}$, and assume that

$$\forall s \in \mathbb{R}, \quad |s\alpha + s^3\beta| \le \gamma(1 + \frac{1}{2}s^2 + \frac{s^4}{24}).$$

Applying that inequality for $s = 1, 3$ gives $|\alpha + \beta| \le \gamma\frac{37}{24}$, $|3\alpha + 27\beta| \le \gamma\frac{213}{24}$ and thus

$$24|\beta| = |3\alpha + 27\beta - 3(\alpha + \beta)| \le \frac{324}{24}\gamma, \quad |\beta| \le \gamma\frac{324}{24^2},$$

$$|\alpha| = |\alpha + \beta - \beta| \le \gamma\frac{37 \times 24 + 324}{24^2} = \gamma\frac{1212}{576}.$$

As a result, from (4.3.15), we get for $\rho(x) > 0$, $\|a'(x)\| \le 3\rho(x)^3$, $\|a^{(3)}(x)\| \le 4\rho(x)$. If $\rho(x) = 0$, we use the inequality (4.3.14) with $h = \epsilon T$ where T is a unit vector and ϵ is a positive parameter, providing $|\epsilon a'(x)T + \epsilon^3\frac{1}{6}a^{(3)}(x)T^3| \le \frac{\epsilon^4}{24}$. Dividing by ϵ and letting it go to zero, we find $a'(x)T = 0$, for all T, i.e., $a'(x) = 0$. Next we find that for all vectors T, $a^{(3)}(x)T^3 = 0$, implying that the symmetric trilinear form $a^{(3)}(x)$ is zero (see Section 4.2.2). The proof is complete. □

Lemma 4.3.13. *Let a, ρ, Ω be as above. The metric $\frac{|dx|^2}{\rho(x)^2}$ is slowly varying on the open set Ω, i.e., there exists $C_0 \ge 1 > r_0 > 0$ such that*

$$x \in \Omega \text{ and } |x - y| \le r_0\rho(x) \implies y \in \Omega, \quad C_0^{-1} \le \frac{\rho(x)}{\rho(y)} \le C_0. \qquad (4.3.16)$$

The constants r_0 and C_0 can be chosen as "universal" fixed constants (independently of the dimension and of the function a, which is normalized by the condition $\|a^{(4)}\|_{L^\infty} \le 1$).

Proof. Using Taylor's formula, one gets, using (4.3.14), Section 4.2.2,

$$\rho(x + h)^4 = a(x + h) + \|a''(x + h)\|^2$$

$$\le a(x) + a'(x)h + \frac{1}{2}a''(x)h^2 + \frac{1}{6}a'''(x)h^3 + \frac{1}{24}|h|^4$$

$$+ 3\|a''(x)\|^2 + 3\|a'''(x)\|^2|h|^2 + 3\frac{1}{4}|h|^4$$

$$\le 2a(x) + a''(x)h^2 + \frac{1}{12}|h|^4 + 3\|a''(x)\|^2 + 3\|a'''(x)\|^2|h|^2 + 3\frac{1}{4}|h|^4$$

$$\le 2\rho(x)^4 + \rho(x)^2|h|^2 + \frac{1}{12}|h|^4 + 3\rho(x)^4 + 3 \cdot 2^4\rho(x)^2|h|^2 + \frac{3}{4}|h|^4$$

$$\le 5\rho(x)^4 + |h|^2\rho(x)^2(1 + 3 \cdot 2^4) + |h|^4(\frac{1}{12} + 3 \cdot 2^{-2})$$

$$\le 3^4\big(\rho(x) + |h|\big)^4.$$

This implies that

$$\rho(x + h) \leq 3\big(\rho(x) + |h|\big). \tag{4.3.17}$$

As a consequence, we have for $\|T\| \leq 1, r \geq 0$, $\rho(x + r\rho(x)T) \leq 3(1 + r)\rho(x)$, and thus

$$|y - x| \leq r\rho(x) \Longrightarrow \rho(y) \leq 3(1 + r)\rho(x).$$

Moreover if $y = x + r\rho(x)T$ with $r \geq 0$ and $|T| \leq 1$, (4.3.17) gives

$$\rho(x) = \rho\big(y - r\rho(x)T\big) \leq 3\big(\rho(y) + r\rho(x)\big)$$

and if $r \leq 1/6$ we find $\frac{1}{2}\rho(x) \leq 3\rho(y) \leq (9 + \frac{3}{2})\rho(x)$ providing the result of the lemma with $C_0 = 1/r_0 = 6$. □

Remark 4.3.14. When the normalisation condition $\|a^{(4)}\|_{L^\infty} \leq 1$ is not satisfied, it is of course possible to divide a by a constant to get back to that normalization condition. When $\|a^{(4)}\|_{L^\infty} \neq 0$, Lemma 4.3.11 provides the inequalities

$$\|a'(x)\|^{4/3} \leq 3^{4/3}\Big(a(x)\|a^{(4)}\|_\infty^{1/3} + \|a''(x)\|^2\|a^{(4)}\|_\infty^{-2/3}\Big), \tag{4.3.18}$$

$$\|a^{(3)}(x)\|^4 \leq 4^4\Big(a(x)\|a^{(4)}\|_\infty^3 + \|a''(x)\|^2\|a^{(4)}\|_\infty^2\Big). \tag{4.3.19}$$

Note that if $\|a^{(4)}\|_{L^\infty} = 0$, i.e., $a^{(4)} \equiv 0$, a is a polynomial of degree ≤ 3, and the nonnegativity implies $a^{(3)} \equiv 0$ so that, if its minimum is realized at 0, a is the sum of a non-negative quadratic form and of a non-negative constant.

Lemma 4.3.15. *Let a, ρ, Ω be as above. Let θ be such that $0 < \theta \leq 1/2$. If $y \in \Omega$ verifies $a(y) \geq \theta\rho(y)^4$, then*

$$|x - y| \leq \theta\rho(y)2^{-3} \Longrightarrow a(x) \geq \theta\rho(y)^4/2.$$

Proof. We note that for $|x - y| \leq r\rho(y)$, using Lemma 4.3.11 and Taylor's formula, we have

$$a(x) \geq a(y) - r\rho(y)3\rho(y)^3 - \frac{1}{2}r^2\rho(y)^2\rho(y)^2 - \frac{1}{6}r^3\rho(y)^34\rho(y) - \frac{1}{24}r^4\rho(y)^4,$$

implying, for $a(y) \geq \theta\rho(y)^4$, that $a(x) \geq \rho(y)^4(\theta - 3r - \frac{r^2}{2} - \frac{2r^3}{3} - \frac{r^4}{24})$ and since for $r \leq 1/2$, we have

$$3r + \frac{r^2}{2} + \frac{2r^3}{3} + \frac{r^4}{24} \leq r(3 + 1/8 + 1/12 + 1/384) \leq 4r,$$

we obtain indeed $a(x) \geq \frac{1}{2}\theta\rho(y)^4$ if $r \leq \theta/8$. □

Lemma 4.3.16. *Let a, ρ, Ω be as above. There exists $R_0 > 0$ such that if $y \in \Omega$ verifies $a(y) < \rho(y)^4/2$, then there exists a unit vector T such that,*

$$|x - y| \leq R_0\rho(y) \Longrightarrow a''(x)T^2 \geq 2^{-1}\rho(y)^2.$$

One can take $R_0 = 10^{-2}$.

Proof. We have $\|a''(y)\| \geq 2^{-1/2}\rho(y)^2$, so with the results of Section 4.2.2, we find a unit vector T such that $|a''(y)T^2| \geq 2^{-1/2}\rho(y)^2$. Then we have, for all real s,

$$0 \leq a(y + s\rho(y)T)$$

$$\leq a(y) + s\rho(y)a'(y)T + \frac{s^2}{2}\rho(y)^2 a''(y)T^2 + \frac{s^3}{6}\rho(y)^3 a'''(y)T^3 + \frac{s^4}{24}\rho(y)^4.$$

The quantity $s\rho(y)a'(y)T + \frac{s^3}{6}\rho(y)^3 a'''(y)T^3$ is odd in the variable s so that

$$a(y) + \frac{s^2}{2}\rho(y)^2 a''(y)T^2 + \frac{s^4}{24}\rho(y)^4 \geq |s\rho(y)a'(y)T + \frac{s^3}{6}\rho(y)^3 a'''(y)T^3| \geq 0,$$

and in particular,

$$\forall s \neq 0, \ a''(y)T^2 \geq -\frac{s^2}{12}\rho(y)^2 - s^{-2}2a(y)\rho(y)^{-2} \implies a''(y)T^2 \geq -2 \times 6^{-1/2}a(y)^{1/2}.$$

Since $|a''(y)T^2| \geq 2^{-1/2}\rho(y)^2$, this implies

$$a''(y)T^2 \geq 2^{-1/2}\rho(y)^2, \tag{4.3.20}$$

otherwise we would have $-2 \times 6^{-1/2}a(y)^{1/2} \leq a''(y)T^2 \leq -2^{-1/2}\rho(y)^2$ and thus

$$a(y)^{1/2} \geq 6^{1/2}2^{-3/2}\rho(y)^2 \implies a(y) \geq \frac{3}{4}\rho(y)^4$$

which is incompatible with $a(y) < \rho(y)^4/2$. Using the Taylor expansion for $x \mapsto a''(x)T^2$ yields the following; we write, for $|x - y| \leq \rho(y)s$,

$$a''(x)T^2 \geq a''(y)T^2 - |s|\rho(y)4\rho(y) - \frac{s^2}{2}\rho(y)^2$$

$$\geq \rho(y)^2\left(\frac{1}{\sqrt{2}} - 12|s| - 3s^2/2\right) \geq \rho(y)^2/2,$$

provided $|s| \leq 10^{-2}$.   □

Lemma 4.3.17. *Let $a, \rho, \Omega, C_0, r_0, R_0, T$ be as above. There exists a positive constant θ_0 such that if $0 < \theta \leq \theta_0$ and $y \in \Omega$ is such that $a(y) < \theta\rho(y)^4$, the following property is true. For all x such that $|x - y| \leq \theta^{1/2}\rho(y)$, the function $\tau \mapsto a'(x + \tau\rho(y)T)T$ has a unique zero on the interval $[-\theta^{1/4}, \theta^{1/4}]$. The constant θ_0 is a universal constant that will be chosen also $\leq \min(1/2, r_0^2, R_0^4)$.*

Proof. From the previous lemma, we know that for $y \in \Omega$ such that $a(y) < \rho(y)^4/2$, there exists a unit vector T such that,

$$|x - y| \leq R_0\rho(y) \implies a''(x)T^2 \geq 2^{-1}\rho(y)^2.$$

The second-order Taylor's formula gives, for $|t| \leq r_0$, using (4.3.16),

$$0 \leq a(y + t\rho(y)T) \leq a(y) + t\rho(y)a'(y)T + \frac{\rho(y)^2 t^2}{2} C_0^2 \rho(y)^2$$

and thus $|t|\rho(y)|a'(y)T| \leq a(y) + C_0^2 \rho(y)^4 t^2/2 \leq \theta\rho(y)^4 + C_0^2\rho(y)^4 t^2/2$. As a result choosing $t = \theta^{1/2}$ (which is indeed smaller than r_0), we get

$$|a'(y)T| \leq \rho(y)^3(\theta^{1/2} + C_0^2 \theta^{1/2}/2). \qquad (4.3.21)$$

We have for s real

$$a'(y + s\rho(y)T)T = a'(y)T + s\rho(y)a''(y)T^2 + \frac{s^2}{2}\rho(y)^2 a'''(y)T^3$$
$$+ \int_0^1 \frac{1}{2}(1 - t)^2 a^{(4)}(y + ts\rho(y)T)T^4 dt s^3 \rho(y)^3,$$

so that, using (4.3.21), we have

$$a'(y + s\rho(y)T)T \leq \rho(y)^3\theta^{1/2}\overbrace{(1 + C_0^2/2)}^{=C_1} + s\rho(y)a''(y)T^2 + \frac{s^2}{2}\rho(y)^3 4 + \frac{1}{6}|s|^3\rho(y)^3$$
$$\leq \rho(y)^3\left(\theta^{1/2}C_1 + s\frac{a''(y)T^2}{\rho(y)^2} + 2s^2 + \frac{|s|^3}{6}\right).$$

The coefficient of s inside the bracket above belongs to the interval $[2^{-1/2}, 1]$. For $s = -\theta^{1/4}$, we get that

$$a'(y - \theta^{1/4}\rho(y)T)T \leq \rho(y)^3\left(\theta^{1/2}C_1 - \theta^{1/4}2^{-1/2} + 2\theta^{1/2} + \frac{|\theta|^{3/4}}{6}\right) < 0,$$

if θ is small enough with respect to a universal constant. Since we have also the inequality

$$a'(y + s\rho(y)T)T \geq -\rho(y)^3\theta^{1/2}C_1 + s\rho(y)a''(y)T^2 - \frac{s^2}{2}\rho(y)^3 4 - \frac{1}{6}|s|^3\rho(y)^3$$
$$\geq \rho(y)^3\left(-\theta^{1/2}C_1 + s\frac{a''(y)T^2}{\rho(y)^2} - 2s^2 - \frac{|s|^3}{6}\right),$$

the choice $s = \theta^{1/4}$ shows that $a'(y + \theta^{1/4}\rho(y)T)T > 0$. As a result the function ϕ defined by $\phi(\tau) = a'(y + \tau\rho(y)T)T$ vanishes for some τ with $|\tau| \leq \theta^{1/4} \leq R_0$. Moreover, from Lemma 4.3.16, its derivative ϕ' satisfies

$$\phi'(\tau) = a''(y + \tau\rho(y)T)T^2\rho(y) \geq 2^{-1}\rho(y)^3 > 0,$$

so that ϕ is monotone increasing in τ on the interval $[-\theta^{1/4}, \theta^{1/4}]$, with a unique zero on that interval. Considering now for $|y - x| \leq \theta^{1/2}\rho(y)$ the function

$$\psi(\tau, x) = a'(x + \tau\rho(y)T)T,$$

we get that

$$\phi(\tau) - \theta^{1/2}\rho(y)C_0^2\rho(y)^2 \le \psi(\tau,x) \le \phi(\tau) + \theta^{1/2}\rho(y)C_0^2\rho(y)^2$$

so that with the same reasoning as before, we find that for all x such that $|x - y| \le \theta\rho(y)$, the function $\tau \mapsto a'(x + \tau\rho(y)T)T$ has a unique zero on the interval $[-\theta^{1/4}, \theta^{1/4}]$, provided that θ is smaller than a positive universal constant. $\qquad\square$

Remark 4.3.18. Let $a, \rho, \Omega, r_0, C_0, R_0, \theta_0$ be as in Lemma 4.3.17 and $0 < \theta \le \theta_0$. Let y be a point in Ω such that $a(y) < \theta\rho(y)^4$. We may choose the linear orthonormal coordinates such that the vector T given by Lemma 4.3.17 is the first vector of the canonical basis of $\mathbb{R}^m$. Then a consequence of Lemma 4.3.17 is that, for all $x' \in B_{\mathbb{R}^{m-1}}(y', \theta^{1/2}\rho(y))$ the map $\tau \mapsto \partial_1 a(\tau, x')$ has a unique zero $\alpha(x')$ on the interval $[-\theta^{1/4}\rho(y) + y_1, \theta^{1/4}\rho(y) + y_1]$. We have thus

$$|x' - y'| \le \theta^{1/2}\rho(y) \implies \partial_1 a(\alpha(x'), x') \equiv 0, \quad |\alpha(x') - y_1| \le \theta^{1/4}\rho(y). \quad (4.3.22)$$

From Lemma 4.3.16, we get also

$$\partial_1^2 a(\alpha(x'), x') \ge \rho(y)^2/2. \quad (4.3.23)$$

Since the function $\partial_1 a$ is $C^{2,1}$, the implicit function theorem entails that the function α is C^2; let us show in fact that α is $C^{2,1}$. Denoting by ∂_2 the x' derivative, with a and its derivatives always evaluated at $x_1 = \alpha(x')$, we obtain by differentiating the identity $\partial_1 a(\alpha(x'), x') \equiv 0$,

$$\alpha' \partial_1^2 a + \partial_1 \partial_2 a = 0, \quad (4.3.24)$$

$$\alpha'' \partial_1^2 a + \alpha'^2 \partial_1^3 a + 2\alpha' \partial_1^2 \partial_2 a + \partial_1 \partial_2^2 a = 0. \quad (4.3.25)$$

The identities (4.3.24) – (4.3.25) give for $|x' - y'| \le \theta^{1/2}\rho(y)$, using (4.3.23),

$$\begin{cases} |\alpha(x') - y_1| \le \theta^{1/4}\rho(y), \\ |\alpha'(x')| \le 2\rho(y)^{-2}\rho(\alpha(x'), x')^2 \le 2C_0^2 \lesssim 1, \\ |\alpha''(x')| \le 2\rho(y)^{-2}\big(4^2C_0^4 + 4^2C_0^2 + 12\big)\rho(\alpha(x'), x') \lesssim \rho(y)^{-1}. \end{cases} \quad (4.3.26)$$

We have also the identity, using (4.3.25),

$$\alpha''(x') = -(\partial_1^2 a(\alpha(x'), x'))^{-1}\Big(\alpha'^2 \partial_1^3 a(\alpha(x'), x') + 2\alpha' \partial_1^2 \partial_2 a(\alpha(x'), x')$$
$$+ \partial_1 \partial_2^2 a(\alpha(x'), x')\Big), \quad (4.3.27)$$

so that the function α'' is Lipschitz-continuous. Applying formally the chain rule from (4.3.25) would give the identity

$$\alpha''' \partial_1^2 a + 3\alpha'' \alpha' \partial_1^3 a + 3\alpha'' \partial_1^2 \partial_2 a + \alpha'^3 \partial_1^4 a + 3\alpha'^2 \partial_1^3 \partial_2 a + 3\alpha' \partial_1^2 \partial_2^2 a + \partial_1 \partial_2^3 a = 0.$$

However the meaning of the last four terms above is not clear since the fourth derivative of a is only L^∞, so to restrict it to the hypersurface $x_1 = \alpha(x')$ does not make sense. In fact, we do not need that, but only the fact that the composition of Lipschitz-continuous functions gives a Lipschitz-continuous function with the obvious bound on the Lipschitz constant. We start over from (4.3.27) and we write the duality products with a smooth compactly supported test function χ, a_ϵ a regularized a,

$$\langle a''', \chi \rangle = -\int a'' \chi' dm = \int \chi'(a'^2 \partial_1^3 a + 2\alpha' \partial_1^2 \partial_2 a + \partial_1 \partial_2^2 a)(\partial_1^2 a)^{-1} dm$$

$$= \lim_{\epsilon \to 0} \int \chi'(\alpha'^2 \partial_1^3 a_\epsilon + 2\alpha' \partial_1^2 \partial_2 a_\epsilon + \partial_1 \partial_2^2 a_\epsilon)(\partial_1^2 a)^{-1} dm$$

$$= -\lim_{\epsilon \to 0} \int \chi \Big((\alpha'^2 \partial_1^3 a_\epsilon + 2\alpha' \partial_1^2 \partial_2 a_\epsilon + \partial_1 \partial_2^2 a_\epsilon)(\partial_1^2 a)^{-1}\Big)' dm.$$

The computation of the derivative between the parenthesis above, with uniform bounds with respect to ϵ, gives indeed

$$|a'''(x')| \lesssim \rho(y)^{-2}. \tag{4.3.28}$$

4.3.6 From discrete sums to finite sums

At the end of the proof of Theorem 2.5.19, we established that

$$a(x) = \sum_{1 \le j \le 1 + N_{m-1}} \sum_{\nu \in \mathbb{N}} b_{\nu,j}(x)^2 \varphi_\nu(x)^2 \tag{4.3.29}$$

with (φ_ν) satisfying the properties of Lemma 2.5.20 and the $b_{\nu,j}$ are $C^{1,1}$ functions such that

$$|b_{\nu,j}^{(l)}| \le c_0 \rho_\nu^{2-l}, \quad 0 \le l \le 2, \quad |(b_{\nu,j}' b_{\nu,j}'')'| \le c_0 \quad |(b_{\nu,j} b_{\nu,j}'')''| \le c_0, \tag{4.3.30}$$

where c_0 is a universal constant. We keep the normalization assumption

$$\|a^{(4)}\|_{L^\infty(\mathbb{R}^m)} \le 1.$$

We want to write a as a finite sum with similar properties, using the slow variation of the metric $|dx|^2/\rho(x)^2$. We are given a positive number $r \le r_0'$, where r_0' is defined in Lemma 2.5.20. We define a sequence (x_ν) and balls U_ν as in that lemma.

- $\mathcal{N}_1 =$ maximal subset of $\mathbb{N}$ containing 0 such that for $\nu' \ne \nu''$ both in $\mathcal{N}_1$, $U_{\nu'} \cap U_{\nu''} = \emptyset$. Let $\nu_2 = \min \mathcal{N}_1^c$.

- $\mathcal{N}_2 =$ maximal subset of $\mathcal{N}_1^c$ containing ν_2 such that for $\nu' \ne \nu''$ both in $\mathcal{N}_2$, $U_{\nu'} \cap U_{\nu''} = \emptyset$. Let $\nu_3 = \min(\mathcal{N}_1 \cup \mathcal{N}_2)^c$.

- ... Let $\nu_{k+1} = \min(\mathcal{N}_1 \cup \cdots \cup \mathcal{N}_k)^c$.

- $\mathcal{N}_{k+1}$ = maximal subset of $(\mathcal{N}_1 \cup \cdots \cup \mathcal{N}_k)^c$ containing ν_{k+1} such that for $\nu' \neq \nu''$ both in $\mathcal{N}_{k+1}$, $U_{\nu'} \cap U_{\nu''} = \emptyset$. Let $\nu_{k+2} = \min(\mathcal{N}_1 \cup \cdots \cup \mathcal{N}_{k+1})^c$.

- $\ldots$

We observe the following.

- The sets $\mathcal{N}_j$ are two-by-two disjoint.

- For all j, k such that $1 \leq j \leq k$, there exists $\nu \in \mathcal{N}_j$ so that $U_\nu \cap U_{\nu_{k+1}} \neq \emptyset$: otherwise, we could find $1 \leq j \leq k$ so that for all $\nu \in \mathcal{N}_j$, $U_\nu \cap U_{\nu_{k+1}} = \emptyset$, so that the set $\mathcal{N}_j \cup \{\nu_{k+1}\}$ would satisfy the property that the maximal $\mathcal{N}_j$ should satisfy.

- For k large enough, we have $\mathcal{N}_1 \cup \cdots \cup \mathcal{N}_k = \mathbb{N}$: otherwise ν_{k+1} is always well-defined and using the property above, we get that one can find $\mu_j \in \mathcal{N}_j, 1 \leq j \leq k$, so that $U_{\mu_j} \cap U_{\nu_{k+1}} \neq \emptyset$. As a consequence, for $1 \leq j \leq k$, we find $y_j \in U_{\mu_j}$ such that $|x_{\mu_j} - y_j| \leq r\rho(x_{\mu_j}) \leq C_0 r\rho(y_j)$,

$$|x_{\nu_{k+1}} - y_j| \leq r\rho(x_{\nu_{k+1}}) \leq C_0 r\rho(y_j) \leq C_0^2 r\rho(x_{\nu_{k+1}})$$

and thus

$$|x_{\nu_{k+1}} - x_{\mu_j}| \leq (C_0^2 r + r)\rho(x_{\nu_{k+1}}), \qquad (4.3.31)$$

with distinct μ_j (they belong to two-by-two disjoint sets). On the other hand, we know by construction (see Lemma 1.4.9 in [71]) that there exists a positive r_1 such that , for $\nu' \neq \nu''$,

$$|x_{\nu'} - x_{\nu''}| \geq r_1 \rho(x_{\nu'}),$$

so that, with a fixed $r_2 > 0$, the balls $\big(B(x_{\mu_j}, r_2\rho(x_{\mu_j}))\big)_{1 \leq j \leq k}$ are two-by-two disjoint as well as $\big(B(x_{\mu_j}, r_3\rho(x_{\nu_{k+1}}))\big)_{1 \leq j \leq k}$ with a fixed positive r_3. Thanks to (4.3.31), they are also all included in $B(x_{\nu_{k+1}}, r_4\rho(x_{\nu_{k+1}}))$ with a fixed positive r_4 so that $k \leq r_4^m/r_3^m$ and thus k is bounded. We can thus write, with $M_m = \lambda_0^m$, since the balls $U_\nu (\supset \operatorname{supp} \varphi_\nu)$ are two-by-two disjoint for ν running in each $\mathcal{N}_k$,

$$a = \sum_{1 \leq j \leq 1+N_{m-1}} \sum_{1 \leq k \leq M_m} \left(\sum_{\nu \in \mathcal{N}_k} b_{\nu,j} \varphi_\nu \right)^2$$

and defining $B_{j,k} = \sum_{\nu \in \mathcal{N}_k} b_{\nu,j} \varphi_\nu$ we get

$$a = \sum_{1 \leq j \leq 1+N_{m-1}} \sum_{1 \leq k \leq M_m} B_{j,k}^2 \qquad (4.3.32)$$

with $|B_{j,k}''| \leq \sum_{\nu \in \mathcal{N}_k} c_0 \psi_\nu \lesssim 1$. Moreover the identities

$$(B_{j,k}' B_{j,k}'')' = \sum_{\nu \in \mathcal{N}_k} \big((b_{\nu,j}\varphi_\nu)'(b_{\nu,j}\varphi_\nu)''\big)' \psi_\nu, \qquad (4.3.33)$$

$$(B_{j,k} B_{j,k}'')'' = \sum_{\nu \in \mathcal{N}_k} \big((b_{\nu,j}\varphi_\nu)(b_{\nu,j}\varphi_\nu)''\big)'' \psi_\nu \qquad (4.3.34)$$

yield the sought estimates on the derivatives. As a final question, one may ask for some estimate of the Pythagorean number, i.e., the number of squares necessary for the decomposition. From the formula (4.3.32), we have the estimate

$$N_m \leq (1 + N_{m-1})\lambda_0^m, \quad \lambda_0 \text{ universal constant,}$$

which gives $N_m \leq \mu_0^{m^2}$, which is probably a very crude estimate, compared to the exponential bound known for the Artin theorem of decomposition as a sum of squares of non-negative rational fractions. As a matter of fact, a recent paper of J.-M. Bony [17] provides the equality $N_1 = 2$, which is optimal in view of Glaeser's counterexample ([48]); however his proof is much more involved than our argument as exposed above with our set of indices $\mathcal{N}_k$.

4.3.7 On families of rapidly decreasing functions

Lemma 4.3.19 ([18]). *Let $(f_k)_{k \geq 1}$ be a sequence of functions from $\mathbb{R}_+$ to itself such that*

$$\forall k, N \in \mathbb{N}^*, \quad \sup_{t \geq 0}\big(f_k(t)(1+t)^N\big) = \alpha_{k,N} < +\infty. \tag{4.3.35}$$

Then there exists some positive constants $\beta_N, \beta_{k,N}$, a decreasing function F from $\mathbb{R}_+$ to $\mathbb{R}_+^*$, depending only on the sequence $(\alpha_{k,N})$, such that

$$\forall N \in \mathbb{N}, \forall t \geq 0, \quad F(t) \leq \beta_N(1+t)^{-N}, \tag{4.3.36}$$

$$\forall k, N \in \mathbb{N}^*, \forall t \geq 0, \quad f_k(t) \leq \beta_{k,N}F(t)^N. \tag{4.3.37}$$

Proof. Defining for $t \geq 0$, $g_k(t) = \inf_N \alpha_{k,N}(1+t)^{-N}$, we have $f_k(t) \leq g_k(t)$. We have for $m, p \in \mathbb{N}^*$, $g_k(t)(1+t)^{mp} \leq (1+t)^{mp-N}\alpha_{k,N} \leq (1+t)^{-1}\alpha_{k,mp+1}$ so that

$$t \geq \alpha_{k,mp+1} \implies g_k(t)(1+t)^{mp} \leq 1, \quad \text{i.e., } g_k(t)^{1/m}(1+t)^p \leq 1,$$

and thus $t \geq \alpha_p = \max_{k+m \leq p} \alpha_{k,mp+1}$ implies that $\forall k, m \in \mathbb{N}^*$ with $k + m \leq p$, $g_k(t)^{1/m}(1+t)^p \leq 1$, so that $\sup_{t \geq \alpha_p, k+m \leq p} g_k(t)^{1/m}(1+t)^p \leq 1$. We set for $q \geq 1$, $T_q = q + \max_{1 \leq p \leq q} \alpha_p$. The sequence $(T_q)_{q \geq 1}$ is increasing and tends to $+\infty$. We define (with $T_0 = 0$),

$$F(t) = \sum_{q \geq 0} \mathbf{1}_{[T_q, T_{q+1}[}(t)(1+t)^{-q}.$$

The function F is positive-decreasing and since

$$F(t)(1+t)^N \leq \sum_{0 \leq q < N} \mathbf{1}_{[T_q, T_{q+1}[}(t)(1+t)^{N-q} + \sum_{N \leq q} \mathbf{1}_{[T_q, T_{q+1}[}(t),$$

we have $\sup_{t\geq 0} F(t)(1+t)^N = \beta_N < +\infty$, proving (4.3.36). Moreover, for $k, m \in \mathbb{N}^*$, we have for all $t \geq 0$,

$$g_k(t)^{1/m} F(t)^{-1}$$

$$= \sum_{0\leq q<k+m} g_k(t)^{1/m}(1+t)^q \mathbf{1}_{[T_q,T_{q+1}[}(t) + \sum_{q\geq k+m} g_k(t)^{1/m}(1+t)^q \mathbf{1}_{[T_q,T_{q+1}[}(t)$$

$$\leq \sum_{0\leq q<k+m} g_k(t)^{1/m}(1+t)^q \mathbf{1}_{[T_q,T_{q+1}[}(t) + \sum_{q\geq k+m} \mathbf{1}_{[T_q,T_{q+1}[}(t)\mathbf{1}\{t \geq \alpha_q\}$$

$$+ \sum_{q\geq k+m} g_k(t)^{1/m}(1+t)^q \mathbf{1}_{[T_q,T_{q+1}[}(t)\mathbf{1}\{t < \alpha_q\}.$$

Since $\alpha_q \leq T_q$, the product $\mathbf{1}_{[T_q,T_{q+1}[}(t)\mathbf{1}\{t < \alpha_q\} = 0$ implying

$$g_k(t)^{1/m} F(t)^{-1} \leq \sup_{[0,T_{k+m}]} g_k(t)^{1/m}(1+t)^{k+m-1} + 1 = \widetilde{\beta}_{k,m},$$

proving (4.3.37), $f_k(t) \leq g_k(t) \leq \beta_{k,m} F(t)^m$.   □

Lemma 4.3.20. *Let γ be a positive-definite quadratic form on $\mathbb{R}^n$, let $g = \gamma \oplus \gamma$ be the positive-definite quadratic form on $\mathbb{R}^n \times \mathbb{R}^n$ defined by $g(x,\xi) = \gamma(x) + \gamma(\xi)$, let U_r (resp.$\mathbf{b}_r$) be the the g-ball (resp. γ-ball) with center 0 and radius r, and let $A \in \mathrm{Conf}(g, U_r)$ with $r \leq 1$ (cf. Definition 2.3.1). Then for all $\epsilon > 0$, there exists $b \in \mathrm{Conf}(g, U_r)$ and a smooth function a verifying*

$$\forall k, N \in \mathbb{N}, \quad \sup_{\gamma(t)\leq 1} |a^{(k)}(x)t^k|\big(1 + \gamma^{-1}(x - \mathbf{b}_{r+\epsilon})\big)^N < +\infty, \qquad (4.3.38)$$

such that $A(x,\xi) = a(x)b(x,\xi)$. The semi-norms of b in $\mathrm{Conf}(g, U_r)$ as well as the quantities (4.3.38) depend only on the semi-norms of A in $\mathrm{Conf}(g, U_r)$.

Proof. According to (4.4.18), we have $g^\sigma = \gamma^{-1} \oplus \gamma^{-1}$. We have

$$\sup_{T,g(T)\leq 1} |A^{(k)}(X)T^k| \leq \|A\|_{g,U_r}^{(k,2N)}\big(1 + g^\sigma(X - U_r)\big)^{-N},$$

so that defining for $t \in \mathbb{R}_+$,

$$f_k(t) = \inf_{N\geq 0} \|A\|_{g,U_r}^{(k,2N)}(1+t)^{-N},$$

we see that the hypothesis (4.3.35) is fulfilled with $\alpha_{k,N} = \|A\|_{g,U_r}^{(k,2N)}$. According to Lemma 4.3.19, there exists a decreasing positive function F defined on $\mathbb{R}_+$ such that $\forall N \geq 0$, $\beta_N = \sup_{t\geq 0}(1+t)^N F(t) < +\infty$ and

$$\sup_{T,g(T)\leq 1} |A^{(k)}(X)T^k| \leq f_k\big(g^\sigma(X - U_r)\big) \leq \beta_{k,N} F\big(g^\sigma(X - U_r)\big)^N.$$

The function F and the constants $\beta_N, \beta_{k,N}$ depend only on the sequence $\|A\|_{g,U_r}^{(k,2N)}$. We define for $\epsilon > 0$, $x \in \mathbb{R}^n$, $\rho \in C_c^\infty(\mathbb{R}; \mathbb{R}_+)$, supp $\rho = [-1, 1]$, $\int_{\mathbb{R}^n} \rho(|z|^2)dz = 1$,

$$a(x) = \int F\big(\gamma^{-1}(x - y - \mathbf{b}_{r+\epsilon})\big)|\gamma|^{1/2}\epsilon^{-n}\rho\big(\gamma(\epsilon^{-1}y)\big)dy.$$

We note that the triangle inequality for the norm $\|x\|_\gamma = \gamma(x)^{1/2}$ implies that

$$\forall y \in \mathbf{b}_\epsilon, \mathbf{b}_{r+\epsilon} + y \supset \mathbf{b}_r,$$

since $\|s\|_\gamma \leq r, \|y\|_\gamma \leq \epsilon$, gives $s = y + s - y$ and $\|s - y\|_\gamma \leq r + \epsilon$. As a result $\forall y \in \mathbf{b}_\epsilon, \|x - y - \mathbf{b}_{r+\epsilon}\|_{\gamma^{-1}} \leq \|x - \mathbf{b}_r\|_{\gamma^{-1}}$ and since F is decreasing (and $\rho \geq 0$), we obtain

$$a(x) \geq F\big(\|x - \mathbf{b}_r\|_{\gamma^{-1}}^2\big) \int |\gamma|^{1/2}\epsilon^{-n}\rho\big(\gamma(\epsilon^{-1}y)\big)dy = F\big(\gamma^{-1}(x - \mathbf{b}_r)\big) > 0. \quad (4.3.39)$$

On the other hand, we have $\forall y \in \mathbf{b}_\epsilon, \mathbf{b}_{r+\epsilon} + y \subset \mathbf{b}_{r+2\epsilon}$ and thus

$$\forall y \in \mathbf{b}_\epsilon, \quad \|x - y - \mathbf{b}_{r+\epsilon}\|_{\gamma^{-1}} \geq \|x - \mathbf{b}_{r+2\epsilon}\|_{\gamma^{-1}},$$

entailing that $a(x) \leq F\big(\gamma^{-1}(x - \mathbf{b}_{r+2\epsilon})\big)$. For $\gamma(t) \leq 1$, we have also

$$|a^{(k)}(x)t^k| \leq \int_{\|y\|_\gamma \leq \epsilon} F\big(\gamma^{-1}(x - y - \mathbf{b}_{r+\epsilon})\big)|\gamma|^{1/2}\epsilon^{-n}dy C_k(\rho)\epsilon^{-k}$$

$$\leq C_k(\rho, n)\epsilon^{-k}F\big(\gamma^{-1}(x - \mathbf{b}_{r+2\epsilon})\big), \quad (4.3.40)$$

proving (4.3.38) (with 2ϵ replacing ϵ). We consider now the function b defined on $\mathbb{R}^n \times \mathbb{R}^n$ by $b(x, \xi) = A(x, \xi)/a(x)$, which makes sense from (4.3.39). For $t \in \mathbb{R}^n$, we have (see the Faà de Bruno formula in (4.3.1))

$$(\partial_x^k b)(x, \xi)t^k = \sum_{\substack{k_1+k_2=k \\ 0 \leq l \leq k_2, \sum m_j = k_2}} C(k_1, k_2, l)(\partial_x^{k_1} A)(x, \xi)t^{k_1} a(x)^{-1-l}a^{(m_1)}t^{m_1} \ldots a^{(m_l)}t^{m_l},$$

so that, for $(t, \tau) \in \mathbb{R}^n \times \mathbb{R}^n$, with $\gamma(t), \gamma(\tau) \leq 1$, we have with $X = (x, \xi)$,

$$|(\partial_\xi^j \partial_x^k b)(x, \xi)t^k \tau^j|$$

$$\leq C(j, k, \epsilon)\beta_{j+k,k+2}$$

$$\times \sum_{l \leq k} F\big(g^\sigma(X - U_r)\big)^{k+2}F\big(\gamma^{-1}(x - \mathbf{b}_r)\big)^{-1-l}F\big(\gamma^{-1}(x - \mathbf{b}_{r+2\epsilon})\big)^l$$

$$\leq C(j, k, \epsilon)\beta_{j+k,k+2}$$

$$\times \sum_{l \leq k} F\big(\gamma^{-1}(x - \mathbf{b}_r)\big)^{k+1-1-l}F\big(\gamma^{-1}(x - \mathbf{b}_{r+2\epsilon})\big)^l F\big(g^\sigma(X - U_r)\big)$$

$$\leq C'(j, k, \epsilon)\beta_{j+k,k+2}F\big(g^\sigma(X - U_r)\big)$$

and b belongs to $\mathrm{Conf}(g, U_r)$ with semi-norms depending only on those of A. We have thus $A(x, \xi) = a(x)b(x, \xi)$ and the lemma is proven. $\qquad\square$

4.3.8   Abstract lemma for the propagation of singularities

We want to prove here an elementary lemma in a Hilbertian framework, hopefully helping the reader to understand the theorem of propagation of singularities in a simple setting as well as providing a simple expanation for the orientation of the propagation by the sign of the imaginary part. Let $\mathbb{H}$ be a complex Hilbert space, I an interval of $\mathbb{R}$ and $I \ni t \mapsto R(t) \in \mathscr{L}(\mathbb{H})$ be a continuous mapping. We shall define

$$\operatorname{Re} R(t) = \frac{1}{2}\big(R(t) + R^*(t)\big), \qquad \operatorname{Im} R(t) = \frac{1}{2i}\big(R(t) - R^*(t)\big), \qquad (4.3.41)$$

and assume that $\inf_{t \in I} \operatorname{Im} R(t) > -\infty$, i.e.,

$$\exists \mu_0 \in \mathbb{R}, \quad \operatorname{Im} R(t) + \mu_0 \geq 0. \qquad (4.3.42)$$

We calculate for $u \in C_c^1(I, \mathbb{H})$, $t_0 < t_1$ in I,

$$2 \operatorname{Re}\langle \frac{1}{i}\partial_t u + R(t)u(t), i\mathbf{1}_{[t_0,t_1]}(t)u(t)\rangle_{L^2(I;\mathbb{H})}$$

$$= |u(t_0)|_{\mathbb{H}}^2 - |u(t_1)|_{\mathbb{H}}^2 + 2 \int_{t_0}^{t_1} \langle \operatorname{Im} R(t)u(t), u(t)\rangle_{\mathbb{H}} dt$$

$$\geq |u(t_0)|_{\mathbb{H}}^2 - |u(t_1)|_{\mathbb{H}}^2 - 2\mu_0 \int_{t_0}^{t_1} |u(t)|_{\mathbb{H}}^2 dt,$$

proving with $L = \frac{1}{i}\partial_t + R(t)$ that

$$|u(t_0)|_{\mathbb{H}}^2 \leq |u(t_1)|_{\mathbb{H}}^2 + 2 \operatorname{Re}\langle Lu, iu\rangle_{L^2([t_0,t_1];\mathbb{H})} + 2\mu_0 \int_{t_0}^{t_1} |u(t)|_{\mathbb{H}}^2 dt. \qquad (4.3.43)$$

We claim that the previous inequality contains all the qualitative information for a propagation-of-singularities theorem: assuming for simplicity that $\mu_0 = 0$ (which means $\operatorname{Im} R(t) \geq 0$), and $Lu = 0$, we see that $|u(t_0)|_{\mathbb{H}}^2 \leq |u(t_1)|_{\mathbb{H}}^2$ whatever are $t_0 \leq t_1$ in I. In particular if $|u(t_0)|_{\mathbb{H}}^2 \gg 1$ (some type of singular behaviour) then $|u(t_1)|_{\mathbb{H}}^2 \gg 1$, meaning that this "singular behaviour" did actually propagate forward in time. Note that the "regularity" propagates backward in time: if $|u(t_1)|_{\mathbb{H}}^2 \ll 1$ then $|u(t_0)|_{\mathbb{H}}^2 \ll 1$. If we change the hypothesis (4.3.42) into

$$\operatorname{Im} R(t) + \mu_0 \leq 0, \qquad (4.3.44)$$

the calculation above gives

$$|u(t_1)|_{\mathbb{H}}^2 \leq |u(t_0)|_{\mathbb{H}}^2 - 2 \operatorname{Re}\langle Lu, iu\rangle_{L^2([t_0,t_1];\mathbb{H})} - 2\mu_0 \int_{t_0}^{t_1} |u(t)|_{\mathbb{H}}^2 dt.$$

Obviously the "propagation of singularities" is now backward in time. Finally we see that we must expect that the sign of the imaginary part is actually orienting the direction of propagation, essentially forward for a positive imaginary part and backward for a negative imaginary part, a feature rather easy to memorize. The inequality (4.3.43) is simple to state and to prove, but we want to prove now a sharper version of it.

Lemma 4.3.21. *Let $\mathbb{H}$ be a Hilbert space, I an interval of $\mathbb{R}$ and $I \ni t \mapsto R(t) \in \mathscr{L}(\mathbb{H})$ be a continuous mapping such that (4.3.42) holds. Then for all $t_0 \le t_1$ in I, we have with $L = \frac{d}{idt} + R(t)$,*

$$\sup_{t_0 \le s \le t_1} |u(s)|_{\mathbb{H}} e^{-\mu_0(t_1-s)} \le |u(t_1)|_{\mathbb{H}} + \int_{t_0}^{t_1} |e^{-\mu_0(t_1-t)} Lu(t)|_{\mathbb{H}} dt. \qquad (4.3.45)$$

Proof. We calculate for $u \in C_c^1(I, \mathbb{H})$, $t_0 < t_1$ in I, $v(t) = u(t)e^{\mu_0 t}$,

$$2\operatorname{Re}\langle \frac{1}{i}\partial_t u + R(t)u(t), i\mathbf{1}_{[t_0,t_1]}(t)e^{2\mu_0 t}u(t)\rangle_{L^2(I;\mathbb{H})}$$

$$= 2\operatorname{Re}\langle \frac{1}{i}\partial_t v + (R(t) + i\mu_0)v(t), i\mathbf{1}_{[t_0,t_1]}(t)v(t)\rangle_{L^2(I;\mathbb{H})}$$

and since $\mu_0 + \operatorname{Im} R(t) \ge 0$, we get, using (4.3.42) with $\mu_0 = 0$ and (4.3.43)

$$|e^{\mu_0 t_0}u(t_0)|_{\mathbb{H}}^2 \le |e^{\mu_0 t_1}u(t_1)|_{\mathbb{H}}^2 + 2\operatorname{Re}\langle e^{\mu_0 t} Lu, ie^{\mu_0 t}u(t)\rangle_{L^2([t_0,t_1];\mathbb{H})},$$

and thus with

$$\sigma(t) = 2\int_t^{t_1} |e^{\mu_0(s-t_1)} Lu(s)|_{\mathbb{H}} |e^{\mu_0(s-t_1)}u(s)|_{\mathbb{H}} ds + |u(t_1)|_{\mathbb{H}}^2$$

we get $-\dot{\sigma}(t) = 2|e^{\mu_0(t-t_1)} Lu(t)|_{\mathbb{H}} |e^{\mu_0(t-t_1)}u(t)|_{\mathbb{H}} \le 2|e^{\mu_0(t-t_1)} Lu(t)|_{\mathbb{H}} \sqrt{\sigma(t)}$ so that

$$\frac{d}{dt}\left(-\sqrt{\sigma(t)}\right) \le |e^{\mu_0(t-t_1)} Lu(t)|_{\mathbb{H}}$$

$$\implies \text{for } t \le t_1, \quad \sigma(t)^{1/2} \le \sigma(t_1)^{1/2} + \int_t^{t_1} |e^{\mu_0(s-t_1)} Lu(s)|_{\mathbb{H}} ds.$$

As a result, we get

$$\sup_{t_0 \le s \le t_1} |u(s)|e^{-\mu_0(t_1-s)} \le \sup_{t_0 \le s \le t_1} \sigma(s)^{1/2} \le |u(t_1)| + \int_{t_0}^{t_1} |e^{-\mu_0(t_1-t)} Lu(t)|_{\mathbb{H}} dt.$$

$\square$

Remark 4.3.22. It may be useful to write the expression of that result for the evolution equation in the more familiar form $\frac{d}{dt} + Q(t)$, under the assumption

$$\mu_0 + \operatorname{Re} Q(t) \ge 0. \qquad (4.3.46)$$

In fact we have $e^{-\mu_0 t}\left(\frac{d}{dt} + Q(t)\right)e^{\mu_0 t} = i\left(\frac{d}{idt} + \operatorname{Im} Q(t) - i\operatorname{Re} Q(t) - i\mu_0\right)$ and using the hypothesis (4.3.44) and the same method as above, we get

$$\sup_{t_0 \le s \le t_1} |u(s)|e^{-\mu_0(s-t_0)} \le |u(t_0)| + \int_{t_0}^{t_1} |e^{-\mu_0(t-t_0)}\left(\frac{du}{dt} + Q(t)u(t)\right)|_{\mathbb{H}} dt. \qquad (4.3.47)$$

4.4   On the symplectic and metaplectic groups

4.4.1   The symplectic structure of the phase space

We are given a finite-dimensional real vector space E (dimension n) and we consider the phase space attached to E, namely the $2n$-dimensional $E \oplus E^* = \mathcal{P}$. On the latter, we introduce the symplectic form σ (which is a bilinear alternate form) defined by

$$\sigma(X, Y) = \langle \xi, y \rangle_{E^*, E} - \langle \eta, x \rangle_{E^*, E}, \quad \text{for } X = (x, \xi), Y = (y, \eta). \qquad (4.4.1)$$

We can identify σ with the isomorphism[7] of $\mathcal{P}$ to $\mathcal{P}^*$ such that $\sigma^* = -\sigma$, with the formula $\sigma(X, Y) = \langle \sigma X, Y \rangle_{\mathcal{P}^*, \mathcal{P}}$. A symplectic mapping Ξ is an endomorphism of $\mathcal{P}$ such that

$$\forall X, Y \in \mathcal{P}, \quad \sigma(\Xi X, \Xi Y) = \sigma(X, Y), \quad \text{i.e., such that } \Xi^* \sigma \Xi = \sigma. \qquad (4.4.2)$$

Note that the symplectic mappings make a subgroup of $Gl(E \oplus E^*)$: if Ξ_1, Ξ_2 are symplectic, then $\Xi_1 \Xi_2, \Xi_j^{-1}$ are also symplectic:

$$(\Xi_1 \Xi_2)^* \sigma \Xi_1 \Xi_2 = \Xi_2^* \Xi_1^* \sigma \Xi_1 \Xi_2 = \Xi_2^* \sigma \Xi_2 = \sigma,$$

$$\sigma(X, Y) = \sigma(\Xi_1 \Xi_1^{-1} X, \Xi_1 \Xi_1^{-1} Y) = \sigma(\Xi_1^{-1} X, \Xi_1^{-1} Y).$$

Definition 4.4.1. Let E be a finite-dimensional real vector space and σ be given by (4.4.1). The symplectic group $Sp(E \oplus E^*)$ is the group of symplectic mappings of the phase space $E \oplus E^*$. If $n = \dim E$, it is isomorphic to the symplectic group $Sp(2n)$, the group of $2n \times 2n$ matrices Ξ such that $\Xi^* \left(\begin{smallmatrix} 0 & I_n \\ -I_n & 0 \end{smallmatrix} \right) \Xi = \left(\begin{smallmatrix} 0 & I_n \\ -I_n & 0 \end{smallmatrix} \right)$.

It is interesting to see that this example of "polarized " symplectic space is in fact the general case, as shown by the following result.

Proposition 4.4.2. *Let F be a finite-dimensional real vector space equipped with a non-degenerate bilinear alternate form σ. Then the dimension of F is even (say $2n$) and F has a symplectic basis, i.e., a basis $(e_1, \ldots, e_n, \epsilon_1, \ldots, \epsilon_n)$ such that $\sigma(e_j, e_k) = \sigma(\epsilon_j, \epsilon_k) = 0, \sigma(\epsilon_j, e_k) = \delta_{j,k}$, and (F, σ) is thus isomorphic to $\mathbb{R}_x^n \times \mathbb{R}_\xi^n$ with the symplectic form $\xi \cdot y - \eta \cdot x$ (with the canonical dot product on $\mathbb{R}^n$).*

Proof. Let ϵ_1 be a non-zero vector in F (assumed to be not reduced to $\{0\}$) and let $V = \{x \in F, \sigma(\epsilon_1, x) = 0\}$. Since σ is non-degenerate alternate, V is a hyperplane which contains ϵ_1. It is thus possible to find $e_1 \notin V$ such that $\sigma(\epsilon_1, e_1) = 1$. We have

$$F = \mathbb{R} e_1 \oplus V, \quad V = \mathbb{R}\epsilon_1 \oplus F', \quad F' = \{x \in V, \sigma(x, e_1) = 0\},$$

since $e_1 \wedge \epsilon_1 \neq 0$. The restriction of σ to F' is non-degenerate since, if $x \in F'$ is such that $\forall y \in F', \sigma(x, y) = 0$, then $F' \subset V$ and its definition give $\sigma(x, \epsilon_1) = \sigma(x, e_1) = 0$ and thus $x = 0$. We can conclude by an induction on the dimension of F. $\qquad \square$

---

[7] $\sigma((x, \xi)) = (\xi, -x)$ is an isomorphism, identified with the $2n \times 2n$ matrix $\begin{pmatrix} 0 & I_n \\ -I_n & 0 \end{pmatrix}$.

Theorem 4.4.3. *Let E be a finite-dimensional real vector space and $Sp(E \oplus E^*)$ the symplectic group attached to the phase space $E \oplus E^*$ (see Definition 4.4.1). With obvious block notation for linear mappings from $E \oplus E^*$ to itself we have the following properties. The group $Sp(E \oplus E^*)$ is included in $Sl(E \oplus E^*)$ and generated by the following mappings:*

$$\begin{pmatrix} I_E & 0 \\ A & I_{E^*} \end{pmatrix}, \quad A : E \longrightarrow E^*, \quad A = A^*,$$

$$\begin{pmatrix} B^{-1} & 0 \\ 0 & B^* \end{pmatrix}, \quad B : E \longrightarrow E, \ invertible,$$

$$\begin{pmatrix} I_E & -C \\ 0 & I_{E^*} \end{pmatrix}, \quad C : E^* \longrightarrow E, \quad C = C^*.$$

For A, B, C as above, the mapping

$$\Xi_{A,B,C} = \begin{pmatrix} B^{-1} & -B^{-1}C \\ AB^{-1} & B^* - AB^{-1}C \end{pmatrix} = \begin{pmatrix} I_E & 0 \\ A & I_{E^*} \end{pmatrix} \begin{pmatrix} B^{-1} & 0 \\ 0 & B^* \end{pmatrix} \begin{pmatrix} I_E & -C \\ 0 & I_{E^*} \end{pmatrix}$$

belongs to $Sp(E \oplus E^*)$. Moreover, we define on $E \times E^*$ the generating function S of the symplectic mapping $\Xi_{A,B,C}$ by the identity

$$S(x, \eta) = \frac{1}{2}(\langle Ax, x \rangle + 2\langle Bx, \eta \rangle + \langle C\eta, \eta \rangle) \quad so \ that \quad \Xi\left(\frac{\partial S}{\partial \eta} \oplus \eta\right) = x \oplus \frac{\partial S}{\partial x}. \quad (4.4.3)$$

For a symplectic mapping Ξ, to be of the form above is equivalent to the assumption that the mapping $x \mapsto \pi_E \Xi(x \oplus 0)$ is invertible from E to E; moreover, if this mapping is not invertible, the symplectic mapping Ξ is the product of two mappings of the type $\Xi_{A,B,C}$.

Proof. The expression of Ξ above as well as (4.4.3) follow from a simple direct computation left to the reader. The inclusion of the symplectic group in the special linear group follows from the statement on the generators. We consider now Ξ in $Sp(E \oplus E^*)$: we have

$$\Xi = \begin{pmatrix} P & Q \\ R & S \end{pmatrix}, \quad where \begin{cases} P : E \to E, & Q : E^* \to E, \\ R : E \to E^*, & S : E^* \to E^*, \end{cases}$$

and the equation

$$\Xi^* \sigma \Xi = \sigma \tag{4.4.4}$$

is satisfied with $\sigma = \begin{pmatrix} 0 & I_{E^*} \\ -I_E & 0 \end{pmatrix}$, which means

$$P^* R = (P^* R)^*, \quad Q^* S = (Q^* S)^*, \quad P^* S - R^* Q = I_{E^*}. \tag{4.4.5}$$

We can note also that the mapping $\Xi \mapsto \Xi^*$ is an isomorphism of $Sp(E \oplus E^*)$ with $Sp(E^* \oplus E)$ since $\Xi \in Sp(E \oplus E^*)$ means

$$\Xi^* \sigma \Xi = \sigma \implies \Xi^{-1} \sigma^{-1} (\Xi^*)^{-1} = \sigma^{-1} \implies \Xi^{-1}(-\sigma^{-1})(\Xi^*)^{-1} = (-\sigma^{-1}),$$

and since $(-\sigma^{-1}) = \begin{pmatrix} 0 & I_E \\ -I_{E^*} & 0 \end{pmatrix}$, we get that $\Xi^* \in Sp(E^* \oplus E)$. As a result,

$\Xi = \begin{pmatrix} P & Q \\ R & S \end{pmatrix}$ belongs to $Sp(E \oplus E^*)$ is also equivalent to

$$PQ^* = (PQ^*)^*, \quad RS^* = (RS^*)^*, \quad PS^* - QR^* = I_E. \qquad (4.4.6)$$

Let us assume that the mapping P is invertible, which is the assumption in the last statement of the theorem. We define then the mappings A, B, C by

$$A = RP^{-1}, \quad B = P^{-1}, \quad C = -P^{-1}Q$$

so that we have $\quad A^* = P^{*-1}R^*PP^{-1} = P^{*-1}P^*RP^{-1} = RP^{-1} = A$

as well as $C^* = -Q^*P^{*-1} = -P^{-1}PQ^*P^{*-1} = -P^{-1}QP^*P^{*-1} = -P^{-1}Q = C$

and $\quad P = B^{-1}, \quad R = AB^{-1}, \quad Q = -B^{-1}C,$

$$S = P^{*-1}(I_{E^*} + R^*Q) = B^*(I_{E^*} - B^{*-1}A^*B^{-1}C) = B^* - AB^{-1}C.$$

We have thus proven that any symplectic matrix Ξ as above such that P is invertible is indeed given by the product appearing in Theorem 4.4.3. Let us now consider the case where $\det P = 0$; writing $E = \ker P \oplus N$ we have that P is an isomorphism from N onto $\operatorname{ran} P$. Let $B_1 \in Gl(E)$ such that $B_1 P$ is the identity on N. We have

$$\begin{pmatrix} B_1 & 0 \\ 0 & B_1^{*-1} \end{pmatrix} \begin{pmatrix} P & Q \\ R & S \end{pmatrix} = \begin{pmatrix} B_1 P & B_1 Q \\ B_1^{*-1}R & B_1^{*-1}S \end{pmatrix}. \qquad (4.4.7)$$

If $p = \dim(\ker P)$, we have for the $n \times n$ matrix $B_1 P$ the block decomposition

$$B_1 P = \begin{pmatrix} 0_{p,p} & 0_{p,n-p} \\ 0_{n-p,p} & I_{n-p} \end{pmatrix}, \qquad (4.4.8)$$

where $0_{r,s}$ stands for an $r \times s$ matrix with only 0 as an entry. On the other hand, we know from (4.4.5) that the mapping

$$(B_1 P)^* B_1^{*-1} R = P^* R$$

is symmetric. Writing $B_1^{*-1}R = \begin{pmatrix} \tilde{R}_{p,p} & \tilde{R}_{p,n-p} \\ \tilde{R}_{n-p,p} & \tilde{R}_{n-p,n-p} \end{pmatrix}$, where $\tilde{R}_{r,s}$ stands for an $r \times s$ matrix, this gives the symmetry of

$$\begin{pmatrix} 0_{p,p} & 0_{p,n-p} \\ 0_{n-p,p} & I_{n-p} \end{pmatrix} \begin{pmatrix} \tilde{R}_{p,p} & \tilde{R}_{p,n-p} \\ \tilde{R}_{n-p,p} & \tilde{R}_{n-p,n-p} \end{pmatrix} = \begin{pmatrix} 0_{p,p} & 0_{p,n-p} \\ \tilde{R}_{n-p,p} & \tilde{R}_{n-p,n-p} \end{pmatrix}$$

implying that $\tilde{R}_{n-p,p} = 0$. The symplectic matrix (4.4.7) is thus equal to

$$\begin{pmatrix} 0_{p,p} & 0_{p,n-p} & & B_1 Q \\ 0_{n-p,p} & I_{n-p} & & \\ \tilde{R}_{p,p} & \tilde{R}_{p,n-p} & & B_1^{*-1}S \\ 0_{n-p,p} & \tilde{R}_{n-p,n-p} & & \end{pmatrix}, \quad \text{where } B_1 Q \text{ and } B_1^{*-1}S \text{ are } n \times n \text{ blocks.}$$

The invertibility of (4.4.7) implies that $\tilde{R}_{p,p}$ is invertible. We consider now the $n \times n$ symmetric matrix

$$C = \begin{pmatrix} I_{p,p} & 0_{p,n-p} \\ 0_{n-p,p} & 0_{n-p,n-p} \end{pmatrix}$$

and the symplectic mapping

$$\begin{pmatrix} I_E & C \\ 0 & I_E \end{pmatrix} \begin{pmatrix} B_1 & 0 \\ 0 & B_1^{*-1} \end{pmatrix} \begin{pmatrix} P & Q \\ R & S \end{pmatrix} = \begin{pmatrix} I_E & C \\ 0 & I_E \end{pmatrix} \begin{pmatrix} B_1 P & B_1 Q \\ B_1^{*-1} R & B_1^{*-1} S \end{pmatrix} \quad (4.4.9)$$

which is a symplectic mapping $\begin{pmatrix} P' & Q' \\ R' & S' \end{pmatrix}$ with

$$P' = B_1 P + C B_1^{*-1} R$$

$$= \begin{pmatrix} 0_{p,p} & 0_{p,n-p} \\ 0_{n-p,p} & I_{n-p} \end{pmatrix} + \begin{pmatrix} I_{p,p} & 0_{p,n-p} \\ 0_{n-p,p} & 0_{n-p,n-p} \end{pmatrix} \begin{pmatrix} \tilde{R}_{p,p} & \tilde{R}_{p,n-p} \\ 0_{n-p,p} & \tilde{R}_{n-p,n-p} \end{pmatrix}$$

$$= \begin{pmatrix} \tilde{R}_{p,p} & \tilde{R}_{p,n-p} \\ 0_{n-p,p} & \tilde{I}_{n-p} \end{pmatrix},$$

which is an invertible mapping. From the equation (4.4.9) and the first part of our discussion, we get that $\begin{pmatrix} P' & Q' \\ R' & S' \end{pmatrix} = \begin{pmatrix} I_E & 0 \\ A' & I_{E*} \end{pmatrix} \begin{pmatrix} B'^{-1} & 0 \\ 0 & B'^* \end{pmatrix} \begin{pmatrix} I_E & -C' \\ 0 & I_{E*} \end{pmatrix}$ with A', C' symmetric and B' invertible and

$$\Xi = \begin{pmatrix} B_1^{-1} & 0 \\ 0 & B_1^* \end{pmatrix} \begin{pmatrix} I_E & -C \\ 0 & I_E \end{pmatrix} \begin{pmatrix} I_E & 0 \\ A' & I_{E*} \end{pmatrix} \begin{pmatrix} B'^{-1} & 0 \\ 0 & B'^* \end{pmatrix} \begin{pmatrix} I_E & -C' \\ 0 & I_{E*} \end{pmatrix},$$

proving that the $\Xi_{A,B,C}$ generate the symplectic group and more precisely that every Ξ in the symplectic group is the product of at most two mappings of type $\Xi_{A,B,C}$. The proof of Theorem 4.4.3 is complete. □

Proposition 4.4.4. *Let $n \geq 1$ be an integer; the symplectic group $Sp(2n)$ (Definition 4.4.1), is included in $Sl(2n, \mathbb{R})$ and $Sp(2) = Sl(2, \mathbb{R})$. A set of generators of $Sp(2n)$ is*

(i) $(x, \xi) \mapsto (B^{-1}x, B^*\xi)$, where $B \in Gl(n, \mathbb{R})$,

(ii) $(x, \xi) \mapsto (\xi, -x)$,

(iii) $(x, \xi) \mapsto (x, Ax + \xi)$, where A is $n \times n$ symmetric.

Another set of generators of $Sp(2n)$ is

(j) $(x, \xi) \mapsto (B^{-1}x, B^*\xi)$, where $B \in Gl(n, \mathbb{R})$,

(jj) $(x, \xi) \mapsto (\xi, -x)$,

(jjj) $(x, \xi) \mapsto (x - C\xi, \xi)$, where C is $n \times n$ symmetric.

The mapping $\Xi \mapsto \Xi^*$ is an isomorphism of $Sp(2n)$.

Proof. The last statement follows from the previous theorem. The inclusion in $Sl(2n, \mathbb{R})$ is a consequence of the previous theorem and the equality with $Sl(2, \mathbb{R})$ when $n = 1$ follows from the defining equations (4.4.5) of the symplectic group: the first two are satisfied for p, q, r, s scalar and the remaining one is the definition of $Sl(2, \mathbb{R})$. Moreover we note that, with a symmetric $n \times n$ matrix A,

$$\begin{pmatrix} 0 & -I_n \\ I_n & 0 \end{pmatrix} \begin{pmatrix} I_n & -A \\ 0 & I_n \end{pmatrix} \begin{pmatrix} 0 & I_n \\ -I_n & 0 \end{pmatrix} = \begin{pmatrix} 0 & -I_n \\ I_n & -A \end{pmatrix} \begin{pmatrix} 0 & I_n \\ -I_n & 0 \end{pmatrix} = \begin{pmatrix} I_n & 0 \\ A & I_n \end{pmatrix}$$

so that

$$\underbrace{\begin{pmatrix} I_n & 0 \\ A & I_n \end{pmatrix}}_{\text{(iii)}} = \underbrace{\begin{pmatrix} -I_n & 0 \\ 0 & -I_n \end{pmatrix}}_{\text{(j)}} \underbrace{\begin{pmatrix} 0 & I_n \\ -I_n & 0 \end{pmatrix}}_{\text{(jj)}} \underbrace{\begin{pmatrix} I_n & -A \\ 0 & I_n \end{pmatrix}}_{\text{(jjj)}} \underbrace{\begin{pmatrix} 0 & I_n \\ -I_n & 0 \end{pmatrix}}_{\text{(jj)}}.$$

Since the previous theorem gives that the symplectic group is generated by matrices of type (j), (jjj) and (iii), this equality proves that (j), (jjj) and (jj) generate the symplectic group. Similarly, we prove that (i), (iii) and (ii) generate the symplectic group, e.g., using that for an $n \times n$ symmetric matrix C, we have

$$\underbrace{\begin{pmatrix} I_n & -C \\ 0 & I_n \end{pmatrix}}_{\text{(jjj)}} = \underbrace{\begin{pmatrix} 0 & I_n \\ -I_n & 0 \end{pmatrix}}_{\text{(ii)}} \underbrace{\begin{pmatrix} I_n & 0 \\ C & I_n \end{pmatrix}}_{\text{(iii)}} \underbrace{\begin{pmatrix} 0 & I_n \\ -I_n & 0 \end{pmatrix}}_{\text{(ii)}} \underbrace{\begin{pmatrix} -I_n & 0 \\ 0 & -I_n \end{pmatrix}}_{\text{(i)}},$$

completing the proof.   □

Remark 4.4.5. The symplectic matrix

$$\begin{pmatrix} 0 & I_n \\ -I_n & 0 \end{pmatrix} = 2^{-1/2} \begin{pmatrix} I_n & I_n \\ -I_n & I_n \end{pmatrix} 2^{-1/2} \begin{pmatrix} I_n & I_n \\ -I_n & I_n \end{pmatrix}$$

is not of the form $\Xi_{A,B,C}$ but is the square of such a matrix. It is also the case of all the mappings $(x_k, \xi_k) \mapsto (\xi_k, -x_k)$ with the other coordinates fixed. Similarly the symplectic matrix

$$\begin{pmatrix} 0 & -I_n \\ I_n & I_n \end{pmatrix} = \begin{pmatrix} I_n & -I_n \\ 0 & I_n \end{pmatrix} \begin{pmatrix} I_n & 0 \\ I_n & I_n \end{pmatrix}$$

is not of the form $\Xi_{A,B,C}$ but is the product $\Xi_{0,I,I} \Xi_{I,I,0}$.

Proposition 4.4.6. *The Lie algebra $sp(2n)$ of the Lie group $Sp(2n)$ is the $n(2n+1)$-dimensional real vector space of $2n \times 2n$ real matrices such that*

$$(\sigma M)^* = \sigma M, \quad \text{with } \sigma = \begin{pmatrix} 0 & I_n \\ -I_n & 0 \end{pmatrix}.$$

Proof. The Lie algebra $sp(2n)$ is the vector space of $2n \times 2n$ real matrices such that for all $t \in \mathbb{R}$, $\exp tM \in Sp(2n)$, according to the general definition of the Lie algebra of a matrix group such as $Sp(2n)$. In fact, according to (4.4.4), getting $\exp tM \in Sp(2n)$, is equivalent to have for all real t $(\exp tM)^* \sigma \exp tM = \sigma$, which means

$$\sigma = \sum_{k,l \geq 0} \frac{t^{k+l}}{k!l!} M^{*k} \sigma M^l, \tag{4.4.10}$$

implying $M^* \sigma + \sigma M = 0$, i.e., $\sigma M = M^* \sigma^* = (\sigma M)^*$. Conversely, the latter condition is enough to get (4.4.10): using the identity $\sigma M^l = M^{*l} \sigma(-1)^l$, easily proven by induction on l, it implies

$$\sum_{k,l \geq 0} \frac{t^{k+l}}{k!l!} M^{*k} \sigma M^l = \sum_{k,l \geq 0} \frac{t^{k+l}}{k!l!} M^{*k} M^{*l} \sigma(-1)^l = \exp tM^* \exp -tM^* \sigma = \sigma.$$

Note that, for $n \times n$ matrices $\alpha, \beta, \gamma, \delta$ we have

$$\begin{pmatrix} 0 & I_n \\ -I_n & 0 \end{pmatrix} \begin{pmatrix} \alpha & \beta \\ \gamma & \delta \end{pmatrix} = \begin{pmatrix} \gamma & \delta \\ -\alpha & -\beta \end{pmatrix}$$

so that the condition σM symmetric is equivalent to γ, β symmetric, $\delta^* = -\alpha$, that is $2\frac{n(n-1)}{2} + n^2 = 2n^2 - n$ independent linear conditions, ensuring that

$$\dim sp(2n) = (2n)^2 - 2n^2 + n = n(2n+1).$$

Incidentally, this proves that the Lie group $Sp(2n)$ (defined as a subgroup of $Sl(2n, \mathbb{R})$ by the equations (4.4.5)) has the same topological dimension $n(2n+1)$ and that the $2\frac{n(n-1)}{2} + n^2$ equations (4.4.5) are independent. Note also that the exponential map is not onto, even if $n = 1$: in particular

$$\begin{pmatrix} -1/2 & 0 \\ 0 & -2 \end{pmatrix} \neq \exp M, \quad \text{for any } 2 \times 2 \text{ real matrix } M.$$

Otherwise, we would find $M = \begin{pmatrix} \alpha & \beta \\ \gamma & \delta \end{pmatrix}$ a real 2×2 matrix such that

$$\exp M = \begin{pmatrix} -1/2 & 0 \\ 0 & -2 \end{pmatrix},$$

implying that $\text{trace} M = 0$, i.e., $\delta = -\alpha$. The characteristic polynomial of M is thus $X^2 - \alpha^2 - \beta\gamma$ and $M^2 = (\alpha^2 + \beta\gamma)I$ implying that

$$\exp M = \sum_{k \geq 0} \frac{(\alpha^2 + \beta\gamma)^{2k}}{(2k)!} I + \sum_{k \geq 0} \frac{(\alpha^2 + \beta\gamma)^{2k}}{(2k+1)!} M \tag{4.4.11}$$

and since $\exp M$ and I are symmetric, we get that M is symmetric as well (the factor of M above cannot be 0, otherwise $\exp M$ would be proportional to I). As a result, $M = \begin{pmatrix} \alpha & \beta \\ \beta & -\alpha \end{pmatrix}$ with $\rho = (\alpha^2 + \beta^2)^{1/2} > 0$ and from (4.4.11), we get

$$\begin{pmatrix} -1/2 & 0 \\ 0 & -2 \end{pmatrix} = \exp M = I \cosh \rho + M \frac{\sinh \rho}{\rho}$$

which implies $\beta = 0$ and thus M diagonal, which is obviously impossible. Using for $n \geq 1$ the symplectic mapping $(x_1, \xi_1) \mapsto (-x_1/2, -2\xi_1)$, the other coordinates fixed, yields the same result. $\qquad\square$

Lemma 4.4.7. *Let $n \in \mathbb{N}^*$. Then we have*

$$Sp(2n) \cap O(2n) = Sp(2n) \cap GL(n, \mathbb{C}) = O(2n) \cap GL(n, \mathbb{C}) = U(n), \quad (4.4.12)$$

where $O(2n)$ is the orthogonal group in $2n$ dimensions, defined by $A^*A = \mathrm{Id}$, $GL(n, \mathbb{C})$ is the group of invertible $n \times n$ matrices with complex entries, identified with the group of real $2n \times 2n$ matrices such that $\sigma A = A\sigma$.

N.B. If $A = \begin{pmatrix} P & Q \\ R & S \end{pmatrix}$ is a real $2n \times 2n$ matrix given in $n \times n$ blocks such that $\sigma A = A\sigma$, it means that

$$\begin{pmatrix} 0 & I_n \\ -I_n & 0 \end{pmatrix} \begin{pmatrix} P & Q \\ R & S \end{pmatrix} = \sigma A = \begin{pmatrix} R & S \\ -P & -Q \end{pmatrix}$$

$$= A\sigma = \begin{pmatrix} P & Q \\ R & S \end{pmatrix} \begin{pmatrix} 0 & I_n \\ -I_n & 0 \end{pmatrix} = \begin{pmatrix} -Q & P \\ -S & R \end{pmatrix},$$

i.e., $S = P, R = -Q$. The mapping

$$\begin{aligned} \Phi: \quad GL(n, \mathbb{C}) &\longrightarrow GL(2n, \mathbb{R}) \\ V &\mapsto \begin{pmatrix} \mathrm{Re}\, V & \mathrm{Im}\, V \\ -\mathrm{Im}\, V & \mathrm{Re}\, V \end{pmatrix} \end{aligned}$$

is an injective homomorphism of groups since

$$\Phi(V_1)\Phi(V_2) = \begin{pmatrix} \mathrm{Re}\, V_1 & \mathrm{Im}\, V_1 \\ -\mathrm{Im}\, V_1 & \mathrm{Re}\, V_1 \end{pmatrix} \begin{pmatrix} \mathrm{Re}\, V_2 & \mathrm{Im}\, V_2 \\ -\mathrm{Im}\, V_2 & \mathrm{Re}\, V_2 \end{pmatrix}$$

and $\Phi(V_1 V_2) = \begin{pmatrix} \mathrm{Re}(V_1 V_2) & \mathrm{Im}(V_1 V_2) \\ -\mathrm{Im}(V_1 V_2) & \mathrm{Re}(V_1 V_2) \end{pmatrix}$

$$= \begin{pmatrix} \mathrm{Re}\, V_1 \, \mathrm{Re}\, V_2 - \mathrm{Im}\, V_1 \, \mathrm{Im}\, V_2 & \mathrm{Re}\, V_1 \, \mathrm{Im}\, V_2 + \mathrm{Im}\, V_1 \, \mathrm{Re}\, V_2 \\ -\mathrm{Re}\, V_1 \, \mathrm{Im}\, V_2 - \mathrm{Im}\, V_1 \, \mathrm{Re}\, V_2 & \mathrm{Re}\, V_1 \, \mathrm{Re}\, V_2 - \mathrm{Im}\, V_1 \, \mathrm{Im}\, V_2 \end{pmatrix}$$

which means that $GL(n, \mathbb{C})$ is isomorphic to $\Phi(GL(n, \mathbb{C}))$, which is a subgroup of $GL(2n, \mathbb{R})$. Note also that iI_{2n} is identified with

$$\begin{pmatrix} 0 & I_n \\ -I_n & 0 \end{pmatrix} = \sigma.$$

We shall identify $GL(n, \mathbb{C})$ with $\Phi(GL(n, \mathbb{C}))$, i.e., with the $2n \times 2n$ invertible matrices of the form

$$\begin{pmatrix} P & Q \\ -Q & P \end{pmatrix},$$

that is to the $2n \times 2n$ invertible matrices A such that $\sigma A = A\sigma$. Similarly, the unitary group $U(n)$ will be identified with the $2n \times 2n$ matrices

$$\begin{pmatrix} P & Q \\ -Q & P \end{pmatrix}, \quad \text{such that } P^*P + Q^*Q = I_n, \quad P^*Q - Q^*P = 0,$$

since it is equivalent to $(P^* - iQ^*)(P + iQ) = I_n$.

Proof of the lemma. If $A \in Sp(2n) \cap O(2n)$, then $\sigma A = A^{*-1}\sigma = A\sigma$ and thus $A \in Sp(2n) \cap GL(n, \mathbb{C})$. If $A \in Sp(2n) \cap GL(n, \mathbb{C})$, then

$$A^*A = -A^*\sigma\sigma A = -A^*\sigma A\sigma = -\sigma^2 = \text{Id}$$

and thus $A \in GL(n, \mathbb{C}) \cap O(2n)$. If $A \in GL(n, \mathbb{C}) \cap O(2n)$, then $A^*A = I_{2n}$ and

$$A = \begin{pmatrix} P & Q \\ -Q & P \end{pmatrix}, \quad \text{with } P^*P + Q^*Q = \text{Id}, \quad P^*Q - Q^*P = 0, \qquad (4.4.13)$$

so that $A \in U(n)$. If $A \in U(n)$, then (4.4.13) holds and $A^*A = I_{2n}$ along with (4.4.5), proving that $A \in Sp(2n) \cap O(2n)$. The proof of the lemma is complete. $\square$

Lemma 4.4.8. *Let $A \in Sp(2n)$ symmetric positive-definite. Then for all $\theta \in \mathbb{R}$, $A^\theta \in Sp(2n)$.*

Proof. There is an orthonormal basis $(e_j)_{1 \leq j \leq 2n}$ such that $Ae_j = \lambda_j e_j$ with some positive λ_j. Moreover we have $\lambda_j\lambda_k\langle\sigma e_j, e_k\rangle = \langle\sigma Ae_j, Ae_k\rangle = \langle\sigma e_j, e_k\rangle$ and thus, if $\langle\sigma e_j, e_k\rangle \neq 0$, it implies $\lambda_j\lambda_k = 1$, so that

$$\langle\sigma A^\theta e_j, A^\theta e_k\rangle = \langle\sigma e_j, e_k\rangle\lambda_j^\theta\lambda_k^\theta = \begin{cases} \langle\sigma e_j, e_k\rangle & \text{if } \langle\sigma e_j, e_k\rangle \neq 0, \\ 0 = \langle\sigma e_j, e_k\rangle & \text{if } \langle\sigma e_j, e_k\rangle = 0, \end{cases}$$

implying that the matrix A^θ is symplectic. $\square$

Proposition 4.4.9. *The manifold $Sp(2n)$ is diffeomorphic to the Cartesian product of the group $U(n)$ with a convex open cone of a vector space of dimension $n(n+1)$. The symplectic group is path connected and the injection of $U(n)$ in $Sp(2n)$ induces an isomorphism of $\mathbb{Z} = \pi_1(U(n))$ onto $\pi_1(Sp(2n))$.*

Proof. Let $\Xi \in Sp(2n)$: then $\Xi = (\Xi\Xi^*)^{1/2}(\Xi\Xi^*)^{-1/2}\Xi$ and the matrix $(\Xi\Xi^*)^{1/2}$ is symplectic from Lemma 4.4.8 as well as the matrix $(\Xi\Xi^*)^{-1/2}\Xi$ which is also orthogonal since $\Xi^*(\Xi\Xi^*)^{-1/2}(\Xi\Xi^*)^{-1/2}\Xi = \Xi^*(\Xi\Xi^*)^{-1}\Xi = I_{2n}$. From Lemma 4.4.7, we have a mapping

$$
\Psi: \quad
\begin{array}{ccc}
Sp(2n) & \longrightarrow & U(n) \times \overbrace{\big(Sp(2n) \cap \{\text{symmetric positive-definite matrices}\}\big)}^{\text{denoted by } Sp_+(2n)} \\
\Xi & \longmapsto & \big((\Xi\Xi^*)^{-1/2}\Xi, (\Xi\Xi^*)^{1/2}\big).
\end{array}
$$

$$(4.4.14)$$

This mapping is bijective since on the one hand it is obviously one-to-one and also for $V \in U(n)$, S symplectic and positive-definite, the matrix SV is symplectic and, since $VV^* = I_{2n}, S = S^* \gg 0$, we have

$$
\Psi(SV) = \big((SVV^*S^*)^{-1/2}SV, (SVV^*S^*)^{1/2}\big) = (V, S).
$$

We have $\Psi^{-1}(V, S) = SV$ and thus $Sp(2n)$ is diffeomorphic to $U(n) \times Sp_+(2n)$. Let $\Xi \in Sp_+(2n)$: according to Theorem 4.4.3, it implies

$$
\Xi = \begin{pmatrix} B^{-1} & B^{-1}A \\ AB^{-1} & B + AB^{-1}A \end{pmatrix}, \ B \text{ symmetric positive-definite, } A \text{ symmetric.}
$$

Note also that the same theorem gives as well the more transparent

$$
\begin{pmatrix} B^{-1} & B^{-1}A \\ AB^{-1} & B + AB^{-1}A \end{pmatrix} = \begin{pmatrix} I & 0 \\ A & I \end{pmatrix} \begin{pmatrix} B^{-1} & 0 \\ 0 & B \end{pmatrix} \begin{pmatrix} I & A \\ 0 & I \end{pmatrix},
$$

with B symmetric positive-definite, A symmetric. With $Sym(n, \mathbb{R})$ standing for the $n \times n$ symmetric matrices and $Sym_+(n, \mathbb{R})$ for the $n \times n$ positive-definite symmetric matrices, the mapping

$$
\Gamma: \quad
\begin{array}{ccc}
Sym(n, \mathbb{R}) \times Sym_+(n, \mathbb{R}) & \longrightarrow & Sp_+(2n) \\
(A, B) & \longmapsto & \begin{pmatrix} B^{-1} & B^{-1}A \\ AB^{-1} & B + AB^{-1}A \end{pmatrix}
\end{array}
$$

is a diffeomorphism and $Sym(n, \mathbb{R}) \times Sym_+(n, \mathbb{R})$ is an open cone in the vector space $Sym(n, \mathbb{R}) \times Sym(n, \mathbb{R})$, which is $2\frac{n(n+1)}{2}$-dimensional. Since the fundamental group of the unitary group $U(n)$ is $\mathbb{Z}$ (whose dimension is n^2, see the next lemma) and $Sym(n, \mathbb{R}) \times Sym_+(n, \mathbb{R})$ is contractible, we get also the last statement. We note also that

$$
\dim Sp(2n) = 2\frac{n(n+1)}{2} + n^2 = n(2n+1). \qquad \square
$$

Lemma 4.4.10. *Let $n \in \mathbb{N}^*$. The unitary group $U(n)$ is compact connected with fundamental group $\mathbb{Z}$. The special unitary group is compact connected and simply connected.*

Proof of the lemma. The unitary group is the subgroup of the group of invertible $n \times n$ complex matrices such that $U^*U = I$. It is a real Lie group with dimension n^2, since we have $n(n-1) + n$ independent constraints in a space of dimension $2n^2$. The compactness is obvious and the path connectivity is due to the spectral reduction

$$U = S^* \operatorname{diag}(e^{i\theta_1}, \dots, e^{i\theta_n})S, \quad S \in U(n)$$

so that $U_t = S^* \operatorname{diag}(e^{it\theta_1}, \dots, e^{it\theta_n})S$, $t \in [0,1]$ is a path in $U(n)$ to I. The same arguments give the compactness and the path connectivity of the special unitary group. On the other hand, the equivalence relation in $SU(n)$ given by "having the same first column" induces a fibration for $n \geq 2$,

$$SU(n) \longrightarrow \mathbb{S}^{2n-1}, \quad \text{with fibers } SU(n-1).$$

As a result, since $SU(1) = \{1\}$, we get by induction that $SU(n)$ is simply connected. The homomorphism $\det U(n) \longrightarrow \mathbb{S}^1$ identifies $\mathbb{S}^1$ with $U(n)/SU(n)$, proving that the fundamental group of $U(n)$ is $\mathbb{Z}$, since $SU(n)$ is simply connected. $\square$

Remark 4.4.11. The symplectic group $Sp(2n)$ can be identified with the set of $(C_1, \dots, C_n, C_{n+1}, \dots, C_{2n})$ where the C_j belong to $\mathbb{R}^{2n}$ and are such that (the brackets are the symplectic form),

$$[C_{j+n}, C_j] = 1, \ 1 \leq j \leq n, \quad \text{all other brackets } [C_k, C_l] = 0.$$

Considering the equivalence relation $(C_l)_{1 \leq l \leq 2n} \sim (C_l')_{1 \leq l \leq 2n}$ meaning $C_1 = C_1'$ and $C_{n+1} = C_{n+1}'$, we see that for $n \geq 1$,

$$\left(Sp(2n)/\sim\right) = (\mathbb{R}^{2n} \backslash \{0\}) \times \mathbb{R}^{2n-1} (\text{diffeomorphic to } \mathbb{S}^{2n-1} \times \mathbb{R}^{2n}),$$

since $C_1 \in \mathbb{R}^{2n} \backslash \{0\}$ and C_{n+1} belongs to the hyperplane $[X, C_1] = 1$. Since the symplectic orthogonal of $\mathbb{R}C_1 \oplus \mathbb{R}C_{n+1}$ is

$$\operatorname{Vect}(C_l)_{\substack{2 \leq l \leq n \\ n+2 \leq l \leq 2n-1}},$$

which is a $2n-2$-dimensional symplectic space, the fibers of the canonical projection $Sp(2n) \longrightarrow (Sp(2n)/\sim)$ are isomorphic to $Sp(2n-2)$, showing as well that the fundamental group of $Sp(2n)$ is isomorphic to the one of $Sp(2, \mathbb{R}) = SL(2, \mathbb{R})$; since $SL(2, \mathbb{R})$ is diffeomorphic[8] to $\mathbb{S}^1 \times \mathbb{R}^2$, this gives a sketch of another proof of $\pi_1(Sp(2n)) = \mathbb{Z}$. Note also that the dimension of the fibers adds up correctly since

$$\sum_{1 \leq k \leq n} ((2k-1) + 2k) = 2(n(n+1)) - n = 2n^2 + n = n(2n+1)$$

which is indeed the dimension of $Sp(2n)$.

---

[8]We have $SL(2, \mathbb{R}) = \{(a,b,c,d) \in \mathbb{R}^4, \text{ s.t. } ad - bc = 1\}$ and the mapping

$$\begin{array}{ccc} \mathbb{S}^1 \times \mathbb{R}_+^* \times \mathbb{R} & \longrightarrow & SL(2, \mathbb{R}) \\ (e^{i\theta}, r, t) & \mapsto & (r\cos\theta, r\sin\theta, t\cos\theta - r^{-1}\sin\theta, t\sin\theta + r^{-1}\cos\theta) \end{array}$$

is a diffeomorphism.

The Maslov index of a loop in $Sp(2n)$

We consider the diffeomorphism Ψ given by (4.4.14) ; for $\Xi \in Sp(2n)$, the matrix $(\Xi\Xi^*)^{-1/2}\Xi = \Psi_0(\Xi) = \begin{pmatrix} P & Q \\ -Q & P \end{pmatrix}$ can be identified to the $n \times n$ unitary matrix $P + iQ$. We define

$$
\begin{aligned}
\rho : Sp(2n) &\longrightarrow \quad \mathbb{S}^1 \\
\Xi &\longmapsto \quad \rho(\Xi)= \det(P + iQ).
\end{aligned}
\tag{4.4.15}
$$

Definition 4.4.12. Let $\gamma : \mathbb{R}/\mathbb{Z} \longrightarrow Sp(2n)$ be a loop. The Maslov index of γ is defined as $\mu(\gamma) = \deg(\rho \circ \gamma)$.

With $\Gamma = \rho \circ \gamma : \mathbb{R}/\mathbb{Z} \longrightarrow \mathbb{S}^1$, we define $\alpha(t) = \frac{1}{2i\pi} \int_0^t \dot{\Gamma}(s)\Gamma(s)^{-1}ds$, and we have

$$
\frac{d}{dt}(e^{-2i\pi\alpha(t)}\Gamma(t)) = e^{-2i\pi\alpha(t)}\left(-\dot{\Gamma}(t)\Gamma(t)^{-1}\Gamma(t) + \dot{\Gamma}(t)\right) = 0
$$

and thus $\Gamma(t) = e^{2i\pi\alpha(t)}\Gamma(0)$ with $\deg\Gamma = \alpha(1) - \alpha(0)$, which belongs to $\mathbb{Z}$.

Proposition 4.4.13. *Two loops in $Sp(2n)$ are homotopic if and only if they have the same Maslov index. For γ_1, γ_2 loops in $Sp(2n)$, we have $\mu(\gamma_1\gamma_2) = \mu(\gamma_1) + \mu(\gamma_2)$. The Maslov index provides an explicit isomorphism between $\pi_1(Sp(2n))$ and $\mathbb{Z}$.*

Proof. We have seen previously (Proposition 4.4.9) that the map ρ induces an isomorphism of the fundamental groups, and the homotopy property follows. The product law is obvious for loops of unitary matrices and since every loop in $Sp(2n)$ is homotopic to a unitary loop (in Proposition 4.4.9, we have seen that $Sp(2n)$ is diffeomorphic to the Cartesian product of a contractible space with $U(n)$), this gives the result. $\qquad\square$

4.4.2   The metaplectic group

Let E be a real finite-dimensional vector space (dimension n). In the previous paragraph, we have studied the symplectic group $Sp(E \oplus E^*)$, which is a subgroup of the special linear group of the phase space $\mathcal{P} = E \oplus E^*$. We want now to *quantize* the symplectic group, that is to find a group of unitary transformations of $L^2(E)$, $Mp(E)$, such that the formula (2.1.14) is satisfied.

The phase space $\mathcal{P}$ can be identified to the quantization of linear forms, via the mapping $\mathcal{P} \ni Y \mapsto L_Y^w$, where $L_Y(X) = \sigma(X, Y)$ so that with $Y = (y, \eta)$ $L_Y^w = y \cdot D_x - \eta \cdot x$, a linear operator continuous on $\mathscr{S}(E), \mathscr{S}'(E)$.

Proposition 4.4.14. *The set G of automorphisms M of $\mathscr{S}'(E)$ such that, for all $Y \in \mathcal{P}$, there exists $Z \in \mathcal{P}$ such that, $ML_Y^w M^{-1} = L_Z^w$ is a subgroup of the group of automorphisms of $\mathscr{S}'(E)$. For each $M \in G$, there exists a unique $\chi_M \in Sp(E \oplus E^*)$ such that, for all $Y \in \mathcal{P}$,*

$$
ML_Y^w M^{-1} = L_{\chi_M Y}^w.
\tag{4.4.16}
$$

The mapping $\Phi : G \ni M \mapsto \chi_M \in Sp(E \oplus E^*)$ *is a surjective homomorphism whose kernel is* $\mathbb{C}^*$ Id.

Proof. We may note first that, for each $M \in G$, this defines a linear mapping $Y \mapsto Z = \chi_M Y$ from $\mathcal{P}$ into itself: first of all, Z is uniquely determined by the identity $ML_Y^w M^{-1} = L_Z^w$ and the mapping $Y \mapsto L_Y^w$ is obviously linear. Moreover χ_M belongs to the symplectic group; we note first that for $Y = (y, \eta)$, $Z = (z, \zeta) \in \mathcal{P}$, we have

$$[L_Y^w, L_Z^w] = [y \cdot D_x - \eta \cdot x, z \cdot D_x - \zeta \cdot x] = -\frac{1}{2i\pi}\langle \zeta, y \rangle + \frac{1}{2i\pi}\langle \eta, z \rangle = \frac{1}{2i\pi}\sigma(Y, Z).$$

For $M \in G$, we get

$$\frac{1}{2i\pi}\sigma(\chi_M Y, \chi_M Z) = [L_{\chi_M Y}^w, L_{\chi_M Z}^w] = [ML_Y^w M^{-1}, ML_Z^w M^{-1}] = M[L_Y^w, L_Z^w]M^{-1}$$

which is equal to $\frac{1}{2i\pi}\sigma(Y, Z)$ implying $\chi_M \in Sp(E \oplus E^*)$. Moreover G is a subgroup of the automorphisms of $\mathscr{S}'(E)$ since for $M_1, M_2 \in G, Y \in \mathcal{P}$, we have

$$(M_1^{-1}M_2)L_Y^w(M_1^{-1}M_2)^{-1} = M_1^{-1}L_{\chi_{M_2}Y}^w M_1 = L_{\chi_{M_1}^{-1}\chi_{M_2}Y}^w,$$

so that $M_1^{-1}M_2 \in G$ and this proves also the homomorphism property. To obtain the surjectivity, we shall use Theorem 4.4.3 and check, using the notation of this theorem, that $\Xi_{A,B,C}$ is in the image of the homomorphism; we calculate for $Y = (y, \eta) \in \mathcal{P}$, the product of operators (with A, B, C as in Theorem 4.4.3)

$$e^{i\pi\langle Ax,x\rangle}L_Y^w e^{-i\pi\langle Ax,x\rangle} = e^{i\pi\langle Ax,x\rangle}(y \cdot D_x - \eta \cdot x)e^{-i\pi\langle Ax,x\rangle}$$

$$= y \cdot D_x - y \cdot Ax - \eta \cdot x = y \cdot D_x - (Ay + \eta) \cdot x = L_{\Xi_{A,I,0}Y}^w,$$

$$e^{i\pi\langle CD_x,D_x\rangle}L_Y^w e^{-i\pi\langle CD_x,D_x\rangle} = e^{i\pi\langle CD_x,D_x\rangle}(y \cdot D_x - \eta \cdot x)e^{-i\pi\langle CD_x,D_x\rangle}$$

$$= y \cdot D_x - \eta \cdot x - \eta \cdot CD_x = (y - C\eta) \cdot D_x - \eta \cdot x = L_{\Xi_{0,I,C}Y}^w,$$

and with $(T_B u)(x) = u(Bx)|\det B|^{1/2}$,

$$T_B L_Y^w T_B^{-1} = T_B(y \cdot D_x - \eta \cdot x)T_B^{-1} = (B^{-1}y) \cdot D_x - B^*\eta \cdot x = L_{\Xi_{0,B,0}Y}^w,$$

so that the surjectivity is fulfilled. We prove now the statement on the kernel of Φ: if $M \in \ker \Phi$, we have for all $Y \in \mathcal{P}$ that $L_Y^w M = ML_Y^w$, i.e., if $\mu(x, x')$ stands for the distribution kernel of M, we have for all $(y, \eta) \in \mathcal{P}$,

$$y \cdot \frac{\partial \mu}{\partial x}(x, x') = -y \cdot \frac{\partial \mu}{\partial x'}(x, x') \Longrightarrow \forall j, \quad \frac{\partial \mu}{\partial x_j} + \frac{\partial \mu}{\partial x'_j} = 0,$$

$$\eta \cdot x\mu(x, x') = \eta \cdot x'\mu(x, x') \Longrightarrow \forall j, \quad (x_j - x'_j)\mu = 0,$$

which implies that $\mu(x, x') = c\delta(x - x')$, i.e., $M = c$ Id: in fact, the change of variables $y = x - x', z = x + x'$, gives $\mu = 1 \otimes \nu(y), \nabla \hat{\nu} = 0$, so that $\hat{\nu} = C^{te}$ and $\nu = c\delta$. $\qquad \square$

Incidentally, we have also proven the following

Lemma 4.4.15. *Let A, B, C be as in Theorem 4.4.3, and S be the generating function of $\Xi_{A,B,C}$. We define the operator $M_{A,B,C}$ on $\mathscr{S}(E)$ by*

$$(M_{A,B,C}v)(x) = \int_{E^*} e^{2i\pi S(x,\eta)}\hat{v}(\eta)d\eta|\det B|^{1/2}. \qquad (4.4.17)$$

This operator is an automorphism of $\mathscr{S}'(E)$ and of $\mathscr{S}(E)$ which is unitary on $L^2(E)$, belongs to G and $\Phi(M_{A,B,C}) = \Xi_{A,B,C}$.

Definition 4.4.16. Let E be a real finite-dimensional vector space. The metaplectic group $Mp(E)$ is the subgroup of the $M \in G$ (defined in Proposition 4.4.14), such that M is also a unitary isomorphism of $L^2(E)$.

Proposition 4.4.17. *The metaplectic group is generated by the $M_{A,B,C}$ given in (4.4.17) and by the multiplication by complex numbers with modulus 1. The group $Mp(E)$ is isomorphic to $G/\mathbb{R}_+^* \operatorname{Id}$.*

Proof. $Mp(E)$ is $G \cap \mathscr{U}(E)$, where $\mathscr{U}(E)$ is the group of unitary isomorphisms of $L^2(E)$. Note that the metaplectic group contains the $M_{A,B,C}$ given in (4.4.17). Moreover, since the mapping Φ is a surjective homomorphism with kernel $\mathbb{C}^* \operatorname{Id}$, $Sp(\mathcal{P})$ is isomorphic to $G/\mathbb{C}^*$; as a result $G/\mathbb{C}^*$ is generated by the $M_{A,B,C}$ given in (4.4.17) and from Theorem 4.4.3, we know that any element in $G/\mathbb{C}^*$ can be written as a product $M_{A_1,B_1,C_1}M_{A_2,B_2,C_2}$ so that any element of G can be written as

$$zM_{A_1,B_1,C_1}M_{A_2,B_2,C_2}, \quad \text{with } z \in \mathbb{C}^*,$$

implying that the group $Mp(E)$ is generated by the $zM_{A,B,C}$ with $|z| = 1$ and also that the group $G/\mathbb{R}_+^* \operatorname{Id}$ is isomorphic to $Mp(E)$: in fact the canonical mapping $Mp(E) \longrightarrow G \longrightarrow G/\mathbb{R}_+^* \operatorname{Id}$ is proven onto by the previous remarks on the generators and it is also one-to-one since if M_1, M_2 belong to $Mp(E)$ are such that $M_2 = \rho M_1$ with $\rho > 0$, the fact that both M_j are unitary implies $\rho = 1$.   □

We have also proven the following result.

Proposition 4.4.18. *The following diagram of group homomorphims is commutative and the horizontal lines are exact sequences,*

$$
\begin{array}{ccccccccc}
1 & \longrightarrow & \mathbb{C}^* & \overset{j}{\longrightarrow} & G & \overset{\Phi}{\longrightarrow} & Sp(\mathcal{P}) & \longrightarrow & 1 \\
 & & \downarrow{\scriptstyle p_1} & & \downarrow{\scriptstyle p_2} & & \downarrow{\scriptstyle \operatorname{Id}} & & \\
1 & \longrightarrow & \mathbb{S}^1 & \underset{j}{\longrightarrow} & Mp(E) & \underset{\Pi}{\longrightarrow} & Sp(\mathcal{P}) & \longrightarrow & 1
\end{array}
$$

where $j(z) = z \operatorname{Id}$ (the canonical injection), Φ is defined in Proposition 4.4.14,

$p_1(z) = z/|z|$, and p_2 is the canonical projection of G onto $Mp(E)$ since $Mp(E) = G/\mathbb{R}_+^*$. Since Φ maps $\mathbb{C}^* \operatorname{Id}$ to the identity of $Sp(\mathcal{P})$, the mapping Π can be taken as the restriction of Φ to the subgroup $Mp(E)$. We have for $\theta \in \mathbb{R}$,

$$\Pi(e^{i\theta} M_{A,B,C}) = \Xi_{A,B,C}, \qquad \text{for } M_{A,B,C} \text{ given in } (4.4.17).$$

Remark 4.4.19. Note also that for $M \in Mp(E), Y \in \mathcal{P}$, the formula (4.4.16) means

$$M\sigma(X,Y)^w M^{-1} = \sigma\big(X, \Pi(M)Y\big)^w = \sigma\big(\Pi(M)^{-1}X, Y\big)^w = \sigma\big(\Pi(M^{-1})X, Y\big)^w$$

so that, changing M in M^{-1}, we get $\sigma\big(\Pi(M)X, Y\big)^w = M^{-1}\sigma(X,Y)^w M$, which is (2.1.14) for linear forms a. As a consequence, for a real linear form L on $\mathcal{P}$, L^w is unitary equivalent to the multiplication by x_1 acting on $\mathscr{S}(E)$, so that the operator $\exp(iL^w)$ has the Weyl symbol e^{iL}. Writing now for $a \in \mathscr{S}'(\mathbb{R}^{2n}), \chi \in Sp(2n), \Pi(M) = \chi$, the inverse Fourier transform formula, we obtain[9]

$$(a \circ \chi)^w = \int \big(e^{2i\pi\langle \chi X, \Xi\rangle}\big)^w \hat{a}(\Xi) d\Xi = \int e^{2i\pi(\langle \chi X, \Xi\rangle)^w} \hat{a}(\Xi) d\Xi$$

$$= \int e^{2i\pi(M^*\langle X, \Xi\rangle^w M)} \hat{a}(\Xi) d\Xi = M^* \int e^{2i\pi\langle X, \Xi\rangle^w} \hat{a}(\Xi) d\Xi M = M^* a^w M$$

which is (2.1.14).

Remark 4.4.20. We have followed the terminology adopted by J. Leray in [87] for the definition of the metaplectic group. When $E = \mathbb{R}^n$, we shall note $Mp(E)$ by $Mp(n)$. However the fibers of the mapping Π in Proposition 4.4.18 are isomorphic to $\mathbb{S}^1$ whereas the metaplectic group introduced by A. Weil in [149] is a two-fold covering of $Sp(2n)$. That group $Mp_2(n)$ is the group generated by the

$$M_{A,B,C}^{[\mu]} = e^{i\frac{\pi}{2}\mu} M_{A,B,C}$$

for a fixed choice of $\mu(B) = \arg(\det B)/\pi$; $Mp_2(n)$ is a proper subgroup of $Mp(E)$ and the image by Π of $Mp_2(n)$ is the whole $Sp(2n)$, while the fibers have two elements. The restriction to $Mp_2(n)$ of Π is a two-fold covering with kernel $\{\pm \operatorname{Id}_{L^2}\}$.

4.4.3 A remark on the Feynman quantization

Going back to Remark 2.1.4 on the Feynman quantization, let us prove that this quantization is *not* invariant by the symplectic group: we assume that $n = 1$ and consider the symplectic mapping $\chi(x, \xi) = (x, \xi + Sx)$ where S is a non-zero real number. We shall prove now that one can find some $a \in \mathcal{S}(\mathbb{R}^{2n})$ such that

$$(a \circ \chi)^F \neq M^* a^F M,$$

---

[9]The integrals should be interpreted weakly, e.g., for $a \in \mathscr{S}'(\mathbb{R}^{2n}), u, v \in \mathscr{S}(\mathbb{R}^n)$, with the Wigner function $\mathcal{H}(u, v)$ given by (2.1.3),

$$\langle a^w u, v\rangle_{\mathscr{S}^*(\mathbb{R}^n), \mathscr{S}(\mathbb{R}^n)} = \prec a, \mathcal{H}(u, v) \succ_{\mathscr{S}'(\mathbb{R}^{2n}), \mathscr{S}(\mathbb{R}^{2n})} = \prec \hat{a}, \widecheck{\mathcal{H}(u, v)} \succ_{\mathscr{S}'(\mathbb{R}^{2n}), \mathscr{S}(\mathbb{R}^{2n})}.$$

where M is the unitary transformation of $L^2(\mathbb{R})$ given by $(Mu)(x) = e^{i\pi Sx^2}u(x)$. We compute

$$2\langle (a \circ \chi)^F u, v \rangle = \int e^{2i\pi(x-y)\xi}\big(a(x, \xi + Sx) + a(y, \xi + Sy)\big)u(y)\overline{v(x)}dy dx d\xi$$

$$= \int e^{2i\pi(x-y)(\xi-Sx)}a(x, \xi)u(y)\overline{v(x)}dy dx d\xi$$

$$+ \int e^{2i\pi(x-y)(\xi-Sy)}a(y, \xi)u(y)\overline{v(x)}dy dx d\xi$$

$$= \int e^{2i\pi(x-y)\xi}\big(a(x, \xi)e^{-i\pi S(x-y)^2} + a(y, \xi)e^{i\pi S(x-y)^2}\big)(Mu)(y)\overline{(Mv)(x)}dy dx d\xi$$

so that $(a \circ \chi)^F = M^*KM$ where the kernel k of the operator K is given by

$$2k(x, y) = \widehat{a}^2(x, y - x)e^{-i\pi S(x-y)^2} + \widehat{a}^2(y, y - x)e^{i\pi S(x-y)^2}.$$

On the other hand the kernel l of the operator a^F is given by

$$2l(x, y) = \widehat{a}^2(x, y - x) + \widehat{a}^2(y, y - x).$$

Checking the case $S = 1, a(x, \xi) = e^{-\pi(x^2+\xi^2)}$, we see that

$$2k(1, 0) = -e^{-2\pi} - e^{-\pi}, \quad 2l(1, 0) = e^{-2\pi} + e^{-\pi},$$

proving that $K \neq a^F$ and the sought result.

4.4.4 Positive quadratic forms in a symplectic vector space

Lemma 4.4.21. *Let V be a Euclidean vector space with even dimension $2n$. Let S be an isomorphism of V such that ${}^tS = -S$ (i.e., $\langle Su, v\rangle_V = -\langle u, Sv\rangle_V$). Then there exists an orthonormal basis $(e_1, \ldots, e_n, \epsilon_1, \ldots, \epsilon_n)$ of V such that*

$$Se_j = -\mu_j\epsilon_j, \quad S\epsilon_j = \mu_j e_j, \quad 1 \leq j \leq n, \quad \mu_j > 0.$$

Proof. The isomorphism S has at least an eigenvector $X + iY \neq 0$ $(X, Y \in V)$, corresponding to a non-zero eigenvalue $\lambda + i\mu$ $(\lambda, \mu \in \mathbb{R})$ so that

$$SX = \lambda X - \mu Y, \quad SY = \lambda Y + \mu X.$$

Writing $0 = \langle SX, X\rangle_V = \langle SY, Y\rangle_V$, we obtain $\lambda = 0$ (since $\|X\|^2_V + \|Y\|^2_V > 0$) and $\langle X, Y\rangle_V = 0$, so that $X \wedge Y \neq 0$. Moreover we have

$$\mu\|Y\|^2_V = -\langle SX, Y\rangle_V = \langle X, SY\rangle_V = \mu\|X\|^2_V$$

so that we may assume that X and Y are both unit vectors. The plane (X, Y) is stable by S and its matrix in the basis (X, Y) is $\begin{pmatrix} 0 & \mu \\ -\mu & 0 \end{pmatrix}$; we may assume that

$\mu > 0$ since otherwise we can exchange the rôle of X and Y. Let us note now that $W = (\mathbb{R}X \oplus \mathbb{R}Y)^{\perp}$ is stable by S since, if $Z \in W$,

$$\langle SZ, X \rangle_V = -\langle Z, SX \rangle_V = \mu \langle Z, Y \rangle_V = 0, \quad \text{and similarly } \langle SZ, Y \rangle_V = 0.$$

We can then proceed by induction on n. The proof of the lemma is complete. $\square$

Definition 4.4.22. Let Q be a positive-definite quadratic form on a symplectic vector space $(\mathcal{P}, \sigma)$, i.e., a linear mapping from $\mathcal{P}$ to $\mathcal{P}^*$ such that $Q^* = Q$ and $\mathcal{P} \ni X \mapsto \langle QX, X \rangle_{\mathcal{P}^*, \mathcal{P}}$ is positive for $X \neq 0$. We define

$$Q^{\sigma} = \sigma^* Q^{-1} \sigma \tag{4.4.18}$$

and we note that Q^{σ} is a positive-definite quadratic form on $\mathcal{P}$.

In fact, we have $(\sigma^* Q^{-1} \sigma)^* = \sigma^* Q^{-1} \sigma$ and for $0 \neq X \in \mathcal{P}$,

$$\langle \sigma^* Q^{-1} \sigma X, X \rangle_{\mathcal{P}^*, \mathcal{P}} = \langle Q^{-1} \sigma X, \sigma X \rangle_{\mathcal{P}, \mathcal{P}^*} = \langle Q Q^{-1} \sigma X, Q^{-1} \sigma X \rangle_{\mathcal{P}^*, \mathcal{P}} > 0.$$

Remark 4.4.23. The Cauchy-Schwarz inequality can be written as (with brackets of duality as above)

$$\forall \Xi \in \mathcal{P}^*, \forall X \in \mathcal{P}, \quad \langle \Xi, X \rangle^2 \leq \langle \Xi, Q^{-1} \Xi \rangle \langle QX, X \rangle \tag{4.4.19}$$

proven by the standard method of checking the discriminant of the always non-negative second-degree polynomial $t \mapsto \langle Q(X + tQ^{-1}\Xi), X + tQ^{-1}\Xi \rangle$ (the equality in (4.4.19) occurs iff $\Xi \wedge QX = 0$). We infer from this,

$$\forall X, Y \in \mathcal{P}, \quad \langle \sigma X, Y \rangle^2 \leq \langle Q^{\sigma} X, X \rangle \langle QY, Y \rangle \tag{4.4.20}$$

and

$$\langle Q^{\sigma} X, X \rangle = \sup_{0 \neq T \in \mathcal{P}} \frac{\langle \sigma X, T \rangle^2}{\langle QT, T \rangle}, \tag{4.4.21}$$

since $\langle \sigma X, Q^{-1} \sigma X \rangle^2 = \langle Q^{\sigma} X, X \rangle \langle Q(Q^{-1} \sigma X), Q^{-1} \sigma X \rangle$, we have (for $X \neq 0$),

$$\langle Q^{\sigma} X, X \rangle = \frac{\langle \sigma X, T \rangle^2}{\langle QT, T \rangle} \quad \text{with } T = Q^{-1} \sigma X. \tag{4.4.22}$$

Remark 4.4.24. Let Q_1, Q_2 be a pair of positive-definite quadratic forms on a symplectic vector space such that $Q_1 \leq Q_2$; then we get immediately from (4.4.21) that $Q_2^{\sigma} \leq Q_1^{\sigma}$.

Lemma 4.4.25. *Let $(\mathcal{P}, \sigma)$ be a symplectic vector space of dimension $2n$ and let Q be a positive-definite quadratic form on $\mathcal{P}$. There exists a symplectic basis $(e_1, \ldots, e_n, \epsilon_1, \ldots, \epsilon_n)$ of $\mathcal{P}$ (in the sense of Proposition 4.4.2) which is also an orthogonal basis for Q. The matrix of σ in this basis is $\begin{pmatrix} 0 & I_n \\ -I_n & 0 \end{pmatrix}$. The matrix of Q in this basis is $\begin{pmatrix} \mu & 0 \\ 0 & \mu \end{pmatrix}$, $\mu = \text{diag}(\mu_1, \ldots, \mu_n)$, $\mu_j > 0$. The matrix of Q^{σ} in this basis is $\begin{pmatrix} \mu^{-1} & 0 \\ 0 & \mu^{-1} \end{pmatrix}$.*

Proof. The mapping $Q^{-1}\sigma$ is antiadjoint on the Euclidean space $\mathcal{P}$ equipped with the dot product $\ll X, Y \gg_Q = \langle QX, Y \rangle_{\mathcal{P}^*, \mathcal{P}}$:

$$\ll Q^{-1}\sigma X, Y \gg_Q = \langle \sigma X, Y \rangle = -\langle X, \sigma Y \rangle = - \ll X, Q^{-1}\sigma Y \gg_Q .$$

Applying Lemma 4.4.21 to $Q^{-1}\sigma$ gives a basis $(e_1, \ldots, e_n, \epsilon_1, \ldots, \epsilon_n)$ of $\mathcal{P}$, Q-orthonormal such that

$$Q^{-1}\sigma e_j = -\mu_j \epsilon_j, \quad Q^{-1}\sigma \epsilon_j = \mu_j e_j, \quad 1 \le j \le n, \quad \mu_j > 0,$$

so that $\langle \sigma e_j, \epsilon_j \rangle_{\mathcal{P}^*, \mathcal{P}} = \ll Q^{-1}\sigma e_j, \epsilon_j \gg_Q = -\mu_j \|\epsilon_j\|_Q^2 = -\mu_j$. We also have

$$\langle \sigma e_j, e_k \rangle_{\mathcal{P}^*, \mathcal{P}} = \ll Q^{-1}\sigma e_j, e_k \gg_Q = 0, \quad \text{and } \langle \sigma \epsilon_j, \epsilon_k \rangle_{\mathcal{P}^*, \mathcal{P}} = 0 ,$$

as well as $\langle \sigma \epsilon_j, e_k \rangle_{\mathcal{P}^*, \mathcal{P}} = \delta_{j,k} \mu_j$. As a result the basis

$$(\mu_1^{-1/2} e_1, \ldots, \mu_n^{-1/2} e_n, \mu_1^{-1/2}\epsilon_1, \ldots, \mu_n^{-1/2}\epsilon_n)$$

is at the same time an orthogonal basis for Q and a symplectic basis in the sense of Proposition 4.4.2. To get the statement on Q^σ we calculate

$$\begin{pmatrix} 0 & -I_n \\ I_n & 0 \end{pmatrix} \begin{pmatrix} \mu^{-1} & 0 \\ 0 & \mu^{-1} \end{pmatrix} \begin{pmatrix} 0 & I_n \\ -I_n & 0 \end{pmatrix} = \begin{pmatrix} \mu^{-1} & 0 \\ 0 & \mu^{-1} \end{pmatrix}.$$

The proof of the lemma is complete. $\square$

Let Q_1 and Q_2 be two positive-definite quadratic forms on a finite-dimensional real vector space $\mathbb{V}$. Let us recall the definition[10] of the geometric mean, denoted by $\sqrt{Q_1 \cdot Q_2}$. We consider the mappings

$$\kappa_C = \begin{pmatrix} Q_1 & C \\ C & Q_2 \end{pmatrix}$$

from $\mathbb{V} \oplus \mathbb{V}$ into $\mathbb{V}^* \oplus \mathbb{V}^*$ with a non-negative symmetric $C : \mathbb{V} \to \mathbb{V}^*$. The geometric mean of Q_1 and Q_2 is the largest C such that $\kappa_C \ge 0$. Considering $(\mathbb{V}, Q_1)$ as a Euclidean space, we can identify Q_1 with the identity matrix and Q_2 with a positive-definite matrix B: in that situation, the geometric mean is simply $B^{1/2} = \exp \frac{1}{2} \text{Log } B$ as given by (4.1.9). To get a more symmetrical matrix definition, identifying Q_1, Q_2 with positive-definite matrices, we define

$$\sqrt{Q_1 \cdot Q_2} = Q_1^{1/2}(Q_1^{-1/2} Q_2 Q_1^{-1/2})^{1/2} Q_1^{1/2}$$

and we verify that $\sqrt{Q_1 \cdot Q_2} = \sqrt{Q_2 \cdot Q_1}$: we must check

$$Q_1^{1/2}(Q_1^{-1/2} Q_2 Q_1^{-1/2})^{1/2} Q_1^{1/2} = Q_2^{1/2}(Q_2^{-1/2} Q_1 Q_2^{-1/2})^{1/2} Q_2^{1/2},$$

---

[10]See, e.g., the articles [3],[11] and the references therein.

which is equivalent to

$$(Q_1^{-1/2} Q_2 Q_1^{-1/2})^{1/2} = Q_1^{-1/2} Q_2^{1/2} (Q_2^{-1/2} Q_1 Q_2^{-1/2})^{1/2} Q_2^{1/2} Q_1^{-1/2},$$

which is equivalent to

$$Q_1^{-1/2} Q_2 Q_1^{-1/2} =$$
$$Q_1^{-1/2} Q_2^{1/2} (Q_2^{-1/2} Q_1 Q_2^{-1/2})^{1/2} Q_2^{1/2} Q_1^{-1/2} Q_1^{-1/2} Q_2^{1/2} (Q_2^{-1/2} Q_1 Q_2^{-1/2})^{1/2} Q_2^{1/2} Q_1^{-1/2}$$

and the latter matrix is equal to

$$Q_1^{-1/2} Q_2^{1/2} (Q_2^{-1/2} Q_1 Q_2^{-1/2})^{1/2} Q_2^{1/2} Q_1^{-1} Q_2^{1/2} (Q_2^{-1/2} Q_1 Q_2^{-1/2})^{1/2} Q_2^{1/2} Q_1^{-1/2}$$
$$= Q_1^{-1/2} Q_2^{1/2} (Q_2^{-1/2} Q_1 Q_2^{-1/2})^{1/2} (Q_2^{-1/2} Q_1 Q_2^{-1/2})^{-1} (Q_2^{-1/2} Q_1 Q_2^{-1/2})^{1/2} Q_2^{1/2} Q_1^{-1/2}$$
$$= Q_1^{-1/2} Q_2^{1/2} Q_2^{1/2} Q_1^{-1/2} = Q_1^{-1/2} Q_2 Q_1^{-1/2}, \text{ qed.}$$

Definition 4.4.26. Let Q_1 and Q_2 be two positive-definite quadratic forms on a finite-dimensional real vector space $\mathbb{V}$. We define

$$Q_1 \wedge Q_2 = 2(Q_1^{-1} + Q_2^{-1})^{-1}, \quad \text{the harmonic mean,} \tag{4.4.23}$$

$$\sqrt{Q_1 \cdot Q_2} \quad \text{the geometric mean,} \tag{4.4.24}$$

$$Q_1 \vee Q_2 = \frac{1}{2}(Q_1 + Q_2), \quad \text{the arithmetic mean.} \tag{4.4.25}$$

Considering the Euclidean space $(\mathbb{V}, Q_1)$, we can identify Q_2 with a symmetric positive-definite matrix: that matrix has positive real eigenvalues in an orthonormal basis (for Q_1) of eigenvectors. We can thus consider that Q_1 is the identity matrix and Q_2 is a diagonal matrix $\text{diag}(\lambda_1, \ldots, \lambda_d), d = \dim \mathbb{V}, \lambda_j > 0$. Then we identify

$$Q_1 \wedge Q_2 = 2 \, \text{diag}(1 + \lambda_j^{-1})^{-1}, \quad \sqrt{Q_1 Q_2} = \text{diag}(\lambda_j^{1/2}), \quad Q_1 \vee Q_2 = \text{diag}(\frac{1 + \lambda_j}{2})$$

and we obtain readily

$$Q_1 \wedge Q_2 \leq \sqrt{Q_1 Q_2} \leq Q_1 \vee Q_2, \quad Q_1 \wedge Q_2 \leq 2Q_j \; (j = 1, 2). \tag{4.4.26}$$

Also for a positive-definite quadratic form Q on the vector space $\mathbb{V}$, $T \in \mathbb{V}$, we use the notation $Q(T) = \langle QT, T \rangle_{\mathbb{V}^*, \mathbb{V}}$ and we have

$$(Q_1 \wedge Q_2)(T) = 2 \inf_{T' + T'' = T} (Q_1(T') + Q_2(T'')). \tag{4.4.27}$$

In fact, for $T \in \mathbb{V}$, we define $\Xi = (Q_1^{-1} + Q_2^{-1})^{-1} T, \; T_j = Q_j^{-1} \Xi$ and (note $T_1 + T_2 = T$) and

$$2(Q_1^{-1} + Q_2^{-1})^{-1}(T) = 2 \langle (Q_1^{-1} + Q_2^{-1})^{-1} T, T \rangle$$
$$= 2 \langle \Xi, (Q_1^{-1} + Q_2^{-1}) \Xi \rangle = 2 \langle \Xi, Q_1^{-1} \Xi \rangle + 2 \langle \Xi, Q_2^{-1} \Xi \rangle$$
$$= 2 \langle Q_1 T_1, T_1 \rangle + 2 \langle Q_2 T_2, T_2 \rangle.$$

Moreover, we have for $S \in \mathbb{V}$, T_j as above,

$$2\langle Q_1(T_1 + S), (T_1 + S)\rangle + 2\langle Q_2(T_2 - S), (T_2 - S)\rangle$$
$$= \underbrace{2\langle Q_1 T_1, T_1\rangle + 2\langle Q_2 T_2, T_2\rangle}_{=2(Q_1^{-1}+Q_2^{-1})^{-1}(T)}$$
$$+ \underbrace{4\langle Q_1 T_1, S\rangle - 4\langle Q_2 T_2, S\rangle}_{=0} + \underbrace{2\langle Q_1 S, S\rangle + 2\langle Q_2 S, S\rangle}_{\geq 0},$$

giving (4.4.27).

Definition 4.4.27. Let $(\mathcal{P}, \sigma)$ be a symplectic vector space of dimension $2n$ and let Q be a positive-definite quadratic form on $\mathcal{P}$. We define

$$Q^\natural = \sqrt{Q \cdot Q^\sigma} \tag{4.4.28}$$

where Q^σ is defined by (4.4.18). According to Lemma 4.4.25, there exists a symplectic basis $(e_1, \ldots, e_n, \epsilon_1, \ldots, \epsilon_n)$ of $\mathcal{P}$ (in the sense of Proposition 4.4.2) which is also an orthogonal basis for Q. From Lemma 4.4.25, in that basis, the matrix of $Q^\natural$ is I_{2n}. As a result, we have always $Q^\natural = (Q^\natural)^\sigma$ and the implication

$$Q \leq Q^\sigma \implies Q \leq Q^\natural = (Q^\natural)^\sigma \leq Q^\sigma. \tag{4.4.29}$$

Remark 4.4.28. If Q_1, Q_2 are positive-definite quadratic forms on a symplectic vector space $\mathcal{P}$, we have

$$Q_1^\sigma \wedge Q_2^\sigma = 2\big((Q_1^\sigma)^{-1} + (Q_2^\sigma)^{-1}\big)^{-1} = 2\big(\sigma^{-1}Q_1\sigma^{*-1} + \sigma^{-1}Q_2\sigma^{*-1}\big)^{-1}$$
$$= \sigma^* 2\big(Q_1 + Q_2\big)^{-1}\sigma,$$

so that
$$Q_1^\sigma \wedge Q_2^\sigma = \big(Q_1 \vee Q_2\big)^\sigma. \tag{4.4.30}$$

Remark 4.4.29. Similarly with $Q_1, \ldots, Q_k$ positive-definite quadratic forms on a symplectic vector space $\mathcal{P}$, we can verify that

$$Q_1^\sigma \wedge \cdots \wedge Q_k^\sigma = (Q_1 \vee \cdots \vee Q_k)^\sigma, \tag{4.4.31}$$

where $Q_1 \vee \cdots \vee Q_k = k^{-1}(Q_1 + \cdots + Q_k)$ is the arithmetic mean and

$$Q_1 \wedge \cdots \wedge Q_k = k(Q_1^{-1} + \cdots + Q_k^{-1})^{-1} \tag{4.4.32}$$

is the harmonic mean. We have also, when $k \geq 2$,

$$\left(\frac{1}{k-1}(Q_1 \wedge \cdots \wedge Q_{k-1})\right) \wedge Q_k = \frac{2}{k}(Q_1 \wedge \cdots \wedge Q_k). \tag{4.4.33}$$

In fact, we have

$$\left(\left(\frac{1}{k-1}(Q_1 \wedge \cdots \wedge Q_{k-1})\right) \wedge Q_k\right)^{-1} = ((Q_1^{-1} + \cdots + Q_{k-1}^{-1})^{-1} \wedge Q_k)^{-1}$$

$$= 2^{-1}(Q_1^{-1} + \cdots + Q_{k-1}^{-1} + Q_k^{-1}) = \frac{k}{2}(Q_1 \wedge \cdots \wedge Q_k)^{-1}.$$

As a consequence, from (4.4.32), (4.4.27), we have for $k \geq 2$, $T \in \mathcal{P}$,

$$\frac{2}{k}(Q_1 \wedge \cdots \wedge Q_k)(T) = \inf_{T'+T''=T}\left\{\frac{2}{k-1}(Q_1 \wedge \cdots \wedge Q_{k-1})(T') + 2Q_k(T'')\right\}, \quad (4.4.34)$$

implying that

$$\frac{1}{k}(Q_1 \wedge \cdots \wedge Q_k)(T' + T'') \leq \frac{1}{k-1}(Q_1 \wedge \cdots \wedge Q_{k-1})(T') + Q_k(T''). \quad (4.4.35)$$

Using the notation

$$\Delta_{(j)}^{(j+1)} = \sup_{T \neq 0} \frac{Q_{j+1}(T)}{Q_j(T)}, \quad 1 \leq j < k, \quad (4.4.36)$$

we have, since the function $t \mapsto t^{-1}$ is monotone-matrix[11] (decreasing) on the positive-definite quadratic forms,

$$\Delta_{(1)}^{(2)}Q_1 \geq Q_2 \implies Q_1^{-1} \leq Q_2^{-1}\Delta_{(1)}^{(2)}$$

$$\implies Q_1^{-1} + \cdots + Q_k^{-1} \leq (1 + \Delta_{(1)}^{(2)})(Q_2^{-1} + \cdots + Q_k^{-1})$$

and thus $(Q_1^{-1} + \cdots + Q_k^{-1})^{-1} \geq (1 + \Delta_{(1)}^{(2)})^{-1}(Q_2^{-1} + \cdots + Q_k^{-1})^{-1}$. This implies readily

$$(Q_1^{-1} + \cdots + Q_k^{-1})^{-1} \geq (1 + \Delta_{(1)}^{(2)})^{-1}(1 + \Delta_{(2)}^{(3)})^{-1} \ldots (1 + \Delta_{(k-1)}^{(k)})^{-1}Q_k$$

and

$$\frac{1}{k}(Q_1 \wedge \cdots \wedge Q_k) \geq Q_k \prod_{1 \leq j < k}(1 + \Delta_{(j)}^{(j+1)})^{-1}. \quad (4.4.37)$$

Remark 4.4.30. If Q_1, Q_2 are conformal positive-definite quadratic forms on a symplectic vector space, we have with a scalar quantity μ that $Q_2 = \mu Q_1$ so that $Q_2^\sigma = \mu^{-1}Q_1^\sigma$ and $Q_2^\natural = Q_1^\natural$.

---

[11]We use only here that, for Q_1, Q_2 positive-definite quadratic forms, the inequality $Q_1 \leq Q_2$ implies $Q_2^{-1} \leq Q_1^{-1}$, a fact that can be proven directly by simultaneous diagonalization. The book [39] provides much more refined information on the topic of monotone matrix functions.

4.5   Symplectic geometry

4.5.1   Symplectic manifolds

Let (M, σ) be a C^∞ symplectic manifold, i.e., a C^∞ differentiable manifold equipped with a closed 2-form σ. For a smooth function f defined on M, we define the Hamiltonian vector field H_f of f by the identity

$$\sigma \lrcorner H_f = -df, \qquad (4.5.1)$$

where $\lrcorner$ is the interior product. We define also the Poisson bracket of two functions f, g by

$$\{f, g\} = \langle \sigma, H_f \wedge H_g \rangle \qquad (4.5.2)$$

and we note that $H_f(g) = \{f, g\}$ since

$$H_f(g) = \langle dg, H_f \rangle = -\langle \sigma \lrcorner H_g, H_f \rangle = -\langle \sigma, H_g \wedge H_f \rangle = \{f, g\}.$$

We have also, with $\mathscr{L}_X$ standing for the Lie derivative,

$$\mathscr{L}_{H_f}(\sigma) = d\sigma \lrcorner H_f + d(\sigma \lrcorner H_f) = 0, \quad \text{since } \sigma \text{ is closed and } \sigma \lrcorner H_f \text{ is exact,}$$

so that with the volume form given by $\omega = \sigma^{\wedge n}/n!$, we have

$$0 = \mathscr{L}_{H_f}(\omega) = (\operatorname{div} H_f)\omega,$$

implying the Liouville theorem $\operatorname{div} H_f = 0$. Let us prove the Jacobi identity,

$$[H_f, H_g] = H_{\{f,g\}}. \qquad (4.5.3)$$

We have, using $\mathscr{L}_{H_f}(\sigma) = 0$ and the commutation of the Lie derivative with the exterior derivative,

$$\sigma \lrcorner [H_f, H_g] = \sigma \lrcorner \mathscr{L}_{H_f}(H_g) = \mathscr{L}_{H_f}(\sigma \lrcorner H_g) = -\mathscr{L}_{H_f}(dg) = -d(\mathscr{L}_{H_f}(g))$$
$$= -d(H_f(g)) = -d\{f, g\},$$

so that (4.5.1) gives (4.5.3). Another way of writing the Jacobi identity is

$$\{f, \{g, h\}\} + \{g, \{h, f\}\} + \{h, \{f, g\}\} = 0, \qquad (4.5.4)$$

for f, g, h differentiable functions on M. In fact (4.5.3) implies

$$\{f, \{g, h\}\} = H_f H_g(h) = [H_f, H_g](h) + H_g H_f(h) = H_{\{f,g\}}(h) + \{g, \{f, h\}\}$$

entailing $\{f, \{g, h\}\} = \{\{f, g\}, h\} + \{g, \{f, h\}\}$ which is (4.5.4).

The canonical symplectic structure on the phase space $M = \mathbb{R}^n_x \times \mathbb{R}^n_\xi$ is given by (2.1.1), (2.1.2), i.e., by

$$\sigma = \sum_{1 \le j \le n} d\xi_j \wedge dx_j \qquad (4.5.5)$$

and the Hamiltonian vector field of a function f defined on (an open subset of) $\mathbb{R}^{2n}$ is

$$H_f = \sum_{1 \le j \le n} \frac{\partial f}{\partial \xi_j} \frac{\partial}{\partial x_j} - \frac{\partial f}{\partial x_j} \frac{\partial}{\partial \xi_j}, \tag{4.5.6}$$

since

$$(\sigma \lrcorner H_f) \cdot X = \sigma(\partial_\xi f \cdot \partial_x - \partial_x f \cdot \partial_\xi, a \cdot \partial_x + b \cdot \partial_\xi) = -\partial_x f \cdot a - \partial_\xi f \cdot b = -df \cdot (a \cdot \partial_x + b \cdot \partial_\xi).$$

The Poisson bracket $\{f, g\}$ coincides with (1.1.23) with $\{f, g\} = H_f(g)$.

4.5.2 Normal forms of functions

Proposition 4.5.1. *Let p be a C^∞ complex-valued function positively-homogeneous of degree 1 in a conic-neighborhood of $(x_0, \xi_0) \in \dot{T}^*(\mathbb{R}^n)$ such that $p(x_0, \xi_0) = 0, \partial_\xi p(x_0, \xi_0) \ne 0$. Then there exist C^∞ canonical coordinates (y, η) in a conic-neighborhood Γ_0 of (x_0, ξ_0) with y (resp. η) homogeneous of degree 0 (resp. 1) so that*

$$for \ (x, \xi) \in \Gamma_0, \quad p(x, \xi) = e(x, \xi)(\eta_1 + if(y, \eta)), \tag{4.5.7}$$

with $e \in C^\infty$, positively-homogeneous of degree 0, non-vanishing on Γ_0, $f \in C^\infty$, homogeneous of degree 1, real-valued and independent of η_1 and such that at (x_0, ξ_0), $\eta_1 = 0 = f(y, \eta)$.

Proof. We may assume that $\partial_{\xi_1} \operatorname{Re} p(x_0, \xi_0) \ne 0$, $x_0 = 0, \xi_0 = e_n$ and using the Malgrange-Weierstrass theorem (see Theorem 7.5.5 in [71]), we can find a conic-neighborhood of (x_0, ξ_0) in which

$$p(x, \xi) = \big(\xi_1 + a(x_1, x', \xi') + ib(x_1, x', \xi')\big) e_0(x, \xi)$$

where a, b are real-valued positively-homogeneous of degree 1 and e is non-vanishing, homogeneous of degree 0. Noting that the Poisson bracket

$$\{\xi_1 + a, x_1\} \equiv 1,$$

we use the homogeneous version of the Darboux theorem (see Theorem 21.1.9 in [73]), we take $\eta_1 = \xi_1 + a(x_1, x', \xi'), y_1 = x_1$, and we find $y_2, \ldots, y_n$ homogeneous of degree 0, $\eta_2, \ldots, \eta_n$ homogeneous of degree 1, such that (y, η) is a local system of symplectic coordinates, i.e.,

$$d\eta_1 \wedge \cdots \wedge d\eta_n \wedge dy_1 \wedge \cdots \wedge dy_n \ne 0, \quad \{\eta_j, y_k\} = \delta_{j,k}, \quad \{\eta_j, \eta_k\} = \{y_j, y_k\} = 0.$$

We have

$$p(x, \xi) = e_0(x, \xi)\big(\eta_1 + if(y, \eta)\big)$$

with $f(y, \eta) = (b \circ \chi)(y, \eta)$ and $(x, \xi) = \chi(y, \eta)$ a local symplectomorphism. We note that

$$-\partial_{\eta_1} f = \{y_1, f\} = \{x_1, b\} \circ \chi = 0,$$

so that f does not depend on η_1, completing the proof. $\qquad \square$

4.6   Composing a large number of symbols

Let us begin by revisiting the composition formula in the Weyl quantization.

Theorem 4.6.1. *Let $(a_j)_{1 \le j \le 2k+1}$ be some $L^1(\mathbb{R}^{2n})$ functions. Then for $X \in \mathbb{R}^{2n} = \mathcal{P}$, we have*

$$(a_1 \natural \ldots \natural a_{2k})(X)$$
$$= 2^{2nk} \int_{\mathcal{P}^{2k}} a_1(Y_1) \ldots a_{2k}(Y_{2k}) \exp 4i\pi \sum_{1 \le j < l \le 2k} (-1)^{j+l}[X - Y_j, X - Y_l] dY,$$

$$(4.6.1)$$

and

$$(a_1 \natural \ldots \natural a_{2k} \natural a_{2k+1})(X)$$
$$= 2^{2nk} \int_{\mathcal{P}^{2k}} a_1(Y_1) \ldots a_{2k}(Y_{2k}) a_{2k+1}\Big(X + \sum_{1 \le j \le 2k} (-1)^j Y_j\Big)$$
$$\times \exp 4i\pi \sum_{1 \le j < l \le 2k} (-1)^{j+l}[X - Y_j, X - Y_l] dY. \qquad (4.6.2)$$

N.B. We can note that, thanks to (2.1.13), the operators a_j^w are bounded on $L^2(\mathbb{R}^n)$, so that the composition $a_1^w \ldots a_k^w$ makes sense as a bounded operator on $L^2(\mathbb{R}^n)$. The formulae above provide $a_1 \natural \ldots \natural a_{2k+1}$ as an $L^1(\mathbb{R}^{2n})$ function and $a_1 \natural \ldots \natural a_{2k}$ as an $L^\infty(\mathbb{R}^{2n})$ function and their Weyl quantization continuously sends $\mathscr{S}(\mathbb{R}^n)$ into $\mathscr{S}'(\mathbb{R}^n)$ (see (2.1.8)).

Proof. Let us first prove by induction on k that, for Y_j in $\mathbb{R}^{2n}$,

$$\sigma_{Y_1} \ldots \sigma_{Y_{2k+1}} = e^{4i\pi(\sum_{1 \le j < l \le 2k+1}(-1)^{j+l}[Y_j, Y_l])} \sigma_{\sum_{1 \le j \le 2k+1}(-1)^{j+1}Y_j}. \qquad (4.6.3)$$

We note first that, from Lemma 2.1.3, we have

$$\sigma_{Y_1} \sigma_{Y_2} \sigma_{Y_3} = \tau_{2Y_1 - 2Y_2} \sigma_{Y_3} e^{4i\pi[Y_2, Y_1]} = e^{4i\pi\big([Y_1, Y_3] - [Y_1, Y_2] - [Y_2, Y_3]\big)} \sigma_{Y_1 - Y_2 + Y_3},$$

which is the above formula for $k = 1$. Assuming that formula true for some $k \ge 1$ gives

$$S = \sigma_{Y_1} \ldots \sigma_{Y_{2k+1}} \sigma_{Y_{2k+2}} \sigma_{Y_{2k+3}}$$
$$= e^{4i\pi(\sum_{1 \le j < l \le 2k+1}(-1)^{j+l}[Y_j, Y_l])} \sigma_{\sum_{1 \le j \le 2k+1}(-1)^{j+1}Y_j} \sigma_{Y_{2k+2}} \sigma_{Y_{2k+3}},$$

so that, using the proven formula for $k = 1$ gives the sought result

$$S = \sigma_{2 \sum_{1 \le j \le 2k+3}(-1)^{j+1}Y_j} e^{4i\pi(\sum_{1 \le j < l \le 2k+3}(-1)^{j+l}[Y_j, Y_l])},$$

since

$$\sum_{1\le j<l\le 2k+3}(-1)^{j+l}[Y_j,Y_l] = \sum_{1\le j<l\le 2k+1}(-1)^{j+l}[Y_j,Y_l] + \sum_{1\le j\le 2k+1}(-1)^{j+1}[Y_j,Y_{2k+3}]$$
$$- \sum_{1\le j\le 2k+1}(-1)^{j+1}[Y_j,Y_{2k+2}] - [Y_{2k+2},Y_{2k+3}].$$

With (4.6.3), we get

$$2^{-n(2k+1)}a_1^w\ldots a_{2k+1}^w$$
$$=\int_{\mathbb{R}^{2n(2k+1)}}\prod_{1\le j\le 2k+1}a_j(Y_j)e^{4i\pi(\sum_{1\le j<l\le 2k+1}(-1)^{j+l}[Y_j,Y_l])}\sigma_{\sum_{1\le j\le 2k+1}(-1)^{j+1}Y_j}dY$$
$$=\iint_{\mathbb{R}^{4nk}\times\mathbb{R}^{2n}}\prod_{1\le j\le 2k}a_j(Y_j)a_{2k+1}\Big(X+\sum_{1\le j\le 2k}(-1)^j Y_j\Big)$$
$$\times e^{4i\pi(\sum_{1\le j<l\le 2k}(-1)^{j+l}[Y_j,Y_l]-\sum_{1\le j\le 2k}(-1)^j[Y_j,X+\sum_{1\le l\le 2k}(-1)^l Y_l])}dY\sigma_X dX.$$
$$(4.6.4)$$

The factor Φ of $4i\pi$ in the phase in the previous integral is

$$\Phi = \sum_{1\le j<l\le 2k}(-1)^{j+l}[Y_j,Y_l] - \sum_{1\le j\le 2k}(-1)^j[Y_j,X+\sum_{1\le l\le 2k}(-1)^l Y_l],$$

so that

$$\Phi = -\sum_{1\le l<j\le 2k}(-1)^{j+l}[Y_j,Y_l] - \sum_{1\le j\le 2k}(-1)^j[Y_j,X]$$
$$= -\sum_{1\le l<j\le 2k}(-1)^{j+l}[Y_j,Y_l] - \sum_{1\le j\le 2k}(-1)^j[Y_j,X]$$
$$= -\sum_{1\le l<j\le 2k}(-1)^{j+l}[Y_j-X,Y_l-X] - \sum_{1\le l<j\le 2k}(-1)^{j+l}\big([X,Y_l]+[Y_j,X]\big)$$
$$- \sum_{1\le j\le 2k}(-1)^j[Y_j,X].$$

We remark that

$$0 = \sum_{1\le l<j\le 2k}(-1)^{j+l}\big([X,Y_l]+[Y_j,X]\big) + \sum_{1\le j\le 2k}(-1)^j[Y_j,X],$$

since with $Z_j = (-1)^j Y_j$,

$$\sum_{1 \leq l < j \leq 2k} (-1)^{j+l}(Y_j - Y_l) + \sum_{1 \leq j \leq 2k} (-1)^j Y_j$$

$$= \sum_{1 \leq l < j \leq 2k} Z_j (-1)^l - \sum_{1 \leq l < j \leq 2k} (-1)^j Z_l + \sum_{1 \leq j \leq 2k} Z_j$$

$$= \sum_{1 \leq l < j \leq 2k} Z_j (-1)^l - \sum_{1 \leq j < l \leq 2k} (-1)^l Z_j + \sum_{1 \leq j \leq 2k} Z_j$$

$$= \sum_{1 \leq j \leq 2k} Z_j \Big(1 + \sum_{l,\, 1 \leq l < j} (-1)^l - \sum_{l,\, j < l \leq 2k} (-1)^l \Big)$$

$$= \sum_{1 \leq j \leq 2k} Z_j \sum_{l,\, 0 \leq l \leq 2k-1} (-1)^l = 0.$$

As a result, we get from the formula (4.6.4) and the expression of Φ above, that

$$a_1^w \ldots a_{2k+1}^w = 2^{2nk} \iint \prod_{1 \leq j \leq 2k} a_j(Y_j) a_{2k+1}\Big(X + \sum_{1 \leq j \leq 2k} (-1)^j Y_j \Big)$$

$$\times e^{-4i\pi(\sum_{1 \leq l < j \leq 2k} (-1)^{j+l}[Y_j - X, Y_l - X])} dY 2^n \sigma_X dX,$$

yielding the formula (4.6.2) (note the different rôles of the indices j, l here and in the statement of the theorem). To obtain (4.6.1), we choose $a_{2k+1} = 1$ in (4.6.2) (note that the previous calculations use only $(a_j \in L^1(\mathbb{R}^{2n}))_{1 \leq j \leq 2k}$, $a_{2k+1} \in L^\infty(\mathbb{R}^{2n})$, $a_{2k+1}^w \in \mathcal{L}(L^2(\mathbb{R}^n))$). The proof of the theorem is complete. $\square$

We want now to introduce the notion of multiconfinement, generalizing the confinement and biconfinement estimates of Definition 2.3.1 and Theorem 2.3.2.

Theorem 4.6.2. *Let $n \geq 1$ be a given integer. There exists a positive constant κ_n such that the following property is satisfied. Let $\nu \geq 2$ be an integer. Let $g_1, \ldots, g_\nu$ be ν positive-definite quadratic forms on $\mathbb{R}^{2n}$ such that for all $j \in \{1, \ldots, \nu\}$, $g_j \leq g_j^\sigma$ and let a_j be g_j-confined in U_j, a g_j-ball of radius ≤ 1 (see Definition 2.3.1 and (4.4.18)). Then for all $X \in \mathbb{R}^{2n}, N \in \mathbb{N}$,*

$$|(a_1 \sharp \ldots \sharp a_\nu)(X)| \leq \kappa_n^\nu 2^{\nu 2N} \|a_1\|_{g_1, U_1}^{(N+2n+1)} \ldots \|a_\nu\|_{g_\nu, U_\nu}^{(N+2n+1)}$$

$$\times \Big(1 + \nu^{-1} \max_{1 \leq j \leq \nu} (g_1^\sigma \wedge \cdots \wedge g_\nu^\sigma)(X - U_j) \Big)^{-N/2}, \quad (4.6.5)$$

where the semi-norms $\|a\|_{g,U}^{(l)}$ are defined in (2.3.3).

N.B. The point of this result is the precise control of the constants and the semi-norms: the constant κ_n depends only on the dimension n, and for each N, to achieve the required decay, we need only to control $N + 2n + 1$ semi-norms of each a_j, a "shift" which *does not depend on* ν.

Proof. We check first

$$(a_1 \sharp a_2)(X) = 2^{2n} \iint e^{-4i\pi[X-Y,X-Z]} a_1(Y) a_2(Z) dY dZ.$$

For X, Y given in $\mathbb{R}^{2n}$, we choose θ such that $[X - Y, \theta] = g_2^\sigma(X - Y)^{1/2}$, $g_2(\theta) = 1$. The vector θ depends on $X - Y$, but not on Z, so that, with $N' \in \mathbb{N}$,

$$(i + \frac{1}{2}\theta \cdot D_Z)^{N'} (e^{-4i\pi[X-Y,X-Z]}) = (i + g_2^\sigma(X-Y)^{1/2})^{N'} e^{-4i\pi[X-Y,X-Z]}$$

and

$$(a_1 \sharp a_2)(X) = 2^{2n} \iint e^{-4i\pi[X-Y,X-Z]}$$
$$\times (i + g_2^\sigma(X-Y)^{1/2})^{-N'} a_1(Y) ((i - \frac{1}{2}\theta \cdot D_Z)^{N'} a_2)(Z) dY dZ,$$

so that with $N_1, N_2 \in \mathbb{N}$,

$$|(a_1 \sharp a_2)(X)| \le 2^{2n} \iint (1 + g_2^\sigma(X-Y))^{-N'/2} \|a_1\|_{g_1, U_1}^{(0, N_1)} (1 + g_1^\sigma(Y - U_1))^{-N_1/2}$$
$$\times (1 + \frac{1}{4\pi})^{N'} \|a_2\|_{g_2, U_2}^{(N', N_2)} (1 + g_2^\sigma(Z - U_2))^{-N_2/2} dY dZ.$$

We take now $N' = N_1 + N_2$, and we get

$$|(a_1 \sharp a_2)(X)| \le 2^{2n} (1 + \frac{1}{4\pi})^{N_1+N_2} \|a_1\|_{g_1, U_1}^{(0, N_1)} \|a_2\|_{g_2, U_2}^{(N_1+N_2, N_2)}$$
$$\times (1 + (g_1^\sigma \wedge g_2^\sigma)(X - U_1))^{-N_1/2} 2^{N_1/2}$$
$$\times \iint (1 + g_2^\sigma(X-Y))^{-N_2/2} (1 + g_2^\sigma(Z - U_2))^{-N_2/2} dY dZ.$$

Using the assumption $g_2 \le g_2^\sigma$ and that U_2 is a g_2-ball with radius ≤ 1, we obtain, with X_2 the center of U_2,

$$1 + g_2(Z - X_2)^{1/2} \le 2 + g_2(Z - U_2)^{1/2} \le 2(1 + g_2^\sigma(Z - U_2)^{1/2}),$$

so that

$$|(a_1 \sharp a_2)(X)| \le 2^{2n} (1 + \frac{1}{4\pi})^{N_1+N_2} \|a_1\|_{g_1, U_1}^{(0, N_1)} \|a_2\|_{g_2, U_2}^{(N_1+N_2, N_2)} 2^{N_1/2 + N_2/2}$$
$$\times (1 + (g_1^\sigma \wedge g_2^\sigma)(X - U_1))^{-N_1/2}$$
$$\times \iint (1 + g_2^\sigma(X-Y))^{-N_2/2} (1 + g_2(Z - X_2)^{1/2})^{-N_2} dY dZ.$$

Choosing $N_2 = 2n + 1$ gives, with

$$c(n) = 2^{3n+\frac{1}{2}}(1+\frac{1}{4\pi})^{2n+1}\iint(1+g_2^\sigma(X-Y))^{-n-\frac{1}{2}}(1+g_2(Z-X_2)^{1/2})^{-2n-1}dYdZ,$$

$$(4.6.6)$$

which depends only on n, that

$$|(a_1\sharp a_2)(X)| \le c(n)\|a_1\|_{g_1,U_1}^{(N_1)}\|a_2\|_{g_2,U_2}^{(N_1+2n+1)}2^{N_1}\left(1+(g_1^\sigma\wedge g_2^\sigma)(X-U_1)\right)^{-N_1/2}.$$

Using that $\overline{a_1\sharp a_2} = \overline{a_2}\sharp\overline{a_1}$, we get

$$|(a_1\sharp a_2)(X)| \le c(n)\|a_1\|_{g_1,U_1}^{(N_1+2n+1)}\|a_2\|_{g_2,U_2}^{(N_1+2n+1)}2^{N_1}$$

$$\times\left(1+\max((g_1^\sigma\wedge g_2^\sigma)(X-U_1),(g_1^\sigma\wedge g_2^\sigma)(X-U_2))\right)^{-N_1/2}$$

which gives (4.6.5) for $\nu = 2$, provided

$$\kappa_n \ge c(n)^{1/2}. \qquad (4.6.7)$$

We start now an induction on ν: the theorem is proven true for $\nu = 2$ and we assume its conclusion for some $\nu \ge 2$. We write

$$(a_1\sharp\ldots\sharp a_\nu\sharp a_{\nu+1})(X) = 2^{2n}\iint e^{-4i\pi[X-Y,X-Z]}(a_1\sharp\ldots\sharp a_\nu)(Y)a_{\nu+1}(Z)dYdZ,$$

and following exactly the previous proof, for X, Y given in $\mathbb{R}^{2n}$, we choose θ such that $[X-Y,\theta] = g_{\nu+1}^\sigma(X-Y)^{1/2}, g_{\nu+1}(\theta) = 1$. The vector θ depends on $X-Y$, but not on Z, so that, with $N' \in \mathbb{N}$,

$$(i+\frac{1}{2}\theta\cdot D_Z)^{N'}(e^{-4i\pi[X-Y,X-Z]}) = (i+g_{\nu+1}^\sigma(X-Y)^{1/2})^{N'}e^{-4i\pi[X-Y,X-Z]}$$

and

$$(a_1\sharp\ldots\sharp a_\nu\sharp a_{\nu+1})(X) = 2^{2n}\iint e^{-4i\pi[X-Y,X-Z]}$$

$$\times(i+g_{\nu+1}^\sigma(X-Y)^{1/2})^{-N'}(a_1\sharp\ldots\sharp a_\nu)(Y)((i-\frac{1}{2}\theta\cdot D_Z)^{N'}a_{\nu+1})(Z)dYdZ,$$

so that with $N_1, M_1, N_2 \in \mathbb{N}$, using the induction hypothesis, we get for $j_0 \in \{1,\ldots,\nu\}$,

$$|(a_1\sharp\ldots\sharp a_\nu\sharp a_{\nu+1})(X)| \le$$

$$2^{2n}\iint(1+g_{\nu+1}^\sigma(X-Y))^{-N'/2}\kappa_n^\nu 2^{\nu 2N_1}\prod_{1\le j\le\nu}\|a_j\|_{g_j,U_j}^{(N_1+2n+1)}$$

$$\times\left(1+\nu^{-1}(g_1^\sigma\wedge\cdots\wedge g_\nu^\sigma)(Y-U_{j_0})\right)^{-N_1/2}$$

$$\times(1+\frac{1}{4\pi})^{N'}\|a_{\nu+1}\|_{g_{\nu+1},U_{\nu+1}}^{(N',N_2)}(1+g_{\nu+1}^\sigma(Z-U_{\nu+1}))^{-N_2/2}dYdZ.$$

We take now $N' = N_1 + N_2$, and we get

$$|(a_1 \sharp \ldots \sharp a_\nu \sharp a_{\nu+1})(X)| \leq 2^{2n} \kappa_n^\nu 2^{\nu 2 N_1} \prod_{1 \leq j \leq \nu} \|a_j\|_{g_j, U_j}^{(N_1 + 2n + 1)}$$

$$\times \iint (1 + g_{\nu+1}^\sigma(X - Y))^{-N_2/2} (1 + g_{\nu+1}^\sigma(X - Y))^{-N_1/2}$$

$$\times \left(1 + \nu^{-1}(g_1^\sigma \wedge \cdots \wedge g_\nu^\sigma)(Y - U_{j_0})\right)^{-N_1/2}$$

$$\times (1 + \frac{1}{4\pi})^{N_1 + N_2} \|a_{\nu+1}\|_{g_{\nu+1}, U_{\nu+1}}^{(N_1 + N_2, N_2)} (1 + g_{\nu+1}^\sigma(Z - U_{\nu+1}))^{-N_2/2} dY \, dZ.$$

We check now

$$\left(1 + g_{\nu+1}^\sigma(X - Y)\right)\left(1 + \nu^{-1}(g_1^\sigma \wedge \cdots \wedge g_\nu^\sigma)(Y - U_{j_0})\right)$$

$$\geq \left(1 + \nu^{-1}(g_1^\sigma \wedge \cdots \wedge g_\nu^\sigma)(Y - U_{j_0}) + g_{\nu+1}^\sigma(X - Y)\right)$$

(from (4.4.35)) $\geq \left(1 + (\nu + 1)^{-1}(g_1^\sigma \wedge \cdots \wedge g_\nu^\sigma \wedge g_{\nu+1}^\sigma)(X - U_{j_0})\right),$

so that

$$|(a_1 \sharp \ldots \sharp a_\nu \sharp a_{\nu+1})(X)| \leq 2^{2n} \kappa_n^\nu 2^{\nu 2 N_1} \prod_{1 \leq j \leq \nu} \|a_j\|_{g_j, U_j}^{(N_1 + 2n + 1)}$$

$$\times \left(1 + (\nu + 1)^{-1}(g_1^\sigma \wedge \cdots \wedge g_\nu^\sigma \wedge g_{\nu+1}^\sigma)(X - U_{j_0})\right)^{-N_1/2}$$

$$\times (1 + \frac{1}{4\pi})^{N_1 + N_2} \iint (1 + g_{\nu+1}^\sigma(X - Y))^{-N_2/2}$$

$$\times \|a_{\nu+1}\|_{g_{\nu+1}, U_{\nu+1}}^{(N_1 + N_2, N_2)} (1 + g_{\nu+1}^\sigma(Z - U_{\nu+1}))^{-N_2/2} dY \, dZ.$$

Using the assumption $g_{\nu+1} \leq g_{\nu+1}^\sigma$ and that $U_{\nu+1}$ is a g_ν-ball with radius ≤ 1, we obtain, with $X_{\nu+1}$ the center of $U_{\nu+1}$,

$$1 + g_{\nu+1}(Z - X_{\nu+1})^{1/2} \leq 2 + g_{\nu+1}(Z - U_{\nu+1})^{1/2} \leq 2\left(1 + g_{\nu+1}^\sigma(Z - U_{\nu+1})^{1/2}\right),$$

so that choosing $N_2 = 2n + 1$ gives, with the same $c(n)$ as above in (4.6.6), that

$$|(a_1 \sharp \ldots \sharp a_\nu \sharp a_{\nu+1})(X)| \leq c(n) \kappa_n^\nu 2^{\nu 2 N_1} \left(\prod_{1 \leq j \leq \nu} \|a_j\|_{g_j, U_j}^{(N_1 + 2n + 1)}\right) \|a_{\nu+1}\|_{g_{\nu+1}, U_{\nu+1}}^{(N_1 + 2n + 1)}$$

$$\times \left(1 + (\nu + 1)^{-1} \max_{1 \leq j \leq \nu} (g_1^\sigma \wedge \cdots \wedge g_\nu^\sigma \wedge g_{\nu+1}^\sigma)(X - U_j)\right)^{-N_1/2} 2^{N_1}.$$

Using that $\overline{a_1 \sharp \ldots \sharp a_{\nu+1}} = \overline{a_{\nu+1}} \sharp \ldots \sharp \overline{a_1}$, we get that

$$|(a_1 \sharp \ldots \sharp a_\nu \sharp a_{\nu+1})(X)| \leq c(n) \kappa_n^\nu 2^{\nu 2 N_1} 2^{N_1} \prod_{1 \leq j \leq \nu+1} \|a_j\|_{g_j, U_j}^{(N_1 + 2n + 1)}$$

$$\times \left(1 + (\nu + 1)^{-1} \max_{1 \leq j \leq \nu+1} (g_1^\sigma \wedge \cdots \wedge g_\nu^\sigma \wedge g_{\nu+1}^\sigma)(X - U_j)\right)^{-N_1/2},$$

so that with $N_1 = N$, we obtain

$$|(a_1 \sharp \ldots \sharp a_\nu \sharp a_{\nu+1})(X)| \le c(n) \kappa_n^\nu 2^{\nu 2N} 2^N \prod_{1 \le j \le \nu+1} \|a_j\|_{g_j, U_j}^{(N_1 + 2n + 1)}$$

$$\left(1 + (\nu+1)^{-1} \max_{1 \le j \le \nu+1} (g_1^\sigma \wedge \cdots \wedge g_\nu^\sigma \wedge g_{\nu+1}^\sigma)(X - U_j)\right)^{-N/2}$$

which concludes the induction, since we may assume that $\kappa_n \ge \max(c(n)^{1/2}, c(n))$ so that (4.6.7) holds and

$$c(n) \kappa_n^\nu 2^{\nu 2N} 2^N \le \kappa_n^{\nu+1} 2^{(\nu+1)2N}.$$

The proof of the theorem is complete.   □

N.B. The reader may be puzzled by the occurrence of the term ν^{-1} near the max in the formula (4.6.5). The induction proof relies on the formula (4.4.35) and, although it is possible afterwards to replace the ν^{-1} in (4.6.5) by 1 and change the $2^{2\nu N}$ into $2^{2\nu N} \nu^{N/2}$ (a harmless modification), it is indeed the precise (4.6.5) that we are able to prove by induction on ν. After all, a version of this theorem will allow us to define analytic functions of an operator with symbol in $S(1, g)$, so it is quite natural to meet this type of difficulty, related to the control of derivatives of an analytic function. Trying to prove directly the formula (4.6.5) with ν replaced by 1, leads to a polynomial term in ν in the induction step: that term would trigger an unacceptable $\nu!$ in the final formula.

Remark 4.6.3. Looking at the statement of Theorem 2.3.2, we see that for g_1, g_2 two positive-definite quadratic forms on $\mathbb{R}^{2n}$ such that $g_j \le g_j^\sigma$ and for $a_j, j = 1, 2$ g_j-confined in U_j, a g_j-ball of radius ≤ 1, for all k, N, for all $X, T \in \mathbb{R}^{2n}$, we have

$$|(a_1 \sharp a_2)^{(k)} T^k|$$

$$\le A_{k,N} (g_1 + g_2)(T)^{k/2} \left(1 + (g_1^\sigma \wedge g_2^\sigma)(X - U_1) + (g_1^\sigma \wedge g_2^\sigma)(X - U_2)\right)^{-N/2},$$

with $A_{k,N} = \gamma(n, N, k) \|a_1\|_{g_1, U_1}^{(m)} \|a_2\|_{g_2, U_2}^{(m)}, \quad m = 2n + 1 + k + N$. With

$$\rho_{1,2} = 1 + \sup_{T \ne 0} \frac{g_2(T)}{g_1(T)} + \sup_{T \ne 0} \frac{g_1(T)}{g_2(T)}, \tag{4.6.8}$$

the previous estimate implies that, with $N', N'' \in \mathbb{N}$,

$$|(a_1 \sharp a_2)^{(k)} T^k|$$

$$\le \gamma(n, N' + N'', k) 2^{\frac{k + N' + N''}{2}} \|a_1\|_{g_1, U_1}^{(2n+1+k+N'+N'')} \|a_2\|_{g_2, U_2}^{(2n+1+k+N'+N'')}$$

$$\times \rho_{1,2}^{\frac{k+N'}{2}} \left(1 + (g_1^\sigma \wedge g_2^\sigma)(U_2 - U_1)\right)^{-N''/2} g_1(T)^{k/2} \left(1 + g_1^\sigma(X - U_1)\right)^{-N'/2}$$

so that $a_1 \sharp a_2$ is indeed g_1-confined in U_1 (and also g_2-confined in U_2). Assuming now that

$$\rho_{1,2} \leq C_0 \big(1 + (g_1^\sigma \wedge g_2^\sigma)(U_2 - U_1)\big)^{N_0}, \tag{4.6.9}$$

we get, for $k, N, N'' \in \mathbb{N}$,

$$\sup_{g_1(T)=1} |(a_1 \sharp a_2)^{(k)} T^k| \big(1 + g_1^\sigma(X - U_1)\big)^{N/2}$$

$$\leq \gamma(n, N+N'', k) 2^{\frac{k+N+N''}{2}} \|a_1\|_{g_1,U_1}^{(2n+1+k+N+N'')} \|a_2\|_{g_2,U_2}^{(2n+1+k+N+N'')}$$

$$\times C_0^{\frac{k+N}{2}} \big(1 + (g_1^\sigma \wedge g_2^\sigma)(U_2 - U_1)\big)^{-\frac{N''}{2} + \frac{N_0(k+N)}{2}},$$

so that with $N'' = N_0(k+N) + M$, we obtain, for $k, N, M \in \mathbb{N}$,

$$\sup_{g_1(T)=1} |(a_1 \sharp a_2)^{(k)} T^k| \big(1 + g_1^\sigma(X - U_1)\big)^{N/2}$$

$$\leq \Gamma(n, N_0, C_0, k, N, M) \|a_1\|_{g_1,U_1}^{(2n+1+(k+N)(N_0+1)+M)}$$

$$\times \|a_2\|_{g_2,U_2}^{(2n+1+(k+N)(N_0+1)+M)} \big(1 + (g_1^\sigma \wedge g_2^\sigma)(U_2 - U_1)\big)^{-\frac{M}{2}},$$

with

$$\Gamma(n, N_0, C_0, k, N, M) = \gamma(n, N(1+N_0) + kN_0 + M, k) 2^{\frac{(k+N)(N_0+1)+M}{2}} C_0^{\frac{k+N}{2}}.$$

This implies that for $j = 1, 2$, for all $l \in \mathbb{N}$,

$$\|a_1 \sharp a_2\|_{g_j,U_j}^{(l)} \leq \Gamma(n, N_0, C_0, l, l, M) \|a_1\|_{g_1,U_1}^{(2n+1+l(2N_0+2)+M)}$$

$$\times \|a_2\|_{g_2,U_2}^{(2n+1+l(2N_0+2)+M)} \big(1 + (g_1^\sigma \wedge g_2^\sigma)(U_2 - U_1)\big)^{-\frac{M}{2}}. \tag{4.6.10}$$

A consequence of the previous remark is the following result.

Theorem 4.6.4. *Let $n \geq 1$ be a given integer. Let $\nu \geq 2$ be an integer and $g_1, \ldots, g_\nu$ be ν positive-definite quadratic forms on $\mathbb{R}^{2n}$ such that for all $j \in \{1, \ldots, \nu\}, g_j \leq g_j^\sigma$ and let a_j be g_j-confined in U_j, a g_j-ball of radius ≤ 1 (see Definition 2.3.1 and (4.4.18)). With*

$$\delta_{j,j+1} = 1 + (g_j^\sigma \wedge g_{j+1}^\sigma)(U_{j+1} - U_j), \tag{4.6.11}$$

we assume that, for $j \in \mathbb{N}, 1 \leq j < \nu$,

$$\rho_{j,j+1} = 1 + \sup_{T \neq 0} \frac{g_{j+1}(T)}{g_j(T)} + \sup_{T \neq 0} \frac{g_j(T)}{g_{j+1}(T)} \leq C_0 \delta_{j,j+1}^{N_0}. \tag{4.6.12}$$

Then for all $X \in \mathbb{R}^{2n}, M \in \mathbb{N}$,

$$|(a_1 \sharp \ldots \sharp a_\nu)(X)| \leq \omega(n, C_0, N_0, M)^\nu$$

$$\times \|a_1\|_{g_1,U_1}^{(M+(2n+1)(2N_0+2))} \ldots \|a_\nu\|_{g_\nu,U_\nu}^{(M+(2n+1)(2N_0+2))} \prod_{1 \leq j < \nu} \delta_{j,j+1}^{-M/4}. \tag{4.6.13}$$

where the semi-norms $\|a\|_{g,U}^{(l)}$ *are defined in* (2.3.3), *and* $\omega(n, C_0, N_0, M)$ *depends only on* n, C_0, N_0, M.

N.B. Here also, one must point out that the number of semi-norms used to control the lhs depends on the required decay (expressed by M) but is *independent of* ν.

Proof. We assume first that ν is even and we apply Theorem 4.6.2 for $N = 0$. Writing $a_1 \natural \ldots \natural a_\nu = (a_1 \natural a_2) \natural \ldots \natural (a_{\nu-1} \natural a_\nu)$, we get

$$|(a_1 \natural \ldots \natural a_\nu)(X)| \le \kappa_n^{\nu/2} \|a_1 \natural a_2\|_{g_1, U_1}^{(2n+1)} \ldots \|a_{\nu-1} \natural a_\nu\|_{g_{\nu-1}, U_{\nu-1}}^{(2n+1)}.$$

Using now the estimates (4.6.10) in the previous remark, we obtain

$$|(a_1 \natural \ldots \natural a_\nu)(X)| \le \kappa_n^{\nu/2} \Gamma(n, N_0, C_0, 2n+1, 2n+1, M)^{\nu/2}$$
$$\times \prod_{1 \le j \le \nu/2} \|a_{2j-1}\|_{g_{2j-1}, U_{2j-1}}^{((2n+1)(2N_0+3)+M)} \|a_{2j}\|_{g_{2j}, U_{2j}}^{((2n+1)(2N_0+3)+M)} \delta_{2j-1, 2j}^{-M/2}. \quad (4.6.14)$$

We have also, writing $a_1 \natural \ldots \natural a_\nu = a_1 \natural (a_2 \natural a_3) \natural \ldots \natural (a_{\nu-2} \natural a_{\nu-1}) \natural a_\nu$,

$$|(a_1 \natural \ldots \natural a_\nu)(X)|$$
$$\le \kappa_n^{1+\frac{\nu}{2}} \|a_1\|_{g_1, U_1}^{(2n+1)} \|a_2 \natural a_3\|_{g_2, U_2}^{(2n+1)} \ldots \|a_{\nu-2} \natural a_{\nu-1}\|_{g_{\nu-2}, U_{\nu-2}}^{(2n+1)} \|a_\nu\|_{g_\nu, U_\nu}^{(2n+1)},$$

and we obtain (still with (4.6.10)),

$$|(a_1 \natural \ldots \natural a_\nu)(X)| \le \kappa_n^{1+\frac{\nu}{2}} \Gamma(n, N_0, C_0, 2n+1, 2n+1, M)^{\frac{\nu}{2}-1} \|a_1\|_{g_1, U_1}^{(2n+1)} \|a_\nu\|_{g_\nu, U_\nu}^{(2n+1)}$$
$$\times \prod_{2 \le j \le \frac{\nu}{2}-1} \|a_{2j}\|_{g_{2j}, U_{2j}}^{((2n+1)(2N_0+3)+M)} \|a_{2j+1}\|_{g_{2j+1}, U_{2j+1}}^{((2n+1)(2N_0+3)+M)} \delta_{2j, 2j+1}^{-M/2}. \quad (4.6.15)$$

We take now the geometric mean of the (4.6.14), (4.6.15) to get

$$|(a_1 \natural \ldots \natural a_\nu)(X)| \le \kappa_n^{\frac{1+\nu}{2}} \Gamma(n, N_0, C_0, 2n+1, 2n+1, M)^{\frac{\nu-1}{2}}$$
$$\times \prod_{1 \le j \le \nu} \|a_j\|_{g_j, U_j}^{((2n+1)(2N_0+3)+M)} \prod_{1 \le j < \nu} \delta_{j, j+1}^{-M/4}.$$

When ν is odd, the reasoning is similar and we get (4.6.13) with

$$\omega = \max\big(\kappa_n, \Gamma(n, N_0, 2n+1, 2n+1, M)\big),$$

since we may assume that κ_n and $\Gamma(n, N_0, 2n+1, 2n+1, M)$ are larger than 1. $\square$

We want now to extend the results of Theorems 4.6.4, 4.6.2 to the derivatives of the composition; we claim that this extension should be easy since the derivative of the "product" $a_1 \natural \ldots \natural a_\nu$ is simply given by the standard non-commutative formula expressed in the following lemma.

Lemma 4.6.5. *Let $(a_j)_{1 \le j \le \nu}$ be ν functions in $\mathscr{S}(\mathbb{R}^{2n})$. Then for all $T, X \in \mathbb{R}^{2n}$, $\nu, k \in \mathbb{N}$, we have*

$$\frac{1}{k!}(a_1 \sharp \ldots \sharp a_\nu)^{(k)}(X)T^k = \sum_{k_1 + \cdots + k_\nu = k} \frac{1}{k_1! \ldots k_\nu!}(a_1^{(k_1)}T^{k_1} \sharp \ldots \sharp a_\nu^{(k_\nu)}T^{k_\nu})(X).$$

Proof. Let us prove first that the formula holds true for $\nu = 2$ and all $k \in \mathbb{N}$; using the formula (2.1.18), we see that

$$(a_1 \sharp a_2)(X) = 2^{2n} \iint_{\mathbb{R}^{2n} \times \mathbb{R}^{2n}} e^{-4i\pi[Y,Z]} a_1(Y + X)a_2(Z + X)dY\,dZ$$

so that by the usual Leibniz' formula, we get

$$\frac{1}{k!}(a_1 \sharp a_2)^{(k)}(X)T^k$$

$$= 2^{2n} \iint_{\mathbb{R}^{2n} \times \mathbb{R}^{2n}} e^{-4i\pi[Y,Z]} \sum_{k_1 + k_2 = k} \frac{1}{k_1!}\frac{1}{k_2!} a_1^{(k_1)}(Y + X)T^{k_1} a_2^{(k_2)}(Z + X)T^{k_2}dY\,dZ$$

$$= \sum_{k_1 + k_2 = k} \frac{1}{k_1!}\frac{1}{k_2!}(a_1^{(k_1)}T^{k_1} \sharp a_2^{(k_2)}T^{(k_2)})(X). \tag{4.6.16}$$

Assuming that the formula is true for some $\nu \ge 2$, we write

$$\frac{1}{k!}(a_1 \sharp \ldots \sharp a_\nu \sharp a_{\nu+1})^{(k)}(X)T^k = \sum_{k_1 + k_2 = k} \frac{1}{k_1!}\frac{1}{k_2!}((a_1 \sharp \ldots \sharp a_\nu)^{(k_1)}T^{k_1} \sharp a_{\nu+1}^{(k_2)}T^{k_2})(X)$$

and we get immediately the result from the induction[12] hypothesis. $\square$

The previous lemma implies the following corollary to Theorem 4.6.2.

Corollary 4.6.6. *Let $n \ge 1$ be a given integer. There exists a positive constant κ_n such that the following property is satisfied. Let $\nu \ge 2$ be an integer. Let $g_1, \ldots, g_\nu$ be ν positive-definite quadratic forms on $\mathbb{R}^{2n}$ such that for all $j \in \{1, \ldots, \nu\}, g_j \le g_j^\sigma$ and let a_j be g_j-confined in U_j, a g_j-ball of radius ≤ 1. Then for all $X, T \in \mathbb{R}^{2n}, k, N \in \mathbb{N}$,*

$$|(a_1 \sharp \ldots \sharp a_\nu)^{(k)}(X)T^k| \le \kappa_n^\nu 2^{\nu 2N} \left(\max_{1 \le j \le \nu} g_j(T)\right)^{k/2}$$

$$\times \left(1 + \nu^{-1} \max_{1 \le j \le \nu} (g_1^\sigma \wedge \cdots \wedge g_\nu^\sigma)(X - U_j)\right)^{-N/2}$$

$$\times \sum_{k_1 + \cdots + k_\nu = k} \frac{k!}{k_1! \ldots k_\nu!} \prod_{1 \le j \le \nu} \|a_j\|_{g_j, U_j}^{(k_j + N + 2n + 1)},$$

$$\tag{4.6.17}$$

where the semi-norms $\|a\|_{g,U}^{(l)}$ are defined in (2.3.3).

---

[12] A direct proof is possible using the formulas (4.6.1), (4.6.2) in Theorem 4.6.1, but it would be a very complicated way to get that simple lemma, much easier to prove than that theorem.

We obtain also from Lemma 4.6.5 the following corollary to Theorem 4.6.4.

Corollary 4.6.7. *Let $n \geq 1$ be a given integer. Let $\nu \geq 2$ be an integer and $g_1, \ldots, g_\nu$ be ν positive-definite quadratic forms on $\mathbb{R}^{2n}$ such that for all $j \in \{1, \ldots, \nu\}, g_j \leq g_j^\sigma$ and let a_j be g_j-confined in U_j, a g_j-ball of radius ≤ 1 (see Definition 2.3.1 and (4.4.18)). Assuming (4.6.12), we get that for all $X \in \mathbb{R}^{2n}, M \in \mathbb{N}$,*

$$|(a_1 \sharp \ldots \sharp a_\nu)^{(k)}(X)T^k| \leq \omega(n, C_0, N_0, M)^\nu \big(\max_{1 \leq j \leq \nu} g_j(T)\big)^{k/2} \prod_{1 \leq j < \nu} \delta_{j,j+1}^{-M/4}$$

$$\times \sum_{k_1 + \cdots + k_\nu = k} \frac{k!}{k_1! \ldots k_\nu!} \prod_{1 \leq j \leq \nu} \|a_j\|_{g_j, U_j}^{(k_j + M + (2n+1)(2N_0+2))} \quad (4.6.18)$$

where the semi-norms $\|a\|_{g,U}^{(l)}$ are defined in (2.3.3), and $\omega(n, C_0, N_0, M)$ depends only on n, C_0, N_0, M.

4.7   A few elements of operator theory

4.7.1   A selfadjoint operator

Let L be a real-valued linear form on $\mathbb{R}^{2n}$; we define

$$D = \{u \in L^2(\mathbb{R}^n), L^w u \in L^2(\mathbb{R}^n)\}$$

and we consider the unbounded operator L^w with domain D. Note that D contains $\mathscr{S}(\mathbb{R}^n)$ and thus is dense in $L^2(\mathbb{R}^n)$. On the other hand, if $\chi \in \mathscr{S}(\mathbb{R}^{2n})$ is real-valued and such that $\chi(0) = 1$, the operator $\chi(h^{1/2}x, h^{1/2}\xi)^w$ converges strongly to the identity of $L^2(\mathbb{R}^n)$ when $h \to 0_+$: with $\chi_h(X) = \chi(h^{1/2}X)$, the simplest symbolic calculus of Section 1.1 shows that

$$\chi_h \sharp \chi_h = \chi_h^2 + h^2 \omega_h, \quad \sup_{0 < h \leq 1} \|\omega_h\|_{\mathcal{L}(L^2)} < +\infty,$$

and also $(\chi_h^2)^{\text{Wick}} = (\chi_h^2)^w + h\Omega_h$, $\sup_{0 < h \leq 1} \|\Omega_h\|_{\mathcal{L}(L^2)} < +\infty$. As a result, we have, for $u \in L^2(\mathbb{R}^n)$,

$$\|(\chi_h^w - \text{Id})u\|^2 = \langle ((\chi_h - 1)\sharp(\chi_h - 1))^w u, u\rangle$$

$$= \int_{\mathbb{R}^{2n}} (\chi_h(X) - 1)^2 \overbrace{|Wu(X)|^2}^{\in L^1(\mathbb{R}^{2n})} dX + O(h\|u\|_{L^2(\mathbb{R}^n)}^2),$$

so that the Lebesgue dominated convergence theorem provides the answer. As a consequence, we have, for $u, v \in D$,

$$\langle L^w u, v\rangle_{L^2(\mathbb{R}^n)} = \lim_{h \to 0} \langle L^w u, \chi_h^w v\rangle_{\mathscr{S}'(\mathbb{R}^n), \mathscr{S}(\mathbb{R}^n)} = \lim_{h \to 0} \langle u, L^w \chi_h^w v\rangle_{\mathscr{S}'(\mathbb{R}^n), \mathscr{S}(\mathbb{R}^n)}$$

$$= \lim_{h \to 0} \big(\langle u, [L^w, \chi_h^w]v\rangle_{\mathscr{S}'(\mathbb{R}^n), \mathscr{S}(\mathbb{R}^n)} + \langle u, \chi_h^w L^w v\rangle_{\mathscr{S}'(\mathbb{R}^n), \mathscr{S}(\mathbb{R}^n)}\big)$$

$$= \lim_{h \to 0} \big(\langle u, [L^w, \chi_h^w]v\rangle_{\mathscr{S}'(\mathbb{R}^n), \mathscr{S}(\mathbb{R}^n)}\big) + \langle u, L^w v\rangle_{L^2(\mathbb{R}^n)}.$$

The operator $[L^w, \chi_h^w]$ has a symbol in $\mathscr{S}(\mathbb{R}^{2n})$ with an $\mathcal{L}(L^2(\mathbb{R}^n))$ norm which is an $O(h^{1/2})$ from the symbolic calculus so that $\langle L^w u, v \rangle = \langle u, L^w v \rangle$ and the operator L^w is symmetric. If we consider now $u_0 \in$

$$D^* = \{u \in L^2(\mathbb{R}^n), \exists C > 0, \forall v \in D, |\langle u, L^w v \rangle| \leq C \|v\|_{L^2(\mathbb{R}^n)}\}$$

we get, for $v \in \mathcal{S}(\mathbb{R}^n)$,

$$|\langle L^w u_0, v \rangle_{\mathscr{S}'(\mathbb{R}^n), \mathscr{S}(\mathbb{R}^n)} = \langle u_0, L^w v \rangle_{\mathscr{S}'(\mathbb{R}^n), \mathscr{S}(\mathbb{R}^n)} = \langle u_0, L^w v \rangle_{L^2(\mathbb{R}^n)}| \leq C \|v\|_{L^2(\mathbb{R}^n)}$$

so that $L^w u_0 \in L^2(\mathbb{R}^n)$ and $u_0 \in D$, making L^w a selfadjoint operator with domain D.

4.7.2 Cotlar's lemma

We present here a precise version of the celebrated Cotlar's lemma which is Lemma 4.2.3 in [20].

Lemma 4.7.1. *Let $\mathbb{H}$ be a Hilbert space and $(\Omega, \mathcal{A}, \nu)$ a measured space such that ν is a σ-finite positive measure. Let $(A_y)_{y \in \Omega}$ be a measurable family of bounded operators on $\mathbb{H}$ such that*

$$\sup_{y \in \Omega} \int_{\Omega} \|A_y^* A_z\|^{1/2} \, d\nu(z) \leq M, \quad \sup_{y \in \Omega} \int_{\Omega} \|A_y A_z^*\|^{1/2} \, d\nu(z) \leq M. \tag{4.7.1}$$

Then, for all $u \in \mathbb{H}$, we have

$$\iint_{\Omega \times \Omega} |\langle A_y u, A_z u \rangle_{\mathbb{H}} \, |d\nu(y) d\nu(z) \leq M^2 \|u\|_{\mathbb{H}}^2 \tag{4.7.2}$$

which implies the strong convergence of $A = \int_{\Omega} A_y d\nu(y)$ and $\|A\| \leq M$.

Remark 4.7.2. We use this result in our Section 2.5.1 on L^2 boundedness of pseudo-differential operators with $\Omega = \mathbb{R}^{2n}$ and $d\nu(Z) = |g_Z|^{1/2} dZ$. The novelty of this lemma does *not* rely on the generalization to a measured space, but on the estimate (4.7.2), which is optimal. In the more familiar discrete framework, the lemma implies that for a sequence (A_j) of bounded operators on $\mathbb{H}$, the conditions

$$\sup_j \sum_k \|A_j^* A_k\|^{1/2} \leq M, \quad \sup_j \sum_k \|A_j A_k^*\|^{1/2} \leq M$$

imply $\sum_{j,k} |\langle A_j u, A_k u \rangle| \leq M^2 \|u\|^2$ and the strong convergence of $A = \sum_j A_j$ with $\|A\| \leq M$.

Proof. From the σ-finiteness of ν, we know that Ω is a countable union of measurable sets ω with $\nu(\omega) < \infty$. Moreover since $y \mapsto \|A_y\|$ is everywhere finite, we have $\cup_{k \in \mathbb{N}} \{y \in \omega, \|A_y\| \leq k\} = \omega$. Then Ω is a countable union of measurable sets

K such that $\nu(K) < \infty$ and $\sup_{y \in K} \|A_y\| < \infty$. We consider then such a K and for a measurable function $\theta(y, z)$ with modulus smaller than 1, we define

$$T_{K,\theta} = \int_{K \times K} A_y^* A_z \theta(y, z) d\nu(y) d\nu(z).$$

We have $\|T\|^2 = \|T^*T\|$, $\|T\|^4 = \|T^*TT^*T\|$, $\ldots$, $\|T\|^{2m} = \|(T^*T)^m\|$ for m (e.g., a power of 2), so that

$$\|T\|^{2m} = \left\| \int_{K^{4m}} A_{z_1}^* A_{y_1} A_{y_1'}^* A_{z_1'} \ldots A_{z_m}^* A_{y_m} A_{y_m'}^* A_{z_m'} \prod_{1 \leq j \leq m} \bar{\theta}(y_j, z_j)\theta(y_j', z_j') \right.$$
$$\left. \times d\nu(y_j) d\nu(z_j) d\nu(y_j') d\nu(z_j') \right\|.$$

Using then the standard Cotlar's trick, we take the above operators two-by-two, beginning with the first or the second, we get inside the integral the estimates from above

$$\prod_{1 \leq j \leq m} \|A_{z_j}^* A_{y_j}\| \|A_{y_j'}^* A_{z_j'}\|, \qquad \|A_{z_1}^*\| \prod_{1 \leq j < m} \|A_{y_j} A_{y_j'}^* A_{z_j'} A_{z_{j+1}}^*\| \|A_{y_m} A_{y_m'}^* A_{z_m'}\|$$

so that, taking the geometric mean, we have

$$\|T_{K,\theta}\|^{2m} \leq \sup_K \|A_y\| \int_{K^{4m}} \prod_{1 \leq j \leq m} \|A_{z_j}^* A_{y_j}\|^{1/2} \|A_{y_j'}^* A_{z_j'}\|^{1/2}$$
$$\times \prod_{1 \leq j < m} \|A_{y_j} A_{y_j'}^*\|^{1/2} \|A_{z_j'} A_{z_{j+1}}^*\|^{1/2} \|A_{y_m} A_{y_m'}^*\|^{1/2}$$
$$\times d\nu(y_j) d\nu(z_j) d\nu(y_j') d\nu(z_j').$$

Integrating with respect to z_m', will give a quantity depending on y_m' but $\leq M$,

integrating with respect to y_m' will give a quantity depending on y_m but $\leq M$,

integrating with respect to y_m will give a quantity depending on z_m but $\leq M$,

integrating with respect to z_m will give a quantity depending on z_{m-1}' but $\leq M$,

and iterating this process up to $j = 2$, we get

$$\|T_{K,\theta}\|^{2m} \leq \sup_K \|A_y\| M^{4m-4}$$
$$\times \int_{K^4} \|A_{z_1}^* A_{y_1}\|^{1/2} \|A_{y_1'}^* A_{z_1'}\|^{1/2} \|A_{y_1} A_{y_1'}^*\|^{1/2} d\nu(y_1) d\nu(z_1) d\nu(y_1') d\nu(z_1')$$
$$\leq \sup_K \|A_y\| M^{4m-1} \nu(K),$$

and taking the $2m$-th root, letting m go to infinity, we obtain $\|T_{K,\theta}\| \leq M^2$. We consider now $u_0 \in \mathbb{H}$ and $\theta(y, z)$ mesurable with modulus 1 such that

$$\langle A_y u_0, A_z u_0 \rangle \theta(y, z) = |\langle A_y u_0, A_z u_0 \rangle|.$$

We have thus

$$M^2\|u_0\|^2 \geq \langle Tu_0, u_0\rangle = \int_{K^2} \langle A_y u_0, A_z u_0\rangle \theta(y, z) d\nu(y) d\nu(z)$$

$$= \int_{K^2} |\langle A_y u_0, A_z u_0\rangle| d\nu(y) d\nu(z)$$

and then (4.7.2). Considering now an increasing sequence $(K_n)_{n\geq 0}$ of measurable sets with union Ω, such that $\nu(K_n) < +\infty, \sup_{y\in K_n} \|A_y\| < +\infty$, we may consider, for $u \in \mathbb{H}$, the sequence $v_n = \int_{K_n} A_y u d\nu(y)$ which belongs to $\mathbb{H}$ and is a Cauchy sequence: with $n, l \in \mathbb{N}$,

$$\|v_{n+l} - v_n\|_{\mathbb{H}}^2 = \iint_{(K_{n+l}\setminus K_n)^2} \langle A_y u, A_z u\rangle d\nu(y) d\nu(z)$$

$$\leq \iint_{(K_n^c)^2} |\langle A_y u, A_z u\rangle| d\nu(y) d\nu(z) \underset{n\to+\infty}{\longrightarrow} 0 \quad \text{from (4.7.2).}$$

Defining Au as the limit of the Cauchy sequence (v_n), we see that A is linear and we get also from (4.7.2) the estimate $\|A\| \leq M$. This definition of Au does not depend on the choice of the sequence (K_n) satisfying the above properties. If (L_n) is such a sequence and $w_n = \int_{L_n} A_y u d\nu(y)$, we have

$$v_n - w_n = \int_{K_n \cap L_n^c} A_y u d\nu(y) - \int_{K_n^c \cap L_n} A_y u d\nu(y)$$

and

$$\|v_n - w_n\|_{\mathbb{H}}^2 \leq 2 \iint_{(K_n^c)^2} |\langle A_y u, A_z u\rangle| d\nu(y) d\nu(z) + 2 \iint_{(L_n^c)^2} |\langle A_y u, A_z u\rangle| d\nu(y) d\nu(z)$$

which goes to 0 as n tends to infinity as already seen above. The proof of the lemma is complete. $\qquad\square$

Remark 4.7.3. We shall see that Cotlar's lemma will give pretty refined results of L^2-boundedness and in particular will be much better than the so-called Schur criterion: let us recall here that for a measured space M with a positive measure μ, a Schur kernel $k(x, y)$ defined on $M \times M$ is such that

$$\kappa_1 = \sup_{x\in M} \int_M |k(x, y)| d\mu(y) < \infty, \quad \kappa_2 = \sup_{y\in M} \int_M |k(x, y)| d\mu(x) < \infty.$$

Then, for $u, v \in L^2(M, d\mu)$, K the operator with kernel k, we have

$$|\langle Ku, v\rangle| \leq \iint |k(x, y)|^{1/2} |u(y)| |k(x, y)|^{1/2} |v(x)| d\mu(y) d\mu(x)$$

$$\leq \left(\iint |k(x, y)| |u(y)|^2 d\mu(y) d\mu(x)\right)^{1/2} \left(\iint |k(x, y)| |v(x)|^2 d\mu(y) d\mu(x)\right)^{1/2}$$

$$\leq \kappa_2^{1/2} \|u\|_{L^2(M, d\mu)} \kappa_1^{1/2} \|v\|_{L^2(M, d\mu)}$$

and the operator K is bounded on $L^2(M, d\mu)$ with a norm $\leq \sqrt{\kappa_1\kappa_2}$. The Hilbert-Schmidt condition $k \in L^2(M \times M, d\mu \otimes d\mu)$ gives rise to a compact operator (we assume that μ is σ-finite) with an $\mathcal{L}(L^2)$ norm smaller than $\|k\|_{L^2(M \times M, d\mu \otimes d\mu)}$:

$$|\langle Ku, v\rangle| \leq \iint |k(x,y)||u(y)||v(x)|d\mu(y)d\mu(x)$$

$$\leq \|k\|_{L^2(M \times M, d\mu \otimes d\mu)}\|u\|_{L^2(M,d\mu)}\|v\|_{L^2(M,d\mu)}. \qquad (4.7.3)$$

Remark 4.7.4. It was remarked by A. Unterberger in [143] that the key hypothesis (4.7.1) can be relaxed by assuming that

$$(y,z) \mapsto \|A_y^* A_z\|^{1/2} \text{ and } (y,z) \mapsto \|A_y A_z^*\|^{1/2}$$

are the kernels of bounded operators on $L^2(\Omega, d\nu)$. Moreover in [144], the same author obtains the boundedness of the operator $A = \int B_y^* B_y d\nu(y)$ under the assumption that $k(y,z) = \|B_y B_z^*\|$ is the kernel of a bounded operator K on $L^2(\Omega, d\nu)$: the proof goes as follows. We have

$$\|Au\|_{\mathbb{H}}^2 = \iint \langle B_z^* B_z u, B_y^* B_y u\rangle_{\mathbb{H}} d\nu(y)d\nu(z) = \iint \langle B_y B_z^* B_z u, B_y u\rangle_{\mathbb{H}} d\nu(y)d\nu(z)$$

so that

$$\|Au\|_{\mathbb{H}}^2 \leq \iint k(y,z)\|B_z u\|_{\mathbb{H}}\|B_y u\|_{\mathbb{H}} d\nu(y)d\nu(z) = \langle K\|B.u\|_{\mathbb{H}}, \|B.u\|_{\mathbb{H}}\rangle_{L^2(\Omega, d\nu)}$$

which gives

$$\|Au\|_{\mathbb{H}}^2 \leq \|K\|_{\mathcal{L}(L^2(\Omega, d\nu))} \int_\Omega \|B_y u\|_{\mathbb{H}}^2 d\nu(y) = \|K\|_{\mathcal{L}(L^2(\Omega, d\nu))}\langle Au, u\rangle_{\mathbb{H}}$$

and $\|A\|_{\mathcal{L}(\mathbb{H})} \leq \|K\|_{\mathcal{L}(L^2(\Omega, d\nu))}$.

Examples 4.7.5. Let us consider the Hilbert space $\ell^2(\mathbb{Z})$ and the Hilbert operator given by the infinite matrix $A = (a_{j,k})_{j,k\in\mathbb{Z}}$,

$$a_{j,k} = \frac{1}{\pi(j-k)} \quad \text{for } j \neq k, \qquad a_{jj} = 0.$$

The Schur criterion fails for this operator with a logarithmic divergence (needless to say, the Hilbert-Schmidt condition is not satisfied). Writing now that for a function $u \in L^2_{\text{loc}}(\mathbb{R})$, 1-periodic,

$$u(x) = \sum_{\xi \in \mathbb{Z}} \hat{u}(\xi)e^{2i\pi x\xi}, \quad (\hat{u}(\xi))_{\xi \in \mathbb{Z}} \in \ell^2(\mathbb{Z}),$$

the operator A is identified with $\widehat{Au}(\xi) = \sum_{\eta \in \mathbb{Z}, \eta \neq \xi} \frac{\hat{u}(\eta)}{\pi(\xi - \eta)}$ so that for $\varphi, u \in C^\infty(\mathbb{R})$, 1-periodic,

$$\int_0^1 (Au)(x)\varphi(x)dx = \sum_{\xi \in \mathbb{Z}} \hat{\varphi}(-\xi) \sum_{\eta \in \mathbb{Z}, \eta \neq \xi} \hat{u}(\eta)\frac{1}{\pi(\xi - \eta)} = \sum_{\xi \in \mathbb{Z}^*, \eta \in \mathbb{Z}} \hat{\varphi}(-\xi - \eta)\hat{u}(\eta)\frac{1}{\pi\xi}$$

and using that $\sum_{\xi \geq 1} \frac{\sin(2\pi x \xi)}{\pi \xi} = \frac{1}{2} - x$, for $x \in (0,1)$, we obtain that $(Au)(x) = i(1 - 2x)u(x)$ so that

$$\|A\|_{\mathcal{L}(\ell^2(\mathbb{Z}))} = 1. \tag{4.7.4}$$

The continuous version $\mathcal{H}$ of the Hilbert transform is the convolution with pv $\frac{1}{\pi x}$, whose Fourier transform is $-i\,\text{sign}\,\xi$, so that (say for $u \in \mathscr{S}(\mathbb{R})$),

$$\lim_{\epsilon \to 0_+} \int_{|y| \geq \epsilon} \frac{u(x - y)}{\pi y} dy = (\mathcal{H}u)(x) = -i(\text{sign}\,Du)(x) \Longrightarrow \|\mathcal{H}\|_{L^2(\mathbb{R})} = 1, \tag{4.7.5}$$

meaning that $\mathcal{H}$ can be extended to a bounded operator on $L^2(\mathbb{R})$ with norm 1. The Hardy operator is defined, in its continuous version, by its distribution kernel

$$\frac{H(x)H(y)}{\pi(x + y)}, \quad H = 1_{\mathbb{R}_+} \text{ is the Heaviside function.}$$

Section 4.1.5 contains a proof of the L^2-boundedness of that operator as well as the exact computation of its $\mathcal{L}(L^2)$ norm (which is 1). The discrete version of the Hardy operator is defined by the infinite matrix

$$\left(\frac{1}{\pi(j + k)} \right)_{j,k \in \mathbb{N}^*}$$

acting on $\ell^2(\mathbb{N}^*)$ and can be proven to have $\mathcal{L}(\ell^2(\mathbb{N}^*))$ norm equal to 1. It is interesting to note that all the operators mentioned in this series of examples are as well L^p bounded for $1 < p < \infty$.

4.7.3 Semi-classical Fourier integral operators

Let us consider the standard semi-classical metric $|dx|^2 + h^2|d\xi|^2$ on $\mathbb{R}^{2n}$ where $h \in (0, 1]$. We want to quantize a canonical transformation χ given by a generating function in a semi-classical framework, namely we wish to find an operator M, L^2-bounded such that $M^* a^w M$ is close to $(a \circ \chi)^w$ in a sense made precise below for $a \in S^0_{scl}$. We shall follow the "sharp Egorov principle" as given by C. Fefferman and D.H. Phong in Lemma 3.1 of [45].

Let $(x_0, \xi_0) \in \mathbb{R}^{2n}$, V_0 be a neighborhood of (x_0, ξ_0) and $\chi : W_0 \longrightarrow V_0$ be a canonical transformation, where W_0 is a neighborhood of a point $(y_0, \eta_0) \in \mathbb{R}^{2n}$. We assume that this canonical transformation is given by a generating function $S(x, \eta)$ such that

$$\chi(\partial_\eta S(x, \eta), \eta) = (x, \partial_x S(x, \eta)), \quad \text{with } \det\left(\frac{\partial^2 S}{\partial x \partial \eta}\right) \neq 0.$$

We consider $\omega \in C_c^\infty(\mathbb{R}^{2n})$, supported near (x_0, η_0) and the operator M given by

$$(Mv)(x) = \int e^{2i\pi h^{-1} S(x, h\eta)} \omega(x, h\eta) \hat{v}(\eta) d\eta, \tag{4.7.6}$$

whose kernel is $\mu(x,y) = \int e^{2i\pi(h^{-1}S(x,h\eta)-y\eta)}\omega(x,h\eta)d\eta$. We consider a smooth function $a \in C_c^\infty(\mathbb{R}^{2n})$ (depending also on h) and its standard semi-classical quantization, $a(x,hD,h)$ with kernel $\kappa_a(x,x') = \int e^{2i\pi(x-x')\xi}a(x,h\xi,h)d\xi$ so that

$$a(x,h\xi,h) = \int e^{2i\pi x'\xi}\kappa_a(x,x'+x)dx'.$$

We calculate $M^*a(x,hD,h)M$ which has the kernel $k(y,y')$ and we have

$$k(y,y'+y) = \int \overline{\mu(x,y)}\kappa_a(x,x')\mu(x',y+y')dxdx',$$

so that

$$
\begin{aligned}
k(y,y'+y) = \int & e^{2i\pi(-h^{-1}S(x,h\eta'')+y\eta'')}\bar\omega(x,h\eta'')e^{2i\pi(x-x')\xi}a(x,h\xi) \\
& \times e^{2i\pi(h^{-1}S(x',h\eta')-y'\eta'-y\eta')}\omega(x',h\eta')d\eta''d\eta'd\xi dx dx'.
\end{aligned}
$$

Now the (standard) symbol b of $M^*a(x,hD,h)M$ is given by

$$b(y,h\eta,h) = \int k(y,y'+y)e^{2i\pi y'\eta}dy'$$

and we have

$$
\begin{aligned}
b(y,h\eta,h) &= \int e^{2i\pi y'\eta}e^{2i\pi(-h^{-1}S(x,h\eta'')+y\eta'')}\bar\omega(x,h\eta'')e^{2i\pi(x-x')\xi}a(x,h\xi,h) \\
& \qquad \times e^{2i\pi(h^{-1}S(x',h\eta')-y'\eta'-y\eta')}\omega(x',h\eta')d\eta''d\eta'd\xi dx dx'dy' \\
&= \int \exp 2i\pi\big[h^{-1}S(x',h\eta) - h^{-1}S(x,h\eta'') + y(\eta''-\eta) + (x-x')\xi\big] \\
& \qquad \times \bar\omega(x,h\eta'')\omega(x',h\eta)a(x,h\xi,h)d\eta''d\xi dx dx',
\end{aligned}
$$

so that

$$
\begin{aligned}
b(y,\eta,h) = \int & \exp 2i\pi\big[h^{-1}S(x',\eta) - h^{-1}S(x,\eta'') + h^{-1}y(\eta''-\eta) + (x-x')\xi h^{-1}\big] \\
& \times \bar\omega(x,\eta'')\omega(x',\eta)a(x,\xi,h)d\eta''d\xi dx dx'h^{-2n}
\end{aligned}
$$

and changing the name of the integration variables we get

$$
\begin{aligned}
b(y,\eta,h) = \int_{\mathbb{R}^{4n}} & \exp 2i\pi h^{-1}\big[S(x',\eta) - S(x'',\eta'') + y(\eta''-\eta) + (x''-x')\xi''\big] \\
& \times \bar\omega(x'',\eta'')\omega(x',\eta)a(x'',\xi'',h)d\eta''d\xi''dx''dx'h^{-2n}. \quad (4.7.7)
\end{aligned}
$$

The following lemma summarizes the above calculations.

Lemma 4.7.6. *Let* $a(\cdot, h), \omega \in C_c^\infty(\mathbb{R}^{2n})$. *Let* W *be a neighborhood of* $\operatorname{supp}\omega$, $S :$ $W \longrightarrow \mathbb{R}$ *be a smooth function and let* M *be the operator given by* (4.7.6). *The operator* $M^* a(x, hD_x, h)M = b(y, hD_y, h)$ *where* b *is given by*

$$b(y, \eta, h) = h^{-\frac{4n}{2}} \int_{\mathbb{R}^{4n}} e^{2i\pi h^{-1}\Phi_{y,\eta}(x'', x', \xi'', \eta'')} \bar{\omega}(x'', \eta'')\omega(x', \eta)$$

$$\times a(x'', \xi'', h)d\eta''d\xi''dx''dx', \quad (4.7.8)$$

with $\Phi_{y,\eta}(x'', x', \xi'', \eta'') = S(x', \eta) - S(x'', \eta'') + y(\eta'' - \eta) + (x'' - x')\xi''$. (4.7.9)

Let $(x_0, \eta_0) \in \mathbb{R}^n \times \mathbb{R}^n$ be given such that $\det\left(\frac{\partial^2 S}{\partial x \partial \eta}\right)(x_0, \eta_0) \neq 0$, and let us define $y_0 = (\partial_\eta S)(x_0, \eta_0)$, $\xi_0 = (\partial_x S)(x_0, \eta_0)$. There exists a neighborhood of (x_0, η_0, y_0) on which the equation $y = (\partial_\eta S)(x, \eta)$ is equivalent to $x = f(y, \eta)$. The mapping

$$(y, \eta) \mapsto \big(f(y, \eta), (\partial_x S)(f(y, \eta), \eta)\big) = \chi(y, \eta)$$

is a symplectomorphism from a neighborhood of (y_0, η_0) to a neighborhood of (x_0, ξ_0). With $(y, \eta) \in \mathbb{R}^{2n}$ as parameters, the phase in the above integral on $\mathbb{R}^{4n}$ is

$$\Phi_{y,\eta}(x'', x', \xi'', \eta'') = S(x', \eta) - S(x'', \eta'') + y(\eta'' - \eta) + (x'' - x')\xi''$$

and we have

$$\partial_{x''}\Phi = -(\partial_{x''}S)(x'', \eta'') + \xi'', \qquad \partial_{x'}\Phi = (\partial_{x'}S)(x', \eta) - \xi'',$$
$$\partial_{\xi''}\Phi = x'' - x', \qquad \partial_{\eta''}\Phi = -(\partial_{\eta''}S)(x'', \eta'') + y,$$

so that the critical point of $\Phi_{y,\eta}$ satisfies

$$\chi(y, \eta'') = \chi((\partial_{\eta''}S)(x'', \eta''), \eta'') = (x'', (\partial_{x''}S)(x'', \eta''))$$
$$= (x'', \xi'') = (x', (\partial_{x'}S)(x', \eta)) = \chi((\partial_\eta S)(x', \eta), \eta)$$

and thus setting $(x, \xi) = \chi(y, \eta)$, we get that the critical point (x'', x', ξ'', η'') is such that

$$\eta'' = \eta, \quad y = (\partial_\eta S)(x'', \eta) = (\partial_\eta S)(x', \eta), \quad x'' = x' = x, \quad \xi'' = \xi,$$

and the phase vanishes there. Moreover, assuming that the Hessian of $S(x, \eta) - \langle x, \eta \rangle$ is vanishing at (x_0, η_0), we get that the Hessian of $\Phi_{y,\eta}/2$ at the critical point $(x_0, x_0, \xi_0, \eta_0)$ is

$$(\mathbb{R}^n)^4 \ni (t'', t', \tau'', \sigma'') \mapsto -t''\sigma'' + (t'' - t')\tau''$$

which has the same signature as $-t''\sigma'' - t'\tau''$ which has signature 0. As a result, there is a neighborhood of (y_0, η_0) in which the signature of the Hessian of $\Phi_{y,\eta}$ at its critical point is 0.

Remark 4.7.7. An application of the stationary phase formula with parameters (see, e.g., Theorem 7.7.6 in [71]) gives readily with $(x, \xi) = \chi(y, \eta)$ (i.e., $x = f(y, \eta), \xi = \varphi(y, \eta) = (\partial_x S)(f(y, \eta), \eta))$

$$b(y, \eta, h) = |\det \Phi''_{y, \eta}(x, x, \xi, \eta)|^{-1/2} \Big[|\omega(x, \eta)|^2 a(x, \xi, h) + \sum_{1 \leq j < N} i^j h^j c_j(x, \xi, h) \Big]$$

$$+ h^N r_N(y, \eta, h), \quad c_j \in C_c^\infty(V_0), r_N(\cdot, h) \in C_b^\infty(\mathbb{R}^{2n}), \quad (4.7.10)$$

with $c_j = L_j \alpha$ where L_j is a differential operator with real coefficients of order $2j$ in the variables x'', x', ξ'', η'' and $\alpha = \bar{\omega}(x'', \eta'')\omega(x', \eta)a(x'', \xi'', h)$. Note in particular that all c_j are real-valued if a and ω are real-valued. Also the previous formula ensures that M^*M is a bounded operator on $L^2(\mathbb{R}^n)$.

We have the identity

$$\Phi''_{y, \eta}(x, x, \varphi(y, \eta), \eta) = \Phi''_{(\partial_\eta S)(x, \eta), \eta}(x, x, \varphi((\partial_\eta S)(x, \eta), \eta), \eta)$$

and we may thus choose

$$\omega(x, \eta) = \omega_0(x, \eta) = |\det \Phi''_{(\partial_\eta S)(x, \eta), \eta}(x, x, \varphi((\partial_\eta S)(x, \eta), \eta), \eta)|^{1/4} \quad (4.7.11)$$

on a neighborhood Ω of (x_0, η_0) such that

$$\mathrm{supp}\big[(x, \eta) \mapsto a(x, (\partial_x S)(x, \eta))\big] \subset \Omega,$$

a condition which can be achieved by shrinking the support of a.

Theorem 4.7.8. *Let χ be a canonical transformation from a neighborhood of a point $(y_0, \eta_0) \in \mathbb{R}^n \times \mathbb{R}^n$ to a neighborhood of (x_0, ξ_0) so that χ is given by a generating function $S(x, \eta)$ such that the Hessian of $S(x, \eta) - \langle x, \eta \rangle$ is vanishing at (x_0, η_0). There exists a neighborhood V_0 of (x_0, ξ_0) and a neighborhood Ω_0 of (x_0, η_0), $\omega_0 \in C_c^\infty(\Omega_0)$ such that for all $a(\cdot, h) \in C_c^\infty(V_0)$, we have*

$$M^*a(x, hD, h)M = b(y, hD_y, h), \quad \text{with the operator } M \text{ given by (4.7.6)},$$

where b is given by (4.7.8). We have $b(y, \eta, h) = a(\chi(y, \eta), h)$ modulo $hC_b^\infty(\mathbb{R}^{2n})$ and with Weyl quantization

$$M^*a(x, h\xi, h)^w M = b(y, h\eta, h)^w, \quad (4.7.12)$$

$$b(y, \eta, h) = a(\chi(y, \eta), h) \mod h^2 C_b^\infty(\mathbb{R}^{2n}). \quad (4.7.13)$$

Proof. Lemma 4.7.6 gives the first statement. Formula (4.7.10) with the choice of ω_0 given by (4.7.11) provides

$$b(y, \eta, h) = a(\chi(y, \eta), h) + ihc_1(\chi(y, \eta), h) + h^2 r_2(y, \eta, h).$$

If a is real-valued, since our choice (4.7.11) of ω_0 is also real-valued, we get that c_1 is real-valued, according to Remark 4.7.7. Using now the Weyl symbols, in the case a real-valued, we have with $\rho_2(\cdot, h) \in \dot{C}_b^\infty(\mathbb{R}^{2n})$,

$$M^* a(x, h\xi, h)^w M = M^* \frac{1}{2}\left(a(x, hD, h) + a(x, hD, h)^*\right)M + h^2 \rho(y, hD_y, h)$$
$$= \beta(y, h\eta, h)^w$$

and with β real-valued (since $M^* a(x, h\xi, h)^w M$ is selfadjoint), we get, modulo $h^2 C_b^\infty$,

$$\beta + i\pi h D_y \cdot D_\eta \beta \equiv a + ihc_1 + ih\pi D_x \cdot D_\xi a,$$

so that taking the real parts we obtain $\beta(y, \eta, h) \equiv a(\chi(y, \eta), h)$. The statement on Weyl quantization follows by separating real and imaginary parts. $\qquad\square$

Remark 4.7.9. For $a, b \in C_b^\infty(\mathbb{R}_{x_1} \times \mathbb{R}_{x', \xi'}^{2n-2}; \mathbb{R})$, we may consider the function

$$f(x, \xi) = \xi_1 + a(x_1, x', \xi') + ib(x_1, x', \xi'),$$

obtained as a normal form of the principal symbol of a principal-type pseudo-differential operator (see Proposition 4.5.1), up to a non-vanishing multiplicative factor. Defining the functions

$$\eta_1 = \xi_1 + a(x_1, x', \xi'), \quad y_1 = x_1$$

we have $\{\eta_1, y_1\} \equiv 1$, so that we may find, say in a neighborhood of 0 in $\mathbb{R}^{2n}$, some symplectic coordinates $(y_1, y', \eta_1, \eta') \in \mathbb{R} \times \mathbb{R}^{n-1} \times \mathbb{R} \times \mathbb{R}^{n-1}$, completing y_1, η_1, i.e., such that $\{\eta_j, y_k\} = \delta_{j,k}$, $\{\eta_j, \eta_k\} = \{y_j, y_k\} = 0$. We have then with a real-valued function q,

$$f(x, \xi) = \eta_1 + iq(y, \eta), \quad q\big(y(x, \xi), \eta(x, \xi)\big) = b(x_1, x', \xi'),$$

and since $(x, \xi) \mapsto (y, \eta)$ is a local symplectomorphism, we get

$$\frac{\partial q}{\partial \eta_1} = \{q, y_1\} = \{b, x_1\} \equiv 0$$

and $q(y, \eta)$ does not depend on η_1. Assuming as we may that $da(0) = 0$, we see that our symplectic transformation is tangent to the identity at 0, and we can apply the quantization Theorem 4.7.8 to get

$$M^*\big(hD_{x_1} + a(x_1, x', h\xi')^w + ib(x_1, x', h\xi')^w\big)M$$
$$= hD_{y_1} + iq(y_1, y', h\eta')^w \quad \text{mod } h^2 \text{op}(S_{scl}^0). \quad (4.7.14)$$

4.8   On the Sjöstrand algebra

The Sjöstrand algebra is defined in Proposition 2.5.6. We give here some more properties of that algebra.

Lemma 4.8.1. *Let b be a function in $\mathcal{A}$ and $T \in \mathbb{R}^{2n}, t \in \mathbb{R}$. Then the functions $\tau_T b$, b_t defined by $\tau_T b(X) = b(X - T), b_t(X) = b(tX)$ belong to $\mathcal{A}$ and*

$$\sup_{T \in \mathbb{R}^{2n}} \|\tau_T b\|_{\mathcal{A}} \leq C\|b\|_{\mathcal{A}}, \qquad \|b_t\|_{\mathcal{A}} \leq (1 + |t|)^{2n} C\|b\|_{\mathcal{A}}, \qquad (4.8.1)$$

where C depends only on the dimension.

Proof. We check, using that $T = S + j_0, j_0 \in \mathbb{Z}^{2n}, S \in [0,1]^{2n}$,

$$\mathcal{F}(\chi_j \tau_T b)(\Xi) = \int e^{-2i\pi X\Xi} \chi_j(X) b(X - T) dX$$

$$= e^{-2i\pi T\Xi} \int e^{-2i\pi X\Xi} \chi_{j-j_0}(X + S) b(X) dX$$

$$= e^{-2i\pi T\Xi} \int e^{-2i\pi X\Xi} \chi_{j-j_0}(X + S) \Big(\sum_{|k| \leq R_0} \chi_{j-j_0+k}(X) b(X) \Big) dX$$

$$= e^{-2i\pi T\Xi} \sum_{|k| \leq R_0} \mathcal{F}\big((\tau_{-S} \chi_{j-j_0})(\chi_{j-j_0+k} b) \big)(\Xi)$$

$$= e^{-2i\pi T\Xi} \sum_{|k| \leq R_0} \big(\mathcal{F}(\tau_{-S} \chi_{j-j_0}) * \mathcal{F}(\chi_{j-j_0+k} b) \big)(\Xi)$$

so that

$$|\mathcal{F}(\chi_j \tau_T b)(\Xi)| \leq C_0 \int |\mathcal{F}(\tau_{-S} \chi_{j-j_0})(\Xi - \Xi')| \omega_b(\Xi') d\Xi'$$

$$= C_0 \int |\mathcal{F}(\chi_0)(\Xi - \Xi')| \omega_b(\Xi') d\Xi',$$

entailing

$$\int \sup_{j \in \mathbb{Z}^{2n}} |\mathcal{F}(\chi_j \tau_T b)(\Xi)| d\Xi \leq C_0 \|\widehat{\chi_0}\|_{L^1} \|b\|_{\mathcal{A}}$$

and the first part of (4.8.1).

The second part is obvious if $t = 0$ since $\mathcal{A}$ is continuously embedded in $C^0 \cap L^\infty$ (Proposition 2.5.6). Assuming $t \neq 0$, we look now at

$$\mathcal{F}(\chi_j b_t)(\Xi) = \int e^{-2i\pi X\Xi} \chi_0(X - j) b(tX) dX$$

$$= \sum_{k \in \mathbb{Z}^{2n}} \int e^{-2i\pi X\Xi} \chi_0(X - j) \chi_k(tX) b(tX) dX,$$

and since on the support of the integrand, we have

$$|X - j| \le R_0, \quad |k - tX| \le R_0$$

and thus

$$|k - tj| \le R_0 + |t|R_0,$$

we get

$$\mathcal{F}(\chi_j b_t)(\Xi) = \int e^{-2i\pi X\Xi} \chi_0(X - j) b(tX) dX$$

$$= \sum_{\substack{k \\ |k-tj|\le R_0(1+|t|)}} \int e^{-2i\pi X\Xi} \chi_0(X - j) b(tX) \chi_k(tX) dX$$

$$= \sum_{|k-tj|\le R_0(1+|t|)} \iint e^{-2i\pi t^{-1}X\Xi} \widehat{\chi_0}(N) e^{2i\pi N(t^{-1}X - j)} (\chi_k b)(X) dX dN |t|^{-2n}$$

$$= \sum_{|k-tj|\le R_0(1+|t|)} \iint e^{-2i\pi t^{-1}X\Xi} \widehat{\chi_0}(tN) e^{2i\pi N(X - tj)} (\chi_k b)(X) dX dN$$

$$= \sum_{|k-tj|\le R_0(1+|t|)} \iint \widehat{\chi_0}(tN) e^{-2i\pi Ntj} e^{2i\pi X(N - t^{-1}\Xi)} (\chi_k b)(X) dX dN$$

$$= \sum_{|k-tj|\le R_0(1+|t|)} \iint \widehat{\chi_0}(tN + \Xi) e^{-2i\pi Ntj} e^{-2i\pi \Xi j} e^{2i\pi XN} (\chi_k b)(X) dX dN$$

$$= \sum_{|k-tj|\le R_0(1+|t|)} \int \widehat{\chi_0}(-tN + \Xi) e^{2i\pi Ntj} e^{-2i\pi \Xi j} \left(\int e^{-2i\pi XN} (\chi_k b)(X) dX \right) dN,$$

so that

$$|\mathcal{F}(\chi_j b_t)(\Xi)| \le \sum_{|k-tj|\le R_0(1+|t|)} \int |\widehat{\chi_0}(-tN + \Xi)| \left| \int e^{-2i\pi XN} (\chi_k b)(X) dX \right| dN$$

$$\le C_n R_0^{2n} \int |\widehat{\chi_0}(-tN + \Xi)| \omega_b(N) dN (1 + |t|)^{2n},$$

and finally the sought result $\int \sup_j |\mathcal{F}(\chi_j b_t)(\Xi)| d\Xi \le C_n R_0^{2n} (1+|t|)^{2n} \|\widehat{\chi_0}\|_{L^1} \|b\|_{\mathcal{A}}$. $\square$

4.9 More on symbolic calculus

4.9.1 Properties of some metrics

Lemma 4.9.1. *Let Γ be a positive-definite quadratic form on $\mathbb{R}^{2n}$ such that $\Gamma = \Gamma^\sigma$ and let $g_X = \lambda(X)^{-1}\Gamma$ be a metric conformal to Γ such that g is slowly varying* and $\inf_X \lambda(X) \geq 1$. Then the metric g satisfies $g_X(T) \leq Cg_Y(T)\big(1+\Gamma(X-Y)\big)$, i.e.,

$$\frac{\lambda(Y)}{\lambda(X)} \leq C\big(1 + \Gamma(X-Y)\big), \tag{4.9.1}$$

implying that g is admissible.

Proof. Since g is slowly varying, we may assume, with a positive r_0, $g_Y(Y-X) \geq r_0^2$, which means $\Gamma(Y-X) \geq r_0^2\lambda(Y)$ and using $\lambda(X) \geq 1$ we get $\lambda(Y)/\lambda(X) \leq r_0^{-2}\Gamma(Y-X)$. $\qquad\square$

Lemma 4.9.2. *Let Γ be a positive-definite quadratic form on $\mathbb{R}^{2n}$ such that $\Gamma = \Gamma^\sigma$ and let $g_X = \lambda(X)^{-1}\Gamma$ be a metric conformal to Γ. Assume that $\lambda(X) = d(X)^2 + \lambda_1(X)$ with a function d uniformly Lipschitz-continuous (with respect to Γ) and $\lambda_1^{-1}\Gamma$ slowly varying with $\lambda_1 \geq 1$. Then the metric g is slowly varying.*

Proof. Let us assume that $|X-Y|^2 \leq r^2\big(d(X)^2 + \lambda_1(X)\big)$. If $d(X)^2 \leq \lambda_1(X)$, using the fact that $\lambda_1^{-1}\Gamma$ is slowly varying, we can choose r small enough so that $\lambda_1(X) \leq C_1\lambda_1(Y)$ and thus

$$\lambda(X) \leq 2C_1\lambda_1(Y) \leq 2C_1\lambda(Y).$$

If $d(X)^2 > \lambda_1(X)$, we have, with L standing for the Lipschitz constant of d,

$$2^{-1/2}\lambda(X)^{1/2} < d(X) \leq d(Y) + L|X-Y| \leq \lambda(Y)^{1/2} + Lr\lambda(X)^{1/2}$$

so that, for $r \leq \frac{1}{2^{3/2}L+1}$ we get $\lambda(X) \leq 8\lambda(Y)$. This concludes the proof since, according to Remark 2.2.2, the slow variation property of a metric g is satisfied whenever there exists $r_0 > 0, C_0 > 0$ such that for all $X, Y, T \in \mathbb{R}^{2n}$, $g_X(Y-X) \leq r_0^2$ implies $g_Y(T) \leq C_0 g_X(T)$. $\qquad\square$

4.9.2   Proof of Lemma 3.2.12 on the proper class

All norms in this proof are taken with respect to the constant quadratic form Γ, so we omit the index everywhere and denote $\|\cdot\|_\Gamma$ by $|\cdot|$. Since for all $j \in \mathbb{N}$, $|f^{(j)}(X)| \leq \gamma_j\Lambda^{m-\frac{j}{2}}$, we get

$$1 \leq \lambda(X) \leq 1 + \Lambda \max_{\substack{0 \leq j < 2m \\ j \in \mathbb{N}}} \gamma_j^{\frac{2}{2m-j}} = 1 + \gamma\Lambda \leq (1+\gamma)\Lambda$$

and (3.2.24). For $0 \leq j < 2m$, we have from the definition of λ, the estimate $|f^{(j)}(X)| \leq \lambda(X)^{m-\frac{j}{2}}$, and for $j \geq 2m$, we can use

$$|f^{(j)}(X)| \leq \gamma_j\Lambda^{m-\frac{j}{2}} = \gamma_j\Lambda^{-\frac{(j-2m)}{2}} \leq \gamma_j\lambda^{-\frac{(j-2m)}{2}}(1+\gamma)^{\frac{(j-2m)}{2}},$$

so that $f \in S(\lambda^m, \lambda^{-1}\Gamma)$ with a j-th semi-norm less than 1 for $j < 2m$ and less than $\gamma_j(1+\gamma)^{\frac{(j-2m)}{2}}$ for $j \geq 2m$.

Let us now prove that $\lambda^{-1}\Gamma$ is slowly varying. Let us assume that $|X - Y|^2 \leq r^2\lambda(X)$. Using Taylor's formula, we get for the smallest integer $N \geq 2m$ ($N = -[-2m]$) and $0 \leq j < 2m$,

$$|f^{(j)}(X)| \leq \sum_{l, j+l < 2m} |f^{(j+l)}(Y)| \frac{r^l}{l!} \lambda(X)^{l/2} + \gamma_N \Lambda^{m-\frac{N}{2}} \frac{r^{N-j}}{(N-j)!} \lambda(X)^{(N-j)/2},$$

so that $|f^{(j)}(X)| \leq \sum_{l, j+l < 2m} \lambda(Y)^{\frac{2m-j-l}{2}} \lambda(X)^{\frac{1}{2}} \frac{r^l}{l!} + \gamma_N \Lambda^{\frac{2m-N}{2}} \lambda(X)^{\frac{N-j}{2}} \frac{r^{N-j}}{(N-j)!}$,

and

$$|f^{(j)}(X)|$$

$$\leq \sum_{l, j+l < 2m} (\lambda(Y)^{\frac{2m-j}{2}})^{\frac{2m-j-l}{2m-j}} (\lambda(X)^{\frac{2m-j}{2}})^{\frac{l}{2m-j}} \frac{r^l}{l!} + \gamma_N \overbrace{\Lambda^{\frac{2m-N}{2}}}^{\leq 0} \lambda(X)^{\frac{N-j}{2}} \frac{r^{N-j}}{(N-j)!}$$

$$\leq \sum_{l, j+l < 2m} \frac{2m-j-l}{2m-j} \lambda(Y)^{\frac{2m-j}{2}} \frac{r^l}{l!} + \frac{l}{2m-j} \lambda(X)^{\frac{2m-j}{2}} \frac{r^l}{l!}$$

$$+ \gamma_N (1+\gamma)^{\frac{N-2m}{2}} \lambda(X)^{\frac{2m-j}{2}} \frac{r^{N-j}}{(N-j)!},$$

implying

$$|f^{(j)}(X)| \leq \lambda(Y)^{\frac{2m-j}{2}} \overbrace{\sum_{l, j+l < 2m} \frac{2m-j-l}{2m-j} \frac{r^l}{l!}}^{= p(r) \text{ a polynomial in } r}$$

$$+ \lambda(X)^{\frac{2m-j}{2}} \underbrace{\left(\sum_{1 \leq l, j+l < 2m} \frac{l}{2m-j} \frac{r^l}{l!} + \gamma_N (1+\gamma)^{\frac{N-2m}{2}} \frac{r^{N-j}}{(N-j)!} \right)}_{= \epsilon(r) \text{ goes to zero with } r}.$$

Assuming then that j was chosen so that $\lambda(X) = 1 + |f^{(j)}(X)|^{\frac{2}{2m-j}}$, we get

$$\lambda(X) \leq 1 + \left(\lambda(Y)^{\frac{2m-j}{2}} p(r) + \lambda(X)^{\frac{2m-j}{2}} \epsilon(r) \right)^{\frac{2}{2m-j}},$$

so that there exist $r_0 > 0, C_0 \geq 1$, depending only on the N first semi-norms of f, such that for $r \leq r_0$, we have

$$|X - Y|^2 \leq r^2\lambda(X) \implies \lambda(X) \leq C_0\lambda(Y),$$

and thus $r \leq r_0, |X - Y|^2 \leq r^2 C_0^{-1}\lambda(X) \implies C_0^{-1}\lambda(X) \leq \lambda(Y) \leq C_0\lambda(X)$, which is the property (2.2.3). The property (2.2.12) is obviously satisfied since $\lambda(X) \geq 1$. Moreover, we get a stronger property than (2.2.16) from Lemma 4.9.1 above in this appendix. $\quad\square$

4.9.3   More elements of Wick calculus

Most of the results of this section are rather technical and useful only for Section 3.7.

Lemma 4.9.3. *Let Γ be a fixed positive-definite symplectic quadratic form. Let M be an admissible weight for Γ (cf. Definition 2.2.15), i.e., a positive function such that $M(X)M(Y)^{-1} \leq C(1+\Gamma(X-Y))^N$. Then if a measurable function a defined on $\mathbb{R}^{2n}$ satisfies for all X, $|a(X)| \leq C_1 M(X)$, the symbol $a * \exp -2\pi\Gamma$ belongs to $S(M,\Gamma)$ with semi-norms depending only on C_1. More generally, for a polynomial p the symbol A defined by*

$$A(X) = \int a(Y)p(X-Y)\exp -2\pi\Gamma(X-Y)dY$$

belongs to $S(M,\Gamma)$.

Proof. We check first

$$(a * 2^n \exp -2\pi\Gamma)^{(k)}(X) = \int a(Y)P_k(X-Y)2^n \exp -2\pi\Gamma(X-Y)dY \quad (4.9.2)$$

with a polynomial P_k, which gives

$$M(X)^{-1}|(a * 2^n \exp -2\pi\Gamma)^{(k)}(X)|$$
$$\leq C_1 \int \frac{M(Y)}{M(X)}|P_k(X-Y)|2^n \exp -2\pi\Gamma(X-Y)dY$$
$$\leq C_1 \int C(1+\Gamma(X-Y))^N |P_k(X-Y)|2^n \exp -2\pi\Gamma(X-Y)dY$$
$$= C_1 C\gamma(k,N,n).$$

Let us examine $A^{(k)}$: it is a sum of terms of type (4.9.2) and thus the above argument works. $\square$

Lemma 4.9.4. *Let g be an admissible metric on $\mathbb{R}^{2n}$ (see Definition 2.2.15) such that, for all $X \in \mathbb{R}^{2n}, g_X^\flat = \Gamma$, where $\Gamma^{1/2}$ is a fixed symplectic norm. Let m be an admissible weight for g such that $m \in S(m,g)$. Then, if $A \in op(S(m,g))$, there exists a semi-norm γ of the symbol of A such that, for all $v \in \mathscr{S}(\mathbb{R}^n)$,*

$$|\langle Av, v\rangle| \leq \gamma\langle m^{Wick}v, v\rangle = \gamma \int_{\mathbb{R}^{2n}} m(Y)\|\Sigma_Y v\|_{L^2}^2 dY. \quad (4.9.3)$$

Proof. Lemma 2.6.18 shows that the space $\mathcal{H}(m^{1/2},g)$ is equal to $\mathcal{H}(m^{1/2},\Gamma)$, provided that $m^{1/2}$ belongs to $S(m^{1/2},g) \cap S(m^{1/2},\Gamma)$, and the latter condition is satisfied since $m \in S(m,g)$, $g_X \leq g_X^\flat = \Gamma$ and $m^{1/2} \in S(m^{1/2},g)$ (see Lemma 2.2.22). Moreover Corollary 2.6.17 and Theorem 2.6.29 imply

$$|\langle Av, v \rangle| \le \|Av\|_{\mathcal{H}(m^{-1/2}, g)} \|v\|_{\mathcal{H}(m^{1/2}, g)}$$

$$\le \gamma \|v\|^2_{\mathcal{H}(m^{1/2}, g)} = \gamma \|v\|^2_{\mathcal{H}(m^{1/2}, \Gamma)} = \gamma \int_{\mathbb{R}^{2n}} m(Y) \|\theta^w_Y u\|^2_{L^2} dY,$$

where (θ_Y) is a partition of unity related to the metric Γ. We have, using $1^{\text{Wick}} = \text{Id}$ (see Proposition 2.4.3), Proposition 2.2.20 and (2.4.9), with $\langle T \rangle^2 = 1 + \Gamma(T)$, for all N_1, N_2,

$$\int m(Y) \|\theta^w_Y u\|^2 dY = \iiint_{\mathbb{R}^{2n}} m(Y) \langle \theta^w_Y \Sigma_{Z_1} \Sigma_{Z_1} u, \theta^w_Y \Sigma_{Z_2} \Sigma_{Z_2} u \rangle dY dZ_1 dZ_2$$

$$\le \iiint m(Z_1)^{1/2} m(Z_2)^{1/2} \|\Sigma_{Z_1} u\| \|\Sigma_{Z_2} u\|$$

$$\times \langle Y - Z_1 \rangle^{-N_1} \langle Z_2 - Z_1 \rangle^{-N_2} dY dZ_1 dZ_2 C_{N_1, N_2}$$

$$\le \iint m(Z_1)^{1/2} m(Z_2)^{1/2} \|\Sigma_{Z_1} u\| \|\Sigma_{Z_2} u\| \langle Z_2 - Z_1 \rangle^{-N_2} dZ_1 dZ_2 C_{N_1, N_2}$$

$$\le \int m(Z) \|\Sigma_Z u\|^2 dZ,$$

which completes the proof of the lemma. $\qquad\square$

Lemma 4.9.5. *Let m_1, m_2 be two weights for Γ (a fixed positive-definite symplectic quadratic form) and a_1, a_2 be two locally Lipschitz-continuous functions such that $|a_1(X)| \le m_1(X)$, $|a'_2(X)| \le m_2(X)$. Then the operator*

$$a_1^{\text{Wick}} a_2^{\text{Wick}} \in (a_1 a_2)^{\text{Wick}} + op(S(m_1 m_2, \Gamma)). \tag{4.9.4}$$

Proof. We use Definition 2.4.1 and Taylor's formula to write

$$a_1^{\text{Wick}} a_2^{\text{Wick}} = \iint a_1(Y) \Big(a_2(Y) + \int_0^1 a'_2(Y + \theta(Z - Y)) d\theta (Z - Y)\Big) \Sigma_Y \Sigma_Z dY dZ$$

$$= (a_1 a_2)^{\text{Wick}} + R^w,$$

with (see (2.4.10))

$$R(X) = \iiint_0^1 a_1(Y) a'_2(Y + \theta(Z - Y))(Z - Y) e^{-\frac{\pi}{2}|Y - Z|^2}$$

$$\times e^{-2i\pi[X-Y, X-Z]} 2^n e^{-2\pi|X - \frac{Y+Z}{2}|^2} dY dZ d\theta. \tag{4.9.5}$$

We have, using (2.2.19),

$$|R(X)| \le \iiint_0^1 m_1(Y) m_2(Y) \frac{m_2(Y + \theta(Z - Y))}{m_2(Y)} |Z - Y| e^{-\frac{\pi}{2}|Y - Z|^2}$$

$$\times 2^n e^{-2\pi|X - \frac{Y+Z}{2}|^2} dY dZ d\theta$$

$$\leq Cm_1(X)m_2(X) \iiint_0^1 (1+|Y-X|^2)^N (1+|Y-Z|^2)^{N+1/2}$$

$$\times e^{-\frac{\pi}{2}|Y-Z|^2} e^{-2\pi|\frac{Y+Z}{2}-X|^2} dY\,dZ\,d\theta$$

$$\leq Cm_1(X)m_2(X) \iint (1+|T/2+S|^2)^N (1+|T|^2)^{N+1/2} e^{-\frac{\pi}{2}|T|^2} e^{-2\pi|S|^2} dT\,dS$$

$$= C'm_1(X)m_2(X).$$

Moreover taking derivatives of R in its defining formula (4.9.5) above leads to the same estimate for $R^{(k)}(X)$. The proof of the lemma is complete. $\qquad\square$

Lemma 4.9.6. *Let (χ_k) be a partition of unity and (ψ_k) be a sequence as in Theorem 4.3.7 for an admissible metric of type $\lambda^{-1}(X)\Gamma$, where λ is a Γ-weight and $\Gamma = \Gamma^\sigma$. Let ω be a locally bounded function such that $|\omega(X)| \leq M(X)$ where M is a Γ-weight. Assume that, for each k, there exist a bounded function ω_k such that $\omega(X) = \omega_k(X)$ for all $X \in \text{supp}\,\chi_k$ and such that for all $X \in \mathbb{R}^{2n}$, $|\omega_k(X)| \leq M(X)\lambda(X)^{N_0}$. Then with $\widetilde{\omega}(X) = \int \omega(Y)2^n \exp{-2\pi\Gamma(X-Y)}dY$, we have*

$$\chi_k(X)\widetilde{\omega}(X) = \chi_k(X)\widetilde{\omega_k}(X) + r_k(X), \quad \sum_k r_k \in S(\lambda^{-\infty}, \Gamma). \qquad (4.9.6)$$

Proof. We already know from Lemma 4.9.3 that

$$X \mapsto \widetilde{\omega}(X) = \int \omega(Y)2^n \exp{-2\pi\Gamma(X-Y)}dY$$

belongs to $S(M,\Gamma)$. We check now

$$\chi_k(X)\widetilde{\omega}(X) = \chi_k(X) \int \omega(Y)2^n \exp{-2\pi\Gamma(X-Y)}dY$$

$$= \chi_k(X) \int \psi_k(Y)\omega(Y)2^n \exp{-2\pi\Gamma(X-Y)}dY$$

$$+ \chi_k(X) \int_{Y,\psi_k(Y)\neq 1} (1-\psi_k(Y))\omega(Y)2^n \exp{-2\pi\Gamma(X-Y)}dY,$$

so that

$$\chi_k(X)\widetilde{\omega}(X) = \chi_k(X) \int \psi_k(Y)\omega_k(Y)2^n \exp{-2\pi\Gamma(X-Y)}dY + r_k(X). \quad (4.9.7)$$

We have

$$\Gamma(U_k - (U_k^*)^c) = \inf_{\Gamma(T)<1\leq\Gamma(S)} \Gamma(X_k + r_0\lambda(X_k)^{1/2}T - X_k - \lambda(X_k)^{1/2}2r_0 S)$$

and thus $\Gamma(U_k - (U_k^*)^c) \geq \lambda(X_k)r_0^2$. Since ψ_k is equal to 1 on U_k^* (notation of Theorem 4.3.7) we obtain from (4.9.7),

$$|r_k^{(j)}(X)|_\Gamma \leq C_j\psi_k(X)\exp{-\pi\Gamma(U_k - (U_k^*)^c)} \leq C_{j,N,r_0}\psi_k(X)\lambda(X)^{-N}$$

and thus $\sum_k r_k \in S(\lambda^{-\infty}, \Gamma)$. We obtain

$$\chi_k \tilde{\omega} = \chi_k \big(\psi_k \omega_k * 2^n \exp -2\pi\Gamma \big) + r_k$$
$$= \chi_k \big(\omega_k * 2^n \exp -2\pi\Gamma \big) + \chi_k \big(\omega_k (\psi_k - 1) * 2^n \exp -2\pi\Gamma \big) + r_k,$$

and applying again the same reasoning to the penultimate term above, we get for $Y \in (U_k^*)^c$ and $X \in U_k$, that $\Gamma(X - Y) \geq \lambda(X_k) r_0^2$ the following estimate for the integrand

$$\exp -\pi\Gamma(X - Y) \exp -\pi\lambda(X_k) r_0^2 \times M(Y)\lambda(Y)^{N_0}$$
$$\leq CM(X)\lambda(X)^{N_0} (1 + \Gamma(X - Y))^{N_0} \exp -\pi\Gamma(X - Y) \exp -\pi\lambda(X_k) r_0^2$$
$$\leq C'M(X)\lambda(X_k)^{N_0} (1 + \Gamma(X - X_k))^{N_0} \exp -\frac{\pi}{2}\Gamma(X - Y) \exp -\pi\lambda(X_k) r_0^2$$
$$\leq C''M(X)\lambda(X_k)^{3N_0} \exp -\frac{\pi}{2}\Gamma(X - Y) \exp -\pi\lambda(X_k) r_0^2$$
$$\leq C'''M(X)\lambda(X_k)^{3N_0} \exp -\frac{\pi}{2}\Gamma(X - Y) \exp -\pi\lambda(X_k) r_0^2$$

which yields the result. $\qquad\square$

Definition 4.9.7. Let Γ be a symplectic quadratic form on $\mathbb{R}^n \times \mathbb{R}^n$, i.e., a positive-definite quadratic form such that $\Gamma = \Gamma^\sigma$. There exists a unique linear symplectic mapping A such that for all $X = (x, \xi)$, $\Gamma(AX) = \sum_{1 \leq j \leq n} x_j^2 + \xi_j^2$. Let U be a metaplectic transformation in the fiber of A. Then for $a \in L^\infty(\mathbb{R}^{2n})$, we define

$$a^{\text{Wick}(\Gamma)} = \int a(Y) 2^n \big(\exp -2\pi\Gamma(\cdot - Y) \big)^w dY = U(a \circ A)^{\text{Wick}} U^*. \qquad (4.9.8)$$

Remark 4.9.8. Note that since U is uniquely determined up to a factor of modulus 1, that definition is consistent. We remark also that, defining for $X \in \mathbb{R}^{2n}$, $\Phi(X) = 2^n \exp -2\pi\Gamma(X)$, we have $\Phi(AX - AY) = 2^n \exp -2\pi|X - Y|^2$, which is the Weyl symbol of Σ_Y (Definition 2.4.1). From the Segal formula (2.1.14), we have, with a metaplectic U in the fiber of A,

$$\Phi(X - Z)^w = U\Phi(AX - Z)^w U^*$$

and thus we can justify the equality in formula (4.9.8) since

$$\int a(Y) 2^n \big(\exp -2\pi\Gamma(X - Y) \big)^w dY = \int a(AY)\Phi(X - AY)^w dY$$
$$= \int a(AY) U\Phi(AX - AY)^w U^* = \int a(AY) U\Sigma_Y U^* dY = U(a \circ A)^{\text{Wick}} U^*.$$

Remark 4.9.9. We can also notice that the definition above is consistent with the fact that Wick and Weyl quantization coincide for linear forms: if a is a linear form, we have

$$a^{\text{Wick}(\Gamma)} = U(a \circ A)^{\text{Wick}} U^* = U(a \circ A)^w U^* = UU^* a^w UU^* = a^w. \qquad (4.9.9)$$

Also, it is easy with the formula (4.9.8) to check that the results of Section 2.4.1 on the Wick quantization can be extended, *mutatis mutandis*, to the Wick(Γ) quantization.

4.9.4 Some lemmas on symbolic calculus

Let g be an admissible metric on $\mathbb{R}^{2n}$ and m be a g-weight (see Definition 2.2.15). Then, at each point $X \in \mathbb{R}^{2n}$, we can define the metric $g_X^{\natural}$ by taking the geometric mean of g_X, g_X^{σ} (see (2.2.21)) so that in particular

$$g_X \leq g_X^{\natural} = (g_X^{\natural})^{\sigma} \leq g_X^{\sigma}. \tag{4.9.10}$$

We define

$$h(X) = \sup_{g_X^{\natural}(T)=1} g_X(T), \tag{4.9.11}$$

and we note that whenever $g^{\sigma} = \lambda^2 g$ we get from Definition 2.2.15 that $g^{\natural} = \lambda_g g$ and $\lambda_g = 1/h$.

Definition 4.9.10. Let l be a non-negative integer. We define the set $S_l(m, g)$ as the set of smooth functions a defined on $\mathbb{R}^{2n}$ such that a satisfies the estimates of $S(m, g)$ for derivatives of order $\leq l$, and the estimates of $S(m, g^{\natural})$ for derivatives of order $\geq l + 1$, which means

$$|a^{(k)}(X)T^k| \leq C_k m(X)$$

$$\times \begin{cases} g_X(T)^{k/2} & \text{if } k \leq l, \\ g_X^{\natural}(T)^{k/2} h(X)^{\frac{l+1}{2}} & \text{if } k \geq l + 1, \text{ with } h(X) = \sup_{g_X^{\natural}(T)=1} g_X(T). \end{cases}$$

Note that since $h \leq 1$ and $g \leq h g^{\natural}$, we get $S(m, g) \subset S_l(m, g)$. If $g = \lambda(X)^{-1}\Gamma_X$, where $\lambda(X)$ is positive (scalar) and $\Gamma_X = \Gamma_X^{\sigma}$, then $g_X^{\natural} = \Gamma_X$ and a belongs to $S_l(m, \lambda^{-1}\Gamma)$ means

$$|a^{(k)}(X)|_{\Gamma_X} \leq C_k m(X) \times \begin{cases} \lambda(X)^{-k/2} & \text{if } k \leq l, \\ \lambda(X)^{-l/2} & \text{if } k \geq l + 1. \end{cases}$$

Moreover, if $g \equiv g^{\natural}$, then for all l, $S(m, g) = S_l(m, g)$.

Lemma 4.9.11. *Let Γ be a positive-definite quadratic form on $\mathbb{R}^{2n}$ such that $\Gamma = \Gamma^{\sigma}$ and λ be a Γ-weight. Let b be a symbol in $S_1(\lambda^m, \lambda^{-1}\Gamma)$, where m is a real number. Then $b \natural b - b^2 \in S(\lambda^{2m-1}, \Gamma)$*

Proof. We have $(b \natural b)(X) = \exp i\pi[D_{X_1}, D_{X_2}](b(X_1) \otimes b(X_2))_{|X_1=X_2=X}$ so that using Taylor's formula with integral remainder for $s \mapsto e^s$ yields

$$(b \natural b)(X) = b(X)^2 + \int_0^1 \exp i\pi\theta[D_{X_1}, D_{X_2}] d\theta i\pi[D_{X_1}, D_{X_2}] b(X_1) \otimes b(X_2)_{|X_1=X_2=X}.$$

Since $b' \in S(\lambda^{m-1/2}, \Gamma)$ and

$$\exp i\pi\theta[D_{X_1}, D_{X_2}](a_1(X_1) \otimes a_2(X_2))$$
$$= \exp i\pi[D_{X_1\theta^{-1/2}}, D_{X_2\theta^{-1/2}}](a_1(\theta^{-1/2}X_1\theta^{1/2}) \otimes a_2(\theta^{-1/2}X_2\theta^{1/2}))_{|X_1=X_2=X}$$
$$= \exp i\pi[D_{Y_1}, D_{Y_2}](a_1(\theta^{1/2}Y_1) \otimes a_2(\theta^{1/2}Y_2))_{|Y_1=Y_2=\theta^{-1/2}X}$$
$$= ((a_1 \circ \theta^{1/2})\sharp(a_2 \circ \theta^{1/2}))(\theta^{-1/2}X),$$

we get that, if $a_j \in S(\lambda^{m_j}, \Gamma)$, we have $a_j \circ \theta^{1/2} \in S(\lambda^{m_j}, \theta\Gamma)$ so that the symbolic calculus for the metric $\theta\Gamma$ (observe that it is admissible for θ bounded) gives

$$(a_1 \circ \theta^{1/2})\sharp(a_2 \circ \theta^{1/2}) \in S(\lambda^{m_1+m_2}, \theta\Gamma)$$

which implies $((a_1 \circ \theta^{1/2})\sharp(a_2 \circ \theta^{1/2})) \circ \theta^{-1/2} \in S(\lambda^{m_1+m_2}, \Gamma)$. Applying this to the integral above gives the result of the lemma. $\qquad\square$

Lemma 4.9.12. *Let Γ be a positive-definite quadratic form on $\mathbb{R}^{2n}$ such that $\Gamma = \Gamma^\sigma$ and λ is a Γ-weight. Let $l \in \mathbb{N}, \mu \in \mathbb{R}$ and a be a locally bounded function defined on $\mathbb{R}^{2n}$ such that*

$$\forall j \in \{0, \dots, l\}, \quad |a^{(j)}(X)| \leq C\lambda(X)^{\mu-\frac{j}{2}}.$$

Then the function $a * \exp -2\pi\Gamma$ belongs to $S_l(\lambda^\mu, \lambda^{-1}\Gamma)$.

Proof. We use the formula $(a * \exp -2\pi\Gamma)(X) = \int a(X-Y) \exp -2\pi\Gamma(Y) dY$ to obtain an estimate for the derivatives of order $\leq l$: we get for $k \leq l$,

$$|(a * \exp -2\pi\Gamma)^{(k)}(X)| \leq C\lambda(X)^{\mu-\frac{k}{2}} \int \frac{\lambda(X-Y)^{\mu-\frac{k}{2}}}{\lambda(X)^{\mu-\frac{k}{2}}} \exp -2\pi\Gamma(Y) dY$$

$$\leq C\lambda(X)^{\mu-\frac{k}{2}} \int (1+\Gamma(Y))^{N|\mu-\frac{k}{2}|} \exp -2\pi\Gamma(Y) dY = C'\lambda(X)^{\mu-\frac{k}{2}},$$

and for $k > l$ we have $(a * \exp -2\pi\Gamma)^{(k)} = (a^{(l)} * (\exp -2\pi\Gamma)^{(k-l)})$ yielding immediately the result. $\qquad\square$

Let us recall the composition formula in the Weyl quantization, with the symplectic form $[,]$ given in (2.1.18). We have $a^w b^w = (a\sharp b)^w$ and, for $X \in \mathbb{R}^{2n}$,

$$(a\sharp b)(X) = 2^{2n} \iint_{\mathbb{R}^{2n} \times \mathbb{R}^{2n}} a(Y)b(Z) \exp -4i\pi[X-Y, X-Z] dY dZ$$

$$= 2^{2n} \iint_{\mathbb{R}^{2n} \times \mathbb{R}^{2n}} a(Y+X)b(Z+X) \exp -4i\pi[Y, Z] dY dZ. \qquad (4.9.12)$$

We note also that

$$(a\sharp b)' = a'\sharp b + a\sharp b'. \qquad (4.9.13)$$

Moreover, if a is a function only of ξ, we have

$$(a\sharp b)(x,\xi) = 2^{2n}\int_{\mathbb{R}^{4n}} a(\eta)b(z,\zeta)e^{-4i\pi(\xi-\eta)(x-z)}e^{4i\pi(x-y)(\xi-\zeta)}\,dy\,d\eta\,dz\,d\zeta$$

$$= 2^n\int_{\mathbb{R}^{2n}} a(\eta)b(z,\xi)e^{-4i\pi(\xi-\eta)(x-z)}\,d\eta\,dz$$

$$= 2^n\int_{\mathbb{R}^{2n}} ((1+D_\eta^2/4)^N a)(\eta)b(z,\xi)(1+|x-z|^2)^{-N}e^{-4i\pi(\xi-\eta)(x-z)}\,d\eta\,dz$$

$$= 2^n\int_{\mathbb{R}^{2n}} ((1+D_\eta^2/4)^N a)(\eta)(1+|\xi-\eta|^2)^{-N}(1+D_z^2/4)^N$$

$$\times\, \big(b(z,\xi)(1+|x-z|^2)^{-N}\big)e^{-4i\pi(\xi-\eta)(x-z)}\,d\eta\,dz,$$

so that with $N \geq E(n/2)+1$

$$|(a\sharp b)(x,\xi)| \leq \max_{j\leq 2N}\|a^{(j)}\|_{L^\infty}\max_{j\leq 2N}\|b^{(j)}\|_{L^\infty}(1+|\xi-\operatorname{supp}a|)^{-N/2}c(n,N). \quad (4.9.14)$$

4.9.5 The Beals-Fefferman reduction

Lemma 4.9.13. *Let* $F:\mathbb{R}\to\mathbb{R}$ *be a* C^2 *function such that*

$$16|F(0)| < F'(0)^2,\quad \|F''\|_{L^\infty(\mathbb{R})}\leq 1. \quad (4.9.15)$$

We set $\rho = |F'(0)|/4$. *Then there exists* $t_0\in[-\rho/2,\rho/2]$ *and* $e\in C^1(\mathbb{R})$ *such that*

$$\text{for } |t|\leq\rho,\ F(t) = (t-t_0)e(t),\quad 8\rho\geq e(t)\geq\rho,\quad \|e'\|_{L^\infty(\mathbb{R})}\leq 1. \quad (4.9.16)$$

Proof. Assume first that $F(0) = 0$ and $F'(0) = 4\rho$. Then, for $|t|\leq 2\rho$,

$$F(t) = te(t),\quad 6\rho\geq e(t)\geq 4\rho-2\rho = 2\rho, \|e'\|_{L^\infty(\mathbb{R})}\leq 1.$$

Now if $F(0) > 0$ and $F'(0) = 4\rho$, $F(-\frac{\rho}{2})\leq\rho^2-\frac{\rho}{2}4\rho+\frac{\rho^2}{4} < 0$, so that, for some $t_0\in]-\rho/2,0[$ we have $F(t_0) = 0$. Using what was done above, we have for $|s|\leq|F'(t_0)|/2$,

$$F(s+t_0) = (s+t_0)e_0(s),\quad 3|F'(t_0)|/2\geq e_0(s)\geq|F'(t_0)|/2,\quad \|e_0'\|_{L^\infty(\mathbb{R})}\leq 1.$$

But since

$$\frac{|F'(t_0)|}{2}\geq\frac{1}{2}(4\rho-\frac{\rho}{2}) = \frac{7\rho}{4}\quad\text{and}\quad\frac{7\rho}{4}-\frac{\rho}{2} = \frac{5\rho}{4}\geq\rho,$$

we have on $[t_0-\frac{7\rho}{4}, t_0+\frac{7\rho}{4}]$ which contains $[-\rho,\rho]$,

$$F(t) = (t-t_0)e(t),\quad |t_0|\leq\rho/2,\quad 8\rho\geq\frac{27\rho}{4}\geq e(t)\geq 7\rho/4\geq\rho, \|e'\|_{L^\infty(\mathbb{R})}\leq 1.$$

$\square$

Lemma 4.9.14. *Let $F : \mathbb{R}^d \to \mathbb{R}$ be a C^2 function such that*

$$2^6 |F(0)| < \|\nabla F(0)\|^2, \quad \|F''\|_{L^\infty(\mathbb{R}^d)} \le 1. \tag{4.9.17}$$

We set $\rho = \|\nabla F(0)\| 2^{-5}$. There exists two C^1 functions $\alpha : \mathbb{R}^{d-1} \to [-5\rho, 5\rho]$ and $e : \mathbb{R}^d \to [7\rho, 70\rho]$, a set of orthonormal coordinates $(x_1, x') \in \mathbb{R} \times \mathbb{R}^{d-1}$ such that for $\max(|x_1|, |x'|) \le \rho$,

$$F(x) = (x_1 + \alpha(x'))e(x), \quad \|e'\|_{L^\infty(\mathbb{R}^d)} \le 1, \quad \|\alpha'\|_{L^\infty(\mathbb{R}^{d-1})} \le 1. \tag{4.9.18}$$

Proof. We can choose the coordinates so that $\nabla F(0) = \frac{\partial F}{\partial x_1}(0)\vec{e_1}$. Then for $|x'| \le \rho$, we have $|F(0, x')| \le 2^{-6+10}\rho^2 + \rho 2^5 \rho + \frac{1}{2}\rho^2 = \rho^2(2^5 + 2^4 + 2^{-1})$ and

$$\left|\frac{\partial F}{\partial x_1}(0, x')\right| \ge \left|\frac{\partial F}{\partial x_1}(0, 0)\right| - \rho = (2^5 - 1)\rho$$

so that

$$\frac{16|F(0, x')|}{\left|\frac{\partial F}{\partial x_1}(0, x')\right|^2} \le \frac{16 \times 48.5}{31^2} < 1.$$

Applying Lemma 4.9.13, we get for all $|x'| \le \rho$ the existence of $\alpha(x')$ such that, when $|x_1| \le 31\rho/4$,

$$F(x_1, x') = (x_1 + \alpha(x'))e(x), \quad |\alpha(x')| \le \frac{33\rho}{8} < 5\rho,$$

$$70\rho \ge 8 \times 33\rho/4 \ge |e(x)| \ge 31\rho/4 > 7\rho.$$

The implicit function theorem guarantees the C^1 regularity of the function α and the Taylor-Lagrange formula with integral remainder provides the regularity of e. $\qquad\square$

Remark 4.9.15. If the function F in Lemma 4.9.14 is C^∞, since the function α is obtained by the implicit function theorem, and e by Taylor's formula with integral remainder, both functions α, e are C^∞. Moreover, the identity $F(-\alpha(x'), x') = 0$ implies that

$$|\alpha^{(k)}(x')| \le C_k \rho^{1-k}, \quad |e^{(k)}(x')| \le C_k \rho^{-k}$$

where C_k are semi-norms of the function F in $\max(|x_1|, |x'|) \le \rho$. In particular, if we apply this result to the function (3.2.34)

$$F(T) = \Lambda^{1/2} q(t, Y + \nu(t, Y)^{1/2}T) \mu(t, Y)^{-1/2} \nu(t, Y)^{-1}$$

we get that $|F^{(k)}|$ is bounded above by $\gamma_k(q)$ and $1/2 \le \rho \le \gamma_1(q)$ as defined in (2.1.1). We get then from Lemma 4.9.14,

$$\Lambda^{1/2} q(t, Y + \nu(t, Y)^{1/2}T) \mu(t, Y)^{-1/2} \nu(t, Y)^{-1} = e_0(T)(T_1 + \alpha_0(T'))$$

so that e_0, α_0 are smooth with fixed bounds and thus

$$
\Lambda^{1/2} q(t, X) \mu(t, Y)^{-1/2} = e_0 \big((X - Y)\nu(t, Y)^{-1/2} \big) \nu(t, Y)^{1/2}
$$
$$
\times \big(X_1 - Y_1 + \alpha_0 \big((X' - Y')\nu(t, Y)^{-1/2} \big) \nu(t, Y)^{1/2} \big),
$$

which corresponds exactly to (3.2.35), (3.2.36), (3.2.37).

4.9.6 On tensor products of homogeneous functions

Let $n \geq 1$ be an integer and $N = n + 1$. Let $(y_0; \eta_0) \in \mathbb{R}^N \times \mathbb{S}^{N-1}$ such that

$$
(y_0; \eta_0) = (t_0, x_0; \tau_0, \xi_0) \in \mathbb{R} \times \mathbb{R}^n \times \mathbb{R} \times \mathbb{R}^n, \quad \text{with } \tau_0 = 0, \ \xi_0 \in \mathbb{S}^{n-1}.
$$

Let $r \in]0, 1/4]$ be given. There exists a function $\chi_0 \in C^\infty(\mathbb{R}^N; [0, 1])$ such that for $\lambda \geq 1$ and $\eta \in \mathbb{R}^N$ with $|\eta| \geq 1$, we have $\chi_0(\lambda \eta) = \chi_0(\eta)$ ("homogeneity of degree zero outside the unit ball') and

$$
\chi_0(\tau, \xi) = \begin{cases} 1 & \text{if } \tau^2 + |\xi|^2 \geq 1 \text{ and } |\tau| \leq r|\xi|, \\ 0 & \text{if } \tau^2 + |\xi|^2 \leq 1/4 \text{ or } |\tau| \geq 2r|\xi|. \end{cases}
$$

There exists a function $\psi_0 \in C^\infty(\mathbb{R}^n; [0, 1])$ such that for $\lambda \geq 1$ and $\xi \in \mathbb{R}^n$ with $|\xi| \geq 1$, we have $\psi_0(\lambda \xi) = \psi_0(\xi)$ and,

$$
\psi_0(\xi) = \begin{cases} 1 & \text{if } |\xi| \geq 1 \text{ and } |\frac{\xi}{|\xi|} - \xi_0| \leq r, \\ 0 & \text{if } |\xi| \leq 1/2 \text{ or } |\frac{\xi}{|\xi|} - \xi_0| \geq 2r\ . \end{cases}
$$

We define the function Φ_0 by

$$
\Phi_0(\tau, \xi) = \chi_0(\tau, \xi)\psi_0(\xi). \tag{4.9.19}
$$

Lemma 4.9.16. *The function Φ_0 is such that for $\lambda \geq 1$ and $\eta \in \mathbb{R}^N$ with $|\eta| \geq 2$, we have $\Phi_0(\lambda \eta) = \Phi_0(\eta)$. Moreover, with $\eta_0 = (0, \xi_0)$, we have*

$$
\Phi_0(\eta) = 1 \text{ for } |\eta| \geq 2 \text{ and } \left| \frac{\eta}{|\eta|} - \eta_0 \right| \leq r/2, \tag{4.9.20}
$$

$$
\Phi_0(\eta) = 0 \text{ for } |\eta| \geq 2 \text{ and } \left| \frac{\eta}{|\eta|} - \eta_0 \right| \geq 4r. \tag{4.9.21}
$$

Proof. The function Φ_0 is such that for $\lambda \geq 1$ and $\eta \in \mathbb{R}^N$ with $|\eta| \geq 2$, we have $\Phi_0(\lambda \eta) = \Phi_0(\eta)$: in fact, if $\tau^2 + |\xi|^2 \geq 4$ and $|\tau| \leq 2r|\xi|$, we get

$$
|\xi|^2 \geq 4(1 + 4r^2)^{-1} \geq 1,
$$

so that $\psi_0(\lambda \xi) = \psi_0(\xi)$ and since we have also in that case $\chi_0(\lambda \eta) = \chi_0(\eta)$, we get the sought property. Now if $\tau^2 + |\xi|^2 \geq 4$ and $|\tau| > 2r|\xi|$, we see that

$\chi_0(\lambda\tau, \lambda\xi) = \chi_0(\tau, \xi) = 0$ so that, $\Phi_0(\lambda\eta) = 0 = \Phi_0(\eta)$. Moreover, if $\tau^2 + |\xi|^2 \geq 4$ and

$$\frac{\tau^2}{\tau^2 + |\xi|^2} + \left|\frac{\xi}{(\tau^2 + |\xi|^2)^{1/2}} - \xi_0\right|^2 \leq r^2/4,$$

we get that $|\tau| \leq r|\xi|(4 - r^2)^{-1/2} \leq r|\xi|$ and thus $\chi_0(\tau, \xi) = 1$; also this implies $|\xi| \geq 2(1 + r^2)^{-1/2} \geq 1$, so that $\psi_0(\xi) = \psi_0(\xi/|\xi|)$. We have then

$$\left|\frac{\xi}{|\xi|} - \xi_0\right| \leq \frac{r}{2} + \left|\frac{\xi}{|\xi|} - \frac{\xi}{(\tau^2 + |\xi|^2)^{1/2}}\right| \leq \frac{r}{2} + |\xi||\tau|^2|\xi|^{-3} \leq \frac{r}{2} + \frac{r^2}{4 - r^2} \leq r,$$

which implies $\psi_0(\xi) = \psi_0(\xi/|\xi|) = 1$, so that Φ_0 is equal to 1 on a conic-neighborhood of $(0, \xi_0)$ in $\mathbb{R}^N$ minus a ball. Similarly, if $\tau^2 + |\xi|^2 \geq 4$ and

$$\frac{\tau^2}{\tau^2 + |\xi|^2} + \left|\frac{\xi}{(\tau^2 + |\xi|^2)^{1/2}} - \xi_0\right|^2 \geq 16r^2,$$

either $|\tau| \geq 2r|\xi|$ and $\chi_0(\tau, \xi) = 0$, entailing $\Phi_0(\tau, \xi) = 0$ or $|\tau| \leq 2r|\xi|$ and then

$$\left|\frac{\xi}{(\tau^2 + |\xi|^2)^{1/2}} - \xi_0\right|^2 \geq 12r^2 \quad \text{and} \quad |\xi| \geq 2(1 + 4r^2)^{-1/2} \geq 1,$$

so that $\psi_0(\xi) = \psi_0(\xi/|\xi|)$. In this case, we have

$$\left|\frac{\xi}{|\xi|} - \xi_0\right| \geq 2\sqrt{3}r - \left|\frac{\xi}{|\xi|} - \frac{\xi}{(\tau^2 + |\xi|^2)^{1/2}}\right| \geq 2\sqrt{3}r - \frac{\tau^2}{|\xi|^2} \geq 2\sqrt{3}r - 4r^2 \geq 2r,$$

implying $\psi_0(\xi) = 0$ and thus $\Phi_0(\tau, \xi) = 0$. Eventually, we have proven that Φ_0 is also supported in a conic-neighborhood of $(0, \xi_0)$ in $\mathbb{R}^N$. $\square$

4.9.7 On the composition of some symbols

Lemma 4.9.17. *Let G, g be metrics on $\mathbb{R}_t \times \mathbb{R}_x^n \times \mathbb{R}_\tau \times \mathbb{R}_\xi^n$ (equipped with its canonical symplectic structure) defined by*

$$G = |dt|^2 + |dx|^2 + \frac{|d\xi|^2 + |d\tau|^2}{1 + |\xi|^2 + \tau^2}$$

$$\leq g = |dt|^2 + |dx|^2 + \frac{|d\xi|^2}{1 + |\xi|^2} + \frac{|d\tau|^2}{1 + |\xi|^2 + \tau^2}. \tag{4.9.22}$$

These metrics are admissible and for all $s \in \mathbb{R}$, $(1 + |\xi| + |\tau|)^s$ is a G-weight and $(1 + |\xi|)^s$ is a g-weight.

Proof. Lemma 2.2.18 provides the admissibility of the classical metric G, whose "Planck function" λ_G is given by $(1 + |\xi| + |\tau|)$ so that $(1 + |\xi| + |\tau|)^s$ is also an admissible weight for G. The metric g satisfies the uncertainty principle (2.2.14) since $\lambda_g = 1 + |\xi| \geq 1$; it is also slowly varying (cf. Definition 2.2.1) since

$$g_{t,x,\tau,\xi}(s-t, y-x, \sigma-\tau, \eta-\xi) \leq r^2 \implies |\xi-\eta|^2 \leq r^2(1+|\xi|^2), \ |\sigma-\tau|^2 \leq r^2(1+|\tau|^2),$$

and since the metric $|d\xi|^2/(1 + |\xi|^2)$ is slowly varying on $\mathbb{R}^n$, we get that the ratios $\langle\xi\rangle/\langle\eta\rangle, \langle\tau\rangle/\langle\sigma\rangle$ are bounded above and below by some positive constants. We have also, with the same notation as above,

$$\frac{1+|\tau|}{1+|\sigma|} + \frac{1+|\tau|+|\xi|}{1+|\sigma|+|\eta|} \leq 2 + 2|\tau - \sigma| + |\xi - \eta|,$$

providing the temperance. $\qquad\square$

Lemma 4.9.18. *Let G, g be the metrics on $\mathbb{R}^{2N}$ defined in (4.9.22) and let s_1, s_2 be two real numbers. Let $a \in S(\langle\xi\rangle^{s_1}, g)$ and $b \in S(\langle\xi,\tau\rangle^{s_2}, G)$ such that*

$$\operatorname{supp} b \subset Z_C = \{(t,x,\tau,\xi) \in \mathbb{R}^{2N}, |\tau| \leq 1 + C|\xi|\}.$$

Then the symbols $a\sharp b, b\sharp a, a \circ b, b \circ a$ belong to $S(\langle\xi,\tau\rangle^{s_1+s_2}, G)$ and are essentially supported in Z_C, i.e., are the sum of a symbol of $S(\langle\xi,\tau\rangle^{s_1+s_2}, G)$ supported in Z_C and of a symbol in $S(\langle\xi,\tau\rangle^{-\infty}, G) = \cap_N S(\langle\xi,\tau\rangle^{-N}, G)$.

Proof. We have

$$(a \circ b)(t,x,\tau,\xi) = \int e^{-2i\pi(s\sigma+y\eta)} a(t,x,\tau+\sigma,\xi+\eta) b(t+s, x+y, \tau, \xi) ds d\sigma dy d\eta,$$

$$(4.9.23)$$

so that, using the standard expansion of the symbols and the fact that b is supported in Z_C,

$$a \circ b = \sum_{|\alpha| < \nu} \frac{1}{\alpha!} \overbrace{D_{\tau,\xi}^\alpha a \ \partial_{t,x}^\alpha b}^{\in S(\langle\tau,\xi\rangle^{1-|\alpha|}, G)}$$

$$+ \int_0^1 \frac{(1-\theta)^{\nu-1}}{(\nu-1)!} e^{-2i\pi(s\sigma+y\eta)} D_{\tau,\xi}^\nu a(t,x,\tau+\theta\sigma, \xi+\theta\eta)$$

$$\times \partial_{t,x}^\nu b(t+s, x+y, \tau, \xi) ds d\sigma dy d\eta d\theta.$$

We define

$$I_\theta(\tau,\xi) = \int e^{-2i\pi(s\sigma+y\eta)} D_{\tau,\xi}^\nu a(t,x,\tau+\theta\sigma, \xi+\theta\eta)$$

$$\times \partial_{t,x}^\nu b(t+s, x+y, \tau, \xi) ds d\sigma dy d\eta, \quad (4.9.24)$$

and integrating by parts, we obtain for all non-negative even integers m that

$$I_\theta(\tau,\xi) = \int e^{-2i\pi(s\sigma+y\eta)} \langle\sigma\rangle^{-m} \langle D_s\rangle^m \langle s\rangle^{-m} \langle D_\sigma\rangle^m \langle y\rangle^{-m} \langle D_\eta\rangle^m \langle\eta\rangle^{-m}$$

$$\times \langle D_y\rangle^m D_{\tau,\xi}^\nu a(t,x,\tau+\theta\sigma, \xi+\theta\eta) \partial_{t,x}^\nu b(t+s, x+y, \tau, \xi) ds d\sigma dy d\eta,$$

and consequently

$$|I_\theta(\tau, \xi)| \lesssim \int \langle \sigma \rangle^{-m} \langle s \rangle^{-m} \langle y \rangle^{-m} \langle \eta \rangle^{-m} (1 + |\xi + \theta \eta|)^{s_1 - \nu} ds d\sigma dy d\eta$$
$$\times (1 + |\xi| + |\tau|)^{s_2} \mathbf{1}(|\tau| \lesssim |\xi|).$$

In the integrand, when $|\eta| \leq |\xi|/2$, we get, since $\theta \in [0, 1]$, $|\xi + \theta \eta| \geq |\xi| - |\eta| \geq |\xi|/2$. As a result, we get for this part of the integral the estimate

$$(1 + |\xi|)^{|s_1| - \nu}(1 + |\xi| + |\tau|)^{s_2} \mathbf{1}(|\tau| \lesssim |\xi|) \lesssim (1 + |\xi| + |\tau|)^{-\nu/2}, \quad \text{for } \nu \text{ large enough.}$$

When $|\eta| > |\xi|/2$, we use the term $\langle \eta \rangle^{-m}$ and the estimate

$$(1 + |\xi|)^{-m/2}(1 + |\xi| + |\tau|)^{s_2} \mathbf{1}(|\tau| \lesssim |\xi|) \lesssim (1 + |\xi| + |\tau|)^{-m/4}, \quad \text{for } m \text{ large enough.}$$

To check that the derivatives of I_θ will satisfy the expected estimates, we differentiate the expression (4.9.24) and repeat the previous proof. We know now that, for all $\nu \geq N_0$,

$$a \circ b = \sum_{|\alpha| < \nu} \frac{1}{\alpha!} D_{\tau, \xi}^\alpha a \, \partial_{t,x}^\alpha b + r_\nu, \quad r_\nu \in S(\langle \tau, \xi \rangle^{-\nu/2}, G).$$

Using the standard Borel argument, we find $c \in S(\langle \tau, \xi \rangle^{s_1 + s_2}, G)$, essentially supported in Z_C such that, for all ν,

$$c - \sum_{|\alpha| < \nu} \frac{1}{\alpha!} D_{\tau, \xi}^\alpha a \, \partial_{t,x}^\alpha b \in S(\langle \tau, \xi \rangle^{s_1 + s_2 - \nu}, G),$$

entailing that, for all $\nu \geq N_0$,

$$a \circ b - c = -c + \sum_{|\alpha| < \nu} \frac{1}{\alpha!} D_{\tau, \xi}^\alpha a \, \partial_{t,x}^\alpha b + r_\nu \in S(\langle \tau, \xi \rangle^{\max(-\nu/2, s_1 + s_2 - \nu)}, G),$$

implying that $a \circ b - c \in S(\langle \tau, \xi \rangle^{-\infty}, G)$, which gives the result of the lemma for $a \circ b$. To get the result for $b \circ a$ is somewhat easier by looking at (4.9.24), to obtain the estimate

$$|I_\theta(\tau, \xi)| \lesssim \int \langle \sigma \rangle^{-m} \langle s \rangle^{-m} \langle y \rangle^{-m} \langle \eta \rangle^{-m} (1 + |\xi + \theta \eta| + |\tau + \theta \sigma|)^{s_2 - \nu}$$
$$\times \mathbf{1}(|\tau + \theta \eta| \lesssim |\xi + \theta \sigma|) ds d\sigma dy d\eta (1 + |\xi|)^{s_1}.$$

When $|\tau| \lesssim |\xi|$ the discussion is the same as for $a \circ b$. When $|\tau| \gg |\xi|$, we split the integral in two parts: the region where $|\sigma| \leq |\tau|/2$, in which we get negative powers of $(1 + |\tau|)$ from the term with the exponent $s_2 - \nu$, and the region where $|\sigma| > |\tau|/2$ in which we use the term $\langle \sigma \rangle^{-m}$. The last part of the discussion is the same. To obtain the result for $a \sharp b$ (which will give also $b \sharp a$ since $\overline{a \sharp b} = \bar{b} \sharp \bar{a}$), we use the group $J^t = \exp 2i\pi t D_x D_\xi$ and the formula $a \sharp b = J^{-1/2}\big(J^{1/2}a \circ J^{1/2}b\big)$. Using the assumptions of the lemma, we see that $J^{1/2}a$ satisfies the same hypothesis as a and $J^{1/2}b$ is essentially supported in Z_C. The proofs above give thus that $J^{1/2}a \circ J^{1/2}b$ satisfies the conclusion of the lemma, which is "stable" by the action of $J^{-1/2}$. The proof of Lemma 4.9.18 is complete. $\square$

Bibliography

[1] S. Alinhac and P. Gérard, *Opérateurs pseudo-différentiels et théorème de Nash-Moser*, InterEditions, Paris, Editions du CNRS, Meudon, 1991.

[2] Hiroshi Ando and Yoshinori Morimoto, *Wick calculus and the Cauthy problem for some dispersive equations*, Osaka J. Math. **39** (2002), no. 1, 123–147. MR1883917 (2003b:35219)

[3] T. Ando, *Concavity of certain maps on positive definite matrices and applications to Hadamard products*, Linear Algebra Appl. **26** (1979), 203–241. MR535686 (80f:15023)

[4] Michael Atiyah and Daniel Iagolnitzer (eds.), *Fields Medallists' lectures*, World Scientific Series in 20th Century Mathematics, vol. 5, World Scientific Publishing Co. Inc., River Edge, NJ, 1997. MR1622945 (99b:00010)

[5] Michèle Audin, *Les systèmes hamiltoniens et leur intégrabilité*, Cours Spécialisés [Specialized Courses], vol. 8, Société Mathématique de France, Paris, 2001. MR1972063 (2004g:37074)

[6] R. Beals, *A general calculus of pseudodifferential operators*, Duke Math. J. **42** (1975), 1–42.

[7] _____, *Characterization of pseudodifferential operators and applications*, Duke Math. J. **44, 1** (1977), 45–57.

[8] R. Beals and C. Fefferman, *On local solvability of linear partial differential equations*, Ann. of Math. (2) **97** (1973), 482–498. MR0352746 (50 #5233)

[9] _____, *Spatially inhomogeneous pseudodifferential operators. I*, Comm. Pure Appl. Math. **27** (1974), 1–24. MR0352747 (50 #5234)

[10] F.A. Berezin, *Quantization*, Math.USSR, Izvest. **8** (1974), 1109–1165.

[11] Rajendra Bhatia and Chandler Davis, *More matrix forms of the arithmetic-geometric mean inequality*, SIAM J. Matrix Anal. Appl. **14** (1993), no. 1, 132–136. MR1199551 (94b:15017)

[12] P. Boggiatto, E. Buzano, and L. Rodino, *Global hypoellipticity and spectral theory*, vol. 92, Akademie Verlag, Mathematical Research, 1996.

[13] J.-M. Bony, *Principe du maximum, inégalite de Harnack et unicité du problème de Cauchy pour les opérateurs elliptiques dégénérés*, Ann. Inst. Fourier (Grenoble) **19** (1969), no. fasc. 1, 277–304 xii. MR0262881 (41 #7486)

[14] _____, *Second microlocalization and propagation of singularities for semilinear hyperbolic equations* , Hyperbolic equations and related topics (Katata/Kyoto, 1984), Academic Press, Boston, MA, 1986, pp. 11–49. MR925240 (89e:35099)

[15] _____, *Caractérisations des opérateurs pseudo-différentiels*, Séminaire sur les Équations aux Dérivées Partielles, 1996–1997, École Polytech., Palaiseau, 1997, pp. Exp. No. XXIII, 17. MR1482829 (98m:35233)

[16] _____, *Sur l'inégalité de Fefferman-Phong*, Séminaire EDP, Ecole Polytechnique (1998-99), Exposé 3.

[17] _____, *Sommes de carrés de fonctions dérivables*, Bull. Soc. Math. France **133** (2005), no. 4, 619–639. MR2233698 (2007e:26004)

[18] _____, *private communications*, , 2006.

[19] J.-M. Bony and J.-Y. Chemin, *Espaces fonctionnels associés au calcul de Weyl-Hörmander*, Bull. Soc. Math. France **122** (1994), 77–118.

[20] J.-M. Bony and N. Lerner, *Quantification asymptotique et microlocalisations d'ordre supérieur. I*, Ann. Sci. École Norm. Sup. (4) **22** (1989), no. 3, 377–433. MR1011988 (90k:35276)

[21] A. Boulkhemair, *Remarks on a Wiener type pseudodifferential algebra and Fourier integral operators*, Math.Res.Lett. **4** (1997), 53–67.

[22] _____, *L^2 estimates for Weyl quantization*, J.Func.Anal. **165** (1999), 173–204.

[23] _____, *On the Fefferman-Phong inequality*, Ann. Inst. Fourier (Grenoble) **58** (2008), no. 4, 1093–1115. MR2427955

[24] Haïm Brezis, *On a characterization of flow-invariant sets*, Comm. Pure Appl. Math. **23** (1970), 261–263. MR0257511 (41 #2161)

[25] R. Brummelhuis, *A counterexample to the Fefferman–Phong inequality for systems*, C. R. Acad. Sci. Paris **310** (1990), série I, 95–98.

[26] Francine Bruyant, *Estimations pour la composition d'un grand nombre d'opérateurs pseudo-différentiels d'ordre 0 et une application* , C.R.Acad.Sci.Paris Sér.A-B **289** (1979), no. 14, A667–A669. MR560329 (80k:58093)

[27] V. S. Buslaev, *Quantization and the WKB method*, Trudy Mat. Inst. Steklov. **110** (1970), 5–28. MR0297258 (45 #6315)

[28] R.D Coifman and Y. Meyer, *Au delà des opérateurs pseudo-différentiels*, Astérisque, vol. 57, Société Mathématique de France, Paris, 1978, With an English summary. MR518170 (81b:47061)

[29] A. Cordoba and C. Fefferman, *Wave packets and Fourier integral operators*, Comm. PDE **3** (1978), (11), 979–1005.

[30] E. B. Davies, *Linear operators and their spectra*, Cambridge Studies in Advanced Mathematics, vol. 106, Cambridge University Press, Cambridge, 2007. MR2359869 (2008k:47001)

[31] _____, *Non-self-adjoint operators and pseudospectra*, Spectral theory and mathematical physics: a Festschrift in honor of Barry Simon's 60th birthday, Proc. Sympos. Pure Math., vol. 76, Amer. Math. Soc., Providence, RI, 2007, pp. 141–151. MR2310201 (2008c:47009)

[32] J.-M. Delort, *F.B.I. transformation, second microlocalization and semilinear caustics*, Lecture Notes in Mathematics, vol. 1522, Springer Verlag, 1992.

[33] N. Dencker, *Estimates and solvability*, Arkiv.Mat. **37** (1999), 2, 221–243.

[34] _____, *The solvability of pseudo-differential operators*, vol. 1, pp. 175–200, Pubbl. Cent. Ric. Mat. Ennio Giorgi, Scuola Norm. Sup., Pisa, 2004.

[35] _____, *The resolution of the Nirenberg-Treves conjecture*, Ann. of Math. **163** (2006), 2, 405–444.

[36] N. Dencker, J. Sjöstrand, and M. Zworski, *Pseudospectra of semiclassical (pseudo-) differential operators*, Comm. Pure Appl. Math. **57** (2004), no. 3, 384–415. MR2020109 (2004k:35432)

[37] M. Dimassi and J. Sjöstrand, *Spectral asymptotics in the semi-classical limit*, London Mathematical Society Lecture Note Series, vol. 268, Cambridge University Press, Cambridge, 1999. MR1735654 (2001b:35237)

[38] _____, *Spectral asymptotics in the semi-classical limit*, LMS lecture note series, vol. 268, Cambridge University Press, 1999.

[39] William F. Donoghue, Jr., *Monotone matrix functions and analytic continuation*, Springer-Verlag, New York, 1974, Die Grundlehren der mathematischen Wissenschaften, Band 207. MR0486556 (58 #6279)

[40] Javier Duoandikoetxea, *Fourier analysis*, Graduate Studies in Mathematics, vol. 29, American Mathematical Society, Providence, RI, 2001, Translated and revised from the 1995 Spanish original by David Cruz-Uribe. MR1800316 (2001k:42001)

[41] Y. V. Egorov, *Subelliptic pseudodifferential operators*, Soviet Math. Dok. **10** (1969), 1056–1059.

[42] Y. V. Egorov and V. A. Kondrat'ev, *The oblique derivative problem*, Mat. sb. (N.S.) **78 (120)** (1969), 148–176. MR0237953 (38 #6230)

[43] G.I. Eskin, *Boundary value problems for elliptic pseudodifferential equations*, Translations of Mathematical Monographs, vol. 52, Soviet Math. Dok., American Mathematical Society, Providence, R.I., 1981.

[44] C. Fefferman and D. H. Phong, *On positivity of pseudo-differential operators*, Proc. Nat. Acad. Sci. U.S.A. **75** (1978), no. 10, 4673–4674. MR507931 (80b:47064)

[45] ———, *The uncertainty principle and sharp Gårding inequalities*, Comm. Pure Appl. Math. **34** (1981), no. 3, 285–331. MR611747 (82j:35140)

[46] G.B. Folland, *Harmonic analysis in phase space*, Annals of Mathematics Studies, vol. 122, Princeton University Press, Princeton, NJ, 1989. MR983366 (92k:22017)

[47] D. Geller, *Analytic pseudodifferential operators for the Heisenberg group and local solvability*, Mathematical Notes, vol. 37, Princeton University Press, Princeton, NJ, 1990. MR1030277 (91d:58243)

[48] Georges Glaeser, *Racine carrée d'une fonction différentiable*, Ann. Inst. Fourier (Grenoble) **13** (1963), no. fasc. 2, 203–210. MR0163995 (29 #1294)

[49] A. Grigis and J. Sjöstrand, *Microlocal analysis for differential operators*, London Mathematical Society Lecture Note Series, vol. 196, Cambridge University Press, Cambridge, 1994, An introduction. MR1269107 (95d:35009)

[50] ———, *Microlocal analysis for differential operators*, London Mathematical Society Lecture Note Series, vol. 196, Cambridge University Press, Cambridge, 1994, An introduction. MR1269107 (95d:35009)

[51] K. Gröchenig and M. Leinert, *Wiener's lemma for twisted convolution and Gabor frames*, J. Amer. Math. Soc. **17** (2004), no. 1, 1–18 (electronic). MR2015328 (2004m:42037)

[52] G. Grubb, *Functional calculus of pseudodifferential boundary problems*, second ed., Progress in Mathematics, vol. 65, Birkhäuser Boston Inc., Boston, MA, 1996. MR1385196 (96m:35001)

[53] Pengfei Guan, C^2 *a priori estimates for degenerate Monge-Ampère equations*, Duke Math. J. **86** (1997), no. 2, 323–346. MR1430436 (98d:35074)

[54] Victor Guillemin and Shlomo Sternberg, *Symplectic techniques in physics*, second ed., Cambridge University Press, Cambridge, 1990. MR1066693 (91d:58073)

[55] B. Helffer, *Semi-classical analysis for the Schrödinger operator and applications*, Lecture Notes in Mathematics, vol. 1336, Springer-Verlag, Berlin, 1988. MR960278 (90c:81043)

[56] ———, *h-pseudodifferential operators and applications: an introduction*, Quasiclassical methods (Minneapolis, MN, 1995), IMA Vol. Math. Appl., vol. 95, Springer, New York, 1997, pp. 1–49. MR1477208 (99a:35284)

[57] _____, *Semiclassical analysis, Witten Laplacians, and statistical mechanics*, Series in Partial Differential Equations and Applications, vol. 1, World Scientific Publishing Co. Inc., River Edge, NJ, 2002. MR1936110 (2003j:58038)

[58] _____, *Semiclassical analysis, Witten Laplacians, and statistical mechanics*, Series in Partial Differential Equations and Applications, vol. 1, World Scientific Publishing Co. Inc., River Edge, NJ, 2002. MR1936110 (2003j:58038)

[59] B. Helffer and F. Nier, *Hypoelliptic estimates and spectral theory for Fokker-Planck operators and Witten Laplacians*, Lecture Notes in Mathematics, vol. 1862, Springer-Verlag, Berlin, 2005. MR2130405 (2006a:58039)

[60] B. Helffer and J. Nourrigat, *Hypoellipticité maximale pour des opérateurs polynômes de champs de vecteurs*, Progress in Mathematics, vol. 58, Birkhäuser Boston Inc., Boston, MA, 1985. MR897103 (88i:35029)

[61] F. Hérau, *Melin-Hörmander inequality in a Wiener type pseudo-differential algebra*, Ark. Mat. **39** (2001), no. 2, 311–338. MR1861063 (2002j:47084)

[62] L. Hörmander, *On the theory of general partial differential operators*, Acta Math. **94** (1955), 161–248. MR0076151 (17,853d)

[63] _____, *Differential equations without solutions*, Math. Ann. **140** (1960), 169–173. MR0147765 (26 #5279)

[64] _____, *Pseudo-differential operators and non-elliptic boundary problems*, Ann. of Math. (2) **83** (1966), 129–209. MR0233064 (38 #1387)

[65] _____, *Pseudo-differential operators and hypoelliptic equations*, Singular integrals (Proc. Sympos. Pure Math., Vol. X, Chicago, Ill., 1966), Amer. Math. Soc., Providence, R.I., 1967, pp. 138–183. MR0383152 (52 #4033)

[66] _____, *A class of hypoelliptic pseudodifferential operators with double characteristics*, Math. Ann. **217** (1975), no. 2, 165–188. MR0377603 (51 #13774)

[67] _____, *Propagation of singularities and semiglobal existence theorems for (pseudo)differential operators of principal type*, Ann. of Math. (2) **108** (1978), no. 3, 569–609. MR512434 (81j:35110)

[68] _____, *Subelliptic operators*, Seminar on Singularities of Solutions of Linear Partial Differential Equations (Inst. Adv. Study, Princeton, N.J., 1977/78), Ann. of Math. Stud., vol. 91, Princeton Univ. Press, Princeton, N.J., 1979, pp. 127–208. MR547019 (82e:35029)

[69] _____, *The Weyl calculus of pseudodifferential operators*, Comm. Pure Appl. Math. **32** (1979), no. 3, 360–444. MR517939 (80j:47060)

[70] _____, *Pseudodifferential operators of principal type*, Singularities in boundary value problems (Proc. NATO Adv. Study Inst., Maratea, 1980),

NATO Adv. Study Inst. Ser. C: Math. Phys. Sci., vol. 65, Reidel, Dordrecht, 1981, pp. 69–96. MR617227 (83m:35003)

[71] _____, *The analysis of linear partial differential operators. I*, Grundlehren der Mathematischen Wissenschaften [Fundamental Principles of Mathematical Sciences], vol. 256, Springer-Verlag, Berlin, 1983, Distribution theory and Fourier analysis. MR717035 (85g:35002a)

[72] _____, *The analysis of linear partial differential operators. II*, Grundlehren der Mathematischen Wissenschaften [Fundamental Principles of Mathematical Sciences], vol. 257, Springer-Verlag, Berlin, 1983, Differential operators with constant coefficients. MR705278 (85g:35002b)

[73] _____, *The analysis of linear partial differential operators. III*, Grundlehren der Mathematischen Wissenschaften [Fundamental Principles of Mathematical Sciences], vol. 274, Springer-Verlag, Berlin, 1985, Pseudodifferential operators. MR781536 (87d:35002a)

[74] _____, *The analysis of linear partial differential operators. IV*, Grundlehren der Mathematischen Wissenschaften [Fundamental Principles of Mathematical Sciences], vol. 275, Springer-Verlag, Berlin, 1985, Fourier integral operators. MR781537 (87d:35002b)

[75] _____, *Notions of convexity*, Progress in Mathematics, vol. 127, Birkhäuser Boston Inc., Boston, MA, 1994. MR1301332 (95k:00002)

[76] _____, *Symplectic classification of quadratic forms, and general Mehler formulas*, Math. Z. **219** (1995), no. 3, 413–449. MR1339714 (96c:58172)

[77] _____, *On the solvability of pseudodifferential equations*, Structure of solutions of differential equations (Katata/Kyoto, 1995), World Sci. Publ., River Edge, NJ, 1996, pp. 183–213. MR1445340 (98f:35166)

[78] _____, *private communications*, september 2002 – august 2004, 2004.

[79] I. L. Hwang, *The L^2-boundedness of pseudodifferential operators*, Trans. Amer. Math. Soc. **302** (1987), no. 1, 55–76. MR887496 (88e:47096)

[80] V. Ivrii, *Microlocal analysis and precise spectral asymptotics*, Springer Monographs in Mathematics, Springer-Verlag, Berlin, 1998. MR1631419 (99e:58193)

[81] Warren P. Johnson, *The curious history of Faà di Bruno's formula*, Amer. Math. Monthly **109** (2002), no. 3, 217–234. MR1903577 (2003d:01019)

[82] J.-L. Journé, *Calderón-Zygmund operators, pseudodifferential operators and the Cauchy integral of Calderón*, Lecture Notes in Mathematics, vol. 994, Springer-Verlag, Berlin, 1983. MR706075 (85i:42021)

[83] O. D. Kellogg, *On bounded polynomials in several variables*, Math. Z. **27** (1928), no. 1, 55–64. MR1544896

[84] H. Kumano-go, *Pseudodifferential operators*, MIT Press, Cambridge, Mass., 1981, Translated from the Japanese by the author, Rémi Vaillancourt and Michihiro Nagase. MR666870 (84c:35113)

[85] Bernard Lascar, *Singularités des solutions d'équations aux dérivées partielles non linéaires*, C. R. Acad. Sci. Paris Sér. A-B **287** (1978), no. 7, A527–A529. MR512097 (80b:35101)

[86] P. D. Lax and L. Nirenberg, *On stability for difference schemes: A sharp form of Gårding's inequality*, Comm. Pure Appl. Math. **19** (1966), 473–492. MR0206534 (34 #6352)

[87] Jean Leray, *Lagrangian analysis and quantum mechanics*, MIT Press, Cambridge, Mass., 1981, A mathematical structure related to asymptotic expansions and the Maslov index, Translated from the French by Carolyn Schroeder. MR644633 (83k:58081a)

[88] N. Lerner, *Sur les espaces de Sobolev généraux associés aux classes récentes d'opérateurs pseudo-différentiels*, C. R. Acad. Sci. Paris Sér. A-B **289** (1979), no. 14, A663–A666. MR560328 (80k:47055)

[89] _____ , *Sufficiency of condition (ψ) for local solvability in two dimensions*, Ann. of Math. (2) **128** (1988), no. 2, 243–258. MR960946 (90a:35242)

[90] _____ , *Wick-Wigner functions and tomographic methods*, SIAM J. Math. Anal. **21** (1990), no. 4, 1083–1092. MR1052888 (92g:35232)

[91] _____ , *An iff solvability condition for the oblique derivative problem*, Séminaire sur les Équations aux Dérivées Partielles, 1990–1991, École Polytech., Palaiseau, 1991, pp. Exp. No. XVIII, 7. MR1131591 (92m:35287)

[92] _____ , *Nonsolvability in L^2 for a first order operator satisfying condition (ψ)*, Ann. of Math. (2) **139** (1994), no. 2, 363–393. MR1274095 (95g:35222)

[93] _____ , *Energy methods via coherent states and advanced pseudo-differential calculus*, Multidimensional complex analysis and partial differential equations (São Carlos, 1995), Contemp. Math., vol. 205, Amer. Math. Soc., Providence, RI, 1997, pp. 177–201. MR1447224 (98d:35244)

[94] _____ , *Perturbation and energy estimates*, Ann. Sci. École Norm. Sup. (4) **31** (1998), no. 6, 843–886. MR1664214 (2000b:35289)

[95] _____ , *When is a pseudo-differential equation solvable?*, Ann. Inst. Fourier (Grenoble) **50** (2000), no. 2, 443–460. MR1775357 (2001e:35188)

[96] _____ , *Second microlocalization methods for degenerate Cauchy-Riemann equations*, Carleman estimates and applications to uniqueness and control theory (Cortona, 1999), Progr. Nonlinear Differential Equations Appl., vol. 46, Birkhäuser Boston, Boston, MA, 2001, pp. 109–128. MR1839170 (2002d:30054)

[97] _____, *Solving pseudo-differential equations*, Proceedings of the International Congress of Mathematicians, Vol. II (Beijing, 2002) (Beijing), Higher Ed. Press, 2002, pp. 711–720. MR1957078 (2004c:35437)

[98] _____, *Cutting the loss of derivatives for solvability under condition* (ψ), Bull. SMF (2006), 68 pages.

[99] _____, *Some facts about the Wick calculus*, Pseudo-differential operators, Lecture Notes in Math., vol. 1949, Springer, Berlin, 2008, pp. 135–174. MR2477145

[100] N. Lerner and Y. Morimoto, *On the Fefferman-Phong inequality and a Wiener-type algebra of pseudodifferential operators*, Publ. RIMS (Kyoto) (2006), 33 pages.

[101] N. Lerner and J. Nourrigat, *Lower bounds for pseudo-differential operators*, Ann. Inst. Fourier (Grenoble) **40** (1990), no. 3, 657–682. MR1091836 (92a:35172)

[102] H. Lewy, *An example of a smooth linear partial differential equation without solution*, Ann. of Math. (2) **66** (1957), 155–158. MR0088629 (19,551d)

[103] André Martinez, *An Introduction to Semi-Classical and Microlocal Analysis*, Universitext, Springer-Verlag, New York, 2002.   MR1872698 (2003b:35010)

[104] V. P. Maslov and M. V. Fedoriuk, *Semiclassical approximation in quantum mechanics*, Mathematical Physics and Applied Mathematics, vol. 7, D. Reidel Publishing Co., Dordrecht, 1981, Translated from the Russian by J. Niederle and J. Tolar, Contemporary Mathematics, 5. MR634377 (84k:58226)

[105] Dusa McDuff and Dietmar Salamon, *Introduction to symplectic topology*, second ed., Oxford Mathematical Monographs, The Clarendon Press Oxford University Press, New York, 1998. MR1698616 (2000g:53098)

[106] Anders Melin and Johannes Sjöstrand, *A calculus for Fourier integral operators in domains with boundary and applications to the oblique derivative problem*, Comm. Partial Differential Equations **2** (1977), no. 9, 857–935. MR0458508 (56 #16708)

[107] S. Mizohata, *Solutions nulles et solutions non analytiques*, J. Math. Kyoto Univ. **1** (1961/1962), 271–302. MR0142873 (26 #440)

[108] R. D. Moyer, *Local solvability in two dimensions: necessary conditions for the principal type case*, Mimeographed manuscript, University of Kansas, 1978.

[109] T. Muramatu and S. Wakabayashi, *On the norm of a symmetric multilinear form*, http://www.math.tsukuba.ac.jp/~wkbysh/note3.pdf (2009), 5p.

[110] Alexander Nagel and E. M. Stein, *Lectures on pseudodifferential operators: regularity theorems and applications to nonelliptic problems*, Mathematical Notes, vol. 24, Princeton University Press, Princeton, N.J., 1979. MR549321 (82f:47059)

[111] _____, *Some new classes of pseudodifferential operators*, Harmonic analysis in Euclidean spaces (Proc. Sympos. Pure Math., Williams Coll., Williamstown, Mass., 1978), Part 2, Proc. Sympos. Pure Math., XXXV, Part, Amer. Math. Soc., Providence, R.I., 1979, pp. 159–169. MR545304 (83d:47053)

[112] L. Nirenberg and F. Treves, *Solvability of a first order linear partial differential equation*, Comm. Pure Appl. Math. **16** (1963), 331–351. MR0163045 (29 #348)

[113] _____, *On local solvability of linear partial differential equations. I. Necessary conditions*, Comm. Pure Appl. Math. **23** (1970), 1–38. MR0264470 (41 #9064a)

[114] _____, *On local solvability of linear partial differential equations. II. Sufficient conditions*, Comm. Pure Appl. Math. **23** (1970), 459–509. MR0264471 (41 #9064b)

[115] _____, *A correction to: "On local solvability of linear partial differential equations. II. Sufficient conditions" (Comm. Pure Appl. Math. 23 (1970), 459–509)*, Comm. Pure Appl. Math. **24** (1971), no. 2, 279–288. MR0435641 (55 #8599)

[116] J. Nourrigat, *Subelliptic systems*, Comm. Partial Differential Equations **15** (1990), no. 3, 341–405. MR1044428 (91c:35040)

[117] _____, *Systèmes sous-elliptiques. II*, Invent. Math. **104** (1991), no. 2, 377–400. MR1098615 (92f:35048)

[118] A. Parmeggiani, *A class of counterexamples to the Fefferman-Phong inequality for systems*, Comm. Partial Differential Equations **29** (2004), no. 9-10, 1281–1303. MR2103837 (2005h:35385)

[119] B. Plamenevskii, *Algebras of pseudodifferential operators with piecewise smooth symbols*, Symposium "Partial Differential Equations" (Holzhau, 1988), Teubner-Texte Math., vol. 112, Teubner, Leipzig, 1989, pp. 204–212. MR1105811

[120] K. Pravda-Starov, *A complete study of the pseudo-spectrum for the rotated harmonic oscillator*, J. London Math. Soc. (2) **73** (2006), no. 3, 745–761. MR2241978 (2007c:34133)

[121] _____, *Etude du pseudospectre d'opérateurs non autoadjoints*, Ph.D. thesis, Université de Rennes 1, 2006.

[122] _____, On the pseudospectrum of elliptic quadratic differential operators, Duke Math. J. **145** (2008), no. 2, 249–279. MR2449947

[123] Karel Pravda-Starov, Pseudo-spectrum for a class of semi-classical operators, Bull. Soc. Math. France **136** (2008), no. 3, 329–372. MR2415346 (2009h:35473)

[124] Jeffrey Rauch, Singularities of solutions to semilinear wave equations, J. Math. Pures Appl. (9) **58** (1979), no. 3, 299–308. MR544255 (83c:35078)

[125] Didier Robert, Autour de l'approximation semi-classique, Progress in Mathematics, vol. 68, Birkhäuser Boston Inc., Boston, MA, 1987. MR897108 (89g:81016)

[126] X. Saint Raymond, Elementary introduction to the theory of pseudodifferential operators, Studies in Advanced Mathematics, CRC Press, Boca Raton, FL, 1991. MR1211419 (94b:47066)

[127] Mikio Sato, Takahiro Kawai, and Masaki Kashiwara, Microfunctions and pseudo-differential equations, Hyperfunctions and pseudo-differential equations (Proc. Conf., Katata, 1971; dedicated to the memory of André Martineau), Springer, Berlin, 1973, pp. 265–529. Lecture Notes in Math., Vol. 287. MR0420735 (54 #8747)

[128] I. E. Segal, Transforms for operators and symplectic automorphisms over a locally compact abelian group, Math. Scand. **13** (1963), 31–43. MR0163183 (29 #486)

[129] M. A. Shubin, Pseudodifferential operators and spectral theory, Springer Series in Soviet Mathematics, Springer-Verlag, Berlin, 1987, Translated from the Russian by Stig I. Andersson. MR883081 (88c:47105)

[130] J. Sjöstrand, An algebra of pseudodifferential operators, Math.Res.Lett. **1** (1994), 2, 189–192.

[131] _____, Wiener type algebras of pseudodifferential operators, Séminaire EDP, École Polytechnique (1994-95), Exposé 4.

[132] E.M. Stein, Harmonic analysis: real-variable methods, orthogonality, and oscillatory integrals, Princeton Mathematical Series, vol. 43, Princeton University Press, Princeton, NJ, 1993, With the assistance of Timothy S. Murphy, Monographs in Harmonic Analysis, III. MR1232192 (95c:42002)

[133] Daniel Tataru, On the Fefferman-Phong inequality and related problems, Comm. Partial Differential Equations **27** (2002), no. 11-12, 2101–2138. MR1944027 (2003m:35259)

[134] M.E. Taylor, Pseudodifferential operators, Princeton Mathematical Series, vol. 34, Princeton University Press, Princeton, N.J., 1981. MR618463 (82i:35172)

[135] _____, *Pseudodifferential operators and nonlinear PDE*, Progress in Mathematics, vol. 100, Birkhäuser Boston Inc., Boston, MA, 1991. MR1121019 (92j:35193)

[136] S. Thangavelu, *An introduction to the uncertainty principle*, Progress in Mathematics, vol. 217, Birkhäuser Boston Inc., Boston, MA, 2004, Hardy's theorem on Lie groups, With a foreword by Gerald B. Folland. MR2008480 (2004j:43007)

[137] Joachim Toft, *Schatten-von Neumann properties in the Weyl calculus, and calculus of metrics on symplectic vector spaces*, Ann. Global Anal. Geom. **30** (2006), no. 2, 169–209. MR2234093 (2007f:35314)

[138] Lloyd N. Trefethen and Mark Embree, *Spectra and pseudospectra*, Princeton University Press, Princeton, NJ, 2005, The behavior of nonnormal matrices and operators. MR2155029 (2006d:15001)

[139] J.-M. Trépreau, *Sur la résolubilité microlocale des opérateurs de type principal*, Conference on Partial Differential Equations (Saint Jean de Monts, 1982), Soc. Math. France, Paris, 1982, pp. Conf. No. 22, 10. MR672289 (84a:58079)

[140] F. Treves, *A new method of proof of the subelliptic estimates*, Comm. Pure Appl. Math. **24** (1971), 71–115. MR0290201 (44 #7385)

[141] _____, *Introduction to pseudodifferential and Fourier integral operators. Vol. 1*, Plenum Press, New York, 1980, Pseudodifferential operators, The University Series in Mathematics. MR597144 (82i:35173)

[142] _____, *Introduction to pseudodifferential and Fourier integral operators. Vol. 2*, Plenum Press, New York, 1980, Fourier integral operators, The University Series in Mathematics. MR597145 (82i:58068)

[143] A. Unterberger, *Sur la continuité L^2 des opérateurs pseudo-différentiels*, C. R. Acad. Sci. Paris Sér. A-B **287** (1978), no. 14, A935–A936. MR520772 (80c:47043)

[144] _____, *Extensions du lemme de Cotlar et applications*, C. R. Acad. Sci. Paris Sér. A-B **288** (1979), no. 4, A249–A252. MR524785 (80d:47055)

[145] _____, *Oscillateur harmonique et opérateurs pseudo-différentiels*, Ann. Inst. Fourier (Grenoble) **29** (1979), no. 3, xi, 201–221. MR552965 (81m:58077)

[146] _____, *Les opérateurs métadifférentiels*, Complex analysis, microlocal calculus and relativistic quantum theory (Proc. Internat. Colloq., Centre Phys., Les Houches, 1979), Lecture Notes in Phys., vol. 126, Springer, Berlin, 1980, pp. 205–241. MR579751 (82j:35142)

[147] _____, *Automorphic pseudodifferential analysis and higher level Weyl calculi*, Progress in Mathematics, vol. 209, Birkhäuser Verlag, Basel, 2003. MR1956320 (2003k:11062)

[148] A. Unterberger and H. Upmeier, *Pseudodifferential analysis on symmetric cones*, Studies in Advanced Mathematics, CRC Press, Boca Raton, FL, 1996. MR1382864 (97g:58166)

[149] A. Weil, *Sur certains groupes d'opérateurs unitaires*, Acta Math. **111** (1964), 143–211. MR0165033 (29 #2324)

[150] H. Weyl, *Gruppentheorie und Quantenmechanik*, second ed., Wissenschaftliche Buchgesellschaft, Darmstadt, 1977. MR0450450 (56 #8744)

[151] H. Widom, *Asymptotic Expansions for Pseudodifferential Operators on Bounded Domains*, Lecture Notes in Mathematics, vol. 1152, Springer-Verlag, Berlin, 1985. MR811855 (87d:35156)

Index

Notation

$D_{x_j} = \frac{1}{2i\pi}\frac{\partial}{\partial x_j}$, 288

$H(m, g)$, 150

$H = \mathbf{1}_{\mathbb{R}_+}$, 295

$J^t = \exp 2i\pi t D_x \cdot D_\xi$, 292

$Mp(E)$, 336

$Mp(n)$, 337

$Q_1 \vee Q_2$, 341

$Q_1 \wedge Q_2$, 341

$S^m_{1,0}$, 12

$S^m_{\rho,\delta}$, 22

$S^m_{\text{loc}}(\Omega \times \mathbb{R}^n)$, 29

S^m_{scl}, 22

$S^{-\infty}_{1,0} = \cap_m S^m_{1,0}$, 14

$Sp(2n)$: symplectic group, 324

$V \Subset \Omega$: $\overline{V}$ compact $\subset \Omega$, 45

WFu, 35

$WF_s u$, 37

$\Delta_r(X, Y)$, 82

Log: the principal determination of the logarithm in $\mathbb{C}\backslash\mathbb{R}_-$, 289

$\Psi^m(\Omega)$, 30

$\Psi^m_{ps}(\Omega)$, 31

Σ^m, 28

Υ_+: the $n \times n$ complex non-singular symmetric matrices with non-negative real part, 290

Υ^*_+: the $n \times n$ complex symmetric matrices with a positive-definite real part, 291

char A, 33

curl, 52

$\delta_r(X, Y)$, 80

$\dot{T}^*(\Omega)$, 32

essupp, 34

λ_g, 72

$\lrcorner$ interior product, 344

$\mathbb{P}$: the Leray projection, 51

$\mathcal{H}(u, v)$, Wigner function, 58

$\mathcal{L}(L^2(\mathbb{R}^n))$, 5

$C^\infty_b(\mathbb{R}^{2n})$, 2

$\omega_k(a, b)$, 65

$\text{op}_t a$, 9

$\text{Op}_\Omega(a)$, 31

σ: symplectic form, 324

$\sigma_{x,\xi}$ phase symmetry, 58

$\sqrt{Q_1 \cdot Q_2}$, 341

$\tau_{x,\xi}$ phase translation, 61

$\langle\xi\rangle = (1 + |\xi|^2)^{1/2}$, 12

$\{a, b\}$ Poisson bracket, 15

$a \boxtimes b$, 64

$a \diamond b$: composition of symbols, 6

$a\natural b$, 62, 65

a^w, 19

$g^\natural$, 78

g^σ, 339

$[X, Y] = \langle \sigma X, Y\rangle_{\mathcal{P}^*, \mathcal{P}}$, 57

$\mathcal{P} = E \oplus E^*$ phase space, 57

$H^s(\mathbb{R}^n)$: Sobolev space, 14

$Sp(2n)$, 57

$Sp_+(2n)$, 332

$Sym(n, \mathbb{R})$, 332

$Sym_+(n, \mathbb{R})$, 332

$U(n)$, 332

$sp(2n)$, 328

adjoint operator, 8
admissible metric, 76
admissible weight, 76
analytic functional calculus, 134
annihilation operator, 162
arithmetic mean, 341

bicharacteristic curves, 38
biconfinement estimates, 84

Calderón-Zygmund, 183
canonical transformation, 361
characteristic points, 33
characterization of ψdos, 140
complex harmonic oscillator, 170
composition formula, 62
condition $(\overline{\Psi})$, 177
condition (Ψ), 177
condition (P), 179
confinement estimates, 84
confinement lemma, 98
conic-neighborhood, 32
Cotlar's lemma, 357
creation operator, 162

Dencker's lemma, 190
drift of a pair, 227

elliptic points, 33
elliptic regularity theorem, 36
essential support, 34
Euler's reflection formula, 296

Faà di Bruno formula, 303
Fefferman-Phong inequality, 115
Feynman quantization, 63
Fock-Bargmann space, 104
Fourier inversion formula, 288
Fourier transform, 287

Gårding inequality, 113
Gamma function, 295
Gauss formula, 296
generating function, 361
geodesic temperance, 152

geometric mean, 341

Hörmander's classes, 22
Hamiltonian vector field, 38
Hans Lewy operator, 170
Hardy operator, 299
harmonic mean, 341
Heaviside function, 295
Helmoltz-Weyl projector, 51
Hilbert transform, 51

interior product, 344
iterated brackets, 206

Jacobi identity, 344

Leray-Hopf projection, 51
Lewy operator, 170
Liouville theorem, 344
local solvability, 42
logarithmic convexity, 306
loss of derivatives, 42

main distance function, 80
Maslov index, 334
metaplectic group, 60
Mizohata-Nirenberg-Treves op., 168
multi-index, 288
multiconfinement, 348

Nirenberg-Treves estimate, 219
normal forms of functions, 345
null bicharacteristic curves, 38

oblique derivative problem, 216

partitions of unity, 308
phase symmetry, 58
Poisson bracket, 15
propagation of singularities, 37
proper class, 183
pseudo-differential calculus, 57

quasi-convexity, 200

real principal type, 49

Riesz operators, 51

Schur criterion, 359
Schwartz space, 287
Segal formula, 61
semi-classical calculus, 22
semi-classical Fourier integral opera-
          tors, 361
semi-global solvability, 225
sharp Egorov principle, 361
sharp Gårding inequality, 19
Shubin's classes, 27
sign function, 295
signature, 291
Sjöstrand algebra, 116
slowly varying metric, 68
Sobolev spaces, 137
solvability with loss, 46
subellipticity, 207
symbolic calculus, 91
symplectic form, 57
symplectic group, 57
symplectic intermediate metric, 78
symplectic metric, 78

temperate metrics, 74
tempered distributions, 288
transversal sign changes, 220

uncertainty principle, 72
uniformly confined family of
          symbols, 98

wave-front-set, 35
Weierstrass formula, 296
weight, 76
Weyl quantization, 58
Wick quantization, 100
Wiener Lemma, 117
Wigner function, 58

Pseudo-Differential Operators (PDO) Theorie and Applications

Edited by
M.W. Wong, York University, Canada

In cooperation with an international editorial board

Pseudo-Differential Operators: Theory and Applications is a series of moderately priced graduate-level textbooks and monographs appealing to students and experts alike. Pseudo-differential operators are understood in a very broad sense and include such topics as harmonic analysis, PDE, geometry, mathematical physics, microlocal analysis, time-frequency analysis, imaging and computations. Modern trends and novel applications in mathematics, natural sciences, medicine, scientific computing, and engineering are highlighted.

Forthcoming

Nicola, F. / Rodino, L.
Global Pseudo-Differential Calculus on Euclidean Spaces
ISBN 978-3-7643-8511-8

de Gosson, M.A.
Bopp Pseudo-Differential Operators and Deformation Quantization
ISBN 978-3-7643-9991-7

Available

PDO 3: Lerner, N.
Metrics on the Phase Space and Non-Selfadjoint Pseudo-Differential Operators (2010)
ISBN 978-3-7643-8509-5

PDO 2: Ruzhansky, M. / Turunen, V.
Pseudo-Differential Operators and Symmetries. Background Analysis and Advanced Topics (2009)
ISBN 978-3-7643-8513-2

This monograph develops a global quantization theory of pseudo-differential operators on compact Lie groups.

Traditionally, the theory of pseudo-differential operators was introduced in the Euclidean setting with the aim of tackling a number of important problems in analysis and in the theory of partial differential equations. This also yields a local theory of pseudo-differential operators on manifolds. The present book takes a different approach by using global symmetries of the space which are often available. First, a particular attention is paid to the theory of periodic operators, which are realized in the form of pseudo-differential and Fourier integral operators on the torus. Then, the cases of the unitary group SU(2) and the 3-sphere are analyzed in extensive detail. Finally, the monograph also develops elements of the theory of pseudo-differential operators on general compact Lie groups and homogeneous spaces.

The exposition of the book is self-contained and provides the reader with the background material surrounding the theory and needed for working with pseudo-differential operators in different settings. The background section of the book may be used for independent learning of different aspects of analysis and is complemented by numerous examples and exercises.

PDO 1: Unterberger, A.
Quantization and Arithmetic (2008)
ISBN 978-3-7643-8790-7

The primary aim of this book is to create situations in which the zeta function, or other L-functions, will appear in spectral-theoretic questions. A secondary aim is to connect pseudo-differential analysis, or quantization theory, to analytic number theory. Both are attained through the analysis of operators on functions on the line by means of their diagonal matrix elements against families of arithmetic coherent states: these are families of discretely supported measures on the line, transforming in specific ways under the part of the metaplectic representation or, more generally, representations from the discrete series of SL(2,R), lying above an arithmetic group such as SL(2,Z).

BIRKHÄUSER